COSPAR COLLOQUIA SERIES

VOLUME 3
SOLAR WIND SEVEN

SOLAR WIND SEVEN

*Proceedings of the 3rd COSPAR Colloquium
held in Goslar, Germany, 16–20 September 1991*

Edited by

E. MARSCH

and

R. SCHWENN

*Max-Planck-Institut für Aeronomie, Postfach 20,
D-W3411 Katlenburg-Lindau, Germany*

PERGAMON PRESS

OXFORD · NEW YORK · SEOUL · TOKYO

U.K.   Pergamon Press Ltd, Headington Hill Hall,
            Oxford OX3 0BW, England

U.S.A.   Pergamon Press, Inc., 660 White Plains Road,
            Tarrytown, New York 10591-5153, U.S.A.

KOREA   Pergamon Press Korea, Room 613 Hanaro Building,
            194-4 Insa-Dong, Chongno-ku, Seoul 110-290, Korea

JAPAN   Pergamon Press Japan, Tsunashima Building Annex,
            3-20-12 Yushima, Bunkyo-ku, Tokyo 113, Japan

First edition 1992

British Library Cataloguing in Publication Data
A catalogue record for this book is available from the British Library.

Library of Congress Cataloging in Publication Data
COSPAR Colloquium (3rd: 1991: Goslar, Germany) Solar wind seven: proceedings of the 3rd COSPAR Colloquium held in Goslar, Germany, 16–20 September 1991/edited by E. Marsch and R. Schwenn – 1st ed.
(COSPAR colloquia series: v. 3)
"Sponsored by Max-Planck-Gesellschaft (MPG), Deutsche Forschungsgemeinschaft (DFG), and the Committee on Space Research (COSPAR)" – Half t.p.
Includes index.
1. Solar wind – Congresses. 2. Sun–Corona – Congresses. 3. Cosmic rays. 4. Space plasmas – Congresses. 5. Heliosphere - Congresses. I. Marsch, E. (Eckart), 1947- . II. Schwenn, R. (Rainer), 1941- . III. Max-Planck-Gesellschaft zur Förderung der Wissenschaften. IV. Deutsche Forschungsgemeinschaft. V. COSPAR. VI. Title. VII. Series.
QB529.C69 1991
523.5'8 – dc20   92-20040

ISBN 0-08-042049-4

In order to make this volume available as economically and as rapidly as possible the author's typescript has been reproduced in its original form. This method unfortunately has its typographical limitations but it is hoped that they in no way distract the reader.

Whilst every effort is made by the publishers and editorial board to see that no inaccurate or misleading data, opinion or statement appears in this journal, they wish to make it clear that the data and opinions appearing in the articles and advertisements herein are the sole responsibility of the contributor or advertiser concerned. Accordingly, the publishers, the editorial board and editors and their respective employers, officers and agents accept no responsibility or liability whatsoever for the consequences of any such inaccurate or misleading data, opinion or statement.

Printed and bound in Great Britain by BPCC Wheatons Ltd, Exeter

CONTENTS

PREFACE

The International Conference Solar Wind Seven was held in Goslar (Germany) from September 15 to 21, 1991. The large number of 214 participating scientists from 22 countries exceeded all our expectations. What attracted them, we think, was the ambitious scientific program set up by some of the leading experts in the various areas of heliospheric physics. Or was it perhaps the tempting prospect of spending a few days in the enchanting atmosphere of the famous Kaiserstadt Goslar?

This Solar Wind conference was held as the third COSPAR colloquium. As organizers we were glad to receive substantial financial support from the Max-Planck-Gesellschaft and the Deutsche Forschungsgemeinschaft. Thanks to their support (and from contributions from all the other participants paying the regular fee) we were able to subsidize more than 40 colleagues, coming mainly from eastern European countries. Otherwise, they could not have come. In fact, some of them we only knew by name and by their work, but had never met before! Many participants and observers of the scene expressed their surprise upon realizing how big and active the solar wind community actually is all over the globe. Also, the attendance of so many young bright students should be viewed as an encouraging sign for the vitality of the field.

The scientific program was supposed to cover the whole field of research on the solar wind, with special emphasis on its source regions. Solar wind interactions with "obstacles" such as planets, comets and their plasma environments as well as the more general issue of stellar winds were not addressed at this time. The program was split into five main topics. For each of them we had nominated an individual board of three conveners. It was the responsibility of these conveners to create the detailed program. Early on it had been decided that oral presentations, both reviews and contributed papers, should be given upon invitation only, and that all the other contributions would be accepted for poster presentations. That resulted in 64 oral presentations and in the exhibition of 153 posters.

After the meeting, the conveners handled the refereeing process for "their" papers from the proceedings. All papers which had arrived in time were submitted to capable referees. We as editors and the conveners alike are very grateful to the referees for their competent and prompt evaluations of the papers. In order not to delay the printing of the proceedings unduly, the review process had to be cut short and we could not admit as many iterations as might have been required in some cases. Therefore, a small number of papers could not be accepted in their submitted form and thus are not included in the proceedings. We ask for the understanding of the respective authors and encourage them to publish their revised work in the established journals. Unfortunately, many contributions (mainly posters) have not been written up and submitted at all. Despite this loss of papers the proceedings in hand encompass an impressive collection of high-quality research and review papers.

The group of conveners consisted of Bruno Bavassano, Mike Bird, Alfred Bürgi, Len Burlaga, Ken Dere, Ruth Esser, Ernie Hilder, Todd Hoeksema, Fred Ipavich, Martin Lee, Andre Mangeney, Eberhard Möbius, Richard Steinolfson, Steve Suess and George Withbroe. The solar wind community owes its sincere thanks to all of them.   Also, we express our thanks to the town of Goslar, and in particular to the tourist office lead by Mrs Sundermeier who helped in arranging local tours and social occasions for the staff of the congress center "Der Achtermann", and last but not least to our conference secretaries, Mrs Ute Spilker and Gerlinde Bierwirth.

We are looking forward to Solar Wind Eight, which our colleagues from JPL and Los Alamos promised to organize in a few years.

Eckart Marsch
Rainer Schwenn

Session 1

Coronal Heating and Solar Wind Acceleration
Conveners: Ken Dere
            Ruth Esser
            George Withbroe

THE ORIGIN OF HIGH SPEED SOLAR WIND STREAMS

W. I. Axford and J. F. McKenzie

Max-Planck-Institut für Aeronomie, 3411 Katlenburg-Lindau, Germany

ABSTRACT

Standard coronal heating mechanisms such as dissipation of acoustic waves generated in turbulent photospheric layers, or dissipative heating by current sheets and resonant absorption (see proceedings of Heidelberg conference /1/) cannot produce high speed solar wind streams. We propose that the source of energy in these streams lies in the regions of strong magnetic field which define the boundaries of the chromospheric supergranulation network pattern. If the magnetic field in these "trunks" is not strictly unipolar but instead also contains closed loops then the available free energy can be readily released in impulsive reconnection events ("microflares") which give rise to high frequency hydromagnetic waves with periods of the order of or less than one second. It appears necessary to invoke a strong source of relatively high frequency waves to allow substantial dissipation in the corona itself and also to heat selectively species of differing charge to mass ratios by cyclotron interactions. Moreover such wave-particle interactions are also necessary to account for the detailed properties of high speed streams in which the perpendicular temperature exceeds the parallel temperature, ionic species tend to have the same temperature per unit mass and heavier species tend to move faster than the protons by about the Alfvén speed. It is noteworthy that the same source mechanism for generating the high frequency waves may also be responsible for an enhancement of low FIP ions if the required closed loops of magnetic flux emerge in the transition region where the temperature is between 10^4 and 10^{5o}K.

DISTINCTION BETWEEN FAST AND SLOW STREAMS

The solar wind may generally be divided into fast and slow streams, the former having a clear association with coronal holes and the latter with current sheets or sector boundaries. Fast streams are characterized not only by their speeds (500 - 800 km/sec) but also by their relatively uniform composition and steady bulk plasma properties, all of which suggest that that these streams are in equilibrium with their coronal base. In contrast the slow streams (300 - 400 km/sec) are often highly variable both in composition and in properties such as temperature and density, suggesting transient variations in the associated coronal conditions are somehow involved and that the streams are not in equilibrium (Axford /2/). For these reasons the first task of theory is to account for the properties of fast streams.

The simple procedure of mapping high speed streams back to their associated coronal holes leads to the conclusions that the average (unipolar) magnetic field strength at the base of the corona must be of the order of 5 - 10 gauss and that the total (including gravitational) energy flux must be of the order of 5 - 10 x 10^5 ergs/cm^2, which is comparable to the energy flux in UV/EUV radiation emitted from the base of the corona in quiet regions outside coronal holes (Axford /3/, Gabriel /4/; see Table 1). We may conclude that the energy flux in the form of hydromagnetic waves which normally leads to coronal heating in magnetically closed regions and is largely dissipated as radiation from the transition region, is converted into solar wind bulk flow energy in long-term

1

magnetically open regions and little is radiated from the transition region, thus accounting for the coronal holes.

It is evident from the detailed properties of high speed solar wind streams that even at relatively large distances (0.3 - 1.0 A.U.) their nature is determined to a great extent by wave-particle interactions involving waves which originate at the sun and by cascading from such waves to higher and lower frequencies. Furthermore:

1) the perpendicular temperature is usually larger than the parallel temperature which is the opposite to what would be expected in the absence of preferential absorption of wave energy into perpendicular motion in the solar wind frame (see Leer and Axford /5/);

2) all ionic species tend to have the same temperature per unit mass;

3) heavier species tend to move faster than protons by an amount which is of the order of the Alfvén speed;

4) at 0.3 A.U. from the sun, waves in the frequency range usually measured (10^2 - 10^5 Hz) are outwards propagating Alfven waves.

It is very difficult to account for these features on the basis of any collision-dominated or totally exospheric model. On the contrary, it appears necessary to invoke a strong source of waves with frequencies high enough to allow substantial dissipation in the corona itself and also to heat selectively species of differing charge-to-mass ratios by interaction through their differing cyclotron frequencies. This can be achieved if the source(s) of the waves are "microflares" which must be be of sufficiently small scale, with large characteristic Alfven speeds so that high frequency waves are emitted, and also sufficiently prevalent that a fairly uniform outflow of high speed solar wind results (Axford /6/, McKenzie /7/). (In this context "micro" is intended as meaning "very small" rather than an indication of a millionth). The "nanoflares" envisaged by Parker /8/ arise from the slow twisting and braiding of coronal flux tubes in closed field regions associated with the motion of their footprints in the photosphere. In contrast the microflares we envisage result from reconnection of closed magnetic loops fairly deep in the network and these give rise to high frequency (periods <1 second) waves in both predominantly open and closed field regions. The suggestion that coronal heating is the result of microflares was to our knowlege first made by Gold in the 1960's.

MICROFLARES IN A COMPLEX NETWORK MODEL

An important step forward in understanding the solar wind was the recognition that the magnetic field that extends out into the corona and interplanetary space is largely rooted in the "network" defining the boundaries of the chromospheric supergranulation (e.g. /4/). The first models were however rather too simple in that they assumed a static configuration with a unipolar magnetic field in the network and some undefined source of coronal heating above it. In fact the network itself must contain the source of the heating and since a unipolar field has no energy available for dissipation other than low frequency (ca. 300 second period) twisting and buffeting from the sides and below, we are lead to assume that the field is not unipolar but has only an excess of say, outwards flux over inwards flux. The field should close over a short distance either in the same channel of the network in the case of a coronal hole, or possibly in a neighbouring channel in the case of quiet closed coronal regions (see Figure 1). The bidirectional field would be a natural consequence of the continual addition of flux at the sides as a result of the spreading of the slow (but super-Alfvénic) upwelling flow in the centre of the supergranulation and its subsequent cooling and downflow at the edges where the network forms.

It is important to note that the network can only form in the presence of downwards cooling motions since the magnetic field strength can only grow at the expense of the internal gas pressure. In general we may assume at most that the magnetic pressure in the network balances the pressure in the external chromospheric gas and therefore that fields of the order of a few hundred gauss can only occur at mid-chromospheric heights and the maximum field at the level of the photosphere is about 1800 gauss. The cooling and sinking of the network gas has the additional effect of enhancing the Alfvén speed which may easily reach 100 km/sec or more (although still much less than the 1000 km/sec or more prevalent in the lower corona).

Supergranules have a typical life time of one or two days and a scale of about 30,000 km. The horizontal flow speeds are of the order of 1 km/sec or less so that the amount of overturning is fairly limited. Nevertheless enough magnetic flux must be disgorged by the upwelling motion to form the network in the time available. This flux may be in the form of weak magnetic field or possibly of relatively strong filaments, perhaps the remnants of a previous network. We may consider then that the energy made available in the network for heating the corona and/or driving the solar wind is obtained from the supergranulation convection being converted to magnetic energy which can be further converted to intense hydromagnetic activity as a consequence of microflares associated with magnetic field reversals in the not strictly unipolar network. It seems necessary that the microflares should be able to occur throughout the network in order that their effects should be uniformly felt in the corona and at the Earth's orbit where it takes several hours for a flux tube from a given network channel to pass an observer. Flaring at the edge of the network as envisaged by Dere /9/, would not have this property and in any case could scarcely be violent in view of the very low Alfvén speeds prevalent in the chromospheric gas. On the other hand, isolated filaments of strong field could produce flaring as has been observed but this should not be the rule even if it leads to more easily visible features such as X-ray bright points.

On the basis of these arguments we envisage the immediate source of energy for coronal heating in quiet closed field regions and for high speed solar wind streams in open field regions as being microflare activity in the network on very small scales (< 100 km) so that the characteristic period of the hydromagnetic waves emitted (scale/Alfvén speed) is less than one second. Such microflares are at present unresolvable spatially and they may also not produce such intense UV/EUV emission that they can be easily observed individually except in extreme cases. However they should exist and we may interpret the existing observations of turbulent events, jets and bright points as evidence for an underlying spectrum of microflare activity in which the integrated energy output is maximum at smaller scales. In this respect a lack of spatial and temporal resolution need not be a disadvantage provided adequate spectral resolution is at hand.

SUMMARY AND CONCLUSIONS

Mapping high speed streams back to their associated coronal holes shows that the energy flux in high speed streams at their base ($\approx 5 - 10 \times 10^5$ ergs/cm^2 sec) is comparable with the energy flux emitted in EUV radiation from the transition region outside coronal holes. In terms of supergranulation, network pattern, occurrence of X-ray bright points and the characteristics of the upper chromosphere, coronal holes and "quiet" non-hole regions are rather similar. Although the magnetic field in high speed streams is unipolar this need not be the case in the corresponding network where there may only be an excess of one polarity over the other. There are a number of implications which follow from these observations:

1) upgoing energy transforms to high speed solar wind in "holes" whereas in closed field regions it results in coronal /transition layer heating;

2) the source of the upgoing energy flux must be the same in holes and in closed field regions, and it must occur throughout the "trunk" of the network, presumably as microflare activity involving closed magnetic loops in the network;

3) the individual microflares may have relatively small dimensions as far as reconnection is concerned and therefore small time scales ($\lesssim 1$ second);

4) since high speed solar wind is uniform across a supergranule (i.e. on timescales of around 4 hours) the source of energy and acceleration should therefore be distributed throughout the network. The network "trunks" must be supplied continously with the addition of new magnetic flux at the sides to maintain the energy supply.

Microflares occurring within the network have the "right" properties in that they generate hydromagnetic waves with high characteristic frequencies (periods < 1 sec) and produce plasma jets and regions of hot, compressed, downwards moving plasma. The high frequency waves so generated are potentially easily damped by cyclotron interactions thus permitting discrimination of species by charge to mas ratio, perpendicular temperatures greater than parallel ones and heavier species moving faster than protons. A fully self-consistent nonlinear theory of gyroresonance effects in a multi-component flowing plasma is not yet available but recent work, /10/, /11/, indicates that preferential heating of minor ions does occur. Figure 2(a) depicts the history of a closed loop undergoing reconnection which launches blue shifted plasma motions and upgoing large amplitude small scale Alfvén waves accompanied by (red shifted) hot plasma trapped on slowly contracting loops. If this process ocurs in the transition region where the temperature is between 10^4 and $10^{5\circ}$K (see Figure 2 (b)), there is a tendency to enrich low FIP ions and deplete He^+ by the combined effect of intense Lyman- alpha radiation together with plasma transport associated with reconnection in a partially ionized medium (e.g. Ip and Axford /12/).

REFERENCES

1. Ulmschneider. P., E.R. Priest, and R. Rosner (Eds.), Mechanisms of Chromospheric and Coronal Heating. Springer-Verlag, Heidelberg (1991).

2. Axford, W.I., Study of Travelling Interplanetary Phenomena 1977, D. Reidel Dordrecht-Holland, 145 (1977).

3. Axford, W.I., Proc. International Symposium on Solar-Terrestrical Physics, Boulder, U.S.A., 1, 270, (1976).

4. Gabriel, A.H., Phil. Trans. Roy. Soc., A281, 339 (1976).

5. Leer, E., and W.I.Axford, Solar Phys. 23, 238, (1972).

6. Axford, W.I., Solar Phys. 100, 575, (1985).

7. McKenzie, J.F., J. Geomag. Geoelectr. 43, (1991)

8. Parker, E.N., Adv. Space Res. 10, 17 (1990).

9. Dere, K.P., These Proceedings

10. Gomberoff, L., and R. Elguetta, J. Geophys. Res., 96, 9801 (1991).

11. Gomberoff, L., and R. Hernández, These Proceedings

12. Ip, W.-H., and W.I. Axford, Adv. Space Res. 11 No.1, 247 (1991).

Table I

High Speed Streams (1AU) / Coronal Holes ($1R_o$)

Quantity	Earth's orbit	Coronal base
Number density	$4 cm^{-3}$	$\sim 2 \times 10^8$ cm^{-3}
Particle flux	3×10^8 (cm^2 sec)$^{-1}$	$\sim 10^{14}$ (cm^2 sec)$^{-1}$
Speed	750km/sec	$\sim$ 10km/sec
Proton temperature	$10^{5\circ}$K	$1.5 \times 10^{6\circ}$ K
Electron temperature	$10^{5\circ}$K	$1.5 \times 10^{6\circ}$ K
Energy flux	2 - 3 ergs/cm sec	8×10^5 ergs /cm sec
Radial magnetic field	3×10^{-5} G	10 G
Alfvén speed	50km/sec	1400km/sec
Sound speed	60km/sec	140km/sec

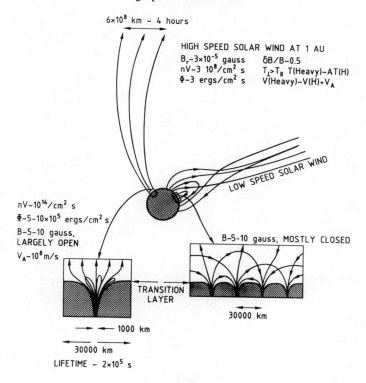

Figure 1: The origin and properties of high speed solar wind from a coronal hole.

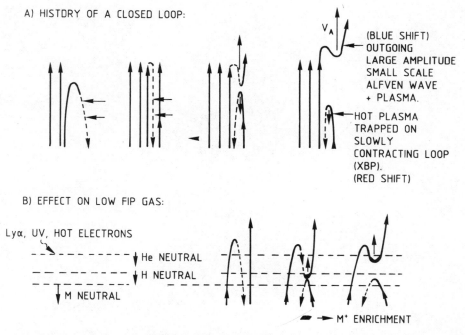

Figure 2(a): The evolution of a closed loop in a predominantly unipolar region depicting reconnection which launches "blue shifted" plasma and outgoing Alfvén waves accompanied by hot plasma trapped on slowly contracting loop (b) Enrichment of low FIP ions by this process combined with $Ly\alpha$ and UV radiation.

A CONSISTENT TREATMENT OF THERMALLY CONDUCTIVE MAGNETOHYDRODYNAMIC FLOWS IN HELMET-STREAMER CORONAL STRUCTURES

S. Cuperman,*,** T. R. Detman,* C. Bruma** and M. Dryer*

* Space Environment Laboratory, ERL, NOAA, Boulder, CO 80303, U.S.A.
** Raymond and Beverly Sackler, Faculty of Exact Sciences, School of Physics and Astronomy, Tel Aviv University, Ramat Aviv, Tel Aviv 69978, Israel

ABSTRACT

The consistent treatment of the acceleration of the solar wind plasma in helmet-streamer coronal magnetic configurations is considered within the framework of the conductive, magnetohydrodynamic model equations. General analytical expressions and numerical procedures for the calculation of the transverse electrical currents entering Maxwell's equations are developed and presented.

INTRODUCTION

In a previous paper (Cuperman, Ofman and Dryer, 1990, hereafter Paper I) we investigated the problem of thermally conductive MHD flows in prescribed helmet-streamer coronal structures with azimuthal symmetry. The conductive solutions obtained were contrasted with those provided by the isothermal model (Pneuman and Kopp, 1971).

The present work is concerned with the consistency of the helmet-streamer configuration and the solar wind-plasma characteristics. Now, the use of an iterative, converging procedure (a generalization of that used by Pneuman and Kopp, 1971 for the isothermal case) requires a very accurate mathematical treatment - especially for the calculation of the curvature radius of the magnetic field lines and of the derivative of the plasma pressure in the direction toward the center of curvature of the field lines. Thus, we here develop a general mathematical formalism aimed to provide the desired accuracy mentioned above. The various parts of the formalism are first tested on an analytical case and then applied to the actual helmet-streamer problem.

THE HELMET-STREAMER MAGNETIC CONFIGURATION

In spherical geometry (r, ϕ, θ), with axisymmetry and $B_\phi = 0$, Maxwell's equations reduce to (see, e.g., Paper I)

$$\nu^2 \frac{\partial^2 f}{\partial \nu^2} + (1 - \mu^2) \frac{\partial^2 f}{\partial \mu^2} = - \frac{(1 - \mu^2)^{1/2}}{\nu^2} j \tag{1}$$

where

$$f \equiv A_\phi \sin\theta / r_0 B_{0,p} \tag{2}$$

$$j \equiv 4\pi r_0 J_\phi / B_{0,p} \tag{3}$$

and

$$\nu \equiv r_0/r , \quad \mu = \cos\theta, \quad \phi = \phi .$$

(4)

Here A_ϕ is the ϕ-component of the vector potential; $B_{0,p}$ and r_0 are reference constants. The field components B_r and B_θ are given by, correspondingly

$$\bar{B}_r \equiv B_r/B_{0,p} = -\nu \left[\frac{\partial f}{\partial \mu}\right]$$

(5)

and

$$\bar{B}_\theta \equiv B_\theta/B_{0,p} = \frac{\nu^2}{(1-\mu^2)^{1/2}} \left[\frac{\partial f}{\partial \nu} - \frac{f}{\nu}\right] .$$

(6)

Now, the helmet-streamer magnetic configuration can be obtained by solving Eqs. (1)-(6), subject to the boundary conditions defining it (see, e.g., Paper I). As already mentioned above, the consistency of the problem requires the calculation and incorporation in Eq. (1) of the transverse plasma currents.

ANALYTICAL FORMALISM FOR THE CALCULATION OF THE TRANSVERSE CURRENTS

Consider a local rectangular coordinate system defined by the orthogonal unit vectors $\underline{e}_\ell \equiv \underline{B}/B$, $\underline{e}_c \equiv R (\partial \underline{e}_\ell/\partial \ell)$ and $\underline{e}_n \equiv \underline{e}_\ell \times \underline{e}_c$. Using the basic methods of the differential geometry, one obtains the following useful results:

a. Transformation from B_r, B_θ to B_ν, B_μ:

$$\begin{pmatrix} B_\nu \\ B_\mu \end{pmatrix} = \begin{pmatrix} -1 & 0 \\ 0 & -1 \end{pmatrix} \begin{pmatrix} B_r \\ B_\theta \end{pmatrix} .$$

(7)

b. The unit vector along the field line :

$$\underline{e}_\ell = \frac{B_\mu}{B} \underline{e}_\mu + \frac{B_\nu}{B} \underline{e}_\nu ,$$

(8)

$$\left[B = \sqrt{B_\mu^2 + B_\nu^2} = \sqrt{B_r^2 + B_\theta^2}\right].$$

c. The unit vector pointing toward the center of curvature of the field line:

$$\underline{e}_c = \frac{\underline{m}}{|\underline{m}|} \left(-\frac{B_\nu}{B} \underline{e}_\mu + \frac{B_\mu}{B} \underline{e}_\nu\right),$$

(9)

where

$$m = \frac{\nu}{r_0} \left\{ B_\mu S \begin{vmatrix} B_\mu & B_\nu \\ \frac{\partial}{\partial\mu} B_\mu & \frac{\partial}{\partial\mu} B_\nu \end{vmatrix} + \nu B_\nu \begin{vmatrix} B_\mu & B_\nu \\ \frac{\partial}{\partial\nu} B_\mu & \frac{\partial}{\partial\nu} B_\nu \end{vmatrix} + B^2 B_\mu \right\}$$

(10)

with $S \equiv (1-\mu^2)^{1/2}$.

d. The radius of curvature, R:

$$\frac{1}{R} = \frac{|m|}{B^3} \,.$$ (11)

e. The derivative along the field line:

$$\frac{\partial}{\partial \ell} = \frac{1}{B} \frac{\nu}{r_0} \left[SB_\mu \frac{\partial}{\partial \mu} + \nu B_\nu \frac{\partial}{\partial \nu} \right].$$ (12)

f. The derivative in the direction of $\underline{e}_c$:

$$\frac{\partial}{\partial c} = \frac{1}{B} \frac{m}{|m|} \frac{\nu}{r_0} \left[-SB_\nu \frac{\partial}{\partial \mu} + \nu B_\mu \frac{\partial}{\partial \nu} \right].$$ (13)

g. An equation replacing eq (1):

$$\bar{B}/\bar{R} - \partial \bar{B}/\partial \bar{c} = j \,,$$ (14)

where $\bar{R} \equiv \dfrac{R}{r_0}$, $\bar{c} = \dfrac{c}{r_0}$, $\bar{B} = \dfrac{B}{B_{0,p}}$ and j is defined in eq. (3).

IMPLEMENTATION OF THE FORMALISM

The accuracy of the solution method of eq. (1), as well as of the expression (7)–(14), were checked on the case of the analytical model presented in Paper I. Thus, (i) the numerical evaluation of $j(\mu,\nu)$ from eq. (1) with analytically given f-distribution, (ii) the alternative expression (14) involving the eqs. (8)–(13) based on differential geometry and (iii) the analytical current distribution were found to provide, within less than 10^{-4} relative error, the same values.

As known (see Paper I), the helmet-streamer structure is basically determined by the imposition of specific boundary conditions on the function f during the solution of eq. (1); the actual distribution of currents, $j(r,\theta)$, will only modify the characteristics of the basic helmet-structure (it will change the location of the neutral point, etc.). Then, for the purpose of illustrating the basic procedures developed in this paper we discarded the plasma current distribution. Thus, Figure 1 gives the spatial dependence of the flux function $F(\mu.\nu) \equiv f(\mu,\nu)/\nu$ with $f(\mu,\nu)$ being the solution of eq. (1) for the helmet-streamer with the neutral point at $\nu \equiv r_0/r = 0.4$ $(r/r_0 = 2.5)$. Figure 2 gives the corresponding sheet current, extending between $\bar{r} = 2.5$ and $\bar{r} = \infty$ $(\nu = 0)$, in the equatorial plane.

The accuracy of the calculations is remarkable. To see this better, we selected several representative field lines, labelled a–e (see Fig. 3) and presented separately the contribution of the terms entering equation (14) for the currents, for one of them. Thus, Fig. 4 gives the ℓ-dependence, along the corresponding field line, of the following quantities: the magnetic field $\bar{B}$, the curvature $1/R$, the ratio $\bar{B}/\bar{R}$ and the field derivative $\partial \bar{B}/\partial \bar{c}$. As it is seen, the last two curves are almost indistinguishable – their difference is practically zero. This explains the overall excellent result presented in Fig. 2, indicating that indeed only the sheet current is non-zero, and, therefore, both analytical and numerical procedures used here work fine.

SUMMARY

We developed and tested the mathematical techniques required for the consistent treatment of the thermally conductive magnetohydrodynamic flows in helmet-streamer

S. Cuperman *et al.*

coronal structures.

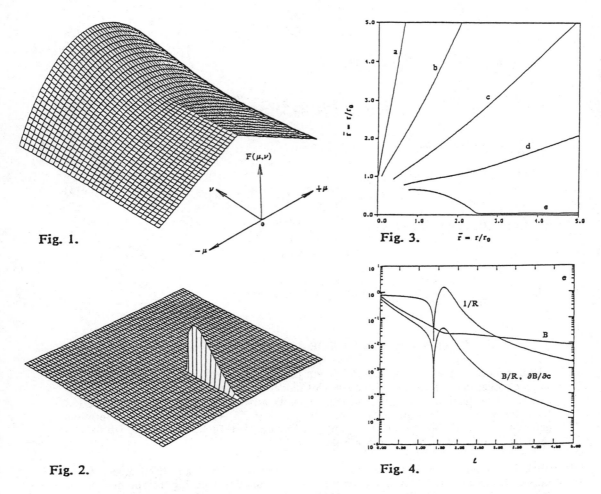

Fig. 1.

Fig. 3. $\bar{r} = r/r_0$

Fig. 2.

Fig. 4.

Fig. 1. The spatial dependence of the magnetic flux function $F(\mu,\nu) \equiv f(\mu,\nu)/\nu$, where
$f(\mu,\nu)$ is the solution of eq. (1) for the helmet-streamer case with the neutral
point at $\nu \equiv r_0/r = 0.4$ $(r/r_0 = 2.5)$, where r_0 is the sun's radius.

Fig. 2. The spatial dependence of the transverse electrical current $j(\mu,\nu)$ for the case
defined in the caption to Fig. 1, as calculated by the formula (14).

Fig. 3. Five magnetic field lines of the helmet-streamer configuration described in the
caption to Fig. 1, in the r/r_0 and θ space. They were selected such that at
$r/r_0 = 65$, the corresponding θ-values are 7.5^0(a), 22.5^0(b), 45^0(c), 67.5^0(d) and
88^0(e).

Fig. 4. The "ℓ" dependence of the quantities entering the transverse currents j, eq. (14),
for the field line labelled e in Fig. 3. For simplicity, the bars on the symbols B,
R, c, ℓ were omitted.

REFERENCES

S. Cuperman, L. Ofman, and M. Dryer, Ap.J., 350, 846 (1990).
S. Cuperman, T. Detman, and M. Dryer, Ap.J., 330, 446 (1988) (Paper I) .
G. W. Pneuman, and R. A. Kopp, Solar Phys. 18, 258 (1971).

EXPLOSIVE EVENTS AND MAGNETIC RECONNECTION IN THE SOLAR ATMOSPHERE

K. P. Dere

*E. O. Hulburt Center for Space Research, Naval Research Laboratory,
Code 4163, Washington, DC 20375, U.S.A.*

ABSTRACT

Signatures of explosive events are prominent in transition region spectra where they display large (100 km s^{-1}) in small areas (1500 km). Their physical properties, deduced from previous analyses of HRTS spectra and more recent work, are summarized here. It now appears that the explosive events are associated with the process of magnetic cancellation and HRTS data obtained during the most recent rocket flight in 1990 provide direct evidence for this assertion. Flux cancellation, which tends to proceed with time-scales of hours, most likely involved magnetic reconnection in a very bursty manner which explains the small, short-lived explosive events.

INTRODUCTION

In 1975, the first rocket flight of the NRL High Resolution Telescope and Spectrograph took place. The ability of this instrument to simultaneously record high resolution spectra over a large field of view had an immediate impact on our view of the 'quiet' solar transition zone. These observations made it clear that the quiet transition zone was a highly dynamic environment. One of the more obvious examples of this sort of dynamic behavior were the newly discovered coronal jets and explosive events /1/. The coronal jets were characterized by velocities as high as 500 km s^{-1} , were roughly 10" in size, and showed repeated accelerations at the same site. The explosive events were characterized by velocities of 100 km s^{-1} and spatial scales around 1500 km. They were much more numerous than the coronal jets. The deduced mass and energy associated with these events indicated that they potentially provided the source of the solar wind mass outflow and the heating for the quiet corona. Over the years, there have been 7 rocket flights of the HRTS and a Spacelab 2 mission in 1985. The analysis of these data sets have caused our ideas about coronal jets and explosive events to evolve. For one thing, it is now clear that the coronal jets are relatively rare events and were only fortuitously observed in the first two rocket flights. It seems unlikely that they occur sufficiently frequently to play a major role in the global balance of mass and energy in the solar corona. These later HRTS rocket flights have allowed a comparison of the locations and occurrences of the explosive events with photospheric magnetic fields and their evolution and this has brought about a much greater understanding of their basic nature. In this paper, I will then describe the basic properties of explosive events and the observations which point to their relationship with the reconnection of magnetic fields in the Sun.

PROPERTIES OF EXPLOSIVE EVENTS

Explosive events are observed as small-scale, impulsive features in the spectra of chromospheric, transition region, and coronal lines. They appear to be most prominent in transition region lines formed near 10^5 K where they most commonly appear in the HRTS spectra. Analyses of their properties as they appear in 'the HRTS spectra have been published previously /1,2,3/. In the HRTS spectra, the explosive events are detected as small features with large doppler shifts. Porter *et al.* /4/ have reported small scale short period brightenings with large doppler signatures in SMM UVSP data that are probably closely related to explosive events seen in the HRTS spectra. Typical maximum velocities for the explosive events are 100 km s^{-1}. This can be compared with a sound speed of 50 km s^{-1} at 10^5 K. The profiles are usually asymmetric with the outflowing (blueshifted) and downflowing (redshifted) plasmas often displaced along the slit by 1000 km. Some of the profiles seem to be symmetric, perhaps indicating locations of high turbulence, but this is not usually the case. The redshifted profile is often different from the blueshifted profiles and in many examples, one of these components is often quite weak. On average, the velocities inferred from explosive event profiles show no dependence on their position with respect to either sun center or the limb. This implies that, in a statistical sense, the horizontal velocities (observed at the solar limb) and the vertical velocities (observed at disk center) are roughly equal. The velocities show no preferred direction and are in a sense isotropic. Times series of C IV spectra are available which show that over a time period of 200 s, no apparent motion of the explosive events along the slit is observed. If the explosive event were a single plasmoid moving at a velocity of 100 km s^{-1}, one would expect to detect motions of the spectral features along the slit. The observations place an upper limit of 5 km s^{-1} for the component of the plasmoid velocity across the solar surface.

The size of the explosive events tend to be around 1500 km. This is commensurate with the spatial displacements observed between blueshifted and redshifted components in the spectra. Structure on smaller spatial scales is also evident. It is clear that these are not just point-like explosions but three-dimensional objects whose structure is probably directly related to their dynamics.

In the HRTS spectra, the highest time resolution observations are found in a raster sequence with 20 s intervals between spectra at the same spatial location. This data set shows that the typical lifetime of an explosive event is 60 s but with strong changes often taking place with 20 s. Porter et al./5/ find transition brightenings in active regions with similar lifetimes and with some as short as 20 s. There is some evidence for repeatability of explosive events at the same site with some residual high nonthermal velocities between events.

Explosive events observed in the quiet solar atmosphere are generally found at heights below typical transition region structures. This is demonstrated by the observations shown in Figure 1. These data were obtained on December 11, 1987 when two sounding rockets were launched, one carrying the AS&E X-ray telescope and the other carrying the NRL HRTS instrument for its fifth rocket flight (HRTS-5). Full disk magnetograms and He I λ10830 spectroheliogram were also recorded by the Kitt Peak National Observatory. The data set obtained by this dual rocket flight is quite comprehensive, includes an image of the photospheric fields together with images spanning a full range of temperatures from the chromosphere, through the transition region and corona. The HRTS performed a raster sequence over a region of mostly quiet features, including the limb. The spectrograph was centered on the spectral region around 1550 Å which contains the C IV transition region lines. Images of the intensity, velocity in the C IV lines were reconstructed from the spectra and coaligned with the full disk images. The locations of the explosive events found in the spectra were also placed onto a map for correlation with various images where they are indicated by small rectangles. Portions of these various images near the limb are shown in Figure 1. Each image is 177" x 220". Directly at the limb, a number of explosive events are found at a height less than 2000 km just above the photosphere and below the limb brightening of the C IV transition region lines.

There has been general speculation that X-ray bright points, He I dark points, microflares and explosive events are related phenomena. This idea can be tested with the HRTS-5 data set. The locations of explosive events with respect to structures seen on the solar disk are shown in Figure 2. A number of explosive events seem to occur in the vicinity of the three X-ray bright points seen in this figure but the overall appearance of their distribution appears to be nearly random with regard to the X-ray bright points. Likewise, a number

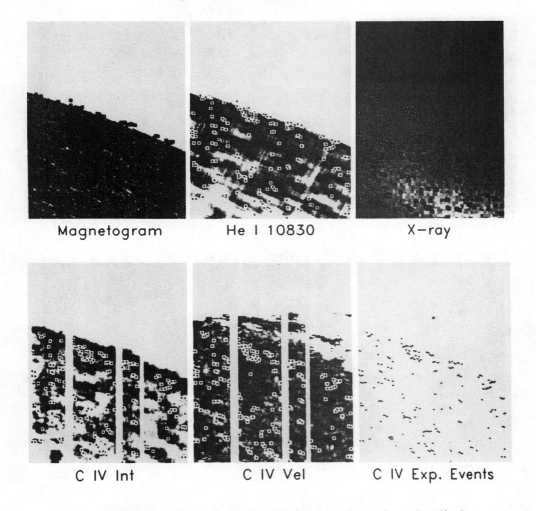

Figure 1: Location of explosive events at the solar limb

of explosive events occur inside He I dark points but there is no general correlation between the two phenomena. There is an indication that the explosive events are related to the supergranular network evident in the photospheric magnetogram, the He I spectroheliogram and the C IV intensity image.

In the quiet sun, a birthrate for the explosive events of 10^{-20} cm^{-2} s^{-1} has been derived from a sequence of 11 rasters obtained during the third HRTS rocket flight /3/. The birthrate in a polar coronal hole observed in the same flight was about half this number. A single large area raster obtained during the HRTS-7 rocket flight indicated that the explosive event birthrate in the quiet sun and in a coronal hole on the disk were equal /8/.

One of the most important parameters for determining the relevance of explosive events on the global mass and energy balance of the solar coronal is their density. One method of calculating the density is to determine the emission measure ($\int n^2 dV$). The density follows when the volume of the feature is measured. The main problem with this method is that it is possible that the filling factor of the emitting elements of the structure is much smaller than its envelope. This is the general case for the quiet and active transition zone /6/. A more accurate measurement involves the use of density sensitive line intensity ratios, where

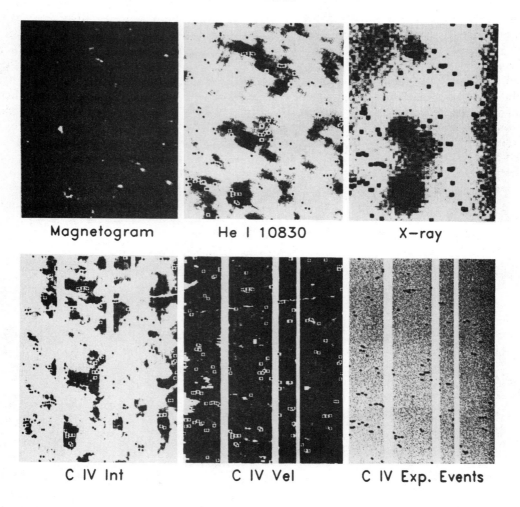

Figure 2: **Location and environment of explosive events on the solar disk.**

the intensities of different lines from the same ion exhibit a different dependence on the density. This happens when metastable levels are present whose population does not vary linearly with density which is the general case with allowed lines directly excited from the ground level. The best line ratios in the HRTS spectral region for performing such an analysis for explosive events arise from the O IV ion formed near 1.7×10^5 K. Calculations /7/ show that line ratios involving these lines are sensitive indicators of electron density. Unfortunately, they are not extremely intense lines and explosive events are not especially bright. One useable example where the O IV lines are bright enough to derive reliable densities has been found in the HRTS Spacelab 2 data. The line ratios in this event are consistent with a density of 7×10^{10} cm^{-3}. The average density obtained using the emission measure is 1×10^9 cm^{-3}. The difference in the two density values indicates a volumetric fill factor of about 10^{-4}. The explosive event analyzed is an especially bright one but if the fill factors derived from this event apply to most explosive events then the masses and kinetic particle energies associated with explosive events much be reduced by this factor of 10^{-4} from the nominal values previously derived. As a consequence, the explosive events do not appear to play a direct role in the mass and energy balance of the corona. Nevertheless, they are probably a direct indicator of important processes occurring in the solar atmosphere that are directly connected to the mass and energy balance of the corona.

Explosive events have typically been in observed in transition region lines formed at a temperature near 10^5 K. There are also a number of strong chromospheric lines in the HRTS spectra such as C I λ1561 which should be sensitive to explosive event plasmas near 10^4 K. However, explosive events are only rarely seen in these lines. Although we have not attempted to determine the exact frequency, a rough estimate is that less that 1% of the explosive events seen in transition zone lines are also seen in chromospheric lines such as C I. They are seen weakly in C II λ1335 formed at 4×10^4 K. Dynamic events very similar to the transition zone explosive events have been detected in coronal and upper transition zone lines observed with the NRL slitless spectroheliograph on Skylab /9,10/. Velocities are on the order of 200-500 km s^{-1}. Most of these events occur in active regions although some occur in quiet regions but their maximum temperatures are less than about 10^6 K. There are many fewer XUV streaks than transition region explosive events, even though the Skylab XUV spectroheliograph simultaneously observed the complete solar disk. One reason may be that only in the active regions is there enough plasma heating to generate the coronal temperatures.

Properties of Explosive Events	
Velocities	100 km s^{-1} Asymmetric profiles Apparent motions < 5 km s^{-1} No center-limb variation: statistically isotropic
Spatial scales	1500 km Evidence for smaller scales
Time scales	Average lifetime = 60 s Evidence for repeatability Residual high nonthermal velocities
Height	1000-2000 km above photosphere Below quiet transition zone
Density	7×10^{10} cm^{-3} (one event) Low fill factor
Temperature	Maximum brightness at 10^5 K Rarely seen in chromospheric lines Occasionally seen in coronal lines
Birthrate	10^{-20} cm^{-2} s^{-1} (quiet sun and coronal holes)

THE ASSOCIATION OF EXPLOSIVE EVENTS WITH EMERGING AND CANCELING FLUX

It has generally been assumed that explosive events are driven by magnetic forces. For a variety of reasons, photospheric magnetic field data has been lacking until the last several rocket flights so that it has not been possible to delve into this question. The first clues came from HRTS C IV spectra over an active region obtained during Spacelab 2. These showed unusually high velocities and brightness over an extended region that was associated with an emerging flux region seen in Hα spectroheliograms obtained at the Hida observatory /11/. The association of explosive events with emerging flux was further demonstrated with data obtained during the sixth HRTS rocket flight (HRTS-6). The primary target for this flight was a coronal hole on the disk but the slit raster sequence also included a small emerging active region. This active region was a prolific producer of explosive events, most of which were aligned along the boundaries of strong flux regions with many on the neutral line /8/. These data led us to conclude that the explosive events were produced by the emergence of new magnetic flux into the solar atmosphere. In particular, this would proceed by the reconnection of magnetic flux as in the Heyvaerts, Priest and Rust flare model /12/. The choice of reconnection is supported by the bipolar nature of the reconnection process which would drive material out of the reconnection region in opposite directions as observed in the explosive events.

It does not appear to be possible to associate most of the explosive events with magnetic flux emergence. Under the assumption that the explosive events are associated with magnetic reconnection, the best candidate for the production of the explosive events are the magnetic flux cancellation events. Flux cancellation was discovered by Martin /13/ and is described by Martin et al. /14/ and Livi et al. /15/. The basic picture is that small magnetic bipoles emerge in the supergranular cell centers and are transported to the boundaries where opposite polarity flux elements are driven together by the photospheric flows and forced to cancel.

The identification of flux cancellation events as the source of the explosive events in the quiet sun is supported by the fact that many explosive events occur on the edges of the supergranular network defined both by the C IV line intensities and by the photospheric fields. This can be seen in the HRTS-5 data seen above and also in the HRTS-6 data set /8/. At first glance, the explosive events appear to be randomly distributed over the quiet sun. After some study, their distribution can be described by several general characteristics. They tend to avoid the regions of highest photospheric magnetic fields strength. They frequently occur on the borders of the high field regions, often just outside the high field area where the observed field strength appears to be negligible. Many other are just scattered about the solar surface in places where there appears to be no observable field, probably because the combined resolution and sensitivity of the available observations during the HRTS flights have not been able to detect the weaker bipolar fields that are generally present in the absence of strong fields.

In order to establish a one-to-one observational correlation between explosive events and canceling magnetic flux, the launch of the sixth HRTS rocket, together with the AS&E rocket, was coordinated with photospheric field observations at Kitt Peak and the Big Bear observatory. The latter provided time series images of the photospheric field that could be correlated with the rocket observations. The field of view included a complex, flaring active region together with large regions of quiet sun. In Figure 3 is shown a composite of HRTS C IV profiles, a Big Bear magnetograph image, and a HRTS ultraviolet continuum slit jaw image. The latter provides excellent coregistration of the HRTS spectra to the magnetograms. The position of the slit is shown superposed on the magnetograph image together with X's denoting the positions of the explosive events seen above in the HRTS spectra. The large one to the right of center shows an explosive event along a tongue of magnetic polarity jutting into a large region of opposite polarity. Flaring was observed along this tongue throughout the day. Consequently, it should be possible to attribute the explosive events along this configuration with dynamical changes in the magnetic field, probably due to magnetic reconnection. To the left of center can be seen the largest explosive event observed in the HRTS-6 spectra. Figure 4 shows a time series of Big Bear magnetograms in the vicinity of this large explosive event. It does seem possible to associate this explosive event with a small clump of flux that moves outward from the active region and coalesces with a large preexisting patch of magnetic flux. This data then

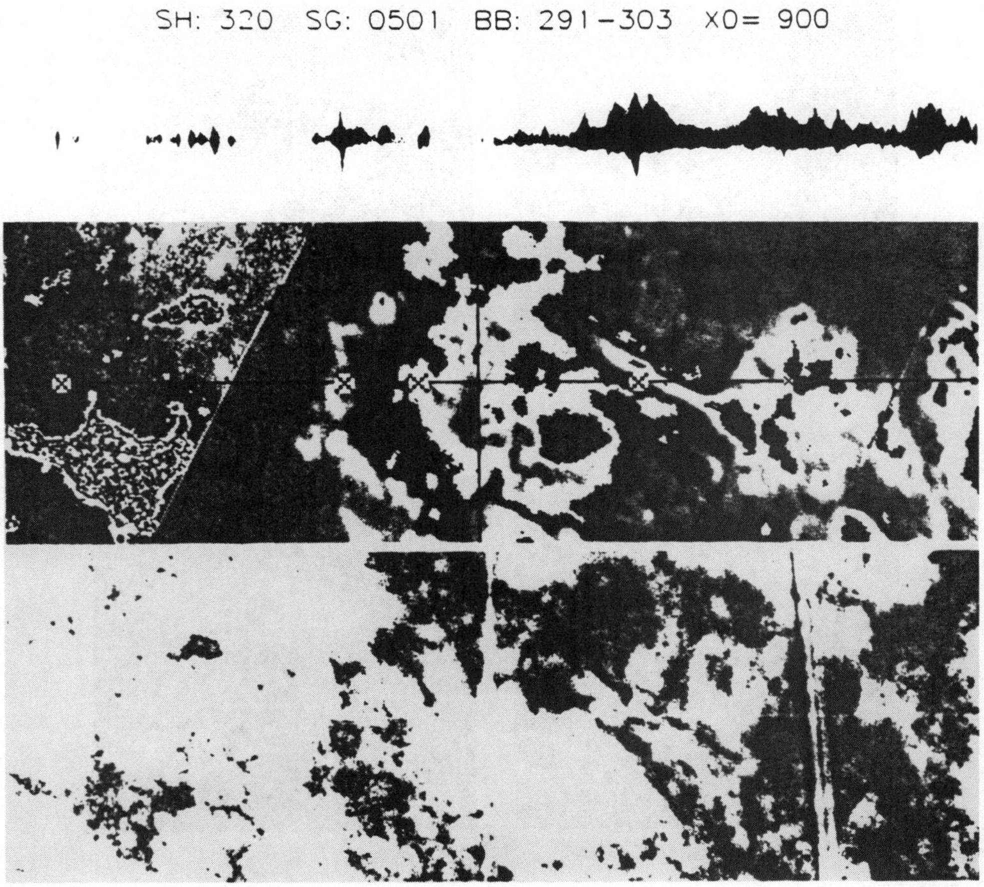

SH: 320 SG: 0501 BB: 291–303 XO= 900

Figure 3: Location of explosive events relative to a BBSO magnetogram.

provides several examples of explosive events probably associated with demonstrable dynamic changes in the field pattern. However, they are still a significant number of explosive events that are situated on the edge of magnetic network but without any corresponding opposite polarity flux element. Many cannot be associated with any observable magnetic flux. At this point, we must assume that this is a question of the sensitivity of current instrumentation.

DIAGNOSTICS OF MAGNETIC RECONNECTION

The characteristics of magnetic cancellation and explosive events can be used to shed some light on the nature of magnetic reconnection as it occurs in the quiet sun outside of flares. For example, if the velocity observed in an explosive event is equated with the local Alfvén speed, then a knowledge of the plasma density leads to a value for the magnetic field strength in the reconnection region. For the single event where the electron density has been measured, a magnetic field strength of 20 gauss is derived. This value seems to be roughly comparable with photospheric field strengths measured with BBSO magnetograms of canceling intranetwork flux elements /15/.

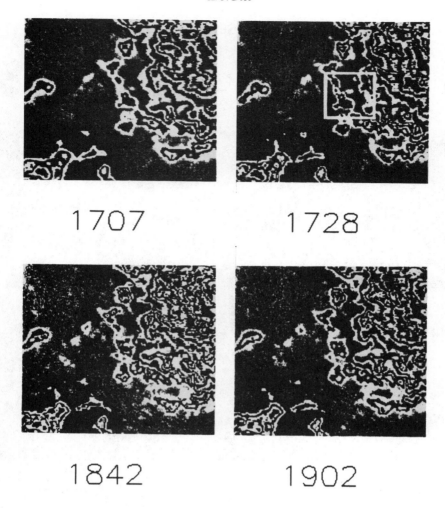

Figure 4: Time series of BBSO magnetograms at the site of a large explosive event.

The cancellation of photospheric flux is observed to occur across boundaries on the order of 6000 km and on time-scales of several hours. A simple model suggests that cancellation then occurs as one flux element is driven into another at a velocity around 0.3 km s^{-1}, very close to the velocity observed for the convection of flux elements through the supergranular cell interior /16/. The explosive events are characterized by size scales of 1500 km and time scales of 60 s, much smaller and shorter than observed for the process of flux cancellation. This suggests that the actual reconnection occurs in short bursts at discrete locations along the overall interaction boundary. This is graphically summarized in Figure 5. The rate at which reconnection proceeds is quite close to the rate predicted by the Petschek mechanism /17/.

The net effect of these flux cancellation/reconnection events on the overall structure and energetics of the solar atmosphere is not immediately clear. With regard to mass transport in the corona, the spectra certainly show that jets directed outward from the solar surface are produced. Some of the jets will occur on field lines that are newly opened to the corona by the reconnection event. The material in these jets can escape into the solar wind. Because the densities are high (at least in one case), the total mass associated with an explosive event at 10^5 K is not high enough for them to play a strong role in the supply for

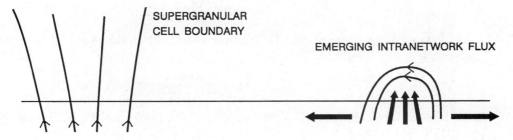

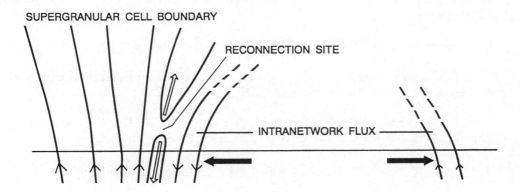

Figure 5:   **Schematic of reconnection in explosive events during magnetic flux cancellation.**

the solar wind. One possibility is that the mass deduced from C IV observations can account for the total mass in the event. Plasma heating does result from the reconnection that accelerates the is involved since 10^5 K material is formed at a height where much cooler material previously existed. The observations reported by Cheng and Kjeldseth-Moe /10/ suggest that in most events, the material is not heated to coronal temperatures. The energy losses in the quiet sun are dominated by the high density structures in the strong network field regions but the explosive events mainly occur on the boundaries of these strong field regions. Parker /18/ suggests that magnetic reconnection in coronal holes may be responsible for the heating in these regions and for the generation of Alfvén waves which accelerate the high speed wind streams. Dere /19/ has deduced the existence of considerable power at wavelengths short enough to be dissipated through classical viscosity and joule heating. It is possible that this power is the result of reconnection processes occurring over a range of spatial scales. Another result of the network boundary reconnection events would be to enhance the rate at which magnetic complexity is built up by the shuffling of the magnetic field lines that Parker /20/ has suggested as the root source of free energy in the corona.

REFERENCES

1.   G. E. Brueckner and J.-D. F. Bartoe, <u>Astrophys J</u>. **272**, 329 (1983).

2.   J. W. Cook, P. A. Lund, J.-D. F. Bartoe, G. E. Brueckner, K. P. Dere, and D. G. Socker, in <u>Cool Stars, Stellar System, and the Sun</u>, eds. J. L. Linsky and R. E. Stencel, Lecture Notes in Physics **291**, Springer-Verlag (1987) p. 150.

3.   K. P. Dere, J.-D. F. Bartoe, G. E. Brueckner, <u>Solar Phys.</u> 123, 41 (1989).

4.   J. G. Porter, R. L. Moore, E. J. Reichmann, O. Engvold, and K. L. Harvey, <u>Astrophys. J.</u> 323, 380 (1987).

5.   J. G. Porter, U. Toomre, and K. B. Gebbie, <u>Astrophys. J.</u> 283, 879 (1984).

6.   K. P. Dere, J.-D. F. Bartoe, G. E. Brueckner, J. W. Cook, and D. G. Socker, <u>Solar Phys.</u> **114**, 223 (1987).

7.   K. P. Dere, J.-D. F. Bartoe, G. E. Brueckner, <u>Astrophys. J.</u> **259**, 366 (1982).

8.   K. P. Dere, J.-D. F. Bartoe, G. E. Brueckner, J. Ewing, and P. Lund, <u>J.G.R.</u>**96**(A6), 9399 (1991).

9.   G. E. Brueckner, N. P. Patterson, and V. E. Scherrer, <u>Solar Phys.</u> **47**, 127 (1976).

10.   C.-C. Cheng and O. Kjeldseth-Moe, Dynamics of Solar Flares, eds. B. Schmieder and E. Priest, Observatoire de Paris, DASOP 1991, p. 101.

11.   G. E. Brueckner, J.-D. F. Bartoe, J. W. Cook, K. P. Dere, D. G. Socker, H. Kurokawa, M. McCabe, <u>Astrophys J.</u> **335**, 986 (1988).

12.   J. Heyvaerts, E. R. Priest, and D. M. Rust, <u>Astrophys. J.</u> **216**, 123 (1977).

13.   S. F. Martin, <u>Small-Scale Dynamical Processes in Quiet Stellar Atmosphere</u>, ed. S. L. Keil, National Solar Observatory, Sunspot, 1984, p. 30.

14.   S. F. Martin, S. H. B Livi, and Wang, J., <u>Aust. J. Phys.</u> **38**, 929 (1985).

15.   S. H. B. Livi, J. Wang, and S. F. Martin, <u>Aust. J. Phys.</u> **38**, 855 (1985).

16.   H. Zirin, <u>Aust. J. Phys.</u> **38**, 961 (1985).

17.   H. E. Petschek, <u>The Physics of Solar flares</u>, ed. W. N. Hess, NASA SP-50, 425 (1964).

18.   E. N. Parker, <u>Astrophys. J.</u> **372**, 719 (1991).

19.   K. P. Dere, <u>Astrophys. J.</u> **340**, 599 (1989).

20.   E. N. Parker, <u>Astrophys. J.</u> **330**, 474 (1988).

COMPRESSIONAL INSTABILITY IN THE SOLAR WIND DRIVEN BY DISSIPATIVE HEATING OF ALFVÉN WAVES

M. K. Dougherty* and J. F. McKenzie**

*Space and Atmospheric Physics, Imperial College, London, SW7 2BZ, U.K.
** Max-Planck-Institut für Aeronomie, Postfach 20, 3411 Katlenburg-Lindau, Germany

ABSTRACT

This paper examines the stability of the solar wind dissipatively heated by saturated sound or Alfvén waves. The amplitude of the waves is assumed to be saturated at some given level by non-linear processes. It is shown that long wavelength compressional modes in the solar wind can be driven unstable by the saturated Alfvén waves whereas in the case of saturated sound waves the system remains stable. The fastest growth times for the instability are shown to be about seven times longer than a characteristic Alfvén wave period, so that in the supersonic region of the flow, where the instability is likely to develop, the growth time can be of the order of minutes to hours. The implication is that this instability can play a significant role in the large scale heating and dynamics of the solar wind.

INTRODUCTION

Magnetoacoustic and Alfvén waves propagating outward from the solar corona undergo amplification arising from WKB growth in the expanding solar wind. It is therefore to be expected that non-linear effects will set in to limit their relative amplitude. If we make the simplest assumption of saturation at some given fixed level this implies dissipative wave heating of the plasma. This is analyzed in the following sections for the case of both saturated sound and Alfvén waves whose wavelengths are "short" compared with the characteristic flow length scale. It is shown that dissipative Alfvén waves can drive "long" wavelength compressional waves unstable whereas their sound wave counterpart cannot by virtue of stabilizing pressure gradients. The marginal stability condition shows that for a fixed saturated wave amplitude instability occurs if the plasma β is less than a critical value which increases dramatically with wave normal obliquity. Growth rate calculations for various saturation levels indicate that maximum growth rates of the order of 1/7 of Alfvén frequencies are obtainable for oblique propagation. These unstable waves may play an important role in the solar wind dynamics and energetics in the supersonic region.

THEORETICAL DEVELOPMENT

The question of the stability of a plasma flow dissipatively heated by saturated Alfvén waves is of some importance, since an instability, if it arises, is likely to result in further plasma heating. This problem has already been investigated by /1/ in a more general framework which included the effects of cosmic rays and self-excited Alfvén waves. Provided the plasma β is not too small, it was found that instabilities can occur. Here we describe a simpler version of that work specifically within the context of a solar wind flow.

First, we stress that the equations used to describe the mutual interaction between the waves and the plasma flow are valid only if the wavelength of the waves λ_w (or period) is small (or short) compared with the length scale L (or flow time) characteristic of the background plasma flow velocity u. In other words we must have

$$\lambda_w \ll L \; , \quad \frac{1}{L} = \frac{1}{u}\frac{du}{dr} \tag{1}$$

for the averaging process over a wavelength (or period) to hold. The unsteady versions of the equations of conservation of mass, momentum, and energy then take the form

$$\frac{D\rho}{Dt} + \rho \, \mathrm{div}\, \mathbf{u} = 0 \tag{2}$$

$$\rho \frac{D\mathbf{u}}{Dt} = -\nabla \left(p_g + p_w \right) - \frac{G\mu\rho}{r^2} \tag{3}$$

$$\frac{Dp_g}{Dt} - c^2 \frac{D\rho}{Dt} = (\gamma - 1) Q \tag{4}$$

where

$$\frac{D}{Dt} = \frac{\partial}{\partial t} + \mathbf{u} \cdot \nabla$$

and ρ and p_g are the plasma density and pressure, p_w is the wave pressure and Q the dissipative heating function. The system is completed by the wave energy exchange equation, namely,

$$\frac{\partial E_w}{\partial t} + \mathrm{div}\, \mathbf{S}_w = \mathbf{u} \cdot \nabla p_w - Q \tag{5}$$

where

$$E_w = \begin{cases} 2p_w & - \text{ Alfvén waves} \\ \dfrac{2}{(\gamma + 1)} p_w & - \text{ sound waves} \end{cases} \tag{6}$$

$$\mathbf{S}_w = \begin{cases} p_w \left(3\mathbf{u} + 2\mathbf{v} \right) & - \text{ Alfvén waves} \\ E_w \left(\mathbf{u} + \mathbf{c} \right) & - \text{ sound waves} \end{cases} \tag{7}$$

The form of the function Q depends on the dissipation processes involved but in general we may write

$$Q = f \left(\rho, \mathbf{u}, p, \nabla\rho, \mathrm{div}\, \mathbf{u}, \nabla p, \nabla^2 \mathbf{u}, \nabla^2 T \right). \tag{8}$$

For example in the case of heat conduction $Q = \mathrm{div}\,(\kappa \nabla T)$, whereas in radiative cooling $Q = -\rho^2 \Lambda(T)$ where $\Lambda(T)$ is some function of the temperature T. As we shall see subsequently the dissipation implied by assuming that the waves are saturated at some given level involves spatial gradients of the density, flow speed and pressure.

For our purposes it will be sufficient to study the stability of the system to compressional disturbances whose wavelength λ is very much less than the background plasma flow length scale L, and in as much as the waves are treated as contributing a single pressure, we require λ to be very much greater than λ_w, and so

$$\lambda_w \ll \lambda \ll L. \tag{9}$$

A uniform background plasma state given by $(\rho_o, p_{go}, p_{wo}, u_o, Q_o)$ is assumed and linearization of equations (2)-(4) for one-dimensional unsteady compressive perturbations $(\rho, p_g, p_w, u, \delta Q)$ yields

$$\frac{D\rho}{Dt} + \rho_o \frac{\partial u}{\partial x} = 0 \tag{10}$$

$$\rho_o \frac{Du}{Dt} = -\frac{\partial p_g}{\partial x} \tag{11}$$

$$\frac{Dp_g}{Dt} - c_o^2 \frac{D\rho}{Dt} = (\gamma - 1) \delta Q \tag{12}$$

where the perturbed dissipative function δQ for a fixed, saturated level of p_{wo} or E_{wo} is given by,

$$\delta Q = \begin{cases} -p_{wo} \dfrac{\partial}{\partial x} (3u + 2v) \ , \ v = -\dfrac{v_o \rho}{2\rho_o} & - \text{ Alfvén waves} \\[2mm] -E_{wo} \dfrac{\partial}{\partial x} (u + c) \ , \ c = \dfrac{c_o}{2} \left(\dfrac{p}{p_o} - \dfrac{\rho}{\rho_o} \right) & - \text{ sound waves} \end{cases} \tag{13}$$

It is straightforward to show that the spatial gradients of density appearing in δQ are destabilizing whereas pressure gradients are stabilizing. In the following sections we examine in detail the stability of a plasma dissipatively heated by sound and Alfvén waves.

SATURATED SOUND WAVES

Equations (10)-(13) can be manipulated to yield the dispersion relation;

$$v_p^2 = c_d^2 \frac{(v_p - v_o)}{(v_p - v_\infty)} \tag{14}$$

where

$$c_d^2 = c_o^2 + \frac{(\gamma - 1) E_{wo}}{\rho_o} \,, v_o = \frac{(\gamma - 1) E_{wo} c_o}{2 \rho_o c_d^2} \,, v_\infty = \frac{(\gamma - 1) E_{wo} c_o}{2 p_o}.$$

The graphical solution of this dispersion relation can be seen in figure 1(a) for $v_o < v_\infty$ and 1(b) for $v_o > v_\infty$. In figure 1(a), there is no possibility of an instability occuring. In figure 1(b) instability can arise if $(\gamma - 1)\left[c_o^2 + \gamma E_{wo}/\rho_o\right] < 0$ requiring $\gamma < 1$ which cannot be satisfied in the case of sound waves. Hence, for a plasma dissipatively heated by saturated sound waves, the pressure perturbation in δQ provides a stabilizing influence on the system and no instability can occur.

SATURATED ALFVÈN WAVES

Equations (10)-(12) can be used together with (13) to derive the dispersion relation for the Alfvén wave case :

$$v_p^2 - \left[c_o^2 + \frac{3 (\gamma - 1) p_{wo}}{\rho_o}\right] = -\frac{(\gamma - 1) p_{wo} v_o/\rho_o}{v_p}. \tag{15}$$

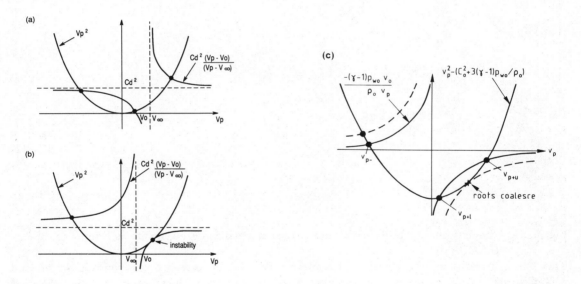

Fig. 1. Graphical solutions of the dispersion relation for the sound wave case in (a) and (b) and the Alfvén wave case in (c).

The graphical solution of this dispersion relation is shown in figure 1(c), where it can be seen that if v_o is sufficiently small, three roots result. The larger of the two positive roots v_{p+u} corresponds to a sound wave propagating at a speed less than $\sqrt{c_o^2 + 3 (\gamma - 1) p_{wo}/\rho_o}$; the smaller positive root, v_{p+l}, corresponds to an entropy wave which because of the dissipation now propagates forward relative to the flow, and it is the coalesence of these two forward propagating modes which leads to instability. This result is different to that obtained in the sound wave case essentially because the phase speed of the short wavelength waves does not contain the stabilizing pressure perturbation which is present in the sound wave case. This instability is anisotropic in nature and needs an imbalance of waves in one direction in order to occur. Generalizing the formulation in order to account for Alfvén waves propagating in both directions leads to the disappearance of the instability when the wave pressures in opposite directions are equal.

MARGINAL STABILITY CONDITION AND GROWTH RATES

The marginal stability condition and growth rates for this instability can be found analytically in the low β approximation and solved numerically for the full MHD dispersion relation. This analysis has been carried out in some detail in /2/, and the results will merely be summarized here.

The marginal stability condition in (β, β_w) space where $\beta = c_o^2/v_o^2$, $\beta_w = u_{wo}^2/v_o^2$ and $p_{wo}^2 = 1/2\rho_o u_{wo}^2$, is given for various propagation angles in figure 2(a), where we have considered perturbations propagating at an angle θ to the background direction. The regions below the curves correspond to instability with oblique propagation greatly extending the unstable region in the (β, β_w) plane. Figure 2(b) displays the normalized growth rate $M_i = \omega_i/kv_o$ as a function of wave normal angle θ for a range of values of β at $\beta_w = 1/2$, where β_w is the saturated relative Alfvén wave power. Results for a range of β_w values can be seen in /2/. The general effect of increasing β is to decrease the growth rate which shifts the maximum to more oblique angles. The fastest normalized growth rates are of the order of 0.15, which corresponds to growth times 7 times as long as the characteristic period of the saturated Alfvén waves.

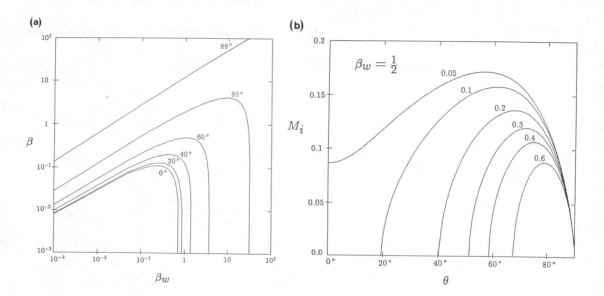

Fig. 2. (a) The marginal stability curve and (b) the normalized growth rate as a function of θ.

DISCUSSION

This paper investigates the stability of a plasma flow in the presence of saturated sound and Alfvén waves. Instability arises in the case of Alfvén waves but does not in the case of sound waves essentially because of the presence of a stabilizing pressure perturbation in the latter case. The numerical results reveal that the fastest growth times are about seven times longer than a characteristic Alfvén wave period and so in the supersonic region of the flow can be of the order of between seven seconds and seven hours (for Alfvén wave periods ranging from one second to one hour). Since these growth times are very much less than the few days transit time of the solar wind from the coronal base to the Earth's orbit this instability may be of significance in the large scale heating and dynamics of the solar wind expansion. This analysis with its wave saturation assumption does however need to be extended to include a more realistic dissipative function and an analysis using a ρu_w^3 type dissipation, corresponding to a Kolmogorov rate /3/, does indicate that such compressional instabilities indeed occur in more realistic models.

REFERENCES

1 J.F. McKenzie, and G.M. Webb, Magnetohydrodynamic plasma instability driven by Alfvén waves excited by cosmic rays, J. Plasma Phys. 31, 275-299 (1984)

2 M.K. Dougherty, and J.F. McKenzie, Compressional instability in the solar wind driven by wave dissipation, J. Geophys. Res. 96, 179-185 (1991)

3 K. Naidu, and J.F. McKenzie, Unpublished manuscript (1990)

DENSITY AND TEMPERATURE STRUCTURE IN A CORONAL HOLE

B. N. Dwivedi

Department of Applied Physics, Institute of Technology, Banaras Hindu University, Varanassi 221 005, India

ABSTRACT

The density and temperature structure in the quiet Sun and coronal holes have been discussed using a new calculation of several EUV diagnostic line intensity ratios. EUV emission lines from some of the selected ions of the boron (Mg VIII and Si X) and nitrogen (Mg VI, Si VIII, and S X) isoelectronic sequences have been considered. In this paper some new work on the determination of theoretical electron density has been presented using computed line intensities from several ions based on the Kopp and Orrall model. This work will be extremely useful in analyzing observational data from Solar Ultraviolet Measurements of Emitted Radiation (SUMER) and Coronal Diagnostic Spectrometer (CDS) instruments aboard SOHO satellite scheduled to be launched by ESA.

INTRODUCTION

The atomic physics of excitation of highly ionized species provides information on the electron density N_e and the electron temperature T_e structure of the source from the measurements of intensity ratios of certain selected line pairs. Density diagnostic techniques generally rely on identifying transitions in which one or both of the levels are metastable. The ratios of emissivities for two optically thin, allowed transitions excited from the same level is dependent on temperature when the difference between the energies of the two transitions is of the order of kT. A number of authors (e.g. Vernazza and Reeves' [1], Doschek *et al.*[2], Dwivedi [3] *etc.*) have attempted to determine the quiet Sun and coronal hole densities using EUV line ratios. However, scanty observations of EUV emission lines from coronal holes pose a serious problem to understand density and temperature or pressure structure of the emitting source. It is expected that the SOHO Mission will provide a wealth of information on EUV emission lines from various solar emitting regions such as quiet Sun, coronal holes, sunspots, prominences and active regions etc. The Solar Ultraviolet Measurements of Emitted Radiation (SUMER) and the Coronal Diagnostic Spectrometer (CDS) instruments on SOHO will respectively cover a wavelength range of 500 Å to 1600 Å and 150 Å to 800 Å. Keeping this in view, a number of density diagnostic line ratios have been studied. Theoretical line intensities have been estimated using the Kopp and Orrall [4] model for the quiet Sun and coronal holes. The ions selected for this study (Mg VI, Mg VIII, Si X, and S X) have maximum ionic concentrations at temperatures ranging from 4×10^5 to 1.6×10^6 K.

LINE EMISSION

The line emissivity (per unit volume, per unit time) for an optically thin spectral line is given by

$$\epsilon(\lambda_{ij}) = N_j \, A_{ji} \, hc/\lambda_{ij}$$

where A_{ji} is the spontaneous transition probability and N_j the number density of the upper level j parametrised as

$$N_j = \frac{N_j(X^{+p})}{N(X^{+p})} \; \frac{N(X^{+p})}{N(X)} \; \frac{N(X)}{N(H)} \; \frac{N(H)}{N_e} \; N_e$$

Here, $N(X^{+p})$ is the ionization ratio of the ion X^{+p} relative to the total number density of the element X and is primarily a function of temperature; $N(X)/N(H)$ is the element abundance which may or may not be constant in the solar atmosphere; $N(H)/N_e$ is the hydrogen abundance which is usually assumed to be about 0.8; N_e is the electron number density. $N_j(X^{+p})/N(X^{+p})$ is the population of level j relative to the total number density of the ion X^{+p} which is a function of electron density and temperature and is determined by solving the detailed balance equations for the ion. The physical processes and atomic data taken here are the same as used by Raju and Dwivedi /5/ for Mg VI, Dwivedi /6/ for Si VIII and S X and Dwivedi /7/ for Mg VIII and Si X.

The integrated line intensity at the Sun is given by

$$I\,(\lambda_{ij}) = \int \varepsilon(\lambda_{ij})\; dh \quad (cm) \quad erg\; cm^{-2} s^{-1} sr^{-1}$$
$$= 7.93 \times 10^9 \int \varepsilon^*(\lambda_{ij})\; N_e\; dh \quad (km)$$

$$\varepsilon^*(\lambda_{ij}) = 1.59 \times 10^{-8} \; \frac{A_{ji}}{\lambda_{ij}} \; \frac{N_j(X^{+p})}{N(X^{+p})} \; \frac{N(X^{+p})}{N(X)} \; \frac{N(X)}{N(H)}.$$

$N(X^{+p})/N(X)$ values have been taken from Jordan /8/ and $N(X)/N(H)$ from Kato /9/. Electron density N_e and temperature T_e as a function of height have been provided by Kopp /10/ in a tabular form.

RESULTS AND DISCUSSION

The SUMER instrument on SOHO will primarily determine the characteristics of emitted radiation with the aim of deriving plasma parameters of the solar atmosphere from atoms or ions. The CDS instrument will extend the wavelength range to shorter values, providing the opportunity to measure additional line pairs for density and temperature determines. These two instruments together will cover a wavelength range of 150 Å to 1600 Å. It is thus essential to determine appropriate lines or line pairs for plasma diagnostics.

In Figures 1 - 4, we have shown some of the density diagnostic line ratios from Mg VI, Si VIII, and S X ions. The values of temperature indicated in these Figures are those at which the respective elements have maximum ionic concentrations. These line ratios are insensitive to temperature variation. We measure the electron density of the solar emitting regions from these theoretical line ratio curves using the observed line intensity ratios. However, we do not have observations of these EUV emission lines from coronal holes. In order to have an idea of these line ratios for density measurement from coronal holes, we have estimated line intensities using a model atmosphere of Kopp and Orrall /4/. The model dependent line intensity ratio values are holes. The theoretical electron densities thus derived are listed in Table 1. Mg VIII and Si VIII have maximum ionic concentration at 8×10^5 K whereas S X and Si X at 1.6×10^6 K. Dwivedi /3/ has reported the Mg VIII $\lambda 430/\lambda 436$ line ratio as a good density monitor from the quiet Sun and coronal holes. Many other line ratios from Mg VIII are discussed by Dwivedi /11/ for density diagnostics. Similarly Si X $\lambda258/\lambda272$, $\lambda347/\lambda356$ and $\lambda638/\lambda653$ line ratios are good candidates for density determinations from the quiet Sun and coronal holes.

Using the ionization equilibrium calculations of Jordan /8/, we find that Si VIII and Mg VIII curves overlap on each other, thereby giving the physical insight that EUV emission lines from these ions emanate from the same source. Similar is the case for S X and Si X. Assuming the maximum ionic temperature as the temperature of emitting source (isothermal structure) and studying many more ions, we can study the density and temperature structure of the emitting regions. In Table 2 we give the average electron densities at temperatures 4×10^5 K, 8×10^5 K and 1.6×10^6 K obtained from the EUV emission lines for the quiet Sun and coronal holes.

In conclusion, solar EUV observation at higher spectral and spatial resolution from SOHO is awaited to study the density and temperature structure of various solar features; and also to test the present theoretical investigation.

Acknowledgements

I am very grateful to Dr. E. Marsch and the organizers of Solar Wind 7 for extending me a most generous financial support and to the CSIR, New Delhi for partial travel support. Helpful comments by Dr. K. Dere are gratefully acknowledged.

REFERENCES

1.   Vernazza, J.E. and Reeves, E.M.: 1978, Astrophys. J. Suppl. 37, 485.

2.   Doschek, G.A., Feldman, U., Bhatia, A.K., and Mason, H.E.: 1978, Astrophys. J. 226, 1129.

3.   Dwivedi, B.N.: 1988, Solar Phys. Lett. 116, 405.

4.   Kopp R.A.: and Orrall, F.Q.: 1976, Astron. Astrophys. 53, 363.

5.   Raju, P.K. and Dwivedi, B.N.: 1990, Astrophys. Space Sci. 173, 13.

6.   Dwivedi, B.N.: 1991, Solar Phys. 131, 49.

7.   Dwivedi, B.N.: 1992, Paper in preparation.

8.   Jordan, C.: 1969, Monthly Notices Roy. Astron. Soc. 142, 501.

9.   Kato. T.: 1976, Astrophys. J. Suppl. 30, 397.

10.   Kopp, R.A.: 1978, Private Communication.

11.   Dwivedi, B.N.: 1989, Solar Phys. Lett. 124, 185.

Table 1: Computed line intensity ratios and inferred theoretical electron densities using the Kopp and Orrall model.

Line Pair	Quiet Sun		Coronal Hole	
	Intensity Ratio	N_e (cm^{-3})	Intensity Ratio	N_e (cm^{-3})
Mg VI T = 4×10^5 K				
$\lambda349/\lambda399$	2.50	2.1×10^9	2.14	1.0×10^9
$\lambda349/\lambda403$	0.85	2.0×10^9	0.71	8.3×10^8
$\lambda387/\lambda400$	0.19	2.2×10^9	0.17	1.2×10^9
$\lambda387/\lambda403$	0.13	2.9×10^9	0.11	1.2×10^9
Si VII T = 10^5 K				
$\lambda277/\lambda319$	0.39	9.6×10^8	0.32	3.8×10^8
$\lambda216/\lambda319$	0.36	9.3×10^8	0.29	4.0×10^8
$\lambda276/\lambda319$	0.18	9.6×10^8	0.12	4.5×10^8
$\lambda214/\lambda319$	0.15	9.6×10^8	0.09	4.1×10^8
$\lambda445/\lambda440$	19.0	8.9×10^8	13.4	4.3×10^8
$\lambda944/\lambda445$	1.96	1.0×10^9	1.29	4.0×10^8
S X T = 8×10^5 K				
$\lambda226/\lambda257$	0.46	5.4×10^8	0.30	2.5×10^8
$\lambda226/\lambda257$	0.15	5.6×10^8	0.09	2.4×10^8
$\lambda226/\lambda257$	0.01	5.6×10^8	.004	2.5×10^8
$\lambda226/\lambda257$	2.78	5.6×10^8	1.73	2.3×10^8

Table 2: Average electron densities derived from EUV line intensity ratios.

Ion	Temperature	Electron Density (cm^{-3})	
		Quiet Sun	Coronal Hole
Mg VI	4×10^5 K	2.3×10^8	1.1×10^8
Si VIII	8×10^5 K	9.5×10^8	4.1×10^8
Mg VIII	8×10^5 K	9.7×10^8	4.8×10^8
S X	1.6×10^6 K	5.6×10^8	2.4×10^8
Si X	1.6×10^6 K	5.1×10^8	1.8×10^8

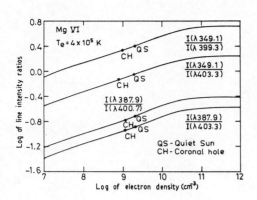

Figure 1: Mg VI intensity ratios

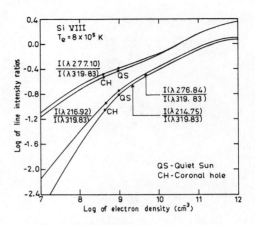

Figure 2: Si VIII intensity ratios

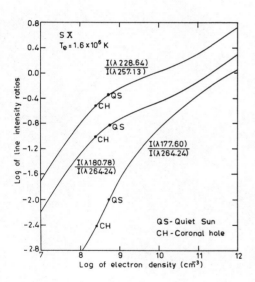

Figure 3: S X intensity ratios

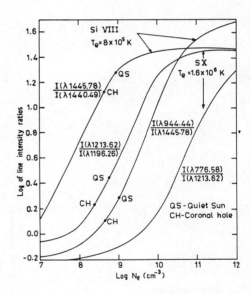

Figure 4:   Si VIII and S X
intensity ratios

INTERPRETING OBSERVATIONS IN THE SOLAR WIND ACCELERATION REGION

R. Esser

The Auroral Observatory, University of Tromsö, Tromsö, Norway

INTRODUCTION

Remote and in situ observations show that the solar wind is highly variable in space and time. Presently it is impossible to develop models that describe all these variations. Most of the models therefore concentrate on the steady solar wind which is assumed to expand along infinitesimal flow tubes. To gain insight into the expansion of the plasma from the models, it is necessary to compare the input parameters and the results of the models to observations.

Measurements particularly important for such comparisons are remote observations in the solar wind acceleration region of large coronal holes and quiet regions which are often observed to be steady over several solar rotations. The interpretation of such measurements is, however, not a straightforward task. To convert observed quantities into plasma parameters, assumptions about the measured plasma have to be made. The most severe of these assumptions is probably the distribution along the line-of-sight of the plasma contributing to the measured signal. The uncertainties in the derived plasma parameters should thus not only include uncertainties due to the instrument and data analysis, but also uncertainties due to the interpretation of the data. Theoretical studies should be designed to investigate the possible effects of varying the specifications of the contributing plasma on the measured quantities.

Difficulties in the comparison between models and observations arise also from the fact that the boundary conditions used in addition to the measurements in the acceleration region (e.g. mass flux and velocity at 1 AU, temperature and density at the coronal base), are usually not measured simultaneously for the same solar wind flow tube. The effect of not including proper uncertainties and treating the boundary conditions as known quantities, on the conclusions drawn from observations can be seen from the interpretation of the coronal hole measured by Munro and Jackson /1/. Assuming a fixed mass flux at 1 AU and a well defined density profile, the authors concluded that a significant energy addition to the flow is needed in the range 2 to 5 R_S to explain the observations. A study by Lallement et al. /2/, assuming different values for the mass flux at 1 AU and including the uncertainties in the density measurements, showed that the same observations could also be explained by the opposite conclusion, namely that no significant energy is needed below 5 R_S. This example not only demonstrates the importance of estimating "decent" error bars but also the close link between observations and the role models play in their interpretation. Many of the solar wind parameters given in the literature as observed quantities, and used as such by many authors are actually parameters derived in conjunction with solar wind models (e.g. /3/)!

In the present paper I want to give a few examples for the kind of problems that might be inherent in the conversion process from the observations to plasma parameters.

HYDROGEN LYMAN-ALPHA LINE PROFILE

Measurements of the resonantly scattered hydrogen Ly-α line (1216 Å) provide an important tool to derive plasma parameters in the inner solar wind acceleration region as a function of distance from the sun. This line was measured during three sounding rocket flights at distances 1.5 to 4 R_S (e.g. /4,5/). Observations of the Ly-α line are also planned from SPARTAN in 1993, and SOHO in 1995. The primary plasma parameter that can be derived from the width of the line is the kinetic temperature of the scattering particles, where the kinetic temperature is due to both thermal and non-thermal motions. A line profile measured at 1.5 R_S during a 1982 rocket flight is shown in figure 1 /6/. The line width is usually defined by fitting a Gaussian profile to the measured line (solid line in figure 1). In principal, profiles like the one in figure 1 can be fit by a range of Gaussians leading to a range of line widths and kinetic temperatures. The temperature range derived from the observations places limits on the proton temperature only under two conditions 1) the scattering neutral hydrogen atoms are closely coupled to the protons, and 2) the scattering particles have a Maxwellian velocity distribution.

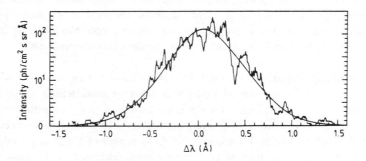

Fig. 1. Resonantly scattered HI Ly-α line profile measured during a 1980 rocket flight together with a Gaussian profile /6/.

By comparing solar wind expansion times with charge exchange times between protons and neutrals, Withbroe et al. /7/ showed that the coupling between the particle species may be questionable as close as 3 R_S from sun center for a low density coronal hole. Therefore the coupling should be carefully examined for each structure measured, especially for low density regions and at large distances from the sun.

The width of a spectral line reflects the temperature of the scattering particles only if the particles have a Maxwellian velocity distribution. If the particles are not Maxwell distributed the resulting line will be non-Gaussian. The reverse is, however, not necessarily true since there are several reasons why the line might be non-Gaussian. One example is shown in figure 2 where the Ly-α spectral line was calculated at 3.0 R_S. For the theoretical expression of the Ly-α line profile see e.g. Withbroe et al. /7/. The line-of-sight passes through a low density coronal hole which is 120° wide, and is surrounded by a high density quiet region. The density ratio between the coronal hole and the quiet region was chosen to be fairly extreme but still inside observed limits (see /8/ for details). The resulting profile is clearly non-Gaussian.

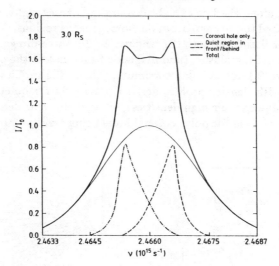

Fig. 2. HI Ly-α line profiles calculated for a coronal hole (thin solid line), a quiet region in front (dash-dotted line) and a quiet region behind (dashed line) the coronal hole, and when all three regions contribute along the line-of-sight (thick solid line). The normalisation factor I_0 is the intensity of the quiet regions at $\nu = 2466 \cdot 10^{15}$ s^{-1} /8/.

Also the presence of waves can give non-Gaussian profiles if the particles participate in the wave motion, depending on the distribution of the waves in the scattering region (e.g. /9/).

On the other hand there is a possibility that a deviation from a Gaussian, if not too profound, could go unnoticed due to the irregularities in the measured line profile (figure 1). In this case the derived temperature could deviate significantly from the 'true' temperature of the protons in the region.

In future observations from SOHO there will be a possibility of getting insight into the distribution of the plasma along the line-of-sight for the long lived coronal structures so that many of these uncertain factors arising from the line-of-sight integration could be removed. The distribution of the particles and waves in the scattering region will always pose some problems. Insight in their effects on the measured lines can to a certain degree be gained by theoretical studies.

CORONAL WHITE LIGHT MEASUREMENTS

The solar photospheric continuum radiation is scattered by coronal electrons and interplanetary dust. To calculate the polarisation brightness the intensity of the scattered light is measured through different polarisers. Assuming that the dust scattered component is unpolarised the electron density can be derived from the polarisation brightness. Two similar methods were developed by Van de Hulst /10/ and Saito et al. /11/ to derive the electron density, both assume an unstructured spherically symmetric corona. Few attempts have been made to model the electron density under the assumption that the corona is structured (e.g. /12/, and references therein). Habbal et al. /13/ therefore made some rough calculations for a given density distribution in a coronal hole. The coronal hole is assumed to have denser structures (plumes) in it like it is often seen in white light photographes. The density ratio between the regions is shown in figure 3a. The assumed geometry is shown in figure 3b. Each structure is 2° wide, and integrating along the line-of-sight the density changes from low to high every 2°. The polarisation brightness calculated from the formula by Van de Hulst /10/ is plotted in figure 3c (solid line). This polarisation brightness is compared to the polarisation brightness when no plumes are present in the coronal hole (dashed line), and when the whole integrated region has the higher density (dotted line). As one might expect, averaging along the line of sight favours the denser regions, it also changes the steepness of the curve compared to the curve from the homogeneous regions. This is also consistent with a recent study by Bagenal and Gibson /12/ who use a fully three-dimensional magnetostatic model to find the distribution of the plasma (and magnetic field) that fit the observed polarisation brightness at distances 1.4 to 2.4 R_S at solar maximum. The radial profiles that they derive at the equator are consistent with density profiles derived previously for quiet regions (which can be considered unstructured due to their high densities), while their polar density profiles are considerably steeper than those derived for polar coronal holes using traditional models assuming an unstructured corona.

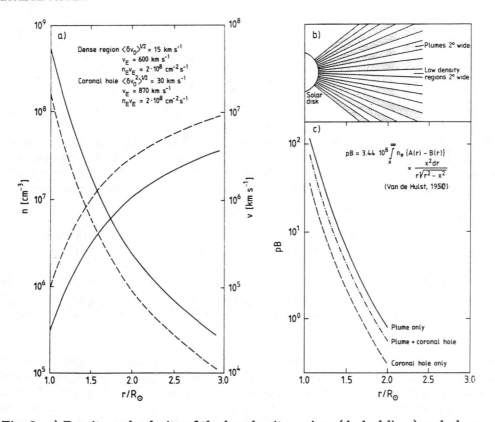

Fig. 3. a) Density and velocity of the low density regions (dashed lines) and plumes (solid lines), b) distribution of the low and high density regions along the line-of-sight, and c) polarisation brightness calculated when only the high density region is present along the line-of-sight (solid line), when only the low density region is present (dashed line), and when the distribution of high and low density regions is as in b) (dash-dotted line) /13/.

As can be seen from figure 3c and also from the paper by Bagenal and Gibson /12/ averaging along the line-of-sight can have a significant effect on the density profiles derived from the polarisation brightness, which can cause problems if the derived densities are used in comparison with models (see introduction).

INTERPLANETARY SCINTILLATION MEASUREMENTS

IPS observations can be used to derive the outflow speed of the solar wind at distances from the sun where other techniques are not presently available. The basic idea is that radio waves from a distant radio source are scattered by electron density fluctuations in the solar wind. The diffraction pattern produced by the scattered waves moves across the earth and the wave intensity is measured at two locations. The time lag is estimated by calculating the cross correlation function from the two intensity time series (e.g. /14/). The derived velocity is an average over the line-of-sight of the velocity component perpendicular to the line-of-sight.

To compare the velocities derived from the observations with velocities calculated from theoretical models it is not only necessary to gain insight into the effect of the line-of-sight integration on the derived velocity, but also to separate the variation of the flow speed due to changes in longitude/latitude and time from variations with radial distance. The few existing observations inside 10 R_S have not met this requirement. Outside of 10 R_S so far only one study has been published with the attempt to eliminate variations due to latitude/longitude /15/.

Information on the distribution of the plasma along the line-of-sight will not be available for the IPS technique, unless the IPS measurements are carried out in conjunction with white light measurements and UV spectral line measurements from SPARTAN or SOHO. Some insight into the effect of the line-of-sight integration on derived velocities could be gained by theoretical studies. For the integration through a homogeneous region the effect is expected to be of the order of a few percent /16/. The consequences assuming an inhomogeneous distribution of the contributing plasma has so far not been investigated in detail. Figure 4 shows one example how the presence of a well defined fast stream can effect the auto and cross correlation function /17/. The correlation functions were calculated for the EISCAT antennas in Northern Scandinavia assuming that the stream contributes 0, 30, and 50%. The velocity of the stream was twice as fast as the background solar wind. Note that the intersection points between the auto and cross correlation functions (usually used to estimate the time lag) do not change in this case. This is however a highly idealized case since the variation of the stream and background wind with distance from the coronal base was not taken into account. The effect of a more realistic integration along the line-of-sight might be that the cross correlation function appears skewed instead of having a well defined hump /16/.

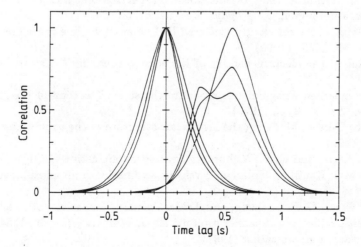

Fig. 4. Auto and cross correlation functions calculated for the EISCAT antenna system when a beam of plasma is present along the line-of-sight. The beam contributes 0, 30 and 50% and has twice the velocity of the background plasma /17/.

CONCLUSION

Observations of the solar wind acceleration region are necessary to improve our understanding of the solar wind expansion. The inversion of remote observations is, however, not a straightforward task, and requires assumptions on the measured plasma. While the line-of-sight assumption is inherent in all remote measurements, others are specific to the technique used such as the assumption about the coupling between protons and neutrals for the HI Ly-α observations, and on the dust scattered component for the electron scattered white light observations. Even though analysis and instrumental errors are often small, the examples given above show that these underlying assumptions might significantly increase the uncertainties in the observed parameters. In many cases theoretical studies will be necessary to gain insight into the consequences that follow from various assumptions. The plasma parameters derived from the observations are never better than the assumptions that go into their inversion. Therefore the resulting plasma parameters should, like the results of model calculations, always be taken with a grain of salt.

REFERENCES

1. R. H. Munro and B. V. Jackson, Physical properties of a polar coronal hole from 2 to 5 $R_\odot$, *Astrophys. J.* 213, 874 (1977)

2. R. Lallement, T. E. Holzer, and R. H. Munro, Solar wind expansion in a polar coronal hole: Inferences from coronal white light and interplanetary Lyman alpha observations, *J. Geophys. Res.* 91, 6751 (1986)

3. G. L. Withbroe, The temperature structure, mass, and energy flow in the corona and inner solar wind, *Astrophys. J.* 325, 442 (1988)

4. J. L. Kohl, H. Weiser, G. L. Withbroe, R. W. Noyes, W. H. Parkinson, E. M. Reeves, R. M. MacQueen, and R. H. Munro, Measurements of coronal kinetic temperatures from 1.5 to 3 solar radii, *Astrophys. J.* 241, L117 (1980)

5. G. L. Withbroe, J. L. Kohl, H. Weiser, and R. H. Munro, Coronal temperatures, heating, and energy flow in a polar region of the Sun at solar maximum, *Astrophys. J.* 297, 324 (1985)

6. L. Strahan, Jr., *Measurement of Outflow Velocities in the Solar Corona*, Ph. D. Thesis, Harvard University (1990)

7. G. L. Withbroe, J. L. Kohl, H. Weiser, and R. H. Munro, Probing the solar wind acceleration region using spectroscopic techniques, *Space Sci. Rev.* 33, 17 (1982).

8. R. Esser and G. L. Withbroe, Line-of-sight effects on spectroscopic measurements in the inner solar wind acceleration region, *J. Geophys. Res.* 94, 6886 (1989)

9. R. Esser, Broadening of the resonantly scattered HI Lyman-alpha line caused by Alfvén waves, *J. Geophys. Res.* 95, 10261 (1990)

10. H. C. Van de Hulst, The electron density of the solar corona, *Bull. Astron. Inst. Neth.* 11, 135 (1950).

11. K. Saito, A non-spherical axisymmetric model of the solar K corona of the minimum type, *Ann. Tokyo Obs., Ser. 2* 12, 53 (1970)

12. F. Bagenal and S. Gibson, Modeling the large-scale structure of the solar corona, *J. Geophys. Res.* 96, 17663 (1991)

13. S. R. Habbal, R. Esser, and G. L. Withbroe, private communication (1991)

14. W. A. Coles and J. J. Kaufman, Solar wind velocity estimation from multi-station IPS, *Radio Science* 13, 591 (1978)

15. W. A. Coles, R. Esser, U.-P. Løvhaug, and J. Markkanen, Comparison of solar wind velocity measurements with a theoretical acceleration model, *J. Geophys. Res.* 96, 13849 (1991)

16. W. A. Coles, private communication (1991)

17. W. A. Coles, unpublished manuscript (1990)

THE GENERATION OF PLASMA WAVES IN THE ACCELERATION REGION OF THE SOLAR WIND

H. Fichtner and H. -J. Fahr

*Institut für Astrophysik und Extraterrestrische Forschung der Universität Bonn,
Auf dem Hügel 71, D-5300 Bonn, Germany*

Abstract

The generation of plasma waves in the acceleration region of the solar wind is investigated. In the first part of the paper a new magnetohydrodynamical model of the solar wind expanding from coronal magnetic field configurations is constructed. The model takes into account especially the two-fluid character of the plasma by describing the thermal properties of each constituent with two temperature equations for the temperature parallel and perpendicular to the magnetic field. The detailed knowledge of the temperature profiles and in particular the temperature anisotropies allows for an analysis of plasma wave generation.

To describe this in detail, in the second part appropriate distribution functions (represented by polynomial expressions) for the main constituents of a plasma are introduced and new algebraic dispersion relations valid for wave propagation in the plasma parallel to an external magnetic field are derived from the theory of waves in a hot plasma. The solution of the dispersion relations is equivalent to the determination of the roots of a complex polynomial of degree 10 for a two-component (and of degree 14 for a three-component) plasma.

In the third part this formalism is applied to the case of the solar wind by use of the above mentioned two-fluid model. The status of the distribution functions reached in a specific step of a simultaneous integration of the solar wind equations is expressed in terms of rational (polynomial) expressions. The theory then enables the computation of the growth rates of the self-generated wave modes in the expanding solar wind. Finally, the results are compared with observations.

Theory

(a) The *solar wind* is described as an expanding two-fluid plasma that is guided by a prescribed coronal magnetic field $\vec{B}$, i.e. we make the assumption $\beta = \frac{2\mu_0 n k_b T}{B^2} < 1$, which is valid in the region $r \leq 50 r_\odot$. For an investigation of the self-generation of plasma waves due to inherent temperature anisotropies it is necessary to treat each plasma constituent as a thermally anisotropic species. This scenario is represented by the following set of differential equations (ds denotes the line element along the magnetic field lines, $\alpha = e, p$):

$$\frac{du}{ds} = \{-\frac{T_p^\perp + T_e^\perp}{B}\frac{dB}{ds} - \frac{h}{r^2}\frac{dr}{ds} - \frac{dT_p^\|}{ds} - \frac{dT_e^\|}{ds}\}/\{u - \frac{T_p^\| + T_e^\|}{u}\} \tag{1}$$

$$\frac{dT_\alpha^\|}{ds} = 3(\eta_\alpha^\| - 1)\{\frac{T_\alpha^\|}{3(\eta_\alpha^\| - 1)^2}\frac{d\eta_\alpha^\|}{ds} - \frac{T_\alpha^\|}{u}\frac{du}{ds} + H_{alpha}^\|\} \tag{2}$$

$$\frac{dT_\alpha^\perp}{ds} = \frac{3}{2}(\eta_\alpha^\perp - 1)\{\frac{2T_\alpha^\perp}{3(\eta_\alpha^\perp - 1)^2}\frac{d\eta_\alpha^\perp}{ds} + \frac{T_\alpha^\|}{B}\frac{dB}{ds} + H_\alpha^\perp\} \tag{3}$$

$$n = \frac{n_0 u_0 B}{B_0 u} \qquad h = \frac{G M_\odot m_p r_\odot}{k_b T_{ref}} \tag{4}$$

with the usual notation of the normalized quantities: $u(s)$ = solar wind velocity, $n(s)$ = number density, $T_{e,p}^{\|,\perp}(s)$ = electron and proton temperatures parallel and perpendicular to the magnetic field, $\eta_{e,p}^{\|,\perp}(s)$ = variable polytropic indices describing nonthermal heating and heat conduction (see

/1/,/2/), $H_{e,p}^{\|,\perp}(s)$ = energy exchange by collisions and wave particle interactions. T_{ref} is an arbitrary reference temperature.

An iterative solution procedure that is explained in detail in /3/ yields profiles of the velocity, density, temperature as well as of the nonthermal heating both for protons and electrons (Figure 1).

(b) The *particle distribution functions* (not only for reasons of mathematical tractability, see below) are represented by:

$$f_\alpha(v_\|, v_\perp) = \frac{N_\alpha}{\underbrace{1 + a_\alpha(\frac{v_\|}{v_{t,\alpha}})^2 + b_\alpha(\frac{v_\|}{v_{t,\alpha}})^4}_{polynomial\ part\ \|\vec{B}}} \underbrace{\frac{m_\alpha}{2\pi k_b T_\alpha^\perp} exp(-\frac{v_\perp^2}{2k_b T_\alpha^\perp/m_\alpha})}_{Maxwellian\ part\ \perp\vec{B}} \tag{5}$$

where N_α is a normalization constant, a_α, b_α are parameters determining the form of the polynomial distribution and $v_{t,\alpha}$ is the "thermal" velocity of the species α parallel to $\vec{B}$.

The distribution of the velocities parallel to the magnetic field (named "polynomial part", because it is determined completely by a polynomial in the denominator) takes account for the non-relaxed state of the distribution functions associated with the absence of collisions. This results in so-called "high energy tails" or "halos" especially of the velocity distributions parallel to the magnetic field, which are observed as a characteristic feature of the solar wind protons (e.g. /4/,/5/). If the resulting anisotropy of a distribution function is developed up to overcritical levels, various wave modes may be generated. These plasma waves propagate mainly along the magnetic field lines.

(c) For the *plasma waves* we have derived by use of the above mentioned distribution functions the following new algbraic dispersion relations from the general theory of waves in hot plasmas (for details see /6/):

$$1 - 4\pi k_b k_\| i \sum_\alpha \frac{\omega_{p,\alpha}^2}{\omega^2} \frac{T_\alpha^\perp N_\alpha}{m_\alpha b_\alpha^2} \sum_{j=2,4} s_{\alpha,j}^\mp - \frac{k_\|^2 c^2}{\omega^2} = 0 \qquad (transversal) \tag{6}$$

$$1 - 4\pi i \sum_\alpha \frac{\omega_{p,\alpha}^2}{\omega} \frac{N_\alpha}{b_\alpha^2} \sum_{j=2,4} t_{\alpha,j} = 0 \qquad (longitudinal) \tag{7}$$

with $k_\|$ = real ("parallel") wave number, ω = complex wave frequency, $\omega_{p,\alpha}$ = plasma frequency of plasma constituent α and $s_{\alpha,j}^\mp(\omega, k_\|, \Omega_\alpha, T_\alpha^{\|,\perp})$, $t_{\alpha,j}(\omega, k_\|, T_\alpha^{\|,\perp})$ being elements of the conductivity tensor.

The left hand sides of the dispersion relations can be transformed into polynomials with complex coefficients. The solutions then can be determined as the roots of these polynomials, which is a mathematically simple and with respect to the minimized numerical efforts an advantegeous procedure. For the transversal dispersion relation there results an equivalent polynomial of degree 10 for a two- and of degree 14 for a three-component plasma. For the former case it reads:

$$\omega^{10} + c_9\omega^9 + c_8\omega^8 + c_7\omega^7 + c_6\omega^6 + c_5\omega^5 + c_4\omega^4 + c_3\omega^3 + c_2\omega^2 + c_1\omega + c_0 = 0$$

$$\omega \neq v_{\alpha,j} k_\| \mp \Omega_\alpha \ ; \quad j \in \{1,2,3,4\} \tag{8}$$

with $v_{\alpha,j}$ being the zeros of the denominator of the polynomial part of (5), with $Im(v_{\alpha,2})$ and $Im(v_{\alpha,4})$ being negative. Explicit analytical forms of the coefficients c_0 through c_9 have been obtained with computer-algebraic methods and can be found in /6/.

Results

The solar wind solution given in Figure 1 has been computed with the coronal boundary condition of isotropic temperatures $T_p^\| = T_p^\perp = T_e^\| = T_e^\perp = 1.6 \cdot 10^6 K$ and a density $n = 6.1 \cdot 10^5 cm^{-3}$ at $r = 2r_\odot$. This corresponds to standard values of the plasma parameters in a coronal hole region. The influences of field geometry (Fig. 1a) and nonthermal heating (Fig. 1b), which itself is not independent of the geometry, typically result in high velocities near the axis and in low velocities at the fringes of a coronal hole (Fig. 1c). The velocity values as well as the corresponding density- and temperature-profiles (Fig. 1d,e) are in good agreement with observations (summaries of relevant observational data: /7/,/8/). The explicit profile of the nonthermal heating is nearly the same as that used by /9/.

These plasma parameters are taken as an input to the kinetic theory, and the application of the outlined theoretical procedure then yields the dispersion curves and damping or growth rates of the different wave modes. For the case of the coronal hole presented above we find only one self-generated wave mode (i.e. $Im(\omega) > 0$), namely the fast magnetosonic wave. Figure 2 gives the growth rates of this mode for the different positions within the coronal hole region indicated in the upper left part. From this we conclude:

The evolving temperature anisotropies in the acceleration region of the expanding solar wind generate plasma waves of short wavelengths ($\lambda < 1000 km$):

- due to the temperature anisotropy $T_p^\|/T_p^\perp > 1$ and the " high energy tail" of the proton distribution functions the *fast magnetosonic mode* is generated. All other modes are damped;
- the maximum growth rates occur near the axis of a coronal hole and increase with increasing heliocentric distances;
- with increasing heliocentric distance the maximum is shifted to lower wave numbers, i.e. higher wavelengths.

If there exists a selective heating of the plasma perpendicular to the magnetic field as suggested by e.g. /10/ to explain the characteristic anisotropy $T_p^\perp/T_p^\| > 1$ of the high speed wind streams, different wave modes can be generated. From a similar calculation with a correspondingly altered nonthermal heating (via prescribing appropriate variations of the polytropic indices) we find that ion-cyclotron waves are generated in high speed streams due to the $T_p^\perp/T_p^\| > 1$ anisotropy. For further investigations of different solar wind types as well as an analysis of measured distribution functions see /6/.

References

1. Fichtner, H., Fahr, H.J. [1989]: Planet. Space Sci. **37**, 987-999

2. Fahr, H.J., Fichtner, H. [1991]: Space Sci. Rev. **58**

3. Fichtner, H., Fahr, H.J. [1991]: Astron. Astrophys. **241**, 187-196

4. Marsch, E., Mühlhäuser, K.H., Schwenn, R., Rosenbauer, H., Pilipp, W., Neubauer, F.M. [1982]: J. Geophys. Res. **87**, 52-72

5. Marsch, E. [1984]: in "Plasma astrophysics", esa-sp-207, 33-40

6. Fichtner, H. [1991]: Ph.D. thesis, University of Bonn

7. Withbroe, G.L. [1988]: Astrophys. J. **325**, 442-467

8. Bird, M.K., Edenhofer, P. [1990]: in "Physics of the inner heliosphere", eds. R. Schwenn, E. Marsch, Springer, Berlin-Heidelberg-New York, 13-97

9. Leer, E., Axford, W.I. [1972]: Sol. Phys. **23**, 238-250

10. Schwartz, S.J., Feldman, W.C., Gary, P.S. [1978]: J. Geophys. Res. **86**, 541-546

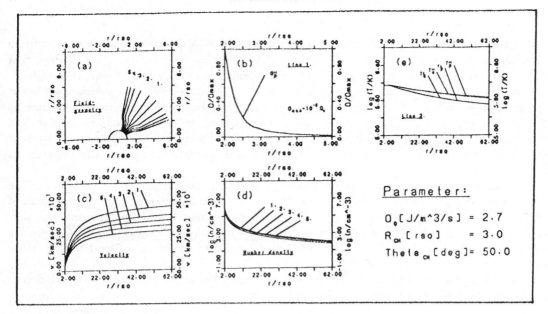

Figure 1: With a prescribed magnetic field geometry (a) and variable polytropic indices the integration of equs. (1) - (4) yields the explicit form of the nouthermal heating (b) and the profiles of the solar wind plasma parameters: (c) bulk velocity, (d) number density and (e) temperatures parallel and perpendicular to the magnetic field. The numbers indicate the corresponding fieldline as labeled in (a).

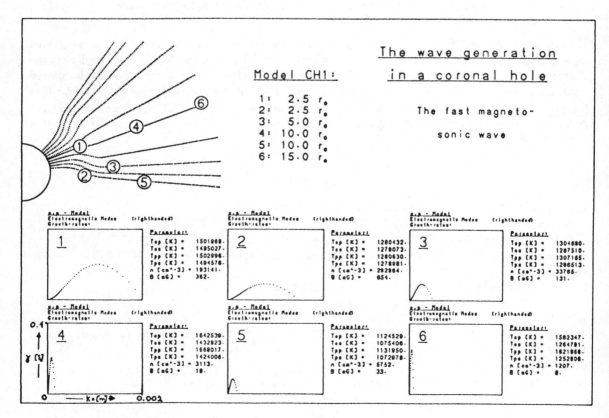

Figure 2: Shown are the variations of the growth rates of the fast magnetosonic mode in a coronal hole region. The plasma parameters for the different positions sketched in the upper left part of the figure have been calculated with the two-fluid solar wind model illustrated in Figure 1.

VARIABLE EUV EMISSION IN THE QUIET SUN AND CORONAL HEATING

S. R. Habbal

Harvard-Smithsonian Center for Astrophysics, Cambridge, MA 02138, U.S.A.

ABSTRACT

We review the characteristics of the variable emission from the small scale structure in the quiet Sun, in view of the recent theoretical proposals that microflares are responsible for the heating of the corona and the solar wind (/1,2/). We consider the observational properties of the variable emission in quiet regions and coronal holes. Our results will be based primarily on simultaneous multiwavelength EUV observations, supplemented by combinations of simultaneous cm radio, He I 10830 Å, x-ray and line of sight photospheric magnetic field measurements. We show that the variable emission from the small scale structure has surprisingly well-defined properties. Yet, within the limit of the temporal and spatial resolution of data currently available, the radiative losses, from this component of the emission, are a factor of ten smaller than that in the quiet Sun. Hence, the theory of microflare heating cannot rely on these observations for support.

INTRODUCTION

Within the past decade or so, it has become clear that solar activity occurs not only in large scale magnetic field regions associated with sunspots, but also in the more prevalent small scale structures that abound on the solar surface (/3/). It has been recently suggested that the continuous activity in the photospheric magnetic field (/4/) produces microflares which are the source of the coronal heating (/1,2,5/). In this presentation, we explore the characteristics of the variable emission from the small scale structure in the quiet Sun which we define as that part of the solar surface which includes quiet regions and coronal holes, but excludes large scale active regions. The small scale structures considered are typically less than 60" in spatial extent. Our goal is to determine the observational characteristics of this emission and their implication for the coronal heating process.

OBSERVATIONAL CHARACTERISTICS OF THE VARIABLE EMISSION

We consider primarily the data acquired by the EUV experiment on *Skylab* (/6/). These observations were made simultaneously, through the same instrumental slit, in six different wavelengths typical of emission spanning the temperature range from the chromosphere to the corona. They are the most comprehensive data available at present. Because of the scarcity and low spatial resolution of simultaneous magnetic field measurements taken with these data, they will be supplemented by multiwavelength observations including line of sight observations of the photospheric magnetic field with simultaneous 20 cm radio and He I 10830 Å, and x-ray observations. Typical time resolutions in all these data are a few minutes, and the spatial resolution ranges from 3" to 5".

To study the characteristics of the small scale closed magnetic structures, we consider two regions with differing overlying large scale field: a quiet region (Fig. 1) and a coronal hole, both at disk center. An example of coronal bright points as they appear in the EUV/*Skylab* data in the coronal line of Mg X and the lower transition region line of C III is shown in Fig. 1. The bright points

41

S. R. Habbal

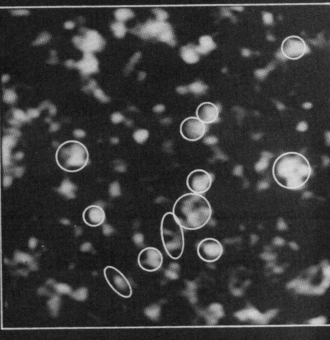

Fig. 1. EUV spectroheliograms, taken on 11 November 1973 at 1:52 UT, with the *Skylab*/EUV spectrometer, of a 5′ × 5′ area in a quiet region at Sun center in the lines of Mg X (625 Å) at 1.4 10⁶ K, and C III (977 Å) at 9 10⁴ K. The spatial resolution is 5″. There is a small active region visible in Mg X to the right of the field of view. The bright points are circled, and some are labelled for reference purposes. (From [1]).

are circled and some are labelled for reference purpose. The preferential location of bright points in the network boundary is clearly visible in the C III emission.

One of the distinguishing characteristics of bright points is their variable emission (/3,7/). Recently, Habbal and Grace (/8/) have shown that, in fact, all the variable emission throughout the quiet Sun is localized in coronal bright points. An example of the variability of the emission in bright points is shown in Fig. 2, in the Mg X and C III lines, for the bright points labelled A_1 and A_2 in Fig. 1. The frames are 5.5 min apart. The variability of the emission is random in nature, ranging anywhere from a factor of two to over an order of magnitude change in the intensity within a few minutes, with no obvious correlation between the temporal variability at the different temperatures of emission within the structure of the bright point (see /3,7/). Despite some claims that the variability of the emission within bright points is associated with explosive events (/5/), which are jets of chromospheric material, with average velocities of 100 km s^{-1}, best seen between 2×10^4 and 2×10^5 K (see, e.g. /9/), it has been recently shown by Porter and Dere (/10/) that the correspondence is not well-founded. (For more details on the nature of explosive events, see the article by K. Dere in these proceedings).

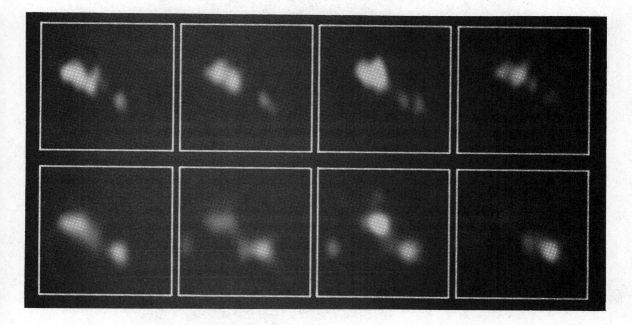

Fig. 2. A time sequence showing bright points A1 and A2 in the quiet region of Fig. 1. The upper panels are in Mg X, the lower ones in C III. The scale is magnified over that shown in Fig. 1. The frames are 40" × 40", and are 5.5 min apart, starting at 1:47 UT. (From /7/).

Habbal and Harvey (/11/) have also shown that bright points are associated with magnetic activity in the boundaries of the network, more readily with regions of magnetic flux cancellation than emerging bipoles (see Fig. 3). The most intriguing finding, however, is that despite the continuous motion of magnetic bipoles and their predominant activity in the network boundary, only a few bipoles give rise to observable emission. These results were also confirmed by simultaneous x-ray and line-of-sight photospheric magnetic field observations (/12/).

Since remnants of structure in the outer solar wind have been reported (/13/), we also consider the small scale structure observed in the extended corona, at the limb, most prominently in polar coronal holes (Fig. 4). The two outstanding structures seen in Fig. 4 are macrospicules (as outlined by the black contours of C III emission) visible in cool lines below 5×10^5 K, and plumes (defined by the white Mg X contours) visible above 10^6 K. These structures are not cospatial. While plumes are distinguishable in white light coronagraph pictures out to several solar radii (/14/), macrospicules reach a maximum height of 60" (/15,16,17/). There is strong evidence that bright points are rooted at the base of plumes (see, e.g., the concentrated Mg X emission at the base of the right-most plume in Fig. 4). Reports of pulse-like propagation in plumes have been made by Withbroe (/18/), with no correlation, however, with the temporal variability of the emission from the underlying bright points. The variability in macrospicules differs observationally from that in plumes or bright points. Throughout their short lifetime (typically 30-45 min), macrospicules undergo structural changes, with evidence of pinching off at their tip, possibly leading to the escape of some plasma (/16,17/), as is clear in the time series of Fig. 4.

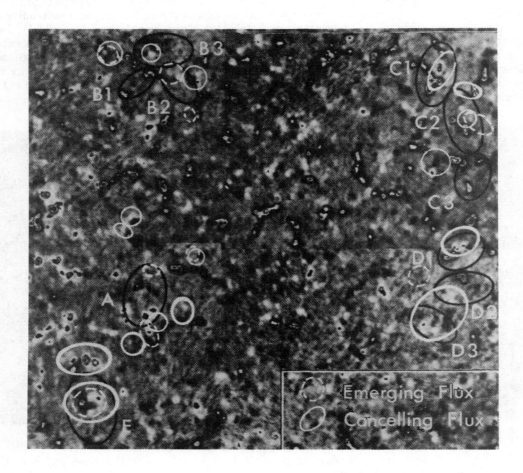

Fig. 3. A 563" × 444" magnetogram (courtesy F. Tang and S. Martin, Big Bear Solar Observatory) acquired over 20 minutes on 1985 September 8, starting at 17:40 UT. The polarity of the magnetic fields is identified by the lowest black (negative) and white (positive) flux levels. Within a given polarity region, the colors reverse whenever the flux levels double. The 20 cm radio sources observed are drawn in black. The white dashed contours encompass areas of emerging flux, while the solid white contours delineate areas of cancelling flux. Although there are many other sites of magnetic flux cancellation and emergence, only those in the vicinity of the radio sources have been indicated. (From /11/).

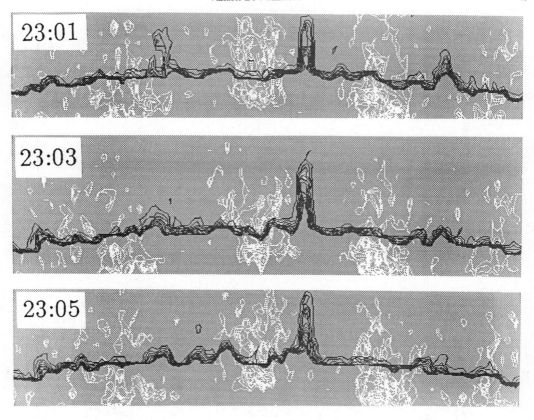

Fig. 4. A six minute time sequence (from 23:01 to 23:06 UT) of solar limb observations in a polar coronal hole made on 11 December 1973, with the *Skylab*/EUV spectrometer. The white contours correspond to the Mg X coronal emission in which polar plumes are visible. The black contours trace out the C III emission from macrospicules and the limb. Note the absence of cospatiality between these two structures. The spatial scale is the same in both directions. The extent of the field of view in the horizontal direction is 5 arcmin.

ENERGY BUDGET OF THE VARIABLE EMISSION

We discuss first the contribution of the variable EUV emission, from the small scale structures on the disk, to the observed energy budget of the quiet Sun. Given that the variable emission is discretely localized in bright points in the quiet Sun (/8/), we measure the spatial density of, or fraction of the solar surface covered by, this component of the emission, as a function of temperature. The results, shown in Fig. 5, were computed for three different regions: a coronal hole at Sun center, a coronal hole at the limb, and a quiet region at Sun center (/8/). We also show the spatial density of the enhanced emission or bright points (top panel) in which the variable emission is embedded. (The difference in the spatial density of the enhanced emission between the coronal hole at the limb and the disk could be due to line of sight effects. This difference does not appear in the variable emission, because the line of sight effects are eliminated in the technique used to deduce the variable emission. See /8/ for details). We note that the fraction of the surface covered by variable emission is a function of temperature, with a marked maximum around 5×10^4 K. The average spatial density over this temperature range is 0.1 in the three regions. This spatial density should be taken into account when computing the contribution of this variable emission to the overall background emission.

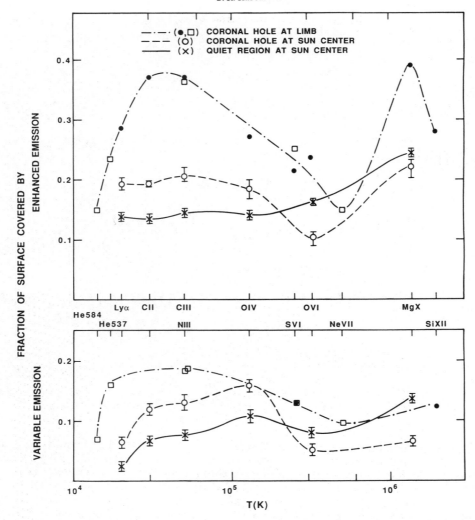

Fig. 5. Plots of the fraction of the solar surface covered by enhanced emission or bright points (top panel) and variable emission (lower panel) for three different regions on the Sun: a coronal hole at the limb, a coronal hole at sun center, and a quiet region at sun center (see legend for corresponding symbols). For the coronal hole at the limb the two sets of data points correspond to two different grating positions in the spectrometer, thus providing information at the standard (Ly α, C II, C III, O IV, O VI and Mg X) or the non standard (He 584 Å, He 537 Å, N III, S VI, Ne VII and Si XII) grating position. The other two regions at Sun center were observed in the standard grating position. For the coronal hole at the limb, the size of the data points corresponds to the empirical spread of the data; this spread is indicated by the vertical lines at each data point in the other two regions. (From /8/).

The contribution, δR, of the variable EUV emission to the coronal heating budget can be inferred directly from the data. We use the technique developed by Withbroe (/19/), where the intensity of the emission from different temperature lines can be combined to deduce the total radiative losses, whether it be in the background quiet Sun or the variable structures. This is the same technique used to obtain the commonly quoted estimates of the energy budget in different regions of the Sun, as given, for example, by Withbroe and Noyes (/20/). Taking into account the respective spatial densities of the variable and background emission, we find that δR in coronal holes is 1.2×10^4 erg cm^{-2} s^{-1}, as compared to the background emission of 3.2×10^5 erg cm^{-2} s^{-1}. In the quiet region, $\delta R = 5 \times 10^4$ erg cm^{-2} s^{-1}, while the background is 6×10^5 erg cm^{-2} s^{-1}. These values yield a ratio of the radiative losses in the variable to the background emission which is 0.04 and 0.08 in the coronal hole and quiet region respectively. Within the structure of the bright point, however, the ratio of the radiative losses in the variable to the total enhanced emission is the same in these two regions and equals 0.27.

We estimate next the kinetic energy content of the plasma blobs escaping from the macrospicules. Using the data shown in Fig. 4, we derive an average upward velocity of 140 km s^{-1} from the time it takes the blob to disappear from the field of view (compare for example the frames at 23:01 and 23:03 UT). Assuming a spherical shape with an observed diameter of 5", and a gas density of 10^{10} cm^{-3} (see /17/), the kinetic energy per blob per macrospicule is 7×10^{25} erg. With a spatial density of 0.04 for macrospicules in a coronal hole, and if the whole process of dissociation takes 8 min, then the energy density delivered to the solar wind environment is 2×10^4 erg cm^{-2} s^{-1}.

Data on the variability observed in polar plumes is very sparse. Computational models are under way to estimate the magnitude of the forces that could produce the observed pulse-like phenomena propagating along the plume structure, and their effect on the solar wind flow.

CONCLUSIONS

The characteristics of the variable emission in the small scale structures observed on the disk in the quiet Sun are well-defined and are the same in coronal holes and quiet regions. Hence, they are independent of the topology of the overlying large scale magnetic field. The variable emission in the small scale structure observed on the disk of the quiet Sun, is localized in coronal bright points, in the boundaries of supergranular cells. The variable emission is discretely localized on the solar surface, with a spatial density that is temperature dependent, averaging about 10% of the quiet Sun area. This variability is associated with the activity in the underlying photospheric bipolar magnetic regions. However, magnetic activity does not always produce observable emission, in the visible, EUV or cm wavelengths.

Structures observed in the extended corona (predominantly in large polar coronal holes because of the reduction of the contribution from possible denser regions along the line of sight) are in the form of plumes and macrospicules. These structures are not cospatial and have two distinct temperature ranges in which they are observed. Estimates of energy input to the solar wind from the dissociation of plasma blobs from macrospicules is 20% the estimates of energy input to coronal holes. Despite the observational evidence that plumes are rooted in bright points at the coronal base, the impact of the observed variable emission close to the coronal base on the solar wind flow has not been made.

It is plausible, on theoretical grounds, that the variable emission is a signature of microflares going off in the small scale structure. Estimates of the energy output in the form of radiation from the variable component is at least a factor of 10 lower than the output from the quiet Sun as a whole. Within the limits of the spatial (a few arcsec) and temporal (a few minutes) resolution, the data available at present indicates that the variability cannot account for the energy budget of the corona.

ACKNOWLEDGMENTS

I would like to thank Martina Arndt for her help in the production of Fig. 4. Funding for this work was provided in part by NASA grant NAGW-249 and Air Force grant AFOSR-91-0244.

REFERENCES

1. E. N. Parker, *Astrophys. J.* 330, 474 (1988)

2. E. N. Parker, *Astrophys. J.* 372, 719 (1991)

3. S. R. Habbal, and G. L. Withbroe, *Solar Phys.* 69, 77 (1981)

4. S. F. Martin, *Solar Phys.* 117, 243 (1988)

5. J. G. Porter, R. L. Moore, E. J. Reichmann, O. Engvold, and K. L. Harvey, *Astrophys. J.* 323, 380 (1987)

6. E. M. Reeves, M. C. E. Huber, and J. G. Timothy, *Applied Optics* 16, 837 (1977)

7. S. R. Habbal, J. F. Dowdy, Jr., and G. L. Withbroe, *Astrophys. J.* 352, 333 (1990)

8. S. R. Habbal, and E. Grace, *Astrophys. J.* 382, December (1991)

9. K. P. Dere, J.-D. F. Bartoe, and G. E. Brueckner, *Solar Phys.* 123, 41 (1989)

10. J. G. Porter, and K. P. Dere, *Astrophys. J.* 370, 775, 1991

11. S. R. Habbal, and K. L. Harvey, *Astrophys. J.* 326, 988 (1988)

12. L. Golub, K. L. Harvey, M. Herant, and D. F. Webb, *Solar Phys.* 124, 211 (1989)

13. K. M. Thieme, E. Marsch, and R. Schwenn, *Ann. Geophysicae* 8, 713 (1990)

14. S. Koutchmy, *Solar Phys.* 51, 399 (1977)

15. J. D. Bohlin, S. N. Vogel, J. D. Purcell, N. R. Sheeley, Jr., R. Tousey, and M. E. VanHoosier, *Astrophys. J. Lett.* 197, L133 (1975)

16. G. L. Withbroe, D. T. Jaffe, P. V. Foukal, M. C. E. Huber, R. W. Noyes, E. M.Reeves, E. J. Schmahl, J. G. Timothy, and J. E. Vernazza, *Astrophys. J.* 203, 528 (1976)

17. S. R. Habbal, and R. D. Gonzalez, *Astrophys. J. Lett.* 326, L25 (1991)

18. G. L. Withbroe, *Solar Phys.* 89, 77 (1983)

19. G. L. Withbroe, *Solar Phys.* 45, 301 (1975)

20. G. L. Withbroe, and R. W. Noyes, *Ann. Rev. Astron. Astrophys.* 15, 363 (1977)

A NEW INTERPRETATION OF THE RED-SHIFT OBSERVED IN OPTICALLY THIN TRANSITION REGION LINES

V. Hansteen

Institute of Theoretical Astrophysics, University of Oslo, PB 1029 Blindern , 0315 Oslo 3, Norway

ABSTRACT

We propose that the pervasive red-shift observed in transition region spectral lines is caused by propagating acoustic (slow-mode) waves. The waves are assumed generated in the corona by the nano-flare (or other episodic) heating mechanism. Numerical hydrodynamic calculations show that the net line shift depends on the characteristic timescale for ionization of the radiating ion. This suggests methods of discriminating various coronal heating mechanisms from observations of transition region spectral lines.

INTRODUCTION

The solar chromosphere–corona transition region consisting of a $2 \times 10^4 K$ to $10^6 K$ plasma is highly structured, both spatially and temporally. Indeed intensity observations (e.g. /1/, /2/) indicate that the dominant portion of transition region radiation arises in areas of variable emission. These observations have been taken as evidence of episodic or nano-flare heating mechanisms. In the nano-flare mechanism as presented by Parker /3/, coronal heating is achieved by the dissipation of the many tangential discontinuities arising spontaneously in coronal loops as a consequence of random continuous motion of the footpoints of the field in the photospheric convection. Releasing a large amount of energy in the course of a few seconds will excite several MHD modes in the low β coronal plasma. The signature of the nano-flare heating mechanism in transition region lines will thus consist of the transition region response to these downwardly propagating waves. This signature will be modulated by the global heating and cooling properties of the loop if the characteristic time between reconnection events, or series of reconnection events, is on the order of the loop cooling time ($10^2 - 10^3 s$).

Analysis of spectral line profiles show that transition region lines display a net red-shift (e.g. /4/, /5/). The observed mean red-shift is typically $5 km/s$ for C IV $\lambda 1548$ formed at $10^5 K$ but varies considerably, the amplitude of the shift is reduced towards lower temperatures and chromospheric lines display little or zero net shift. There are few observations of higher temperature lines but there is evidence that the net red-shift is reduced or disappears in the upper transition region. Doschek et al./4/ report no substantial downflow in O V $\lambda 1218$ formed at $2.5 \times 10^5 K$ while, perhaps more significantly, Hassler et al./5/ report $0 \pm 4 km/s$ for Ne VIII $\lambda 770$ formed at $8 \times 10^5 K$. Center to limb variation of the red-shift seem to indicate that motions are primarily vertical. The only clear correlation between intensity, line width and shift is between intensity and line width (Dere et al./6/). This may be evidence of wave motion.

In this paper we propose that the net red-shifts may be explained as a consequence of nano-flare coronal heating. Propagating compressive waves will cause spectral line shifts and asymmetries in the 'direction' of wave propagation /7/,/8/. A numerical approach to resolving this issue is chosen; energy is released at random intervals into the model corona, the hydrodynamic and radiative response of the atmosphere is followed in a time-dependent calculation. A more detailed discussion of the numerical method and model loop is under preparation.

RESULTS AND DISCUSSION

The loop into which the nano-flares are introduced consists of a semi-circular geometry of length l and with constant cross section, containing a plasma of hydrogen, helium, carbon and oxygen at solar (photospheric) abundances. The evolution of the nano-flare is followed by solving the time-dependent equations of mass, momentum, and energy conservation to obtain the run of density (ρ), the population density (n_{ij}) of elements i and ionization state j, the velocity (u), and the temperature (T_e) along the loop. The ionization state and radiative loss rate are computed consistently with the hydrodynamic equations.

The nano-flare model requires that tangential discontinuities introduced by the random motions of the loop footpoints dissipate explosively, releasing $10^{23} - 10^{25} ergs$ within a few seconds in a small region. The model loop temperature structure is maintained by releasing energy in bursts within randomly located regions near the loop apex. This energy release produces acoustic waves (as well as Alfvén and fast -modes, not modeled here); β is assumed small enough to constrain acoustic perturbations to along the field. The Alfvén waves generated will cause line broadening, but have less impact on net-shifts. We assume that the energy of dissipation is — at the spatial scales we are considering — introduced as heat rather than as momentum.

The (slow-mode) acoustic waves generated by releasing energy in the course of a few seconds are of large amplitude ($\Delta p/p \approx 1$) and have periods on the order of $P_w \approx 10s$. In addition heat is spread throughout the corona by thermal conduction. Both these effects cause motions in the transition region as well as perturbing the temperature gradient and density, leading to variations in line shifts and intensities.

The large amplitude waves propagate towards the transition region where the rapidly decreasing sound speed causes partial reflection of the wave and thus displacement of the transition region. On a slightly longer timescale the net result of several nano-flares is to heat the corona, increasing the coronal pressure. This will also tend to drive the transition region into the chromosphere. At the same time the temperature gradient in the transition region increases in order to support the enhanced thermal conduction from the heated corona. A period of no flares will allow the corona to cool, causing the corona to condense onto the chromosphere — the transition region will in this case move upwards with steadily decreasing temperature gradient.

The intensity contribution for an optically thin transition region resonance line is proportional to the electron density squared (n_e^2), the ionic concentration (n_{ij}/n_i), and a temperature dependent function ($C_{lu} \propto exp(-h\nu_{ul}/kT_e)T_e^{-1/2}$). The emission profile is assumed to be thermally broadened, the resultant line profile will be further broadened by wave motion. The observed intensity results from a region of line formation (Δz) where the intensity contribution is significantly different from zero. The response of the optically thin transition region line profile and intensity will depend on how these factors correlate with the velocity in the presence of a compressive perturbation. Figures 1 and 2 show the run of these variables with time for the resonance lines of C IV λ1548 and O IV λ789.

The nano-flare heating mechanism leads necessarily to preferentially red-shifted emission lines in the transition region. The correlation between velocity and intensity is present but with a large amount of scatter, as in the observations. Coronal lines will behave differently as they are characterized by larger regions of line formation and primarily horizontal motions. According to the results shown in this paper there is a correlation between the characteristic time of achieving ionizational balance and observed red-shift. This correlation should extend also to lines with different excitational characteristic times, i.e. metastable transitions versus allowed transitions. Discriminating between coronal heating mechanisms and especially between episodic and steady-state models can thus be done by observing lines with different characteristic timescales.

REFERENCES

1. J.G. Porter, R.L. Moore, E.J. Reichmann, O. Engvold, & K.L. Harvey, Ap. J. 323, 380 (1987)

2. S. Habbal, in: Mechanisms of Chromospheric and Coronal Heating, Heidelberg, 5-8 June 1990, eds. P. Ulmschneider, E.R. Priest, and R. Rosner, p. 127.

3. E.N. Parker, Ap. J. 330, 474 (1988)

4. J.A. Doschek, U. Feldman, & J.D. Bohlin, Ap. J. 205, L177 (1976)

5. Hassler, D.M., Rottman, G.J., and Orrall, G.Q., Ap. J. 372, 710 (1991)

6. K.P. Dere, J.-D.F. Bartoe, & G.E. Brueckner, Ap. J. 281, 870 (1984)

7. G. Eriksen & P. Maltby, Ap. J. 148, 833 (1967)

8. P. Gouttebroze, Ap. J. 337, 536 (1989)

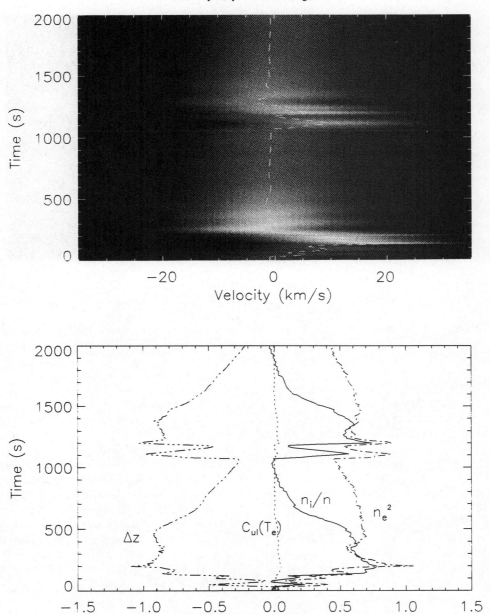

Fig. 1. Relationship between C IV $\lambda 1548$ intensity, line shift and elements of the contribution function $\propto n_e^2 (\frac{n_{ij}}{n_i}) C_{lu}$, the latter plotted on a logaritmic scale in the lower panel along with the region of line formation (Δz). The total line intensity $\int I_\nu d\nu$ is approximately given by the product of the contribution function and the region of line formation. In the displayed model a $200s$ period of several nano-flares is followed by a cooling period of $800s$, thereafter the cycle is repeated. The intensity shown is the total from the entire loop as seen from above. Note that C IV formed at $10^5 K$ is far from ionization equilibrium during most of this period, while O IV (Figure 2) formed at $1.4 \times 10^5 K$ is close to equilibrium at all times. Note also that the variation in temperature of line formation ($< 50\%$ for C IV, $< 20\%$ for O IV) matters little for both lines.

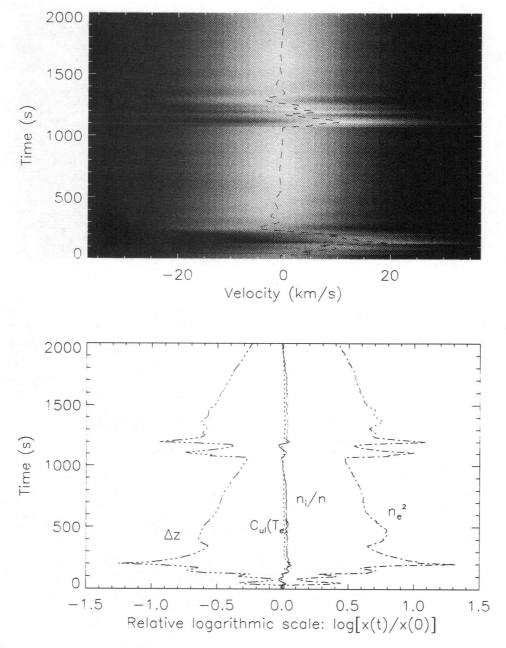

Fig. 2. Relationship between O IV $\lambda789$ intensity, line shift and elements of the contribution function $\propto n_e^2(\frac{n_{ij}}{n_i})C_{lu}$, the latter plotted on a logaritmic scale in the lower panel along with the region of line formation ($\Delta z \propto 1/\nabla T_e$). Variations in the density are compensated by variations in the region of line formation; as the transition region is pushed into the denser chromosphere by the rising coronal pressure, the enhanced conductive flux from the heated corona forces the temperature gradient to rise, reducing the region of line formation. Thus it is the difference in characteristic times for establishing ionization equilibrium that determines the behavior of the total intensity. See Figure 1 for comparison with C IV. O IV has a fairly complicated term structure and the calculation here should be taken more as indicative of the behavior of a line with short characteristic times than an accurate portrayal of O IV.

STATUS OF SOLAR WIND MODELING FROM THE TRANSITION REGION OUTWARDS

J. V. Hollweg

Physics Department and Institute for the Study of Earth, Oceans and Space, University of New Hampshire, Durham, NH 03824, U.S.A.

ABSTRACT

In recent years, solar wind modeling has to some extent undergone a shift of emphasis, from attempts to produce high-speed streams far from the sun, to investigations of what conditions must exist in the solar corona and transition region in order to produce the observed conditions in both the solar wind and low corona. Thus there has been an increased awareness that the solar wind should really be treated as part of the solar atmosphere, and that the problems associated with heating and accelerating the solar wind should be treated in concert with the coronal (and perhaps chromospheric) heating problems. We will discuss several models which take this point of view but we will place particular emphasis on some outstanding problems, viz. the mass flux problem, some puzzling recent IPS data, our persistent difficulties with electron heat conduction, and observational uncertainties about the coronal and transition region boundary conditions which should be put into the models. We will conclude by suggesting some possible alternatives for future models.

INTRODUCTION

In a classic paper, Hartle and Sturrock [1] pointed out three major problems with thermally-driven solar wind models: the predicted flow speed is much slower than is usually observed; the protons are predicted to be much colder than is observed; and the electrons are predicted to be hotter and to carry much more heat flux than is usually observed.

Following the landmark discovery of Alfvén waves in the solar wind by Belcher, Davis, and Smith [2,3], in the 1970's much emphasis was placed on using those waves to directly accelerate the wind and heat the protons; see [4,5,6] for reviews. Those models were partially successful, but there were still important uncertainties about the nature of the wave damping, and the correct description of the heat flux carried by the weakly-collisional electrons.

In the 1980's emphasis shifted from waves to MHD turbulence; see the excellent recent review by Marsch [7]. Although much progress has been made in understanding solar wind turbulence, both observationally and theoretically, there has been very little work done to incorporate turbulence into global models of the solar wind. Tu [8,9,10] has emphasized the evolution of power spectra in the wind, but his models start much further from the sun than traditional models (which start within one solar radius above the solar surface). The questions concerning the energetics of the high-speed solar wind streams have in fact received rather little attention in recent years.

Instead, there has been renewed interest in understanding how the solar wind is linked to the solar corona and underlying layers, such as the 'transition region' (TR) where the temperature rises from 10^4 in the chromosphere to several million Kelvins in the corona. This is a refreshing development, since it is an explicit attempt to treat the corona and solar wind as a unified physical phenomenon. This makes sense not only because the wind flows out of the corona, but also because both the corona and solar wind are undergoing heating (which is not yet fully explained in either region). Perhaps the heating of the corona and solar wind are due to the same physical process, in which case it would be foolish not to treat them together. Moreover, the corona and solar wind are closely coupled via an electron heat flux which can be inward or outward, and by the fact that the wind drains energy from the corona. This viewpoint is to be contrasted with 'traditional' solar wind models which simply assume an already heated corona as an inner boundary condition.

In the following sections we will review some recent models, discuss their shortcomings, and offer some alternatives for the future.

THREE MODELS

Including the corona and TR in solar wind models requires three modifications to the traditional models: 1. Radiative losses must be included. The radiation is taken to be optically thin. The radiation is collisionally produced, scaling as n_e^2 where n_e is electron concentration. 2. Coronal heating must be included. 3. A *downward* electron heat flux, q_e, from the hot corona must be included. This has been taken to be classical, scaling as $T_e^{5/2} \nabla_{||} T_e$, where T_e is electron temperature. As discussed by Shoub /11/, this formulation is difficult to justify since it leads to TR's which are thin compared to the intracollisional path of the heat-carrying electrons. On the other hand, Shoub's fully kinetic calculations of the heat flux /12/ lead to values which are not strongly different from the classical values. Moreover, the classical expression for q_e leads to a predicted behavior for the differential emission measure in the TR which agrees with observations at temperatures above about 10^5K, although the behavior below 10^5K is not well understood (see, for example, Antiochos and Noci /13/). On the other hand, the TR models assume a static TR, while in reality the TR is dynamic, so the correctness of the classical heat conduction formula is really very uncertain.

Withbroe /14/ produced a series of 1-fluid models which include the above features. The coronal heating is modelled by a postulated mechanical energy flux (erg cm^{-1}) varying as $\exp\{-(r-r_{sun})/H_m\}$. In addition he includes the wave pressure from WKB Alfvén waves, and a collisionless expression for q_e /15/ scaling as $p_e V$ beyond 5-10r_{sun} (p_e is electron pressure, and V is solar wind flow speed). Withbroe's work is distinguished by his forcing his models to satisfy a number of empirical constraints: 1. $n_e(r)$; 2. the base pressure in the TR (obtained from spectral lines), which is used to determine the downward heat flux in the TR; 3. V(r) (obtained from Doppler shifts, interplanetary scintillation (IPS) data, and *in situ* measurements); 4. nV (obtained from *in situ* and IPS data); 5. T(r) in the low corona (obtained from spectral lines and the freezing-in of charge states); 6. non-radial expansion; 7. the 'turbulent' velocities in the corona (obtained from Doppler broadening of spectral lines, and from IPS data), which are used to constrain the Alfvén wave power.

Withbroe considers models for several types of coronal regions. We will only discuss his model for an equatorial coronal hole, which leads to the high-speed streams observed by spacecraft near the ecliptic. His model has a mechanical flux density of 4.2×10^5 erg cm^{-2} s^{-1} at the sun; 25% of this is radiated away, and the rest goes into the wind. An additional 2×10^5 erg cm^{-2} s^{-1} is needed in the Alfvén waves to drive a high-speed stream. The value of H_m is about $0.6 r_{sun}$; this is an important example of how solar wind models help to constrain coronal heating mechanisms! The model departs from traditional models inside $r \approx 2 r_{sun}$; models without explicit coronal heating should not be compared with data inside this distance.

Withbroe finds that larger values of H_m lead to less downward q_e (because the heat is spread out over greater heights above the TR). This results in a lower base pressure. More rapid non-r^2 expansion has the same effect, and Withbroe suggests that the combination of large H_m and rapid expansion is the cause of the coronal holes. His models suggest that $0.1 < H_m < 0.5 r_{sun}$ leads to slow wind, while $0.5 < H_m < 1.0 r_{sun}$ combined with Alfvén waves leads to fast wind. Larger values of H_m lead to too low base pressures. Finally, comparing his models with empirical values for base pressure suggests that $H_m \propto p_{base}^{-1.5}$; models for coronal heating have to satisfy this constraint. (Paradoxically, the chromospheric network looks the same in quiet regions and coronal holes /16/, but the coronas in these regions are different. Withbroe's models suggest that the difference arises because the H_m's are different. Does this imply that the coronal heating is not a result of network activity, as argued by Parker /17/?}

Results generally similar to Withbroe's were obtained previously by Hammer /18,19/.

The second model which we will discuss is a 2-fluid model by Hollweg and Johnson /20/. Their goal was to explore the possibility that the corona and solar wind are heated by the same mechanism, viz. Alfvén waves dissipating at the rate given by a Kolmogorov turbulent cascade. In that case, the volumetric heating is $\rho <\delta V^2>^{3/2} L_c^{-1}$, where ρ is density, L_c is a correlation length, and δV refers to the Alfvén waves; uncertain numerical factors are absorbed into L_c. Hollweg and Johnson guessed that L_c scales as the distance between magnetic field lines. (They also took classical heat conduction close to the sun, and the collisionless prescription at greater distances.)

The results of their work are qualitatively similar to Withbroe's. The model succeeds in producing a corona and TR, and high-speed flows. Large dissipation scales, i.e. large L_c, produce fast flows and *too low* base pressures ($n_e T_e \approx 8 \times 10^{13}$ cgs); small dissipation scales produce higher base pressures but slow wind. But the worst result of their model was proton temperatures of $3-4 \times 10^6$K near $r \approx 3r_{sun}$. These temperatures are ruled out by the Lyman-α coronal scattering observations (e.g. /21/). The model *assumed* that the heat went into the *protons*, since that is what is observed in the more distant solar wind /10/. Isenberg /22/ found that putting some energy into helium does not eliminate the problem, and there is no good reason for assuming that the Alfvén wave energy ends up in the electrons. Apparently, the Kolmogorov prescription for coronal heating as employed by Hollweg and Johnson is not fully correct. Perhaps a more sophisticated treatment of the turbulence can save the day /23,24/. Perhaps another coronal heating mechanism is operative. (In any event, Hollweg and Johnson's assumption of WKB Alfvén waves is not consistent with fully-developed turbulence, since the nonlinear terms which drive the turbulence then cancel out. Departures from WKB close to the sun could allow turbulence, but a proper treatment is needed.)

The third model we will mention is from Wang and Sheeley /25/. Their 1-fluid model is very simple, but it is again motivated by a desire to tie what we observe in the wind to what is going on at the sun. They begin with observations /26/ showing that fast solar wind at 1AU is associated with coronal magnetic field lines which are slowly diverging. Conversely, rapid divergence leads to slow flows. Their model is designed to test the hypothesis that the energy source for the wind has a constant flux density (erg cm^{-2} s^{-1}) at the sun; then slower field line divergence implies that more square centimeters of the sun are able to supply the wind, and a faster wind results. Their model uses Alfvén waves to drive the wind, but the same idea could probably be applied to other energy sources. Their model yields a relationship between speed and expansion factor which fits the data.

However, as E. Leer remarked at this meeting, the flow speed at 1AU is also controlled to a large extent by the solar wind mass flux, which is a sensitive function of coronal temperature. By varying the temperature a little bit, one can get a whole range of flow speeds, and it is not clear whether Wang and Sheeley's model really proves their hypothesis since they took a fixed value for the coronal temperature (unless the coronal temperature is indeed very constant). This brings us to some major problems.

OUTSTANDING PROBLEMS

The recent attempts to unify coronal and solar wind modeling represent a welcome change, but all is still not well. We still do not understand the energetics of the high-speed streams, or of the hotter and faster heavy ions, but there seems to be some agreement that the Alfvén waves can fill the bill, with turbulence governing the wave evolution and dissipation. But these are old problems; we will instead concentrate on other problems more germane to the corona-wind coupling.

The mass flux problem. Close to the sun the solar wind is subsonic, and therefore close to hydrostatic equilibrium. The coronal density falls off approximately exponentially with a scale height proportional to temperature. Thus the density at the sonic point, and ultimately the mass flux in the wind, is a sensitive (exponential) function of temperature. Leer and Holzer /27/ have given a simple example in which a coronal temperature change from 10^6 to 2×10^6K leads to a hundredfold mass flux increase. The problem is that the solar wind mass flux is observed to vary only by a factor of 2-10 (depending on whom you ask; see /28/), even though the 'observed' coronal temperature variations should lead to much greater mass flux variations. For example, some of the Ulysses SWICS data reported at this meeting /e.g. 29/ shows ion freezing-in temperatures changing by a factor of 2 (or more!) with no dramatic changes in the overall solar wind properties at Ulysses!

Withbroe /30/ has presented a scheme which works in part by keeping the base temperature and density nearly constant. More recent work has explored the possible effects of helium in the low corona, and in particular the possibility that the helium abundance close to the sun might be much larger than the abundance at 1AU. Leer and Holzer /27/ considered n_α/n_p as large as 30% at the base, compared to the typical value of 4% observed at 1AU (n_α and n_p are helium and proton concentrations). The basic idea behind this approach is that the near exponential fall-off of $n_p(r)$ in the subsonic region is controlled by gravity, Coulomb friction with the helium, and the electrostatic force which exists in order to preserve quasineutrality; the latter two forces are affected by the helium, and the proton mass flux comes into play because the Coulomb friction depends on $n_p V_p$. Using a simple isothermal spherically symmetric model which omits coronal heating and wave forces, Leer and Holzer find that a large helium abundance at the base can decrease the temperature sensitivity of $n_p V_p$. For example, with $n_\alpha/n_p = 0.3$ at the base, the proton flux at 1AU increases only fifteenfold when the temperature is doubled from 10^6 to 2×10^6K, compared to a hundredfold increase in the absence of helium. But their model has a major shortcoming: it predicts a strong positive correlation between n_α/n_p and $n_p V_p$ at 1 AU, whereas the observed correlation is negative /31, and references therein/.

Bürgi /32/ has recently presented a much more sophisticated 3-fluid model to investigate how the helium controls the proton flux. His model includes non-r^2 expansion, classical heat conduction (downward and outward), and a coronal heating function similar to Withbroe's /14/. The heat is assumed to be deposited solely in the electrons, resulting in a strongly peaked T_e near $1.3r_{sun}$. But the electrons are thermally coupled to the protons and helium, and the temperatures of the latter peak also, but they are cooler than the electrons. Because the frictional coupling between helium and protons contains a $T^{-3/2}$ dependence, the helium tends to accumulate near the temperature maximum, and very large values of n_α/n_p can be obtained there, even for modest helium abundances at the base (which is at the top of the TR). The coronal electron temperature is varied by changing the energy flux which heats the corona, and Bürgi finds that $n_p V_p$ at 1AU has a much weaker dependence on T_e than in models without helium. However, the model contains the same defects as Leer and Holzer's: it leads to n_α/n_p at 1AU being a very sensitive function of coronal temperature, and it leads to a positive correlation between n_α/n_p and $n_p V_p$ at 1AU (the opposite of what is observed). {Bürgi's model has another defect: the omission of radiation. Radiation plays a crucial role in models such as Withbroe's /14/and Hollweg and Johnson's /20/. Withbroe shows that radiation leads to a direct relationship between base pressure and the downward heat flux. This feature is missing in Bürgi's model, and he instead simply picks the base density and temperature, and thus the pressure, as an inner boundary condition. He fixes n_e and T_e to be 10^8 cm^{-3} and 10^6K, and the response of base pressure to the varying downward heat flux is lost. But his model probably correctly represents the accumulation of helium at the temperature maximum, which was not contained in Leer and Holzer's work, and the consequent effects on the proton mass flux. As a further development of the model, it would be interesting to ask what happens if the coronal heating were to go into the protons and/or helium; this could affect the results considerably, since the Coulomb friction depends on the temperatures of those two species, and not on the electron temperature.}

On the whole, the mass flux problem has to be viewed in the context of two other unanswered questions: 1. How does helium get out of the corona? 2. Why is n_α/n_p negatively correlated with $n_p V_p$ at 1AU?

IPS velocities. Coles et al. /33/ have recently used IPS to measure the velocities of density irregularities, which are assumed to be approximately equal to the solar wind flow speed. They used simultaneous measurements of nearly radially-aligned radio sources, which gives an instantaneous measurement of the velocity gradient; the dataset is enhanced by making measurements of the same sources on several successive days. They find that some half of the velocity profiles V(r) fit Alfvén-wave-driven models reasonably well. The other half spell trouble. About half of the troublesome cases show slow flows ($\approx$ 100-150 km s^{-1}) near 0.1AU, with rapid acceleration beyond that distance; the other half show constant speed or even deceleration beyond about 0.1AU. Coles et al. believe they are not observing time-dependent effects. Unfortunately, no existing steady models of the solar wind reproduce these troublesome data.

Electron temperature profiles. There is no justification for using classical electron heat conduction in the solar wind, corona, or TR. Out of desperation, some models (e.g. /5,14,20/) use the classical prescription inside 5-10r_{sun} and the collisionless prescription /15/ at greater distances. This results in T_e(r) being fairly flat close to the sun, with a steeper decline further out. However, this behavior is the opposite of the behavior inferred by Marsch et al. /34/. They noted that the Helios data give a rather flat T_e profile in 0.3 < r < 1AU. If these profiles are extrapolated (as power laws) back to the sun, then the inferred coronal electron temperatures end up much too small, especially in the high-speed streams. So Marsch et al. infer that T_e(r) falls off steeply close to the sun, with a flatter fall-off further out. No existing models give this behavior. After two decades of puzzlement, we still do not know what to do with the electrons! But it is worth noting that if T_e(r) does fall off steeply close to the sun, then the electron pressure gradient contributes less to the wind's acceleration than in models with a flatter temperature profile. This is particularly disconcerting since Marsch et al. infer the steepest fall-off to occur in the high-speed streams. This loss of electron pressure has to be made up somehow, perhaps via hot ions or wave pressure. {Marsch et al. possibly overdid their extrapolation a bit, assuming T_e = 2x10^{6}K at r = 3r_{sun}, which is probably too hot there, but their general conclusion seems valid.}

Boundary conditions in the TR. Steady solar wind models with coronal heating /14,20,32/ have to reproduce the 'observed' base pressure (in the TR). However, the TR probably consists of a mix of open and closed magnetic flux tubes /35/. Solar wind models are really only concerned with the pressures on the open flux tubes. The problem is that the observations are unable to cleanly separate the open from the closed regions, and the quoted TR pressures probably represent a weighted average of the pressures in the open and closed regions, with the closed regions having the higher pressures. So the base pressures which have been used may not correctly represent the open regions. It is possible that the open regions have lower pressures than the values which have been used to constrain the models /36/. Thus the model base pressures which Hollweg and Johnson regarded as being "too low" might in fact be reasonable. On the other hand, the coronal hole data may not be heavily contaminated by closed flux tubes, since the holes can be strongly unipolar /37/. In any event, it is not clear how well we know the base pressures.

Another problem is that the TR is dynamic /17,38,39,40/, and it is not clear whether extending *steady* models down into the TR has any meaning at all.

Why do fluctuations have time scales of hours in the solar wind *and* corona? In situ observations of Alfvénic fluctuations in the solar wind show that long time scales, of the order of hours, contain most of the power /e.g. 3/. The corona too seems to be dominated by fluctuations at long time scales; faraday rotation fluctuations /41/ and phase-sensitive radio observations /42,43,44/ indicate that the magnetic field and density fluctuations in the corona have hour time scales, as close to the sun as one solar radius above the surface. The riddle is: what on the sun is responsible for producing time scales of hours? Only two candidates come to mind: the solar mesogranulation /45/, and the flux cancellation events in the photosphere /e.g. 38/ in which several hours are required for two flux tubes of opposite polarity to cancel each other. We believe it is important to understand the solar origin of the coronal and solar wind fluctuations, since they are probably the key to understanding what is responsible for heating the corona and driving the solar wind.

POSSIBLE FUTURE DIRECTIONS

The desire of the solar wind community to include coronal and TR physics in solar wind models is a happy development. (We wish the solar community would be equally eager to recognize the solar wind as being part of the sun.) Withbroe's work /14/ has already pointed to several constraints which coronal heating theories must satisfy. However, we have not yet been successful in integrating a *physical* coronal heating mechanism into solar wind models. The principal attempt in this direction was made by Hollweg and Johnson /20/ but their results disagreed with the Lyman-α data. Their work invoked MHD turbulence, which was handled roughly; perhaps one simply needs a better treatment of the turbulence.

Or perhaps one needs to consider other coronal heating mechanisms. Resonance absorption of MHD waves /46/ or heating by MHD shocks which form in the chromosphere /47/ are possible candidates deserving further study. Parker's suggestion /17/, that reconnection in the network can heat coronal holes, deserves both further theoretical and observational study; network reconnection can provide direct heating and launch waves. Unfortunately, there is as yet no clear evidence that either waves /48/ or reconnection /39,40,49/ provide the required energy flux. Perhaps coronal heating is really the sum of many mechanisms.

It has been suggested /50,51/ that polar plumes are sources of solar wind in coronal holes. This idea has not yet been seriously studied. However, recent ion abundance data from the SWICS instrument on Ulysses make this idea less attractive. Geiss /52/ remarked that the 'FIP effect' is weaker in fast flows (presumably coming from coronal holes). But Feldman and Widing /53/, using optical data, conclude that plumes exhibit a strong FIP effect. As T.E. Holzer remarked at this meeting, the conclusion would seem to be that fast flows do not come from the plumes.

Of course, we may have to altogether abandon the conventional modeling approach, which simply calculates the steady flow along individual streamlines. Perhaps all that is needed is to worry about non-steady flows along the individual magnetic flux tubes. But more radical alternatives may be needed. Only two will be mentioned here. Several authors /51,54,55,56/ have suggested that the solar wind is driven by the expulsion of plasmoids or magnetic bubbles. This idea has not yet been developed into detailed wind models which can be fully tested against data. However, Bochsler /57/ has pointed out that the SWICS data do not show the presence of oxygen with low charge states, which should appear if the corona contains expanding and cooling magnetic bubbles. The second radical approach is to treat the corona and solar wind as an exosphere, containing non-Maxwellian distribution functions with strong excesses of particles with energies greater than a few times the average thermal energy. This view has been strongly argued by Scudder /58/. We believe this view deserves serious consideration. But we wish to mention the following caveat: The title of Scudder's paper /58/ suggests that the problem with the corona is obtaining the high temperature. But the real problem is replenishing the radiative losses (and the solar wind losses) from the corona. Scudder has to show that the exospheric approach can in fact lead to an adequate energy flux to replenish the losses; however, this will probably turn out to be the case since a non-Maxwellian electron distribution function can easily carry a substantial heat flux. It should also be recalled that the radiation is collisionally produced, and it is not clear how far one can push the exospheric approach.

ACKNOWLEDGEMENTS

The author is grateful to Drs. K. Dere, R. Esser, U. Feldman, S. Habbal, E. Leer, and K. Widing for valuable discussions. This work has been supported in part by the NASA Space Physics Theory Program under grant NAG5-1479, and in part by NASA grant NSG-7411, to the University of New Hampshire.

REFERENCES

1. R.E. Hartle, and P.A. Sturrock, Astrophys. J. 151, 1155 (1968)

2. J.W. Belcher, L. Davis Jr., and E.J. Smith, J. Geophys. Res. 74, 2302 (1969)

3. J.W. Belcher, and L. Davis Jr., J. Geophys. Res. 76, 3534 (1971)

4. J.V. Hollweg, Rev. Geophys. 16, 689 (1978)

5. J.V. Hollweg, J. Geophys. Res. 91, 4111 (1986)

6. E. Leer, T.E. Holzer, and T. Fla, Space Sci. Rev. 33, 161 (1982)

7. E. Marsch, in: Physics of the Inner Heliosphere, ed. R. Schwenn, and E. Marsch, Springer-Verlag, Berlin 1991, p. 159.

8. C.-Y. Tu, Z.-Y. Pu, and F.-S. Wei, J. Geophys. Res. 89, 9695 (1984)

9. C.-Y. Tu, Solar Phys. 109, 149 (1987)

10. C.-Y. Tu, J. Geophys. Res. 93, 7 (1988)

11. E.C. Shoub, Astrophys. J. 266, 339 (1982)

12. E.C. Shoub, SUIPR Rep. 946 Institute for Plasma Res., Stanford Univ., Stanford, Calif., (1982)

13. S.K. Antiochos, and G. Noci, Astrophys. J. 301, 440 (1986)

14. G.L. Withbroe, Astrophys. J. 325, 442 (1988)

15. J.V. Hollweg, J. Geophys. Res. 81, 1649 (1976)

16. S.R. Habbal, private communication (1991)

17. E.N. Parker, Astrophys. J. 372, 719 (1991)

18. R. Hammer, Astrophys. J. 259, 767 (1982)

19. R. Hammer, Astrophys. J. 259, 779 (1982)

20. J.V. Hollweg, and W. Johnson, J. Geophys. Res. 93, 9547 (1988)

21. G.L. Withbroe, J.L. Kohl, H. Weiser, G. Noci, and R.H. Munro, Astrophys. J. 254, 361 (1982)

22. P.A. Isenberg, J. Geophys. Res. 95, 6437 (1990)

23. Y. Zhou, and W.H. Matthaeus, J. Geophys. Res. 95, 10291 (1990)

24. Y. Zhou, and W.H. Matthaeus, J. Geophys. Res. 95, 14881 (1990)

25. Y.-M. Wang, and N.R.J. Sheeley, Astrophys. J. Lett. 372, L45 (1991)

26. Y.-M. Wang, and N.R.J. Sheeley, Astrophys. J. 355, 726 (1990)

27. E. Leer, and T.E. Holzer, Ann. Geophys. 9, 196 (1991)

28. M. Neugebauer, this issue.

29. G. Gloeckler, this issue.

30. G.L. Withbroe, Astrophys. J. Lett. 337, L49 (1989)

31. M. Neugebauer, Fundam. Cosmic Phys. 7, 131 (1981)

32. A. Bürgi, Habilitationsschrift University of Bern, (1991)

33. W.A. Coles, R. Esser, U.-P. Løvhaug, and J. Markkanen, J. Geophys. Res., in press (1991)

34. E. Marsch, W.G. Pilipp, K.M. Thieme, and H. Rosenbauer, J. Geophys. Res. 94, 6893 (1989)

35. J.F.J. Dowdy, Solar Phys. 105, 35 (1986)

36. U. Feldman, and K.G. Widing, private communication (1991)

37. K. Dere, private communication (1991)

38. S.H.B. Livi, J. Wang, and S.F. Martin, Aust. J. Phys. 38, 855 (1985)

39. S.R. Habbal, and E. Grace, Astrophys. J., in press (1991)

40. K.P. Dere, J.D.F. Bartoe, G.E. Brueckner, J. Ewing, and P. Lund, J. Geophys. Res. 96, 9399 (1991)

41. J.V. Hollweg, M. Bird, H. Volland, P. Edenhofer, C. Stelzried, and B. Seidel, J. Geophys. Res. 87, 1 (1982)

42. P.S. Callahan, Astrophys. J. 199, 227 (1975)

43. R. Woo, F.-C. Yang, K.W. Yip, and W.B. Kendall, Astrophys. J. 210, 568 (1976)

44. R. Woo, and J.W. Armstrong, J. Geophys. Res. 84, 7288 (1979)

45. A. Nordlund, Solar Phys. 100, 209 (1985)

46. J.V. Hollweg, and G. Yang, J. Geophys. Res. 93, 5423 (1988)

47. J.V. Hollweg, Astrophys. J. 254, 806 (1982)

48. D.M. Hassler, G.J. Rottman, E.C. Shoub, and T.E. Holzer, Astrophys. J. 348, L77 (1990)

49. J.G. Porter, and R.L. Moore, in: Solar and Stellar Coronal Structure and Dynamics, ed. R.C. Altrock, National Solar Observatory, Sunspot 1988, p. 125.

50. I.A. Ahmad, and G.L. Withbroe, Solar Phys. 397 (1977)

51. D.J. Mullan, and I.A. Ahmad, Solar Phys. 75, 347 (1982)

52. J. Geiss, this issue.

53. U. Feldman, and K.G. Widing, this issue.

54. P.J. Cargill, and G.W. Pneuman, Astrophys. J. 276, 369 (1984)

55. P.J. Cargill, and G.W. Pneuman, Astrophys. J. 307, 820 (1986)

56. D.J. Mullan, Astron. Astrophys. 232, 520 (1990)

57. P. Bochsler, this issue.

58. J.D. Scudder, this issue.

NUMERICAL MODEL FOR CORONAL SHOCK WAVE FORMATION IN TWO-FLUID APPROXIMATION

A. G. Kosovichev*,** and T. V. Stepanova***

Institute of Astronomy, University of Cambridge, Cambridge, U.K.
*** Crimean Astrophysical Observatory, Crimea, Ukraine*
*** Department of General Physics and Astronomy of the Russian Academy of Sciences, Moscow, Russia*

ABSTRACT

We present results of modeling of shock waves, generated by coronal transients and expanding flare loops. We consider loss of hydrostatic equilibrium of a magnetic loop as a result of an increase of the magnetic field under the loop and as a result of a flare-induced heating of plasma inside the loop. The expanding magnetic loops act as a piston on coronal plasma and produce shock waves propagating in the solar wind. The shock waves are computed in one-dimensional, two-fluid approximation, by taking into account the processes of turbulent dissipation. The corresponding gas-dynamic equations are solved simultaneously with equations of motion of the magnetic tube to provide a self-consistent picture of the shock formation by the transients.

BASIC EQUATIONS

Following the model of coronal transients /1/ we consider a magnetic flux tube bounded by field lines denoted by indices 1 and 2 and lying at distances r_1 and r_2 from the Sun's center. If it is assumed that the magnetic field is frozen into the plasma and that all parameters inside the tube vary smoothly and monotonically, then the equations of motion for the magnetic tube boundaries can be written as:

$$\rho \frac{dV_1}{dt} = \frac{2(P - P_1)}{D} + \frac{B^2 - B_1^2}{4\pi D} - \frac{B^2}{4\pi R_c} - \frac{GM\rho}{S^2},$$

$$\rho \frac{dV_2}{dt} = -\frac{2(P - P_2)}{D} - \frac{B^2 - B_2^2}{4\pi D} - \frac{B^2}{4\pi R_c} - \frac{GM\rho}{S^2},$$

where $D = (r_2 - r_1)$ is the diameter of the tube, $S = 0.5(r_1 + r_2)$ is its mean radius, R_c is the radius of curvature of the tube, B_1 and B_2 are the magnetic field strengths above and beneath the magnetic tube. Inside the magnetic tube the density ρ is determined from the equation of conservation of mass; the magnetic field B is determined from the condition of conservation of magnetic flux; and the gas-dynamic pressure P is inferred from the adiabaticity equation. The pressure P_1 above the loop is determined from solution the gas-dynamic equations, governing the motion of the overlying plasma. Thus, the motion of the loop depends on the ambient reaction.

Evolution of a shock wave formed by a rising loop is described by the following system of Navier-Stokes equations:

$$\rho \frac{dV}{dt} + \frac{\partial}{\partial r}\left[\frac{k}{M}\rho(T_i + T_e)\right] = \frac{4}{3}\frac{\partial}{\partial r}\left(\eta \frac{\partial V}{\partial r}\right) +$$
$$\frac{4}{3}\left[2\eta \frac{\partial}{\partial r}\left(\frac{V}{r}\right) - \frac{V}{r}\frac{\partial \eta}{\partial r}\right] - \frac{\rho GM}{r^2},$$

$$\frac{3}{2}\rho \frac{k}{M}\frac{dT_e}{dt} - \frac{k}{M}T_e \frac{d\rho}{dt} = \frac{1}{r^2}\frac{\partial}{\partial r}\left(r^2 \kappa_e \frac{\partial T_e}{\partial r}\right) + Q,$$

$$\frac{3}{2}\rho\frac{k}{M}\frac{dT_i}{dt} - \frac{k}{M}T_i\frac{d\rho}{dt} = \frac{4}{3}\eta\left[\left(\frac{\partial V}{\partial r}\right)^2 - \frac{2v}{r}\left(\frac{\partial V}{\partial r} - \frac{1}{2}\frac{V}{r}\right)\right] +$$

$$\frac{1}{r^2}\frac{\partial}{\partial r}\left(r^2\kappa_i\frac{\partial T_i}{\partial r}\right) - Q,$$

$$\frac{d\rho}{dt} + \rho\frac{1}{r^2}\frac{\partial}{\partial r}\left(r^2 V\right) = 0, \qquad \frac{dr}{dt} = V,$$

where

$$\eta = l_1\mu\left(\frac{kT_i}{M}\right)^{5/2}, \quad \kappa_e = 75.5\, l_2\mu\left(\frac{kT_e}{M}\right)^{5/2}, \quad \kappa_i = 3.1\, l_3\mu\left(\frac{kT_i}{M}\right)^{5/2},$$

$$Q = 0.126\, l_4^{-1}\mu^{-1}\rho^2\frac{k}{M}\left(T_i - T_e\right)\left(\frac{kT_e}{M}\right)^{-3/2}, \quad \mu = 1.08\frac{M^3}{e^4\Lambda}.$$

Here $\rho = nM$, V, T_e and T_i are the density, the velocity, and the electron and ion temperatures in the solar wind, respectively; n is the electron and ion density; η is the coefficient of ion viscosity; κ_e and κ_i are the coefficients of electron and ion heat conduction, respectively; Q is the energy exchange rate between electrons and ions; $l_1, ..., l_4$ are the parameters characterizing the turbulence influence upon the transport coefficients; k is the Boltzmann's constant; Λ is the Coulomb logarithm; and M is the mass of an ion.

The effects of the interplanetary magnetic field have not been studied yet in our model. The observed quasi-circular shock shape /2/ suggests that the magnetic effects seem not to be very significant for the global structure of the shocks. However, the magnetic field leads to local distortions of shock fronts, and it also plays an important role in turbulent processes determining the structure of the fronts and plasma parameters in the flow behind the fronts. The model we develop might be useful for studies of the influence of the different models of the transport coefficients on the parameters of the shock waves. Here for the transport coefficients we use an empirical model, similar to that suggested in /3/, which differ from Coulomb coefficients by the terms dependent on distance: $l_j = R_\odot/a_j r$, where a_j are constants from 10 to 10^2 and $R_\odot$ is the radius of the Sun. In the calculations to follow they have the following values: $a_1 = a_2 = a_3 = 10^2$ and $a_4 = 20$. At distances $r \gtrsim 0.1$ A.U. large-scale turbulence due to the Kelvin-Helmholtz instability, which also modifies the transport coefficients, has been taken into account, according to /4/.

RESULTS

We have investigated the behavior of the processes in the corona and the solar wind in the case of disturbances of two types: (i) with an increase of magnetic field strength B_2 below the loop, which corresponds to the model of coronal transients, and (ii) with increasing temperature T_0 and density ρ_0 inside the loop, which believe to occur due to solar flares. The following initial parameters of the loop have been chosen: $S_0 = 1.2 R_\odot$, $D_0 = 0.24 R_\odot$, $\rho_0 = 1.7 \times 10^{-15}$ g/cm^{-3}, $T_0 = 10^5$ K, and $B_0 = 5$ G.

Figure 1 presents the results for an increasing magnetic field B_2 beneath the loop from 7 G to 10 G. At initial moments of time there occurs a fast acceleration of the loop to a constant velocity $\simeq 600$ km/s. As the loop rises, its diameter increases nearly linearly with time. At the beginning of the process magnetic forces predominate over the gas-dynamic and gravitational pressures. However, when $t \gtrsim 10$ h, this relationship of the forces reverses, and at subsequent moments of time the loop decelerates gradually. However, this occurs at distances exceeding $30 R_\odot$ where all external forces are no longer sufficient to substantially change the velocity of the loop. Therefore, its subsequent motion occurs virtually under its own momentum. In front of the rising loop there forms a shock wave which propagates in the solar wind with the velocity $\simeq 800$ km/s and reaches the Earth's orbit in $\simeq 57$ h. At this moment the upper boundary of the abruptly expanded magnetic loop is at the distance of $\simeq 0.9$ A.U. The electron temperature is several times lower than the ion temperature in a wide relaxation zone behind the shock front, of an extent of $\simeq 10 R_\odot$. The disturbed solar

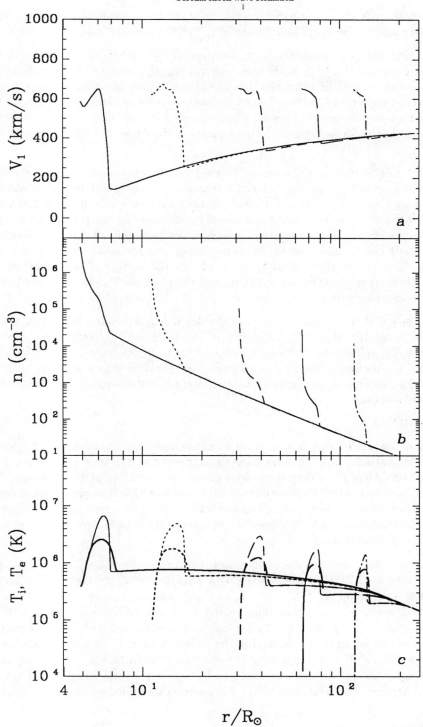

Fig. 1. The dependencies on distance from the solar center of (a) velocity V, (b) density $n = \rho/M$ and (c) ion T_i and electron T_e (heavy lines) temperatures at successive moments of time: 1.64 h (solid lines); 3.64 h (dotted lines); 9.64 h (lines with short dashes); 19.64 h (lines with long dashes); 35.64 h (dash-dotted lines).

wind velocity $\simeq 600$ km/s. Near the upper boundary of the loop, a spheric expansion results in a decrease of density and temperature and "heaping-up" of the material occurs.

If the magnetic field strength below the loop increases to 15 G, then the processes in the outer solar atmosphere occur substantially more intensively. Thus, the transient velocity reaches $\simeq 1200$ km/s; the plasma flow velocities behind the shock front are $\simeq 1500$ km/s, and the ion temperature is $\simeq 4 \times 10^7$ K. The structure of the shock front in this case is determined by large-scale turbulent viscosity. An important role here is played by the viscous heating behind the shock front, which does not result, unlike the preceding case, in the formation of a dense and cold layer near the upper loop boundary.

The development of disturbances of the second ('flare') type has been simulated by the increase in temperature of the loop to $T_0 = 10^7$ K, without an increase of background magnetic field $B_2 = 6$ G. In this case we have also found a fast acceleration of the loop to the velocity of $\simeq 700$ km/s which is accompanied, however, by a more rapid increase of its diameter as compared with the preceding cases. Besides, a deceleration of the rise of the loop starts immediately after the acceleration phase. Processes of formation of the shock wave and its propagation in the solar wind are similar to those shown on Fig. 1. However, at the earliest moments of time the heating of the gas behind the front is stronger, the ion temperature reaches $\simeq 10^7$ K, and therefore the shock front structure at these moments is determined by turbulent viscosity.

Another case of the gas-dynamic disturbance has simulated a simultaneous increase in temperature up to 10^7 K and in density from 1.7×10^{-15} g/cm^{-3} to 2.7×10^{-14} g/cm^{-3} in the flare loop. In this case a substantial role in the resulting balance of forces in the loop is played by the gravity force. As a result, the motion of the loop has a pulsating character with a slow rise with the mean velocity of $\simeq 60$ km/s. This is accompanied by the formation of a series of relatively weak shock waves in the solar corona.

CONCLUSIONS

Equilibrium loss of magnetic loops as a result of an enhancement of the background magnetic field or due to a plasma heating inside the loops leads to their rapid acceleration (in about 1-2 hours) to supersonic velocities. After that, they move with about a constant velocity. However, flare loops are characterized by a more rapid increase of their diameter and by a more well-defined deceleration after the acceleration phase. The very much expanded loops could be observed as magnetic clouds in the interplanetary medium. In both cases the rise of loops leads to the generation of shock waves which propagate in the solar wind.

Shock waves in the solar wind can be divided into two types, depending on their intensity. In the case of relatively weak shock waves moving with velocities of $\lesssim 10^3$ km/s, the plasma behind the shock front is heated up to $\lesssim 5 \times 10^6$ K, and the electron temperature is lower than the ion temperature in a narrow relaxation zone behind the front. A typical feature in this case is the formation of a dense and cold layer ("piling-up" of material) near the upper boundary of the loop. In the case of strong shock waves, having the velocities of $\gtrsim 10^3$ km/s, the ion temperature can increase to $\simeq 5 \times 10^7$ K and can exceed the electron temperature by more than an order of magnitude. An important role in this case is played by large-scale turbulence behind the front and by the viscous heating in the relaxation zone. No appreciable piling-up of plasma occurs in this case.

REFERENCES

1. G.W.Pneuman, Eruptive prominences and coronal transients, *Solar Phys.* **65**,369-385 (1980)

2. R.Schwenn, Relationship of coronal transients to interplanetary shocks: 3D aspects, *Space Science Rev.* **44**, 139-168 (1986)

3. C.W.Wolff, J.C.Brandt, and R.C.Southwick, A two-component model of the quiet solar wind with viscosity, magnetic field, and reduced heat conduction, *Astrophys.J.* **165**, 181-194 (1971)

4. N.P.Korzhov,V.V.Mishin,and V.M.Tomozov,On the role of plasma parameters and Kelvin-Helmholtz instability in a viscous interaction of solar wind streams, *Plan.Sp.Sci.* **32**,1169-1178(1984)

ON THE POSSIBLE ROLE OF PLASMA WAVES IN THE HEATING OF CHROMOSPHERE AND CORONA

E. Marsch

Max-Planck-Institut für Aeronomie, W-3411 Katlenburg-Lindau, Germany

ABSTRACT

The possible importance of kinetic plasma waves at frequencies near and above the ion gyrofrequencies in the transport of energy through the chromosphere and corona and the wave energy deposition in these layers is hardly known and has little been investigated. This paper intends to give a cursory review of some plasma waves that may perhaps be relevant for anomalous transport. Some kinetic instabilities that might arise due to spatial inhomogeneity and associated velocity space anisotropies are briefly discussed. Atmospheric heating rates are then estimated on the basis of quasilinear theory and compared against radiative and conductive losses.

INTRODUCTION

Kinetic plasma waves have as yet received little attention in the theoretical concepts for heating the solar chromosphere and corona. Whereas ample literature exists, that addresses various heating mechanisms of the solar atmosphere by MHD waves (see e.g. the recent proceedings of the Heidelberg conference /1/), the possible role of high-frequency plasma waves in the energy transport and deposition has hardly been evaluated. Of course, only by direct in-situ measurements, such as are possible in the solar wind or in planetary magnetospheres, can the existence of plasma waves and their effects on the underlying particle velocity distribution functions conclusively be established. But given the present experimental techniques and means of observing the sun, this seems impossible for the solar atmosphere. We therefore have to rely on indirect evidence for plasma waves (as in relation to solar radio activity, see e.g. the book reference /2/), or we may infer their existence by analogy with better explored physical situations in other solar system plasmas.

QUASILINEAR HEATING RATES

Kinetic plasma waves have not been directly observed in the solar atmosphere. But they do, in our opinion, certainly exist there, since we are dealing with an inhomogeneous, non-stationary and non-LTE plasma through which a considerable amount of mechanical energy has to be fed in order to support the different atmospheric layers. Plasma waves will play a prominent role in the energy dissipation and coronal heating (which ultimately requires small-scale kinetic processes anyway), given they occur at sizable relative amplitudes, and given there is a kind of a turbulent energy cascade of MHD fluctuations to the kinetic scales. They are thus expected to largely determine the non-collisional ionic transport in the chromosphere and corona.

Basic properties of the solar atmosphere as a plasma medium can be found for example in the book by Krüger /2/, which contains a compilation of the important plasma parameters and of their variations with the height and solar magnetic field and dispersion diagrams demonstrating the main modes in a warm magnetoplasma. For the present purpose we only give a few parameters scaled in a way convenient for the sun. The thermal speed is $v_j = (k_B T_j/m_j)^{1/2} = 91 T_{j6}^{1/2} A_j^{-1/2}$ [km s^{-1}], the gyroradius $r_j = v_j/\Omega_j = 0.95 T_{j6}^{1/2} A_j^{1/2} Z_j^{-1} B_1^{-1}$ [m]. The isothermal scale height reads $H_j = v_j^2/g_\odot = 30223 T_{j6} A_j^{-1}$ [km]. The collision free path is $\lambda_j = 1100 n_{j8}^{-1} T_{j6}^2 Z_j^{-4}$ [km] with a Coulomb logarithm of 20. The Alfvénspeed is $v_A = B(4\pi n_p m_p)^{-1/2} = 2200 B_1 n_{p8}^{-1/2}$ [km s^{-1}]. Here we used the definitions $T_6 = T/10^6$ K, $B_1 = B/10$ $gauss$, $n_8 = n/10^8$ cm^{-3} for the temperature, magnetic field and number density. The surface gravitational acceleration is $g_\odot$ = 274 m s^{-2} and the solar radius $R_\odot$ is 696260 km.

The book by Melrose /3/ covers the fundamentals of different kinetic instabilities in space plasmas, including the elements of weak turbulence and quasilinear theory. According to this theory the microturbulent heating rates H_j for any species j are given by

$$H_j = < \delta \mathbf{J}_j \cdot \delta \mathbf{E} > \tag{1}$$

where $\delta \mathbf{J}_j$ denotes the current fluctuations associated with particles of the kind j, and $\delta \mathbf{E}$ is the microscopic fluctuation electric field. H_j can be evaluated by help of the quasilinear diffusion equation for the distribution function $f_j(\mathbf{v})$. Quasilinear theory ensures conservation of the overall wave-particle energy and momentum (e.g. Davidson /4/). The dissipation is determined by the properties of the plasma as a linear dielectric medium with tensor $\underline{\epsilon}_j$ that is a functional of the particles' velocity distributions. We restrict ourselves here to electromagnetic waves propagating along the magnetic field. The waves we have in mind are the ion-cyclotron waves and magnetoacoustic waves which can for example be driven unstable by a beam-plasma configuration or temperature anisotropies. For these waves the heating rates may be expressed as

$$H_j = n_j m_j v_A^2 \Omega_j X_j = \frac{B^2}{4\pi} \Omega_j \hat{\rho}_j X_j \tag{2}$$

by means of the background Alfvén speed v_A, gyrofrequency Ω_j, density n_j, fractional mass density $\hat{\rho}_j$, and mass m_j of species j. The unknown dimensionless quantity X_j contains the detailed information on the magnetic power spectrum and the local damping characteristics of the plasma. Following the nomenclature of Marsch et al. /5/ and Dum et al. /6/ we may write:

$$X_j = \frac{1}{2} \sum_{+,-} \int_{-\infty}^{\infty} dk_\parallel \frac{8\pi B_{k_\parallel}^\pm}{B^2} 2Im \left[\frac{(\omega_{k_\parallel}')^*}{\Omega_j} \hat{\epsilon}_j^\pm \right] \left(\frac{\Omega_j}{k_\parallel v_A} \right)^2 \tag{3}$$

$$Im \hat{\epsilon}_j^\pm = \sqrt{\frac{\pi}{2}} \left(\frac{\omega_{k_\parallel}'}{k_\parallel v_j} \right) \exp \left[-\frac{1}{2} \left(\frac{\omega_{k_\parallel}' \pm \Omega_j}{k_\parallel v_j} \right)^2 \right] \tag{4}$$

Here $B_{k_\parallel}^\pm$ means the spectral density of the magnetic field fluctuations with left $(-)$ and right $(+)$ hand polarization and the symbol $\omega_{k_\parallel}'$ denotes the wave frequency in the rest frame of species j that drifts with speed u_j relative to the center-of-mass frame. Note that for a local Maxwellian as assumed in (4), the imaginary part is directly proportional to the number of resonant particles fulfilling $\omega_{k_\parallel}' \pm \Omega_j - k_\parallel v_\parallel = 0$. For waves with phase speed comparable to v_A the formula (4) is extremely sensitive to the plasma beta $\beta_j = (v_j/v_A)^2$.

For electrostatic waves we can derive an expression similar to (3). The waves, with frequency ω_k and wave vector $\mathbf{k}$, that are relevant in this case are the electron plasma oscillations and more importantly the ion acoustic waves, which are ubiquitously observed in the interplanetary solar wind (see the review by Gurnett /7/). The heating rate can be written as:

$$H_j = n_j m_j v_j^2 \omega_j Y_j = p_j \omega_j Y_j \tag{5}$$

with the partial pressure p_j and the dimensionless quantity Y_j reading

$$Y_j = \int d^3 k \left(\frac{\mathcal{E}_k}{n_j k_B T_j} \right) 2Im \left[\frac{\omega_k}{\omega_j} \hat{\epsilon}_j \right] \left(\frac{k_j}{k} \right)^2 \tag{6}$$

Here $\mathcal{E}_k$ denotes the spectral density of electric field fluctuations and $\hat{\epsilon}_j$ is the unmagnetized dieletric susceptibility. The temperatures of species j is denoted by T_j and k_j means the Debye wave vector $k_j = \omega_j/v_j$ with the plasma frequency $\omega_j = (4\pi e_j^2 n_j/m_j)^{1/2}$.

Estimates of the chromospheric and coronal heating rates can now be obtained by balancing the rates (2) and (5) against the radiative and mechanical losses in these atmospheric layers. The dimensionless quantities X_j and Y_j require the knowledge of the relative fluctuation level of the waves involved and the velocity distributions of the ions and electrons, for which only in local thermodynamic equilibrium a Maxwell distribution may be adopted. This assumption is of course questionable almost everywhere in the weakly collisional parts of the solar atmosphere. The detailed solar wind observations provide many examples of nonthermal distributions occuring in a collisionless space plasma (Marsch /8/). Before we actually calculate numbers for (2) and (5) we shall shortly discuss the possible effects of spatial inhomogeneity and weak collisionality on the shape of the velocity distributions.

SPATIAL INHOMOGENEITY AND COLLISIONAL EFFECTS

There is observationally no doubt that in the case of the chromosphere, transition region and lower corona we are dealing with very small-scale inhomogeneous layers and magnetically structured media. For example, the magnetic field lines, in the form of bundles rooted in the chromospheric network boundaries, rapidly diverge across the transition layer and then strongly spread out with height to fill ultimately the whole space above the supergranular cells in the corona. In such magnetic mirror configurations the gyrofrequencies strongly decline with height along field lines. Also the collisional plasma properties strongly vary with distance above the solar surface, whereby the plasma becomes increasingly collisionless with atmospheric height. Particle distribution are under such circumstances expected to deviate strongly from local Maxwellians and will thus develop sufficient free energy in phase space to drive microinstabilities and to excite various kinds of plasma waves.

Such a behaviour of electron and ion velocity distribution functions has often been observed and also modelled for polar wind electrons (e.g. Yasseen et al. /9/) and for solar wind protons (Livi and Marsch /10/). These references give illustrations of the kinetic effects we have in mind. The particle distributions are found to be shaped simultaneously by the magnetic mirror force, that conserves their magnetic moments, and by Coulomb collisions which scatter the particles and tend to produce isotropic pitch-angle distributions. Due to spatial inhomogeneity nonthermal distributions will result which are again prone to plasma instabilities. For the solar wind such processes are extensively described in the review by Marsch /7/. Of course, for the corona the situation is more complicated than in the mentioned solar or polar wind scenarios. For instance, the picture needs to be complemented by the possibility that plasma waves are injected from the bottom of the solar atmosphere into the overlying lower corona and are thus added to the waves generated locally. This energy influx could in turn lead to additional wave-particle interactions affecting the shape of the distributions and their transport properties. Unfortunately, without solar probe measurements /11/ the spectral characteristics of the coronal plasma waves will remain elusive and unknown, and so will the real features of the coronal particles.

ESTIMATES OF HEATING RATES

To estimate the heating rates by plasma waves according to (2-6) we require the spectral energy densities of the electromagnetic or electrostatic waves involved and the knowledge of the dielectric properties and damping characteristics of the plasma in form of $Im\hat{\epsilon}_j$. This may exponentially vary with the ratio of resonant over thermal speed, and it can attain an upper limit of order unity, corresponding to strong Landau or cyclotron damping. Since we do not know $Im\hat{\epsilon}_j$ we will replace it by one.

The question then is how large X_j or Y_j, i.e. the relative wave amplitudes near Ω_j or ω_j, have to be in order to balance radiative cooling or heat conduction losses. We have chosen these two loss terms since for them local rates can be given, whereas for other mechanisms merely energy fluxes can be quoted that still must be transformed into rates by assuming an unknown dissipation length. The radiative loss function (e.g. Cook et al. /12/) is well known to be $C_R = n_e^2 \Lambda(T_e)$, where for the atomic part Λ we simply use 10^{-22} [erg cm^3 s^{-1}]. Consider a hydrogen plasma with $n_e = n_p$ and $T_e = T_p$ and densities of $n_e = 10^{10}(10^8)$ cm^{-3} and temperatures of $T_e = 10^4(10^6)$ K for the chromosphere/transition region interface and the corona, respectively. For the corona we find $C_R \approx 10^{-6}$ [erg cm^{-3} s^{-1}]. Then the equilibrium requirement on X_j for protons is $H_p = C_R$:

$$X_p \approx \frac{\Lambda n_p}{m_p v_A^2 \Omega_p} \; , \quad X_p \approx \Lambda_{-22} n_{p8}^2 B_1^{-3} 1.3 \times 10^{-12} \; , \quad \Lambda_{-22} = \Lambda/[10^{-22} \text{erg cm}^3 \text{s}^{-1}] \qquad (7)$$

For conductive losses the cooling rate is $C_Q = \nabla \cdot \mathbf{Q}$, with the electronic heat flux $\mathbf{Q} = -\kappa \nabla T_e$. Using this classical expression for $\mathbf{Q}$ and the temperature gradient scale length ℓ, we have with $\ell_8 = \ell/10^8$ cm the result: $C_Q \approx 3 \times 10^{-2} T_{e6}^{7/2} \ell_8^{-2}$ [erg cm^{-3} s^{-1}]. The equilibrium requirement $H_p = C_Q$ puts a lower limit on the wave fluctuation level. Here we restrict ourselves to radiative losses. Given the value of C_R and the typical plasma parameters of the different layers, the equations (2) and (5) provide the estimates $X_p = 1.3 \times 10^{-8}$, $Y_p = 5.6 \times 10^{-9}$ for the upper chromosphere and $X_p = 1.3 \times 10^{-9}$, $Y_p = 5.6 \times 10^{-12}$ for the lower corona. For cyclotron waves this corresponds to relative wave amplitudes of $\delta B/B \gtrsim 1.2 \times 10^{-4}$ and $\delta B/B \gtrsim 3.6 \times 10^{-5}$, values which are small as compared with solar wind values for Helios at 0.3 AU: $B = 66$ nT, $\delta B = 1$ nT near 1 Hz yields $\delta B/B \approx 0.015$; for Mariner 10 at 0.95 AU: $B = 4.7$ nT, $\delta B = 0.32$ nT gives $\delta B/B \approx 0.067$. For ion acoustic waves (see /6/) the maximum fluctuation level gives $Y_p \lesssim 10^{-7}$, with corresponding electric fields $\delta E \lesssim 1$ mV/m. The coronal turbulence levels for strong damping are given by $X_p \approx < (\delta B/B)^2 >$ and $Y_p \approx < \delta E^2/8\pi n_p T_p >$. Our above numerical estimates required to balance the losses are not unreasonable and indeed much smaller than those usually observed in the inner solar wind. Of course a weaker damping, due to wave-particle interactions occurring only in the tails of the velocity distributions, would after (3) or (6) require much stronger turbulence amplitudes.

CONCLUSIONS

The solar chromosphere/transition region/corona plasmas are strongly inhomogeneous and weakly collisional. Under such conditions highly non-Maxwellian velocity distributions are to be expected and as a result plasma microinstabilities. Thus kinetic plasma wave turbulence should prevail. If energy is somehow cascaded from large scales to the kinetic regime, then wave amplitudes moderate by comparison with the observed solar wind turbulence would suffice to ensure the heating rates required to balance locally the radiative and conductive losses. Therefore plasma waves should play a major role in solar atmospheric heating and in the dissipation of the larger-scale mechanical energy of various forms.

REFERENCES

1. Ulmschneider, P., E.R. Priest, and R. Rosner (Eds.), *Mechanisms of Chromospheric and Coronal Heating*, Springer-Verlag, Heidelberg (1991)
2. Krüger, A., *Introduction to solar radio astronomy and radio physics*, D. Reidel Publishing Company, Dordrecht (1979)
3. Melrose, D.B., *Instabilities in Space and Laboratory Plasma*, Cambridge University Press (1986)
4. Davidson, R.C., *Methods in Nonlinear Plasma Theory*, Academic Press, New York (1972)
5. Marsch, E., C.K. Goertz, and K. Richter, *J. Geophys. Res.*, 87, 5030 (1982)
6. Dum, C.T., E. Marsch, and W.G. Pilipp, *J. Plasma Physics*, 23, 91 (1980)
7. Gurnett, D.A., in *Physics of the Inner Heliosphere*, Vol. II, R. Schwenn and E. Marsch (Eds.), Springer-Verlag, Heidelberg, 135 (1991)
8. Marsch, E., in *Physics of the Inner Heliosphere*, Vol. II, R. Schwenn and E. Marsch (Eds.), Springer-Verlag, Heidelberg, 45 (1991)
9. Yaseen, F., J.M. Retterer, T. Chang, and J.D. Winningham, *Geophys. Res. Lett.*, 9, 1023 (1989)
10. Livi, S., and E. Marsch, *J. Geophys. Res.*, 92, 7255 (1987)
11. Solar Probe, A report of the 1989 Solar Probe Science Study Team, JPL D-6797, California Institute of Technology (1989)
12. Cook, J.W., C.C. Cheng, V.L. Jacobs, and S.K. Antiochos, *Astrophys. J.*, 338, 1176 (1989)

KNOWLEDGE OF CORONAL HEATING AND SOLAR-WIND ACCELERATION OBTAINED FROM OBSERVATIONS OF THE SOLAR WIND NEAR 1 AU

M. Neugebauer

Mail Stop 169-506, Jet Propulsion Laboratory, California Institute of Technology, Pasadena, CA 91109, U.S.A.

ABSTRACT

Clues to the nature of the mechanisms responsible for heating the corona and accelerating the solar wind can be obtained by contrasting the properties of the quasi-stationary and transient states of the solar wind. Substantial differences exist in the proton temperatures and anisotropies, the entropy, the field strength, β, the Alfvénicity of fluctuations in the field, the distribution of MHD discontinuities, and the helium abundance of the two types of flow. Those differences are displayed as a function of the solar wind speed. Several signals of wave acceleration can be found in the data for quasi-stationary flows. The relatively smooth velocity dependences of proton temperature, helium abundance, and frequency of occurrence of rotational discontinuities suggest that the acceleration mechanisms for flow from coronal holes, coronal streamers, and the quasi-stationary low-speed flows between them may be basically the same, differing only in degree. The properties of the transient flows indicate a lesser influence of wave acceleration and a greater expansion relative to the quasi-stationary flows. There are differences between the two assumed signatures of transient flows (bidirectional electron streaming and helium abundance enhancements), however, that suggest that low-speed intervals of bidirectional electron streaming might not be associated with transient flows.

APPROACH

Observations of the solar wind obtained near 1 AU provide boundary conditions for theoretical models of the heating of the corona and the acceleration of the solar wind. Data from 1 AU may also give clues to what the heating and acceleration mechanisms might be. The mechanisms are probably different for quasi-stationary and transient solar wind flows. Furthermore, there may be distinctions between different types of quasi-stationary flow such as the flows from coronal holes, from sector boundaries, and from the background, low-speed solar wind (if it exists as a distinct type of flow). Therefore, in this paper I examine the properties of the solar wind for limited intervals of time that can be assigned to one or another type of flow with a reasonable level of certainty.

The results presented here are largely based on analysis of plasma and field parameters measured at the ISEE-3 spacecraft between August, 1978 and February, 1980. The plasma data were obtained by the solar wind spectrometer /1/ for which S. J. Bame is the Principal Investigator, while the field data were obtained by the vector helium magnetometer /2/ for which E. J. Smith is the Principal Investigator. For the purposes of this study, I have defined five types of solar wind flow associated with different source regions on the Sun or different acceleration mechanisms. The types are: CH = quasi-stationary flow from coronal holes; PS = quasi-stationary, high-density plasma sheets in which the heliospheric current sheet (or sector boundary) is embedded; IS = quasi-stationary interstream flows associated with neither coronal holes nor coronal streamers; BES = bidirectional electron streaming events identified by Gosling et al. /3/; and HAE = helium abundance enhancements in which $N_a/N_p > 0.08$, where N_a and N_p are the alpha-particle and proton densities, respectively. Both BES and HAE are believed to be signatures of transient flows initiated by coronal mass ejections (CMEs). Further discussion of the selection criteria for the different types of flow, the start and stop times of the intervals selected, and some illustrative examples are given by Neugebauer and Alexander /4/. The ISEE-3 data are supplemented with data for very fast flow from large coronal holes observed by the IMP 6-8 spacecraft in 1971-4 as summarized by Feldman et al. /5/. The sources of data used in this analysis are summarized in Table 1.

TABLE 1. Summary of data used

Type of data	Spacecraft	No. of intervals	Hours
Quasi-stationary flows (Qs)			
CH = Coronal holes	ISEE 3	9	222
	IMP 6-8	19	1060
IS = Interstream	ISEE 3	6	214
PS = Plasma sheet	ISEE 3	7	92
Transient flows (Tr)			
BES = Bidirectional electron streaming	ISEE 3	17	146
HAE = Helium abundance enhancements	ISEE 3	15	235

RESULTS

Most of the analyses presented here are in the form of scatter plots of various plasma parameters versus the solar wind speed, v, after sorting by type of flow. The top row of Figure 1 shows plots of proton temperature T_p versus v, with the quasi-stationary CH, IS, and PS flows on the left, the transient BES and HAE flows in the center panel, and all quasi-stationary and transient flows superimposed in the panel on the right. The left-hand panel exhibits the well-known increase of T_p with v. Note that the data from the three types of quasi-stationary flow all lie close to a common curve, perhaps indicating a common acceleration mechanism. The center and right-hand panels show a different relationship for the transient flows. T_p is roughly independent of v for the HAE intervals, while there is an apparent correlation between T_p and v for the BES flows, but the slope is less steep than for the quasi-stationary wind. The HAE plasma is usually hotter than the BES plasma, but at high velocities, the temperatures of both types of transient flow are very significantly below the temperatures observed in quasi-stationary flows at the same speed. A strong departure from the T_p-v relationship exhibited by the quasi-stationary flow has often been used as a marker of transient plasma /6/. The lower temperature of the high-speed transient flow has been interpreted as evidence of either decreased thermal conduction from the corona due to magnetic disconnection (i.e., a closed plasmoid configuration) /6/ and/or greater than average expansion between the Sun and 1 AU /7/. A quasi-stationary stream is confined to expand within a well defined channel between its neighboring streams, whereas a transient flow can push aside the plasma in its path and expand in three dimensions.

The second row of plots in Figure 1 shows similar data for electron temperature T_e. Despite the discovery of plasma with anomalously low electron temperature following interplanetary shocks /8, 9/, the temperature-velocity relation in Figure 1 is quite different for protons and electrons. A possible interpretation is that T_p is controlled by interplanetary wave-particle interactions whereas T_e is controlled by conduction from the hot corona.

Another fundamental difference between the quasi-stationary and the transient flows is the greater strength of the magnetic field in the latter /3, 7/. Figure 2 illustrates this point. On the left is a

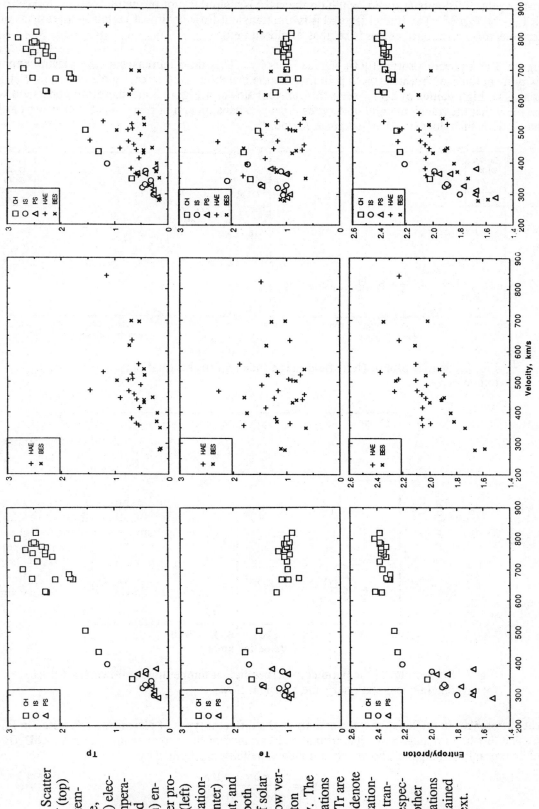

Fig. 1. Scatter plots of (top) proton temperature, (middle) electron temperature, and (bottom) entropy per proton for (left) quasi-stationary, (center) transient, and (right) both types of solar wind flow versus proton velocity. The abbreviations Qs and Tr are used to denote quasi-stationary and transient, respectively; other abbreviations are explained in the text.

scatter plot of |B| versus speed, while on the right is a similar plot of the plasma $\beta = 8\pi k(N_pT_p + N_aT_a + N_eT_e)/B^2$. The higher |B| and β in the transient flow is almost certainly a remnant of its previous magnetic confinement in regions of closed field.

Figure 3 is a plot of proton $T_\parallel/T_\perp$ versus velocity. This ratio increases as the plasma expands away from the Sun, but its growth is limited by the instability of the resulting anisotropic distribution. The high values of $T_\parallel/T_\perp$ seen for transient flows in Figure 3 are therefore consistent with the view that the transient wind undergoes greater expansion and/or that it has fewer wave-particle interactions than does the quasi-stationary wind.

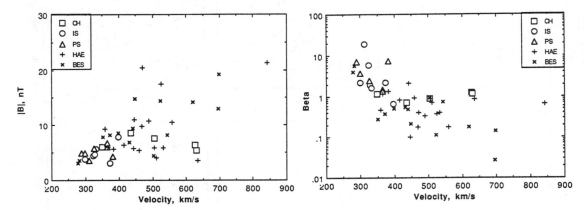

Fig. 2. Scatter plot of (left) field magnitude and (right) total plasma β versus proton velocity.

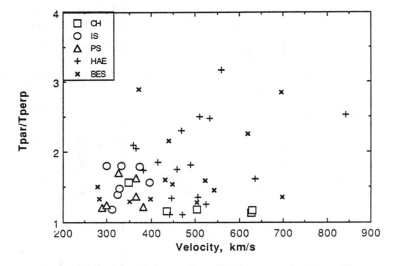

Fig. 3. Scatter plot of the parallel to perpendicular temperature ratio versus velocity for protons in quasi-stationary and transient flows.

The differences in the level of wave activity in the different types of plasma are addressed in Figure 4. The left-hand panel shows the normalized variance of the vector magnetic field, $<\delta\mathbf{B}^2/\mathbf{B}_0^2>$, while the right-hand panel shows the normalized helicity σ_c, defined by

$$\sigma_c = \frac{2 <\delta\mathbf{v} \cdot \delta\mathbf{B}>}{<\delta v^2 + \delta B^2>}$$

Both parameters were calculated from 5-minutes averages of the ISEE-3 field and velocity vectors. The square symbols in the left-hand panel show that the wave intensity in coronal hole flow, as measured by $<\delta\mathbf{B}^2/\mathbf{B}_0^2>$, is correlated with the speed, suggesting that wave acceleration may be

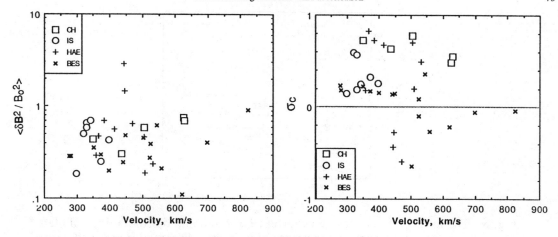

Fig. 4. Scatter plots of (left) the relative variance in the vector magnetic field and (right) the normalized cross helicity versus velocity for two types of quasi-stationary and transient flow.

important in coronal holes. The crosses in that panel indicate that the wave level is often quite low for BES intervals, as previously pointed out by Gosling et al. /3/. For HAE intervals, the relative wave intensity is variable, sometimes being very low and sometimes exceeding unity. The right side of Figure 4 indicates the degree of Alfvénicity of the fluctuations. The sector structure of the interplanetary field was taken into account such that $\sigma_c = +1$ corresponds to outward propagating Alfvén waves while $\sigma_c = -1$ corresponds to waves moving toward the Sun. For the quasistationary wind, both the CH and the IS intervals show outward propagating waves, with the coronal hole intervals being more Alfvénic than the interstream intervals. σ_c is smallest for the BES intervals, indicating that either the fluctuations are not Alfvénic or they consist of an equal number of waves travelling towards and away from the Sun. Only the transient flows exhibit negative values of σ_c, consistent with the hypothesis that the magnetic fields in those flows consist of closed field loops with waves traveling in both directions, in some cases with more inward than outward propagating waves.

An independent indication that turbulent wave fields may have played a role in the acceleration of the fast, quasi-stationary wind is the entropy data shown in the bottom panel of Figure 1. The entropy per proton S is defined as $k \ln(T_p^{3/2}/N_p)$. If it is assumed that in the photosphere all protons have the same entropy, Figure 1 can be interpreted as indicating that the faster wind has acquired more entropy by the time it reaches 1 AU than has the slower wind, and (from the right-hand panel), that for the same speed, the quasi-stationary wind has greater entropy/proton than does the transient wind.

Parker /10,11/ has suggested that convection-driven, random shuffling of the foot points of closed field lines necessarily leads to the creation of tangential discontinuities (TDs) and that the magnetically closed corona is heated by "nanoflares" resulting from bursts of magnetic reconnection across those TDs On field lines that are open to the solar wind, however, the disturbances created by the shuffling foot points would be carried away by waves. Some of the energy and momentum of the waves may contribute to the acceleration of the solar wind. There may be an additional source of waves, as well as particle momentum, in the network microflares that Parker /12/ has suggested as the mechanism responsible for heating the corona in coronal holes. Some of the waves may steepen into rotational discontinuities (RDs) /13, 14, 15/ that propagate away from the Sun in the wind. If these scenarios are correct, one would expect to find more TDs in transient than in quasi-stationary flow and more RDs in quasi-stationary than in transient flow. Figure 5 is a test of that hypothesis. It is a plot of the number of RDs observed per hour by ISEE 3 plotted versus the number of TDs per hour for the five different categories of solar-wind flow. The details of how the discontinuities were identified and classified are given in Neugebauer and Alexander /4/.

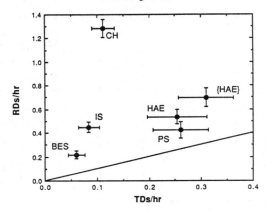

Fig. 5. The average frequency of occurrence of rotational discontinuities plotted versus the average frequency of occurrence of tangential discontinuities for different types of solar wind flow.

The slanted line in Figure 5 indicates an equal number of TDs and RDs. We suspect that the fact that we found more RDs than TDs for all types of flow, whereas some other investigators have found the opposite, can perhaps be explained by the difference in the criteria used to select the discontinuities. The sample used in our study was limited to large, isolated discontinuities for which $|\Delta B|/B > 0.5$, whereas the studies that found more TDs than RDs had no limitation on this ratio. We intend to investigate this discrepancy further in the future.

For the discontinuities that meet our selection criteria, we find that the flow from coronal holes has many more RDs than do other types of flow. This is consistent with the hypothesis stated above. Furthermore, Figure 6 shows that the number of RDs/hr increases with the solar-wind speed in quasi-stationary flow, thus indicating that wave acceleration may play a role. The right side of Figure 6 shows an inverse, if any, correlation between RDs/hr and speed for the transient flows. Similarly, there is no correlation of the number of TDs/hr and speed for any type of flow.

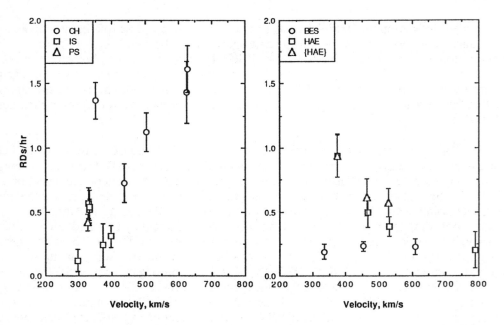

Fig. 6. Scatter plots of the frequency of occurrence of rotational discontinuities versus solar-wind velocity for different types of (left) quasi-stationary and (right) transient flows.

Figure 5 also shows that the two types of flow with the highest frequency of TDs are the transient HAE flows and the quasi-stationary PS flows. The large number of TDs in the PS flows is probably caused by partial and complete crossings of the interplanetary current sheet which is included in PS flows by definition. The hypothesis presented above leads to the expectation that both HAE and BES flows would have a high frequency of TDs. According to Figure 5, this is true for HAEs, but not for BESs. In fact, BES flows are unique in that the fields in them are unusually strong and steady and β is unusually low, as has been pointed out previously by several people and as evident in Figure 2. Neugebauer and Alexander /4/ showed a general relation between the number of discontinuities of any type and the value of proton β and suggested that under conditions of very low β there are not enough proton current carriers to sustain large discontinuities in the field. The difference between the {HAE} and HAE data in Figures 5 and 6 is that intervals that exhibit both high helium abundance and bidirectional electron streaming have been included in intervals designated as HAE, but not in those designated as {HAE}. We conclude that our results on the frequency of occurrence of TDs in different types of flow is consistent with Parker's nanoflare model except for those intervals in which β is so low that there are no discontinuities at all.

Consider next Figure 7, which shows the dependence of the helium abundance, expressed as the ratio N_a/N_p, on speed for the different types of flow. The data are all from ISEE 3 except that the highest velocity point in the left-hand panel represents the average abundance observed in a series of very-high-speed coronal hole flows observed in 1971-4 by IMP 6-8 /16/. With the exception of one point, the left-hand panel of Figure 7 shows a continuous variation of N_a/N_p with speed. Theories of solar-wind abundances /17/ place the region where the composition of the solar wind is determined in the upper chromosphere, below the location of most of the coronal heating and wind acceleration. But even though they occur at very different altitudes, there clearly must be a link between the mass fractionation and the acceleration processes.

The right-hand panel of Figure 7 shows similar data for the transient flows. Note the difference of vertical scale for the two panels. Each of the HAE intervals has $N_a/N_p > 0.08$ by definition, but no relation between N_a/N_p and v is apparent. The situation is, surprisingly, quite different for the BES flows, for which N_a/N_p does appear to increase with increasing v. The abundances in Figure 7 may be compared with recent results from helioseismology /18/ which indicate that N(He)/N(H) = 0.08. The quasi-stationary flows are always depleted in helium, whereas the transient flows may have either more or less helium than the solar average.

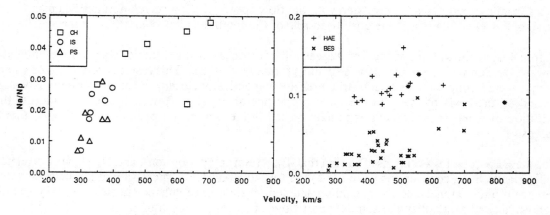

Fig. 7. The abundance of alpha particles as a function of velocity for different types of (left) quasi-stationary and (right) transient flows.

Finally, quite a bit of attention has been paid in the past /19/ to the unexpected observation that the momentum flux density, $N_p m_p v^2$, of the solar wind is nearly constant, independent of the type of flow. It is my view that the momentum flux density is, in fact, no more a constant than is the number flux, $N_p v$, or the energy flux density, $N_p v(m_p v^2/2 + m_p MG/R_s)$, where the second term accounts for the work done against gravity between the solar surface and 1 AU. This conclusion is based on plots such as the ones shown in Figure 8, where the number, momentum, and energy fluxes for both the ISEE-3 and IMP intervals are plotted versus velocity; there is no qualitative difference between the spreads shown in the three plots.

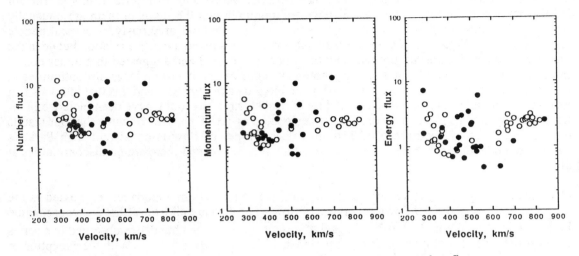

Fig. 8. Scatter plots of the velocity dependence of the proton number flux, momentum flux, and energy flux in the solar wind for quasi-stationary (open circles) and transient (closed) solar wind flows.

DISCUSSION

Most of the relationships presented above have been known for quite some time. I have merely used the ISEE-3 data to present them in a slightly different format after organizing or sorting the data in a slightly different way than has been done in the past. A few new insights did emerge from the exercise, however.

For example, I don't think it was previously realized that the alpha-particle abundance in the solar wind from coronal holes is correlated with the flow speed. Such a correlation is perhaps not surprising, however, in view of the postulated role of Coulomb drag between different ion species.

Figure 4 is, so far as I know, the first suggestion of an occasional net flux of waves travelling back toward the Sun within the magnetic structures formed by CMEs. This result is certainly consistent with the prevailing view that such structures are magnetically closed, as had been previously suggested on the basis of many of the other properties of transient flows. It does not address the question of whether the structures remain rooted in the Sun or are plasmoids closing on themselves.

For the BES data, I was surprised to find that T_p, S (entropy per proton), and N_a/N_p are correlated with speed. Such correlations were not found for the HAE data. There should not be such a difference if bidirectional electron streaming and helium abundance enhancements are simply two diagnostics used to identify the same type of flow. For high-speed transient flows, the HAE and BES intervals are often close together in time and sometimes overlap. Figures 1 and 7 show that it is only at low speeds that the HAE and BES intervals have different values of T_p, S, and N_a/N_p. Perhaps low-speed BES events are not transient flows.

In summary, several signals of wave acceleration can be found in the data for quasi-stationary flows. The relatively smooth velocity dependences of proton temperature, helium abundance, and frequency of occurrence of rotational discontinuities suggest that the acceleration mechanisms for flow from coronal holes (CH), coronal streamers (PS), and the relatively low-speed quasi-stationary flows between them (IS) may be basically the same, differing only in degree. The properties of the transient flows suggest a lesser influence of wave acceleration and a greater expansion relative to the quasi-stationary flows.

There are other differences between quasi-stationary and transient solar wind flows that could not be illustrated with the ISEE-3 plasma and field data on which this paper is based. Important distinctions have been found in energetic particles and in a wide variety of ions heavier than alphas. The remarkable heavy-ion data from Ulysses are discussed in another session of this conference. They contain different kinds of clues to the processes occurring in the region where the solar wind is accelerated, and the two data sets must eventually be considered together.

Acknowledgements: I thank Silvia DaCosta for her assistance in preparing much of the material on which this paper is based. The paper presents the results of one phase of research conducted at the Jet Propulsion Laboratory, California Institute of Technology, under contract with the National Aeronautics and Space Administration.

REFERENCES

1. S. J. Bame, J. R. Asbridge, H. E. Felthauser, J. P. Glore, H. L. Hawk, and J. Chavez, ISEE-C solar wind plasma experiment, IEEE Trans. Geosci. Electron., GE-16, 160 (1978).

2. A. M. A. Frandsen, B. V. Connor, J. van Amersfoort, and E. J. Smith, The ISEE-C vector helium magnetometer, IEEE Trans. Geosci. Electron., GE-16, 195 (1978).

3. J. T. Gosling, D. N. Baker, S. J. Bame, W. C. Feldman, R. D. Zwickl, and E. J. Smith, Bidirectional solar wind electron heat flux events, J. Geophys. Res., 92, 8519 (1987).

4. M. Neugebauer and C. J. Alexander, Shuffling foot points and magnetohydrodynamic discontinuities in the solar wind, J. Geophys. Res., 96, 9409 (1991).

5. W. C. Feldman, J. R. Asbridge, S. J. Bame, and J. T. Gosling, High-speed solar wind flow parameters at 1 AU, J. Geophys. Res., 81, 5054 (1976).

6. J. T. Gosling, V. Pizzo, and S. J. Bame, Anomalously low proton temperatures in the solar wind following interplanetary shock waves: Evidence for magnetic bottles?, J. Geophys. Res., 78, 2001 (1973).

7. L. W. Klein and L. F. Burlaga, Interplanetary magnetic clouds at 1 AU, J. Geophys. Res., 87, 613 (1982).

8. M. D. Montgomery, J. R. Asbridge, S. J. Bame, and W. C. Feldman, Solar wind electron temperature depressions following some interplanetary shock waves: Evidence for magnetic merging?, J. Geophys. Res., 79, 2324 (1974).

9. R. D. Zwickl, J. R. Asbridge, S. J. Bame, W. C. Feldman, J. T. Gosling, and E. J. Smith, Plasma properties of driver gas following interplanetary shocks observed by ISEE-3, in: Solar Wind Five, NASA Conf. Proc. CP-2280, ed. M. Neugebauer, NASA, Washington 1983, p. 711.

10. E. N. Parker, Magnetic reorientation and the spontaneous formation of tangential discontinuities in deformed magnetic fields, Astrophys. J., 318, 876 (1986).

11. E. N. Parker, Intrinsic magnetic discontinuities and solar x-ray emission, Geophys. Res. Lett., 17, 2055 (1990).

12. E. N. Parker, Heating solar coronal holes, <u>Astrophys. J.</u>, 372, 719 (1991).

13. J.-I. Sakai and B. U. O. Sonnerup, Modulational instability of finite amplitude dispersive Alfvén waves, <u>J. Geophys. Res.</u>, 88, 9069 (1983).

14. C. F. Kennel, B. Buti, T. Hada, and R. Pellat, Nonlinear, dispersive, elliptically polarized Alfvén waves, <u>Phys. Fluids</u>, 31, 1949 (1988).

15. B. Buti, Stochastic and coherent processes in space plasmas, in: <u>Cometary and Solar Plasma Physics</u>, Ed. B. Buti, World Scientific, Singapore, 1988, p. 221.

16. S. J. Bame, J. R. Asbridge, W. C. Feldman, and J. T. Gosling, Evidence for a structure-free state at high solar wind speeds, <u>J. Geophys. Res.</u>, 82, 1487 (1977).

17. R. von Steiger and J. Geiss, Supply of fractionated gases to the corona, <u>Astron. Astrophys.</u>, 225, 222 (1989).

18. W. A. Dziembowski, A. A. Pamyatnykh, & R. Sienkiewicz, Helium content in the solar convective envelope from helioseismology, <u>Mon. Not. R. Astr. Soc.</u>, 249, 602 (1991).

19. R. Steinitz and M. Eyni, Global properties of the solar wind. I. The invariance of the momentum flux density, <u>Astrophys. J.</u>, 241, 417 (1980).

HEATING CORONAL HOLES AND ACCELERATING THE SOLAR WIND

E. N. Parker

*Enrico Fermi Institute and Departments of Physics and Astronomy,
University of Chicago, Chicago, IL 60637, U.S.A.*

ABSTRACT

The special energy requirements of a coronal hole combined with current knowledge of the limited dissipation of Alfven and fast mode MHD waves in the solar corona suggest a unique source of heat for the coronal hole. The near coronal hole ($r \lesssim 3R_\odot$) requires approximately $3 - 4 \times 10^5$ ergs/cm^2 sec, which can come only from the fluid jets, fast particles, and short period ($\lesssim 10$ sec) MHD waves from the network activity. The high speed streams of solar wind from coronal holes show that there is substantial heating, of $1 - 2 \times 10^5$ ergs/cm^2 sec, beyond the sonic point ($r \gtrsim 5R_\odot$) in the wind, which can come only from the dissipation by thermal conduction of long period ($\gtrsim 10^2$ sec) MHD waves from subphotospheric convection. Although the Alfven wave flux from the photosphere is generally taken for granted in the literature, we point out that it is a crucial phenomenon that has yet to be established on either a theoretical or observational scientific basis. The same is true for the energy output of the network activity.

Both the steady quiet solar wind at 400 km/sec and the steady fast solar wind at 500–800 km/sec originate in regions of open magnetic field. Open fields develop in regions where the field is weak and particularly in regions where the field lines extend far out to low field densities. It is the heat input to the coronal gas in the open fields that provides the steady wind, which sweeps out the solar system to create the heliosphere. The X–ray coronal holes are unique in that they provide the fast streams in the wind /1–3/. This indicates a special form for the heat input, different from the quiet wind. Withbroe's /4/ analysis of conditions in a coronal hole leads to an estimated total heat input of 5×10^5 ergs/cm^2 sec, with most of the heat introduced in the first $1 - 2R_\odot$ above the surface of the Sun, to maintain the temperature (1.5×10^6 K) of the slowly rising expanding gas. It is this heat input that largely determines the mass flux of the wind. The gas velocity is of the order of 1 km/sec at the base of the corona (where $N \cong 10^8$ H atoms/cm^3) and accelerates gradually out through the solar gravitational field over a period of a few days to the speed of sound (~ 200 km/sec) at the critical point in the wind at a distance of $2 - -6R_\odot$. Beyond the critical point the gas is essentially free of the gravitational field and expands freely into the vacuum at infinity. If no further heat is added, the final wind velocity is of the order of 300 km/sec /5–8/, providing the quiet solar wind. To produce the fast streams at two or more times 300 km/sec, an additional heat source is required beyond the sonic point /9,10/. Thus, for instance, a heat source that maintains an isothermal corona at 1.5×10^6 K yields a wind velocity of about 600 km/sec at the orbit of Earth. It appears that an introduction of momentum, in addition to heat, is required beyond the critical point to produce wind velocities above 600 km/sec /11,12/.

The essential point is that the fast streams in the solar wind, from the coronal holes, require a special form of heat input, with more than half the heat added near the Sun but with a heat input (an "afterburner") beyond the critical point ($r > 3 - 7R_\odot$). Since coronal holes appear always to produce fast wind, this two part character of the heat input is evidently an automatic property of the coronal hole. That is to say, the theory of coronal heating must provide both parts in a natural way. We shall find what appears to be a unique scenario within the limits of present theory of dissipation of Alfven waves /13/.

Note, then, that photospheric convection provides waves with periods of 100–500 sec, and there is no known theoretical wave dissipation mechanism that damps such waves within the first $5R_\odot$. Therefore, it is not currently possible to base a theory for heating the inner corona ($r < 5R_\odot$) on the dissipation of Alfven waves and fast mode waves from the photosphere. The only alternative of which we are aware is the suggestion that the heat requirements for the near coronal hole can be supplied by the network activity /14–19/. Quantitative studies of the rapid sputtering and popping of microflares, jets, and eruptions in the network magnetic fields /18/ suggests an energy release of the order of 5×10^5 ergs/cm^2 sec, although Esser (these Proceedings) finds substantially less. The characteristic times are of the general order of 1–5 sec (associated with a characteristic scale of a few thousand km or less and coronal Alfven speeds of the order of 10^8 cm/sec), so the motions and fields produced by the activity are expected to dissipate quickly into heat. There is no evident alternative to this idea for the near coronal hole, although it remains to be verified that the total energy output of the network activity is $3 - 4 \times 10^{15}$ ergs/cm^2 sec. So we proceed to the remaining portion of the problem, viz the heat supply beyond the sonic point, i.e. $r \gtrsim 5R_\odot$, that is necessary to provide the observed high velocity of the solar wind from coronal holes.

The heat supply beyond $5R_\odot$ is presumably /13/ the result of Alfven waves and fast mode waves with periods of the general order of magnitude of 10^2 sec, so that they do not damp close to the Sun but survive to $5R_\odot$ and beyond /20/. Of course, the network activity may provide a small fraction of its power at time scales of 10^2 sec, but $1 - 2 \times 10^5$ ergs/cm^2 sec is needed so we must look elsewhere for the major portion. The obvious and traditional source of power is the photospheric and subphotospheric convection.

There are several considerations here. The first is the point already made, that there is no known mechanism that has been shown to damp these waves and prevent their arrival in $r > 5R_\odot$. This has not been properly appreciated in some of the literature, unfortunately, so we provide in the paragraphs below a brief listing and discussion of each of the currently known dissipation processes. Then there are the questions of the generation and of the transmission of a sufficient energy flux ($1 - 2 \times 10^5$ ergs/cm^2 sec) from the photosphere into the corona. As we shall see, it is not entirely obvious that the photospheric convection provides a sufficient wave flux in the weak (10 gauss) fields of the coronal hole (see observational evidence in /21/). The high speed of the wind suggests that, somehow, an adequate wave flux is produced, but that statement by itself does not constitute a theory for the phenomenon.

Consider, then, the rate of damping of Alfven waves and fast mode waves in the corona of the Sun, about which so many words have been written but so few quantitative examples have been supplied. There are a variety of effects that provide damping of Alfven waves under coronal conditions and some of the effects are novel and interesting in their own right. These include Landau damping, surface resonance, hyperresistivity, phase mixing, stochastic field lines, plasma turbulence, and mode coupling, in addition to the classical resistivity, viscosity, and thermal conduction. A detailed evaluation of the efficacy of each effect has been presented elsewhere

/13,22/ and only a few brief remarks on each will be presented here. We begin simply by noting that the Alfven speed C in the coronal hole, where $B = 10$ gauss and $N = 10^8$ atoms/cm^3, is 2×10^8 cm/sec, and the sound speed is 2×10^7 cm/sec, where $T = 1.5 \times 10^6$ K. Hence an Alfven wave energy flux $I = \rho \langle v^2 \rangle C$ of 1×10^5 ergs/cm^2 sec involves $\langle v^2 \rangle^{\frac{1}{2}} = 17$ km/sec. This represents waves of small amplitude, with $\langle v^2 \rangle^{\frac{1}{2}}/C \cong 0.01$. The rms magnetic field fluctuation $\langle b^2 \rangle^{\frac{1}{2}}$ is $(4\pi \rho \langle v^2 \rangle)^{\frac{1}{2}} = 0.002B = 0.08$ gauss. The wavelength for a period of 10^2 sec is $\lambda = 2 \times 10^{10}$ cm. The rms current density is $\langle j^2 \rangle^{\frac{1}{2}} = c \langle b^2 \rangle^{\frac{1}{2}}/2\lambda$ or 0.06 esu. The rms electron conduction velocity is $\langle w^2 \rangle^{\frac{1}{2}} = \langle j^2 \rangle^{\frac{1}{2}}/Ne = 1.2$ cm/sec. It is evident that there is no plasma turbulence generated by the wave, nor are there any appreciable nonlinear effects such as wave steepening or hyperresistivity. The excess of the Alfven speed C over the sound speed of $0.1\, C$ reduces collisionless damping /23–25/ to negligible levels. The kinematic viscosity ν is approximately 2×10^{15} cm^2/sec, so the characteristic damping time $1/k^2\nu$ is 5×10^3 sec (for $\tau = 10^2$ sec, $\lambda = 2 \times 10^{10}$ cm, and $k = 3 \times 10^{-10}$/cm), during which the wave propagates a distance $C/k^2\nu = 10^{12}$ cm $= 14\, R_\odot$. The resistive diffusion coefficient is 10^{11} times smaller and entirely negligible. Thermal conductivity provides the most effective diffusion of fast mode waves and has been studied in detail by Habbal and Leer /20/, taking account of the semi-collisionless nature of the coronal electrons in waves with periods as small as 10^2 sec. They estimate damping over a characteristic length of $5R_\odot$.

Phase mixing has been a popular idea for dissipating Alfven waves in the solar corona /26,27/ but it is expected to be effective only over distances of the order of $10R_\odot$. In fact phase mixing is possible only if there is an ignorable coordinate transverse to the field along which the Alfven wave is propagating /22/. The extreme horizontal inhomogeneity of the transition region (Orrall et al. 1990) suggests that an ignorable horizontal coordinate is a rarity. It also suggests that for almost all ω/k there is one, or more, surfaces in the region where the Alfven speed C is equal to ω/k, providing surface resonance. Lee and Roberts /27/, Davilla /29/, Hollweg and Yang /30/; and Hollweg /31/, among others, have developed the theory of surface resonance, showing the conversion of the transverse Alfven wave to a longitudinal mode at the surface. The longitudinal oscillations are eventually damped by viscosity and thermal conductivity so that the wave energy is converted to heat in distances of the order of $5 - 20R_\odot$ /27/.

The dissipation of Alfven waves propagating along a mean field B with a small stochastic component has been pointed out and developed by Similon and Sudan /32/, with application to the closed bipolar magnetic fields of active regions. Now the field of a coronal hole is stochastic as a consequence of the random displacement of the footpoints of the field in the photospheric convection /33,34/. However, the outer "end" of the field is not anchored anywhere, so any fluctuations about the mean field are themselves linearly propagating Alfven waves, rather than a fixed wandering of the field lines. They are, in fact, the waves whose damping we are discussing. So the stochastic damping is inapplicable to the coronal hole.

This survey of the dissipation rates indicates that Alfven and fast mode waves with periods of the order of 10^2 sec have precisely the properties needed to supply the necessary heat beyond the sonic point in the wind. They penetrate into the corona, and then survive to distances beyond the sonic point where they damp gradually to provide heat and momentum to the wind. The principal dissipation mechanism is thermal conductivity. It follows that Alfven waves with substantially shorter period ($\tau < 50$ sec) are dissipated too near the Sun to supply heat beyond the sonic point. There is no known mechanism besides the Alfven waves with periods of the order of 10^2 sec for supplying a substantial amount of heat beyond the sonic point /35,36/.

As a matter of fact it should be understood that the small transverse scale (horizontal scale) of the photospheric granules compared to the wavelength of 2×10^{10} cm (for a wave period of 10^2 sec) means that the wave vectors are nearly perpendicular to the vertical magnetic field /22/. Hence there are probably no pure Alfven waves, the waves generally being fast mode. Note too that the viscosity applicable to the transverse wave number is enormously reduced by the magnetic field, so that it does not provide significant dissipation.

To take up the next question, then, consider the rate of generation of fast mode MHD waves by the photospheric convection (granules) of scale $\ell_0 \cong 5 \times 10^7$ cm and characteristic velocity $\langle v^2 \rangle^{\frac{1}{2}} \cong 1$ km/sec at the visible surface where $\rho \cong 2 \times 10^{-7}$ gm/cm^3 in a mean vertical magnetic field $B = 10$ gauss. The sound speed is approximately 10^6 km/sec and the Alfven speed $C = B/(4\pi\rho)^{\frac{1}{2}}$ is 0.6×10^4 cm/sec. The characteristic granule turnover time is of the order of $\ell/\langle v^2 \rangle^{\frac{1}{2}} \cong 500$ sec, so that we expect waves of comparable period to be generated in the granules. In fact one expects waves over a broad range of periods $10^2 - 10^3$ sec. The high Reynolds number $(\ell \langle v^2 \rangle^{\frac{1}{2}}/\nu \cong 10^8)$ suggests that the convection is strongly turbulent, but it must be appreciated that a Kolmogoroff spectrum provides a kinetic energy density that varies in direct proportion to the characteristic period. Thus the power in photospheric waves of small period ($<< 10^2$ sec) is small.

On the other hand, theoretical studies of wave propagation upward from the photosphere show, first of all, that only the Alfven mode survives into the corona, where it appears as a fast mode wave because $\mathbf{k}$ is nearly perpendicular to $\mathbf{B}$ /37–40/, and, second, that Alfven waves are strongly reflected at the transition zone /41/. The reflection coefficient is estimated at 0.7 for Alfven waves with periods of 10^2 sec. An /42/ show that reflection is nearly complete for wave periods of 10^3 sec or more, providing trapping and resonance (see also Moore /43/). It appears, then, that the principal power lies in Alfven waves with periods of the order of 10^2 sec with longer periods strongly reflected.

As a matter of fact, Collins /44,45/ pointed out that the Alfven waves are not generated strongly in the photospheric convection because the Alfven speed is so small, $C \cong 0.06\langle v^2 \rangle^{\frac{1}{2}}$, and it is not obvious that Alfven waves are generated with sufficient strength to supply even a modest portion of the heat in the coronal hole. To illustrate the difficulty note that the Alfven speed in a mean field of 10 gauss is only about one sixteenth of the characteristic fluid velocity $\langle v^2 \rangle^{\frac{1}{2}} = 1$ km/sec. So there is a poor impedance match between the granule and the Alfven waves. Thus we note that an Alfven wave train with an rms fluid velocity $\langle u^2 \rangle^{\frac{1}{2}}$ transmits an energy flux

$$I = \rho \langle u^2 \rangle C \quad \text{ergs/cm}^2 \text{ sec.}$$

At the photosphere, an energy flux $I = 1 \times 10^5$ ergs/cm^2 sec involves a fluid motion $\langle u^2 \rangle^{\frac{1}{2}} \cong 10^4$ cm/sec. This may be thought of as a transverse velocity $\langle u^2/2 \rangle^{\frac{1}{2}} = 0.7 \times 10^4$ cm/sec in each of the two directions perpendicular to the field. Such waves are strong, with $\langle u^2/2 \rangle^{\frac{1}{2}} \cong C$, representing hairpin loops in the field. If A is the amplitude of the displacement in any one direction perpendicular to the field, then $\langle u^2/2 \rangle = C^2$ implies that $kA \cong 2$, where the wave number k is ω/C. The wavelength is 6 km and it is not clear how a granule can produce such narrow (6 km) hairpin loops in the field, like a river meandering over a broad plane.

The fact is of course, that the field is not 10 gauss in the photosphere but rather it is 10^3 gauss confined to widely separate intense magnetic fibrils that fill approximately 10^{-2} of the total area. The Alfven speed within each fibril is 6 km/sec, considerably larger than the rms fluid velocity of 1 km/sec in the granule. Thus the fibril is relatively unaffected by the motion at the surface of the granule. We may speculate that motion below the surface causes the individual fibrils to lash back and forth, with a large amplitude at a height of a few times 10^3 km above the photosphere. But that is only a conjecture that would have to be substantiated by observation /21/.

An ideal field for generating Alfven waves would be 2×10^2 gauss filling one fifth of the space, so that the $C \cong 1$ km/sec, providing a good impedance match between the granule and the Alfven waves. This matter should be given close observational scrutiny.

There is a gain in the efficiency of generation of Alfven waves in going to the small scales in the turbulence spectrum in the granules, but, of course, the total available energy is greatly diminished. Thus, with the rms turbulent velocity $v(\ell)$ at a scale ℓ proportional to $\ell^{\frac{1}{3}}$, so that

$$v(\ell) = v(\ell_0)(\ell/\ell_0)^{\frac{1}{3}},$$

it follows that the characteristic period is

$$\tau(\ell) = \tau(\ell_0)(\ell/\ell_0)^{\frac{2}{3}},$$

where $\tau(\ell_0) = \ell_0/v_0(\ell_0)$. There is a good impedance match to Alfven waves at the scale ℓ_A where $v(\ell_A) = C$. It follows that

$$\ell_A = \ell_0[C/v(\ell_0)]^3.$$

For a mean field of 10 gauss with $C = 0.6 \times 10^4$ cm/sec and the granule velocity $v(\ell_0) = 10^5$ cm/sec, it follows that $C/v(\ell_0) = 0.06$ and $\ell_A = 2 \times 10^{-4}\ell_0 = 0.1$ km for $\ell_0 = 500$ km. Then $(\ell_A/\ell_0)^{\frac{1}{3}} = 0.06$ and $\tau \cong 2$ sec. The kinetic energy density of these waves is smaller by the ratio of the periods. Thus with $\ell_0 = 500$ km and $v(\ell_0) = 1$ km/sec again, it follows that $\tau_0 = 500$ sec so that $\tau/\tau_0 = 4 \times 10^{-3}$. For the gas density $\rho \cong 2 \times 10^{-7}$ gm/cm^3 at the visible surface, the kinetic energy density of the largest eddies is $\frac{1}{2}\rho v(\ell_0)^2 = 10^3$ ergs/cm^3, with a kinetic energy density of 4 ergs/cm^3 at $\tau = 2$ sec. If this kinetic energy density is transported at the Alfven speed the energy flux is 2×10^4 ergs/cm^2 sec, which is without significant effect.

In conclusion, there appears to be a unique prescription for heating the coronal hole, based on the near heat input of $3 - 4 \times 10^5$ ergs/cm^2 sec and the distant heat input of $1 - 2 \times 10^5$ ergs/cm^2 sec. The near input arises from the network activity, for which precise observational estimates of the total energy are needed to establish the validity of the theory. The distant heating arises from the dissipation, largely by thermal conduction, of the Alfven or fast mode waves produced by the photospheric convection with periods of 10^2 sec or more. Precise estimates (both observational and theoretical) of the energy in these waves is an essential step in establishing the theory, because it is not obvious how the granule motions can produce a sufficient wave flux. Unfortunately the existence of any desired level of wave flux is generally taken for granted in the literature so this question has not been properly addressed so far.

The high speed solar wind suggests that somehow the necessary wave generation is accomplished (or else there is some unknown and quite different heat source for the distant coronal hole). In this connection we note the anomalous heating of the alpha particles in the solar wind /46/ indicating extensive turbulence and waves in the solar wind /47,48/.

Finally, an interesting effect was noted recently by Moore, et al /49/, that there is a low frequency cutoff for Alfven waves propagating vertically in a stratified atmosphere, just as there is for vertically propagating sound waves. Alfven waves with periods longer than about $3\Lambda/C$, where Λ is the pressure scale height, are evanescent. They show that the maximum period for propagation is of the order of 300 sec in a diverging field in an isothermal atmosphere of

$1 - 2 \times 10^6 K$, with shorter periods for lower temperatures. They note that if the principal heat source to the near coronal is Alfven waves from the photospheric convection, with characteristic times of the order of 300 sec, the effect would act as a thermostat, retaining Alfven waves for eventual dissipation and coronal heating if the temperature falls, and releasing Alfven waves to escape if the temperature rises.

In the context of the scenario outlined in the present paper, wherein the near corona is heated by the short period agitation of the network activity, the process would work differently, with the photospheric Alfven waves prevented from reaching the distant wind if the network activity is too weak and the near corona too cool, but free to escape under ordinary conditions. Unfortunately, we know so little of the quantitative aspects of the network heating, and of the period and intensity of the Alfven waves from the photosphere, that little can be said.

This work was supported in part by the National Aeronautics and Space Administration under NASA grant NAGW-2122.

REFERENCES

1. A.S. Krieger, A.F. Timothy, and E.C. Roelof, Solar Phys. 29, 505 (1973). .

2. L. Svalgaard, J.M. Wilcox, and T.L. Duvall, Solar Phys. 37, 157 (1974).

3. V.G. Eselevich, Solar Phys. 137, 179 (1992).

4. G.L. Withbroe, Ap. J. 325, 442 (1988).

5. E.N. Parker, Ap. J. 139, 93 (1964).

6. E.N. Parker, , Space Sci. Rev. 4, 666 (1964).

7. E. Leer and T.E. Holzer, Solar Phys. 63, 143 (1979).

8. E. Leer, T.E. Holzer and T. Fla, Space Sci. Rev. 33, 161 (1982).

9. E. Leer, T.E. Holzer, J. Geophys. Res. 85, 4681 (1980).

10. T.E. Holzer and E. Leer, J. Geophys. Res. 85, 4665 (1980).

11. J.V. Hollweg and W. Johnson, J. Geophys. Res. 93, 9547 (1988).

12. P. Isenberg, J. Geophys. Res. 95, 6437 (1990).

13. E.N. Parker, Ap. J. 372, 719 (1991).

14. S.F. Martin, in: Dynamical processes in quiet solar atmospheres, ed. S.L. Keil, National Solar Observatory/Sacramento Peak, Sunspot, NM 1984, p. 36.

15. S.F. Martin, Solar Phys. 117, 243 (1988).

16. S.F. Martin, Solar photosphere: Structure, convection and magnetic fields, in: IAU Symposium 138, ed. J.O. Stenflo, D. Reidel, Dordrecht 1990, p. 129.

17. J.G. Porter, R.L. Moore, E.J. Reichmann, O. Engvold, and K.L. Harvey, Ap. J. 323, 380 (1987).

18. J.G. Porter and R.L. Moore, in:Proc. Ninth Sacramento Peak Summer Symposium, ed. R.C. Altrock, National Solar Observatory/Sacramento Peak, Sunspot, NM 1988, p. 30.

19. H. Wang, H. Zirin, and G. Ai, Solar Phys. 131, 53 (1991).

20. S.R. Habbal and E. Leer, Ap. J. 253, 318 (1982).

21. J.V. Hollweg, M.K. Bird, H. Volland, P. Edenhofer, C.T. Stelzreid, and B.L. Seidel, J. Geophys. Res. 87, 1 (1982).

22. E.N. Parker, Ap. J. 376, 355 (1991).

23. A. Barnes, Phys. Fluids 9, 1483 (1966).

24. A. Barnes, Ap. J. 155, 311 (1969).

25. A. Barnes in: Solar system plasma physics, Vol. 1, ed. E.N. Parker, C.F. Kennel and L.J. Lanzerotti, Amsterdam: North Holland 1979, p. 249.

26. L. Nocera, B. Leroy and E.R. Priest, Astron. Astrophys. 133, 387 (1984).

27. M. Lee, B. Roberts, Ap. J. 301, 430 (1986).

28. F.Q. Orrall, G.J. Rottman, R.R. Fisher and R.H. Munro, Ap. J. 349, 656 (1990).

29. J.M. Davila, Ap. J. 317, 514 (1987).

30. J.V. Hollweg and G. Yang, _J. Geophys. Res._ 93, 5423 (1988).

31. J.V. Hollweg, G. Yang, V.M. Cadez, and B. Gakovic, _Ap. J._ 349, 335 (1990).

32. P. Similon and R. Sudan, _Ap. J._ 336, 442 (1989).

33. J.R. Jokipii and E.N. Parker, _Phys. Rev. Letters_ 21, 44 (1968).

34. J.R. Jokipii and E.N. Parker, _Ap. J._ 155, 799 (1969).

35. R.F. Stein, _Ap. J._ 154, 297 (1968).

36. R.F. Stein and R.A. Schwartz, _Ap. J._ 177, 807 (1972).

37. R.F. Stein and R.A. Schwartz, _Ap. J._ 186, 1083 (1973).

38. R.F. Stein and J. Leibacher, _Ann. Rev. Astron. Astrophys._ 12, 407 (1974).

39. M. Kuperus, J.A. Ionson, and D.S. Spicer, _Ann. Rev. Astron. Astrophys._ 19, 7 (1981).

40. E.R. Priest, _Solar magnetohydrodynamics_, D. Reidel, Dordrecht 1982, Chapt. 6.

41. J.V. Hollweg, _Ap. J._ 277, 392 (1984).

42. C.H. An, Z.E. Musielak, R.L. Moore, and S.T. Suess, _Ap. J._ 345, 597 (1989).

43. R.L. Moore, Z.E. Musielak, S.T. Suess, and C.H. An, _Ap. J._ 378, 347 (1991).

44. W. Collins, _Ap. J._ 343, 499 (1989).

45. W. Collins, _Ap. J_ 384, 319 (1992).

46. K.M. Thieme, E. Marsch, and H. Rosenbauer, _J. Geophys. Res._ 94, 2673 (1989).

47. P.A. Isenberg and J.V. Hollweg, _J. Geophys. Res._ 88, 3923 (1983).

48. C.V. Tu and E. Marsch, _J. Geophys. Res._ 95, 4337 (1990).

49. R.L. Moore, Z.E. Musielak, S.T. Suess, and C.H. An, _Ap. J._ 378, 347 (1991).

MODELLING OF OPEN AND CLOSED CORONAL STRUCTURES: COMPARISON WITH DETAILED EUV OBSERVATIONS

G. Peres and D. Spadaro

Osservatorio Astrofisico di Catania, Catania, Sicily, Italy

ABSTRACT

We consider the modelling and EUV diagnostics of plasma in steady state motion both within closed coronal structures (siphon flows) and outflowing from coronal holes toward interplanetary space. We take into account non-equilibrium ionization in the synthesis of emission lines originating from the modelled closed structures and compare the computed line intensities with detailed EUV observations, in order to constrain significantly the model. We evaluate the importance of non-equilibrium ionizations effects for some published coronal hole models.

INTRODUCTION

We are working on the detailed modelling of the transition region and lower corona portion of closed solar structures as well as solar wind source regions. Past experience on these studies has shown the validity and effectiveness of the comparison between the emission synthesized from realistic models and detailed observations (cf., for instance, Pallavicini et al. /1/). In its more general form our approach is based on the formulation of accurate hydrodynamic models of the atmosphere, either open to the interplanetary space or confined in magnetically closed loops, and on the synthesis of the ensuing emission, either in bands or lines, which we compare with accurate observations of the relevant structures. Whenever appropriate, we take into account non–equilibrium effects in calculating the EUV line emission originating from the transition region of loop siphon flows or wind outflows; in fact the plasma flow through steep temperature gradients can give rise to significant departures from ionization equilibrium /2/. Here we present some results on closed structures and coronal holes.

SIPHON FLOW MODEL

Our first analysis pertains to models of steady siphon flows in coronal loops, in order to compare the intensities of some EUV transition region emission lines synthesized from these models with representative and rather detailed observations of typical solar regions reported by Pallavicini et al. /1/. We have adopted the steady state siphon flow model formulated by Antiochos /3/, and based on the following assumptions: fully ionized plasma, with solar elemental abundances; constant loop cross-section; negligible gravity (the pressure scale height is larger than the loop height for all the cases here considered); constant heating per unit volume along the loop; subsonic flow along the whole loop. We have also assumed that the axis of the loop is semicircular, with the center of the circle on the solar surface, and that all the physical variables change only along the length of the loop. (We refer to /3/ for a detailed description of the steady flow model.)

We have solved the fluid equations using the same procedure described in Noci et al. /2/; the footpoints temperature is 3×10^4 K. By choosing three different sets of values of density, velocity, and temperature gradients at the footpoints, we have computed three loop models – whose physical characteristics are reported in Table 1 –, aimed at reproducing as closely as possible the overall physical parameters (i.e., total loop length, base pressure, maximum temperature at the loop apex) of the static loop models discussed by Pallavicini et al. /1/ which are pertinent to the observed structures. The final fitting is aimed at the line intensities, as explained below.

TABLE 1   Characteristics of the Loop Models

Loop	T_{top} $(10^6\ K)$	$N_e^{(a)}$ $(10^9\ cm^{-3})$	$v^{(a)}$ $(10^5\ cm\ s^{-1})$	Total Length $(10^9\ cm)$
1	3.34	109. − 0.976	0.055 − 6.12	16.9
2	3.34	362. − 3.24	0.132 − 14.8	4.04
3	2.17	14.5 − 0.200	0.165 − 12.0	32.1

$(^a)$ Range from base to top.

Note: The base temperature of all the loop models is 3×10^4 K.

LINE SYNTHESIS

From the solutions of the fluid equations (electron density, flow velocity, and plasma temperature along the loop), we have determined the distributions of the number density of all the carbon and oxygen ions within each loop model, both with and without the approximation of ionization equilibrium. Deviations from ionization equilibrium are indeed important in all the three models here considered (for more details, cf. Peres, Spadaro and Noci /4/). We refer the reader to Noci et al. /2/ and Spadaro et al. /5/ for the detailed description of the calculation of the ion densities. We have determined the emissivity along the upflowing and the downflowing legs for each loop model for the emission lines: C III (977.0 Å), O IV (554.4 Å), O V (629.7 Å), O VI (1031.9 Å); using Withbroe's expression for the total emissivity /6/. We obtain the total intensity of a spectral line emitted by one leg of the loop structure, $I_{(leg)}$, by integrating the line emissivity along the loop length from the lower boundary to the apex. We have considered the non−equilibrium ion densities in the calculation of carbon and oxygen line emissivities from loop models.

The experimental data of Pallavicini et al. /1/ are values averaged over the brightest data points of EUV rasters, and are representative of typical EUV line intensities from the observed structures. Therefore, in order to have an immediate comparison with such average values of intensity, we have also computed the quantity $I_{(loop)}$ for each loop model and for each line analyzed, calculating the arithmetic mean between the values of $I_{(leg)}$ obtained for both the legs of the loop.

LOOP MODEL RESULTS

Table 2 reports, for each model and line, the total intensities in the upflowing leg and in the downflowing one, as well as their average, calculated considering non−equilibrium ionization. The intensity from the upflowing leg is always higher than that from the downflowing one for two reasons: first, in the upflowing leg the emitting ions here considered are overpopulated while they are underpopulated in the downflowing one; second, the temperature gradient is lower in the upflowing leg (cf., also, /2/, /3/) and therefore the integral of the emissivity, which increases with the width of the transition region, is larger for the upflowing leg of the loop.

We incidentally note that siphon flow loop models with uniform heating predict a prevalence of blue−shifted radiation (e.g., /3/), at variance from the observed prevalence of red−shifts seen in transition region lines (for a review, cf. /7/). Our approach, instead, aims only at exploring whether siphon flow models, with proper attention to non−equilibrium ionization, can reproduce well observed total line intensities. We leave more refined analyses for future work.

In Table 3 the values of line intensity calculated from the loop models are compared with the data reported by Pallavicini et al. /1/, here presented together with their errors; there we report also the line intensities synthesized by Pallavicini et al. from static loop models. The observational data are pertinent to the structures identified by Pallavicini et al. as "loop structures extending over a typical solar active region" (for comparison with the results of loop 1), "compact high−pressure loops in the core of the region" (loop 2) and "large−scale loops interconnecting active regions" (loop 3). The comparison shows that the average values of the total intensity here calculated are in general agreement with the observational data reported by Pallavicini et al. /1/; in fact they

TABLE 2 Total Line Intensities ($erg\ cm^{-2}\ s^{-1}sr^{-1}$)

	Line	$I_{(upflowing\ leg)}$	$I_{(downflowing\ leg)}$	$I_{(loop)}$ [a]
Loop 1				
	C III	1.11×10^4	5.42×10^2	5.82×10^3
	O IV	1.35×10^4	1.81×10^3	7.63×10^3
	O V	1.98×10^4	2.84×10^3	1.13×10^4
	O VI	6.26×10^3	1.90×10^3	4.08×10^3
Loop 2				
	C III	4.07×10^4	7.07×10^1	2.04×10^4
	O IV	4.42×10^4	1.18×10^3	2.27×10^4
	O V	7.10×10^4	2.96×10^3	3.70×10^4
	O VI	2.32×10^4	3.52×10^3	1.34×10^4
Loop 3				
	C III	1.68×10^3	1.32	8.39×10^2
	O IV	1.73×10^3	4.22×10^1	8.84×10^2
	O V	2.80×10^3	1.05×10^2	1.45×10^3
	O VI	9.39×10^2	1.49×10^2	5.44×10^2

[a] Average between upflowing and downflowing leg.

TABLE 3 Loop Total Line Intensities ($erg\ cm^{-2}\ s^{-1}\ sr^{-1}$)

	Line	This Work [a]	Observed ($\mp$Error)	Static Model [b]
Loop 1				
	C III	5.82×10^3	$6.11(-2.20; +1.69) \times 10^3$	1.07×10^4
	O IV	7.63×10^3	$1.12(-0.34; +0.36) \times 10^4$	9.15×10^3
	O VI	4.08×10^3	$3.59(-0.75; +1.51) \times 10^3$	4.26×10^3
Loop 2				
	C III	2.04×10^4	$1.64(-0.57; +0.62) \times 10^4$	3.64×10^4
	O IV	2.27×10^4	$3.64(-1.38; +1.65) \times 10^4$	2.95×10^4
	O VI	1.34×10^4	$1.56(-0.74; +0.58) \times 10^4$	1.48×10^4
Loop 3				
	C III	8.39×10^2	$1.11(-0.52; +1.01) \times 10^3$	1.38×10^3
	O IV	8.84×10^2	$2.21(-1.48; +2.57) \times 10^3$	1.11×10^3
	O VI	5.44×10^2	$4.25(-2.45; +3.00) \times 10^2$	5.04×10^2

[a] Average between upflowing and downflowing leg.   [b] From Pallavicini et al. /1/.

always fall within the error bars, except for the O IV line in loop 1 which is marginally out of the error bars, and on the average fall closer to observational data than those of the static model. If we take into account the root mean square deviation of results from observational data (each deviation being divided by the respective observational error), it turns out that, on the whole, loop 1 and 2 of this work fit observations better than the corresponding static models; loop 3, instead, fits the observations a little worse than the static model but it is still quite acceptable.

Therefore it appears that, as for low temperature EUV line intensities, steady siphon flow models of compact active region loops in general agree with the observational data of Pallavicini et al. /1/ better than the static loop models there reported, while for the large interconnecting loop (loop 3) the static loop model works slightly, but not significantly, better. Loop 3 is the longest one and its height is smaller than, yet comparable to, the pressure scale height; since the code used in this work does not consider solar gravity, the discrepancy could be ascribed also to a slightly inexact vertical stratification of plasma. In alternative we can speculate that these results might reflect an important characteristic of loop physics, i.e., that small, recently emerged and highly dynamic loops in active regions might be the site of marked plasma flows while old, larger loops, might be closer to a static configuration.

EXPLORATORY CORONAL HOLE STUDIES

We are developing an accurate model for flows in open coronal regions. This model shall be used, among other things, to perform detailed comparison of emission synthesized in appropriate lines and bands with relevant on-disk observations. As an exploratory analysis we have studied some physical implications derived from published models. In particular we have computed from results on hydrodynamic coronal hole models (Rosner and Vaiana /8/; Withbroe /9/) the characteristic recombination and ionization time scales of seven ions emitting prominent lines which we report in Table 4. We have found that everywhere in the transition region and lower corona of such models the recombination and ionization time (τ_i) is lower, at least by an order of magnitude, than the characteristic variation time of plasma temperature due to the steady outward flow (τ_f). (See /2/, for a definition of τ_i and τ_f.) As an example, we report in Table 4 the characteristic times obtained at several temperatures for the model of Rosner and Vaiana /8/.

TABLE 4   Characteristic Times at Different Temperatures[a]

Ion		C IV	O V	O VI	Ne VII	Ne VIII	Mg X	Si XII
T (K)	τ_f (s)				τ_i (s)			
1.02×10^5	222	13.75	4.57	7.95	4.68	5.90	4.68	4.17
2.75×10^5	607	8.37	11.07	29.0	9.16	23.04	18.28	12.95
3.42×10^5	898	6.27	12.42	36.47	12.98	36.64	25.96	20.62
6.49×10^5	3297	3.49	9.06	32.58	27.73	75.24	86.48	77.63
7.94×10^5	5382	3.45	7.83	30.02	32.99	82.52	111.83	115.08
1.03×10^6	24584	4.16	9.41	32.88	45.23	117.42	181.3	190.55
1.15×10^6	37370	5.35	9.65	32.67	55.80	145.77	250.63	305.9
1.27×10^6	1.58×10^5	9.76	17.6	59.65	101.89	266.16	457.61	558.

[a] Values obtained for the coronal hole model of Rosner and Vaiana /8/.

Therefore the approximation of ionization equilibrium can be safely applied to the line synthesis from such flow models over the whole temperature range examined. An analogous estimate will be performed on the flow models under development in order to obtain a realistic line synthesis.

We acknowledge financial support by Agenzia Spaziale Italiana (contract No. ASI–91–RS–91).

REFERENCES

1. R. Pallavicini, G. Peres, S. Serio, G.S. Vaiana, L. Golub and R. Rosner, Astrophys. J. 247, 692 (1981)

2. G. Noci, D. Spadaro, R.A. Zappalà and S.K. Antiochos, Astrophys. J. 338, 1131 (1989)

3. S.K. Antiochos, Astrophys. J. 280, 416 (1984)

4. G. Peres, D. Spadaro and G. Noci, Astrophys. J., in press (1992)

5. D. Spadaro, R.A. Zappalà, S.K. Antiochos, G. Lanzafame and G. Noci, Astrophys. J. 362, 370 (1990)

6. G.L. Withbroe, in: Activity and Outer Atmospheres of the Sun and Stars, ed. A.O. Benz, Y. Chmielewski, M.C.E. Huber, and H. Nussbaumer, Sauverny, Observatoire de Geneve 1981, p. 1.

7. R.G. Athay and K.P. Dere, Astrophys. J. 346, 514 (1989)

8. R. Rosner and G.S. Vaiana, Astrophys. J. 216, 141 (1977)

9. G.L. Withbroe, Astrophys. J. 325, 442 (1988)

VARIATION OF THE GREEN CORONA IRRADIANCE OVER THE SOLAR CYCLE

V. Rušin and M. Rybanský

Astronomical Institute, Slovak Academy of Sciences, 059 60 Tatranska Lomnica, Czechoslovakia

ABSTRACT

The green corona irradiance varies on a very wide range of scales: from days to cycles (of 1.74 to 27.02 x 10^{-16} W/sr in its daily flux). Short-term intensity variations, of about 20 %, periodic (waves or oscillations from 5.7 s to 5 minutes) or non-periodic (from 2 s to 20 min), are practically observed at any time in the green and red emission corona. Stable and isolated coronal holes or coronal streamers can not be responsible for sudden geomagnetic disturbances or highly geomagnetic activity. An additional source, e.g. flare, eruptive prominence or any dynamic process is required.

INTRODUCTION

Spots, prominences and the solar corona were one of the first features to show their variations over solar cycle. This was already known at the beginning of our century /1/. The structure of the white-light corona and its form (shape) during a solar eclipse were used for such a type of study. Later on, several new measurements in the solar corona were introduced such as the integral brightness of the white-light corona, relative or absolute intensities of emission spectral lines etc. to study its variations over solar cycles, and to compare these variations with similar solar phenomena. Rybanský /1/ proposed a new index of solar activity, called coronal index or CI. This index, expressed in W/sr, represents an energy output (irradiance) in the green corona (530.3 nm), emitted by the whole solar corona to the Earth. The CI may be very easily compared with similar full-disk solar indices, obtained at present in the whole range of the electromagnetic spectrum. In this article we present very briefly: (1) the temporal variations of the CI over the period 1964-1990, (2) its comparison with other indices, and (3) the green corona intensity distribution during the last solar minimum (1986) when high geomagnetic activity was occuring.

RESULTS AND DISCUSSION

Monthly mean variations of the CI from January 1964 to December 1990 are shown in Figure 1, where for a comparison, the monthly mean Wolf·s number and cosmic ray data are shown too. Cycle-to-cycle variability, as well as other periodic or non-periodic variabilities are clearly seen. Results of periodic variations, as derived from the CI are given in Table 1 /3/. In the crudest sense, the agreement between courses of individual phenomena exists. However, differences exist in the details.

A relation between the Wolf·s number (R),the cosmic ray,the 2800
MHz radio flux and the CI,is not unambiguous.For example, if the
CI is increasing, it means that an intensity of the green coro-
(the ion Fe XIV) radiation is growing above the active region.
However, if there are not sunspots in the active region or their
number is nearly stable, the R is not growing. Apart from this,
the R is derived from the visible part of the Sun, and the CI
from observations above the sun·s limb only. A long-term decreasing of cosmic ray is caused by the increasing of the entire heliosphere volume.The heliosphere volume is not unambiguously given in relation to the radiation from the inner corona emission
lines, as the CI is derived. The interpretation of the 2800 MHz
radio flux is very complicated,because its intensity depends not
only on the intensity source on the Sun but from propagation conditions as well (it depends on the matter distribution between
the source and observer). The time shift between both the R and
CI in the maximum cycle 22 is not fully explained. Nevertheles,
the full-disk total magnetic flux or the X-ray indices show nearly the same course than the CI in cycle 21 /4/.

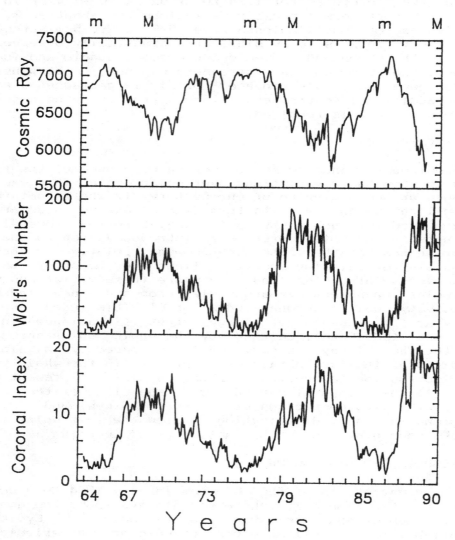

Fig. 1. Monthly means of CI 1964-1990, the Wolf's number
         and cosmic ray; m - minimum, M - maximum.

On the other hand, the observations of short-term intensity variations which have been made with the TV-photometer, have confirmed not only the existence of 5 min oscillations in the green
and red corona, but showed the existence of shorter ones, e.g.
30, 12 and 7.5 s /5/. Moreover, many non-periodic intensity enhancements (up to 20 % of intensity) with duration of 2-20 s are
observed in the solar corona. While the intensity oscillations
are proposed to be connected with the existence of different type of waves in the corona, occurrence of non-periodic intensity
enhancements, both in the corona and background, might be connected with the existence of nanoflares, as discussed by Parker
/6/. Both features, oscillations and non-periodic variations may
play very important role in the corona heating.

The best conditions for the solar-terrestrial study are being
expected over solar minimum. Usually, a few isolated active regions are distributed on the solar surface, and involved geomagnetic activity should be unambiguously related to these solar
activity phenomena.

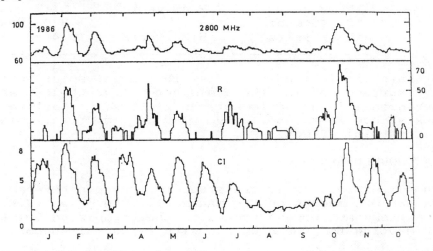

Fig. 2. The 2800 MHz radio flux, the Wolf·s number and
the CI in 1986 (a minimum of cycle 21-22).

The last solar cycle minimum has occurred in the year 1986 (September), even if the minimum values of the CI were observed in
February 1987 /7/. Daily plots of the Wolf·s number, the 2800 MHz
radio flux, and the CI in the year 1986 are shown in Figure 2. It
seems that this last solar minimum differed from earlier minima.
As was already pointed out by Rušin and Rybanský /7/, green coronal intensities, observed over this period, reached very high values above the photospheric active regions as compared with earlier minima. The green corona brightness at the beginning of the
year 1986 was connected with highly geoeffective photospheric
regions ARs 4711 and 4713. A space between these active regions,
similarly as for the surrounding areas in the north-south direction showed very low intensities, usually compared with coronal
holes. The smallest green coronal intensities occurred in August
and September 1986. A new increase of the green corona intensities was observed to the end of 1986 and was connected with ARs
4755 and 4754, still belonging to the cycle 21, and with AR 4750,
belonging to the new solar cycle 22. However, the extremely high
geomagnetic activity occurred only in February 1986, when a series of flares originated in AR 4711 with following CMEs.

One may confirm earlier results /7,8/, that high geomagnetic activity in February 1986 was connected with a series of flares
(an impulse) which arose in the active region connected with
a dense corona (a material to the CMEs) rather than isolated coronal holes /9/. Existence of coronal holes or dense streamers
without an original impulse,e.g. flare or eruptive prominence in
their surrouding or directly in them, can not probably involve
sudden geomagnetic disturbances or highly geomagnetic activity.

TABLE 1 Significant Intermediate and Long-term
Periodicities as Derived from CI

Cycle 20	Cycle 21	1964-1987
88.8 days	75.0 days	-
-	185.8 days	182.3 days
-	-	345 days
2.19 years	2.49 years	2.17 years
3.045 years	-	3.03 years
5.07 years	5.24 years	10.44-11.49 years

We note that time-latitudinal distribution of solar activity on
the solar surface and in the atmosphere, e.g. the N-S asymmetry
/10/, meridional large-scale motions /11,12/, an "active longitude" existence /13/ etc., may play an important role in the
solar-terrestrial effect.

ACKNOWLEDGMENTS

The authors express their gratitude to unknown referee for his
(her) valuable comments on the manuscript.One of us (V.R.) would
like to express my deepest thanks to the Organizing Comittee of
the SW 7 meeting to providing him with financial support.

REFERENCES

1. V. Rušin and M. Rybanský, Slnečná koróna, VEDA, Bratislava,
   (1990)
2. M. Rybanský, Bull. Astron. Inst. Czechosl. 26, 367 (1975)
3. V. Rušin and J. Zverko, Sol. Phys. 128, 261 (1990)
4. V. Rušin and M. Rybanský, in Proceedings of SOLERS 22 Workshop, ed. R.F. Donnelly, Boulder, in press (1991)
5. V. Rušin and M. Minarovjech, in:Mechanisms of Chromospheric
   and Coronal Heating, ed. P. Ulmschneider, E.R. Priest and
   R. Rosner, Springer-Verlag, Berlin 1991,p.30.
6. E.N. Parker, this issue.
7. V. Rušin and M. Rybanský, Bull. Astron. Inst. Czechosl. 41,
   263 (1990)
8. H.A. Garcia and M.Dryer, Sol. Phys. 109, 119, (1987)
9. A. Hewish, Sol. Phys. 116, 195, (1988)
10. V. Rušin and E. Dzifčáková, Bull. Astron. Inst. Czechol.
    41, 69, (1990)
11. H.B. Snodgrass, Sol. Phys. 110, 35, (1987)
12. V. Bumba, V. Rušin and M. Rybanský, Bull. Astron. Inst.
    Czechosl. 41, 69 (1990)
13. V. Bumba and R. Howard, Astrophys. J. 141, 1502 (1965)

A TWO-FLUID MODEL OF THE SOLAR WIND

Ø. Sandbæk,* E. Leer* and T. E. Holzer**

*Institute of Theoretical Astrophysics, University of Oslo, PO Box 1029,
Blindern N-0315 Oslo, Norway*
***High Altitude Observatory, National Center for Atmospheric Research,
PO Box 3000, Boulder, CO 80307, U.S.A.***

ABSTRACT

A two-fluid model of the solar wind was introduced by Sturrock and Hartle [1] (see also [2]). In these studies the heat conduction term was neglected in the proton energy equation. The full two-fluid solar wind equations with proton heat conduction were solved by Durney [3], Couturier [4], and Cuperman et al. [5], but most of the results are for relatively low coronal base densities and high coronal base temperatures. Here we present a method for integrating the two-fluid solar wind equations, which is applicable to a broad range of coronal base densities and temperatures.

INTRODUCTION

The two-fluid solar wind model introduced by Sturrock and Hartle [1] showed that the one-fluid model, where $T_e = T_p = T$, could not be justified assuming that electrons and protons interact by Coulomb collisions in the solar wind plasma. In their works (i.e. [1], and [2]) the divergence of the proton heat conduction flux was neglected in the proton energy equation. Later several authors have integrated the full two-fluid model equations (e.g. [3], [4], [5]). In these works dimensionless variables were used in a simultaneous integration of the two-fluid model equations. The integrations were started at the critical point and the equations were integrated (simultaneously) outward to infinity and toward the coronal base imposing the boundary conditions that the gas pressure vanishes at infinity and that the flow speed equation has a subsonic-supersonic solution passing continously through the critical point.

Here we present a method for integrating the two-fluid model equations. We use the dimensional (ordinary) form of the equations and solve the equations for a given solar wind proton flux. An iteration procedure is used where the integration of each equation is carried out separately. First the energy balance equation for the flow is integrated inwards from an outer boundary and an electron temperature profile is found. Then the flow speed equation is integrated from the coronal base, through a critical point, to the outer boundary. Finally the proton energy equation is integrated inwards, and we find a proton temperature profile. This procedure is continued in an iterative process until the solutions satisfy certain convergence criteria.

In this paper we present a mathematical description of the model and the numerical method we use. We also present some results.

MATHEMATICAL DESCRIPTION

We consider a steady, radial, spherically symmetric flow of electrons and protons. The equations for conservation of mass and momentum can be written as ([6], [1], [2])

$$mnur^2 = (mnur^2)_E \tag{1}$$

and

$$nmu\frac{du}{dr} = -\frac{dp}{dr} - nm\frac{GM_\odot}{r^2} \tag{2}$$

where r is heliocentric distance, $n = n_e = n_p$ is the electron (proton) number density, u is the radial flow speed, $p = p_e + p_p$ is the gas pressure, where $p_e = nkT_e$ and $p_p = nkT_p$, m is the proton mass, $M_\odot$ is the solar mass, and G is the gravitational constant. Subscripts "e" and "p" refer to electrons and protons respectively and "E" refers to 1 AU. k is the Boltzmann constant and T_e and T_p are the electron and proton temperatures.

The energy equations for electrons and protons can be written as (cf. /1/ and /2/)

$$\frac{n^\gamma}{\gamma - 1}u\frac{d}{dr}\left(\frac{p_e}{n^\gamma}\right) = \frac{1}{r^2}\frac{d}{dr}\left(r^2\kappa_{eo}T_e^{5/2}\frac{dT_e}{dr}\right) + \frac{3}{2}\nu nk(T_p - T_e) \tag{3}$$

$$\frac{n^\gamma}{\gamma - 1}u\frac{d}{dr}\left(\frac{p_p}{n^\gamma}\right) = \frac{1}{r^2}\frac{d}{dr}\left(r^2\kappa_{po}T_p^{5/2}\frac{dT_p}{dr}\right) + \frac{3}{2}\nu nk(T_e - T_p) \tag{4}$$

We use transport coefficients κ_{eo}, κ_{po}, and ν as given by Braginskii /7/ and $\gamma = \frac{5}{3}$. By adding equations (3) and (4) and neglecting the proton heat conduction flux, as compared to the electron heat conduction flux, combining equations (1) and (2), and keeping equation (4) we obtain the equations we integrate numerically:

$$nu\left[\frac{3}{2}k\left(\frac{dT_e}{dr} + \frac{dT_p}{dr}\right) + k(T_e + T_p)\left(\frac{1}{u}\frac{du}{dr} + \frac{2}{r}\right)\right] = \frac{1}{r^2}\frac{d}{dr}\left(r^2\kappa_{eo}T_e^{5/2}\frac{dT_e}{dr}\right) \tag{5}$$

$$\frac{1}{u}\frac{du}{dr}\left(u^2 - \frac{k(T_e + T_p)}{m}\right) = \frac{2}{r}\frac{k(T_e + T_p)}{m} - \frac{k}{m}\left(\frac{dT_e}{dr} + \frac{dT_p}{dr}\right) - \frac{GM_\odot}{r^2} \tag{6}$$

$$nu\left[\frac{3}{2}k\frac{dT_p}{dr} + kT_p\left(\frac{1}{u}\frac{du}{dr} + \frac{2}{r}\right)\right] = \frac{1}{r^2}\frac{d}{dr}\left(r^2\kappa_{po}T_p^{5/2}\frac{dT_p}{dr}\right) + \frac{3}{2}\nu nk(T_e - T_p) \tag{7}$$

where $nu = (nur^2)_E/r^2$.

METHOD OF SOLUTION

Equations (5)-(7) are integrated numerically using a fourth-order Runge-Kutta method. To solve equations (5)-(7) we use the following iteration procedure:

1. We specify the coronal base temperature, $T_e(r_o) = T_p(r_o) = T_o$, and the proton flux density at 1 AU, $(nu)_E$. The proton flux is kept constant during the iterations. A value of $(nu)_E$ corresponds to a value of the coronal base electron density, $n_o = n(r_o)$, for a given T_o.

2. We guess profiles for $T_p(r)$ and $u(r)$ and integrate equation (5) inwards from a heliocentric distance $r_1 \approx 1100R_\odot$ to the coronal base to find the electron temperature, $T_e(r)$. We adjust $T_e(r_1)$ until the calculated base temperature differs with less than 0.01% from the specified base temperature T_o. (Notice that we take $\frac{dT_e}{dr} = -\frac{2}{7}\frac{T_e}{r}$ at the outer boundary. If heat conduction is not a dominant term in equation (5) the temperature gradient is quickly adjusted.)

3. We integrate equation (6) for the flow speed, $u(r)$, outward from $r = 1R_\odot$ through a critical point using the calculated T_e-profile and the assumed T_p-profile. We apply a "splitting-shooting" method (cf. /4/) to pass the critical point.

4. We integrate equation (7) for the proton temperature, $T_p(r)$, inward from $r = r_1$ using the calculated u- and T_e-profiles. We adjust the value of $T_p(r_1)$ and use a "splitting-shooting" method to integrate into the collision dominated region above the coronal base. We can integrate equation (7) to a distance r_x before the proton temperature profiles diverge. The collision term in equation (7) causes the proton temperature to be unstable for $r < r_x$; i.e. the proton temperature diverges "up" in $r < r_x$ if $T_p(r_x) > T_e(r_x)$, and it diverges "down" in $r < r_x$ if $T_p(r_x) < T_e(r_x)$. For $T_p(r_x) > T_e(r_x)$ the collisional heat transfer to the electrons is consistent with a large proton heat flux from the Sun and $T_p(r)$ increases with decreasing values of r. For $T_p(r_x) < T_e(r_x)$ the collisional heat transfer to the protons is consistent with a large (and increasing) proton heat flux toward the Sun and $T_p(r)$ decreases rapidly with decreasing values of r. The solutions diverging "up" and "down" follow the same $T_p(r)$ profile in $r_x < r < r_1$. We iterate until $(T_e(r_x) - T_p(r_x))/T_p(r_x) < 2\%$. $T_p(r)$ is interpolated between r_x and the coronal base. r_x lies typically between 1.0 and $2.0R_\odot$, depending on the coronal base density. (We take $\frac{dT_p}{dr} = -\frac{2}{7}\frac{T_p}{r}$ at the outer boundary, when heat conduction is an important term in equation (7)).

This procedure is repeated, i.e. **a)** we integrate equation (5) to find $T_e(r)$ using the calculated $u(r)$ and $T_p(r)$ profiles, **b)** we integrate equation (6) to find $u(r)$ using the calculated $T_e(r)$ and $T_p(r)$ profiles, and **c)** we integrate equation (7) to find $T_p(r)$ using calculated $T_e(r)$ and $u(r)$ profiles. The process is continued until the value for the flow speed at the base of the corona changes by less than 0.1% between the iterations. When this accuracy is obtained the values of flow speed and temperatures at 1 AU vary by less than 0.2% between the iterations. The desired convergence is obtained after 10-15 iterations. For values of n_o and T_o for which the total energy per unit mass in the flow is small and the asymptotic flow speed is very low the convergence is not as good as described above.

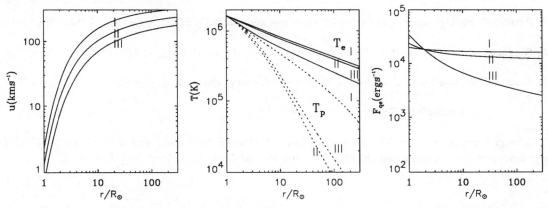

Fig. 1. The flow speed, u, temperatures, T_e and T_p, and electron heat conductive flux from an area $A_o = 1cm^2$ at the coronal base, F_{qe}, versus heliocentric distance, r, in a spherically symmetric thermally driven electron proton solar wind model. The coronal base temperature is $T_o = 1.65 \cdot 10^6 K$. Curves I, II, and III are for proton flux densities at 1 AU $(nu)_E = 0.3 \cdot 10^8$, $1.0 \cdot 10^8$, and $3.0 \cdot 10^8 cm^{-2}s^{-1}$ respectively (1 AU=$215R_\odot$).

SOME RESULTS

Figure 1 shows profiles for flow speed, u, temperatures, T_e and T_p , and electron heat conduction flux from an area of $1cm^2$ at the coronal base, $F_{qe} = -\left(\frac{r}{r_o}\right)^2 \kappa_{eo} T_e^{5/2} \frac{dT_e}{dr}$. The solutions shown are for a coronal base temperature $T_o = 1.65 \cdot 10^6 K$. Curves I, II, and III correspond to proton flux densities at 1 AU of $(nu)_E = 0.3 \cdot 10^8$, $1.0 \cdot 10^8$, and $3.0 \cdot 10^8 cm^{-2}s^{-1}$ respectively. These proton flux densities correspond to coronal base electron densities of $n_o = 6.0 \cdot 10^6$, $3.5 \cdot 10^7$, and $2.0 \cdot 10^8 cm^{-3}$. The figure shows that for low proton fluxes the electron temperature varies as $r^{-2/7}$ and that the proton temperature is high throughout the flow. By increasing the proton flux density (i.e. the base density) the temperatures decrease more rapidly with heliocentric distance. Hence, the flow speed also decreases. Adiabatic cooling is much stronger in the proton gas than in the electron gas due to a much lower heat conduction flux, and for higher proton fluxes (case II) the proton temperature at 1 AU is very low. In case III, however, the high coronal base density leads to an increase of the collisional heating of the protons and the proton temperature at 1 AU increases with increasing proton fluxes in this parameter range. Note that in case III adiabatic cooling of the solar wind electrons is significant also in the subsonic region of the flow. Hence, the increase in the proton flux with increasing n_o is slower than linear (cf. /8/).

CONCLUSION

The method presented here for integration of the two-fluid solar wind equations , with proton heat conduction, can be used for coronal base conditions for which subsonic-supersonic solar wind solutions exists. The method is relatively simple to implement.

REFERENCES

1. Sturrock, P. A. and Hartle, R. E., Two-fluid model of the solar wind, Phys. Rev. Letters 16, 628 (1966)

2. Hartle, R. E. and Sturrock, P. A., Two-fluid model of the solar wind, Astrophys. J. 151, 1155 (1968)

3. Durney, B. R., Solar wind properties at the Earth as predicted by the two-fluid model, Solar Phys. 30, 223 (1973)

4. Couturier, P., A complete numerical solution of the two fluid solar wind model using the shooting-splitting method of integration, Astron. Astrophys. 59, 239 (1977)

5. Cuperman, S., Detman, T., and Dryer, M., Effect of coupled electron-proton thermal conductivities on the two-fluid solutions for the quiet solar wind, Astrophys. J. 330, 466 (1988)

6. Parker, E. N., Dynamics of the interplanetary gas and magnetic fields, Astrophys. J. 128, 664 (1958)

7. Braginskii, S. I., Transport processes in plasmas, Rev. of Plasma Phys. 1, 205 (1965)

8. Sandbæk, Ø., and Leer, E., Adiabatic cooling of solar wind electrons, J. Geophys. Res., in press (1992)

Acknowledgement:

This work was supported by the Norwegian Research Council for Science and the Humanities (NAVF), under contracts D.428.91/012 and D.428.91/052 and by NASA under contract W-17016.

STRUCTURE OF SOLAR CORONAL STREAMERS

C. G. Schultz

Space Research Laboratory, University of Turku, Tykistökatu 4, SF-20520 Turku, Finland

ABSTRACT

We present here a new numerical method for solving generalized potential problems. The method is a direct method, effectively a shooting method, working outwards from an O-point, under the assumption that flux surfaces exist. The standard difficulties encountered in calculating equilibria with vanishing flux gradients, such as neutral sheets in solar coronal streamers and internal separatrices in tokamaks, are traded for a different kind of difficulties.

INTRODUCTION

The model for a large helmet streamer calculated by Pneuman and Kopp (1971, /1/) is a current sheet above the closed flux tube, distinguishing it from the fully developed flare. Adding more physics gives a better understanding of the quasi-steady state solar wind plasma /2/. However, a time-dependent calculation of secondary phenomena such as reconnection becomes prohibitively expensive unless a good time-independent equilibrium is available, in analogy to problems tackled by the fusion community /3/.

The mathematical problem consists of solving the steady state plasma equilibrium across the separatrix between the open and closed field lines. The iterative numerical scheme used by P-K is a way of avoiding the singularities arising from the vanishing field gradient, $\nabla \psi$, at the separatrix. These singularities present formidable challenges to Poisson solvers, which additionally require a fixed-boundary approach. For the Grad-Shafranov-Schlüter (GS) equation for a toroidal plasma with generalized pressure* and poloidal current profiles prescribed as a function of flux ψ, an analytic solution is available. Thus the quality of the solution can readily be checked against the Solovev-solution traditionally used to verify tokamak fusion reactor equilibrium solvers.

A direct solution of an equation of the GS-type is here presented to construct the equilibrium. The generalized shooting method is particularly designed for difficult boundary conditions (separatrices, O-points, line-tying, convoluted equilibria). We have not yet explored the limits of this approach, nor done the perturbation analysis. We hope eventually to see a breakup of the current sheet in the coronal streamer, which is characteristic of the tearing instability, leading to a series of expanding current loops, ejecting packets of plasma in well-defined bunches. Unlike the inverse flux method, the underlying spatial grid appears as part of the solution. The main drawback is that all possible solutions will compete with the desired one, requiring extensive numerical trickery to filter out ballooning behaviour on the scale of the gridding used. All attempts to tie the solution method described below to a predescribed grid have failed.

---

* The outward velocity field of the solar wind is thus subject to restrictions.

Only smoothly varying quantities such as the local distance between adjacent flux surfaces, the direction of the normal to the flux surface and the inverse curvature radius are used to build up the solution (effectively, a set of Frenet coordinates).

Although we have chosen to program only a two-dimensional equation, nothing restricts the method to azimuthal symmetry – the bookkeeping only becomes more cumbersome in the three-dimensional case with true two-dimensional surfaces . As no matrix inversion or iteration is necessary, the method is extremely fast. The only iteration is on a small number of global parameters, such as the total flux enclosed and the geometrical parameters describing the boundary.

GENERAL DESCRIPTION

Time-independent incompressible fluid dynamics as well as ideal MHD in two dimensions amounts to solving

$$\nabla^2 \psi = f(\bar{x}, \psi, \nabla\psi) \tag{1}$$

where a position dependent source function also depends on the value of the flux and its gradient. Conceptually, cylindrical (R,Z)-coordinates with azimuthal symmetry ϕ are implied, and topologically toroidal flux surfaces are centered at an O-point located at $R = R_0$. We construct each flux surface on top of the previous ones. Two key ingredients are now used.

First, the Laplacian is expressed, using normal derivatives only, as

$$\frac{\partial^2 \psi}{\partial n^2} + \frac{\partial \psi}{\partial n}(\alpha_1 + \alpha_2) \tag{2}$$

where the two local inverse curvature radii at a given position of a flux surface are taken at cuts perpendicular to each other (no other restriction); $\alpha_i = 1/r_{ci}$ is signed. Making the obvious directional choices, the curvature radius in the azimuthal direction is simple (although non-trivial, see below); the perpendicular one (ϕ=constant) must be calculated with great care. The scaling of ψ is a complicated unknown nonlinear relation. Fortunately, it makes sense to define the shapes of the pressure and current profiles in terms of flux normalized to unity. We can then chose a predescribed one-dimensional division for the flux function from zero to one, leaving the scaling of ψ to be equivalent to the strength of the pressure- and current-dependent terms of the source function. This now allows us to use the local gradient in the normal direction $\partial\psi/\partial n$ as one of the variables we march outwards from its starting value of 0 at the O-point, and the local value of the distance between these predescribed surfaces as the other.

Second, the innermost flux surface around the O-point has a special role. It is easy to see that specifying the normal gradient along this surface will completely determine the solution. For reasonably smooth source functions, elongation appears in the limit of the O-point, but no information from higher moments. Hence, a good approximation for the shape of this innermost flux surface is an ellipse. On the other hand, the solution at the O-point is an asymptotic one (which becomes painfully clear if one tries to integrate from the outside inwards). Thus, given a value of the flux ψ, the position of the innermost flux surface is known better than its radial derivative at this surface. Or, viewed as a one-point boundary condition on a second-order differential equation, we can specify a value for the flux and its radial derivative. The best initial guess for the normal derivative along the ellipse is of course given by the value derived from a constant-source approximation, but this is the local "free parameter" we change to achieve a desired boundary (in addition to the position of the O-point, the central elongation of the ellipse, the total value of flux (the strength of the source), and the geometric scale factor). The equation solver turns out to be an effective amplifier for anything done to the normal derivative on this innermost elliptical flux surface, including numerical noise due to finite difference techniques and finite numerical accuracy. By construction, the equation will always be *locally* satisfied.

THE GRAD-SHAFRANOV EQUATION

The Grad-Shafranov equation in cylindrical (R,Z)-coordinates with azimuthal symmetry ϕ is

$$\nabla \cdot (\frac{\nabla \psi}{R^2}) = -\frac{j}{R} \tag{3}$$

and gives the solution to a magnetic field B from a current density j:

$$B = \nabla \phi \times \nabla \psi + T \nabla \phi \qquad j = R\frac{dp}{d\psi} + \frac{T}{R}\frac{dT}{d\psi} \tag{4}$$

where j contains two free functions, the pressure profile $p(\psi)$ and the poloidal current profile $T(\psi)$.

For the special case of both $dp/d\psi$ and $TdT/d\psi$ constant, the analytic Solovev-solution is

$$\psi = \frac{(R^2 - R_0^2)^2}{4R_0^2} + \frac{Z^2}{E^2}(1 + h\frac{(R^2 - R_0^2)}{R_0^2}) \tag{5}$$

where E is the elongation, R_0 is the O-point and h is related to the pressure constant. A peculiarity of this solution is that every flux surface qualifies as the plasma boundary. Adding solutions corresponding to the corresponding homogenous equation - vacuum solutions caused by external current distributions – enables creation of a whole family of shapes useful for testing equation solvers.

NUMERICAL IMPLEMENTATION

The general description above needs to be complemented by a working prescription to be useful. The innermost flux surface is parametrised with an angle-like variable, with high resolution in directions where the surface-normal lines fan out, such as corners of the outside boundary. The inverse radius of the flux surface is the local derivative of the normal direction with respect to flux surface length element, as described below. The inverse radius in the perpendicular direction then becomes $\alpha_2 = \cos(\theta)/R$, where θ is the angle the normal makes with the Z=constant plane. The normal marching prescription is a predictor-corrector method for the normal gradient, due to the source being known initially only on the old surface. The distance between the flux surfaces dn is calculated from an ordinary second-order equation, arising from the source equation and from $\psi_{new} - \psi_{old} = dn(\psi'_{new} + \psi'_{old})/2$. The differencing and choice of divisions is made to be error-free under the assumption that ψ has a quadratic dependence on the normal distance from the O-point. At each flux surface a Fourier-filter is used, with respect to the flux surface length variable, to prevent numerical error from being amplified on the scale of the grid size (which shows up in the form of adjacent normal directions trying to run into each other – the grid folds upon itself, since everything else is held in place by construction). The filtering is applied to the spacing between flux surfaces. The value of the inverse curvature radius and the normal direction are filtered with a special procedure to avoid effects of local wrinkles: the Fourier filter cutoff defines a minimum angle over which the normal direction changes; the corresponding points on each side of the desired one are identified along the previous flux surface. For the corresponding number of surface length variables on the new surface $(dl_{i+\frac{1}{2}})$ around each point i a weighted summation is used for the new normal direction θ_i

$$\tan(\theta_i) = \frac{\sum \sin(\theta_{j+\frac{1}{2}})/dl_{j+\frac{1}{2}}}{\sum \cos(\theta_{j+\frac{1}{2}})/dl_{j+\frac{1}{2}}} \tag{6}$$

– summing atleast two chords, which defines a circle. The resulting directions θ_i are then filtered with the Fourier filter. The inverse radius α_{1i} is subsequently calculated at each point from the chord joining the two extreme directions used in the sum. For the Fourier filter itself, a singular value decomposition method is preferable, since Gaussian elimination has difficulties with overcomplete base function choices when increasing the gridding. A standard fitting was used /4/. The sophisticated numerics becomes necessary when going beyond 32 angular divisions, 16 radial divisions on ψ, and a few Fourier modes in the filtering.

It is in the nature of the problem that $\partial\psi/\partial n$ decreases monotonically going outwards. It is trivial to program a check for negative values somewhere along the surface, or the passing of the intended boundary, and subsequently to take corrective action in an iterative fashion. All $\psi(l, n)$-matrices are solutions to the equation, but not necessarily with the correct boundary or the correct amount of total current enclosed. The iterative procedure needed to handle a Pneumann-Kopp coronal streamer Y-shaped separatrix is yet to be tried out.

SOME CALCULATED EQUILIBRIA

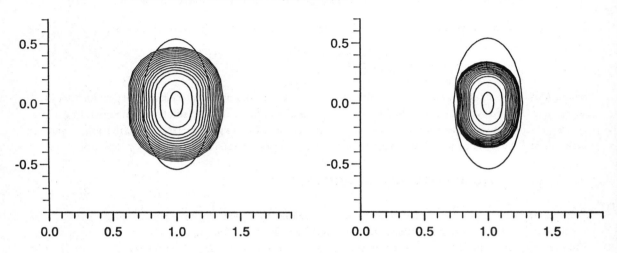

Fig. 1.
The ellipse is an enlargement of the central flux surface at a radius corresponding to the geometrical scale factors set in the problem. The difference between the two cases is the amount of current flowing in the torus.

ACKNOWLEDGEMENTS

Turun yliopistosäätiö, the Academy of Finland and the J. and A. Wihuri foundation are thanked for financial support.

REFERENCES

1. G.W. Pneuman and R.A. Kopp: Solar physics 18, 258 (1971)

2. S. Cuperman, T.R. Detman, C. Bruma and M. Dryer: these proceeedings;
   S. Cuperman, L. Ofman and M. Dryer: The Astrophysical Journal 350, 846 (1990)

3. R. Gruber and J. Rappaz: Finite Element Methods in Linear Ideal Magnetohydrodynamics, Springer-Verlag 1985, and references therein.

4. W.H. Press, B.P. Flannery, S.A. Teukolsky and W.T. Vetterling: Numerical Recipes, Cambridge University Press 1986, chapter 14.3.

THE CAUSE OF THE CORONAL TEMPERATURE INVERSION OF THE SOLAR ATMOSPHERE AND THE IMPLICATIONS FOR THE SOLAR WIND

J. D. Scudder

NASA Goddard Space Flight Center, Code 692, Greenbelt, MD 20771, U.S.A.

ABSTRACT

The energy contained in suprathermal tails at the base of the transition region is shown to be transformed into the rarefied, but hotter, transition region and low corona *WITHOUT* any further addition of energy to the gas above the base of the transition region. Quantitative agreement with the scale length of the transition region and the temperature of the coronal base at $1.03 R_o$ are demonstrated, regardless of the magnetic topology. Possible critical point location and asymptotic wind speed are shown to be controlled by the suprathermal tail strength parameter used to model possible suprathermal velocity distribution functions at the base of the transition region. This process shows promise for producing temperature profiles that peak near, but outside of, the fluid critical point without *ad hoc* energy deposition. The coronal temperature inversion above the solar photosphere is argued to be a generic feature around all stars with non-thermal distributions at the heights where the atmosphere last becomes mostly ionized.

INTRODUCTION

The solar wind was theoretically suggested as the consequence of the known million degree temperature of the coronal base at $1.03 R_o$ /1/. At the hydrodynamic level the known high temperature at $1.03 R_o$ and density profile determine a strong outward gas pressure gradient which overcomes the attraction of gravity, yielding an organized outward average acceleration to supersonic flow. Such flows have come to be known as "thermally" driven, as contrasted with "radiation/wave" accelerated, winds. A crucial factor in the theory is the determinant of the coronal temperature of the gas; if it is too high the wind does not go supersonic; if it is too small the asymptotic wind speed is uninteresting. The cause of the observed, inverted multi-million degree corona above the 5500° photosphere and, hence, the supply of thermal energy for the solar wind has remained a mystery. The possible MHD wave source(s) for direct acceleration within the transition region of the incipient solar wind also remains ill-defined.

After the experimental verification of the existence of the supersonic solar wind, theoretical efforts focussed on finding magnetized fluid wind solutions supported by thermal conduction that predicted the observed bulk speed and radial variation of species temperature as a function of source regions on the sun. The coronal base temperature of the inner boundary of the solutions at $1.03 R_o$ remained a given, constrained by the observations, but not explained. These models have only been partially successful /2-4/. The principal success is the recovery of a wind with 300-400km/s speed at 1AU. The predictions of the radial temperature profiles for electrons and ions and the strong correlation of the later with wind speed were not well described. The observations prompted considerable efforts to construct wave driven theories that deposit energy above the critical point. Wave scenarios for providing heating/momentum addition to the coronal plasma above $1.03 R_o$ remain in vogue /5/, although there are serious questions whether such waves 1) can be adequately transmitted into the low transition region after the strong attenuation/reflection that takes place below this distance /6/; 2) can deposit their energy or momentum at the necessary locations in the corona; 3) can be consistent with the observed, but low, levels witnessed at the orbit of Helios /7,8/; or 4) play any role in determining the coronal temperature inversion above the low TR.

The conflict between solar wind observations and theory is especially severe in the high speed wind over coronal holes. Most theorists, except Olbert /9/, have discounted the sufficiency of the internal energy density at the base of the transition region for such high speed winds in favor of some form of radiation pressure to produce the high speed wind. Alfven wave models are superficially attractive since they can provide a "direct" acceleration without building a thermal pressure gradient in the presence of the high thermal conductivity of the plasma as occurs with magnetosonic waves. The *ad hoc* nature of the wave amplitudes assumed is rarely discussed; neither is their possible origin. Occasionally, support for the appropriate wave amplitudes is sought in the excess Doppler widths of transition region lines. The wave driven wind modelers presuppose the temperature of the corona is given at $1.03 R_o$ and then rely on the accuracy of polytrope closure, or the Spitzer conduction law (or its "saturated version") for its contributions to the extension of the solar wind. Since the Spitzer-Braginskii-Fourier heat loss down the transition region is a sink to the build up of the coronal temperature, significant depositions of energy above $1.03 R_o$ are required by the wave treatments /10/ that explain a non-monotonic profile above $1.03 R_o$.

While the low energy density content of the solar wind may be viewed as a possible perturbation to that of the transition region (TR), the TR determines which particles comprise *all* the transmitted mass, momentum, and energy included in the low corona. However, unlike fluid flow in a pipe, not *all* the particles that comprise the initial number density, mass flux, and energy flux at the base of the TR are transmitted across the TR into the low corona: the TR is a high pass, speed dependent strainer, or filter /11,18/ for particles enroute to the low corona. The boundary conditions suitable for a separate model of the corona must reflect and not stifle the kinetic signatures impressed by the energy dependent filter of the TR /11/. Of particular interest is the appropriate size of the higher order kinetic moments of the velocity distribution function at the usual inner boundary of the compartmentalized domain of solar wind solutions above $1.03 R_o$. Puzzling are the suggestions /12,13/ that the solar wind temperature profile above a recurrent coronal hole is consistent with being non-monotonic above $1.03 R_o$. At the very least such possibilities together with the Fourier heat description appear to require a strong deposition of energy somewhere at the outermost extremity of the transition region; at their worst two distinct heat sources are required: one to form the corona above the transition region and another to produce the distended temperature maximum at 2-$6 R_o$. Recent work /13/ questions the data and uniqueness of the Munro and Jackson conclusion about an energy source *above* $1.03 R_o$, without addressing of the heat deposition required *near* $1.03 R_o$.

It is the identity of this seemingly inescapable, localized energy source, disjoint from the mechanical power implied by the hydrogen convection zone (HCZ) and its attenuation /6/, that has eluded description for the 50 years since the temperature inversion of the corona was established. This paper suggests a promising, new way to retrieve the inversion above the low TR to multi-million degree coronal values, regardless of the magnetic topology; the same process explains the temperature profile rising outwards above $1.03 R_o$.

The internal energy equation contains an effective heat source when $\nabla \cdot \underline{Q} > 0$. Spitzer's heat flux relation $(Q = -\varkappa(T)\underline{\nabla}T)$ is *not* model independent; it is not a valid synthesis of the kinetic physics if the microscopic distributions are not perturbatively close to local Maxwellians /14/. Spitzer's Fourier-like form for the heat flux closure suggests that a temperature profile requires a heat source at local maxima and other locations of strong negative curvature in the temperature profile as near $1.03 R_o$. It has long been known that Fourier's heat law may be derived from a Taylor series perturbation expansion of the solution of the Boltzmann equation of transport for *neutral* gases, where the small parameter is the mean free path over the scale height and the lowest order distribution function is a spatially uniform Maxwellian. This scheme can work because the mean free path is nearly synonymous with the free path of particles of any given energy, whether thermal or extremely suprathermal. When the velocity dependent and unbounded free path in a *fully ionized plasma*, such as the low TR, is considered, it is now widely agreed /9,15-17/ that the Fourier-Spitzer-Braginskii estimate of the heat flux is not on solid theoretical grounds immediately at and above $1.03 R_o$ and even lower in the TR /11, 16/. Even the possibility that the heat law can be formulated as a local functional relationship between moment variables or their gradients in the same place has been in question for the solar wind at least since 1979 /19/. Coulomb collisions by themselves are not sufficiently frequent, at a sufficiently broad spectrum of random speeds to make the Knudsen number small enough in an inhomogeneous, fully ionized plasma to suppress non-local effects /15,18/. Because of the strong inverse energy dependence of the free path, it is extremely difficult at finite density to have sufficient collisions to preempt non-local effects from "booming" through the locally scattered part of the velocity distribution function, f(v). These non-local effects preclude a consistent local description /19/, which in turn, is the foundation of the usual, truncated fluid theory of a plasma that leads to $(\underline{Q} = -\varkappa(T)\underline{\nabla}T$. The theoretical difficulties of the TR temperature profile, the solar wind energy supply at $1.03 R_o$, and solar wind temperature profiles above $1.03 R_o$ are argued in this and related previous work /11,18/ to be connected to our inability to accurately estimate the size or describe the process of the transport of internal energy from one locale to another in such violently inhomogeneous media, where the index of inhomogeneity is so strongly energy dependent and where Chapman-Enskog local gradient expansions are inappropriate.

THE MODEL OF VELOCITY FILTRATION

This paper examines a global transport process, called "velocity filtration", that cannot always be treated by the traditional fluid description /18/; it provides a novel, quantitative, alternate explanation for the temperature inversion of the corona relative to the base of the transition region, but does not involve energy deposition above the base of the transition region. The theoretical mechanism of energy rearrangement *is* a viable way to explain the temperature inversion of *either* "open" or "closed" field structures that traverse the transition region /11/. While providing a detailed TR temperature profile, and suggesting the basic ingredients of a one fluid temperature profile that has its maximum above $1.03 R_o$ and beyond the critical point, it also determines temperatures at critical points that are compatible with those isothermal temperatures Parker presupposed were appropriate in his earlier models. In suggesting what controls the temperature of the solar corona at the critical point, conclusions concerning the types of (supra)thermally driven winds that can be realized and the scaling of temperature inversions around other stars are suggested and compared with the data.

There is wide agreement that whatever remnants of the HCZ power that can get up into the chromosphere are either reflected or damped out by the time they reach the low transition region /6/. This attenuation presents severe difficulties for schemes that would transport or amplify waves from the HCZ to be damped in some other locale above the transition region /20/ in magnetic structures that permit the expanding solar wind to develop. Such waves are thought to somehow provide the increase of the temperature /12/, or the requisite momentum/heat addition outside the critical point thought to be essential for the fluid description of the high speed solar wind /21/.

There is no a priori reason /11,18/ that the fully ionized plasma that accepts the dissipating HCZ wave energy in the low transition region should do so in a self-similar (quasi-Gaussian) way, preserving a Maxwellian velocity distribution function with a single adjustable shape parameter, its variance or temperature. Much more likely, since the H-theorem does not apply in this locale /11/, is that there will be a (random) speed dependent response to the deposition of energy from the wave remnants: the more frequently scattered low energy particles will absorb and share energy in a more self- similar way; the suprathermal particles are relatively ineffective at sharing their acquired energy, yielding a more coherent response to the transfer of energy from the decaying waves of the HCZ /22-23/. This speed dependent response to energy deposition is exacerbated by the rapid decrease of the density (scattering centers) in the outward radial direction at the virtual "edge" of the star, with the higher energy particles sensing their collisional exobase at smaller radii than the lower energy particles.

The implications are now summarized of presupposing a non-thermal electron and/or ion distribution at the base of the low transition region where the plasma is fully ionized. At the outset it should be mentioned that the fluid equation truncated at the energy equation cannot address the consequences of this suggestion /18/. The theoretical approach for the kinetic results summarized here has established /11,18/ that nonthermal distributions, self-similar in an attractive potential *at low energy* to a Maxwellian, approximately zero the collision operator at low energies even when collisions of the rms speed particle in the plasma are frequent. For such a restricted class of boundary functions, the collisional solution of interest reduces in first approximation to that of a collisionless Boltzmann form, even though collisions are not being ignored. It is for a similar reason that Maxwell's exponential, isothermal atmosphere can be interchangeably derived from the Vlasov or Boltzmann equations: the Gaussian boundary function is self- similar in the equivalent potential, remaining everywhere a solution of the collision operator by its special shape regardless of the thermal collision frequency /11/.

In an attractive potential, as at the base of an ionized stellar atmosphere, an equivalent potential well is formed (even with suprathermal f(v)'s) with a depth approximately 1/2 the gravitational one /11/. Conservation of energy prevents, or "filters", certain particles from the boundary distribution function from moving too far in the radial direction. All particles in the boundary function can be divided into positive and negative total energy particles based on their initial kinetic energy in the boundary velocity distribution. A counterintuitive consequence for a non-thermal boundary distribution function, illustrated in the left frame of Figure 1, is that conservation

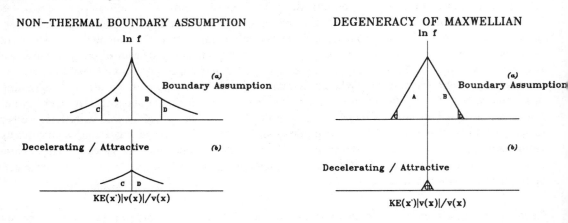

Fig. 1 Signatures of Velocity Filtration (left) for a non-thermal distribution and (right) for a exact Maxwell-Boltzmann distribution. Notice that the Maxwellian image under Velocity Filtration is similar to the boundary distribution and hence is at the *same* temperature; for any suprathermally over populated distribution /18/ Velocity Filtration *increases* the mean energy per particle, its kinetic temperature!

of energy causes the rms spread of the distribution of observable kinetic energies to increase as the over all density is decreased; this results from the exclusion of the low energy particles that cannot get out of the well - a process called "velocity filtration" /11,18/. The rms spread of a distribution controls the kinetic generalization, "T" = P/(nk), of thermodynamic temperature. Conversely, a Maxwellian distribution (the kinetic foundation for the fluid approximation and the premise of Spitzer's heat law) shows no evolution of rms temperature "T" under the same circumstances above (as illustrated in the right frame of Figure 1), even though its density is decreased for a similar reason. The anti-correlation of temperature "T" and density caused by an attractive potential with a non-thermal boundary velocity distribution function is a *generic* property of any non-thermal distribution /11/ under "velocity filtration". While velocity filtration follows directly from conservation of energy, its predictions cannot generally be retrieved from the truncated moment equation /18/; the existence of suprathermal tails are reflected in moment signatures in the deviation of the fourth and higher even moments from their Gaussian based

values, assumed unimportant in the usual truncation of the fluid moment hierarchy expansion /11,18/.
Additionally, no polytrope with a customary polytrope index $\gamma > 1$ can replicate this effect!

Spatial solutions of the Vlasov equation have been found /11/ for a family of non-thermal velocity
distribution functions of ions and electrons at the base of the solar transition region. The modeled non-
thermal distribution is the generalized Lorentzian distribution, called the kappa, κ, distribution introduced
by Olbert /24/. Mathematically defined by the relation

$$f_\kappa \propto \left(1 + \frac{v^2}{\kappa w_c^2}\right)^{-(\kappa+1)} , \tag{1}$$

f_κ resembles a Gaussian at low speeds, and evolves smoothly into an inverse power law with increasing
random speed, v, as controlled by κ. Velocity filtration alone operating on a wide range of non-thermal
boundary distribution functions produces coronal temperatures in excess of several million degrees without
ad hoc heating /11/.

An example of the radial profile of "T" that can be produced considering the effects of velocity filtration
on a non-thermal distribution is illustrated in the left frame of Figure 2. Starting from a base low transition
region temperature $T \simeq 5500°K$, velocity filtration precipitously increases "T(r)", easily achieving values
near $4–6 \times 10^{6}°K$ by $3R_o$. As indicated below in eqtn (4), when there is a mass flux on an "open"
field lines, about half of the available temperature is actually diverted into flow energy by the distance
$R/R_o \simeq \kappa$, yielding estimates of $2–3 \times 10^6$ as the temperatures near the possible critical point. The dashed
curve corresponds to the electron temperature profile and the solid curve to that of the ion; the different
curves result from the slight differences in the velocity distribution function of ions and electrons (κ_+
$\neq \kappa_-$) assumed at the base of the transition region. This temperature "inversion" has been shown to
occur if *either* electrons *and/or* ions are non-thermal at the innermost boundary of the solution in the
low transition region /11/. A significant temperature inversion (greater than a million degrees) can be
obtained with κ between 2 and 8, which corresponds to a rather broad range of boundary distribution
functions ranging from very strong tails (2) to very modest ones (8) /11/. The crucial ingredient is that
κ not be large compared to this range. The logarithmic scale height of the temperature can easily be as
short as 375-450km /11/, again comparable to that inferred from spectral observations /25/.

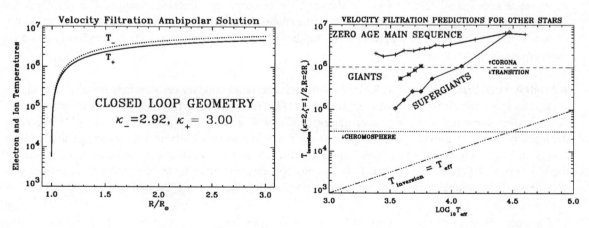

Fig. 2 (Left):Closed magnetic field configuration temperature profile according to Velocity Filtration
solution; (Right): Estimates of stellar inversion temperatures at $2R_*$, assuming that non-thermal
distributions like at the sun occur at the base of these atmospheres when they become fully ionized.

The physical and intrinsically kinetic reason for this sharp increase in "T" derives from the smallness of
the mean kinetic energy $kT(R_*)$ in the TR (0.5eV) in relation to the depth of the equivalent gravitational
plus electrical potential, $|\Delta\Psi(R_*)| \simeq 980eV$. Before the particles move very far (379km, initially) up the

potential hill they must store kT(R_*) worth of kinetic energy as potential energy. It should be emphasized that unlike fluid model heating, where *all* the particles of the velocity space dilute any localized enhancement of the energy density, velocity filtration excludes *part* of the velocity space preferentially, removing in an attractive potential more density than pressure. This kinetic effect fractionally depletes the density of a non-thermal distribution to a larger degree than the pressure, so that the *average* energy of those that defeat the filtration is higher than those that do not (cf Fig 1, left frame). The random speed, v'(R_*), in the transition region boundary distribution that has zero speed at solar radius r = R_*+h is given by the approximate relation v'(R_*,h)/V_{th}(R_*)$\simeq$ 30.7(h/R_*) = 5.3 at the traditional base of the solar corona (h = .03R_*), where V_{th}(R_*) is the thermal speed of the gas at the TR base. Only particles of the boundary distribution function, f(v,R_*), above v'(R_*) = 5.3V_{th}(R_*) can make up the distribution at the traditional coronal base at 1.03R_o. In words, "... the typical particle at 1.03R_o is clearly the atypical, extremely suprathermal particle of the transition region only 2000 km below..." In turn, this implies that significant and rapid "filtration" has returned all the lower kinetic energy particles below v'(R_*) at some intermediate altitude giving the transport physics a *global* character alien to the *local* premise that is the underpinning of the fluid closure.

Analytic forms for the effects of velocity filtration have been derived for a restricted, but representative, subset of analytic, nonthermal boundary distributions given by the kappa function /11,18/. In the isomagnetic flux tube limit, when $\kappa_+ = \kappa_-$, and when there is no mass flux, an especially simple summary of the effect is /18/

$$T(\Delta B = 0, r \to \infty) = T_* \left(1 + \frac{R_*}{(2\kappa - 3)L_D} \right), \tag{2}$$

where T(R_*) is the temperature of the inner boundary condition at the base of the transition region, R_* is the solar radius at the base of the transition region, L_D is the density scale height, and 1/κ is a measure of the non-thermal content of the boundary distribution function at the base of the transition region. The Maxwellian boundary condition corresponds to $\kappa \to \infty$. Mathematically, the temperature inversion owes its existence to the finite size of κ or, more generally, the existence of *any* suprathermal tail (figure 1) /18/; in the Maxwellian limit, velocity filtration reverts back to the well known isothermal exponential atmosphere /26,27/ and, consistently, the temperature inversion disappears in (2). Scudder /11/ develops the arguments that this effect can explain the temperature inversion across the transition region of the closed or open magnetic field regions of the solar corona and many previously reported correlations of heating signatures.

All tightly bound atmospheres (R_*/L_D>>1) with suprathermal boundary distributions should have significant temperature inversions above their surfaces /11/. When the boundary distributions possess a mass flux, as over a coronal hole, the temperature profile will initially rise almost as rapidly as boundary distributions without a mass flux; the peak temperature is somewhat lower due to the opening of new energy flux "exit channels" of mass and conduction flux that are *not* present in the solution illustrated in the left frame of Figure 2. Eventually, as the gas becomes supersonic, the ion temperature profile will cool beyond the general location of the critical point, being then in an accelerating potential /18,28/.

The considerations of this paper that lead to the temperature inversion are not peculiar to the sun: 1) the speed dependence of the coulomb collision frequency; 2) the a strong gravitational potential in relation to the thermal energy of the gas at the stellar "surface" ($\Phi_G/kT_* \gg 1$); and, 3) the ubiquity of non-thermal distribution functions in inhomogeneous, fully ionized plasmas. The form of equation (2) emphasizes that this effect should be commonplace in all stars with "bound" atmospheres. The ratio R_*/L_D is set by the ratio of surface strength of gravity and temperature which is set by the theory of stellar structure. The right frame of figure 2 illustrates the suggested inversion temperature around a subset of stars whose gravity and surface temperature are readily available /29/; all stars have been assigned κ =2 at their bases. All zero age main sequence stars (ZAMS) are calculated to possess super million degree inversion envelopes about their surface, regardless of their surface temperature. Thus, all ZAMS stars are energetically capable of thermal X-ray emission, as observed. Giants and Supergiants with smaller R_*/L_D (less tightly bound atmospheres) cannot achieve such a high inversion temperature. By increasing κ /11/ to give inversion temperatures that do not contradict observations of emissions from a given group of stars, an estimate of the $\frac{U_\infty}{V_{esc}(R_*)} \simeq \left(\frac{2(\kappa-1)}{2\kappa^2-3\kappa} ln20\right)^{1/2}$ /11/ can be made, to illustrate that the Giants and Supergiants have winds that are generally smaller in the dimensionless sense than those on the main sequence. This too, is in accord with the stellar observations /25/.

Within any *possible* critical point the spatially dependent form of the "T" profile of (3) can be written as

$$T(r, U=0) = T_* \left(1 + \frac{2(\Psi(r) - \Psi(r_*))}{(2\kappa - 3)kT_*}\right). \tag{3}$$

If the suprathermal tail strength of the ions and electrons is the same, then with rigor $\Psi = \Phi_G/2$, where Φ_G is the gravitational potential for a proton, and κ is the common suprathermal tail index of electrons and ions. While the flow is still subsonic (3) suggests that the temperature of the plasma and the gravitational potential are strongly coupled at each spatial location, viz

$$v_{th}^2(r) = v_{th}^2(R_*) + \frac{\left(V_{esc}^2(R_*) - V_{esc}^2(r)\right)}{\alpha(r)(2\kappa - 3)}, \tag{4}$$

where $\alpha(r) = 1$ for field lines with no mass flux, but will, in the presence of a mass flux, increases from 1 towards 2 at the critical point r_*. The function $\alpha(r)$ accounts for the diversion of thermal energy into flow energy that will occur when mass flux is allowed at the coronal base. If a critical point is possible in the fluid equations at r_*, with (4) as the equation of state (and $\alpha(r \simeq r_*)$ is slowly varying), it then must occur /11/ at $r_*/R_* \simeq \kappa$. According to velocity filtration, the temperature at the critical point would be

$$T_{VF}(r_*) \simeq 5500 + 1.1 x 10^7 \frac{(\kappa - 1)}{(2\kappa^2 - 3\kappa)} °K, \tag{5}$$

which ranges between $3.6 x 10^6$ and $1.4 x 10^6 °K$ for κ between 2.4 and 4.5, respectively. If Parker's critical point occurred at the radial locations required by velocity filtration above, the *isothermal* temperature required to allow a critical point at that location would be

$$T_{Parker}(r_*) = \frac{MV_{esc}^2(r_*)}{8k_B} = 5.25 x 10^6 \frac{R_*}{r_*} °K, \tag{6}$$

which ranges between 2.2-$1.6 x 10^6 °K$ for critical point locations $r_*(\kappa)$ suggested above for κ between 2.4 and 4.5, respectively. Velocity filtration provides a temperature at the critical point, $T(r_*)$, similar to that isothermal value above $1.03 R_o$ presupposed by Parker. However, *unlike* Parker's isothermal temperature profile, the velocity filtration approach yields 1) a TR like profile below $1.03 R_o$ *and* 2) a temperature profile that is still rising above the critical point $r_* \simeq \kappa R_*$ with an approximate, but positive, power law exponent $\beta = (\kappa - 1)^{-1}$, consistent with the temperature profile of Munro and Jackson /12/, but *without* their inferred deposition of heat! Parker's original work /1/ exploited conservation laws to show that the conditions for the critical point permitted temperature gradients of either sign at the critical point so long as the power law exponent is between -1 and 2 at the critical point. Imposing this condition yields the requirement that $\kappa > 1.5$. Actually $\kappa > 2$ is required /11/ for a finite flux energy supply to the corona, so that $r_* > 2 R_*$. Thus Parker's critical point existence arguments translate into a *lower* bound on the possible value of the critical radius, r_*.

Under velocity filtration a transonic critical point may be realized 1) if the boundary distribution function's non-thermal characteristics are suitable; 2) provided a mass flux exists at the inner boundary condition; and, 3) in the circumstances that the magnetic topology allows particle access above , $r_* = \kappa R_*$. If the magnetic tubes of force reenter the lower TR before extending beyond $2R_*$, no critical point can occur for any physically allowed value of κ at the inner boundary. To achieve supersonic flow the boundary condition must possess a mass flux so that the transonic condition can be realized. Mass flux from a boundary distribution characterized by κ_*, on a magnetic arcade with its apex *inside* of $\kappa_* R_*$ might be a candidate for the (time dependent) spicules, which return most of their upwelling mass flux back to lower altitudes on the returning flux of the arcade. Conversely, the steady supersonic wind should come from boundary f_{κ_*}'s with a mass flux on magnetic tubes of force with loci r(s) that extend *beyond* r(s) $= \kappa_* R_*$.

Thus Parker's parametric freedom of the "coronal" temperature at the critical point is removed; with velocity filtration the temperature $T(r_*)$ at the critical point is calculated from the equilibrium kinetic solution as a reflection of the kinetic boundary conditions at the base of the TR (H $\simeq$ 2090km). Parker showed that the asymptotic wind speed is determined primarily by the temperature near the critical point. In particular, the wind speed would scale like the square root of the temperature at the critical point. The asymptotic wind speed can be estimated /11/ with these kinds of effects to suggest that faster winds would accompany smaller κ's. This situation has already been anticipated, at least for the electrons /30/. Assuming a $20R_o$ isothermal region, and noting that $V_{esc}(R_*) \simeq 616$km/s the following speeds {1066, 841, 542, 458, 346, 215} km/s correspond to κ's of size {2., 2.45, 4.5, 6, 10, 25} respectively. Such a range is consistent with the extremes observed in the solar wind.

SUMMARY

The most striking consequences of postulating a non-thermal electron and/or ion velocity distribution at the mid-transition region of H$\simeq$ 2100km is the recovery of a thin scale (L$\simeq$ 400km) transition region, a temperature profile anti-correlated with the density and a multi-million degree plasma at $1.03R_o$ starting from transition region temperatures ($\simeq 5500°$K) *WITHOUT* distended heat addition to the system above the inner boundary where the non-thermal distribution is postulated. Excess Doppler widths of transition region lines have elsewhere /11/ been suggested to reflect unsurmised non-thermal *random* speed probability distribution functions ($\kappa_+ = 2.48$), rather than unresolved *coherent* motions, as the explanation of the augmented Doppler lines widths reported in the TR. These inferred non-thermal ion distributions can, in turn, explain the inverted coronal temperature profile above either closed or open magnetic topologies. Velocity filtration causes the sound speed and gravitational escape speed at the critical point to be functionally set by the strength of the suprathermal portion of the boundary velocity distribution /11/. Estimates of critical point locations /11/ are encouragingly low 2.4-6RR_o. "Open" topology heating is still significant under velocity filtration as the plasma is accelerated; however, outside the critical point the ion temperature will cool. In the simplest closed loop regimes where there is no net flow, this is not possible and the temperature achieved is largest at the highest point of the loop consistent with observations /31/. When the inner boundary distribution is postulated to be a pure Maxwellian, the critical point recedes to large distance and the asymptotic wind speed becomes small; from the inferred suprathermal strengths required for the alternate interpretation of the the "turbulent" Doppler ion line widths in the TR /11/, the critical point is between 2.4 and 4.R_o, consistent with our current picture of the solar wind expansion. More non-thermal distributions would yield lower critical points, although a minimum critical point radius r_* of approximately 2R$_*$ is suggested. By this same approach all stars should have distended coronal-like envelopes with temperatures limited only by the strength of the suprathermal distribution function where the atmosphere becomes fully ionized. In particular ZAMS stars are capable of X-ray emitting inversion shells, if their κ's near the base are similar to that of the solar case.

The process of "velocity filtration", is not new, but underlies the well known isothermal exponential atmosphere /26,27/. The new macroscopic effects result from considerations of velocity space transport possible when suprathermal distributions are NOT precluded in the kinetic specification of the boundary conditions on the plasma. Velocity filtration represents a very promising approach to the longstanding problem of the origin/maintenance of the corona and of the observed solar wind expansion; it opens the attractive new possibility that the coronal temperature inversion is *imprinted* on the velocity distribution function in the form of a *suprathermal tail* in, or adjoining to, the same locale where the HCZ power is strongly dispersed. Thus, the coronal heating problem comes full circle, with a mechanism for transporting a quantity made by the attenuation of HCZ waves, stored in the microstate's suprathermal tail, and concentrated by the velocity filtration process to raise the mean energy of the particles that survive, thus determining a temperaure increase. This should be contrasted with the early scenario which has only recently been contradicted /6/: a subset of the HCZ waves deposit their energy sufficiently high in the thin overlying atmosphere to raise its mean energy to form the corona. Further ramifications of this idea including correlation of heating with orientation and strength of the magnetic field, scaling of excess "turbulent" Doppler widths with formation temperature, and a possible factor in solar wind minor ions having a temperature proportional to their masses are discussed at length in references /11,18/.

REFERENCES

1. Parker, E.N., Astrophys. J., 128, 664(1958a).

2. Chamberlain, J.W., Astrophys. J., 131, 47(1960); Whang, Y.C. and Chang, C.C., J. Geophys. Res., 70, 4175(1965); Jockers, K. Solar Physics, 3, 603(1968); Hartle, R.E. and Sturrock, P.E., Astrophys. J., 151, 1155(1968); Durney, B. Astrophys. J., 166, 669(1971); Roberts, P.H. and Soward, A.M., Proc. Roy. Soc., A328, 185(1972); Hartle, R.E. and Barnes, A., J. Geophys. Res., 75, 6915(1970); Olbert, S., Solar Wind 5, M. Neugebauer, ed., (Washington: NASA Conf. Pub. 2280), 149(1983).

3. Neugebauer, M. and Snyder, C.W. J. Geophs. Res.,71, 4469(1966); Burlaga, L.F. and Ogilvie, K.W., Astrophys. J.,, 159, 659(1970); Belcher, J.W. and Davis, L., Jr., J. Geophys. Res., 76, 3534(1971); Ogilvie, K.W. and Scudder, J.D.,J. Geophys. Res., 183, A8,3776(1978); E. C. Sittler, Jr. and J. D. Scudder, J. Geophys. Res., 85, 5131(1980); Rosenbauer, H., AGU Boulder, Co.(1976); Hundhausen, A.J., Coronal Expansion and the Solar Wind, (1972) Springer-Verlag, (Heidelberg); Schwenn, R. and Marsch, E., Physics of the Inner Heliosphere, 2. Particles Waves and Turbulence, (1991) Springer-Verlag, (Heidelberg).

4. Withbroe, G., Kohl, J.L., Weiser, H., Noci, G., and Munro, R.H., Astrophys. J., 297, 324 (1985).

5. Belcher, J.W. and Davis, L., Jr., J. Geophys. Res. 76, 3534(1971), and Fla, T., Habbal, S., Holzer, T.E., and Leer, E., Astrophys. J., 280, 382(1984).

6. Narain, U. and Ulmschneider, P., Space Sciences Review, 54, 377(1990); Ulmschneider, P., Rosner, R., and Priest, E. 1991, Mechanisms of Chromospheric and Coronal Heating, Springer-Verlag (Heidelberg).

7. Parker, E.N., Adv. Space Res., 10, (9) 17(1990).

8. Roberts, D.A., J. Geophys. Res., 94, 6899(1989).

9. Olbert, S., Solar Wind 5, M. Neugebauer, ed., (Washington: NASA Conf. Pub. 2280), 149(1983).

10. Withbroe, G., Kohl, J.L., Weiser, H., Noci, G., and Munro, R.H., Astrophys. J., 297, 324 (1985).

11. Scudder, J.D., Astrophys. J., in press(1992b).

12. Munro, R.H. and Jackson, B.V., Astrophys. J., 213 874(1977).

13. Lallement, R., Holzer, T.E., and Munro, R.H.,J. Geophys.Res.,91, 751(1986).

14. deGroot, S.R. and Mazur, P. , Non-Equilibrium Thermodynamics, (New York: Dover) 183(1984).

15. Scudder, J. D. and Olbert, S. J. Geophys.Res., 84, 2755(1979a); Scudder, J.D. and Olbert, S., 1983, in Solar Wind 5, M. Neugebauer, ed., (Washington: NASA Conf. Pub. 2280), 163.

16. Shoub, E.C., Astrophys. J.., 226, 339(1983); Ljepojevic, N.N. and MacNeice, P., Solar Physics, 117, 123(1988); and Ljepojevic, N.N. and MacNeice, P., Pys. Rev. A, 40, 981(1989).

17. Gray, D.R. and Kilkenny, J.D., Plasma Phys., 22, 81(1980).

18. Scudder, J.D., Astrophys. J. in press(1992a).

19. Scudder, J.D. and Olbert, S. J. Geophys. Res., 84, 2755(1979a) cf eqtn 11ff.

20. Parker, E.N., Adv. Space Res., 10, (9)17(1990).

21. Leer, E. and Holzer, T.E., J. Geophys. Res., 85, 4681(1980).

22. Parker, E.N., Astrophys. J.,128, 677(1958b).

23. Dreicer, H., Phys. Rev., 115, 238(1959).

24. Olbert, S., Egidi, A., Moreno, G., and Pai, L.G. Trans AGU, 48 177(1968); Olbert, S., 1969, Physics of the Magnetospheres, ed. R.C. Carovillano, J. F. McClay,and H.R. Radoski, (Dordrecht: Reidel), 641.

25. Cassinelli, J. P. and MacGregor, K.B., Physics of the Sun, III, P. A. Sturrock, ed., (Dordrecht: D. Reidel), 47(1986).

26. Maxwell, J.C., Nature, 8, 537(1873).

27. Boltzmann, L., Wien. Ber., 72, 427(1875).

28. Jockers, K., Astron. and Astrophys., 6, 219(1970).

29. Bowers, R. and Deeming, T., Astrophysics I, Stars, (Boston: Jones and Bartlett)(1984).

30. Fairfield, D.H. and J.D. Scudder, J. Geophys. Res., 90, 4055(1985).

31. Zirin, H. ,Astrophysics of the Sun,(Cambridge UK: Cambridge University Press)(1988).

O VI EXTREME ULTRAVIOLET RADIATION FROM SOURCE REGIONS OF SOLAR WIND

D. Spadaro and R. Ventura

Osservatorio Astrofisico di Catania, Catania, Sicily, Italy

ABSTRACT

We present numerical simulations of intensities of the O VI λ1032 and λ1037 lines expected from several configurations of coronal holes, when observed on the plane of the sky in the range of heliocentric distance 1.1 $R_\odot$ – 10 $R_\odot$. We discuss how the observables considered depend on the physical parameters of the examined structures, in order to evaluate the extent to which they can place constraints on empirical models of the extended solar corona. These results may help in preparing and interpreting SPARTAN and UVCS/SOHO future observations.

INTRODUCTION

It has been recently shown that some EUV emission lines, in particular the H I Ly–α and the O VI resonance doublet at λ1032 and λ1037, can be useful diagnostic tools for the determination of physical parameters, such as density, temperature and outflow speed, in coronal structures where the solar wind originates (e.g., /1,2,3/). Data on intensity and profile of these lines in the range of heliocentric distance 1.2 $R_\odot$ – 10 $R_\odot$ are expected to be obtained in the next future by means of ultraviolet coronagraph spectrometers on board space missions (SPARTAN – UVCS/SOHO). We have investigated the sensitivity of these spectroscopic observables to the physical parameters of regions originating the solar wind. In particular, we have simulated the intensities of both lines of the O VI resonance doublet expected from several configurations of coronal holes. These simulated observations have also been used to evaluate the fractional contributions to observed spectral intensities from the considered coronal structures, when they co-exist with other structures along the simulated lines of sight.

CORONAL STRUCTURES AND MECHANISMS OF LINE FORMATION

We have considered three types of coronal holes:
– an equatorial coronal hole at solar minimum (hereafter EHMI);
– a polar coronal hole at solar minimum (hereafter PHMI);
– a polar coronal hole at solar maximum (hereafter PHMA).
These regions are described as radial flow tubes with a cross–sectional area nonradially increasing with heliocentric distance. The adopted nonradial expansion factor, $f(r)$, is that given by Kopp and Holzer /4/. The radial variations of coronal temperatures, densities and outflow speeds inside these structures are those determined by Withbroe /5/ through a one–fluid steady–state radiative energy balance model. We assume the O VI kinetic temperature and outflow speed to be equal to the plasma temperature and velocity, respectively, calculated in /5/.
The boundaries of the coronal regions considered are shown in Figure 1. For each region, we assume that the line of sight and the axis of symmetry of the coronal structure lie always on the same plane and are perpendicular. ρ is the heliocentric distance, expressed in solar radii, of the point where the selected line of sight intersects the "plane of the disc" (i.e., the plane perpendicular to the line of sight which passes through the Sun center); note that in the chosen geometry the

axis of symmetry of the structures considered passes through that point.

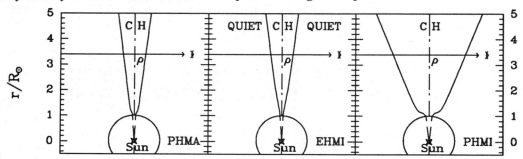

Fig. 1. Geometry of the simulated observations of the three coronal holes surrounded by the quiet corona

For each selected ρ, we have calculated the contribution to the spectral intensity of the O VI lines for volume elements along the line of sight and obtained the expected intensities by adding up all these contributions.

In the case of the polar coronal holes, we have considered only the contribution from the plasma inside the examined structures. For the equatorial coronal hole, we have considered also the contribution from the quiet corona surrounding the coronal hole. The values of temperature, density and outflow velocity adopted for the quiet corona are those calculated by Withbroe /5/ for a quiet, unstructured coronal region. Moreover, we have calculated the fractional contribution from the quiet corona by arbitrarily setting the outflow velocity equal to zero and leaving unchanged the other physical parameters, in order to evaluate the dependence of the fractional contribution on the expansion velocity of the region surrounding the coronal hole.

The emissivity in the O VI lines of a volume element along the line of sight has been calculated taking into account the following mechanisms of line formation:
– resonant scattering of chromospheric $\lambda 1032$ and $\lambda 1037$ photons by the coronal O VI ions;
– collisional excitation by electron impact.

We have taken into account the Doppler dimming for the resonantly scattered component of each line resulting from the solar wind expansion velocity. Moreover, we have considered the excitation of the resonantly scattered component of the O VI line at 1037.613 Å   produced by chromospheric photons of the C II line at 1037.018 Å, when the outflow velocity is large enough to cause red–shift of the incoming C II profile in the coronal ion rest frame by an amount sufficient to make the incoming C II radiation profile and the O VI $\lambda 1037$ scattering profile overlap. All these processes have been extensively described in /1,2,3/.

The intensity and profile of the chromospheric O VI and C II exciting lines considered in this work are those reported by Noci et al. /3/. We assume the lower atmosphere to be uniformly bright in the exciting radiation. We have used the ionization balance of oxygen reported by Noci et al. /6/, calculated under the approximation of ionization equilibrium. Note that this assumption might not be valid for the coronal structures here considered, where the outflowing plasma passes through temperature gradients with velocities of several $km\ s^{-1}$ (10–300 $km\ s^{-1}$, /5/). It is known that mass flows through a temperature gradient can give rise to significant departures from ionization equilibrium (e.g., /7/). We postpone the investigation on the deviations from ionization equilibrium to a future and more extended work.

RESULTS AND DISCUSSION

Table 1 presents total intensities of the two oxygen lines simulated for the three examined coronal structures at different values of ρ. In the case of the equatorial hole, also reported are the intensities calculated considering the contribution from the surrounding quiet corona, either expanding or static. The total intensity varies similarly for all structures, as a function of ρ, for both lines: the intensity of the $\lambda 1032$ line decreases from 1.1 $R_\odot$ to 8 $R_\odot$, like that of the $\lambda 1037$ line; however, while the first line decreases monotonically, the second has a bump in the range of heliocentric distance $3\,R_\odot - 6\,R_\odot$, since there the outflow velocity is large enough to allow the nearby C II line to produce additional excitation of the O VI line. Note that the intensities obtained for PHMI and EHMI are smaller than the corresponding PHMA intensities and exhibit a steeper decrease with ρ. This is due to the higher expansion velocity in the first two structures (see /5/), which determines a more efficient Doppler dimming for the examined lines. Considering the equatorial coronal hole as surrounded by the expanding quiet corona, we notice that the fractional contribution from the hole to the total intensity is 25%–30% for $\rho \le 2\,R_\odot$. For larger heliocentric distances the contribution rapidly decreases to only some percent, although in the $\lambda 1037$ line between $4\,R_\odot$ and $6\,R_\odot$ it is more than 50% (due to the additional excitation by the C II line in the coronal hole, that is not effective in the quiet corona whose outflow velocity is smaller of more than a factor of two /5/). We find a similar situation for $\rho \le 2\,R_\odot$ in the case of the static quiet corona, but for larger values of ρ the contribution from the coronal hole is much smaller (and rather negligible) than in the previous case.

Table I
SIMULATED TOTAL LINE INTENSITIES
(erg cm^{-2} s^{-1} sr^{-1})

ρ (R$_\odot$)		1.1	1.3	1.5	2.0	2.5	3.0	4.0	5.0	6.0	7.0	8.0
STRUCTURE	LINE											
PHMI	λ 1037	18.11	1.73	0.43	2.60E-02	2.62E-03	9.18E-04	8.73E-04	4.07E-04	7.49E-05	7.06E-06	1.13E-06
	λ 1032	62.99	6.43	1.62	9.42E-02	7.52E-03	7.17E-04	2.78E-05	7.39E-06	4.10E-06	2.62E-06	1.76E-06
PHMA	λ 1037	25.30	3.59	1.10	0.13	1.69E-02	2.17E-03	8.58E-04	1.40E-03	6.13E-04	1.20E-04	1.48E-05
	λ 1032	80.87	12.82	3.98	0.46	6.01E-02	7.09E-03	1.94E-03	4.86E-05	2.01E-05	1.03E-05	5.96E-06
EHMI	λ 1037	10.62	1.28	0.39	3.32E-02	3.40E-03	9.26E-04	1.69E-03	9.33E-04	1.93E-04	2.24E-05	3.11E0-06
	λ 1032	35.02	4.61	1.42	0.12	1.08E-02	1.12E-03	8.78E-05	3.03E-05	1.38E-05	6.60E-06	4.11E-06
EHMI+EXP.Q.R.	λ 1037	35.78	4.56	1.3	0.13	2.52E-02	8.31E-03	2.33E-03	1.15E-03	3.99E-04	2.81E-04	3.55E-04
	λ 1032	113.54	16.27	4.84	0.50	9.68E-02	2.92E-02	2.09E-03	4.21E-04	1.71E-04	8.96E-05	5.77E-05
EHMI+STAT.Q.R.	λ 1037			1.34	0.173	5.72E-02	3.48E-02	1.46E-02	8.51E-03		3.14E-03	
	λ 1032			5.02	0.67	0.23	0.14	5.30E-02	3.11E-02		1.29E-02	

Noci et al. /3/ have shown that measurements of the intensity ratio of the two lines of the O VI doublet at several heliocentric distances within a coronal structure can provide sufficient information on the outflow speed inside the structure. We have calculated, for each ρ, the ratio between the $\lambda 1037$ and $\lambda 1032$ emissivity, determined at the point where the line of sight intersects the axis of symmetry of the considered structure, and the ratio between the corresponding intensities integrated along the line of sight. As evidenced by Figure 2, for a given ρ and for all the three structures the two kinds of line ratio differ significantly. However, both ratios exhibit a behaviour similar to that described by Noci et al. /3/ and can be taken as representative indicators of the outflow speed. We notice that the intensity ratios calculated taking into account also the contribution from the expanding quiet corona and from the static corona exhibit considerable differences with respect to the ratio of the intensities obtained considering EHMI only, since at heliocentric distances larger than $5\,R_\odot$ the contribution from the surrounding atmosphere is predominant, so as to control the behaviour of the intensity ratio. In the case of the static corona, in particular, the intensity ratio is always nearly equal to 0.25. The increasing behaviour at larger heliocentric distances noticed for the expanding corona is due to the additional excitation by the C II line, effective in that region at $\rho > 8\,R_\odot$, where the outflow velocity is higher than $120\ km\ s^{-1}$ /5/.

It is worthwhile to point out that in these simulations we have assumed the oxygen velocity to be equal to the velocity of protons. However, several authors have developed solar wind models where the velocity of the oxygen ions is lower than that of the protons by a factor of 2–5 in the range $1.5\,R_\odot$–$10\,R_\odot$ (e.g., Bürgi and Geiss /8/). Adopting this kind of models, we have two consequences: a less efficient Doppler dimming (and probably no C II additional excitation of the $\lambda 1037$ line), and an enhancement of the abundance of oxygen relative to hydrogen in the considered regions. Both effects result in an increase of the intensity of the O VI lines.

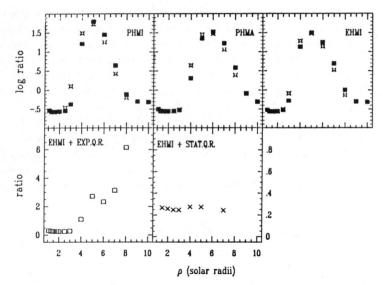

Fig. 2. (Top panels) Ratio of the $\lambda1037$ and $\lambda1032$ line emissivities (filled squares) and intensities (stars) versus the heliocentric distance of the selected line of sight for the three structures considered, and (bottom panels) ratio of the intensities of the same lines calculated in the case of the equatorial hole surrounded by the expanding quiet corona (open squares) and by the static quiet corona (crosses).

CONCLUSIONS

The results obtained in this work show that, within 2–3 $R_\odot$ of heliocentric distance in coronal holes, the intensities of the lines of the O VI resonance doublet exhibit a good sensitivity to the physical parameters of the examined structures, even if these last co–exist with other structures along the line of sight. This does not hold for $\rho > 3\ R_\odot$, where the contribution to the simulated intensities from the surrounding corona is predominant (although we have examined only the case of EHMI, we expect analogous results also for PHMA and PHMI when considered as surrounded by the quiet corona, since these structures are characterized by physical parameters similar to those of EHMI). This conclusion depends on the models of the expanding corona adopted (in particular for the oxygen velocity profile) and therefore we plan to perform this analysis for other models. We believe that even in the present preliminary form these results may give useful information for planning the mission operations of the SPARTAN and UVCS/SOHO space instruments.

We acknowledge financial support by Agenzia Spaziale Italiana (contract No. ASI–91–RS–91).

REFERENCES

1. J.L. Kohl and G.L. Withbroe, Astrophys. J. 256, 263 (1982)

2. G.L. Withbroe, J.L. Kohl, H. Weiser and R.H. Munro, Space Sci. Rev. 33, 17 (1982)

3. G. Noci, J.L. Kohl and G.L. Withbroe, Astrophys. J. 315, 706 (1987)

4. R.A. Kopp and T.E. Holzer, Solar Phys. 49, 43 (1976)

5. G.L. Withbroe, Astrophys. J. 325, 442 (1988)

6. G. Noci, D. Spadaro, R.A. Zappalà and F. Zuccarello, Astron. Astrophys. 198, 311 (1988)

7. D. Spadaro, Adv. Space Res. 11, # 1, 221 (1991)

8. A. Bürgi and J. Geiss, Solar Phys. 103, 347 (1986)

ALFVÉN WAVE REFLECTION AND HEATING IN CORONAL HOLES: THEORY AND OBSERVATION

S. T. Suess,* R. L. Moore,* Z. E. Musielak** and C. -H. An***

* Space Science Lab/ES52, NASA Marshall Space Flight Center, Huntsville, AL 35812, U.S.A.
** Department of Mechanical Engineering, University of Alabama in Huntsville, Huntsville, AL 35899, U.S.A.
*** Applied Research Inc., 6700 Odyssey Dr., Huntsville, AL 35806, U.S.A.

ABSTRACT

We present evidence for significant reflection of Alfvén waves in an isothermal, hydrostatic model corona and that heating in coronal holes is provided by Alfvén waves. For Alfvén waves with periods of 5 min, upward propagating waves are reflected if the temperature is less than 10^6 K, but escape into the solar wind if the temperature is greater than 10^6 K. This sensitive temperature dependence may provide the self-limiting mechanism that has been suspected to exist because the reflected waves result in heating which raises the temperature which, in turn, decreases the reflection. The reflection occurs mostly inside of $\sim 6 R_\odot$, depending on temperature, wave period, and magnetic field strength and geometry. The importance of this process has often been overlooked due to a poor choice of coronal Alfvén speed and temperature. SOHO is well-suited to measure whether the required properties for reflection exist. Solar Probe, however, is the only definitive experiment to show if the waves actually exist to the degree necessary.

1. INTRODUCTION

The basic concept that mechanical energy is transmitted into the atmosphere where it dissipates to provide coronal heating has long been accepted, but the specific mode of energy transmission and dissipation remains largely unknown. Here, we use a model to develop the argument for Alfvén wave reflection playing a major role that is compatible with the linkage between magnetic fields and coronal heating. There are three well-known agents that can generate waves on coronal field lines in quiet regions: (1) granular convection, (2) p-mode oscillations, and (3) spicules and macrospicules in the chromosphere and transition region. We expect each of these sources to produce a spectrum of Alfvén waves having most of its power in the period range $100 - 1000s$ and broadly peaking around $300s$ /7/. The power level of the Alfvén waves generated by these three sources is plausibly high enough for these waves to carry much of the nonthermal energy that heats the corona - Moore, et al. /7/ estimate it to be $\sim 10^6 erg\ cm^{-2}\ s^{-1}$ (see also /2,10/). Furthermore, these mechanisms (especially (3)) can deliver the energy above the transition region where it is more able to propagate outward into the corona.

Fig. 1 Schematic of the region of most important Alfvén wave reflection for $300s$ waves. Superimposed are the regions to be observed by the Ultraviolet Coronagraph and Spectrometer for SOHO (J. Kohl, PI) and the perihelion portion of the proposed Solar Probe orbit.

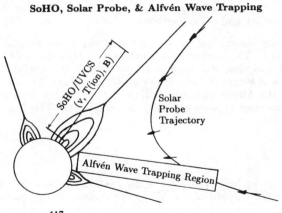

SoHO, Solar Probe, & Alfvén Wave Trapping

Alfvén wave reflection is due to the wavelength becoming too long with respect to the density scale height in the corona - a non-WKB effect. For magnetic field and mass density representative of observed coronal holes, such that the Alfvén speed at $1.5 - 2R_\odot$ is ca. 2500 km/s, Alfvén waves with periods of a few minutes are reflected if the temperature is low enough, but escape into the solar wind if the temperature is too high. The transition to escaping waves occurs at about 1×10^6 K. The reflection itself is a distributed process; the waves undergo continuous partial reflection over a range in height depending on the magnitude, gradient, and second derivative of the Alfvén speed. However, it is generally the case that reflection will be more important inside $6R_\odot$. One suggestion is that the reflection region constitutes a wave cavity, but we believe this will not be the case because the lower boundary (the transition region) is probably not reflective. Rather, the reflective region presents the possibility that the wave can do work on the background medium, undergoing nonlinear dissipation and accelerating the solar wind. This process (without dissipation) has been investigated superficially with a negative result /5/ because, we believe, of a lack of awareness of the strong temperature dependence and consequent choice of too high a coronal temperature.

Two missions will study the energy balance of the solar corona - The Solar and Heliospheric Observatory (SOHO) and the Solar Probe. SOHO is scheduled for launch first and the instrument of interest is the Ultraviolet Coronagraph and Spectrometer (UVCS, J. Kohl PI, G. Noci CoPI). UVCS will measure ion temperature and the velocity vector, mapping these parameters across coronal holes at heights from $1.2R_\odot$ out to $5R_\odot$ /12/. Fig. 1 schematically depicts this in the context of the relative size of coronal streamers, coronal holes, and the heights over which reflection may be important. Solar Probe is still in the planning stage, but would go in to $4R_\odot$. Its primary mission is to measure the role of hydromagnetic waves and other coronal heating and transport mechanisms inside $60R_\odot$, guided by radio scintillation data showing a region of strongly increased scattering generally starting at about $30R_\odot$. Although Alfvén wave reflection is probably not important outside $\sim 6R_\odot$, wave tunneling will cause significant penetration byond $6R_\odot$. Therefore, given a periapsis of $4R_\odot$, Solar Probe should be well suited to diagnose the wave amplitudes and periods in the appropriate regions of the corona to assess the importance of wave reflection. It will be an advantage, in this, if Solar Probe crosses a coronal hole at or near perihelion. Polar passage will be at $5 - 10R_\odot$ which is close enough to test the wave hypothesis /3/. The wavelength involved is, for $v_A = 1000km/s$, $\sim 50,000km$, short enough so that the spacecraft should be able to distinguish a wave field from stationary structures.

2. REFLECTION OF ALFVÉN WAVES IN A MODEL CORONAL HOLE

We discuss wave reflection in terms of the critical frequency - below which waves do not freely propagate, which is a function of Alfvén speed, and which can be computed analytically for simple models of the magnetic field. We do this for an approximation to a coronal hole: a potential magnetic field and a hydrostatic density distribution.

The potential field is assumed to be radial in the plane of a great circle on the Sun with the spreading being symmetric about the plane. Such a field can be described in this plane by

$$\frac{B}{B_S} = \left(\frac{R}{R_S}\right)^{-\sigma} \tag{1}$$

where B_S is the field at the base and σ is the spreading power - a positive, even integer /6/. We will examine the critical frequency for $\sigma = 2, 4$, and 6, corresponding to the observed differences in spreading between the center ($\sigma = 2$) and the edges of polar coronal holes.

For the density, we take an isothermal, hydrostatic distribution given by

$$\frac{\rho}{\rho_s} = e^{-\alpha(1-1/r)} \tag{2}$$

where $r = R/R_S, \alpha = R_S/H_S$, and H_S is the scale height at the base $[H_S = kT/(mg_S)$, where T and m are the temperature and mean particle mass of the plasma and g_S is the acceleration of gravity at the base].

We calculate the critical frequency for Alfvén waves propagating along the radial fieldlines in the plane of symmetry. The results serve to show the dependence of reflection on wave period, field strength and spreading, and plasma density and temperature. To calculate the frequency, we use the results of Musielak, Fontenla, and Moore /9/ who have recently shown that by transforming the spatial coordinate and dependent variables, the Alfvén wave equation can be cast in the form of a Klein-Gordon equation in which the local critical frequency appears explicitly as a coefficient. This frequency is found to be

$$\Omega = \frac{\omega_A}{2}\sqrt{1 + 2\frac{H_{V_A}}{H_{V'_A}}} \tag{3}$$

where $\omega_A = | V'_A | = V_A/H_{V_A}, V'_A$ is the gradient of the Alfvén speed along the magnetic field, $H_{V_A} = V_A/ | V'_A |$ is the scale height of the Alfvén speed along the magnetic field, and $H_{V'_A} = | V'_A | / | V''_A |$ is the scale height of the gradient of the Alfvén speed, V_A'' being the Alfvén speed's second derivative along the magnetic field. Therefore, the local critical frequency depends on the second derivative of the Alfvén speed as well as on the first derivative and the speed itself.

The physical meaning of the local critical frequency is found by considering the wavelength of the wave in comparison to the rate of change of properties in the background medium. In any medium in which Alfvén waves are a natural mode of oscillation, and in which there is a smooth variation of the Alfvén speed, Alfvén waves of sufficiently high frequency (or short wavelength) will propagate freely with negligible reflection because the Alfvén speed changes only slightly over a wavelength; i.e. the WKB approximation is valid. In the opposite extreme the waves will be reflected by the nonuniformity of the medium.

The physics of Alfvén wave propagation through a variable medium has been investigated several times. Early studies determined standing wave solutions, providing no information on local partial reflection (e.g. /4/). Analytic and numerical solutions for the impulse response of a stratified atmosphere show Alfvén waves undergo continuous partial reflection which becomes stronger (weaker) the further the wave frequency is above (below) the critical frequency /1,8,11/. The downward reflection of upward propagating waves is therefore negligible at low heights in the corona where $\omega_A \ll \omega$ and becomes strong above the height at which $\omega_A = \omega$. The waves are mostly reflected in the vicinity of the height at which $\omega_A = \omega$, so we will approximate the overall process by assuming negligible reflection for frequencies below the critical frequency and strong reflection above the critical frequency.

The radial variation of the Alfvén speed in our model is shown in Fig. 2. The radius and Alfvén speed have been normalized by their values at the base of the model so these profiles hold for any choice of R_S, ρ_S, and B_S. Each profile is completely specified by the choice of T and the σ. In each panel the spreading power is fixed and the profiles show the strong effect of changing the temperature through the range of expected coronal temperatures ($T \sim 10^6$ K). The counteracting effects of the power-law magnetic field and the exponential density is demonstrated by the existence of a local maximum above the base of the model for sufficiently small temperatures.

Fig. 2 Variation of the Alfvén speed with radius in our model coronal hole, for increasingly greater magnetic field spreading (increasing σ). Each curve is for a particular coronal temperature and is labelled with the logarithm of that temperature.

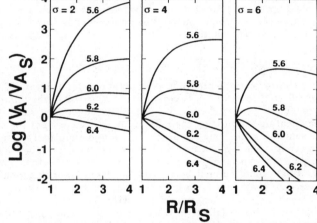

The critical frequency is shown in Fig. 3 for each of the Alfvén speed profiles shown in Fig. 2. Again, because of the normalization of the critical frequency and radius by their values at the base, the profiles in this figure hold for any choice of density, field strength, and base radius, and are completely specified by the choice of temperature and field spreading power. For temperatures below about 10^6 K, the maximum critical frequency occurs above the base, like the Alfvén speed profiles.

Fig. 3 shows whether upward propagating Alfvén waves will pass through the model, the height around which reflection occurs, and how this depends on temperature and spreading power. If ω/Ω_S is less than the maximum Ω/Ω_s, but greater than unity, the waves propagate through the base and up to the height where $\omega = \Omega$, about which height they are reflected. If $\omega/\Omega_s < 1$, waves are immediately reflected in the vicinity of the base. From Fig. 3, there is therefore a maximum critical frequency, Ω_{max}, and a minimum critical period, P_{cmin}, somewhere in the model - either at or above the base. Finally, we note that there is bound to be some tunneling through to the top of the model when P is only slightly greater than P_{cmin}.

The variation of P_{cmin} with temperature and spreading power is shown in Fig. 4. Each of the three curves gives P_{cmin} as a function of temperature for the fixed spreading power σ with which it is labelled. These curves are for an electron density of 3×10^7 cm^{-3} and a magnetic field of 10 gauss at a base radius of 1.15 $R_\odot$. These curves all rise with increasing temperature and all three pass through values of $P_{cmin} \sim 300s$ at $T = 10^{6.0}$ K. Because these are similar to solar parameters, this indicates that solar Alfvén waves with periods much longer than 5 min escape into the solar wind.

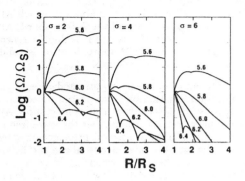

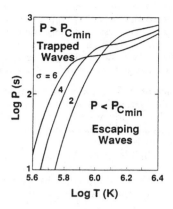

Fig. 3 (left) Variation of the critical frequency for Alfvén wave reflection with radius for each of the σ, T cases. Alfvén waves of *frequency above the maximum critical frequency can propagate through* to the solar wind. In the range of field spreading spanned by the three panels, the maximum critical frequency occurs at a height that decreases with increasing temperature until at temperatures greater than about 10^6 K the maximum occurs at the base of the model.

Fig. 4 (right) Dependence of the minimum critical period, P_{cmin}, on coronal temperature and field spreading, for the three σ-values. *The region above the curve contains reflected waves.* These curves depend on the base radius and are inversely proportional to the Alfvén speed at the base. We have taken $R_S = 1.15R_\odot$, $n_{e_S} = 3 \times 10^7$ cm^{-3}, and $B_S = 10$ gauss.

3. DISCUSSION

Fig. 4 shows for *the center* of large coronal holes (i.e. for σ = 2 and $T = 10^{6.0} - 10^{6.1}$ K) that the transition from reflected to escaping waves occurs at a period in the middle of the expected power spectrum of Alfvén waves. That the σ = 2 curve has a positive slope suggests this occurs because the coronal temperature is set by heating supplied by the reflected Alfvén waves. For the observed temperature, the σ = 2 curve shows that something like half the power (the long-period half) is reflected in the coronal hole. This curve also shows that if the temperature were only slightly below $10^{6.0}$ K, the transition period would be ∼ 100s and nearly all the expected power spectrum would be reflected. Two additional observations fit this hypothesis well. First, small coronal holes with increased spreading result in reduced wave flux per unit area being delivered to the reflection region. Second, the same reasoning applies to the edges of coronal holes - where the spreading is also faster than r^{-2}, with the additional point that the path length is longer so that the evaluation of the gradients in (3) must be done differently. The net result is an even greater reduction in temperature towards the edges of coronal holes.

Acknowledgement: This research has been supported by a grant from the Cosmic and Heliospheric Physics Branch of the Office of Space Science and Applications, U.S. National Aeronautics and Space Administration.

REFERENCES

1. An, C.-H., Musielak, Z. E., Moore, R. L., and Suess, S. T., *Astrophys. J.*, **345**, 597 (1989).
2. Esser, R., Leer, E., Habbal, S. R., and Withbroe, G. L., *J. Geophys. Res.*, **91**, 2950 (1986).
3. Feldman, W., and 22 others, *Solar Probe, Scientific Rationale and Mission Concept*, Jet Propulsion Laboratory **JPL D-6797** (1989).
4. Ferraro, V.C.A., and Plumpton, C., *Astrophys. J.*, **127**, 459 (1958).
5. Heinemann, M., and Olbert, S., *J. Geophys. Res.*, **85**, 1311-1327 (1980).
6. Jahnke, E., and Emde, F., *Tables of Functions*, Dover Publ., New York, p. 108 (1945).
7. Moore, R. L., Musielak, Z. E., Suess, S. T., and An, C.-H., *Astrophys. J.*, **378**, 347 (1991).
8. Musielak, Z. E., An, C.-H., Moore, R. L., and Suess, S. T., *Astrophys. J.*, **344**, 487 (1989).
9. Musielak, Z. E., Fontenla, J. M., and Moore, R. L., *Physics of Fluids B* (Letters), in press (1991).
10. Parker, E. N., in *Solar and Stellar Coronal Structure and Dynamics*, R. C. Altrock ed., Natl. Solar Observatory/Sacramento Peak, Sunspot, New Mexico, p.1 (1988).
11. Rosner, R., Low, B. C., and Holzer, T. E., in *Physics of the Sun*, Vol. II: *The Solar Atmosphere*, P. A. Sturrock, T. E. Holzer, D. M. Mihalas, and R. K. Ulrich, eds., Reidel, Boston, p. 135 (1986).
12. van Beek, H. F., Delache, P., Huber, M. C. E., Malinovsky-Arduini, M., patchett, B., van der Raay, H. B., and Schwenn, R., *Solar and Heliospheric Observatory Assessment Study*, European Space Agency **SCI(83)3** (1983).

GREEN CORONA LOW BRIGHTNESS REGIONS AS THE SOURCE OF THE SOLAR WIND

J. Sýkora

Astronomical Institute of the Slovak Academy of Sciences, 059 60 Tatranská Lomnica, Czechoslovakia

ABSTRACT

The frequency curves of the Green Corona Low Brightness Regions (GCLBRs) manifest the solar cycle but, are displaced considerably with respect to the sunspot cycle. The observed displacement increases with the size of GCLBRs and reaches up to 4-5 years for the largest regions. It is interesting that the displacement in the equatorial zone has opposite trend to that in the higher-latitude zones. An older idea on the physical affinity between GCLBRs and coronal holes led us to study the frequency of GCLBRs and the properties of High Speed Plasma Streams (HSPSs) in the solar wind. Maximum velocity and duration of the coronal-related HSPSs seem to be well correlated with the number and size of GCLBRs located in the N60-N20 and S20-S60 latitudinal zones. This is particularly evident at the end of the solar cycle.

INTRODUCTION

The "Atlas of the Green Corona Synoptic Charts for the Period 1947-1976" /1/, has been used to analyse the long-term distribution on the solar surface of the Green Corona Low Brightness Regions. All the GCLBRs, which we define in our "Atlas" as the areas limited by dashed isolines only, represent certain "concavities" of the brightness profile on the sun's surface, characterized very probably by a quasi-open configuration of the magnetic field lines. We are interested in investigating these regions because they have, so to speak, a certain affinity to coronal holes. This follows from the fact that the intensity of the green coronal line (530.3 nm) decreases greatly in regions of low plasma density and temperature which, of course, also are the most pronounced properties of coronal holes.

In a number of papers (lately, for example, /2/ and /3/) it has been proved beyond doubt that the High Speed Plasma Streams in the solar wind originate in coronal holes on the sun. Hence it may be conceived that the GCLBRs should possess the same property. To prove this assumption the catalogue of HSPSs compiled in three parts /4, 5, 6/ was analysed.

LONG-TERM DISTRIBUTION OF THE GCLBRs

For the purpose of the present study we determined the positions (the heliographic longitude and latitude of the centres of gravity) and size of the GCLBRs on the synoptic charts of the Green Corona Atlas. The regions were classified into 5 groups according to their size. It was then possible to look for trends in the long-term distribution of the GCLBRs on the sun's surface. This was done separately for the equatorial zone N20-S20 and for the higher-latitude zones N60-N20 and S20-S60. Figure 1 depicts the number of GCLBRs versus time for different size- and latitude-groups. At least three conclusions can be drawn from Figure 1:

(1) The number of GCLBRs in the equatorial and higher-latitude zones evidently reflects the solar cycle for all sizes. However, the frequency curves are displaced significantly with respect to the curve of the sunspot cycle. This is simply because the GCLBRs represent an "inactive" feature on the sun's surface while the sunspots define the classical solar activity cycle. The above facts, and also some further statements, are quantitatively presented in Table 1. The table presents the results of a cross-correlation of each GCLBRs curve with the curve of Zürich sunspot number Rz in order to determine the time-lag (in years) at maximum correlation (MC). More exactly, two neighbouring maxima of the cross-correlation are tabulated (indexed by 1 and 2) to see the differences lag(2) - lag(1). These differences in any case should by near to the length of the solar cycle (~11 years) if we are right in affirming that the frequency curves of the GCLBRs exhibit the solar cycle.

(2) The observed systematic displacement of the maxima of the GCLBRs frequency curves (indicated by the slope of the almost vertical dashed lines) suprisingly amounts to 4-5 years between the smallest (class 1) and largest (class 5) sizes of the regions.

(3) A remarkable fact is that the displacement of the GCLBRs curves in the equatorial zone has an opposite trend to that in the higher-latitude zones (see the opposite inclination of the vertical dashed lines connecting in Figure 1 the possitions of the maxima, indicated by black dots at the zero levels of the curves). At this time we have no appropriate explanation

for this observed   trend. We are of the opinion that, in some sense, it resembles the feature of torsional oscillations of the velocity fields found in /7/ and /8/.

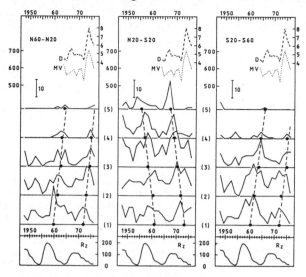

Fig. 1. The frequency curves of GCLBRs, classified into five different sizes, are
plotted separately for three latitudinal zones. The zero levels of the curves are indicated by five horizontal lines and the frequency scale is indicated by vertical bar
above the fifth curve. At the top left the scale of the averaged maximum velocities
MV (in km/sec) and at the top right the scale of averaged duration D of HSPSs (in
days) are given.

TABLE 1 The Quantitative Analysis of Figure 1 (for Explanation See Text)

Size of GCLBRs	MC(1)	Lag(1)	MC(2)	Lag(2)	Lag(2)-Lag(1) (years)
			N60 - N20		
1	0.745	2.6	0.614	13.9	11.3
2	0.466	3.0	0.450	14.8	11.8
3	0.600	4.1	0.633	15.6	11.5
4	0.609	4.8	0.613	16.7	11.9
5	0.487	5.6	0.652	17.6	12.0
			N60 - N20		
1	0.636	2.0	0.755	13.8	11.6
2	0.528	2.1	0.710	12.7	10.6
3	0.560	-0.5	0.724	10.7	11.2
4	0.864	-1.1	0.794	9.9	11.0
5	0.573	-2.7	0.749	8.9	11.5
			S20 - S60		
1	0.612	2.2	0.596	14.2	12.0
2	0.622	2.8	0.558	14.2	11.4
3	0.581	5.8	0.739	17.4	11.6
4	0.733	5.9	0.871	17.7	11.8
5	0.547	8.0	0.551	17.3	9.3

DISTRIBUTION OF THE HSPSs IN LINDBLAD's CATALOGUE

With respect to the introductory part of this paper the question arises if and how the data
on the HSPS in the solar wind can be fitted to the time- and space-distribution of the
GCLBRs. To answer the question it was necessary to analyse the catalogue /4, 5, 6/ in a suitable manner. In this catalogue the individual HSPSs are distinguished by their source regions, i.e. the HSPSs are attributed to the solar flares or coronal holes. Hence, it was possible to draw Figure 2 using the same pattern.

At the top left in Figure 2 the time patterns of the following parameters of coronal-hole-re-
lated HSPSs are plotted from bottom to top: number of streams (N), sum of importances of the
single streams (ΣI), sum of the maximum velocities reached by the streams (ΣMV) and sum of
durations of the HSPSs (ΣD). The importance of a single stream is the product of the difference between its maximum and initial velocities and the duration of the stream in days. Although the three upper curves of the summary quantities are directly proportional to the num-

ber   streams (that is why they also exhibit an increasing trend similar to that of N), it
is useful to present them and to take them into account in any considerations of the solar
wind, because they represent certain "global quantities" affecting the Earth's atmosphere in
a given period of time. Further, it is worth noticing that the relatively flat distribution of
the number of streams (N) in 1973-1977 changes into a particularly pronounced peak of all
three summary parameters in 1974.

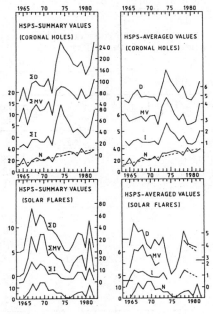

Fig. 2. Trends in the long-term behaviour of some parameters of the HSPSs. The scales
to the left and to the right refer alternately to the curves in their order from bottom to top. Some of the scales are divided by 10 or 100, but it is believed that the
absolute values are not important in the context of the present paper. The physical
meaning of the scales follows from the text.

The mean values of I, MV and D, i.e. the averaged values of these parameters calculated for
one high-spead plasma stream of the solar wind are plotted in the top right-hand part of Figure 2. The cause of the 1974 maximum of the summary quantities can be hopefully derived from
this part of the Figure. There is no doubt that, in the years preceding the minimum of the
solar cycle (1976), MV and D distincly increased (by about 25 % and 65 %, respectively) in
comparison with the other phases of the solar cycle. Logically, also the product of these two
parameters increased, too, thus throwing more light on the solar-terrestrial relations towards the end to the cycle. This result is not, of course, unknown, having been pointed out
in various ways in a number of studies. However, we believe that characteristics of the decreasing part of the 20th solar cycle and the exceptionality of 1974 are shown here in convincing details.

In the lower part of Figure 2 the same summary quantities and the same averaged parameters
are presented as described above, but in this case only the flare-related streams were considered. Evidently, all the curves differ significantly from those derived for the coronal-hole-
-related HSPSs. The time pattern of the number, importance, maximum velocity and duration of
the streams resembles the shape of the 20th solar cycle very much. This is understandable
because the same pattern is characteristic for the frequency of solar flares - the source of
the discussed HSPSs and, at the same time, one of the most outstanding manifestations of solar activity. The curves of the averaged parameters (right-hand bottom) are less smooth owing
to the smaller statistical set. Nevertheless, it is clear that the flare-related streams
reach the highest velocities and the longest durations in the period of maximum solar activity or during the increasing phase of the solar cycle when, as we known, the largest flares
occur more frequently. However, comparing the right-hand upper and lower parts of Figure 2,
it can be seen that the maximum values of the averaged I, MV and D are considerably higher if
the streams originated in coronal holes than if they were flare-related.

RELATION OF GCLBRs TO HSPSs

We now return to the discussion of a possible relation between GCLBRs and HSPSs. It is evident that the averaged HSPSs parameters MV and D (taken from the upper right-hand part of Figure 2 and plotted into all the three parts of Figure 1) cannot be fitted practically to any
of the GCLBRs curves of the equatorial zone N20-S20 (see the middle part of Figure 1 and Ta-

ble 2). On the contrary, it is obvious that, apart from the end of 19th colar cycle, when the regular measurements of the HSPSs only began, the MV and D curves, with their conspicuous maximum in 1974, can be fitted far better to the three middle curves of the higher-latitude zones (see the left- and right-hand sides of Figure 1 and the upper and lower parts of Table 2). It should be said that, in terms of our classification of the GCLBRs we are talking about their second, third and partly fourth magnitudes, which are, on the one hand, sufficiently large to be compared with the coronal holes and, on the other hand, also sufficiently frequent to be the possible source of the HSPSs.

TABLE 2 Correlation of the GCLBRs Frequency Curves with the Data on Maximum Velocity and Duration of the HSPSs

Size of GCLBRs	Maximum velocity				Duration		
	Corr.coeff. at Lag(0)	MC coeff.	Lag (years)		Corr.coeff. at Lag(0)	MC coeff.	Lag (years)
				N60 - N20			
2	0.696	0.814	+0.7		0.623	0.792	+0.9
3	0.644	0.767	-0.7		0.482	0.649	-0.9
4	0.730	0.877	-0.6		0.621	0.800	-0.6
				N20 - S20			
2	0.101	0.722	+4.2		-0.025	0.759	+4.2
3	-0.089	0.574	+5.3		-0.110	0.663	+5.7
4	-0.419	0.672	+6.1		-0.320	0.681	+6.2
				S20 - S60			
2	0.164	0.549	+2.0		-0.086	0.599	+2.3
3	0.307	0.627	-1.3		0.127	0.559	-2.4
4	0.109	0.660	-1.4		0.223	0.517	-2.4

The fact that the pattern of the HSPSs parameters correlates well with the frequency of the most important (sufficiently large and numerous) GCLBRs situated in the N60-N20 and S20-S60 zones seems to be in good agreement with some older post-Skylab era hypotheses claiming that the sources of the solar wind are mainly located at the outer boundaries of the sunspot activity belts, from where the streams of particles move along pear-shaped trajectories to the plane of the ecliptic and cause geoactivity /9/. The values of the correlation coefficients and the lags at maximum correlation (Table 2) show that the southern hemisphere data was somewhat less convincing for the above assumption then the northern hemisphere data. However, one should keep in mind that the supposed mechanism results from the summary effect of both hemispheres.

REFERENCES

1. V. Letfus and J. Sýkora, Atlas of the green corona synoptic charts for the period 1947--1976, Veda, Bratislava (1982).

2. B.A. Lindblad, Coronal sources of high-speed plasma streams in the solar wind during the declining phase of solar cycle 20, Astrophys. Space Sci. 170, 55 (1990).

3. J. Xanthakis, B. Petropoulos and H. Mavromichalaki, Coronal line intensity as an integrated index of solar activity, Astrophys. Space Sci. 164, 117 (1990).

4. B.A. Lindblad and H. Lundstedt, A catalogue of high speed plasma streams in the solar wind, Solar Phys. 74, 197 (1981).

5. B.A. Lindblad and H. Lundstedt, A catalogue of high speed plasma streams in the solar wind 1975-78, Solar Phys. 88, 377 (1983).

6. B.A. Lindblad, H. Lundstedt and B. Larsson, A third catalogue of high speed plasma streams in the solar wind - data for 1978-1982, Solar Phys. 120, 145 (1989).

7. R. Howard and B.J. LaBonte, The sun is observed to be a torsional oscillator with a period of 11 years, Astrophys. J. Lett. 239, L33-36 (1980).

8. B.J. LaBonte and R. Howard, Torsional waves on the sun and the activity cycle, Solar Phys. 75, 161 (1982).

9. J.D. Bohlin, Extreme ultraviolet observations of coronal holes, Solar Phys. 51, 377 (1977).

THE RELATIONSHIP BETWEEN SOLAR WIND SPEED AND THE AREAL EXPANSION FACTOR

Y. -M. Wang and N. R. Sheeley Jr

Code 4172, Naval Research Laboratory, Washington, DC 20375, U.S.A.

ABSTRACT

Empirical studies indicate that the solar wind speed at Earth is inversely correlated with the divergence rate of the coronal magnetic field. This result suggests that the mechanical energy flux at the coronal base (in the form of Alfvén waves, for example) is roughly constant within open field regions.

INTRODUCTION

There has been much confusion in the past concerning the relationship between solar wind speed and the areal expansion rate. In their well-known study of a polar coronal hole, Munro and Jackson /1/ noted that the *total* areal cross-section of the hole at $3R_\odot$ was 7 times greater than if its boundaries had expanded radially from the base of the corona. Because *Skylab* observations indicated that such polar holes give rise to high-speed wind streams, it was perhaps inevitable that rapid areal expansion should thereafter come to be associated with high wind speeds. It was not generally recognized that magnetic flux tubes near the boundary of a polar hole have expansion factors much greater than 7, and that such boundary regions are sources of slow wind at Earth.

Levine, Altschuler, and Harvey /2/, in a study based on two months of *Skylab*-era data, pointed out that high wind speeds are associated with *small* expansion factors. We recently performed a more extensive analysis using spacecraft and magnetograph observations over a 22-yr period /3/; our conclusion was that the solar wind speed measured at Earth is indeed inversely correlated with the divergence rate of the magnetic field near the Sun. This paper discusses the theoretical implications of the empirical relationship.

IMPLICATIONS FOR WIND ACCELERATION MODELS

We begin with a simple and general explanation of our main result. Spacecraft measurements indicate that the solar wind energy flux at Earth, F_E, is dominated by the kinetic energy flux, $\rho_E u_E^3/2$ (where ρ_E denotes the mass density and u_E the wind speed at Earth). If mass and energy are conserved along a flow tube, it follows that /4/

$$u_E^2/2 = F_E/(\rho_E u_E) = F_c/(\rho_c u_c) = F_0/(\rho_0 u_0), \qquad (1)$$

where the subscripts "E," "c," and "0" indicate that the flow variables are evaluated at 1 AU ($r = r_E$), at the critical point ($r = r_c$), and at the coronal base ($r = r_0 = R_\odot$). In most solar wind models, the thermal pressure gradient dominates for $r < r_c$, and $\rho_c u_c$ is mainly a function of the coronal temperature. By continuity along a flow tube, the mass flux $\rho_0 u_0$ at the coronal base must therefore increase with the expansion factor. If we suppose that the energy flux F_0 at the coronal base remains invariant, it then follows that $u_E = [2F_0/(\rho_0 u_0)]^{1/2}$ must be a decreasing function of the expansion factor, as observed.

It is well known that acceleration models that include only the effect of thermal pressure cannot account for the observed high-speed streams, and it is generally believed that some form of mechanical energy must be deposited in the wind somewhere beyond the critical point /5,6/. If we denote this mechanical energy flux by F_m, the empirical relationship between wind speed and expansion factor suggests that F_{m0} must be roughly constant within open field regions, and in particular, that it must be essentially independent of the expansion factor.

In order to make these arguments more quantitative, we now suppose that the mechanical energy is in the form of Alfvén waves /7,8/. We adopt a steady-state, one-fluid model in which the flow, driven by thermal and Alfvén-wave pressure gradients, is confined to an infinitesimal, radially-oriented flux tube; the field strength B, mass density ρ, flow velocity u, and time-averaged Alfvén-wave velocity amplitude $\langle \delta v^2 \rangle^{1/2}$ are then functions of r alone. Our basic equations are similar to those used by Leer, Holzer, and Flå /4/. The conservation of energy along the flux tube may be expressed as

$$F/B = (F_{thc} + F_{Ac})/B_c, \tag{2}$$

where

$$F_{thc} = \rho_c u_c [\frac{1}{2} u_c^2 + (5 + \frac{3}{2}\alpha)\frac{1}{2} v_{Tc}^2 - \frac{GM}{r_c}]; \tag{3}$$

$$F_{Ac} = \rho_c \langle \delta v_c^2 \rangle v_{Ac}(1 + \frac{3}{2} M_{Ac}) = (\frac{\rho_0}{4\pi})^{1/2} \langle \delta v_0^2 \rangle B_c \frac{(1 + \frac{3}{2} M_{Ac})}{(1 + M_{Ac})^2}. \tag{4}$$

Here G is the Gravitational constant, M the Sun's mass, v_T and v_A respectively denote the isothermal sound speed and Alfvén velocity, and $M_A \equiv u/v_A$. The coefficient α determines the magnitude of the collisionless heat flux and will be set equal to 4 /4/. The expressions for the Alfvén-wave energy flux F_A apply in the short-wavelength, dissipation-free approximation /9/; in the second equality of equation (4), we have assumed $M_{A0} \ll 1$. Following /4/, we shall suppose that the coronal temperature T is constant for $r \leq r_c$. Then the momentum equation may be integrated from the coronal base $r = r_0$ to the critical point $r = r_c$ to yield the conditions:

$$r_c = GM/(2u_c^2 \beta_c); \tag{5}$$

$$u_c^2 = v_T^2 + \frac{1}{4}(\frac{\rho_0}{\rho_c})^{1/2} \frac{(1 + 3M_{Ac})}{(1 + M_{Ac})^3} \langle \delta v_0^2 \rangle; \tag{6}$$

$$\rho_c = \rho_0 \exp\{-\frac{GM}{r_0 v_T^2} + (2\beta_c - \frac{1}{2}) + \frac{(u_c^2 - v_T^2)}{v_T^2}[(2\beta_c - \frac{1}{2}) + \frac{2(1 + M_{Ac})(1 + 2M_{Ac})}{(1 + 3M_{Ac})}]\}. \tag{7}$$

Here we have defined $\beta \equiv -(r/2B)dB/dr$ and assumed that $u_0^2 \ll v_T^2$.

In order to obtain u_E, the wind speed at Earth, we determine F_{thc}, F_{Ac}, and $\rho_c u_c$ by solving equations (5)-(7) iteratively for r_c, u_c, and ρ_c, and substitute the results into equation (1). Our objective is to determine how u_E varies as a function of the rate at which the flux tube expands near the Sun. For this purpose, we define the areal expansion factor at any point r along the flux tube to be

$$f(r) = (\frac{r_0}{r})^2 \frac{B_0}{B}. \tag{8}$$

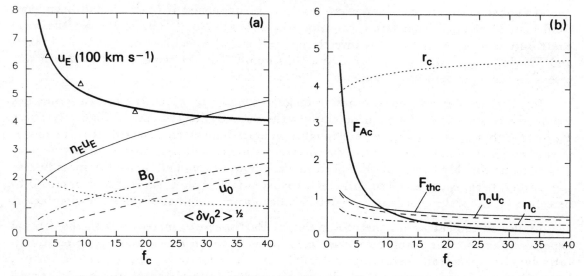

Fig. 1.—Calculated dependence of flow variables on the expansion factor. In (a), u_E (in units of 100 km s^{-1}), $n_E u_E$ (in units of 10^8 cm^{-2} s^{-1}), u_0 and $\langle \delta v_0^2 \rangle^{1/2}$ (in units of 10 km s^{-1}), and B_0 (in units of 10 G) are plotted against f_c. *Triangles* indicate the empirically determined relationship between u_E and f_c /3/. In (b), F_{Ac} and F_{thc} (in units of 10^3 ergs cm^{-2} s^{-1}), r_c (in solar radii), $n_c u_c$ (in units of 10^{12} cm^{-2} s^{-1}), and n_c (in units of 10^5 cm^{-3}) are plotted against f_c.

Assigning a fixed strength $B_E = 3 \times 10^{-5}$ G to the radial field at Earth /10/, we adopt the following magnetic model. The critical point is taken to lie outside the region of rapid expansion near the Sun, and β_c is set equal to 1 (thus $B \propto r^{-2}$ at $r = r_c$). The ratio B_c/B_0 is controlled by the value of f_c, which represents the independent parameter of the model and corresponds roughly to the "expansion factor" used in the empirical studies /2,3/. The field strength at the coronal base, $B_0 = (r_E/r_0)^2 f_E B_E$, is specified by assuming $f_E = 3 f_c^{1/2}$: thus $B_0 = 4.2 f_c^{1/2}$ G. (As discussed below, the choice of f_E affects the qualitative behavior of $\rho_E u_E$ but not that of u_E.)

The coronal temperature will be assigned a value $T = 1.1 \times 10^6$ K; the proton density at the coronal base, $n_0 = \rho_0/m_p$ will be taken as 1.8×10^8 cm^{-3}, so that $n_0 T_0 = 2 \times 10^{14}$ K cm^{-3}, in agreement with observational constraints /11/. In addition, it is necessary to specify either the wave velocity amplitude $\langle \delta v_0^2 \rangle^{1/2}$ or the wave energy flux F_{A0} at the coronal base. In earlier analyses such as that of Leer, Holzer, and Flå /4/, $\langle \delta v_0^2 \rangle^{1/2}$ was assigned a fixed value while $F_{A0} = (\rho_0/4\pi)^{1/2} \langle \delta v_0^2 \rangle B_0$ was implicitly allowed to vary in response to changes in the expansion factor (which affect F_{A0} through its dependence on B_0). Our treatment differs in the crucial respect that F_{A0}, rather than $\langle \delta v_0^2 \rangle^{1/2}$, will be held fixed.

Figure 1 shows the effect of varying f_c (the expansion factor at the critical point), with F_{A0} held constant at a value of 1.5×10^5 ergs cm^{-2} s^{-1}. As f_c decreases from 40 to 2, u_E increases monotonically from 418 km s^{-1} to 776 km s^{-1}, with the slope being steepest for $f_c \lesssim 10$. The $u_E(f_c)$ curve agrees qualitatively with the empirical relation found by Wang and Sheeley /3/, indicated by the *triangles* in Figure 1a. Figure 1b shows that the rapid increase in u_E for small f_c is due to the sharp rise in F_{Ac}, the wave energy flux at the critical point. This behavior follows directly from our assumption that F_{A0} is a constant independent of f_c: from equation (4) (with $M_{Ac} \ll 1$), $F_{Ac} \simeq F_{A0}(B_c/B_0) \propto f_c^{-1}$.

It is instructive to consider what would have happened if $\langle \delta v_0^2 \rangle^{1/2}$ had been kept fixed instead of F_{A0}. Then, from equation (4), $F_{Ac} \propto B_c \propto f_c^{-1} B_0 \propto f_c^{-1} f_E$: the wave contribution to u_E now depends directly on f_E (or B_0) as well as on f_c. Leer, Holzer, and Flå /4/ assumed that $f_c \simeq f_E$, and thus found that the wind speed at Earth was essentially independent of the expansion factor.

In /12/, we describe an alternative magnetic model in which the critical point lies *within* the region of rapid expansion close to the Sun. Unlike the case of Figure 1, where $F_{Ac} \propto f_c^{-1}$ but r_c, n_c, and $n_c u_c$ are roughly independent of f_c, the critical point moves inward and n_c and $n_c u_c$ increase almost linearly with increasing expansion factor, whereas F_{Ac} remains roughly invariant. For both models, however, u_0 and hence the mass flux at the coronal base, $\rho_0 u_0$, increase with the expansion factor; thus, since F_{A0} is assumed to be constant, the wave energy per particle is a decreasing function of the expansion factor.

Figure 1a indicates that the proton flux at Earth, $n_E u_E$, *increases* with the coronal expansion factor. The resulting inverse correlation between $n_E u_E$ and u_E agrees qualitatively with spacecraft measurements /13/. The variation of $n_E u_E$, unlike that of u_E, is sensitive to f_E as well as to f_c. The observed behavior of the proton flux at Earth was matched by assuming a nonlinear relationship between f_E and f_c, such that $f_E > f_c$ for small f_c, but $f_E < f_c$ for large f_c. Such a relationship is suggested by MHD models in which transverse pressure balance results in an increasingly uniform distribution of magnetic flux far from the Sun /10,14/. When a current sheet is used to represent this effect, it is found that the flux tubes that undergo the least expansion near the Sun later fan out more rapidly, whereas those characterized by the largest coronal expansion factors subsequently "reconverge" /3/. Such current sheets are also required to explain the spatial and long-term temporal variation of B_E /10/.

Finally, we note the following observational implications of this model:

1. If Alfvén waves are indeed responsible for boosting the solar wind to high speeds, the requirement that $F_{A0} = (\rho_0/4\pi)^{1/2} \langle \delta v_0^2 \rangle B_0$ be roughly constant implies that the mean-square wave velocity $\langle \delta v_0^2 \rangle$ and the field strength B_0 at the coronal base are inversely correlated (neglecting the variation of $\rho_0^{1/2}$). This prediction could be tested by comparing measurements of spectral line widths and magnetic field strengths in coronal holes.

2. The mass flux $\rho_0 u_0$ and flow speed u_0 at the coronal base should be highest near the boundaries of coronal holes, even though these regions produce the slowest wind at 1 AU.

REFERENCES

1. R. H. Munro and B. V. Jackson, *Astrophys. J.* **213**, 874 (1977)
2. R. H. Levine, M. D. Altschuler, and J. W. Harvey, *J. Geophys. Res.* **82**, 1061 (1977)
3. Y.-M. Wang and N. R. Sheeley, Jr., *Astrophys. J.* **355**, 726 (1990)
4. E. Leer, T. E. Holzer, and T. Flå, *Space Sci. Rev.* **33**, 161 (1982)
5. E. N. Parker, *Space Sci. Rev.* **4**, 666 (1965)
6. E. Leer and T. E. Holzer, *J. Geophys. Res.* **85**, 4681 (1980)
7. J. V. Hollweg, *Rev. Geophys. Space Phys.* **16**, 689 (1978)
8. E. N. Parker, *Astrophys. J.* **372**, 719 (1991)
9. S. A. Jacques, *Astrophys. J.* **215**, 942 (1977)
10. Y.-M. Wang and N. R. Sheeley, Jr., *J. Geophys. Res.* **93**, 11227 (1988)
11. G. L. Withbroe, *Astrophys. J.* **325**, 442 (1988)
12. Y.-M. Wang and N. R. Sheeley, Jr., *Astrophys. J.* **372**, L45 (1991)
13. R. Schwenn, in: *Solar Wind Five*, ed. M. Neugebauer, NASA, Washington, DC 1983, p. 489
14. G. L. Withbroe, *Astrophys. J.* **337**, L49 (1989)

SUMER – SOLAR ULTRAVIOLET
MEASUREMENTS OF EMITTED RADIATION

K. Wilhelm,[a] W. I. Axford,[a] W. Curdt,[a] E. Marsch,[a] A. K. Richter,[a]
M. Grewing,[b] A. H. Gabriel,[c] P. Lemaire,[c] J. -C. Vial,[c] M. C. E. Huber,[d]
M. Kühne,[e] S. D. Jordan,[f] A. I. Poland,[f] R. J. Thomas[f] and J. G. Timothy[g]

[a] *Max-Planck-Institut für Aeronomie, D-3411 Katlenburg-Lindau, Germany*
[b] *Astronomisches Institut, Tübingen, Germany*
[c] *Institut d'Astrophysique Spatiale, Paris, France*
[d] *ESA/ESTEC SSD and Institut für Astronomie, Zürich, Switzerland*
[e] *Physikalisch-Technische Bundesanstalt, Berlin, Germany*
[f] *NASA/Goddard Space Flight Center, Greenbelt, MD, U.S.A.*
[g] *Center for Space Science and Astrophysics, Stanford, CA, U.S.A.*

ABSTRACT

SUMER is a high-resolution normal-incidence UV-spectrometer in the wavelength range from 500 to 1600 Å on SOHO (Solar and Heliospheric Observatory), scheduled for launch in 1995. Its low noise detection system will resolve 700 km on the Sun and plasma velocities down to 1 - 3 km/s can be observed. These characteristics will allow us to perform high-accuracy spectral imaging in many EUV emission lines and to study many physical parameters of the solar atmosphere, especially in the transition region and the low corona. Coronal heating, solar wind acceleration, and the magnetic topology of the upper atmosphere will be investigated. Specific targets will be coronal holes, bright points, magnetic loops, coronal mass ejections and explosive events. Correlative studies with other SOHO instruments and ground-based observers will enhance the scientific return of SUMER.

Here we present only a short summary of the investigation. A full description and an exhaustive scientific report can be found in 'The SOHO mission, ESA SP-1104, 31 (1990)' and 'SUMER Red Book (1991)', respectively. Both documents are available on request.

SCIENTIFIC OBJECTIVES AND DESIGN CHARACTERISTICS

The experiment Solar Ultraviolet Measurements of Emitted Radiation (SUMER) is designed for investigations of plasma flow characteristics, turbulence and wave motions, plasma densities and temperatures, structures and events associated with solar magnetic activity in the chromosphere, the transition zone and the corona. Specifically, SUMER will

- measure profiles and intensities of extreme ultraviolet lines emitted from atoms and ions in the solar atmosphere ranging from the upper chromosphere to the lower corona;

- determine line broadenings, spectral positions and Doppler shifts with high accuracy;

- measure line pairs whose intensity ratios will either be dependent on the plasma density or on the temperature;

- reveal iso-electronic sequences such as Li-like or Be-like ion lines to be used in differential emission measure studies;

- provide stigmatic images of selected areas of the Sun in the EUV with high spatial, temporal and spectral resolution and obtain full images of the Sun and the inner corona in selectable EUV lines, corresponding to a temperature range from 10^4 to more than $1.8 \cdot 10^6$ K.

Figure 1 shows the SUMER optical design which is based upon two parabolic mirrors, a plane mirror and a spherical concave grating. The first off-axis telescope parabola has pointing and

Fig.1. The optical design with four mirrors, made of silicon carbide

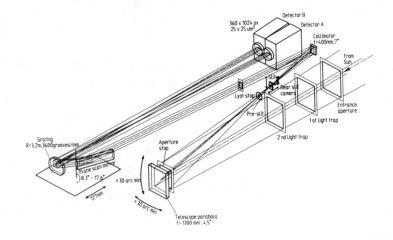

scan capabilities with a range of ± 32 arcmin in two directions and images the Sun on the spectrometer entrance slit. The second parabola collimates the beam leaving the slit which is then deflected by the plane mirror onto the grating. Two 2-dimensional detectors, located in the focal plane of the grating collect the monochromatic images of the spectrometer entrance slit in two orders simultaneously. A baffle system, consisting in particular of an entrance aperture, light traps, an aperture stop, a pre-slit and a Lyot stop, completes the design. The basic characteristics of SUMER are given in Table 1. SUMER is complemented by the other SOHO coronal instruments. Particularly, the Coronal Diagnostic Spectrometer (CDS) which employs a grazing incidence optical system extends the wavelength range down to 154 Å. SOHO will be launched in 1995 and will be placed in a halo orbit around the Lagrangean point L1.

<u>Table 1</u>   Basic Characteristics of the SUMER Normal Incidence Telescope / Spectrometer

The Telescope:	Equivalent f-number	10.66
	Focal length	1300 mm
	Plate scale in slit plane	6.3 µm/arcsec
	Full geometric area	117 cm^2
	Total dynamic field-of-view	64 × 64 arcmin^2
	Step Sizes: Full steps	0.76 arcsec
	Half steps	0.38 arcsec
The Spectrometer:	Wavelength range	500-800 Å (2nd order)
		800-1600 Å (1st order)
	Slit sizes in arcsec^2	1 × 300, 1 × 120, 0.3 × 120, 4 × 300
	Grating ruling	3600 l/mm
	Magnification factor	4.0866 at 800 Å; 25.746 µm/arcsec
	and plate scale	4.4018 at 1600 Å; 27.731 µm/arcsec
The Detectors:	Pixel size	25 × 25 µm^2
	Array size	360 spatial × 1024 spectral pixels

STATUS

The hardware phase has started in May 1991. An engineering model is being integrated now for a delivery in the second half of 1992. However, the team is still open for ideas as long as no re-design of the hardware is required and will appreciate your communication.

FINE STRUCTURE ANALYSIS OF A PROMINENCE IN Hα AND CORONAL LINES

J. E. Wiik,*,** B. Schmieder,* J. C. Noëns*** and P. Heinzel†

* Observatoire de Meudon, DASOP, F-92195 Meudon Principal Cedex, France
** Institute of Theoretical Astrophysics, PO Box 1029 Blindern, N-0315 Oslo 3, Norway
*** OPMT, F-65200 Bagnères-de-Bigorre, France
† Astronomical Institute of the Czechoslovak Academy of Sciences, CS-25165 Ondrejov, Czechoslovakia

Abstract

Prominence observations made simultaneously with the MSDP spectrograph and with the coronagraph at Pic du Midi are presented. The existence of a coronal cavity is discussed. 2D-maps of Hα intensities and the line-of-sight velocities have been obtained. The behaviour of the velocity field suggests the presence of twists along flux ropes or shear motions. Further, using a probabilistic approach to the Hα-line formation, we estimate the range of temperatures, electron densities and optical thicknesses in different parts of the prominence by fitting the observed Hα intensity profiles.

1. Observations

A quiescent prominence has been observed simultaneously in Hα and coronal lines on June 11 1988 respectively with the Multichannel Subtractive Double Pass (MSDP) spectrograph and the 20 cm coronagraph at Pic du Midi. This prominence is edge-on with two parallel legs, one northern leg and one western leg.

The standard data reduction technique for MSDP data, developed by Mein (1977) have been used to obtain intensity and velocity maps from the Hα line profiles (figures 1a and b).

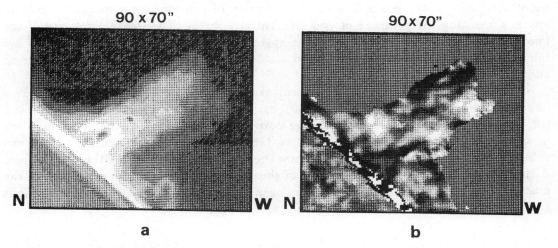

Figure 1 a) Map of the maximum Hα intensities. b) Map of the line-of-sight velocities.

Simultaneous measurements of He I and two iron lines Fe XIII 10747Å and 10798Å are used to determine (i) the position and the shape of the prominence, and (ii) the electron density in the surrounding corona. The results are presented in figures 2a-d.

2. Coronal Cavity

From the ratio of the two observed FeXIII lines we find two regions of low electron density (figure 2d). One north of the prominence and one above the prominence at a high altitude (100 000 km).

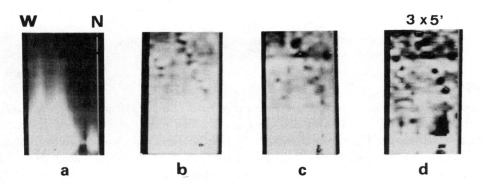

Figure 2 a) He I 10830Å. *b)* I_1=Fe XIII 10747Å. *c)* I_2=Fe XIII 10798Å. *d)* I_2/I_1, i.e. the relative electron density. Note that these images are mirror-images of the MSDP images given in figures 1a and b.

3 Velocity Field

The velocity field is complex, but rather well organized. Looking at the top of the prominence (figure 1b), we see that the northern leg is mainly redshifted, while the western leg mainly blueshifted. Closer to the limb this pattern is reversed. The redshifts are high, reaching -25 km s^{-1}. Even in the redshifted part close to the limb where the Hα line profiles are saturated, we measure redshifts as high as -10 km s^{-1}. The blueshifts are smaller, less than 10 km s^{-1}. In some regions of the prominence the profiles are all double peaked, with the intensity of the two components varying in space, suggesting the presence of two different structures. The relative velocity of the two components is 15-20 km s^{-1}. The systematic blue- and redshifted regions are relatively large, about 100 arcsec^2, and stable on a timescale less than one hour as suggested by Mein *et al.* (1989).

These observations may have different explanations. (i) The prominence behaves like a twisted flux rope (Schmieder 1989), or (ii) the matter is condensing at the top (or bottom) of the loop and then flowing downwards (upwards). If the latter happens in a helical structure as in the Priest *et al.* (1989) model, then we will see red on one side and blue on the other as we look along the direction of the axis of the prominence. A third possibility may be the existence of shear motions along the inversion line of the magnetic field. Similar velocity fields have been reported earlier, both from observations and theory by Malherbe and Priest (1983) and Athay (1985).

4. Non-LTE Hα Line Formation

We use a probabilistic approach to the Hα line formation as developed by Heinzel, see Wiik *et al.* (1992). In one-dimensional slab geometry, a five level hydrogen atom with continuum is considered under the assumption of detailed radiative balance in all Lyman transitions. For a given height above the photosphere, H, thickness of the slab, D, temperature, T, Doppler width, $\Delta\lambda_D$, and a range of values for the electron density, N_e, we derive the source function, S, the maximum intensity, I_{max}, (both in percents of the continuum intensity at disk center), and the optical depth, τ. The calculations are carried out for the values: H = 45 000 km, D = 750 to 3000 km, T = 7500 K, and $\Delta\lambda_D = 0.34$Å.

To demonstrate the results of our method, we plot the electron density versus maximum Hα intensity for various thicknesses D of the slab (figure 3). This parameter is the greatest contributor to the uncertainty in determining the plasma parameters.

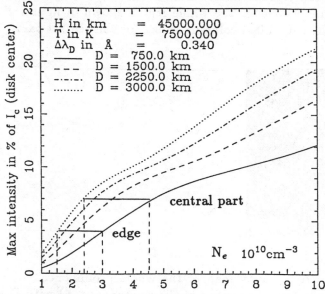

Figure 3 The electron density versus maximum Hα intensity for various thicknesses D. The horizontal bars indicate the range of electron densities for the given regions.

5. Electron Density

By using the non-LTE probabilistic code together with a cloud model method (Mein and Mein 1989), we can fit theoretical Hα profiles to the observed ones. The double-peaked profiles and some high intensity profiles could be fitted using two clouds in the model.

For individual profiles, the electron density can be infered directly from the curves in figure 3. For the low intensity profiles found at the prominence edges (I_{max}/I_c = 2-5%), we get electron densities 1.0-3.5 · 10^{10} cm^{-3}. For profiles in the central parts (I_{max}/I_c >5%), we get electron densities between 2.0-5.5·10^{10} cm^{-3}. Thus, the electron density decreases towards the edges which is in agreement with the Hvar Reference Atmosphere (Engvold *et al.* 1990). However, there is a significant overlap in the range of electron densities for the central parts and the edges, where, for some values of D, the electron density can increase towards the edges (figure 3).

Looking at the top region of the prominence as a whole, we find source functions, S/I_c, between 7.0% and 10.0% and optical depths from 0.3 to 4.0. These values, and those obtained for edges and central parts are listed in table 1. In figures 4b and c we show examples of electron density plots.

	Top region	Edges	Central parts
S/I_c %	7.0-10.0	7.0-8.0	8.0-10.0
τ (Hα)	0.3-4.5	0.3-1.0	1.0-4.5
$N_e 10^{10}$ cm^{-3}	1.0-5.5	1.0-3.5	2.0-5.5

Table 1 Source function, optical thickness and electron density from the observed Hα profiles. First column: values for the whole region shown in figures 4b and c; second column: values for the edges; last column: values for the central parts.

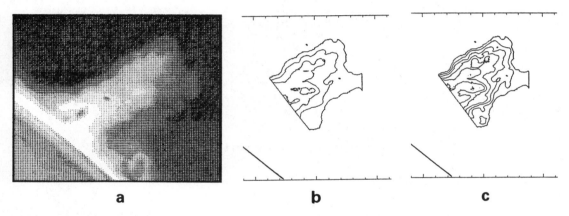

a   b   c

Figure 4 a) MSDP Hα intensity map of the whole prominence. b) and c) Electron density contour plots of the top regions of the prominence, b) with D = 3000 km and c) with D = 750 km. The calculations were done using the input parameters indicated in section 4. The lowest level is 10^{10} cm^{-3}. The level spacing is 1.0×10^{10} cm^{-3}.

References

Athay RG: 1985, *Solar Phys.* **100**, 257

Engvold O, Hirayama T, Leroy JL, Priest ER, Tandberg-Hanssen E: 1990, IAU Coll. 117, *Lecture Notes in Physics* **363**, 294

Malherbe JM, Priest ER: 1983, *Astron. Astrophys.* **123**, 80

Mein P: 1977, *Solar Phys.* **54**, 45

Mein P, Mein N: 1989, *Astron. Astrophys.* **203**, 162

Mein P, Mein N, Schmieder B: 1989, IAU Coll. 117, *Hvar Obs. Bulletin* **13**, 113

Priest ER, Hood AW, Anzer U: 1989, *Astrophys. J.* **344**, 1010

Schmieder B: 1989, "Dynamics and Structure of Quiescent Solar Prominences", 15, Kluwer Academic Publishers, ed. ER Priest

Wiik JE, Heinzel P, Schmieder B: 1992, *Astron. Astrophys.* , Submitted

Session 2

Large-Scale Structure of the Interplanetary Medium

Conveners: Mike Bird
Todd Hoeksema
Steve Suess

MODELING THE LARGE-SCALE STRUCTURE OF THE SOLAR CORONA

F. Bagenal and S. Gibson

*Department of Astrophysical, Planetary and Atmospheric Sciences,
University of Colorado, Boulder, CO 80309, U.S.A.*

ABSTRACT

The aim of this study is to find a quantitative description of both the magnetic field and the distribution of plasma in the lower corona that matches the white light images of the K-corona. We use the magnetostatic model of Bogdan and Low /1/ and data obtained by the High Altitude Observatory Mark III K-Coronameter stationed at Mauna Loa, Hawaii. To start with, we take the simplest, solar minimum case when the corona is approximately longitudinally symmetric. By varying parameters in the Bogdan and Low model we are able to quantitatively match the general characteristics of the lower corona at solar minimum: power law radial profiles of coronal brightness; enhanced brightness at the equator; uniform density depletion at the pole.

INTRODUCTION

While it is evident from inspection of coronal images that the structure is controlled by a magnetic field, there are very few measurements of the coronal magnetic field /2; 3/. On the other hand, the magnetic field in the photosphere has been measured with increasing precision and accuracy since the 1950s revealing a magnetic field of greater and greater complexity. The issue is how to map the photospheric field into the corona. The simplest approach is to assume that there are no electrical currents in the corona so that one can express the magnetic field in the corona as a gradient of a scalar potential which can be expanded as spherical harmonics. One then finds the multipole coefficients that match observations of the photospheric field and extrapolates the magnetic field out into the corona.. The potential field implies that the coronal plasma is spherically uniform with a radial gradient corresponding to hydrostatic equilibrium. The non uniform distribution illustrated by coronal images requires a finite current density in the corona. This raises the difficult issue of finding a density distribution and a magnetic field that are mutually-consistent. In this study, we take the analytical functions for both density and magnetic field that were found to be consistent solutions of the magnetostatic equations for a fully ionized atmosphere by Low /4/ and were developed for the large scale solar corona by Bogdan and Low /1/. We then adjust parameters in these functions to find a model that matches the scattered light in the corona. Although fully three-dimensional, the chosen magnetostatic model relies on some restricting assumptions and is clearly not unique. Nevertheless, it allows a relatively simple quantitative description of the corona based on physical principles which can be tested against observations. For a full description of this work see /5/.

DATA ANALYSIS

In a series of papers, Low has developed mathematical descriptions of three-dimensional structures of magnetostatic atmospheres /4; 1; 6/. Low's objective is to generate realistic, fully three-dimensional solutions to the magnetostatic equation, a mathematically formidable task which was made tractable by imposing the condition that all electrical currents in the corona be perpendicular to gravity /4/. Bogdan and Low /1/ then generated solutions for the magnetic field using a spherical harmonic expansion in terms of Legendre polynomials ($P_n^m(\theta)$ with coefficients g_n^m, h_n^m) and a radial function $\eta(r) \equiv (1 + a/r)^2$ where the parameter a is a length scale for the current density distribution. Setting $a = 0$ gives just the potential field solution with no coronal currents. The situation with $a < 0$ corresponds to a compression of the corona, whereas $a > 0$ corresponds to an expansion of the corona from the potential state. Note that a is assumed to be constant throughout the corona. One of the aims of the study is to see if this simple function is realistic. By taking the radial function of a form such that at a "source surface" /7;8/ $r = b$ the magnetic field becomes purely radial we allow for the fact that at higher altitudes in the corona (somewhere between 1.5 and 3 solar radii) the magnetostatic assumption breaks down as the upper corona evolves into the solar wind. The electron density is expressed in terms of the same parameters g_n^m, h_n^m, a, and b plus parameters c and d to characterize a power law for the background, spherically symmetric density.

The High Altitude Observatory has been measuring pB with a K-coronameter located on Mauna Loa for many years /9/. Each observing day the corona is scanned between approximately 1.1 and 2.4 solar radii (R_S). A radial scan is made every 1° in latitude, sampling at 0.01 R_S intervals in the radial direction. We chose the rotation for May 1986 as typical of solar minimum conditions. The synoptic charts published by Sime et al. /10/ show the three-dimensional structure to be a uniform, bright equatorial disk and dark polar regions, with little variation from rotation to rotation, indicating the temporal stability of these large-scale structures. We initially select only data from the east limb of the image and look at radial profiles of the polarization brightness, pB, for different latitudes. Upon examining the radial profiles we note the strong linear trend of these profiles. The implied power law dependence of pB with radial distance indicates that the coronal structures are not dominated by a particular length scale. By fitting straight lines through each radial profile we preserve information about global structure but remove small scale variations and instrumental noise. We fit each radial profile with the function $\log pB + 2 = \alpha + \beta R$. The linear fit parameters α and β are obtained at 1° intervals for the east limb. We then took 10° averages of α and β to obtain the set of radial profiles, our smoothed data set, shown in Figure 1a, which we aim to match with a theoretical model. Note that these 19 profiles are described by only 38 numbers but they retain the essential information about the large scale radial and latitudinal gradients that comes from the approximately 20,000 raw data points comprising the East limb. Figure 1b shows a contour map of pB that was constructed by linear interpolation of the average values of α and β. With a means of calculating model values of pB over the plane of the sky for a given set of model parameters we then carry out the inverse problem. The equation for pB is non-linear in the model parameters (a, b, c, d, g_n^m, h_n^m). We therefore apply non-linear inversion techniques /11/, which essentially involve linear expansions in a local region of parameter space. The non-linearity of the equations means that we must have reasonable initial values for the parameters and that the inversion is not unique. The only way one can tackle the uniqueness problem is to experiment with the forward calculation, seeing how the model values of pB change when one changes each of the model parameters (a, b, c, d, g_n^m, h_n^m). The aim is not only to find the best fit parameters but also to explore how well the parameters are resolved by this method. To make the inversion manageable, we commence by only fitting a small portion of the data, the smoothed radial profiles of pB shown in Figure 1, and a limited number of parameters with the intention of progressively adding more data and parameters. Since the smoothed radial profiles that we are fitting are strictly linear functions they are fully characterized by taking values of log pB at only two separate radial distances. The model function for log pB, however, is not a linear function of radial distance. When we are fitting the model to the data we therefore calculate log pB at six evenly spaced radial distances for each of the 19 latitudes (at 10° intervals) on the East limb, making a total of 114 data points. Having obtained estimates for the model parameters from the forward calculations we then inserted these as initial values in the data inversion. The resulting parameters from the fit to the data are well-resolved by the data and the formal errors are small. The model that resulted from the inversion is illustrated in Figure 2. Comparison with Figure 1 shows that we have been able to reproduce the general characteristics of the data: approximately linear radial profiles of log pB; the enhanced brightness at the equator; the depletion at the poles. Once we have a set of parameters that best fit the data we can insert them into the equations for the magnetic field and density distribution in the lower corona. In Figure 3a we show magnetic field lines in a meridian plane for the model magnetic field corresponding to the parameters that best fit the coronameter data. Clearly, we would like to compare the magnetic field that we derive from fitting the coronal data with the magnetic field measured in the photosphere. However, it is only the line-of-sight component of the photospheric magnetic field that is measured over large regions of the Sun. This means that the magnetic field in the polar regions is poorly determined by photospheric measurements. Moreover, the synoptic magnetograms /12; 13/ are dominated by small regions of strong magnetic fields, i.e. the high order multipoles, that do not extend up into the corona. Given these very major differences in these two means of determining the solar magnetic field, it is reassuring that the two methods produce values for the g_1^0, g_3^0 and g_5^0 terms that agree in magnitude by a factor of about two /14/. Further detailed comparison of the two data sets is clearly necessary. In Figure 3(b-f) we show the pressure, density, plasma beta, temperature, and current density distribution derived from our model. The temperature was attained from the ratio of pressure to density. Ultimately, one would like to test the degree to which the plasma conditions derived from a magnetostatic model are consistent with an energy equation.

Fig 1. (*left*) Smoothed radial profile and (*right*) contours of log (pB) +2 corresponding to the 10° averages.

Fig 2. (*left*) Radial profile and (*right*) contours of log (pB) +2 that best fit the smoothed data..

Fit parameters:
$a = 0.48$ $b=2.7$
$c=7.9$ $d=6.1$
$g_1^0 = -1.290$
$g_2^0 = -0.252$
$g_3^0 = 1.147$
$g_4^0 = 0.253$
$g_5^0 = -0.647$

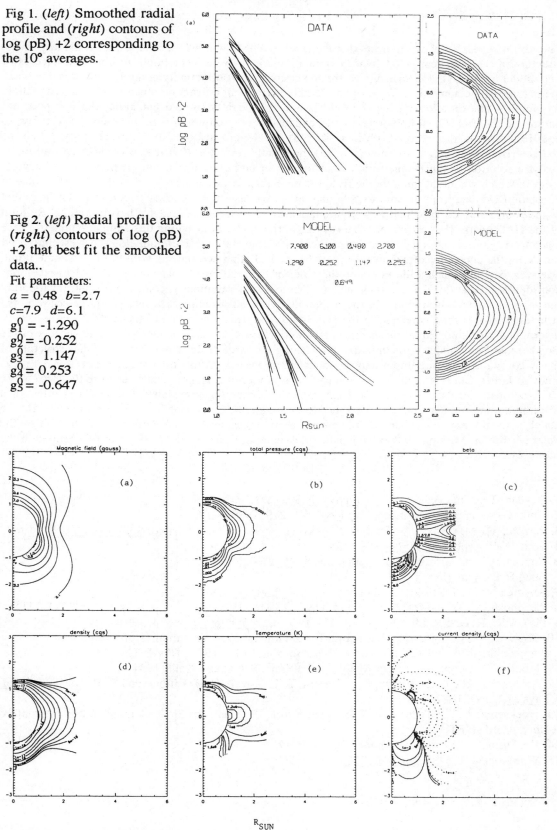

Fig. 3. Plasma parameters derived from the model

CONCLUSIONS

The results discussed above indicate that the Bogdan and Low model is able to match the large-scale structure of the lower corona at solar minimum remarkably well. The implications of the success of this model are that (1) the structure of the lower corona is not strongly controlled by dynamics and the magnetostatic approximation is valid; (2) there are no significant currents in the radial direction: this implies that on a large scale the field is not significantly twisted but rather the non-potential nature of the field is simply due to the expansion or compression of magnetic flux tubes to equilibrate with the coronal plasma /4/; (3) the length scale of the current density can be characterized by a single parameter, a; (4) the simple form for $\eta(r)$ that was chosen to allow an analytic solution to the magnetostatic equations is in fact quite reasonable. There may be several ways in which we might improve the fit to the data. First of all we need to consider the photospheric magnetic field data from the mid to polar latitudes. We also need a means of constraining the height of the source surface, b, which will require looking at a data set that extends beyond the 2.3 R_S limit of the HAO Mark III coronagraph. The success of the model in matching this simple symmetric case leads to the consideration of more complex cases. We need to consider both east and west limbs and removing the constraint of azimuthal symmetry by including coefficients with m≠0. At some point, however, one will reach a limitation on the order of the multipole expansion which arises when one considers spatial scales that are small enough that temporal variations become significant, invalidating the magnetostatic assumption. In the meantime, we wish to emphasize that the large-scale global structure of the corona is related to the very low order structure in the magnetic field that is poorly determined by photospheric measurements. In conclusion, we are encouraged by the success of the Bogdan and Low model in quantitatively matching the large-scale latitudinal and radial structure of the K-corona at solar minimum and we intend to apply the inversion techniques developed in this study to larger data sets in the future. In particular, not only is it important to explore the large Mark III K-coronameter data set from the High Altitude Observatory on Mauna Loa at times other than solar minimum, but it would also be valuable to investigate if the Bogdan and Low model is able to match coronal structures observed at larger radial distances such as those obtained in the past by the Solar Maximum Mission, SOLWIND and Skylab coronagraphs as well as those to be obtained in the future by SOHO.

REFERENCES

1. Bogdan, T. J., and B. C. Low, Astophys. J., 306, 271, 1986.
2. Dulk, G. A., and D. J. McLean, Sol. Phys., 57, 279, 1978.
3. Pätzold, M., M. Bird, H. Volland, G. S. Levy, B. Seidel, and C. Stelzreid, Sol. Phys., 109, 91, 1987.
4. Low, B. C., Astrophys. J., 293, 31, 1985.
5. Bagenal, F., and Gibson, S., J. Geophys.Res. 96, 17663, 1991.
6. Low, B. C., Astrophys. J., 370, 427, 1991.
7. Schatten, K. H., J. M. Wilcox, and N. Ness, Sol. Phys., 6, 442, 1969.
8. Altschuler, M. D. and R. M. Perry, Sol. Phys., 23,410, 1972.
9. Fisher, R., R. Lee, R. MacQueen, and A. I. Poland, Appl. Opt., 20, 1094, 1981.
10. Sime, D. G., C. Garcia, E. Yasukawa, E. Lundin and K. Rack, An atlas of 1986 K-coronameter synoptic charts, Natl. Cent. for Atmos. Res., Boulder, Colo., 1987. NCAR/TN-280+STR.
11. Menke, W., Geophysical Data Analysis, Academic, San Diego, Calif, 1984.
12. Scherrer, P. H., J. M. Wilcox, L. Svalgaard, T. L. Duvall, P. H. Dittmer and E. R. Gustafson, Sol. Phys., 54, 353, 1977.
13. Hoeksema, J. T., Tech. Rep. CSSA-ASTRO-84-07, Cent. for Space Sci. and Astrono, Stanford Univ., Calif.,1984.
14. Hoeksema, J. T., private communtication, 1989.
15. Withbroe, G. L., Astrophys. J., 327, 442, 1988.

THE ULYSSES SOLAR WIND PLASMA INVESTIGATION: EXPERIMENT DESCRIPTION AND INITIAL IN-ECLIPTIC RESULTS

S. J. Bame,* J. L. Phillips,* D. J. McComas,* J. T. Gosling* and
B. E. Goldstein**

*MS D438, Los Alamos National Laboratory, Los Alamos, NM 87545, U.S.A.
** MS 169-506, Jet Propulsion Laboratory, California Institute of Technology,
Pasadena, CA 91109, U.S.A.

ABSTRACT

During the in-ecliptic flight of Ulysses from the Earth toward its encounter with Jupiter, the Los Alamos solar wind plasma experiment has performed well. Briefly described, the instrumentation contains two independent electrostatic analyzers, one for ions and one for electrons. Initial analysis of solar wind electron core temperatures obtained between 1.15 and 3.76 AU yields a gradient of $T \propto R^{-0.7}$ which is flatter than expected for adiabatic expansion of a single-temperature Maxwellian velocity distribution and steeper than that obtained from Mariner-Voyager.

INTRODUCTION AND EXPERIMENT DESCRIPTION

Primary objectives of the Ulysses solar wind plasma investigation include establishing the bulk flow characteristics and internal state conditions of solar wind plasma ions and electrons over the full range of solar latitudes to be reached during the five-year mission. However, the initial phase of the mission presents an opportunity for studying solar wind radial gradients in the ecliptic, since the spacecraft (S/C) must first travel out to Jupiter before it can be gravitationally deflected out of the ecliptic to begin its journey to the high latitude regions over the solar poles. On the way to Jupiter, nearly continuous measurements have been made with the Los Alamos solar wind plasma experiment composed of two independent electrostatic analyzers, described in more detail elsewhere [1, 2]. Two instruments are used to measure the ions and electrons separately because of their widely different characteristics; while the supersonic solar wind ions are always found in a narrow beam flowing nearly radially outward from the sun, the subsonic electrons arrive at the spacecraft from all directions with relatively broad distributions in energy and angle.

The sun-spacecraft-Earth system, illustrated in Figure 1, is constrained by the requirement that the dish antenna point continuously at the Earth, imposing special demands on the ion instrument design. Throughout the mission the sun aspect angle (SAA), defined in the figure, changes continuously as the S/C first travels out to Jupiter in the ecliptic and then climbs over the solar poles. At instrument turn-on, six weeks after launch, the SAA was near 50°. At Jupiter it will be near 0°, and during the polar passages of the sun it will range near 25° - 30°. To accomodate this wide range of SAA while providing the high resolution required for narrow ion distributions, the ion instrument contains 16 channel electron multipliers (CEMs) with central look angles as illustrated in Figure 1. The CEMs are located behind a spherical section analyzer with a 105° bending angle. With 5° spacings, they provide slightly overlapping responses. A set of entrance apertures, selectable by command, provides the dynamic range adjustment necessary for a 1.0 to 5.4 AU mission.

To obtain three-dimensional (3D) ion distributions, counts are accumulated from all 16 CEMs every 5° as the S/C spins and the analyzer voltage is stepped. Since ion distributions lie within a cone of 20° half angle centered on the SAA, only those data from an appropriate group of 6 to 9 contiguous CEMs, taken at appropriate spin angles, are telemetered. The CEM group is selected from a set of ten on the basis of the SAA. Efficient use of telemetry is further enhanced by tracking the E/q position of the proton peak so as to cover only the proton and α peaks. When the S/C is tracked, measurements are obtained every 4 minutes, while during data store periods, measurements are obtained every 8 minutes.

For solar wind electrons, which generally have broad distributions in both energy and angle, a relatively low resolution analyzer with a bending angle of 120° is used. Because the analyzer geometry is relatively open, the analyzer plates are grooved and blackened to inhibit background caused by UV scattering through the analyzer gap to the CEMs. An array of 7 CEMs with 21° spacings is located behind the analyzer to provide overlapping responses and 95% coverage of space. To obtain the dynamic range needed over the mission for electrons, in contrast to the ions which require multiple apertures, it is only necessary to adjust the counting interval by command so as to insure adequate counting statistics without spilling the counters more than once.

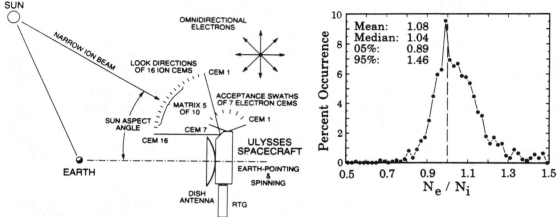

Fig. 1. Sun-spacecraft-Earth system geometry.   Fig. 2. Comparison of particle number densities.

One of two electron energy ranges can be selected: 0.81 to 454 eV or 1.59 to 862 eV. Normally, the higher energy range is used. Although it would be desirable to obtain electron velocity distributions in 3D form only, the time required for transmitting a set of 3D data is prohibitively long, so 3D data sets are mixed with 2D sets formed by on-board summing of counts from the 7 CEMs.

An indicator of the quality of these measurements is the agreement between electron and ion parameters obtained from the separate instruments. Electron density determinations are particularly sensitive to determination of the spacecraft potential, while ion density determinations are strongly driven by the accuracy of fits to the proton and α peaks and by the intrinsic ability of the instrument to resolve cold ion beams. Further, the accuracy of determinations for both instruments relies heavily on laboratory calibrations. Figure 2 shows a statistical comparison of calculated ion and electron densities for all ion spectra and all 3D electron spectra measured between 1.16 and 1.29 AU, with data binned by electron/ion density ratio in bins of 0.02. The densities generally agree within about 5%, which is very good for independent experiments. However, individual cases of larger discrepancies can occur, primarily due to inaccuracies in the determination of spacecraft potential.

IN-ECLIPTIC ELECTRON RESULTS

As a number of earlier investigations have revealed, solar wind electron distributions include a thermal or "core" component and a suprathermal or "halo" component. At 1 AU, a typical core temperature is 1.2×10^5 K with a halo temperature of about 7×10^5 K; the energy breakpoint between core and halo is generally 50-100 eV. A third, colder population of locally generated photoelectrons is bound to the spacecraft by a positive electrostatic potential, typically 2 to 10 volts. Various researchers have estimated widely different radial gradients for the core electron temperature from a number of experimental data sets. The most pertinent estimate for comparison with Ulysses results is based on combined Mariner 10 and Voyager 2 observations between 0.45 and 4.76 AU /3/. This study yielded a polytropic index for electron expansion that is equivalent to a core temperature gradient of $T \propto R^{-0.37}$ for density proportional to R^{-2}. Various other studies have produced widely varying estimates for this gradient; however, all of the gradients have been flatter than the $R^{-4/3}$ expected for adiabatic transport of an isotropic Maxwellian velocity distribution. The Ulysses electron analyzer, which measures 3D velocity distributions, is capable of more definitive sampling of the bi-Maxwellian core distributions than were earlier 1D and 2D instruments. Accordingly, the Ulysses in-ecliptic mission from instrument turn-on at 1.15 AU to Jovian encounter near 5.4 AU provides an unprecedented opportunity for establishing the electron gradient with high accuracy.

In routine data reduction, the 3D Ulysses electron measurements are fit to a convecting bi-Maxwellian core distribution at four energy steps centered at 12.1 to 32.0 eV; these energies are adjusted downward in accordance with an estimate of spacecraft potential. The fit yields core density, the bulk velocity vector, two temperatures, and the orientation of the principal axis of the distribution (ideally, this should be aligned with the magnetic field). The four-step fit was often downgraded to a three energy step fit, and on rare occasions to a two step fit when the core-halo spectral breakpoint fell within the energy passsband centered at 32.0 eV, and/or the photo-core spectral breakpoint was within the passband centered at 12.1 eV.

Figure 3 (top panel) shows an example of a spin-averaged and polar angle-averaged electron spectrum observed at 1.16 AU. In this presentation, a Maxwellian is a straight line. The four-point core fit is plotted as a dashed trace; when subtracted from the observed spectrum (solid trace), the remainder (dotted traces) includes the low-energy photoelectron and suprathermal halo distributions. Compare this distribution, with a core temperature of 1.30×10^5 K, to a similar presentation (bottom panel) of a spectrum observed at 3.11 AU. A three-point core fit to this latter spectrum yielded a

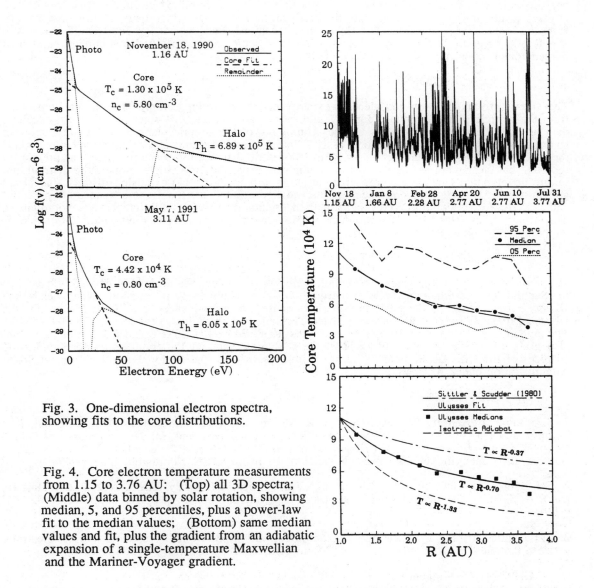

Fig. 3. One-dimensional electron spectra, showing fits to the core distributions.

Fig. 4. Core electron temperature measurements from 1.15 to 3.76 AU: (Top) all 3D spectra; (Middle) data binned by solar rotation, showing median, 5, and 95 percentiles, plus a power-law fit to the median values; (Bottom) same median values and fit, plus the gradient from an adiabatic expansion of a single-temperature Maxwellian and the Mariner-Voyager gradient.

temperature of 4.42×10^4 K. While these two examples were selected to highlight the cooling of the core distribution with increasing heliocentric distance, they are by no means unusual. The top panel of Figure 4 shows core temperature for all 3D Ulysses spectra measured from turn-on on November 18, 1990 through July 29, 1991, a heliocentric range of 1.15 through 3.76 AU. Although the temperature varies widely, the cooling trend is obvious. In order to provide a best estimate of the electron temperature gradient for the quiescent solar wind, three steps were taken in analyzing the data. First, periods of obvious compression, signalled by simultaneous sharp increases in both density and temperature, were eliminated. Second, periods of bidirectional halo streaming, interpreted as passage of topologically distinct coronal mass ejections, were eliminated. Third, the data were divided into ten solar rotations, based on heliographic longitude. Median, 5 percentile, and 95 percentile core temperatures were calculated for each rotation; the results are shown in the middle panel of Figure 4. The solid squares show the median T and R for each rotation, the dashed and dotted traces show the 95 and 5 percentiles, and the solid trace represents the best power law fit to the medians, which yields $T \propto R^{-0.70}$, an intercept value at 1 AU of 1.11 x 10^5 K, and a correlation coefficient of 0.96.

For comparison with other estimates, the bottom panel of Figure 4 shows the same median points and fit, plus two power law fits with the same 1 AU intercept. The upper trace shows the Mariner-Voyager result /3/, while the lower trace represents adiabatic expansion of a single-temperature Maxwellian. The Ulysses fit falls in between the other two curves, suggesting that, while electron transport is clearly not adiabatic, the temperature gradient may not be as flat as determined in the previous study. The Ulysses results more nearly agree with the gradients recently determined using Helios observations from the inner solar system /4/. The Helios determinations show power law fits from $T \propto R^{-0.48}$ to $T \propto R^{-0.84}$, depending on the type of solar wind flow and/or the temperature component, parallel or perpendicular, chosen.

CONCLUSIONS

The Ulysses solar wind ion and electron experiments are returning comprehensive solar wind data from the ecliptic plane enroute to Jupiter. Indications are that the instruments will return excellent data on the heliospheric plasma as Ulysses climbs out of the ecliptic plane to the regions over the solar poles after the Jovian gravitational assist. Initial scientific results show that in the ecliptic plane the core electron temperature decreases with heliocentric distance more sharply than was previously reported, but clearly not as sharply as for an adiabatic expansion of a single-temperature distribution. Future developments in the electron gradient study will include: (1) extension of data analysis to aphelion at 5.4 AU; (2) hand-checking the quality of the electron fit and spacecraft potential calculations; (3) use of observed magnetic field orientation to verify the temperature components; and (4) sorting of solar wind data into different flow regimes to identify any systematic differences in temperature gradients and anisotropies.

ACKNOWLEDGMENTS

The authors wish to acknowledge the engineering support, under the leadership of J.C. Chavez, of the Sandia National Laboratories, Albuquerque. We also acknowledge the support given to us by the Ulysses Project Office at the Jet Propulsion Laboratory, under the leadership of W.G. Meeks, and the Ulysses Project Office at the European Space Technology Center, under the leadership of D. Eaton. The guidance and support of the Project Scientists, K.-P. Wenzel at ESTEC and E.J. Smith at JPL are also gratefully acknowledged. Work at the Los Alamos National Laboratory was done under the auspices of the U.S. Department of Energy with support from the National Aeronautics and Space Administration under W-15,487.

REFERENCES

1. S.J. Bame, J.P. Glore, D.J. McComas, K.R. Moore, J.C. Chavez, T.J. Ellis, G.R. Peterson, J.H. Temple, and F.J. Wymer, The ISPM Solar-Wind Plasma Experiment, in: *The International Solar Polar Mission - Its Scientific Investigations,* esa SP-1050, ed. K.-P. Wenzel, R.G. Marsden, B. Battrick, ESA Scientific and Technical Publications Branch, ESTEC, Noordwijk, 1983, p. 47.

2. S.J. Bame, D.J. McComas, B.L. Barraclough, J.L. Phillips, K.J. Sofaly, J.C. Chavez, B. E. Goldstein, and R.K. Sakurai, The Ulysses solar wind plasma experiment, *Astron. Astrophys. Suppl. Ser.* 92, 237 (1992).

3. E.C. Sittler, Jr, and J.D. Scudder, An empirical polytrope law for solar wind thermal electrons between 0.5 and 4.76 AU: Voyager 2 and Mariner 10, *J. Geophys. Res.* 85, 5131 (1980).

4. W.G. Pilipp, H. Miggenrieder, K.-H. Mühlhäuser, H. Rosenbauer, and R. Schwenn, Large-scale variations of thermal electron parameters in the solar wind between 0.3 and 1 AU, *J. Geophys. Res.* 95 , 6305 (1990).

GLOBAL PROPERTIES OF THE PLASMA IN THE OUTER HELIOSPHERE I. LARGE-SCALE STRUCTURE AND EVOLUTION

A. Barnes,* J. D. Mihalov,* P. R. Gazis,** A. J. Lazarus,***
J. W. Belcher,*** G. S. Gordon Jr*** and R. L. McNutt Jr†

NASA-Ames Research Center, Moffett Field, CA, U.S.A.
**SJSU Foundation, NASA-Ames Research Center, Moffett Field, CA, U.S.A.*
***Physics Department and Center for Space Research, MIT, Cambridge, MA, U.S.A.*
† Visidyne Incorporated, Burlington, MA, U.S.A.

ABSTRACT

Pioneer 10 is now beyond 52 Astronomical Units (AU) from the sun, near the solar equatorial plane; Pioneer 11 and Voyager 2 are at similar heliocentric distances (about 30 AU) and longitudes, but Pioneer 11 is about 17 degrees north of Voyager 2 in latitude. All three spacecraft have working plasma analyzers, so that intercomparison of data from these spacecraft provides important information about the global character of the outer solar wind and its variation over the solar cycle. During the period around the past solar minimum (1985-1987) Pioneer 11 observed fast solar wind characteristic of the high latitude corona, whereas Pioneer 10 and Voyager 2 observed relatively slow wind from near-equatorial regions of the corona. During this period, the inclination of the current sheet associated with the interplanetary magnetic field dropped below the latitude of Pioneer 11. In contrast, after 1987 the strong latitudinal gradient in solar wind speed disappeared, and there have even been occasional signs of a reverse gradient (higher velocities near the solar equator). The solar wind density continues to drop as expected with increasing heliocentric distance. The solar wind ion temperature decreases with increasing heliocentric distance out to 10-15 AU. At larger distances the temperature measurement is more difficult because of the low ion flux, but careful study of these data suggests that (1) the temperature remains fairly constant beyond 15 AU, and (2) there may be a large-scale variation of the temperature with celestial longitude and heliographic latitude. No indication of the presence of a heliospheric terminal shock has been detected to date.

INTRODUCTION

The four spacecraft Pioneers 10-11 and Voyagers 1-2 provide a unique set of observing platforms for study of the global properties of the outer heliosphere. Pioneer 10 is currently beyond 50 AU, near the ecliptic. The other three are on the opposite side of the Sun, at several different heliographic latitudes. Note in particular that Pioneer 11 and Voyager 2 are at similar heliocentric distance and heliographic longitude, but at significantly different heliographic latitude. Table 1 shows the status of the plasma analyzers and magnetometers aboard each spacecraft.

TABLE 1 Outer Heliospheric Spacecraft and Their Instruments

Spacecraft	Plasma Instrument	Magnetometer	Other
Pioneer 10	OK	Failed	
Pioneer 11	OK	OK	Power is critical
Voyager 1	Failed	OK	
Voyager 2	OK	OK	

The present paper is primarily concerned with plasma observations, so that Voyager 1 will not be considered further.

GLOBAL BEHAVIOR OF THE SOLAR WIND IN THE OUTER HELIOSPHERE

The IMP spacecraft provide baseline data from 1 AU, which are useful for distinguishing spatial and temporal variations in the outer heliosphere. Figures 1a and b show the solar wind speed and proton temperature measured at the various outer heliospheric spacecraft as ratios to the corresponding 1 AU quantities measured at IMP. The measurements cover nearly two sunspot cycles. Each entry in the plot represents an average over three solar rotations, and allows for an appropriate delay for time of travel between 1 AU and the distant spacecraft.

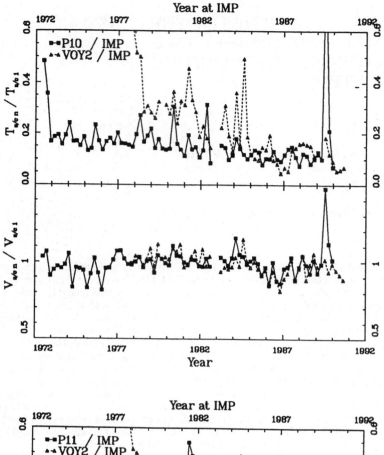

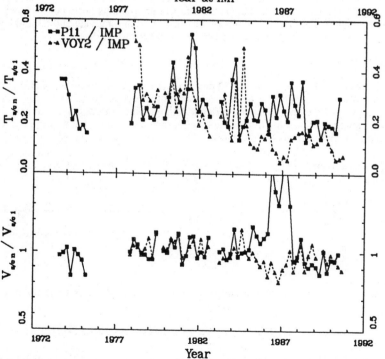

Fig. 1. (a) Three solar rotation averages of the solar wind speed and proton temperature measured at Pioneer 10 and Voyager 2, presented as ratios to the corresponding 1 AU quantities measured at IMP. Each entry in the plot allows for an appropriate delay for time of travel between 1 AU and the distant spacecraft. (b) Same, but for Pioneer 11 and Voyager 2.

Note that the velocities at both Pioneer 10 and Voyager 2 agree very well with the corresponding velocities at 1 AU. Because these spacecraft are nearly in the ecliptic plane, we infer that the mean velocity is independent of heliocentric distance. However, the mean velocity does vary with the sunspot cycle. Figure 2 gives the record of velocity and temperature seen at Pioneer 10. In general the mean speed is higher near sunspot minimum, although near sunspot maximum high-speed flows can occur in association with transient events.

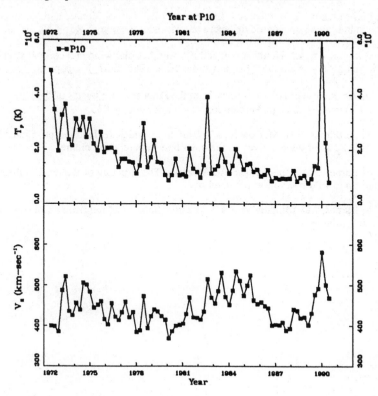

Fig. 2. Three solar rotation averages of velocity and proton temperature measured at Pioneer 10 between 1972 and 1990.

The speed at Pioneer 11 does not always track the IMP velocity as the speeds at the low-latitude outer heliospheric spacecraft do. During the period 1979-1985, a period of moderate to high solar activity, Pioneer 11 did observe about the same mean speed as the lower-latitude spacecraft, about 450 km/s. Evidently during this time there was, on average, little latitude variation in the heliosphere at least in the latitude band 0-17 degrees N. However, during the period around the past solar minimum (1985-1987) Pioneer 11 observed fast solar wind characteristic of the high latitude corona, whereas Pioneer 10 and Voyager 2 observed relatively slow wind from near-equatorial regions of the corona. During this period, the inclination of the current sheet associated with the interplanetary magnetic field dropped below the latitude of Pioneer 11. This period has been discussed elsewhere in detail /1,2,3/. After 1987 the velocities were once again comparable at all three spacecraft, but the picture is complicated by transients, and there were appreciable periods when the mean speed at Pioneer 10 was elevated /4/. The solar-cycle variation in latitudinal morphology of the solar wind speed, discussed above, also appears in the inner heliosphere /5/.

It has been known for a long time that the solar wind proton temperature falls off with increasing heliocentric distance more slowly than the $r^{-4/3}$ dependence expected for adiabatic flow (see /3/ and references cited therein). In fact, beyond 10 AU or so the temperature profile is nearly flat, suggesting that some dissipative process continually heats the outer heliospheric plasma. The temperature appears to vary during the sunspot cycle, being higher near solar minimum.

During the period 1979 to 1985, temperatures observed by Pioneer 11 and Voyager 2 track each other quite well (Fig. 1), being typically ~25,000K. However, in 1985-1986, near solar minimum, Pioneer 11 was immersed in much faster solar wind , as discussed above, and the observed temperature was correspondingly elevated. From 1987 on the situation is not clearcut; the temperature at Pioneer 11 is ~25,000K whereas it remains lower (~15,000K) at Voyager 2. The reason for this apparent latitude gradient is not yet known. On the other side of the Sun at Pioneer 10 the temperature remained fairly constant over this entire period at about ~12,000K. This value is considerably lower than those observed at Pioneer 11 and Voyager 2 (at least through 1985). This apparent asymmetry in celestial longitude could possibly be associated with an extra-heliospheric cause, though we do not regard this as a firm conclusion at present. Further discussion of the temperature measurements and their validity is given in /6/.

REFERENCES

1.   P.R. Gazis, J.D. Mihalov, A. Barnes, A.J. Lazarus, and E.J. Smith, Pioneer and Voyager observations of the solar wind at large heliocentric distances and latitudes, Geophys. Res. Lett., 16, 223 (1989).

2.   J.D. Mihalov, A. Barnes, A.J. Hundhausen, and E.J. Smith, Solar wind and coronal structure near sunspot minimum: Pioneer and SMM observations from 1985-1987, J. Geophys. Res., 95, 8231 (1990).

3.   Barnes, A., Distant solar wind plasma--view from the Pioneers, in: Physics of the Outer Heliosphere, ed. S. Grzedzielski and D.E. Page, Pergamon, New York 1990, p. 235.

4.   Mihalov, J.D., A. Barnes, F.B. McDonald, J.T. Burkepile, and A.J. Hundhausen, Concerning solar sources for cycle 22 solar wind activity in the outer heliosphere, this volume.

5.   P.R. Gazis, A. Barnes, J.D. Mihalov, and A.J. Lazarus, The structure of the inner heliosphere from Pioneer Venus and IMP observations, this volume.

6.   P.R. Gazis, A. Barnes, J.D. Mihalov, and A.J. Lazarus, Solar wind temperature observations in the outer heliosphere, this volume.

CORONAL ALFVÉN WAVES DETECTED BY RADIO SOUNDING DURING THE SOLAR OCCULTATIONS OF THE HELIOS SPACECRAFT

M. K. Bird,* H. Volland,* A. I. Efimov,** G. S. Levy,*** B. L. Seidel*** and C. T. Stelzried***

* Radioastronomisches Institut, Universität Bonn, 5300 Bonn, Germany
** Institute of Radio Engineering and Electronics, Academy of Sciences, Moscow, Russia
*** Jet Propulsion Laboratory, California Institute of Technology, Pasadena, CA 91109, U.S.A.

ABSTRACT

The two Helios spacecraft underwent regular solar occultations during their extended missions from Dec 1974–Feb 1986 (Helios 1) and Jan 1976–Mar 1980 (Helios 2), thereby providing many opportunities for radio propagation experiments in the solar corona. On certain rare occasions over the course of these investigations, Faraday rotation measurements of the linearly polarized Helios signals could be recorded simultaneously at two widely-spaced ground stations. Many of these two-station measurement intervals display clear evidence of wave-like structures with quasi-periods of the order of a few minutes to a few hours. These structures are attributed to coronal Alfvén waves. The radial propagation direction and velocity of these waves are estimated from a cross-correlation analysis of the data between the two stations. The majority of the waves appear to propagate away from the Sun, but about 30% of the cases indicate a propagation direction toward the Sun.

INTRODUCTION

Alfvén waves generated near the solar surface and propagating into interplanetary space have been cited as one of the primary mechanisms for the heating of the solar corona or the acceleration of the solar wind /1/. Alfvénic fluctuations have been measured in situ as close to the Sun as 62 $R_\odot$ with Helios /2/. Remote sensing observations in the inner corona are possible, however, using radio-sounding techniques. Wave-like structures with quasi-periods of the order of a few minutes to a few hours have been detected from Faraday rotation measurements on Helios /3,4/.

In a number of cases, these Faraday rotation measurements could be recorded simultaneously from two widely-spaced ground stations. These data can be used to estimate the propagation direction and speed of coronal inhomogenieties in a statistical sense. This report describes the observations and presents the preliminary results on the implied coronal velocities in the light of our present understanding of the solar corona.

TWO-STATION FARADAY ROTATION OBSERVATIONS

The data for this two-station correlation analysis are available only during times of ground station overlap, which rarely exceeds 5-6 hours over the applicable intercontinental distances. As a result, only a small fraction (ca. 7%) of the total Faraday rotation data base of some 1277 hours between 1975 and 1983 could be used. The observations were made at the MPIfR 100-m Effelsberg radio telescope, FRG, and at the three large NASA Deep Space Network (DSN) tracking stations in California, Australia and Spain. All of the measurements discussed here were taken when the Helios ray path offset distance from the Sun ranged from 3 to 10 $R_\odot$.

The Helios radio signal was linearly polarized with its transmitted electric vector oriented perpendicular to the ecliptic. During the regularly occurring superior conjunctions, the strong magnetic field and high electron density of the solar corona would cause the electric vector to undergo "Faraday rotation" (FR) of up to hundreds of degrees at the S-band downlink frequency of 2.3 GHz. Large-scale changes in coronal FR occur as the source moves through occultation to different solar offsets and the corona changes its orientation via solar rotation. Superimposed on this are ubiquitous oscillations of the FR that presumably arise from smaller-scale coronal inhomogenieties passing through the signal ray path. The two-station observations of coronal FR present an opportunity to determine the velocities associated with these structures.

Figure 1 (left panel) schematically shows the geometical situation at conjunction with the signal ray paths from Helios arriving at Earth-based antennas, in this specific case the DSN ground stations Goldstone (G) and Canberra (C). The right panel of Figure 1 shows one 15-minute segment of Faraday rotation measurements taken on 9 Jan 1983. The Helios 1 ray path was located on the east solar limb at a proximate distance of $R = 4.0\ R_\odot$ (occultation egress). The ray path to the Canberra station was $\Delta R = 2015$ km closer to the Sun than the Goldstone ray path. The FR traces in Figure 1 show only relative changes (i.e. residuals) in Faraday rotation after subtraction of a slowly-varying long-term trend. Variations at time scales longer than 600 sec are eliminated in this process. The data have also been "smoothed", meaning that a running mean filter over ± 20 sec was applied in order to remove high-frequency fluctuations (primarily receiver noise). The wave-like structures are clearly correlated at both stations – the maximum cross-correlation coefficient in this case being 0.87 at a time lag of $\tau = 7.5$ s. The implied projected coronal velocity in this case is $V = 270$ km s^{-1}.

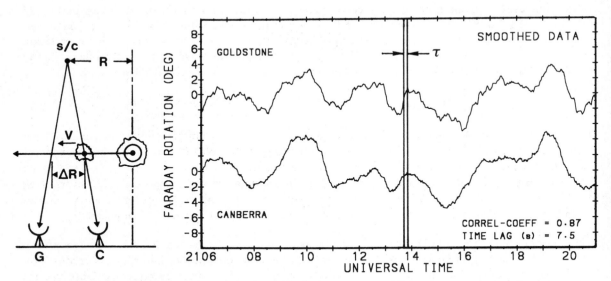

Fig. 1. Left panel: Helios two-station occultation geometry. The solar ray path offset R is the proximate solar distance from the Helios/Earth line. Two ground stations are denoted G (Goldstone) and C (Canberra). The coronal separation between the two ray paths is ΔR. Right panel: Faraday rotation measurements at Goldstone and Canberra on 9 Jan 1983.

CORONAL VELOCITIES IMPLIED BY FR CROSS CORRELATIONS

The FR data in Figure 1 readily display a shift in time between the extrema of the "waves" at both stations. This time lag can be explained as resulting from an arrival of the disturbance at the two coronal ray paths at different times, corresponding to the propagation of a wave structure.

A cross-correlation analysis of all available data was made in order to determine the projected coronal velocities. Each segment of two-station data was divided into time intervals of 15 minutes. The time lag τ of maximum correlation between the two stations was determined for each interval. The correlation was found to be strong (greater than 0.75) for about half of the intervals studied. The observed time lag τ is related to the projected velocity V by $\tau = \Delta R/V$, where ΔR is the radial distance between the two coronal ray paths (Figure 1). Of course, contributions to FR can originate from any point along the radio propagation path. Since the Faraday rotation is an integrated product of the electron density times the magnetic field, both of which decrease rapidly with solar distance, the dominant contributions to FR will come from those ray path segments closest to the Sun.

The radial velocities inferred from the observed τ are plotted in Figure 2 as a function of solar distance. A major surprise of the study was that the implied coronal velocity was sometimes found to be directed toward the Sun. The open (solid) octagons in Figure 2 indicate velocities that are positive (negative). In part because of the magnetic sensitivity of the Faraday rotation, it is suggested that the effective coronal velocity determined from this analysis is actually comprised of two components given by $\vec{V} = <\vec{V}_{sw} + \vec{V}_A>$. The solar wind bulk velocity $<\vec{V}_{sw}>$ is directed radially outward. The velocity of the wave-like fluctuations, assumed to be the Alfvén velocity $<\vec{V}_A>$, is also predominantly directed along a radial. Since both solar wind and the "waves" are expected to propagate radially, the observed time lag τ is thought to be a fair measure of $<V_{sw} + V_A>$.

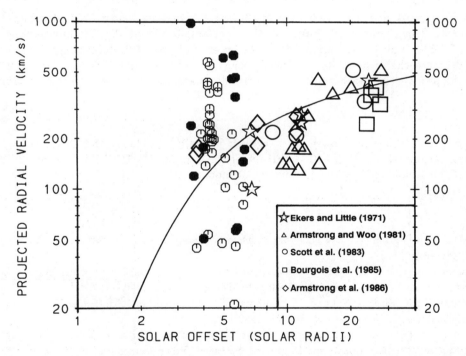

Fig. 2. Projected radial velocities in the corona derived from Faraday rotation time lags. The velocities derived in this work are the octagons that are either open, indicating a radially outward (positive) velocity, or solid, indicating a radially inward (negative) velocity. Only data from 1979–1983 with a correlation coefficient greater than 0.8 are included. Coronal plasma velocity determinations from other cross-correlation studies are given by the various large open symbols. The solid curve is a mean theoretical solar wind velocity profile from /4/.

The data of Figure 2 were restricted in this case to the years 1979–1983, for which the time resolution was sufficient to compute a reasonably accurate correlation time lag. Only those time intervals were used in Figure 2 for which the maximum cross-correlation coefficient was at least 0.8. The velocities vary from 20 km/s to greater than 1000 km/s in the range between 3.5 and 6.3 $R_\odot$, but no significant trend with radial distance is apparent. For comparison, some representative velocities derived by cross-correlation of other radio scintillation data are indicated in Figure 2 by the various large open symbols /5-9/. These earlier IPS data are derived from intensity correlations and are sensitive to much smaller spatial scales than those relevant to the FR correlation studies reported here. The inhomogenieties responsible for the correlation of intensity scintillations are undoubtedly tied to the bulk velocity of the outward flow in the corona, i.e. the solar wind. The solid curve in Figure 2, representative of a solar wind velocity profile (from Figure 2.18 of /4/), supports the "frozen-in" hypothesis for the data plotted with open symbols. A total of 47 two-station FR measurements are plotted in Figure 2, of which 14 (30%) yield velocities that are directed toward the Sun. This would imply that $|V_A| > |V_{sw}|$ with the wave moving toward the Sun.

In trying to explain the negative velocities, it should be noted that there is no reason to expect that coronal fluctuations, which travel at the Alfvén velocity, be restricted to outward propagation. The plasma in the region being probed by the Helios signal is certainly sub-Alfvénic. A "mean" solar wind speed from theoretical models is of the order of 150-300 km/s at 6 $R_\odot$. The Alfvén velocity at this same solar distance, assuming a mean magnetic field B $\simeq$ 50 mG and a mean electron density $N_e \simeq 2{\cdot}10^4$ el·cm^{-3}, is 730 km/s. Another significant difference between these observations and the conventional IPS scintillations, for which only positive velocities are derived, is the correlation time scale. It is surmised that the magnetic field is the agent responsible for maintaining the correlation at characteristic periods of a few minutes (coronal scales: a few thousand kilometers).

The data are consistent with a spectrum of coronal Alfvén waves, up to 70% of which are propagating outward in a statistical sense. The typical amplitudes of the FR fluctuations of a few degrees at S-Band imply wave amplitudes of the order of $\delta B/B \simeq 0.1$. Increasing the time intervals tends to decrease the proportion of negative velocity measurements. This may be attributed to an averaging process that washes out the minority fluctuation component with inward propagation.

ACKNOWLEDGEMENT

This paper presents results of a research project partially funded by the Deutsche Agentur für Raumfahrtangelegenheiten (DARA) GmbH under contract 50 ON 9104. The responsibility for the contents of this publication is assumed by the authors.

REFERENCES

1. G.L Withbroe, The temperature structure, mass and energy flow in the corona and inner solar wind, Ap.J. 325, 442-467 (1988)

2. R. Schwenn, Large-scale structure of the interplanetary medium, in: Physics of the Inner Helio- sphere, eds. R. Schwenn and E. Marsch, Springer-Verlag, Heidelberg, 99-181, 1990.

3. J.V. Hollweg, M.K. Bird, H. Volland, P. Edenhofer, C.T. Stelzried, and B.L. Seidel, Possible evidence for coronal Alfvén waves, J. Geophys. Res. 87, 1-8, 1982.

4. M.K. Bird, and P. Edenhofer, Remote sensing observations of the solar corona, in: Physics of the Inner Heliosphere, eds. R. Schwenn and E. Marsch, Springer-Verlag, Heidelberg, 13-97, 1990.

5. J.W. Armstrong, and R. Woo, Solar wind motion within 30 $R_\odot$: Spacecraft radio scintillation observations, Astron. Astrophys. 103, 415-421, 1981.

6. J.W. Armstrong, W.A. Coles, M. Kojima, and B.J. Rickett, Solar wind observations near the Sun, in: The Sun and the Heliosphere in Three Dimensions, ed. R.G. Marsden, Reidel Publ. Comp., Dordrecht, 59-64, 1986.

7. G. Bourgois, W.A. Coles, G. Daigne, J. Silen, T. Turunen, and P.J., Williams, Measurements of the solar wind velocity with EISCAT, Astron. Astrophys. 144, 452-462, 1985.

8. R.D. Ekers, and L.T. Little, The motion of the solar wind close to the sun, Astron. Astrophys. 10, 310-316, 1971.

9. S.L. Scott, W.A. Coles, and G. Bourgois, Solar wind observations near the sun using interplanetary scintillation, Astron. Astrophys. 123, 207-215, 1983.

MAGNETIC STRUCTURES AT SECTOR BOUNDARIES IN THE INNER HELIOSPHERE

V. Bothmer and R. Schwenn

Max-Planck-Institut für Aeronomie, D-3411 Katlenburg-Lindau, Germany

ABSTRACT

Strong deflections of the IMF out of the ecliptic plane are common at sector boundaries. They may be associated with very different phenomena, such as magnetic clouds, corotating structures, NCDE's or coronal streamers. When their origin is to be explained, the relationships to solar wind features such as corotating interaction regions, stream interfaces, heliospheric current sheet positions have to be properly taken into account. The three dimensional nature of the solar wind and its radial evolution lead to more complicated stream structures and possible superpositions with increasing distance from the sun. With the measurements of the Helios spacecraft from as close as 0.3 AU to the sun the influences of different effects can be more readily separated. A good example is an observation made by the two Helios spacecraft when they were separated by only 5 degrees in heliographic longitude and by a radial distance of 0.2 AU. Results from the analysis of the plasma and magnetic field data for the two spacecraft are presented.

INTRODUCTION

Plots of the one hour averages of solar wind plasma and magnetic field data as measured by Helios 1 and Helios 2 in March, 1976 are shown in figure 1. The large-scale solar wind stream pattern is dominated by two broad high speed streams from coronal holes of opposite magnetic polarity. In the slow wind region between the two high speed streams a sector boundary passage is accompanied by large deflections of the IMF out of the ecliptic plane. The origin of these deflections is the subject of the following investigations.

SECTOR BOUNDARIES AND COMPRESSION REGIONS

An idealized view of a corotating stream interaction region and its evolution from an originally rectangular speed profile at the sun to a more gradual speed increase at 1 AU is shown in figure 2. With increasing distance from the sun the faster plasma with its lesser curved streamlines starts pressing into the slower plasma with the stronger curved streamlines. The characteristic density and speed profiles result from the deflection and compression of both plasma flows. The transition from the slow solar wind to the fast solar wind extends over some 10 to 30 degrees in longitude at 1 AU, while the whole region affected by the interaction amounts to some 30 degrees. Plasma in this region stems from coronal sources originally spanning some 90 degrees in longitude. Sector boundaries are therefore found very often close to stream interfaces at 1 AU although they originally had been well separated from each other at the sun.

Basically, sector boundaries and stream interfaces are completely different features. Measurements of the Helios spacecraft nearer to the sun allow a more precise association of the IMF deflections to sector boundaries, compression regions or transient phenomena. The positions of the two Helios spacecraft for the analyzed event sketched in the figure gives an impression how the observed large-scale heliospheric structures passed the two spacecraft.

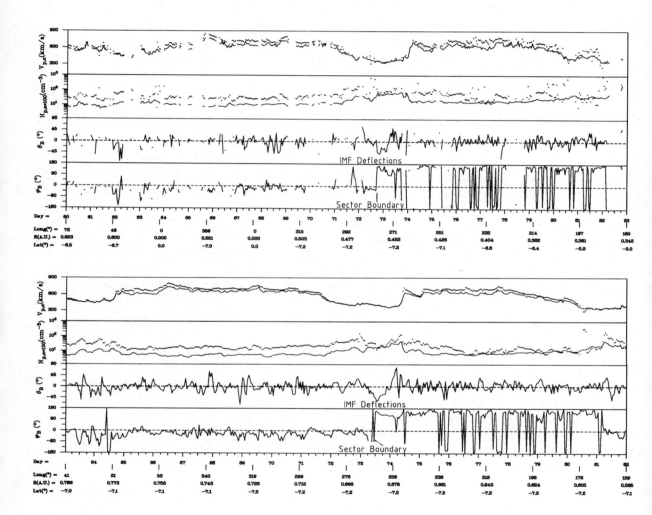

Figure 1. One hour averages of solar wind plasma and magnetic field parameters as measured by Helios 1 (top) and Helios 2 (bottom) during March, 1976. The individual parameters are (dotted curves belong to the α particles): proton bulk speed V_p, proton number density N_p, latitude angle θ and azimuthal angle ϕ of the magnetic field vector. θ is positive for field directions pointing to the north of the ecliptic plane and $\phi = 0°$ is the sunward direction in the ecliptic plane. $-45°$ and $+135°$ correspond to magnetic field directions towards and away from the sun along the nominal Parker spiral. At the bottom the day of the year, the Carrington longitude, the radial distance to the sun in AU and the heliographic latitude are displayed.

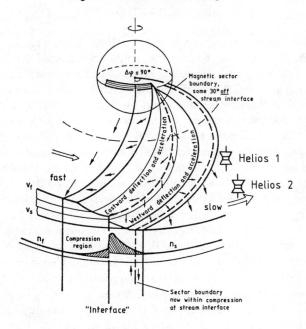

Figure 2. An idealized view of a corotating interaction region and its evolution to 1 AU. The position of Helios 1 and Helios 2 for the analyzed event is indicated.

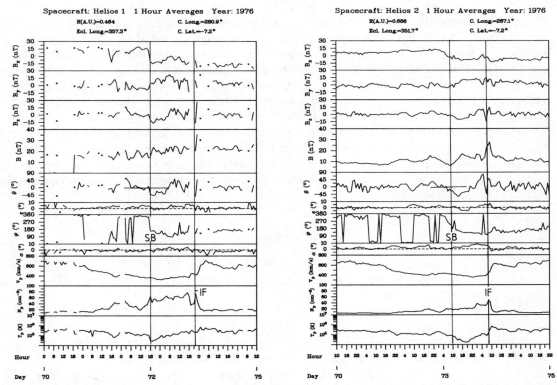

Figure 3. One hour averages of solar wind plasma and magnetic field data for the days around sector boundary passage. The IMF is given in GSE-coordinates. ϕ is measured counterclockwise in the ecliptic plane and the sunward direction corresponds to $\phi = 0°$. ϵ and α denote the plasma flow directions with respect to ecliptic latitude (positive to the north of the ecliptic plane) and longitude (measured counterclockwise in the ecliptic plane, $0°$ is the antisunward direction). SB indicates a sector boundary and IF a stream interface.

HELIOS SPACECRAFT OBSERVATIONS

Cases in which the two Helios spacecraft are nearly radially aligned are outstanding occasions for detailed event studies. In this case the two spacecraft are situated at -7.2 degrees heliographic latitude and separated by only five degrees in heligraphic longitude. The radial distance to the sun is 0.48 AU for Helios 1 and 0.68 AU for Helios 2. Figure 4 shows plots of the individual plasma and magnetic field parameters measured by Helios 1 (left) and Helios 2 (right) for the days around the sector boundary passage. In the slow wind following the first high speed stream the sector boundary (SB) is indicated by a vertical line and the deflections of the IMF out of the ecliptic plane are emphasized by horizontal lines. The sector boundary position was determined from the higher time resolution data (40.5 sec) of the IMF where the decrease of the magnetic field strength right at the sector boundary is best visible. In addition the He++ to H+ ratio showed clear depletions which is in agreement with the results of Borrini et al. [1982] and the density peak and temperature drop in the Helios 1 data right at the first sector boundary may be the signatures of a coronal streamer. The sector boundaries are well separated from the following stream interface (IF). Deflections of the IMF appear again a few hours after the sector boundary passage in relation to a skimming across the heliospheric current sheet. At Helios 2 the deflections of the IMF fall closer to the stream interface for reasons already mentioned. Both spacecraft observe essentially the same solar wind stream pattern. The corotating streams reach Helios 2 later because of a longitudinal offset of plasma flux tubes due to the stronger curvature of the plasma streamlines with increasing distance from the sun.

RESULTS AND CONCLUSIONS

The observed deflections of the IMF are related to the sector boundary, e.g. to the heliospheric current sheet. What might look like the signature of a magnetic cloud at a first glance simply results from a sector boundary crossing and a skimming across the heliospheric current sheet. A pile-up of the IMF deflections at Helios 2 due to compressional effects is likely. Magnetic field lines which tend to remain parallel to the inclined current sheet could explain the observed meridional deflections of the magnetic field at sector boundaries (Rosenberg and Coleman [1980], Schwenn [1990]). The north to south turnings of the IMF observed in this case support this explanation. The changes of the IMF orientation at sector boundaries should depend on the sun's polarity, e.g. the solar cycle and the inclination of the heliospheric current sheet, that is, the phase of the solar cycle. The plasma and magnetic field properties at sector boundaries closer to the sun might reveal interesting features of the coronal streamer belt, emphasizing the importance for further studies on this topic.

REFERENCES

Borrini, G., J. T. Gosling, S. J. Bame, W. C. Feldman, J. M. Wilcox, Solar wind helium and hydrogen structure near the heliospheric current sheet: a signal of coronal streamers at 1 AU, J. Geophys. Res., 86, 4565, 1981.

Gosling, J. T., J. R. Asbridge, S. J. Bame, W. C. Feldman, Solar wind stream interfaces, J. Geophys. Res. 83, 1401, 1978.

Rosenberg, R. L., P. J. Coleman,Jr., Solar cycle-dependent north-south field configurations observed in solar wind interaction regions, J. Geophys. Res., 85, 3021, 1980.

Schwenn, R., Solar wind and its interactions with the magnetoshere: Measured parameters, Adv. Space Res., 1, 3, 1981.

Schwenn, R., Large-Scale Structure of the Interplanetary Medium. In: Physics of the Inner Heliosphere, Vol.1, ed. by R. Schwenn and E. Marsch, Springer-Verlag , Berlin, Heidelberg, New York, 1990.

SOLAR CYCLE CHANGES IN THE TURBULENCE LEVEL OF THE POLAR STREAM NEAR THE SUN

G. Bourgois* and W. A. Coles**

* *Observatoire de Meudon, Meudon 92190, France*
** *ECE Department, University of California, San Diego, CA 92093-0407, U.S.A.*

ABSTRACT

Measurements of the coherence bandwidth of interplanetary scintillations (IPS) near the sun, made at the Nancay observatory from 1982 to 1991, show a clear solar cycle variation. These observations are used to estimate the phase structure function at 10 km, which we denote by D(10km), a measure which is proportional to δN_e^2 the variance of electron density at this scale. We find that there is no statistically significant change in the equatorial variance but there is a strong change in the polar variance. The sense is that the density variance (at scales near 10 km) is reduced by a factor of 5 in the fast polar stream which appears at solar minimum. However, as the polar density N_e is much lower at solar minimum, the level of turbulence $\delta N_e/N_e$ is actually higher in the fast polar stream.

INTRODUCTION

Electron density fluctuations phase modulate electromagnetic waves passing through the solar wind. Thus the radio signal from a distant source, such as a quasar, will fluctuate in phase as a result of passing through the corona. As it continues to propagate to the earth amplitude modulation builds up so the apparent strength of the source fluctuates when measured at the earth. However only the small scale phase fluctuations can cause amplitude fluctuations, so intensity scintillation only measures the density variance on scales smaller than about 100 km. When the scintillation is "weak", far from the sun, the intensity variance is proportional to the density variance. Unfortunately this is not true nearer the sun in "strong" scintillation /1/. However in "strong" scintillation the coherence bandwidth of the intensity fluctuations is inversely proportional to the density variance. Thus the coherence bandwidth is a useful measure of the density variance near the sun /2/.

It has been known for a long time /3,1/ that the density variance δN_e^2 varies with radial distance approximately as does N_e^2, so the ratio $\delta N_e/N_e$ is independent of distance. It is relatively easy to establish the behavior of δN_e^2 versus distance in the equatorial plane by observing a source which is close to the ecliptic plane. However it is much more difficult to establish the latitude variation as both the latitude and distance of non-ecliptic radio sources change simultaneously.

THEORY, OBSERVATIONS AND DATA ANALYSIS

The frequency correlation function $\rho(\delta f)$ of IPS can be derived from the following set of equations. Here equations 3 and 4 are approximations which are valid under the conditions in the solar wind near the sun. More exact expressions are given by Codona et al/2/.

$$D(s) = 4\pi r_e^2 \lambda^2 \int_0^\infty dz \int\int_{-\infty}^\infty d^2\vec{k}[1-cos(\kappa\cdot s)]\Phi_{Ne}(\kappa,z;\kappa_z=0) \tag{1}$$

$$B(k\theta) = F_2[e^{-D(s)/2}] \tag{2}$$

$$I(t) = B(t=L\theta^2/2c) \tag{3}$$

$$\rho(\delta f) = F_1[I(t)] \qquad\qquad (4)$$

Here the line of sight defines the z axis, r_e is the classical electron radius, λ is the electromagnetic wavelength, Φ_{Ne} is the electron density spectrum, $B(\theta)$ is the brightness distribution, I(t) is the impulse response, and $F_n[...]$ denotes an n dimensional fourier transform.

The observations were made with a multi-channel spectrometer. We sampled 16 channels set at different frequencies. This provides estimates of the frequency correlation function at 120 different frequency separations as follows.

$$\rho(f_i - f_j) = <(I(f_i) - <I>)(I(f_j) - <I>)> / <(I - <I>)^2> \qquad\qquad (5)$$

Here the <> are actually replaced by time averages using the assumption of ergodicity. We then estimate D(10km) by model fitting a structure function of the form $D(s) = (s/s_o)^{\alpha}$ to the observed $\rho(\delta f)$. This model fitting shows a systematic error of the order of 10% caused, we believe, by our use of a simple power law model for D(s) when the actual behavior is clearly more complex /6,7/. However for our purposes here this bias is not serious.

Each year we observe two sources as they pass behind the sun (as viewed from the earth). One of these, 3C279, is in the ecliptic plane; and the other, CTA21, is about 6 R_s south of the sun at closest approach. The observations for the two sources are shown in figures 1 and 2. Here the points plotted are daily means and the vertical bars show the daily variation. The estimation error due to receiver noise and the finite observing time is considerably smaller than the scatter.

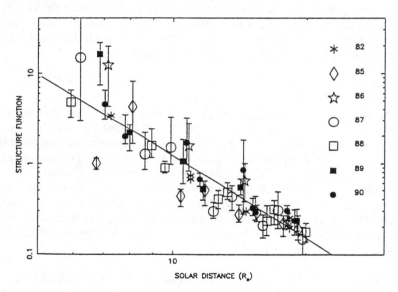

Figure 1. Estimated structure function D(10km) scaled to a wavelength of 20 cm (according to λ^2) from observations of the ecliptic radio source 3C279. The vertical bars show the rms variation of the observations taken on a single day. The abscissa is the minimum distance from the sun to the scattering volume. The solid line drawn through the data points is a best fit power law, $D(10km) = 1.08 (R_s/10)^{-3}$.

DISCUSSION AND INTERPRETATION

One is struck by several obvious features of figures 1 and 2. Although there is considerable daily variation in both figures, the CTA21 data clearly show a bimodal form which apparently depends on the solar activity cycle. At solar maximum the two sources are indistinguishable. However at solar minimum the CTA21 estimates of D(10km) near the sun are considerably lower than those of 3C279 and they show less variability. We interpret this behavior as a latitude variation, in fact as a signal that the density variance in the fast polar stream is low.

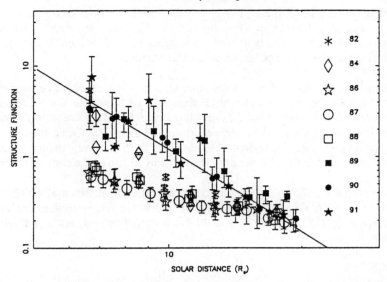

Figure 2. Estimated structure function D(10km) plotted as in figure 1 from observations of the radio source CTA21, which is 6 R_s south of the sun at its closest point of approach.

To establish this we fit a simple power law model to the 3C279 data getting $D(10km)=1.08\,(R_s/10)^{-3.0}$. This behavior is consistent with earlier work; it implies that $\delta N_e \propto R^{-2}$. We then normalize the CTA21 data by this model to estimate the latitude effect. This is plotted in figure 3. Here we see that the density variance is latitude independent at solar maximum (1989, 90 and 91), whereas the polar density variance is reduced by a factor of 5 at solar minimum (1986, 87 and 88). We also see that the density variance was lower in the midlatitudes during the declining phase (1982 and 84). We note that during 1982 and 84 the magnetic field neutral line was not confined near the equator and the center of the fast "polar" stream was more than 30° from the rotational pole /8/. Thus the observed minimum at midlatitudes indicates that the density variance in the fast "polar" stream was already depressed during the declining phase of solar activity.

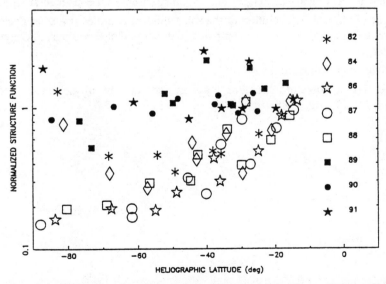

Figure 3. Estimated structure function D(10km) from observations of the radio source CTA21 normalized by the power law model fit to the observations of 3C279. The abscissa is the heliographic latitude of the CTA21 observation.

Another feature that is obvious in all these plots is that the variability from day to day, and the scatter within a day, are considerably smaller in the polar stream. Evidently the density variance does not change much within the stream. Conversely the density variance is quite variable in the equatorial region (even at solar minimum) and also in the polar region at solar maximum.

Although it is clear that the density variance in the polar stream is low, it would be interesting to know the "turbulence level" $\delta N_e/N_e$ in the polar stream. Estimates from Sime /10/ and Withbroe /9/ indicate that the polar coronal hole is between 10 and 50 times less dense than the equatorial plasma at the same solar distance. Even if we take a very conservative estimate of 7 then $\delta N_e/N_e$ is actually 10 times higher in the polar stream than in the equatorial regions. We conclude that the development of compressive turbulence in the polar stream is actually an order of magnitude larger than in the equatorial regions.

It has recently been reported that $\delta N_e^2(R)$ shows an enhancement at the distance at which the solar wind should become supersonic /4,5/. Our study covers the same distance range but does not show a statistically significant change in $\delta N_e^2(R)$ between 15 and 20 R_s. It does, however, reveal a significant change in δN_e^2 at high latitudes.

REFERENCES

1. J. W. Armstrong and W. A. Coles, Interplanetary Scintillations of PSR 0531+21 at 74 MHz, *J. Geophys. Res.* **220**, 346, 1978.

2. J. L. Codona, D. B. Creamer, S. M. Flatte, R. G. Frehlich and F. Henyey, Two-frequency intensity cross-spectrum, *Radio Science* **21**, 805-814, 1986.

3. G. Bourgois, Scintillations interplanetaires de radiosources a 2695 MHz, *Astron. Astrophys.* **2**, 209-217, 1969.

4. N. A. Lotova, D. F. Blums and K. V. Vladimirskii, Interplanetary scintillation and the structure of the solar wind transonic region, *Astron. Astrophys.* **150**, 266-272, 1985.

5. N. A. Lotova, Investigations of the solar wind transonic region, *this issue*.

6. W. A. Coles, W. Liu, J. K. Harmon and C. L. Martin, The solar wind density spectrum near the sun: results from voyager radio measurements, *J. Geophys. Res.* **96**, 1745-1755, 1991.

7. J. K. Harmon and W. A. Coles, The density fluctuation spectrum of the inner solar wind, *this issue*.

8. B. J. Rickett and W. A. Coles, Evolution of the solar wind structure over a solar cycle: interplanetary scintillation velocity measurements compared with coronal observations, *J. Geophys. Res.* **96**, 1717-1736, 1991.

9. D. G. Sime, *private communication*, 1991.

10. G. L. Withbroe, The temperature structure, mass and energy flow in the corona and inner solar wind, *Astrophys. J.* **327**, 442, 1988.

MODELS OF THE SOLAR WIND DRIVEN BY MHD WAVES

I. V. Chashei and V. I. Shishov

P. N. Lebedev Physical Institute, Leninski Prospect 53, Moscow 117924, Russia

When solving the problem of the formation of a hot corona and solar wind, the sources of energy and momentum are of crucial importance. The physical nature of the sources has not yet been ascertained definitively. This paper considers the formation of the corona and solar wind in a model with a wave energy source /1-3/. Attention is focused mainly on open configurations of the magnetic field. We consider the problem for the simple cases of spherically symmetric geometry (SS), isolated coronal hole (CH) and quasidipole structure (QD).

As initial equations, we adopt a stationary MHD system with wave sources of energy and momentum, which for the SS model has the following form:

$$\rho v r^2 = \text{const};\tag{1}$$

$$\rho v \frac{dv}{dr} + \rho g + \frac{d}{dr}\left(p + \frac{W^a}{2}\right) = 0;\tag{2}$$

$$p = \rho v_T^2,\tag{3}$$

$$\frac{1}{r^2}\frac{d}{dr}\, r^2\left\{\rho v\left[\frac{v^2}{2} + \frac{5v_T^2}{2} - \frac{v_g^2}{2}\right] - \kappa\frac{dT}{dr} + H^a + H^s\right\} = -q_{rad};\tag{4}$$

$$H^a\left(\frac{3}{2}v + v_a\right)W^a, \quad H^s = \left(\frac{\gamma+3}{2}v + v_s\right)W^s;\tag{5}$$

$$Q_{rad} = \frac{\rho^2}{M^2}L(T);\tag{6}$$

$$\frac{1}{r^2}\frac{d}{dr}\, r^2\left[\frac{(v+v_{a,s})^2}{v_{a,s}}W_\omega^{a,s}\right] = -\gamma_\omega^{a,s}W_\omega^{a,s},$$

$$W^{a,s} = \int_0^\infty W_\omega^{a,s}\, d\omega;\tag{7}$$

$$Br^2 = \text{const},\tag{8}$$

159

where B is the magnetic field induction, $W^{a,s}$ and $H^{a,s}$ are energy densities and energy fluxes of Alfvén and magnetosonic waves, respectively /4/, $\gamma = 5/3$, $v_{a,s}$ are local Alfvén and sound velocities, v_g is the escape velocity, and Q_{rad} is the radiative loss /5/. We neglect fast magnetosonic waves because of their strong reflection in the transition region; we also neglect the pressure of slow magnetosonic waves which are totally absorbed in the corona where turbulence is weak. Wave attenuation is described by decrements $\gamma^{a,s}$ which include linear absorption by coronal plasma as well as nonlinear, four-wave, absorption of Alfvén waves beyond the confines of the corona.

The boundary conditions are specified in the form:

$$T(r_*) \simeq 10\ K, \quad T(r \to \infty) \to 0, \quad \frac{dT}{dr}(r_*) = 0, \quad V_{a,s}(r_*)\, W^{a,s}_\omega(r_*) = H^{a,s}_* f. \qquad (9)$$

Here $f = f(\omega/\omega_o)$ is the automodel spectrum with a fundamental energy-containing frequency $\omega_o \simeq 2\cdot 10^{-2}\ s^{-1}$ that coincides with the cutoff frequency of acoustic waves; and r_* is the initial level taken as the base of the transition region. Input energy fluxes $H^{a,s}_*$ are formed below the transition region in the regime of strong turbulence, which makes it possible to express $H^{a,s}_*$ as a function of pressure and magnetic field induction B_* in the transition region. Apart from these conditions, one should also use the condition for continuous transition of the flow through the sound velocity at the critical point of the equation of motion (2):

$$v^2 = v_s^2 + v_{s2}^2, \quad \frac{1}{4}v_g^2 = v_s^2 + v_{s2}^2 + \frac{r}{2\rho}\, F_{dis}, \quad [\text{at } r = r_c], \qquad (10)$$

where v_s, v_{s2}, F_{dis} are local sound velocity, second sound velocity, and the Alfvén dissipative force density.

The system (1)–(10) is self-consistent: it includes the mutual influence of waves and plasma, the flow and the corona. The base of the transition region r_* is a sliding boundary; the mass flux is not specified but is determined from the solution. Considering (1)–(10), one can show that the only free parameter which determines the plasma characteristics is the magnetic field induction B_* at the coronal base.

Radiative losses occur mainly in the transition region. The linear damping rates of waves are such that heat sources produced in the corona have a well-defined maximum in the region where the energy-carrying frequency ω_o coincides with the frequency of ion-ion collisions. Slow magnetosonic waves are totally absorbed in the corona. Alfvén waves are only partly absorbed by the corona, and these are largely high-frequency waves, $\omega > \omega_o$; low-frequency waves, $\omega < \omega_o$, propagate through the corona and dissipate in the solar wind via nonlinear interactions, thereby transferring their energy to the plasma near the critical point (located near $r \simeq 10r_*$ according to our estimates).

We specify the corona and the solar wind by the following parameters: the maximal coronal temperature T_m, the fraction of Alfvén wave energy flux ξ^a absorbed in the corona, the pressure in the transition region p_*, and asymptotic outflow velocity v_∞, all of which are dependent on the coronal magnetic field B_*. The variations of these parameters are presented schematically in Figure 1 by the heavy lines.

Two typical regimes are identifiable, according to the magnitude of B_*. One is the weak field regime $B_* < B_{sat}$, in which the main losses are attributable to emission, while the losses for mass lifting are small. In this regime, Alfvén and slow magnetosonic waves make a comparable contribution to the energy balance of the corona. With increasing B_*, the pressure in the transition region varies in such a way that $\beta_* = 4\pi p_*/B_*^2 \simeq 1$ (the transition region descends deep into the atmosphere), and the parameters T_m, and ξ^a increase, reaching values $T_m \simeq 2 \cdot 10^\sigma$ K, and $\xi^a \simeq 0.8$ when $B_* \simeq B_{sat}$. The mass flux starts small and grows exponentially with increasing B_*; the velocity v_∞ decreases, reaching a minimum $v_\infty(B_{sat}) = v_{min} \simeq 300$ km/s when $B_* \simeq B_{sat}$. The saturation value $B_{sat} \simeq 2G$ is determined from the condition that radiative losses are of order of mass lifting losses. In the strong field regime $B_* > B_{sat}$, the energy flux entering into the corona goes mainly into mass lifting and the radiative losses are comparatively small. The pressure in the transition region varies weakly with B_* so that $4\pi p_* < B_*^2$, i.e. the transition region is located above the level where $\beta = 1$. The energy balance of the corona and solar wind is almost totally determined by Alfvén waves. The parameter T_m only weakly depends on B_*, and the same pertains to the mass flux. The asymptotic velocity v_∞ increases at large B_*, reaching an asymptotic value $v_\infty = v_{max} \simeq 1000$ km/s.

The SS model can be valid for times near solar activity maximum, when the global structure of the corona is spherically symmetric. One could thus assume that the background magnetic field in the corona is approximately constant. The solar wind velocity will then be independent of heliolatitude, the situation observed in the IPS measurements at solar maximum /6/. The measured value $V_\infty \simeq 450$ km/s corresponds to an average coronal magnetic field of about 3-4 G. The nondependence on heliolatitude of the solar wind mass flux supports the SS model.

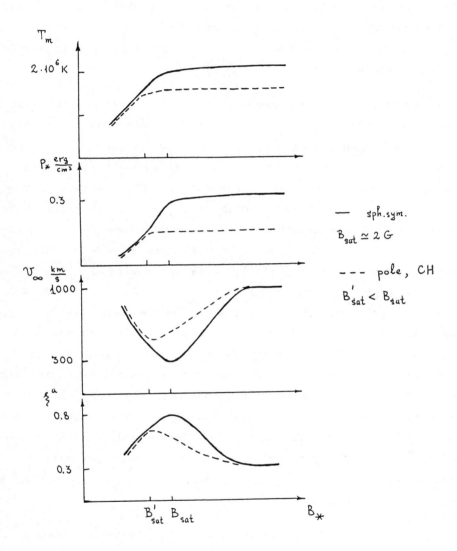

Fig. 1. Dependence of coronal temperature T_m, gas pressure in the transition region p_*, solar wind speed v_∞ and absorbed fraction of Alfvén flux ξ^a on the magnetic field at the coronal base B_* for the SS model (heavy lines), CH model and polar region of QD model (dashed lines).

Considering the corona and solar wind formation in an isolated CH, we assume that the magnetic field at the base of the CH is strong compared with the background magnetic field, and that the spatial magnetic structure of the CH can be represented as a conic tube with the top located at $0 < r < r_*$. Super-radial divergence of the magnetic lines must be taken into account, but the main physical processes of the corona and solar wind formation are the same as for the SS model. The dependence of T_m, p_*, v_∞ and ξ^a for the CH model are presented in Figure 1 by dashed lines. The value of B_{sat} for the CH is smaller than for the SS model. This is the main difference between these models. As a consequence, the coronal energetic source for the CH is mainly Alfvénic and the radiative losses are smaller than mass lifting losses. The value of p_* and T_m in the saturated regime for CH are smaller than the corresponding values for SS. Estimating v_∞, one can show that a velocity of about 700 km/s (typical for fast solar wind streams) corresponds to $B_* \simeq 6-7$ G and to a superradial expansion factor of about 3-4, i.e. the divergence is not very strong.

When considering the configuration corresponding to the QD model, the whole system of MHD equations with wave energy sources was used. We assume that the field B_* at the initial surface $r = r_*$ coincides with the field of the point dipole, $B_* = \frac{1}{2} B_o (1+3\cos^2\theta)^{1/2}$, where θ is the polar angle, and B_o is the polar field. The whole system of equations with the corresponding boundary conditions contains only one free parameter B_o, thereby allowing one to obtain (in principle) the exact solution of the two-dimensional problem. Such a problem, however, is very complicated. In order to obtain approximate solutions, we thus consider closed ($v = 0$) and open ($v \neq 0$) regions separately. The pressure balance condition must be satisfied at the separation boundaries (current surfaces). Unlike similar models considered earlier, the problem is no longer isothermal. Approximately, one may adopt near the equatorial neutral point (cusp, $r = r_o$) $\beta(r_o) = 8\pi P_{intern} / B^2_{extern} \simeq 1$, where the magnetic field B_{extern} is closed to a dipole field. Another possibility for defining the condition near the neutral point is to consider the convective instability of the plasma in the dipole magnetic trap. It can be shown that the condition of marginal stability has the form $\beta(r_o) \simeq 1$, i.e. formally coincides with the condition of pressure balance.

Solving the energy equation for a dipole magnetic arch and taking into account the condition of marginal stability (or pressure balance), we can obtain the dependence of the cusp parameters on the polar magnetic field. If the magnetic field is strong enough, $B_o > 7G$, then the cusp is located near the transition region. The closed near-equatorial coronal belt exists only for a sufficiently weak magnetic field, $B_o < 7G$. The corresponding temperature in the upper parts of the arches will be $T_o < 3 \cdot 10^6$ K. When $B_o < 7G$, a decrease of dipole moment leads to a rapid increase of the cusp height. When $B_o \simeq 2.5G$, the cusp rises as high as $r_o \simeq 2r_*$, and its temperature is $T_o \simeq 2 \cdot 10^6 K$. Decreasing B_o further does not lead to an increase of r_o, but the temperature T_o does decrease.

For the polar region, where the magnetic lines are open and $B \simeq r^{-n}$ with $n > 3$, one can use the general scheme developed for the cases of SS and CH. Strong divergence of the magnetic field leads to an enhanced cooling effect of the corona. The transition from the weak field to the strong field regime occurs at smaller values of B_o than in the SS case: $B_{sat} \simeq 1.5G$, $T_m(B_{sat}) \simeq 1.3 \cdot 10^6 K$, $V_\infty(B_{sat}) = V_{min}^{polar} \simeq 450$ km/s. When $B_o \simeq 6G \simeq 4B_{sat}$ we have: $T_m \simeq 1.4 \cdot 10^6 K$, $V_\infty \simeq 700$ km/s. The latter values of T_m and V_∞ agree with the experimental data on the polar regions of the corona and solar wind /6/ at the minimum of solar activity.

We cannot explain, within the framework of the QD model, the dynamics of the corona and solar wind over the solar cycle by a simple modulation of the value of the dipole moment or orientation. The transition to the global structure typical for solar activity maximum is apparently associated with an increase in the role of small-scale irregular magnetic structures in the solar corona.

REFERENCES

1. I.V. Chashei and V.I. Shishov, Formation of solar wind mass and energy fluxes in the model with the wave source, Geomagnetizm i aeron. 27, 705-711 (1987)

2. I.V. Chashei and V.I. Shishov, Self consistent model of the quiet solar corona with the wave energy source, Sov. Astron. J. 65, 157-166 (1988)

3. I.V. Chashei, On the formation of the corona and fast solar wind streams above the coronal holes, Geomagnetizm i aeron. 28, 190-196 (1988)

4. S.A. Jacques, Momentum and energy transport by waves in the solar atmosphere and solar wind, Astrophys. J. 215, 942-951 (1977)

5. D.P. Cox and W.H. Tucker, Ionization equilibrium and radiative cooling of a low-density plasma, Astrophys. J. 157, 1157-1167 (1987)

6. W.A. Coles, B.J. Rickett, V.H. Rumsey, J.J. Kaufman, D.G. Turley, S. Ananthakrishnan, J.W. Armstrong, J.K. Harmon, S.L. Scott, and D.G. Sime, Solar cycle changes in the polar solar wind, Nature 286, 239-241 (1980)

MAGNETIC FIELD OF THE SOLAR WIND: A STATIONARY KINEMATIC SOLUTION WITH A FINITE ELECTRIC CONDUCTIVITY

A. D. Chertkov

Department of Terrestrial Physics, Institute of Physics, St Petersburg State University, 198904 St Petersburg, Petrodvorets, Russia

ABSTRACT

The stationary models by E.N.Parker for the magnetic field in the solar wind and by K.H.Schatten for the "source surface" in the solar corona are generalized. The configuration of the interplanetary magnetic field, stretched by the expanding corona, depends on the magnitude of the electrical conductivity σ of the solar wind plasma. Knowing the main empirical features of the field configuration, one may estimate the phenomenological value of σ. The estimates show that the electrical conductivity should be approximately 10^{13} times smaller than that calculated by Spitzer.

INTRODUCTION

A theoretical configuration of the magnetic field in the solar wind and in the vicinity of the Sun, stretched by expanding corona, was discussed in /1,2/ and many other papers, see /3/. In all these papers the assumption was made that the electrical resistance of cosmic plasma is negligibly small /4/. However, measurements of the radial dependence of the magnetic field in the solar wind /5/ and studies of the geoefficiency of solar magnetic fields /6/, as well as the proof of a direct link of the solar meridional magnetic field with the solar wind vertical magnetic field near Earth /7/, bring us to another model of the interplanetary magnetic field.

FORMULATION OF THE PROBLEM AND SOLUTIONS OF EQUATIONS

The field equations in a stationary case are:

$$\text{rot } \vec{B} = 4\pi\vec{j}/c; \quad \text{div } \vec{B} = 0; \quad \vec{j} = \sigma(-\text{grad}V + (\vec{v}/c) \times \vec{B}) \qquad (1)$$

where: $\vec{B}$ is magnetic induction; $\vec{j}$ is current density; $\vec{v}$ is plasma velocity relative to field sources: $\vec{v} = v\vec{e}_r - \Omega r\sin\theta\vec{e}_\phi$, $\vec{e}_r$ and $\vec{e}_\phi$ are unit vectors along the radial and azimuthal directions, v is the radial solar wind speed, Ω is angular velocity of the Sun relative to the plasma, V is the electrostatic potential in the rest frame of reference with respect to the field sources. The boundary conditions at $r = R_0$ ($R_0 \geq R_\odot$, where $R_\odot$ is the solar radius) will be discussed below. At $r \to \infty$ all the fields and potentials should tend to zero. We use the right spherical rotating coordinate system (r, θ, ϕ) with the Sun as its center. When $\text{grad}\sigma$, $\text{grad}v$, $\text{grad}\Omega \parallel \vec{e}_r$ (see /7/), one obtains:

$$B_r = \text{Re} \sum_{n=1}^{\infty} \sum_{m=0}^{n} G_{nm} P_{nm} R_{nm}^1 e^{im\phi}$$

$$B_\theta = \text{Re} \sum_{n=1}^{\infty} \sum_{m=0}^{n} \left[G_{nm} \frac{d}{d\theta} P_{nm} \frac{1}{n(n+1)r} \frac{d}{dr}(r^2 R_{nm}^1) + H_{nm} P_{nm} R_{nm}^2 \frac{im}{\sin\theta} \right] e^{im\phi} \qquad (2)$$

$$B_\phi = \mathrm{Re}\sum_{n=1}^{\infty}\sum_{m=0}^{n}\left[G_{nm}P_{nm}\frac{im}{(n(n+1)r\sin\theta)}\frac{d}{dr}(r^2R_{nm}^1) - H_{nm}\frac{d}{d\theta}P_{nm}R_{nm}^2\right]e^{im\phi}$$

$$V = \mathrm{Re}\sum_{n=1}^{\infty}\sum_{m=0}^{n}\left[\frac{\Omega}{c}G_{nm}\frac{\sin\theta}{n(n+1)}\frac{d}{d\theta}P_{nm}r^2R_{nm}^1 + \right.$$

$$\left. + \frac{v}{c}H_{nm}P_{nm}(rR_{nm}^2 - \frac{1}{4\pi\sigma v}c^2\frac{d}{dr}(rR_{nm}^2))\right]e^{im\phi}$$

where n,m are integers; G_{nm} and H_{nm} are complex constants to be calculated from the boundary conditions; $P_{nm}(\cos\theta)=P_{nm}$ are Legendre polynomials; $R_{nm}^1(r)= R_{nm}^1$ and $R_{nm}^2(r)=R_{nm}^2$ are the solutions of the following differential equations:

$$\left[d^2/dr^2 - (4\pi\sigma v/c^2)d/dr - n(n+1)/r^2 + im4\pi\Omega\sigma/c^2\right]r^2R_{nm}^1 = 0 \qquad (3)$$

$$\left[d^2/dr^2 - (4\pi\sigma v/c^2+\frac{\sigma'}{\sigma})d/dr - n(n+1)/r^2 + im4\pi\Omega\sigma/c^2 - v'4\pi\sigma/c^2\right]rR_{nm}^2 = 0 \quad (4)$$

To solve (3,4), we must know the explicit form of $\sigma(r)$, $\Omega(r)$, and $v(r)$. Let us introduce the magnetic Reynolds number: $Re_m=4\pi\sigma vr/c^2=\lambda rf(r)$, $\lambda=const$, $\lambda\equiv Re_m(1A.U.)$. The WKB-method provides asymptotical solutions of (3,4) for any f(r) when f(r) is an arbitrary positive continuous function sufficiently smooth with its derivatives. Let us assume $v=v_1\{1/2+(1/\pi)arctg[(r-r_v)/a_v]\}$, v_1=400km/s, r_v=3-15$R_\odot$ is the position of the solar wind acceleration region, a_v=0.5-2$R_\odot$ is the size of this region; $\Omega=\Omega_s\{1/2+(1/\pi)arctg[(r-r_c)/a_c]\}$, Ω_s=2.7·10^{-6}s^{-1} is the angular velocity of the solar rotation; r_c=5-15$R_\odot$ is the position of the middle of transition region for corotation of the solar wind plasma, a_c is the size of this transition region; $f(r)=r_1^{-1}(r/r_1)^\alpha$, r_1=1A.U., α=const. When $Re_m\ll 1$ (this may occur when $R_\odot\leq r<R_s$, where R_s is the distance where Re_m=1), then $R_{nm}^1=r^{k(n)}$, where $k(n)\approx-n-2+Re_m$. When $Re_m\gg 1$ $(r>R_s)$, then

$$R_{nm}^1 \approx r^{-2}\Xi_{nm}\exp(ims) \qquad (5)$$

where $\Xi_{nm}=\exp\left[-\frac{n(n+1)}{\lambda}h - m^2p/\lambda + O(\lambda^{-2})\right]$, $h\equiv\int r^{-2}f^{-1}dr$, $p\equiv\int\Omega^2/(v^2f)dr$, $s\equiv\int\frac{\Omega}{v}dr$. When $Re_m\ll 1$, then $R_{nm}^2=r^{l(n)}$, where $l(n)\approx-n-1+Re_m$, assuming $\sigma'/\sigma\rightarrow 0$. For $Re_m\gg 1$ one obtains:

$$R_{nm}^2 \approx r^{-1}\Xi_{nm}\exp[im(s-q/\lambda)+g/\lambda] \qquad (6)$$

where $g \equiv \frac{3}{4}\int(f')^2f^{-3}dr - \frac{1}{2}\int f''f^{-2}dr - \lambda\int\frac{v'}{v}dr$, $q \equiv \int f'\Omega/(f^2v)dr$.

It can be shown that (2-6) are the unique solutions of the boundary problem, yielding the magnetic and electric fields created by a rotating magnetized and electrified sphere in the given radial stream of plasma. B_r and V (or, instead of V, $\partial V/\partial r$, or j_r) must be set up on the initial surface r=R_0. The correct solution to the boundary problem should be found only assuming that the electrical conductivity σ is finite. Attempts to solve the problem within the "frozen-in" concept in the nondegenerate cases lead to unresolvable paradoxes stemming from lowering the order of the MHD-system of equations, changing its type to hyperbolic, and violating of the causal chain /7/.

TOPOLOGY OF THE MAGNETIC FIELD

Near the Sun, when $Re_m \rightarrow 0$, the magnetic field configuration is similar to Gaussian. Outside the surface $r > R_s$, where $Re_m(R_s) = 1$, the field is stretched by the solar wind flow. Under the condition $V(R_0, \theta, \phi) = 0$ one obtains:

$$B_r = Re \sum_{n=1}^{\infty} \sum_{m=0}^{n} r^{-2} G_{nm} P_{nm} \Xi_{nm} \Psi_{nm}, \qquad B_\theta \approx -Re_m^{-1} \frac{\partial}{\partial \theta} B_r + O(\lambda^{-2}),$$

$$B_\phi \approx -B_r \, r \sin\theta \frac{\Omega}{V} + O(\lambda^{-1}), \qquad V \approx 0. \tag{7}$$

Here $\Psi_{nm} = \exp[im(s+\phi)]$ describes Parker's spirals, and Ξ_{nm} describes the deviation from them. When $\sigma \rightarrow \infty$, $\sigma'/\sigma < const$, $V = 0$, one obtains Parker's formulae from (7). When $V = 0$, $Re_m(r < R_s) = 0$, $Re_m(r > R_s) = \infty$, one obtains Schatten's model. It should be noted that the condition $V(R_0, \theta, \phi) = 0$, even for the case of finite conductivity, leads to the field structure similar to that of Parker's spiral, i.e. the radial field B_r determines the slowly diminishing azimuthal field B_ϕ and weak meridional field B_θ. In the case when $V \neq 0$, the configuration of magnetic field in space will be much more complicated, since a purely toroidal field slowly diminishing in space will be added to the field (7). The fields B_θ and B_ϕ will then be determined not only by B_r, but mainly by $V \neq 0$. The field configuration would be "convoluted". Hence, the presence of the spiral substructure in the magnetic field of the solar wind /7,8/ directly indicates that the electrostatic field at the initial surface is sufficiently small.

ESTIMATE OF CONDUCTIVITY

Formulae (7) allow us to get a phenomenological estimate of the conductivity when the basic elements of the magnetic field configuration in space are known. Deviations of $B_r(r)$ from r^{-2} and $B_\phi(r)$ from r^{-1} as well as the mean ratio of regular components B_θ/B_r near the Earth then turn out to be important. It follows from (7) that these factors are interrelated. According to /5/, which used data of Mariner 4,5,10 and Pioneer 6,10 in the range 0.4-5A.U., $B_r(r)$ and $B_\phi(r)$ decrease with the distance as $B_r(r) = B_r^1(r/r_1)^{-a}$, $B_\phi(r) = B_\phi^1(r/r_1)^{-b}$, where $B_r^1 = 3.0nT$, $B_\phi^1 = 3.0nT$; $a = 2.13 \pm 0.11$; $b = 1.12 \pm 0.14$. The temporal evolution of B_r^1, B_ϕ^1 was negligible during solar cycle #20 /9/. Using measurements of B_r and B_ϕ on board Pioneers 10 and 11, IMP-8 and ISEE-3 from 1972-1982, the coefficient b was estimated as $b = 1.12 \pm 0.04$. Using data of Helios 1 and 2, b was found to be 1.27 ± 0.06 /10/. The authors found the linear regression coefficients from a best fit to the double logarithmic scale. Using Voyager 1 and 2 data in the range 1 AU$<$r$<$20 AU and applying a linear scale treatment, it was concluded /11/ that Parker's law is valid. Due to the fast decrease of B_r with distance, however, the measurements of $B_r(r)$ at r$>$5AU do not seem to be reliable. The measurements of Pioneer 10, 11 and Voyager 1,2 may be considered as independent and should be treated in the same way (in double logarithmic scale). One then finds a relation $b = 1.10 \pm 0.10$, $a = 2.10 \pm 0.10$. The mean ratio of the regular components near the Earth at r=1A.U. was determined as $\langle B_\theta \rangle / \langle B_r \rangle \simeq 0.1-0.03$ /6,7/. It should be noted that this ratio is observed in the middle and distant solar corona. From (7) it follows that $\langle B_\theta \rangle / \langle B_r \rangle \approx n/\lambda$ when $\alpha = -1$. In this case the magnetic

Reynolds number Re_m is constant in space. Assuming that $\Omega/v \approx 1/r_1$, $\langle P'_{nm}\rangle \approx$ $n\langle P_{nm}\rangle$, n=2-4 (the numbers of dominating components), we estimate: $\lambda\approx100$ /7/. Hence, the conductivity near the Earth's orbit is independent of f(r): $\sigma_\phi=\lambda fc^2/(4\pi v)\approx20s^{-1}$. These estimates give us a perfect description of a and b as well: when $\alpha=-1$, m=0, then a= $2 + n(n+1)/\lambda$, b= $1 + n(n+1)/\lambda$. The harmonic components with m>0 in the range $r>r_1$ diminish faster for larger m because, when $\alpha=-1$, Ξ_{nm} contains the term $\exp(-\frac{1}{\lambda}m^2(r/r_1)^2)$. The sector structure should become simpler beyond 1A.U. Hence, the empirical evaluation $Re_m\approx100$ (instead of the theoretical value $Re_m=10^{15}$ /1/) can be inferred by several independent methods.

CONCLUSION

The stationary models by E.N.Parker for the magnetic field in the solar wind and by K.H.Schatten for the "source surface" in the solar corona are generalized. The magnetic and electric fields are described by continuous functions in the whole space from the lower corona to the very distant solar wind. The plasma flow is taken to be radial. The rotation of the Sun and corotation of plasma is taken into consideration. The magnetic Reynolds number Re_m is variable in space. The electrical conductivity σ and velocity v of the solar wind depend only on solar distance. In the extreme cases ($Re_m=0$, $r<R_s$; $Re_m=\infty$, $r\geq R_s$; $j_r=0$; where R_s is radius of the "source surface"; j_r is radial current density), the model reduces to the Parker-Schatten model. A comparison of the observed and the modelled magnetic field configurations enables one to evaluate the mean Re_m and phenomenological σ values at the distances $\sim$ 1A.U.: $Re_m\approx100$, $\sigma\approx20s^{-1}$. When $\sigma\neq\infty$, the type of the differential equation set is either elliptical or parabolic (in the nonstationary case). Electric fields (static or inductive), which heat the corona and accelerate the solar wind, are possible in the model with $\sigma\neq\infty$.

REFERENCES

1. E.N.Parker, Interplanetary Dynamical Processes, Interscience (1963).

2. K.H.Schatten, J.M.Wilcox, N.F.Ness, Sol.Phys. 6, 442 (1969).

3. M.Dobrovolny and G.Moreno, Space Sci.Rev. 18, 685 (1976).

4. L.Spitzer,Jr., The Physics of Fully Ionized Gases, Interscience (1956).

5. K.W.Behannon, Rev.Geophys.Space Phys. 16, 125 (1978).

6. M.I.Pudovkin and A.D.Chertkov, Solar Phys. 50, 213 (1976).

7. A.D.Chertkov, Solnechnyi Veter i Vnutrennee Stroenie Solnca.(Solar Wind and Internal Structure of the Sun), Moscow, Nauka Publ.(in Russian) (1985).

8. P.H.Scherrer, J.M.Wilcox, R.Howard, Sol.Phys. 22, 418 (1972).

9. J.A.Slavin, E.J.Smith, B.T.Thomas, Geophys.Res.Lett. 11, 279 (1984).

10. J.A.Slavin, G.Jungman, E.J.Smith, Geophys.Res.Lett. 13, 513 (1986).

11. L.W.Klein, L.F.Burlaga, N.F.Ness, J.Geophys.Res. 92, 9885 (1987).

AMBIGUITIES IN IPS G-MAPS: 3D NUMERICAL SIMULATION RESULTS

T. R. Detman, M. Dryer and H. Leinbach

NOAA, Space Environment Laboratory, Boulder, CO 80303, U.S.A.

ABSTRACT

The potential of interplanetary radio scintillation (IPS) for remote sensing of the inner heliosphere is stimulating global research efforts. Like most remote-sensing techniques, IPS provides incomplete information. For example, all-sky "g-maps" of scintillation enhancement factor are two-dimensional (2D) projections onto the plane of the sky. To make use of IPS results it is necessary to make some assumption or inference regarding the three-dimensional (3D) structures and dynamics that produce the observations.

By doing fully 3D, time-dependent MHD solar wind simulations, we are approaching the problem from the opposite direction: beginning with the 3D global dynamics. Then simulated g-maps are produced using "standard" relationships between electron density micro-dynamics and bulk parameters.

We are compiling a catalog, a "Rogues' Gallery," of simulated interplanetary disturbances and synthetic g-maps to address basic issues in the interpretation of g-maps. This modeling will also be discussed in relation to some current controversies and research areas.

INTRODUCTION

All-sky IPS g-maps reveal interplanetary disturbances propagating from the Sun, and thus could, in principle, provide valuable advance warning of sudden commencements and possible geomagnetic storms. The Space Environment Laboratory (SEL) of NOAA is investigating the use of these IPS g-maps to improve its forecasting services. However, many questions remain to be answered.

AMBIGUITIES

G-maps contain inherent ambiguities that must be resolved in order to make optimum use of this valuable information in a forecast. Some of these ambiguities are inherent to an imaging system that scans the sky once each 24 hours with the rotation of Earth: 1) g-maps are plane–of–the–sky projections (i.e., they contain no distance or velocity information about the disturbances they show); and 2) they produce distorted images because the disturbances move and change during the 24 hours required to make the map. G–maps also contain additional ambiguities or uncertainties that arise from the physics of the radio scattering: Amplitude scintillation, under discussion here, is subject to saturation. For the Cambridge array, operating at 81.5 MHz, strong scattering (producing saturation) typically occurs for sources observed at solar elongation angles of 35° or less. Currently, the central area of g–maps (elongation less than 35°) is ignored. The resulting lack of observations within 1/2 AU of the Sun has left room for debate (which is lively) regarding the solar sources of events observed in g-maps. The scattering is caused by micro–scale gradients in electron density (radio refractive index turbulence). Whether this is more a result of waves or turbulence is not fully known, and certainly the relationship to global bulk dynamics is poorly understood.

Leinbach and Grubb /1/ used a geometrical model (an expanding sphere tangent to the solar surface at a fixed point) to gain insight and assess the importance of ambiguities 1 and 2 above. Such a geometrical model gives the outline of a disturbance on a g–map.

The purpose of this work is to further explore the operational impacts of these ambiguities and their resolution using a 3D MHD numerical solar wind model, the 3D IGM (Interplanetary Global Model). Simulated g-maps are constructed by calculating line–of–sight integrals through the model grid using "standard" theoretical expressions for the relation of scattering power to bulk MHD parameters. Tappin et al. /2/ have already indicated the potential of this approach to reproduce the large–scale features of some IPS "events."

THE 3D–IGM

Our 3D model /3,4/ has a computational domain extending from 18 R_s to 330 R_s (1.5 AU), and side boundaries covering ±45° of latitude, centered on the solar equator, and 120° of longitude. Grid resolution is 3° in each angle, and 2 R_s in the helioradial direction. Thus the numerical mesh contains 157 × 31 × 41 grid points. Equations for the conservation of mass, momentum, and energy (without thermal conduction), and the induction equation (ideal MHD) are solved in pseudoconservation form by the two-step Lax-Wendroff method with the addition of artificial viscosity. The model thus satisfies the Rankine-Hugoniot conditions at any shocks that develop.

THE ROGUES' GALLERY

Our basic strategy is to compile a catalog of synthetic g-maps made from simulated disturbances. When a disturbance is observed in a real g-map, the book can be consulted for a matching synthetic g-map. Possible outcomes that may be associated with the observed g-map may then be inferred from the 3D simulations.

Our catalog of simulated disturbances and g-maps presently contains a weak shock (1000 km/s), a strong shock (2000 km/s), and two erupting high-speed streams. High-speed stream one is of 24-hour duration; the momentum flux, NV^2, is held constant while the velocity at the base of the model (18 R_s) is doubled from 250 km/s to 500 km/s. In high-speed stream two, the duration is not limited; the mass flux, NV, is maintained constant while the velocity at the model base is doubled. None of these simulations contains a heliospheric current sheet; this will be added in future work to produce new members of the Rogues' Gallery catalog.

The simulations done thus far lead us to expect "observational" uncertainty due to: 1) the fact that two different disturbances can produce g–maps which are similar, and 2) observed g–maps contain noise. From the flare simulations, it is evident that a *weak* shock observed on day 3 can produce a g-map similar to that produced by a *strong* shock on day 2 (see Fig. 1). This start–time versus speed ambiguity was described by Leinbach and Grubb /1/. For the geometrical model the speed of the disturbance is completely ambiguous from only a single map. For the shocks in Figure 1, one can see the difference in strength. However, considering the "noisy" and variable appearance of observed g-maps, it is questionable whether the two cases in Figure 1 could be distinguished. Identification of the solar origin of the disturbance eliminates this uncertainty. Without identification of the solar origin of the disturbance (which is often impossible) there is also some direction of travel uncertainty. Two shocks that take off from the Sun in different directions and at different times can also produce g-maps which are relatively similar. The direction of travel uncertainty results in simulated maps with a perceptible difference in shape. Whether either of these types of uncertainty will be resolvable in observed g-maps remains to be seen.

CONTROVERSIES AND RESEARCH AREAS

There is lively debate regarding the solar sources of IPS disturbances. Hewish /5/ proposes that IPS events are produced by erupting streams that emanate from coronal holes (regions on the Sun where the magnetic field is open to the interplanetary medium). Others believe that these events come mainly from solar flares (large energy releases), whether the field is open or closed (looping back into the Sun). We plan to examine this issue by simulating specific events (SOLTIP intervals) based on both scenarios to see if either scenario more closely resembles observations.

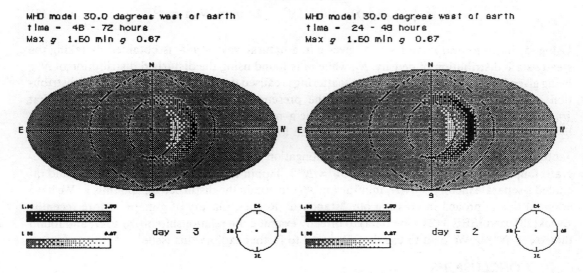

Fig. 1. Comparison of weak shock on day 3 (left) with strong shock on day 2 (right).

Another area of uncertainty involves the relationship of electron density fluctuations (the funda-mental cause of the scintillation) to bulk density. The scintillation index, m is given by:

$$m^2 = 2\pi r_e^2 \lambda^2 \int_{l.o.s.} \int_{k_x} \int_{k_y} \Phi_N(k_x, k_y, k_z = 0) \, F_x F_F dk_x dk_y dz \, , \tag{1}$$

where $l.o.s.$ = line of sight; dz = element of path along $l.o.s.$; k_x, k_y = spatial wave numbers; Φ_N = electron density spatial spectrum; F_s = source size term = $e^{-\frac{1}{2}k^2\theta_0^2 z^2}$; F_F = Fresnel filter term = $4\sin^2(k^2\lambda \, z/4\pi)$; λ = radio wavelength; and r_e = classical electron radius. Note that the total variance of electron density is given by:

$$\sigma_N^2 = \int_{k_x} \int_{k_y} \int_{k_z} \Phi_N(k_x, k_y, k_z) dk^3 \, . \tag{2}$$

Readhead, Kemp, and Hewish /6/, henceforth RKH, took the electron spatial density spectrum to have a Gaussian form and found an analytic solution for the inner wave-space double integral (over k_x and k_y). This greatly simplifies the line-of-sight integration. Their simplified expression for m^2 (see /2,6/) is:

$$m^2 = \int_{l.o.s.} \beta(r) \frac{(4Z)^2}{(1 + h^2)[(1 + h^2)^2 + (4Z)^2]} dz \, , \tag{3}$$

where r = distance from Sun; $Z = \lambda z/2\pi a^2$; $h^2 = 2z^2\theta_0^2/a^2$; θ_0 is e^{-1} diameter of the source; and a = scale length of electron density irregularities.

In this formula $\beta(r)$, the scattering power, must be found such that m will have the proper depen-dence on solar elongation angle. RKH give the following (empirical) formula for $\beta(r)$, which satis-fies this requirement.

$$\beta(r) = 2.73 \times 10^{-3}\lambda^2 r^{-4} \, , \quad \lambda \text{ in m}, \, r \text{ in AU}. \tag{4}$$

The presence of r^{-4} in this equation suggests another possibility: since electron density falls off approximately as r^{-2} ($N \simeq 5 \, r^{-2}$, N in cm^{-3}, r in AU),

$$\beta(r) \simeq 1.09 \times 10^{-4}\lambda^2 N^2 , \quad \lambda \quad \text{in m, } N \text{ in cm}^{-3}. \tag{5}$$

Using (5) in (3) to find m and $<m>$ gives g in a natural way; $<m>$ is obtained by taking the steady-state distribution of density, N_s, while m is found using the disturbed distribution of N, g being given by $g = m / <m>$. Since the scattering is caused by the variance in the spatial distribution of electron density, N, (i.e., $(\sigma_N)^2$), the presence of N^2 in this equation suggests that $(\sigma_N)^2 \propto N^2$. Based on a single pair of synthetic g–maps of the same event, Δg values computed by this method are larger by a factor of two than for the "standard" method described next.

Tappin /7/, by comparing g observed at $90°$ elongation with N observed by Earth-orbiting space-craft, found that $g \propto N^{0.5}$, and hence $\sigma_N \propto N^{0.5}$ Tappin used (4) in (3) to find $<m>$, and included a separate factor $f = N/N_s$ multiplying $\beta(r)$ to obtain the disturbed m, and thus g. We have adopted this approach in creating our "standard" Rogues' Gallery of g–maps. More recently, Zwickl /8/ used ISEE 3 (24 s ion data) to directly examine the relationship of σ_N to N, and found that $\sigma_N \propto N^{0.84}$. We plan to use the 3D model to further explore this issue.

CONCLUSIONS

Due to the "noisy" character of observed g–maps and our limited understanding of the behavior of σ_N, it is apparent that estimation of shock strength and direction of departure from the Sun, based on a single g–map, will have large uncertainty. This is relevant to forecasting since a fast–moving disturbance is likely to be observable in only one g–map from a single station. Additional IPS sites, separated in longitude (and thus observation time), would reduce the uncertainty. However, characterization of the uncertainty remaining when two or more g–maps are obtained (showing the same disturbance) is complex, since both speed and direction are unknown. This requires further research.

We are planning simulations of actual events for which the modeling is well constrained by space-craft observations (SOLTIP intervals). Such detailed comparisons of simulated with observed g–maps could either better delineate the observational uncertainty described above or call into serious question the veracity of the "standard" scattering models. This work thus demonstrates the potential of 3D MHD modeling to address issues of both global dynamics and microphysics.

REFERENCES

1. H. Leinbach and R. Grubb, in: *Solar–Terrestrial Predictions*: Proceedings of a Workshop at Leura, Australia, October 16–20, 1989, ed: R.J. Thompson, D.G. Cole, P.J. Wilkinson, M.A. Shea, D. Smart, and G. Heckman, (1989).

2. S.J. Tappin, M. Dryer, S.M. Han, and S.T. Wu, *Planet. Space Sci.*, **36**, 1155 (1988)

3. S.M. Han, S.T. Wu, and M. Dryer, *Computers and Fluids*, **16**, 81 (1988).

4. S.M. Han, S.T. Wu, and M. Dryer, in *Proceedings of STIP Symposium on Retrospective Analyses*, ed: M.A. Shea and D.F. Smart, Book Crafters Publ. Co., Chelsea, MI, in press (1992).

5. A. Hewish, in *Solar-Terrestrial Predictions*, ed: R.J. Thompson et al. V. 1, 81–101, NOAA Environmental Research Laboratories, Boulder, CO (1990).

6. A.C.S. Readhead, M.C. Kemp, and A. Hewish, *Mon. Not. R. Astr. Soc.*, **185**, 207 (1978).

7. S.J. Tappin, *Planet. Space Sci.*, **34**, 93 (1986)

8. R. Zwickl, private communication (1990).

THREE-DIMENSIONAL STRUCTURING OF THE DISTANT SOLAR WIND BY ANOMALOUS COSMIC RAY PARTICLES

S. Grzedzielski,* H. -J. Fahr** and H. Fichtner**

* Space Research Centre, Ordona 21, 01-237 Warsaw, Poland
** Institut für Astrophysik und Extraterrestrische Forschung der Universität Bonn, Auf dem Hügel 71, D-5300 Bonn, Germany

ABSTRACT

It is well known that both the galactic and the anomalous cosmic rays show positive intensity gradients in the outer heliosphere which are connected with corresponding antiradial pressure gradients. Due to an efficient dynamical coupling between the solar wind plasma and these highly energetic media, by means of a scattering at convected MHD wave turbulences, the latter diffuse through the low energetic solar wind plasma flow, and thus there exists a mutual dynamic interaction of these media. As a prime consequence of this scenario, the diffusion-induced pressure gradients in the cosmic ray distributions influence the distant solar wind expansion. In this paper we concentrate on the interaction of the solar wind with the anomalous cosmic ray component giving a consistent formulation of the system of mutually interacting media. Then we derive numerical solutions in the following steps: First we calculate an aspherical pressure distribution for the anomalous cosmic rays describing their diffusion in an unperturbed radial solar wind. Second we consider the perturbation of the solar wind flow due to these induced anomalous cosmic ray pressure gradients. Within this context we especially take account of the action of an aspherical geometry of the heliospheric shock which may lead to a pronounced upwind/downwind asymmetry in the pressure distribution, and thereby in the resulting solar wind flows. As we can show decelerations of the distant solar wind by between 5 to 11 percent are to be expected, however, deviations of the bulk solar wind flow from the radial directions are only weakly pronounced.

THEORETICAL PROCEDURE

Cosmic ray particles of galactic and anomalous origin diffuse through the expanding solar wind being scattered at intrinsic MHD turbulences that are convected outwards with the wind (see e.g. /1/). Thereby an effective dynamical coupling between the high energy cosmic ray plasma and the low energy solar wind plasma is established giving rise to two main phenomena: (a) The distribution functions of these energetic particles are modulated by diffusion in configuration space and in momentum space and thus their effective pressure varies in space, and (b) the solar wind expansion is modified due to the action of the resulting cosmic ray pressure gradient. Here we concentrate only on the action of the anomalous cosmic ray component - the effect of the galactic cosmic rays is discussed in /2/.

The anomalous cosmic rays are treated here as a hot gas of a negligible mass density, however, representing a considerable energy density. Using an approach for a hydrodynamical diffusion process with an energy-averaged diffusion coefficient $\kappa(r)$ we are led to the following fluid-type energy continuity equation for this gas (see /2/):

$$\nabla \left[\frac{\gamma_c}{\gamma_c - 1} p_c \vec{v} - \kappa \nabla(\frac{p_c}{\gamma_c - 1}) \right] = \vec{v} \cdot \nabla p_c \tag{1}$$

with p_c and γ_c being the anomalous cosmic ray pressure and the polytropic index of this weakly relativistic gas (i.e. $\gamma_c(200MeV/nucleon) \approx 1.6$). The quantity $\vec{v}$ denotes the solar wind bulk flow velocity.

On the other hand, the description of the solar wind expansion is carried out by the use of the conventional set of differential equations describing an one-fluid solar wind dynamics, however, completed here by adequate terms in the momentum and in the energy equation which are connected with the effective pressure gradient $\nabla(\eta_c p_c) \approx \nabla p_c$ of the modulated part of the anomalous cosmic ray distribution function, i.e. we used an efficiency factor (that is defined in /3/) of $\eta_c = 1$, meaning that the whole component is modulated. Thus the following set of equations is used:

$$\nabla(\rho\vec{v}) = 0 \tag{2}$$

$$\rho(\vec{v}\nabla)\vec{v} = -\nabla p - \nabla p_c - \rho\frac{GM_\odot}{r^2}\vec{e_r} \tag{3}$$

$$\nabla\left[\rho\vec{v}(\frac{1}{2}v^2 + \frac{\gamma p}{\gamma - 1})\right] = -\vec{v}\cdot\nabla p_c \tag{4}$$

where ρ and p are the solar wind plasma density and plasma pressure. G is the gravitational constant and $M_\odot$ is the solar mass. The unit vector in the radial direction is denoted by $\vec{e_r}$. γ is the polytropic index of the solar wind plasma.

NUMERICAL PROCEDURE

We start with a solar wind solution obtained for $p_c = 0$ which for large distances simply is given by:

$$\vec{v}(r) = u_E\vec{e_r} \qquad \rho(r) = \rho_E(r_E/r)^2 \qquad p(r) = (p_E/\rho_E^\gamma)\rho^\gamma \tag{5}$$

with u_E, ρ_E and p_E being corresponding reference values at $r = r_E = 1AU$.

We then integrate the cosmic ray equation which in the case of an asymmetric production of anomalous cosmic ray particles at an upwind/downwind asymmetric heliospheric shock boundary causing an asymmetric pressure distribution $p_c = p_c(r, \theta)$ attains the following explicit form:

$$\frac{\gamma_c}{\gamma_c - 1}\frac{u_E}{r^2}\frac{\partial}{\partial r}(r^2 p_c) - \frac{1}{\gamma_c - 1}[\frac{\kappa_E}{r^2}\frac{\partial}{\partial r}(r^2(\frac{r}{r_E})^n\frac{\partial p_c}{\partial r})$$
$$+\frac{\kappa_E(\frac{r}{r_E})^n}{r\sin\theta}\frac{\partial}{\partial \theta}(\sin\theta\frac{1}{r}\frac{\partial p_c}{\partial \theta})] - u_E\frac{\partial p_c}{\partial r} = 0 \tag{6}$$

with r being the radial polar coordinate and θ being the polar inclination angle with respect to the upwind/downwind axis, counted from the upwind direction. In the above equation the energy-averaged diffusion coefficient is treated as a scalar function of the radial coordinate which was taken in the form proposed by /1/ and given by:

$$\kappa(r) = \kappa_E(\frac{r}{r_E})^n \tag{7}$$

In many papers treating the modulation of cosmic ray spectra by shock-induced or genuine turbulences in the solar wind often the value $n = 1$ is used for the above relation (7) (see e.g. /4/,/5/,/6/). Since with this assumption the above equation (6) will become much better tractable, we shall also make use of it here. For $n = 1$ we then obtain from (6) the following simplified equation:

$$2\gamma_c u_E p_c + (u_E - 3\frac{\kappa_E}{r_E})r\frac{\partial p_c}{\partial r} - \frac{\kappa_E}{r_E}r^2\frac{\partial^2 p_c}{\partial r^2}$$
$$-\frac{\kappa_E}{r_E}(\frac{\partial^2 p_c}{\partial \theta^2} + ctg\theta\frac{\partial p_c}{\partial \theta}) = 0 \tag{8}$$

We try to find solutions of (8) by a representation of the function p_c in the following separable form:

$$p_c(r, \theta) = \sum_{\nu=0}^{\infty} \sigma_\nu R_\nu(r) T_\nu(\theta) \tag{9}$$

The constants σ_ν serve to fit the outer boundary conditions. With this representation one obtains the complete solution from the following two equations for the νth part of the series expansion:

$$\lambda^2 R_\nu'' - (\alpha - 3)\lambda R_\nu' + (A_\nu - 2\gamma_c\alpha)R_\nu = 0 \tag{10}$$

and:

$$T_\nu'' + ctg\theta T_\nu' - A_\nu T_\nu = 0 \tag{11}$$

Here the primes on top of the functions R_ν and T_ν in both cases denote derivatives with respect to the respective independent variables, i.e. either r or θ. The quantity $A_\nu = -\nu(\nu+1)$ is the separation constant of this problem. $\lambda = r/r_b \leq 1$ is a newly introduced normalized radial coordinate with $r_b = 80AU$ being the outer boundary of our integration region. In addition the dimensionless quantity α in (10) is introduced by:

$$\alpha = \frac{u_E r_b}{\kappa_b} = \frac{u_E r_E}{\kappa_E} \tag{12}$$

and represents the ratio of the convective and the diffusive transport rates.
In physical connection with an aspherical heliospheric shock front, like the one obtained in the MHD-simulation by /7/, one has to expect an upwind/downwind asymmetric production of anomalous cosmic ray particles. Associated with that an asymmetric pressure distribution $p_{cb} = p_c(r_b, \theta)$ has to be expected at the outer boundary of our integration regime. To restrict ourselves here to the simplest possible form of a nonspherical pressure distribution function (thus of course not to the most general one) that can be made conciliantory with (10) and (11) we start out from the following representation of p_c:

$$p_c(\lambda, \theta) = \sigma_0(R_0(\lambda) + \sigma R_1(\lambda)cos\theta) \tag{13}$$

where $\sigma = \sigma_1/\sigma_0$ denotes the relative amplitude of the nonsymmetric part of the pressure. The free parameters σ_0 and σ_1 were determined by us such that the theoretical pressure profiles calculated by /8/ were exactly matched at a reference point of $20AU$ what concerns the absolute pressure value and the radial gradient at $\theta = 0°$. The special solution obtained under these constraints from (10) and (11) is given by:

$$p_c(\lambda, \theta) = (\gamma_c - 1)[c_1\lambda^{g_1} + c_2\lambda^{g_2} + \sigma(c_3\lambda^{g_3} + c_4\lambda^{g_4})cos\theta] \tag{14}$$

with the following values for the constants c_i and g_i:

$$c_1 = 0.3629 eV/cm^3; c_2 = 2.2461 \cdot 10^{-5} eV/cm^3$$
$$c_3 = 0.3629 eV/cm^3; c_4 = 1.1164 \cdot 10^{-5} eV/cm^3$$

and:

$$g_1 = 4.4564; g_2 = -2.7064; g_3 = 4.7255; g_4 = -2.9755$$

From (14) it is now easy to derive the components of the anomalous cosmic ray pressure in the following form:

$$(\nabla p_c)_r = \frac{\gamma_c - 1}{r_b}[g_1 c_1 \lambda^{g_1 - 1} + g_2 c_2 \lambda^{g_2 - 1}$$
$$+\sigma(g_3 c_3 \lambda^{g_3 - 1} + g_4 c_4 \lambda^{g_4 - 1})cos\theta] \tag{15}$$

and:

$$(\nabla p_c)_\theta = -\frac{\gamma_c - 1}{r_b}\sigma(c_3 \lambda^{g_3} + c_4 \lambda^{g_4})sin\theta \tag{16}$$

where the degree of the relative anisotropy of the anomalous cosmic ray pressure at the integration boundary is expressed by the free coefficient σ varying between $0 \leq \sigma \leq 1$.

NUMERICAL RESULTS

For a specific upwind/downwind production asymmetry, for instance described by the value $\sigma = 1$ in its strongest possible form, we then obtain from (12) and (13) the results shown in Figures 1 and 2 for the anomalous cosmic ray pressure distribution and its radial gradient as functions of r and θ.

As the next logical step within the frame of this mutual interaction scenario of the diffusing anomalous cosmic ray gas and the expanding solar wind, we now include the action of the gradient of the above pressure distribution for the anomalous cosmic ray gas obtained on the 3-dimensional solar wind expansion.

For this purpose the simultaneous integration of the set of 3-dimensional solar wind differential equations has to be carried out with inclusion of the corresponding terms in (3) and (4) that are connected with the pressure gradient $\nabla p_c(r, \theta)$. This numerically very ambitious task we can only fulfill for the solar wind region beyond the critical point when the characteristic equation of the system mathematically has become of a hyperbolic type. Therefore we integrate this system from $1AU$ outwards starting from there with commonly used solar wind parameters ρ_E, u_E, p_E and integrating our 3-dimensional system of equations according to a method explained in detail by /2/ and /9/.

With the pressure distribution shown in Figures 1 and 2 we then calculate those distant solar wind properties shown in Figure 3. There we have plotted both the vectorial flow field of the solar wind bulk velocity and the isocontour lines of the solar wind radial bulk velocity. From Figure 3 it is evident that by means of the action of the anomalous cosmic ray pressure gradient the upwind solar wind flow is decelerated much stronger than the downwind solar wind which close to the downwind axis hardly shows any deceleration. A comparison of these effects with those due to the action of the galactic cosmic rays has been given in /2/.

Thus with these calculations we could show that with any upwind/downwind asymmetric pressure distribution for the anomalous cosmic ray gas at the boundary of our integration volume ($r_b = 80AU$) also a asymmetrical effect on the solar wind expansion is intimately connected. In the case shown by us here we have used the most pronounced case of such an asymmetry with a vanishing pressure at the downwind boundary, i.e. with $p_c(r_b, \theta = 180^0) = 0$. This case would, nevertheless, be of some relevance, since for those highly asymmetric heliospheric shock geometries found by /7/ and /10/ most probably associated with low downwind Mach numbers of $M_{a,down} < 5$, low pick-up ion fluxes there and low conversion factors to high energetic cosmic ray energies (see e.g. /11/) the reality could most likely be placed not far from these assumptions.

REFERENCES

1. J.R. Jokipii, Propagation of Cosmic rays in the Solar Wind, *Rev. Geophys. Space Phys.* 9, 27 (1971)

2. H.J. Fahr, H. Fichtner, S. Grzedzielski, The influence of the Anomalous Cosmic Ray Component on the Dynamics of the Solar Wind, *Solar Physics*, in press (1991)

3. H. Fichtner, H.J. Fahr, W. Neutsch, R. Schlickeiser, A. Crusius-Wätzel, H. Lesch, Cosmic Ray Driven Galactic Winds, *Il Nuovo Cimento* 106 , 909-925 (1991)

4. G.M. Webb, M.A. Forman, W.I. Axford, Cosmic-ray acceleration in stellar wind terminal shocks *Astrophys. J.* 298, 684 (1985)

5. M.A. Lee, W.I. Axford, Model structure of a cosmic-ray mediated stellar or solar wind, Astron. Astrophys. 194, 297 (1988)

6. H. Moraal, 21st *ICRC, Adelaide*, paper SH 6.4-12 (1990)

7. Y. Fujimoto, T. Matsuda, Interaction between the Solar Wind and the magnetized Local Interstellar Medium, preprint (1991)

8. J.R. Jokipii, The anomalous component of cosmic rays, *Proc. of the 1st COSPAR-Colloquium "Physics of the outer heliosphere"*, 169 (1990)

9. H. Fichtner, W. Neutsch, H.J. Fahr, R. Schlickeiser, Three-Dimensional Models of A Galctic Wind Expansion with Ellipsoidal Geometry I., *Astrophys. J.* 371, 98 (1991)

10. T. Matsuda, Y. Fujimoto, E. Shima, K. Sawada, T. Inagushi, Numerical Simulations of Interaction between Stellar Wind and Interstellar Medium, *Prog. Theor. Phys.* 81(4), 810 (1989)

11. H.J. Fahr, Filtration of the Interstellar neutrals at the heliospheric interface and their coupling to the solar wind, *Proc. of the 1st COSPAR-Colloquium "Physics of the outer heliosphere"*, Warsaw (Poland), 327 (1990)

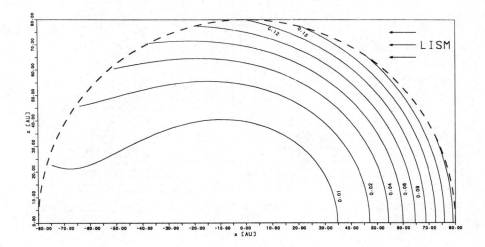

Fig. 1. A contour plot of the anomalous cosmic ray pressure $p_c(r, \theta)$ for an upwind/downwind pressure asymmetry at the boundary ($r_b = 80AU$) given by $\sigma = 1$.

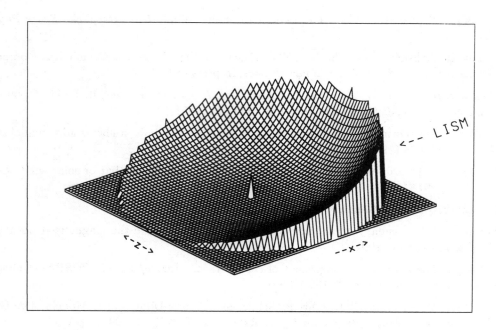

Fig. 2. 3-D plot of the radial component of the anomalous cosmic ray pressure gradient $(\nabla p_c(r, \theta))_r$ based on an upwind/downwind pressure asymmetry characterized by $\sigma = 1$ at the outer boundary of $r_b = 80AU$. The peak in the center indicates the position of the sun. The outer boundary of the circumsolar area represents $r_b = 80AU$.

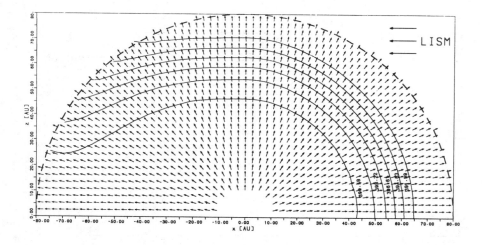

Fig. 3. The resulting solar wind flow field with iso-contours of the solar wind radial bulk velocity. Again the outer circle marks the boundary of our integration volume, a circumsolar spherical shell at $r_b = 80AU$.

SOLAR WIND TEMPERATURE OBSERVATIONS IN THE OUTER HELIOSPHERE

P. R. Gazis,* A. Barnes,** J. D. Mihalov** and A. J. Lazarus***

* SJSU Foundation, NASA Ames Research Center, Moffett Field, CA, U.S.A.
** NASA Ames Research Center, Moffett Field, CA, U.S.A.
*** Center for Space Research, MIT, Cambridge, MA, U.S.A.

ABSTRACT

The Pioneer 10, Pioneer 11, and Voyager 2 spacecraft are now at heliocentric distances of 50, 32 and 33 AU, and heliographic latitudes of 3.5° N, 17° N, and 0° N respectively. Pioneer 11 and Voyager 2 are at similar celestial longitudes, while Pioneer 10 is on the opposite side of the sun. The baselines defined by these spacecraft make it possible to resolve radial, longitudinal, and latitudinal variations of solar wind parameters. The solar wind temperature decreases with increasing heliocentric distance out to a distance of 10-15 AU. At larger heliocentric distances, this gradient disappears. These high solar wind temperatures in the outer heliosphere have persisted for at least 10 years, which suggests that they are not a solar cycle effect. The solar wind temperature varied with heliographic latitude during the most recent solar minimum. The solar wind temperature at Pioneer 11 and Voyager 2 was higher than that seen at Pioneer 10 for an extended period of time, which suggests the existence of a large-scale variation of temperature with celestial longitude, but the contribution of transient phenomena is yet to be clarified.

INTRODUCTION

Since 1980, Pioneer 11 and Voyager 2 have been at similar heliocentric distances – now 32 and 33 AU, respectively – similar celestial longitudes, and headed upstream with respect to the Local Interstellar Medium (LISM). Pioneer 10 is at a greater heliocentric distance – now 50 AU – and on the opposite side of the sun, headed downstream with respect to the LISM. Pioneer 10 and Voyager 2 have remained within 4° of the solar equator, while Pioneer 11 has been at a much higher heliographic latitude. The heliographic latitude of Pioneer 11 has been increasing since 1980 and is now near 17° N. The Pioneer and Voyager spacecraft are thus uniquely placed to conduct observations of the outer heliosphere. The long period during which observations have been available and the fortunate orientation of baselines between these spacecraft make it possible to resolve temporal, radial, longitudinal, and latitudinal variations in the solar wind.

SPATIAL AND TEMPORAL VARIATIONS IN SOLAR WIND TEMPERATURE

Figure 1 shows a comparison of averaged parameters observed at Pioneer 11 (solid line) and Voyager 2 (dashed line), plotted versus time. Each point represents an average over three solar rotations. The upper panel shows temperature while the lower panel shows velocity. To remove the effect of solar cycle variations, Pioneer and Voyager measurements have been divided by measurements made at IMP. The velocity observations are discussed by Barnes /1/.

The average temperature at Pioneer 11 and Voyager 2 dropped rapidly ($\simeq R^{-1}$) between 1 and $\simeq 4$ AU. At larger heliocentric distances, the radial temperature profiles at both spacecraft were almost flat, and the temperature at both spacecraft was $\simeq 25,000$ K. Near solar minimum, a latitudinal gradient developed; the temperature at Pioneer 11 climbed to $\simeq 35,000$ K while the temperature at Voyager 2 dropped to $\simeq 6000$ K. This latitude gradient is consistent with earlier observations /1, 2/.

This latitude gradient persisted after solar minimum; while the temperature at Pioneer 11 dropped back to $\simeq 25,000$ K, the temperature at Voyager only recovered to $\simeq 15,000$ K – significantly less than the temperature at Pioneer 11.

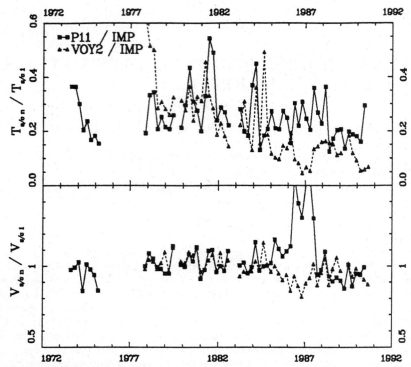

Fig. 1. Averaged parameters observed at Pioneer 11 (solid line) and Voyager 2 (dashed line), plotted versus time. Upper and lower panels show velocity and temperature. Pioneer and Voyager measurements have been divided by measurements made at IMP.

Figure 2 shows a comparison of averaged parameters observed at Pioneer 10 (solid line) and Voyager 2 (dashed line), plotted versus time. The upper panel shows temperature while the lower panel shows velocity. As in Figure 1, these Pioneer and Voyager measurements have been divided by measurements made at IMP.

Prior to the 1986 solar minimum, the average temperature at Pioneer 10 was $\simeq$ 10,000 K; much lower than the temperature at Voyager 2. Since the radial temperature profiles at both spacecraft were nearly flat, this suggests that prior to 1986, the temperature in the outer heliosphere varied with celestial longitude, and was consistently higher in the direction that was upstream with respect to the LISM.

The temperature at Pioneer 10 remained constant throughout the 1986 solar minimum and the rising portion of the next solar cycle. During solar minimum, the temperature at Voyager 2 dropped to a value below that at Pioneer 10. After solar minimum, the temperature at Voyager 2 rose to a value higher than that at Pioneer 10.

SOLAR WIND TEMPERATURES: REAL OBSERVATIONS OR INSTRUMENTAL EFFECTS?

The Pioneer and Voyager temperature observations, particularly the suggestion that temperature may vary with celestial longitude, are so surprising that it is important to exclude the possibility of instrumental effects. It is especially important to exclude the possibility that the different spacecraft have different sensitivities to low temperatures. There are several arguments against this possibility:

1) The determination of temperature is essentially a velocity measurement – i.e. measurement of the width of the distribution function. Since the voltage levels of, and hence the velocity measurements from, the Pioneer and Voyager plasma instruments are well known, the temperature measurements should be reliable as long as the plasma distribution function is resolved, and reasonably close to an isotropic Maxwellian.

2) The Pioneer 10 and 11 instruments and data analysis are almost identical.

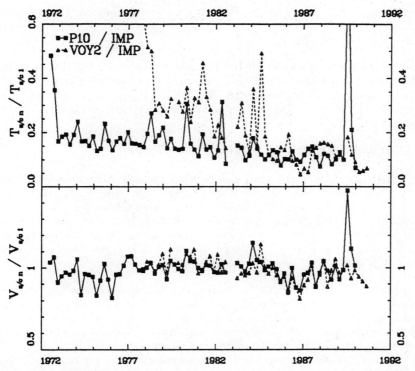

Fig. 2. Averaged parameters observed at Pioneer 10 (solid line) and Voyager 2 (dashed line), plotted versus time. Upper and lower panels show velocity and temperature. Pioneer and Voyager measurements have been divided by measurements made at IMP.

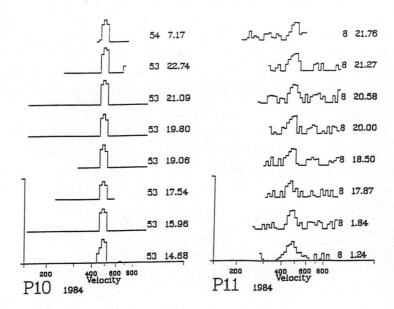

Fig. 3. Spectra from the Pioneer 10 and Pioneer 11 plasma analyzers for corresponding time intervals.

3) The three spacecraft return comparable results when they are in close proximity. Also, temperature measurements made near 1 AU by Pioneer 10, Pioneer 11, and Voyager 2 agreed with temperature measurements made from the IMP spacecraft.

4) Direct examination of spectra from Pioneer 10 and 11 supports the conclusion that temperatures at Pioneer 10 are lower than those at Pioneer 11. Figure 3 shows spectra from the Pioneer 10 and Pioneer 11 instruments for corresponding intervals of solar wind. The Pioneer 10 spectra are significantly narrower than those from Pioneer 11. The Pioneer 10 spectra, while narrow, still occupy several channels, and so are resolved.

SUMMARY: TEMPERATURES IN THE OUTER HELIOSPHERE

The radial profile of the average solar wind temperature between 1 and 50 AU cannot be fit by a simple power law. While the average temperature dropped rapidly ($\simeq R^{-1}$) between 1 and $\simeq 4$ AU, the radial temperature profile was almost flat beyond 10 AU. Solar wind temperatures in the outer heliosphere are much higher than would be expected for simple adiabatic expansion. Whang /4/ suggested that shock heating is inadequate to explain the observed radial temperature profile in the outer heliosphere, but there is more than enough bulk kinetic energy to heat the solar wind if it can be transformed to thermal energy by some other process, such as turbulent decay.

The average temperature appeared to vary with the solar cycle. It was lower near solar maximum and higher near solar minimum – behavior similar to that seen at 1 AU. The solar wind behaved differently before, during, and after the recent solar minimum. Before solar minimum the temperatures at Pioneer 11 and Voyager 2 (upstream with respect to the LISM) were comparable, and higher than the temperature at Pioneer 10 (downstream with respect to the LISM), suggesting that the temperature in the outer heliosphere varied with celestial longitude. During solar minimum, the temperature at Pioneer 11 (high latitude) rose, while the temperature at Voyager 2 (low latitude) dropped below the temperature at Pioneer 10. After solar minimum, the temperature at Pioneer 11 dropped back to the value it had before solar minimum, while the temperature at Voyager 2 rose to a value which, though higher than the temperature at Pioneer 10, was still lower than the temperature at Pioneer 11.

REFERENCES

1. Barnes, A., J. D. Mihalov, P. R. Gazis, A. J. Lazarus, J. W. Belcher, G. S. Gordon, and R. L. McNutt, Global properties of the plasma in the outer heliosphere I. Large-scale structure and evolution, this volume, 1991.

2. Gazis, P. R., J. D. Mihalov, A. Barnes, A. J. Lazarus, and E. J. Smith, Pioneer and Voyager observations of the solar wind at large heliocentric distances and latitudes, *Geophys. Res. Let.*, *3*, 223, 1989.

3. Mihalov, J. D., A. Barnes, A. J. Hundhausen and E. J. Smith, Solar wind and coronal structure near sunspot minimum: Pioneer and SMM observations from 1985-1987, *J. Geophys. Res.*, *95*, 8231, 1990

4. Whang, Y. C., S. Liu, and L. F. Burlaga, Shock heating of the solar wind plasma, *J. Geophys. Res.*, in press, 1990

THE STRUCTURE OF THE INNER HELIOSPHERE FROM PIONEER VENUS AND IMP OBSERVATIONS

P. R. Gazis,* A. Barnes,** J. D. Mihalov** and A. J. Lazarus***

* SJSU Foundation, NASA Ames Research Center, Moffett Field, CA, U.S.A.
** NASA Ames Research Center, Moffett Field, CA, U.S.A.
*** Center for Space Research, MIT, Cambridge, MA, U.S.A.

ABSTRACT

The IMP 8 and Pioneer Venus Orbiter (PVO) spacecraft explore the region of heliographic latitudes between 8° N and 8° S. Solar wind observations from these spacecraft are used to construct synoptic maps of solar wind parameters in this region. These maps provide an explicit picture of the structure of high speed streams near 1 AU and how that structure varies with time. From 1982 until early 1985, solar wind parameters varied little with latitude. During the last solar minimum, the solar wind developed strong latitudinal structure; high speed streams were excluded from the vicinity of the solar equator. Synoptic maps of solar wind speed are compared with maps of the coronal source surface magnetic field. This comparison reveals the expected correlation between solar wind speed near 1 AU, the strength of the coronal magnetic field, and distance from the coronal neutral line.

INTRODUCTION

It has long been known that solar wind parameters can vary with latitude. Recent reports of latitudinal gradients include in situ observations from the inner heliosphere /1/ and outer heliosphere /2/ as well as remote observations using the Interplanetary scintillation (IPS) technique /3/. These observations have established that the solar wind velocity tends to increase with heliographic latitude, and that this gradient extends to high latitudes, is time dependent, has a strong solar cycle dependence, and is stronger near solar minimum. These observations have also established several characteristics of the solar wind stream structure: 1) stream structure can vary strongly with latitude, especially near solar minimum, 2) the IMF sector structure can also vary strongly with latitude near solar minimum – the IMF sector structure vanished at high latitudes in the outer heliosphere during the 1975 and 1986 solar minima – 3) the inclination of the coronal magnetic field drops during solar minimum, which may be related to the exclusion of high speed solar wind from the vicinity of the solar equator during solar minimum. While these observations have provided considerable insight into the latitudinal structure of the solar wind, they do not provide an explicit picture of the structure of solar wind streams from in situ observations. In this report, we describe a method for constructing synoptic maps of solar wind parameters from observations from the Pioneer Venus Orbiter (PVO) and IMP 8 spacecraft. We also describe some of our results.

The IMP 8 and PVO spacecraft are at similar heliocentric distances (0.72 and 1 AU respectively), but the IMP 8 spacecraft explores a somewhat larger range of heliographic latitudes than does PVO (±7° as opposed to ±4°). The data used came from the plasma analyzer and magnetometer aboard PVO, the composite OMNI tape, and coronal source surface magnetic field data (Hoeksema, private communication).

THE METHOD

Six months of hourly averaged data from the IMP 8 and PVO spacecraft are projected back to the coronal source surface, after accounting for solar wind travel time. The data are then binned by Carrington longitude and heliographic latitude. Each bin covers 3.25° in latitude, 20° in longitude – this size was chosen so that each bin would be spanned by multiple tracks from the entire six month period. Data in each bin are averaged, to produce an array of values, and a spline routine is used to fit contour lines to this array. The result is a synoptic map of solar wind parameters. These maps explicitly show the three-dimensional structure of solar wind streams in a narrow band near the solar equator. They have several limitations – 1) latitudinal coverage is restricted to ± 7.25°, 2) spatial resolution is limited, 3) low temporal resolution (each map is a six month average) limits their usefulness when the solar wind undergoes rapid changes, 4) the contouring algorithm can introduce spurious detail, and any features smaller than the bin size (3.25° x 20°) must be treated with caution – but such limitations are also characteristic of IPS measurements.

RESULTS

Figure 1 shows synoptic maps of average solar wind and coronal parameters for six months beginning on day 245 of 1984. The top through bottom panels show plots of solar wind velocity, density, temperature, the component of the interplanetary magnetic field (IMF) parallel to the Parker spiral, and the coronal source surface magnetic field. The well-known correlation between solar wind velocity and temperature, and anti-correlation between solar wind velocity and density can easily be seen. The maps show that these correlations are spatial as well as temporal. There is a strong correlation between the interplanetary and coronal fields, and the interplanetary neutral sheet lies very close to the coronal neutral line. Regions of high density solar wind lie near the neutral sheet, but do not coincide with the neutral sheet itself. Solar wind velocity and temperature appear to correlate with coronal field and distance from the neutral line.

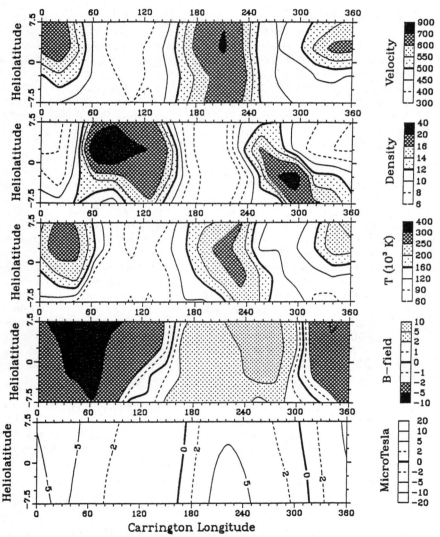

Fig. 1. Average solar wind and coronal parameters for six months beginning day 245 of 1984. Top through bottom panels show solar wind velocity, density, temperature, component IMF parallel to the Parker spiral, and coronal source surface magnetic field.

The solar wind stream structure underwent a marked change during the 1986 solar minimum. Figure 2 shows synoptic maps of average solar wind parameters for the six month period beginning on day 65 of 1987. The top, middle, and bottom panels show velocity, density, and IMF respectively. During the period shown, there were strong latitudinal gradients in solar wind parameters. High speed solar wind was excluded from the vicinity of the solar equator. In contrast, there was a region of high density solar wind near the solar equator. Magnetic field data – not shown in this figure – show that the inclination of the neutral sheet was small and the neutral sheet was close to the solar equator. As was the case prior to solar minimum, there was a strong correlation between solar wind velocity and temperature, and between the IMF and coronal source surface fields.

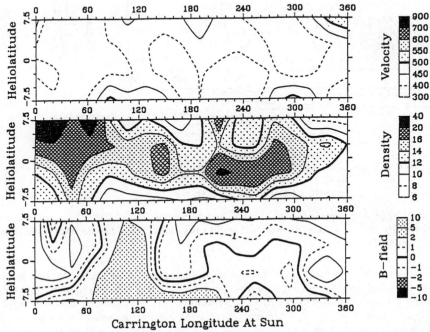

Fig. 2. Average solar wind parameters for six months beginning day 65 of 1987. Top, middle, and bottom panels show velocity, density, and IMF.

After solar minimum, the solar wind stream structure began to recover to the configuration that existed prior to solar minimum. The top and middle panels of Figure 3 show solar wind velocity and density for the six months following day 245 of 1988. During this period, the stream structure was similar to that seen in Figure 1, and high speed streams that stretched across the solar equator with little latitudinal variation. But while this type of structure was typical of the ascending phase of the present solar cycle, there were also many periods when the solar wind was dominated by transients.

CONCLUSIONS

The synoptic maps described above show the expected correlation between solar wind velocity, density, and temperature. They also show a strong correlation between the coronal and interplanetary fields, and the coronal neutral line and the interplanetary neutral sheet. There was an anti-correlation between the solar wind density and the IMF, but regions of high density did not correspond precisely with the interplanetary neutral sheet. Still to be determined are correlations between solar wind parameters and other quantities, such as the coronal field spreading factor, solar wind parameters from IPS measurements, and coronal holes.

The solar wind stream structure was different before, during, and after the latest solar minimum. Prior to the last solar minimum, the inclination of the coronal neutral line was high, latitudinal gradients in the solar wind were small, and high speed streams stretched across the solar equator. During the last solar

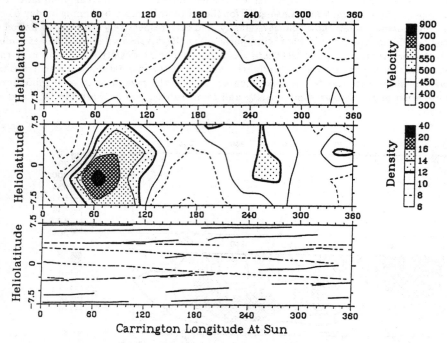

Fig. 3. Average solar wind parameters for six months beginning day 245 of 1988. Top and middle panels show velocity and density. Bottom panel shows IMP (solid lines) and PVO (dashed lines) spacecraft tracks projected back to the sun.

minimum, the inclination of the coronal neutral line dropped, high speed solar wind was excluded from the vicinity of the solar equator, and the region near the solar equator was characterized by low solar wind speeds, low temperatures, and high densities. After solar minimum, the inclination of the coronal neutral line increased, and there were long periods when high speed solar wind returned to the vicinity of the solar equator, but there were intervals when high speed solar wind was absent, as well as numerous transient events.

REFERENCES

1. Miyake, W., T. Mukai, K.-I. Oyama, T. Terasawa, K. Hirag, and A. J. Lazarus., Thin equatorial low-speed region in the solar wind observed during the recent solar minimum, *J. Geophys. Res.*, *94*, 15,359, 1989.

2. Barnes, A., Distant solar wind plasma – view from the Pioneers, *Physics of the Outer Heliosphere*, edited by S. Grzedzielski and D. E. Page, Pergamon Press PLC, Oxford, 235, 1990

3. Coles, W. A., B. J. Rickett, V. H. Rumsey, J. J. Kaufmann, D. G. Turley, S. Anathakrishnan, J. W. Armstrong, J. K. Harmon, S. L. Scott, and D. G. Sime, Solar cycle changes in the polar solar wind, *Nature*, *286*, 239, 1980.

SYNOPTIC MAPS OF HELIOSPHERIC THOMSON SCATTERING BRIGHTNESS FROM 1974–1985 AS OBSERVED BY THE HELIOS PHOTOMETERS

P. Hick,*,** B. V. Jackson* and R. Schwenn***

* Center for Astrophysics and Space Sciences, University of California at San Diego, La Jolla, CA 92093, U.S.A.
** Applied Research Corp., 8201 Corporate Dr., Suite 920, Landover, MD 20785, U.S.A.
*** Max-Planck-Institut für Aeronomie, D-W3411, Katlenburg-Lindau, Germany

ABSTRACT

We display the electron Thomson scattering intensity of the inner heliosphere as observed by the zodiacal light photometers on board the Helios spacecraft in the form of synoptic maps. The technique extrapolates the brightness information from each photometer sector near the Sun and constructs a latitude/longitude map at a given solar height. These data are unique in that they give a determination of heliospheric structures out of the ecliptic above the primary region of solar wind acceleration. The spatial extent of bright, co-rotating heliospheric structures is readily observed in the data north and south of the ecliptic plane where the Helios photometer coverage is most complete. Because the technique has been used on the complete Helios data set from 1974 to 1985, we observe the change in our synoptic maps with solar cycle. Bright structures are concentrated near the heliospheric equator at solar minimum, while at solar maximum bright structures are found at far higher heliographic latitudes. A comparison of these maps with other forms of synoptic data are shown for two available intervals.

INTRODUCTION

The Thomson scattering data obtained with the zodiacal light photometer instruments on board the Helios 1 and 2 spacecraft /1,2/ cover almost a complete solar cycle (1975-1985 for Helios 1, 1976-1980 for Helios 2). These scattering data provide out-of-ecliptic remote-sensing information, covering a wide range of heliographic latitude, about the large-scale density structure of the inner heliosphere (< 1 AU). Hick, Jackson and Schwenn /3/ developed a method of presenting these data in the form of synoptic (Carrington) maps in heliographic longitude and latitude, using data covering a single solar rotation for each map.

The Carrington map has been used extensively for representing solar data in the past, e.g. for Hα, surface magnetic fields, etc. In a recent paper Rickett and Coles analyzed the solar cycle evolution of the solar wind using 16 years of IPS solar wind velocity measurements in a comparison with Mauna Loa K-coronameter brightness data and Stanford magnetic field data. These three data sets were presented as synoptic maps based on 6-month periods (i.e. averages over 6 or 7 solar rotations) and refer to coronal heights of about 2 $R_\odot$, slightly varying for the different data sets. The comparisons confirm the existence of fast (slow) solar wind emanating from low (high) density coronal regions with open (closed) magnetic field structure.

Helios data are available from 1975 until 1985, a subset of the period that was well-sampled by the IPS data used in the Rickett and Coles study (1972-1987). In this paper we extend the method of Hick et al. /3/ and construct synoptic maps, using data for periods as close as possible to those used by Rickett and Coles /4/ and present a preliminary evaluation of the Helios maps in a comparison with the IPS, coronameter and source surface maps as given by Rickett and Coles.

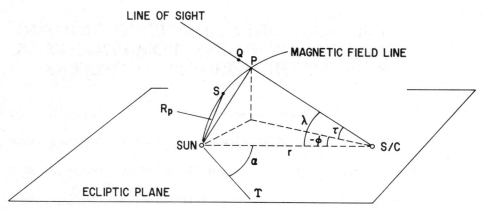

Fig. 1. A sketch of the geometry involved in the production of the synoptic maps. The Helios spacecraft moves in the ecliptic. P is the minimum heliocentric distance location for the line of sight shown. The curved line represents the Archimedean spiral through P. S is the intersection of this field line with the reference surface at a fixed distance from the Sun. The Sun-spacecraft distance and other parameters used in the construction of the synoptic maps are shown as in Hick et al. /3/

SYNOPTIC MAPS

Figure 1 indicates the geometry of the Helios observations for a particular line of sight at elongation λ from the Sun. The observed Thomson scattered brightness is the integrated effect of the scattering electrons along the line of sight. The scattered intensity per unit of length peaks at the point P closest to the Sun, but will have a rather wide half-width (≈ 40 ° for a r^{-2} coronal density). The resulting ambiguity in the actual location of the electrons contributing to the line of sight brightness is removed by assuming that the bulk of the electrons is located near point P. From this point the electrons are traced back along the Archimedean spiral through P to a spherical reference surface (S in Figure 1), assuming a constant, radial solar wind velocity of 400 $km\ s^{-1}$. To minimize the effects of the assumed wind velocity on the backprojection we choose the distance of the reference surface at 0.3 AU. The resulting heliographic location is used in the synoptic maps. For more detailed information we refer to /3/.

It may be noted that the difficulties in interpreting the line of sight measurements are very similar for IPS and Helios observations and are dealt with in similar fashion. Consequently, the Helios maps share with the IPS maps the condition that they are expected primarily to give good results in representing the large-scale, long-lived structure of the inner-heliosphere.

Rickett and Coles /4/ present one synoptic map for each year in their data set, based on observations between DOY 70 and 270, i.e. covering about 7 solar rotations. As far as data coverage allowed we used the same period for constructing the Helios maps. However, Helios data coverage decreased in these six month periods (≈ 1 Helios orbit) in the late phase of the mission (especially for Helios 1 after 1980), as data were collected for decreasing fractions of the heliocentric orbits around perihelion passage at 0.3 AU.

RESULTS AND CONCLUSION

From the total period covered by the Helios observations we selected a few examples for the purpose of this preliminary exposition. Figures 2 and 3 show two Helios maps together with the corresponding IPS, K-coronameter and magnetic source surface maps as given in Rickett and

Coles /4/. In the Helios maps the lowest contour corresponds to 10 S10 units (outermost contour). Values for subsequent contours increase in steps of 10 S10 units. The grey scale for the other plots is a relative scale between maximum and minimum values, where black represents maximum in the IPS and magnetic source surface maps, and minimum in the K-coronameter map (for details see /4/). The south and north hemispheres in Figures 2a and 3a are constructed separately from data obtained from the Helios 1 and Helios 2 photometers, observing south and north of the ecliptic plane, respectively. While IPS, K-coronameter and magnetic field maps refer to a source surface at approximately 2 $R_{\odot}$, the Helios maps are constructed at 0.3 AU (see previous section). At a wind velocity of 400 km s^{-1}, structures in the Helios maps will be shifted about 15° eastward with respect to the corresponding structure in the other maps.

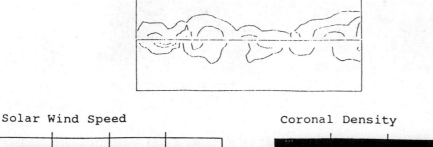

Solar Wind Speed   Coronal Density

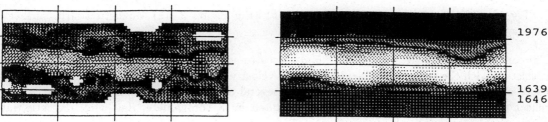

1976

1639
1646

Fig. 2. Synoptic maps for 1976 (Carrington rotations 1639-1646). (*a*) Helios Thomson scattering brightness. (*b*) IPS solar wind velocity. (*c*) K-coronameter map [(b) and (c) taken from /4/].

Figure 2 is representative of solar minimum conditions (1976). The global structure present in the Helios map (Figure 2a) is consistent with the IPS and K-coronameter maps. A region of high brightness (i.e. of higher density relative to the background) and low wind velocity forms a relatively narrow band near the equator, coincident with the flat, equatorial current sheet which existed during this period. The heliosphere is distinctly divided into low speed, high density regions near the equator and high speed, low density poleward regions. Individual large-scale structure may be tentatively identified in all three maps.

Figure 3 is based on data for the rising phase of the solar cycle, close to maximum (1979). The current sheet at this time has developed into several topologically separate structures, extending to high solar latitudes (Figure 3c), reflecting the greater complexity and more transient nature of the inner heliosphere during solar maximum. The IPS map and Helios map do not show any obvious large scale structure matching the magnetic field map. Low velocity regions in the IPS map (Figure 3b) extend over all heliographic latitudes. Similarly, the Helios map (Figure 3a) higher density regions are no longer restricted to the equatorial regions, but extend over essentially all latitudes.

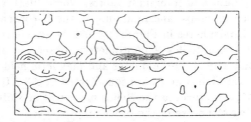

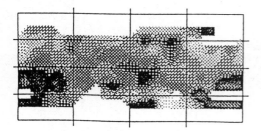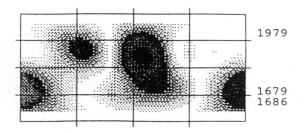

Fig. 3. Synoptic maps for 1979. (*a*) and (*b*) as in Figure 2. (*c*) Magnetic source surface map [(b) and (c) taken from /4/].

The comparison of the Helios maps of Thomson scattered sunlight show a global evolution which is consistent with the evolution evident in the IPS velocity maps. Around solar minimum the maps, based on observations obtained at distances of several tenths of an astronomical unit, accurately reflect the inner heliospheric structure as indicated by K-coronameter data and magnetic data obtained at or near the solar surface.

ACKNOWLEDGEMENTS

We gratefully acknowledge the help of UCSD student Wang Poon in the analysis of the Helios synoptic maps. Part of this work was supported by Air Force contract AFOSR-91-0091 and NASA grant NAGW-2002 to the University of California at San Diego.

REFERENCES

1. C. Leinert, H. Link, E. Pitz, N. Salm and D. Klüppelberg, Helios zodiacal light experiment, Raumfahrtforschung 19, 264 (1975)

2. C. Leinert, E. Pitz, H. Link and N. Salm, Calibration and in-flight-performance of the zodiacal light experiment on Helios, J. Space Sci. Instr. 5, 257-270 (1981)

3. P. Hick, B.V. Jackson and R. Schwenn, Synoptic maps for the heliospheric Thomson scattering brightness as observed by the Helios photometers, Astron. Astrophys. 244, 242-250 (1991)

4. B.J. Rickett and W.A. Coles, Evolution of the solar wind structure over a solar cycle: interplanetary scintillation velocity measurements compared with coronal observations, J. Geophys. Res. 96,#A2,1717-1736 (1991)

LARGE-SCALE STRUCTURE OF THE HELIOSPHERIC MAGNETIC FIELD: 1976–1991

J. T. Hoeksema

Center for Space Science and Astrophysics, Stanford University, Stanford, CA 94305, U.S.A.

ABSTRACT

A uniform set of computed large-scale coronal fields derived from photospheric field observations now exists for the interval 1976 to the present. Using this data base we can begin to compare Solar Cycle 21 with Solar Cycle 22. The character of the field and its evolution are the same, but many of the particulars are different. The polar field was about 25% stronger at solar minimum in 1986 than in 1976, consequently the heliospheric current sheet was confined to a narrower latitude range. In the rising phase of Cycle 21 a strong 4-sector pattern developed; the same pattern was not present in Cycle 22. Other differences include the timing of the polar field reversals and changes in latitudinal extent of the current sheet.

INTRODUCTION

Observations of the photospheric magnetic field with 3 arc minute resolution began at the Wilcox Solar Observatory at Stanford in 1976. The global coronal magnetic fields computed from these large-scale measurements with a potential field – source surface model now span more than 1.5 solar cycles. The field computed at the upper boundary of the model predicts the polarity structure of the interplanetary field quite well. We can now begin to compare the evolution of the three-dimensional heliospheric field structure in Solar Cycle 21 with Cycle 22.

Photospheric observations, similar to those shown for Carrington Rotation 1847 in Figure 1a, provide the inner boundary condition for the coronal field model /1,2,3/. To match the observations high in the corona, the field is constrained to be radial on a sphere 1.5 solar radii above the photosphere called the source surface. If currents are not permitted between the two surfaces, then the field can be described by a scalar potential and a vector solution for the field can be computed from the derivatives of the potential. The potential is described in terms of spherical harmonics. Even though only the line-of-sight photospheric field is measured, the field at 2.5 Rs is sometimes observed to be non-radial, and we know currents must flow low in the corona, these assumptions are good enough to allow for a reasonably accurate model of the large scale coronal field /4,5/. The computed source surface field for CR 1847 is shown in Figure 1b. This polarity pattern is convected outward relatively unchanged by the solar wind. Daily polarity observations taken out to 1 AU in or near the ecliptic match this structure about 80% of the time /6/. The structure itself seems to degrade with increasing distance as dynamic effects become more important /7/.

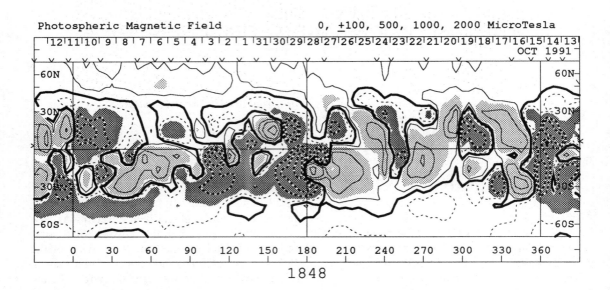

Photospheric Magnetic Field   0, ±100, 500, 1000, 2000 MicroTesla

1848

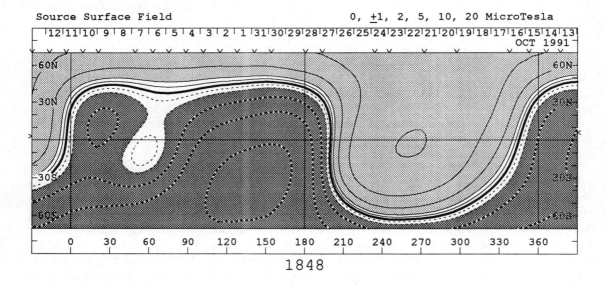

Source Surface Field   0, ±1, 2, 5, 10, 20 MicroTesla

1848

Figure 1: A) The Upper panel shows the photospheric field for CR 1848 in October 1991. Fields greater than 2 Gauss (200 uT) are shaded lightly. Fields below -2 G are shaded more darkly. Central meridian dates are indicated on the upper axis as are times of magnetograms. Time goes from right to left. Earth's latitude is indicated in the left margin. Field values are not well determined above 70°. B) The lower panel shows the computed radial field at the 2.5 Rs source surface, in the same format. Contours and shading values are a factor of 100 smaller and the polarity structure is simpler. This pattern shows the configuration of the heliospheric IMF and generally agrees with observations.

VARIATION WITH THE SOLAR CYCLE

Figure 2 (right) shows how the heliospheric field varies during the solar cycle, from solar maximum at the beginning of 1980 (CR 1690) through minimum in 1986 (CR 1778) and back to maximum in 1989 (CR 1822). Even though only one panel is shown for each 11 rotation interval, it is still possible to see some structures persisting for long periods, e.g. the large negative feature near the center of CR 1690 that lasts through CR 1734 and dies out in CR 1745. The polar fields form after solar maximum and gradually strengthen as the complexity of the field gradually decreases. During most of the declining and minimum phases the field is very stable, alternating between 2 and 4 sectors per rotation near the equatorial plane. At minimum the neutral line is very flat; the Earth never gets very far from the current sheet. The amount of structure rapidly increases during the rising phase of the cycle, reaching the poles again at solar maximum. Notice the sign change of the polar fields between 1989 and 1991.

The two columns of Figure 3 (next page) compare the rising phases of solar cycles 21 and 22. Here successive panels are centered only three rotations apart and it is easier to trace the evolution of individual features. One particularly long-lived example is the negative feature that becomes prominent in CR 1804 near 0 degrees and lasts, with some distortions and gradual movement, through the end of the interval shown in CR 1840. Small isolated regions of opposite polarity can be identified in each cycle.

The general similarity between the structures at comparable times in the two cycles is striking, e.g. CR 1661 & 1801, 1679 & 1819, 1688 & 1828, and 1700 & 1840. Also interesting are some of the differences between the two cycles. Compare the prominent 4-sector structure in CR 1664 - 1670 that did not have a counterpart in Cycle 22.

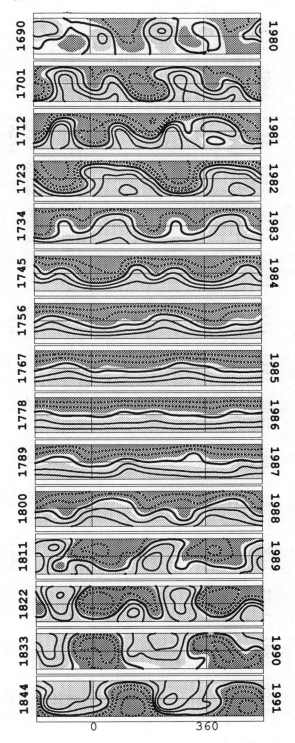

Figure 2: Each panel shows a two Carrington Rotation wide map of the computed coronal magnetic field centered on the rotation indicated at the left. The shading levels are the same as in Fig. 1b. Every 11th rotation is shown. The first rotation shown in a year is labeled on the right.

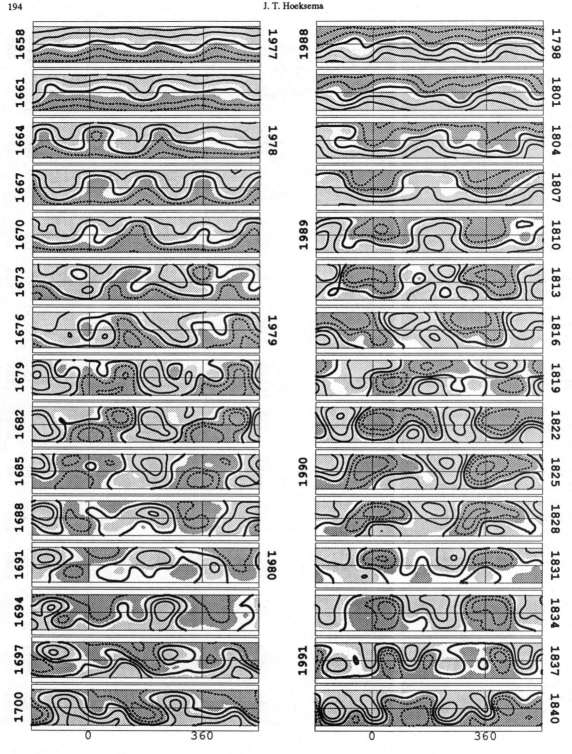

Figure 3: The panels are similar to those in Figure 2 and are centered on every third rotation. The left column shows the rising and maximum phase of Solar Cycles 21; the right column is for Cycle 22. Carrington rotation boundaries are shown at the bottom. The similarities at comparable times during the cycles are striking. The most notable differences occur in 1978 and 1988-1989. Notice the long lifetimes of some features and the rotation rate, shown by the slope. The polar fields reverse near the bottom of each column.

Solar Cycle Variations

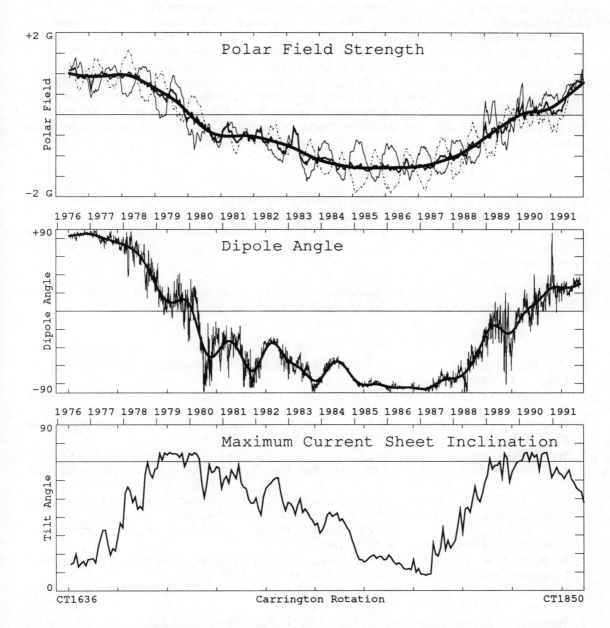

Figure 4: A) The top panel shows the variation of the polar field strength from 1976 to the present. The dark line is a smoothed average of the two poles. The lighter solid line shows the variation of the northern pole; the dashed line shows the southern polar field. The large annual variation is due to the inclination of the Earth's orbit to the solar equator. B) The middle panel shows the dipole angle derived from the spherical harmonic components of the computed coronal field. The darker line is a smoothed average. C) The average inclination of the heliospheric current sheet computed at 2.5 Rs is shown in the bottom panel. The variations of these quantities are all related to changes in the photospheric field. The line at 70° shows the maximum latitude for which the coronal field can be reliably computed.

Various quantities can be defined to characterize the variation of the heliospheric field during the cycle. Figure 4 presents the changing values of of the polar field strength, the dipole angle, and the inclination of the current sheet from 1976 through the present. This includes the solar minima of 1976 and 1986 and the maxima of 1980 and 1989. The top panel shows the variation of the northern and southern polar fields and the smoothed average of the two. The polar field reversal in 1980 was stronger than the 1990 reversal. The field strength was about 25% greater at minimum in 1986 than in 1976. The dipole angle (middle) varies much like the polar field strength, the short term variability is much higher when the polar field strength is low around maximum. The variations in dipole angle are closely related to arrival of flux at the poles. The bottom panel shows how the inclination of the heliospheric current sheet reaches a maximum for two or three years at solar maximum and gradually decreases toward minimum. The smaller inclination at minimum in 1986 is related to the increased polar field strength relative to the previous minimum.

CONCLUSIONS

The structure and evolution of the heliospheric field are determined by the patterns of flux in the photosphere and vary on time scales as long as the solar cycle. A consistent set of measurements of the Sun's magnetic field and the computed coronal field patterns are readily available for the time from 1976 to 1991 /8,9/. These data sets can be useful in characterizing the background state of the interplanetary medium.

Acknowledgement: This work was supported in part by NASA Grant NGR-559, ONR Contract N00014-89-J-1024 and NSF Grant ATM 90-22249.

REFERENCES

1. Schatten, K.H., J.M. Wilcox, and N.F. Ness, A Model of Interplanetary and Coronal Magnetic Fields, *Solar Phys, 6,* 442, 1969.
2. Altschuler, M.D., and G. Newkirk, Jr., Magnetic Fields and the Structure of the Solar Corona, ISolar Phys., **9,** 131, 1969.
3. Hoeksema, J.T., J.M. Wilcox, and P.H. Scherrer, Structure of the Heliospheric Current Sheet in the Early Portion of Sunspot Cycle 21, *J. Geophys. Res.,* **87,** 10331, 1982.
4. Wilcox, J.M., and A.J. Hundhausen, Comparison of Heliospheric Current Sheet Structure Obtained from Potential Magnetic Field Computations and from Observed Maximum Coronal Brightness, *J. Geophys. Res.,* **88,** 8095, 1983.
5. Bruno, R., L.F. Burlaga, and A.J. Hundhausen, K-Coronameter Observations and Potential Field Model Comparison in 1976 and 1977, *J. Geophys. Res.,* **89,** 5381, 1984.
6. Hoeksema, J.T., J.M. Wilcox, and P.H. Scherrer, The Structure of the Heliospheric Current Sheet: 1978-1982, *J. Geophys. Res.,* **88,** 9910, 1983.
7. Behannon, K.W., L.F. Burlaga, J.T. Hoeksema, and L.W. Klein, Spatial Variation and Evolution of Heliosphere Sector Structure, *J. Geophys. Res.,* **94,** 1245, 1989.
8. Hoeksema, J.T. and P.H. Scherrer, The Solar Magnetic Field - 1976 through 1985", WDC-A for Sol-Terr. Phys., *Rpt. UAG-94,* 1986.
9. Hoeksema, J.T., The Solar Magnetic Field 1985 Through 1990, CSSA Rpt. *CSSA-ASTRO-91-01,* 1991.

EVOLUTION OF SPATIAL AND TEMPORAL CORRELATIONS IN THE SOLAR WIND: OBSERVATIONS AND INTERPRETATION

L. W. Klein,* W. H. Matthaeus,** D. A. Roberts*** and
M. L. Goldstein***

* Applied Research Corporation, 8201 Corporate Drive, Landover, MD 20785,
U.S.A.
** Bartol Research Institute, University of Delaware, Newark, DE 19716, U.S.A.
*** NASA, Goddard Space Flight Center, Code 692, Greenbelt, MD 20771,
U.S.A.

ABSTRACT

Observations of solar wind magnetic field spectra from 1-22 AU indicate a distinctive structure in frequency which evolves with increasing heliocentric distance. At 1 AU extremely low frequency correlations are associated with temporal variations at the solar period and its first few harmonics. For periods of 12-96 hours, a 1/f distribution is observed, which we interpret as an aggregate of uncorrelated coronal structures which have not dynamically interacted by 1 AU. At higher frequencies the familiar Kolmogorov-like power law is seen. Farther from the sun the frequency break point between the shallow 1/f and the steeper Kolmogorov spectrum evolves systematically towards lower frequencies. We suggest that the Kolmogorov-like spectra emerge due to *in situ* turbulence that generates spatial correlations associated with the turbulent cascade and that the background 1/f noise is a largely temporal phenomenon, not associated with *in situ* dynamical processes. In this paper we discuss these ideas from the standpoint of observations from several interplanetary spacecraft.

INTRODUCTION

The appearance of Kolmogorov-like spectra in solar wind magnetic field observations has been interpreted as indicative of *in situ* generation of a turbulent cascade by dynamical processes. This concept was first suggested by Coleman /1/ and refined repeatedly since /2-9/. Roberts et al. /6/ observed that in time intervals of less than 10 hours for which these spectra are common, there is an increasing equilibration of inward and outward travelling Alfvénic fluctuations with radial distance; the nonlinear interaction between these modes contributes to the cascade and decay of the turbulence. A spectral evolution consistent with this has been observed in the inner heliosphere /8/. Simulations have been used to suggest that stream shears can be the cause of this evolution. There is growing agreement that in the solar wind much of the structure of field spectra in this high frequency domain is produced by non linear cascades generated in the solar wind.

Matthaeus and Goldstein /7/, using many years of IMP data at 1 AU, observed a 1/f spectrum at wave lengths greater than 10 hours. They suggested that this was due to a hierarchy of reconnecting structures in the low corona. The spectrum produced by these structures is similar to flicker noise observed in other fluid flows due to uncorrelated structures. Matthaeus and Goldstein interpreted the solar wind spectra as signatures of this coronal activity and therefore as largely temporal phenomena rather than due to spatially correlated structures in the flow. Structures with these wavelengths or greater are unlikely to have participated in a turbulent cascade in transit by 1 AU because there is insufficient time for information communicated at either the Alfvén speed or the typical eddy speed to have crossed the structures.

Bavassano et al. /8/ and Klein /9/ have shown that the break point between Kolomogov and the shallower 1/f spectra moves to longer wavelengths with increasing radial distance. We make the interpretation that structures of scale size smaller than those at this break point are spatial correlations produced by *in situ* dynamical processes and that structures larger are remnants of lower coronal activity and are largely temporal phenomena. The size of structures represented at the break point are the largest objects which could interact dynamically in the flow. These concepts were described for isotropic and homogeneous hydrodynamic turbulence theory by Batchelor /10/, who considered the decay of these energy containing structures. In wind tunnel experiments, the size of energy containing structures at scales near the break point has been observed to grow as $R \sim x^{1/2}$, where x measures distance downstream from the grate /11/. In a recent extension of the model to magnetofluids and energy containing eddy transport for solar wind flows, Zhou and Matthaeus /12/ give a prescription for the growth rate of these structures which is slower than that for the hydrodynamic model, or about $R \sim x^{1/3}$. The growth rate is slower for MHD turbulence because of the decorrelation of structures due to Alfvénic propagation; this same effect leads to slower spectral transfer in energy in MHD compared to hydrodynamic turbulence. In this paper, we study this phenomenon using a statistical analysis of many years of interplanetary magnetic field data.

DATA ANALYSIS

We have used hourly averaged magnetic field observations from IMP series, ISEE-3, and Voyager 1 and 2 spacecraft. These data were taken between 1977 and 1984 and represent a heliocentric radial coverage of 1 to 22 AU. The data were bad point edited using sliding filters based on arithmetic means. We computed power spectral densities for subsets of the analysis interval by using the Blackman-Tukey method. We also calculated spectral densities by the method of Fast Fourier Transforms. In our analysis, we have considered intervals long enough and of sufficient quantity so that the spectra are reasonably stationary and also appealed to the notion of frozen-in-flux in order to convert frequency to spatial scales.

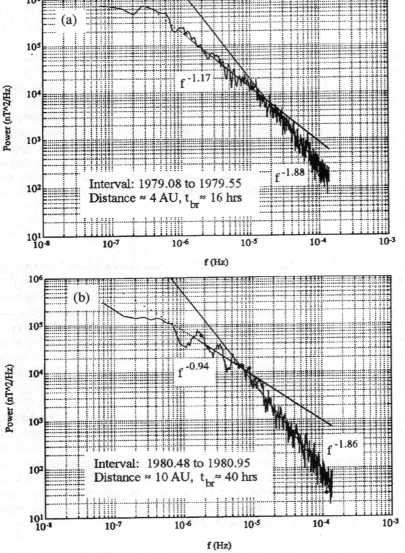

In Figure 1 we show two power spectra of magnetic field taken from the Voyager 1 spacecraft at about 4 and 10 AU from the sun in the ecliptic plane. The spectra were accumulated from 6 months of hour averages. In order to locate the break point in the traces, we did a linear least squares fit to both the higher and lower frequency parts separately. As shown on the plots, the slopes are nearly Kolmogorov-like at the high frequency end and nearly 1/f at the lower end. The calculated intercept of the two lines gives the break point approximately. These are 16 hours and about 40 hours respectively, which correspond to spatial scales of 1/8 and 1/2 AU. In their analysis of 1 AU field data, Matthaeus and Goldstein /7/ found break points at about 10 hours. Analysis of successive six month spectra in the Voyager 2 data set from 1 to 22 AU showed

Fig. 1: Trace of the power spectrum of the magnetic field with fits to two freqeuency regimes for Voyager data near (a) 4 AU and (b) 10 AU.

a similar pattern of movement in the break point although with considerable variability and uncertainty in the identification of the precise location of the break.

Structures existing in the Kolmogorov or inertial range are considered to have largely been created in transit by nonlinear interaction between inward and outward traveling Alfvén waves. We can estimate the largest

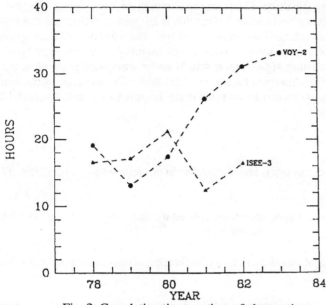

spatial scales in this range by the method above or statistically using the spectral data. This scale—the correlation scale—is defined as $L_c = \int_0^\infty R(\tau)d\tau / R(0)$, where $R(\tau)$ is the correlation function of the field. This method depends on an estimation of the correlation function at infinite wavelength; however, if we use sufficiently long analysis intervals, they can be considered of the Lanczos type and sufficient for our purposes (/9,13/). In Figure 2, we plot the correlation time versus observation time for Voyager 2 as it moves from 1 to about 20 AU. We show the results of the same calculation for ISEE-3 at 1 AU for reference because the data contain many types of solar wind flows, including transients near solar maximum and two and four stream patterns at quiet times. The correlation scale approximately doubles in the flow in the outer heliosphere.

Fig. 2: Correlation time vs. time of observation

We have put these analyses on a more sound statistical basis by accumulating many spectra from the Voyager 1 mission. The interval was broken into 1024 point segments corresponding to about 42 days of data each.

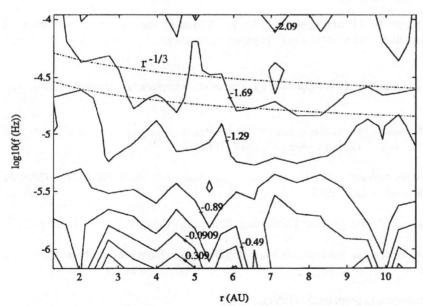

The time series were Fourier transformed and slopes were calculated between 10 logarithmically equally spaced bins in frequency. The segment was advanced 256 points, the binning repeated, and so on through the eight year interval. In Figure 3 we show a contour plot of the resulting spectral slope as a function of frequency and distance. There is a systematic trend in the slopes toward increasing dominance of the Kolmogorov regime, especially between 1 and 7 AU. For refer-

Fig. 3: Spectral slope as a function of frequency and radial distance from the sun to the observation point for Voyager 1 data.

ences we plot two traces of the the predicted growth of the size of the energy containing eddies using the MHD prediction of Zhou and Matthaeus /12/. The growth of the inertial range scales, given by the transition to the -5/3 slope regime, is approximated by the MHD result, with the observations of that slope being roughly bounded by the two curves.

SUMMARY

In this study, we have adopted the picture of the solar wind as an evolving turbulent flow, stirred by stream shears originating in variable sources in the lower corona. The structures leading to the relatively flat spectra at large scales are characterized statistically by a distinctive 1/f power spectrum in the inner heliosphere; these could be associated with a hierarchy of reconnection events /7/, but this is an open question. As the flow evolves in the outer heliosphere, nonlinear interactions among Alfvénic fluctuations produce spatial correlations which cause a cascade to higher wave numbers. These are distinguished by Kolmogorov-like power spectra. Energy containing eddies existing at the largest spatial scales associated with dynamical processes grow and the turbulence destroys the memory of the source of the flow. We have analyzed 8 years of interplanetary magnetic field data in light of this picture and find that it is quite useful and successful in analyzing the solar wind on a statistical basis.

REFERENCES

1. P. J. Coleman, *Astro. Phys. J.*, Turbulence, viscosity, and dissipation in the solar wind plasma, 153, 371 (1968).

2. C.-Y. Tu, Z.-Y. Pu, and F.-S. Wei, The power spectrum of interplanetary fluctuations: derivation of the governing equation and its solution, *J. Geophys. Res.*, 89, 9695 (1984).

3. C.-Y. Tu, The damping of interplanetary Alfvénic fluctuations and the heating of the solar wind, *J. Geophys. Res.*, 93, 7 (1988).

4. Y. Zhou and W. H. Matthaeus, Transport and turbulence modeling of solar wind fluctuations, *J. Geophys. Res*, 95, 10291 (1990).

5. J. V. Hollweg and W. J. Johnson, Transition region, corona and solar wind in coronal holes: some two fluid models, *J.Geophys.Res.*, 93, 9547 (1988).

6. D. A. Roberts, L. W. Klein, W. H. Matthaeus, and M. L. Goldstein, The nature and evolution of magnetohydrodynamic fluctuations in the solar wind: Voyager observations, *J. Geophys. Res.*, 92, 11021 (1987).

7. W. H. Matthaeus and M. L. Goldstein, Low frequency 1/f noise in the interplanetary magnetic field, *Phys. Rev. Lett.*, 57, 495 (1986).

8. B. Bavassano, M. Dobrowolny, F. Mariani, and N. Ness, Radial evolution of power spectra of interplanetary Alfvénic turbulence, *J. Geophys. Res.*, 87, 3617 (1982).

9. L. W. Klein, *Observations of turbulence and fluctuations in the solar wind,* Ph.D. Thesis, The Catholic University of America, Washington, D.C. 1987.

10. C. K. Batchelor, *The Theory of Homogeneous Turbulence,* Cambridge University Press, London 1953.

11. R. W. Stewart and A. A. Townsend, Similarity and self preservation in isotropic turbulence, *Philos. Trans. A,* 243, 359 (1951).

12. Y. Zhou and W. H. Matthaeus, in preparation (1991).

13. W. H. Matthaeus and M. L. Goldstein, Measurement of the rugged invariants of magnetohydrodynamic turbulence, *J. Geophys. Res.*, 87, 6011 (1982).

SOLAR WIND OBSERVED WITHIN 0.3 AU WITH INTERPLANETARY SCINTILLATION

M. Kojima,* H. Washimi,* H. Misawa* and K. Hakamada**

* Solar-Terrestrial Environment Laboratory, Nagoya University, Toyokawa 442,
Japan
** Department of Engineering Physics, Chubu University, Kasugai 487, Japan

ABSTRACT

We report the dependence of the solar wind speed on the heliocentric distance which has been studied with the interplanetary scintillation method. Acceleration was observed beyond 0.1 AU by 0.3 AU both in the low-speed wind and in the high-speed wind. Very slow speed winds were found at distances of 0.18 ~ 0.27 AU, which emanated from around a strong bipolar magnetic region. Origin of the very slow wind is related to low density gradient in the streamer which cannot diverge largely.

INTRODUCTION

Since the high-speed solar wind from the coronal hole was discovered, the solar wind physics has been trying to explain it. If the explanation is successful, it must be able to explain not only why the wind from the coronal hole is fast but also why the wind is slow in a neutral sheet and in a streamer. For these studies it is important to understand the distance dependence of the speed as well as to understand physical properties in the accelerating region such as density, temperature and magnetic field /4,10,11,12/. It is especially important to measure the distance dependence at distances within 0.3 AU where in situ measurements have not been made.
Although only remote sounding methods such as the IPS measurements can be used in this region, their observations were too sporadic to study radial evolution of the solar wind along a stream line from near the sun. Systematic and coordinated observations are necessary for it. Such kind of the IPS observations have started very recently using the EISCAT facility by Coles et al. /3/ and the facility at the Solar-Terrestrial Environment Laboratory, Nagoya University by Kojima and Kakinuma /7/.
We have been carrying out steady IPS observations at 327 MHz using the system consist of three remote stations since 1983. This facility can measure speeds approaching the sun as near as 0.1 AU. We report the distance dependence at distances from 0.1 to 1 AU which have been studied using data obtained with this facility.

HELIOCENTRIC DISTANCE DEPENDENCE OF THE SOLAR WIND SPEED

As there is not enough space in this short paper to write IPS observation, refer papers /1,6,8/ for details of the IPS observations and their reliability.

Rather steady flows are reported from spacecraft and IPS observations at distances of 0.3 ~ 1 AU /2,5,8,9/. However it is not known how far acceleration continues and where it stops within a distance of 0.3 AU. Therefore we examined the distance dependence of the speed in two ways. Firstly we compared the velocity distributions in longitude and latitude coordinates between at distances of 0.1 ~ 0.3 AU and at 0.3 ~ 1 AU. For this comparison we made two velocity distribution maps on the source surface (v-maps): one is derived from the IPS observations at 0.1 ~ 0.3 AU (an inner v-map) and the other is derived from observations at 0.3 ~ 1 AU (an outer v-map).   These two v-maps shown in Figure 1 were observed during Carrington rotations of 1779 ~ 1782. The upper panel is the outer v-map and the lower one is the inner v-map. The year of 1986 was the sunspot minimum. In the minimum phase, the speed is low in a narrow current sheet along the equator and high-speed regions extend widely from the poles towards low latitude. Such structure was observed only at outside of 0.3 AU, and the structure within 0.3 AU was different from it: low-speed winds were observed even at high latitudes and the speed is relatively low all over the surface.

Secondly the distance dependence of the speed is examined for the high-speed and the low-speed winds separately.   It is difficult to guess in the acceleration region which stream will become a high-speed wind or a low-speed one at a large distance because the speed may change largely. Therefore, first of all, the source surface was divided into two source regions of the high-speed wind and the low-speed one as follows: longi-

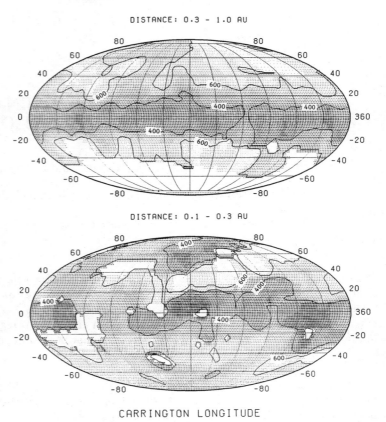

Fig.1. V-maps on the source surface derived from the IPS observations during Carrington rotations of 1779 ~ 1782. The upper panel is the outer v-map and the lower one is the inner v-map.

tude and latitude regions which has speeds faster than 600 km/s in the outer v-map are defined as the high-speed region and those for speeds less than 400 km/s are defined as the low-speed region. Speeds observed in each region during Carrington rotations of 1786 ~ 1794 in 1987 are plotted in Figure 2. The upper panel is for the high-speed wind and the lower one is for the low-speed wind. Although the Figures show large scatter of data, it is obvious that both of the low-speed and the high-speed winds change speeds largely even beyond 0.1 AU and become the steady flows by 0.3 AU. The reason of the large scatter of data within 0.3 AU is the mixture of streams in which acceleration is going on and those which have already been accelerated to a final speed within 0.1 AU. Similar results are reported by Coles and et al. from the IPS observations with the EIS-CAT antennas /3/. We made this analysis for years from 1984 to 1987 and obtained similar results as shown here /7/.

VERY SLOW SPEED WIND

Figure 3 is the inner v-map during the Carrington rotations of 1752 ~ 1756 in 1984. The very slow stream at the longitude of 180° and the latitude of 0° ~ 20° was observed by the three IPS radio sources of 3C273, 3C279 and 1524-13 during September to November. A speed of 207 km/s was observed by 3C273 at the distance of 0.18 AU on 19 September. 3C279 observed 116 km/s at the distance of 0.27 AU on 22 September, and 1524-13 observed 109 ~ 217 km/s at distances around 0.2 AU during three days from 25 to 27 November. This stream has a lower speed even at outside of 0.3 AU than that in the current sheet. A fairly large strong bipolar magnetic region located under this low-speed region, and Hakamada et al./4/ reported that this very low speed wind emanated from the very narrow regions around this magnetic region. Characteristics of the inner v-map should be mentioned here because it does not show exactly the velocity distribution in longitude and latitude coordinates. The inner v-map is derived from data observed at distances of 0.1 ∼ 0.3 AU where the speed is changing largely with a distance. Accordingly the inner v-map represents mixture of the latitude and longitude dependence of the speed and the distance dependence of it.

DISCUSSION

We showed that the acceleration can continue beyond 0.1 AU in both of the low-speed and the high-speed streams but finishes by 0.3 AU. As the velocity change is large at distances of 0.1 ∼ 0.3 AU, the solar wind structure within 0.3 AU is different largely from that at outside of 0.3 AU. It is well known that the solar wind in the current sheet is low. But the solar wind from around the strong bipolar magnetic region was found much slower than it within 0.3 AU. A magnetic flux tube which emanated from a compact region beside the strong bipolar magnetic region converges with the other flux tube which emanated from the opposite polarity side. Then the converged flux tubes make a streamer. If this streamer cannot diverge largely even far from the sun, the density gradient along the radial distance remains small. Conse-

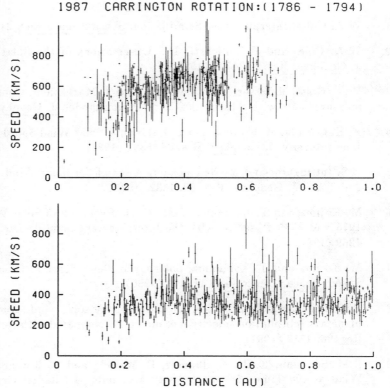

Fig.2. The upper panel plots speeds observed in the high-speed regions during an indicated period and the lower one does speeds for the low-speed regions.

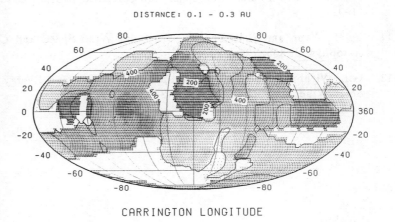

Fig.3. The inner v-map obtained during Carrington rotations of 1752 ∼ 1756 in 1984. The very slow stream at the longitude of 180° and the latitude of 0° ∼ 20° was observed at distances of 0.18 ∼ 0.27 AU.

quently thermal pressure in the streamer is small. This is one of reasons of the low-speed wind.

REFERENCES

1.   W.A. Coles, Interplanetary Scintillation, Space Science Rev. 21, 411 (1978)

2.   W.A. Coles and B.J. Rickett, IPS Observations of the Solar Wind Speed out of Ecliptic, J. Geophys. Res. 81, 4797 (1976)

3.   W.A. Coles, R. Esser, U.-P. Lovhaug, and J. Markkanen, Comparison of solar wind velocity measurements with a theoretical acceleration model, J. Geophys. Res. 96, 13849 (1991)

4.   K. Hakamada, M. Kojima, and T. Kakinuma, Solar Wind Speed and HeI (1083 nm) Absorption Line Intensity, J. Geophys. Res. 96, 5397 (1991)

5.   D.S. Intriligator and M. Neugebauer, A search for Solar Wind Velocity Changes Between 0.7 and 1 AU, J. Geophys. Res. 80, 1332 (1975)

6.   M. Kojima and T. Kakinuma, Solar Cycle Evolution of Solar Wind Speed Structure Between 1973 and 1985 Observed with the Interplanetary Scintillation Method, J. Geophys. Res. 92, 7269 (1987)

7.   M. Kojima and T. Kakinuma, Solar Cycle Dependence of Global Distribution of Solar Wind Speed, Space Science Rev. 53, 173 (1990)

8.   B.J. Rickett and W.A. Coles, Evolution of the solar wind structure over a solar cycle: interplanetary scintillation velocity measurements compared with coronal observations, J. Geophys. Res. 96, 1717 (1991)

9.   R.M. Schwenn, K.-H. Muhlhauser, E. Marsch, and H. Rosenbauer, Two States of the Solar Wind at the Time of Solar Activity Minimum: II. Radial Gradients of Plasma Parameters in Fast and Slow Stream, in: Solar Wind Four, ed. H. Rosenbauer, Max-Planck-Institute fur Aeronomy, Katlenburg-Lindau and Max-Planck-Institute fur Extraterrestrische Physic, Garching, West Germany, 1981, p126.

10. N.R. Sheeley, Jr., E.T. Swanson, and Y.-M. Wang, Out-of- ecliptic tests of the inverse correlation between solar wind speed and coronal expansion factor, J. Geophys. Res. 96, 13861 (1991)

11. Y.-M. Wang and N.R. Sheeley, Jr., Solar Wind Speed and Coronal Flux-Tube Expansion, Astophys. J. 355, 726 (1990)

12. G.L. Withbroe, The Temperature Structure, Mass, and Energy Flow in the Corona and inner Solar Wind, Astrophys. J. 325, 442 (1988)

A NUMERICAL MODEL OF THE LARGE-SCALE SOLAR WIND IN THE OUTER HELIOSPHERE

J. Kóta

Lunar and Planetary Laboratory, University of Arizona, Tucson, AZ 85721, U.S.A.

ABSTRACT

We present a solar wind model developed primarily for the study of cosmic ray modulation in the heliosphere. The model is steady-state in the corotating frame, the wind speed is assumed to vary with *heliomagnetic* latitude, slower wind is taken near the current sheet. A wavy sheet will lead to dynamical effects: corotating interaction regions (CIR) and merged interaction regions (MIR) arise in a natural way. As a simplifying approximation, the wind speed is kept radial, so the calculation is basically 1-dimensional including plasma density, pressure, bulk speed, and the azimuthal component of the magnetic field. Calculations are carried out for various tilt angles of the neutral sheet illustrating various phases of the solar cycle. Possible cosmic-ray effects are discussed.

INTRODUCTION

The dynamical expansion of the solar wind plasma represents a complex phenomenon, involving a number of physical processes. Among these, the present paper addresses one single aspect: the effects connected with the heliomagnetic variation of the solar wind speed are considered. It is generally believed (e.g. Newkirk and Fisk /1/) that the solar wind is slower at the heliospheric current sheet and becomes faster away from the sheet. This, combined with a wavy neutral sheet, will give rise to dynamical processes, corotating interaction regions (CIR) and merged interaction regions appear.

The primary motivation of the present work is to study the effects of CIR-s and MIR-s on the modulation of galactic cosmic rays. In earlier works (e.g. Kóta and Jokipii /2/,/3/,/4/) based on a uniform solar wind speed and (consequently) a uniform Parker-field, particle drifts were found to be extremely important in the heliospheric transport of cosmic rays. It remained, however, to be seen if CIR-s and MIR-s may destroy the global picture of modulation. Motivated by this need we have developed a simple solar wind model that includes CIR-s and MIR-s, although the structures are significantly less complex than reality.

THE MODEL

First, solar wind properties are specified at a base close (but not too close) to the sun. In accord with Newkirk and Fisk /1/ we assume that, near the sun, the solar speed increases from a low value of 350-400 km/sec near the sheet to a plateau of some 600-650 km/sec for high *magnetic* latitudes. The typical half-angle of this transition is 15-20° A tilted dipole field is assumed, thus the current sheet near the sun is in a plane inclined at an angle α to the equator. Solar rotation results in faster wind overtaking previously emitted slower wind, and formation of shocks.

Our approach is basically 1-dimensional and significantly less sophisticated than the 2-dimensional work of Pizzo /5/. As a simplifying assumption, we keep the wind radial, neglect small transverse stresses, and solve the radial mass, momentum, and energy equations, including polytropic pressure and magnetic forces from the azimuthal component of the frozen-in field. We ignore the radial field component which is small in the outer heliosphere. The calculation is followed until it settles into a corotating pattern.

RESULTS AND DISCUSSION

Calculations were carried out for different values of the tilt angle α, illustrating different phases of the solar cycle. The results were found largely insensitive to the initial temperature of the solar wind at the base. Since no heating, except for shock-heating, is included the 5/3 polytropic index leads to a rapid cooloing of the wind in the inner heliosphere.

Figure 1 shows snapshots to illustrate the typical density and velocity variations. CIR-s and MIR-s, similar to those described by Burlaga et al. /6/,/7/, can clearly be identified. Because of the symmetry of our model there are two streams that are symmetrical in the equatorial plane. For a tilt-angle of 30° (left panels) and latitude of 17.5° one of these streams is considerably stronger. The two streams merge to form a one-stream pattern at around 11 A.U. Then, CIR-s merge again in the outer heliosphere around 30–35 A.U. At lower latitudes, closer to the equator, two-stream structures survive longer. It should be noted, however, that the symmetry of the model is not at all an essential feature.

It may be of interest to notice that, as the tilt angle is becoming small, the wind speed is predicted to decrease near the equator while it remains near constant (or might even slightly increase) at higher latitudes. This is similar to that observed by Gazis, Barnes, and Lazarus /8/ around the 1987 solar minimum.

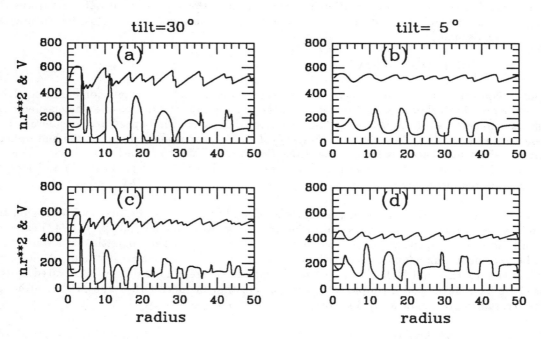

Fig. 1. Snapshots illustrating the radial variation of solar wind density (lower curves) and velocity (upper curves) for latitudes 30° (upper panels: (a) and (b)), and 5° (lower panels: (c) and (d)). Left panels refer to a tilt-angle of $\alpha = 30°$, while right panels are for a small tilt-angle, $\alpha = 5°$. Notice the decrease of the average wind speed at low latitude as α decreases.

Figure 2 shows the typical longitudinal variation of various physical quantities. This approximates what would be seen at a slow moving spacecraft over one 27-day solar rotation. We see the 27-day variation of the solar-wind streams and magnetic field, and cosmic-ray fluxes for both polarity states. Note that the predicted 27-day cosmic-ray variations are quite similar for both polarities. We also see that the cosmic-ray intensity decreases as the observer moves away from the current sheet, as observed by Newkirk and Fisk /1/ for both signs of A.

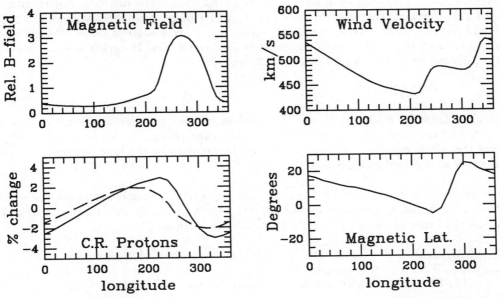

Fig. 2. Longitudinal (or 27-day) variation of various quantities at a latitude of $10°$ and distance of 10 A.U. Shown are magnetic variation relative to the Parker spiral (top left), wind velocity (top right), intensity of 2 GeV protons, lower left (solid line - $A < 0$, dashed line - $A > 0$,) and the magnetic latitude of the observer (lower right).

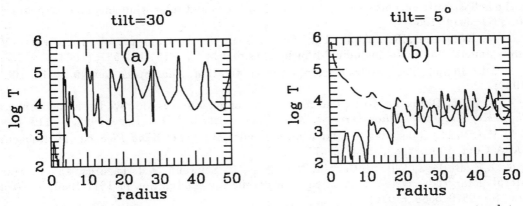

Fig. 3. Snapshots showing the variation of solar wind proton temperature at a latitude of $17.5°$ as obtained for tilt angles $30°$ (left) and $5°$ (right). Solid curves refer to solutions with zero initial temperature, dashed curve on (b) refers to high initial temperature near the sun.

We briefly call attention to one important effect which is seen clearly in Figure 2, and is also present at essentially all radii and at moderate to small latitudes, and for both signs of A. Notice that at the peak of the magnetic-field strength the *rate of decrease* of the cosmic-ray intensity is a maximum. This is quite similar to the effect observed and emphasized by Burlaga, et al. /7/. The rapid decrease of cosmic-ray flux is apparently a result of the smaller diffusion coefficient in the stronger fields of CIR-s. The computations show that this observed relationship is also qualitatively consistent with global model simulations where the global properties are determined to a large extent by the particle drifts. For more details see Kóta and Jokipii /9/.

Finally, Figure 3 shows the variation of ion-temperatures in the wind. Solid curves refer to solutions obtained for cold initial wind, with shock-heating representing the only heat source. Though temperature happens to be rather sensitive to numerical artifacts, our results seem to suggest that shock-heating may be quite important for larger tilt-angles while, not surprisingly, it should become ineffective for small tilt angles i.e. during solar minima. We consider this, at present, as a preliminary result which needs further studies. In earlier works, shock-heating was considered by Goldstein and Jokipii /10/ and Pizzo /5/.

CONCLUSION

A simple solar-wind model has been developed in which CIR-s form and merge in a natural way. Though the structures are much less complex than reality, the model is quite adequate to discuss the effects of CIR-s on cosmic rays. It also reproduces the long-term variation of solar wind speed at different latitudes.

Our simulations suggest that the large-scale 22-year cycle of cosmic-ray modulation is largely controlled by drifts, as proposed earlier /2/,/3/,/4/. Small-scale variations and 27-day variations, on the other hand, are primarily determined by CIR-s.

Acknowledgement. A great part of these results has been obtained in a joint work with Professor J.R. Jokipii. This research was supported, in part by the National Science Foundation under Grant ATM-8922151 and the National Aeronautics and Space Administration under Grant NAGW-2549.

REFERENCES

1. G. Newkirk and L.A. Fisk, Variation of cosmic rays and solar wind properties with respect to the heliospheric current sheet. 1. Five GeV protons and solar wind speed, *J. Geophys. Res.*, **90**, 3391–3414 (1985)

2. J. Kóta and J.R. Jokipii, The effect of drift on the transport of cosmic rays. VI. A three-dimensional model including diffusion, *Astrophys. J.*, **265**, 573–581, (1983)

3. J.R. Jokipii, The physics of cosmic ray modulation, *Adv. Space Res.*, **9**, # 12, 105–119, (1988)

4. J. Kóta, Diffusion, drifts, and modulation of galactic cosmic rays in the heliosphere, in: *Physics of the Outer Heliosphere*, ed. by S. Grzedzielski and D.E. Page, 1990, p. 119.

5. V.J. Pizzo, Quasi-steady solar wind dynamics, in: *Solar Wind Five*, ed. M. Neugebauer, NASA Conf. Publ. 2280, 1983, p. 675.

6. L.F. Burlaga, F.B. McDonald, N.F. Ness, R. Schwenn, A.J. Lazarus, and F. Mariani, Interplanetary flow system associated with cosmic ray modulation in 1977-1980, *J. Geophys. Res.*, **89**, 6579–6589, (1984)

7. L.F. Burlaga, F.B. McDonald, M.L. Goldstein, and A.J. Lazarus, Cosmic ray modulation and turbulent interaction regions near 11 AU. *J. Geophys. Res.*, **90**, 12,027–12,039, (1985)

8. P.R. Gazis, A. Barnes, and A.J. Lazarus, Intercomparison of Voyager and Pioneer plasma observations, in: *Proc. 6th Int. Solar Wind Conf.*, ed. V.J. Pizzo, T.E. Holzer and D.G. Sime, 1988, Vol. II, p. 563.

9. J. Kóta and J.R. Jokipii, The role of corotating interaction regions in cosmic-ray modulation, *Geophys. Res. Lett.*, **18**, 1797–1800, (1991)

10. B.E. Goldstein and J.R. Jokipii, Effects of stream-associated fluctuations upon the radial variation of average solar wind parameters, *J. Geophys. Res.*, **82**, 1095–1105 (1977)

THE INTERPLANETARY H GLOW AS SEEN BY
VOYAGER DURING THE CRUISE

R. Lallement,* J. L. Bertaux,* B. R. Sandel** and E. Chassefière*

Service d'Aéronomie du CNRS, BP No 3, 91371 Verrières le Buisson, France
**Lunar and Planetary Laboratory, University of Arizona, Tucson, AZ 85721,*
U.S.A.

ABSTRACT

Lyman-alpha background maps collected by the Ultraviolet Spectrometer (UVS) instrument on board Voyager 1 and 2 during the whole cruise are compared with current models of the interaction between the Sun and an homogeneous interstellar flow (without any perturbation nor filtration at the heliopause). These data provide precise measurements of the solar wind ionization rate, through the location of the maximum emission region (MER) upwind from the sun. The inferred rate implies a higher interstellar density as compared with previous results. Systematic departures clearly appear, indicating an excess of emission from the downwind ionization cavity as compared with the model. Possible explanations (radiative transfer, heliospheric interface modifications..) are discussed.

INTRODUCTION

Since 1977, the Voyager 1 and 2 UVS have been measuring the resonance glow of H and He atoms due to the scattering of solar photons by interstellar neutral gas. From time to time the Voyager spacecraft executed roll maneuvers (hereafter RM) that allowed a fairly complete sky survey in a limited time (about 2 days), allowing for the first time the mapping of the glow as seen from the outer solar system. Some longer exposures towards some particular directions were also used to derive the glow intensities ("fixed lines of sight", or FL). The flow of the galactic gas is modified by its interaction with the sun (gravitation, radiation pressure, ionization). The study of the glow yields results both on the galactic flow before interaction with the sun, and on the effects of interactions with the Sun. We have compared the measurements with a model which has already been used to fit observations from the inner solar system (0.7 AU < R <1. AU : Mariner 10, Prognoz, Pioneer-Venus). Systematic discrepancies are observed (Lallement et al, 1990, Lallement et al,1991, hereafter LBCS).
During the Roll Maneuvers many UV bright stars crossed the field-of-view and were detected. Thanks to them we were able determine the precise rotation rate, rotation axis direction and the phase by using a systematic search through all rotation parameters and a star catalogue (LBCS). The Lyman-alpha coming from the auxiliary port was taken into account during this exercise, as well as in the RM analysis.

MODEL

Details on the model can be found in Lallement et al (1985 a,b), Lallement & Stewart (1990). The model intensity is computed from the neutral H distribution, assuming optically thin conditions. It is then proportional to n_∞ (interstellar H density) and to the line center solar Lyman-alpha flux Fs (or mu = ratio of radiation pressure over gravitation) while the observed intensity depends on the chosen instrument calibration factor a. In the comparison of computed and measured intensity, n_∞, F_S and a play similar roles and cannot be distinguished from one another. H density distributions were computed for a variety of combinations of solar parameters mu, T_D (= lifetime against ionization at 1 AU, T_D is the inverse of the ionization rate at 1 AU), A_N (solar wind pole/equator anisotropy degree). The solar flux F_S was estimated from the solar radio flux at 10.7 cm and taken into account for the fixed lines of sight data (details are given in Chassefière et al, 1986, Lallement & Stewart, 1990 and LBCS). While the 27-day modulation is of minor importance, the solar cycle variation is very large (50% increase between 1977 and 1982). This solar flux correction was not done for the RM modeling. For a given maneuver, all the measurements are done the same day, and the solar suface inhomogeneity only contributes to scatter the data. As will be seen, the systematic discrepancies from the model exceed largely these small solar line effects.
The Ly-alpha emissivity distribution in a half plane containing the sun-wind axis, is shown in Fig 1 for the case of axial symmetry (isotropic solar ionization) and two different lifetime values. The Lyman-alpha emissivity distribution shows a maximum emissivity region (MER) centered upwind, where the product of the density by the solar flux is maximum. If the ionization is increased by a factor of two, corresponding to a lifetime decrease from $T_D = 2 \times 10^6$ sec (top) down to $T_D = 1 \times 10^6$ sec (bottom), this maximum region moves upwind from 1.1 AU up to 2.6 AU and the maximum brightness decreases drastically. At the same time, the cavity is enlarged. At large solar distances (say, > 20 AU) the density distribution approaches n_∞ and is fairly uniform, therefore does not depend on the solar parameters (mu, T_D, A_N). In contrast, in the inner solar system (say, inside ≈ 10 AU), the H distribution depends

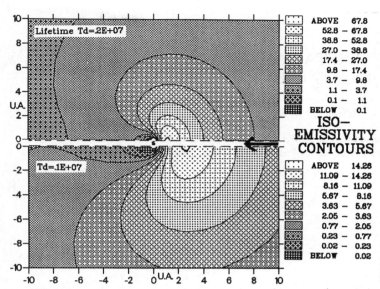

ISO—EMISSIVITY CONTOURS	
ABOVE	67.8
52.8 –	67.8
38.8 –	52.8
27.0 –	38.8
17.4 –	27.0
9.8 –	17.4
3.7 –	9.8
1.1 –	3.7
0.1 –	1.1
BELOW	0.1
ABOVE	14.26
11.09 –	14.26
8.16 –	11.09
5.67 –	8.16
3.63 –	5.67
2.05 –	3.63
0.77 –	2.05
0.23 –	0.77
0.02 –	0.23
BELOW	0.02

Fig. 1: Distribution of H Lyman-alpha emissivity in a half plane containing the wind vector (indicated by an arrow on the right) and the sun for two models with mu = 0.7, T = 8000 K, V= 20 km/s, A_N = 0, and two different lifetimes against ionization. The choice of the unit of emissivity is arbitrary but is the same for the two models. If the lifetime against ionization is decreased by a factor of 2 from 2. 10^6 s (top) to 1. 10^6 s (bottom) at 1 AU (ionization rate multiplied by 2), the region of maximum emissivity (MER) moves further from the sun (from 1.1 AU to 2.6 AU), and becomes dimmer (maximum emissivity divided by 4.5). As a consequence the emissivity gradient along the antisolar direction (between the MER and about 20 AU upwind) becomes much fainter.

sensitively on these solar parameters. The emissivity gradient when moving from the MER towards the upwind antisolar direction is then very dependent on the ionization rate, as can be seen by comparing the two parts of Fig 1. The most important consequence is the relationship between the inferred absolute density far from the sun and the lifetime value used for the modeling, mainly when observing with instruments at small heliocentric distances: as a matter of fact, for the same number of Rayleighs observed from the inner solar system, the use of the model with the smaller lifetime leads to a larger density far from the sun, associated with the weaker emission gradient.

FL DATA/MODEL COMPARISON

At heliocentric distances R<4 AU both V 1 and V 2 measured intensities are well fitted with a model with mu= 0.7, A = 0.40, T_D = 2.10^6 s, n_∞ = 0.06 cm^{-3}, corrected for variations in solar Lyman-alpha and the data/model ratio is independant of the line-of-sight ecliptic longitude, or angle with the wind axis (Fig 2). These last parameters are similar to the values derived from contemporary Prognoz measurements in 1977. If a smaller lifetime is used, the data/model ratio increases significantly when looking towards the downwind side. In fact, given the known solar wind flux ionizing the flow of interstellar H, the best fit value 2 x 10^6s is a fairly large value (Lallement et al,1985). When moving away, this is no longer true: at R= 6 AU, there is a clear modulation of the data/model ratio as a function of the ecliptic longitude of the line-of-sight, with a systematic excess of intensity near 150° longitude, which corresponds to the anti-solar direction (Fig 2). As a conclusion, a model with T_D= 2 x 10^6s fits the distribution of H inside ≈ 5 AU rather well, but not what is seen outside of 5 AU. Indeed, the analysis of the whole set of FL data (Lallement et al, 1990, 1991) shows that decreasing the lifetime T_D, in accordance with solar wind measurements, improves the overall constancy of the data/model ratio for radial distances between 1 to 16 AU, but worsens the fit of data taken in the internal cavity (< 5 AU).

ROLL MANEUVERS (RM)

Each of the RM gives a snapshot of a major fraction of the whole sky at a given time. The problem of the temporal variation of the solar Ly-alpha at line center is then greatly reduced. Here we show the analysis of RM 406 recorded by V 2 when it was at 6.2 AU at ecliptic longitude 158° and ecliptic latitude 2°. In order to estimate the quality of fit of a model, the predicted intensities are first scaled to the "best fit scaling factor" a_1, and a second scaling factor a_2 was computed in such a way that the model is adjusted to the data in the region where the data/model ratio is minimum. Using this factor, a_2, all the differences (data)-(model) have positive values. The relative difference defined as $(I_{data} - a_2 I_{model}) /a_2 I_{model}$ gives an idea of the amplitude of the systematic departures, and is simply derived from the data/model ratio. When using a lifetime of 2. 10^6s, the data are smaller than the model in the maximum intensity region around the sun, and are higher elsewhere. The relative difference reaches 83% in the anti-solar direction and the data/model ratio changes abruptly between the sun and the upwind direction, i.e. around the MER, indicating that the ionization rate is not appropriate. Determining this rate from a limited observed region around the MER is, in our opinion, certainly more reliable than a global fit, which could be biased by secondary effects, e.g. radiative transfer. Fig 3 shows V 2 data and two models around the ecliptic plane. Obviously the model with T_D = 1 x 10^6s gives quite a good fit to the data in all the ionization region, much better than the 2 x 10^6s model. Now if we use T_D = 1 x 10^6s, the shape of the MER is better represented, as shown in Fig 4, but the data now are systematically higher than the model around the downwind direction, by up to 75%. At the same time, all model intensities are decreased, and this is reflected by the best fit scaling factor increase from a_1=4 .5 to a_1=7, and then of the H density. The results are the same for all the maneuvers we have analysed up to now.

As a conclusion, a small lifetime is more appropriate to fit the upwind region and the shape of the maximum emission region, but then the data shows an excess of emission around the downwind direction. Clearly there is a balance between two effects: the downwind large intensity forces the best fit lifetime towards higher values, while the shape of the MER region implies lower values, in agreement with solar wind in situ measurements.

DISCUSSION

What is the nature of the extraneous Lyman-alpha emission which leads inner solar system observations to imply too large a value of T_D ? Possible explanations can be divided in three main types :1) Shortcomings of the model. We have argued that non stationnary effects, certainly present (Fahr & Scherer ,1990), are not the cause of the main type of departures from model which are observed here (LBSC). Elastic collisions with solar wind protons have been proven to have a negligible influence (Chassefière & Bertaux, 1987). The line profile of the integrated disk solar Lyman-alpha line can play a small role, but smaller than what is observed. It is difficult to imagine how stochastic trajectories of the H atoms could produce a filling of the downwind cavity much more efficiently than the smooth trajectories do, since the main effect is the ionization. A serious candidate for the source of discrepancies is the radiative transfer. Keller et al (1981) indicated that the inclusion of radiative transfer increases the intensity in respect to the optically thin approximation, by about 30 % and 10% for downwind and upwind anti-radial directions respectively. There is then a relative downwind increase of about 20%. If radiative transfer is responsible for the downwind enhancement, it means that the excess is at least twice the predicted values. 2) There are perturbations at the heliospheric interface and the flow is inhomogeneous. In this case the relationship between the upwind and downwind velocity distributions can be different from the model (Lallement & Bertaux, 1990). 3) There are some additional sources of H in the solar system like interstellar grains (PAH's molecules ?).

NEUTRAL H DENSITY

Here we discuss the estimate of n_∞ (H) ,the H density far from the sun but still inside the heliopause, maybe smaller than the interstellar density through filtering (Ripken and Fahr, 1983). It is derived from the measured Lyman-alpha intensities by comparison with model intensities, in which n_∞ (H) is a scaling factor. The model intensity will depend crucially of the lifetime T_D, and in order to adjust the model to a given measured intensity, one has to increase n_∞(H) when T_D is decreased (to compensate for the smaller density and emissivity in the MER region). It has to be noticed that, if the downwind intensity increase is due to multiple scattering, at the same time all the intensities are increased. This has the opposite effect to decrease the corresponding derived density (independantly of the lifetime changes). However, this second effect is very probably significantly smaller. If, as we believe now, the lifetime T_D was about 1×10^6 s, the density values derived from Prognoz and Venera 11-12 measurements, which benefit from calibration on stars and on the H corona of Venus respectively (Bertaux et al, 1985, Chassefiere et al,1986), should be multiplied by ≈ 2, yielding n_∞ (H) ≈ 0.10 and 0.13 cm^{-3}. Pioneer Venus measurements by Ajello et al (1987) yeald 0.07 cm^{-3} for T_D=1.5 x 10^6 s, corresponding to 0.11 cm^{-3} when using 1.0 x 10^6 s. These three values 0.1-0.11-0.13 are similar and also correspond to the 0.12 cm^{-3} derived by Shemansky et al (1984) independently of any calibration factor. Now if we refer to the present Voyager calibration factor, n_∞(H) is estimated from the present measurements at 0.3 - 0.4 cm^{-3}, a factor of $\approx$ 3- 4 larger than the previous determinations. Clearly, these discrepant estimates of n_∞(H) are related to different calibration factors of the various instruments.

In principle, the ratio n_∞(H)/n_∞(He) can be used to infer a maximum value for the ionization degree of the local interstellar medium, assuming there is no filtering effect of the heliopause on the neutral He, and helium is not ionized. If the ratio is about 10, equal to the cosmic abundance ratio, then H+ is a minor species, the medium is neutral, and there is no filtering. In a recent analysis, n_∞(He) was found to be 8 ± 4 x 10^{-3} cm^{-3} (Chassefière et al, 1988) yielding n_∞(H)/n_∞(He) = 8.1 for the previous estimate n_∞(H)= 0.065 cm^{-3}. This would imply 20% of H filtering or ionization. This is in agreement with the minimum value derived by Reynolds (1989). Now, with the revised value of n_∞(H) = 0.13 the ratio becomes 16 (+14, -5), and $\approx$ 37 for the Voyager intensity estimate of n_∞(H) = 0.3. The two last estimates exceed the cosmic abundance ratio even without taking into account any possible reduction at the heliopause by filtering of neutral H. This is not totally impossible: it requires however that He be partially ionized in the LISM. Indeed, recent models of the physical state of the ISM show that around warm/hot gas interfaces helium is more strongly ionized than hydrogen, due to enhanced soft X-rays radiation from the interface (Slavin, 1990). Since UV, soft X-rays and optical observations of the local ISM indicate that the sun lies in the vicinity of such an interface (Cox & Reynolds, 1987, Frisch & York, 1986, Lallement et al, 1990), the local interstellar medium could be in this particular physical state.

CONCLUSION

A series of Voyager Lyman-alpha data sets has been analyzed with a standard model of the interstellar H distribution in the solar system and the resulting emission. The lifetime of an H atom at 1 AU has been determined at 1×10^6 s ($\pm$ 20 %) from the shape of the Maximum Emissivity Region. It is clear that inner solar system observations, from Voyager, Prognoz and Pioneer Venus cannot be reconciled with this short lifetime, calling for an additional source of emission from the downwind region. In the absence of detailed calculations of multiple scattering for geometries relevant to Voyager observations, it is not possible to rule out the possibility that multiple scattering is the source of all discrepancies between the standard model (with the optically thin approximation) and observations. However, alternate possibilities as heliopause effects, or additional sources of H in the solar system, mainly in the downwind region, as for example outgassing of cometary grains, also deserve further examination. As a direct consequence of this shorter lifetime, the value of the H density far away from the sun, but still inside the heliopause, has to be revised upward to around around 0.1-0.13 cm^{-3} with the Prognoz calibration factor, $\approx$ 0.3 cm^{-3} with the Voyager calibration factor.

References
Ajello J. M., Stewart A. I., Thomas G. E., Graps A.: Astrophys. J., **317**, 964 (1987)
Chassefière E., Bertaux J.L., Lallement R., Kurt V.G.: Astron. Astrophys. **160**, 229 (1986)
Chassefière E., Bertaux J.L., Astron. Astrophys. **174**, 239 (1987)
Chassefière E. , Bertaux J.L., Lallement R., Sandel B.R., Broadfoot L.: Astron. Astrophys., **199**, 304 (1988)
Cox D.P., Reynolds R.J.: A.R.A.A., **25**, 303 (1987)
Fahr H.J. , Scherer K.: Astron. Astrophys. **232**, 556 (1990)

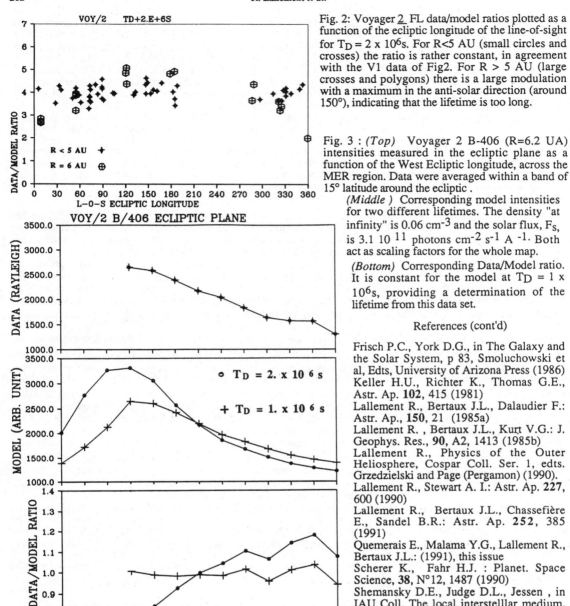

Fig. 2: Voyager 2 FL data/model ratios plotted as a function of the ecliptic longitude of the line-of-sight for $T_D = 2 \times 10^6$s. For R<5 AU (small circles and crosses) the ratio is rather constant, in agreement with the V1 data of Fig2. For R > 5 AU (large crosses and polygons) there is a large modulation with a maximum in the anti-solar direction (around 150°), indicating that the lifetime is too long.

Fig. 3 : *(Top)* Voyager 2 B-406 (R=6.2 UA) intensities measured in the ecliptic plane as a function of the West Ecliptic longitude, across the MER region. Data were averaged within a band of 15° latitude around the ecliptic .

(Middle) Corresponding model intensities for two different lifetimes. The density "at infinity" is 0.06 cm^{-3} and the solar flux, F_s, is 3.1 10^{11} photons cm^{-2} s^{-1} A $^{-1}$. Both act as scaling factors for the whole map.

(Bottom) Corresponding Data/Model ratio. It is constant for the model at $T_D = 1 \times 10^6$s, providing a determination of the lifetime from this data set.

References (cont'd)

Frisch P.C., York D.G., in The Galaxy and the Solar System, p 83, Smoluchowski et al, Edts, University of Arizona Press (1986)
Keller H.U., Richter K., Thomas G.E., Astr. Ap. **102**, 415 (1981)
Lallement R., Bertaux J.L., Dalaudier F.: Astr. Ap., **150**, 21 (1985a)
Lallement R. , Bertaux J.L., Kurt V.G.: J. Geophys. Res., **90**, A2, 1413 (1985b)
Lallement R., Physics of the Outer Heliosphere, Cospar Coll. Ser. 1, edts. Grzedzielski and Page (Pergamon) (1990).
Lallement R., Stewart A. I.: Astr. Ap. **227**, 600 (1990)
Lallement R., Bertaux J.L., Chassefière E., Sandel B.R.: Astr. Ap. **252**, 385 (1991)
Quemerais E., Malama Y.G., Lallement R., Bertaux J.L.: (1991), this issue
Scherer K., Fahr H.J. : Planet. Space Science, **38**, N°12, 1487 (1990)
Shemansky D.E., Judge D.L., Jessen , in IAU Coll. The local interstelllar medium, NASA CP 2345 (1984)
Slavin J. D. : Ap. J., **346**, 718 (1989)

Fig. 4: B-406 relative difference map. Relative differences with a model at $T_D = 1 \times 10^6$s. The model values have been multiplied by the scaling factor a2 indicated in the figure.The MER is well reproduced, but the observed downwind intensity is much larger than expected from the model.

DETECTION OF THE CORONAL PLASMA SHEET ON THE SOLAR DISK

P. Lantos* and C. E. Alissandrakis**

*Observatoire de Paris, F-92195 Meudon, France
**National University of Athens, GR-15783 Athens, Greece

ABSTRACT

The large scale coronal neutral sheet, which is considered to be the source of low speed solar wind, is the low altitude counterpart of the interplanetary heliosheet. It is routinely detected with white light corona observations at the limb. Maps of the sun at meter wavelengths, obtained with the Nançay Radioheliograph, lead to the first detection of this density structure on the disk, in the form of a "plateau" in brightness temperature. The radio observations also show the denser parts of the coronal neutral sheet as discrete emission sources. Some of them may be identified as coronal streamers.

INTRODUCTION

The large scale structure of the solar wind in the inner heliosphere is dominated by the effects of coronal structures, namely the coronal holes and the streamer belt. The streamer belt, also called coronal plasma sheet, is a region of high density which oscillates around the solar equator. It is well observed at the limb with white light coronographs. Its structure derives from the simplification of the large scale solar magnetic fields with altitude (Pneuman et al., 1978, Hoeksema, 1984). As the coronal plasma sheet has only been observed at the limb, with integration along the line of sight, the resolution in longitude has been very limited, which makes difficult the comparison with underlying photospheric and chromospheric structures.

RADIO OBSERVATIONS

Observations with the Nançay Radioheliograph provide daily maps of the corona using earth rotation aperture synthesis, when the sun is sufficiently quiet. The maps are computed on the basis of about 8 hr of one dimensional observations by the East-West (3200 m long) and the North-South (1195 m long) arrays of the Radioheliograph. The instrument operates at five frequencies in the range of 450-150 MHz (wavelength range λ=66 cm to λ=2 m). The flux calibration in done by comparison with the radio source Cygnus A. The instrument, the aperture synthesis method and previous results have been described in Alissandrakis et al. (1985), Lantos et al. (1987) and Lantos et al. (1991).

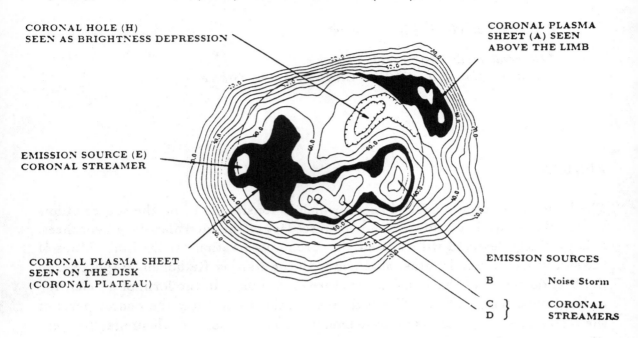

Figure 1: Map of the Sun on July 15, 1984. Isophots in brightness temperature are labelled in 10^5 K. Letters refer to text and to Figure 2.

The map shown in Figure 1 was obtained on July 15, 1984 at λ=1.77 m (169 MHz). The resolution is 1.3 arc min in E-W and 4.8 arc min in N-S. Isophotes are brightness temperature levels in intervals of 50 000 K. The lowest isophote corresponds to 200 000 K. Hatched contours show regions of brightness temperature lower than the background. The circle corresponds to the photospheric limb.

RESULTS

At meter wavelengths, the corona is brighter in regions of higher electron density. The dense coronal plasma sheet is visible above the limb in the north-west quadrant, corresponding to the location (a) on the synoptic white light map (Figure 2) from Mauna Loa Solar Observatory (Sime et al., 1984). Its maximum brightness temperature is 5.5 10^5 K. The main feature in the northern hemisphere is the deep brightness depression (h) associated to a large coronal hole. The brightness temperature in this region is 10^5 K below the surrounding background. The coronal plasma sheet is seen on the disk as a bright *"coronal plateau"*, drawn in black on Figure 1 and corresponding to 6.5- 7 10⁵ K in brightness temperature. It surrounds localized emission sources c, d and e which are identified as coronal streamers. Their brightness temperature ranges from 7 10⁵ K for the source e close to the limb to 8 10⁵ K for the sources near the solar meridian. Source b is of non-thermal origin. It is a faint solar noise storm continuum with a brightness temperature of about 8 10⁵ K. Unlike the other emission sources, source b is close to an active region. Comparison of the radio map with the K corona synoptic map (figure 2) shows a very good agreement of the large scale structure of the coronal plateau with the coronal plasma sheet observed in white light corona.

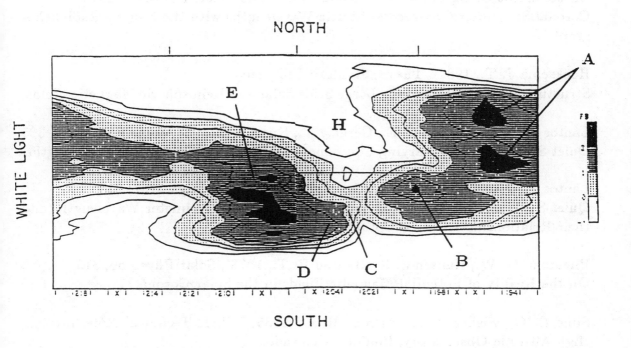

Figure 2: Synoptic map of the white light corona at 1.3 solar radius from Mauna Loa Solar Observatory

CONCLUSION

During 1984, the central part of the dense plasma sheet oscillates in latitude from about 50° North to 50° South. The phase of the solar cycle here is similar to that of the Skylab period in 1973. Thanks to the large oscillation of the coronal plasma sheet, the present result confirms those obtained by Lantos et al. (1991) during the period of cycle minimum with a coronal plasma sheet almost equatorial.

Radio observations at meter wavelengths provide a new tool to study the structure of the coronal plasma sheet. The identification of dense streamers on the disk permits a detailled comparison with the underlying photospheric and chromospheric structures. Daily observations open the possibility of study of temporal evolution over extended periods. Finally, future diagnostics in temperature and density will permit a better knowledge of the regions thought to be the source of slow solar winds.

REFERENCES

Alissandrakis C. E., Lantos P., Nicolaidis E., 1985, Solar Phys., 97, 267
Coronal Structures Observed at Metric Wavelengths with the Nançay Radioheliograph.

Hoeksema J. T., 1984 , Thesis, Stanford University
Structure and Evolution of the Large Scale Solar and Heliospheric Magnetic Fields.

Lantos P., Alissandrakis C. E., Gergely T., Kundu M., 1987, Solar Phys., 112, 325
Quiet Sun and Slowly Varying Component at Meter and Decameter Wavelengths.

Lantos P., Alissandrakis C. E., Rigaud D., 1992, Solar Phys., 137, 225
Quiet Sun Emission and Local Sources at Meter and Decimeter Wavelengths and their Relationship with the Coronal Neutral Sheet.

Pneuman G. W., Hansen S. F., Hansen R. T., 1978 , Solar Phys., 59, 313
On the Reality of Potential Magnetic Fields in the Solar Corona.

Sime D. G., Fisher R. R., Altrock R. C., 1985, NCAR Technical Note TN-251, High Altitude Observatory, Boulder, Colorado.
Solar Coronal White Light, Fe X, Fe XIV and Ca XV Observations During 1984.

INVESTIGATION OF THE SOLAR WIND TRANSONIC REGION

N. A. Lotova

IZMIRAN, Troitsk, Moscow Region 142092, Russia

ABSTRACT

The results of occultation studies of the interplanetary medium near the Sun are presented. Both the subsonic and supersonic regions of the solar wind were involved. Water vapor masers and quasars were used as sources. Consistent with numerous previous experiments, the existence of an enhanced scattering region stretching from about 10 to 30 solar radii is reported. This region was identified as that of sub- to supersonic transition in the solar wind flow. It may be treated as a region of mixed flow where subsonic and supersonic streams coexist and interact. A peculiarity in the structure of the transonic region was found: the wide region of enhanced scattering is preceded by a comparatively narrow zone of diminished scattering, a "predecessor zone". This zone coincides with the local minimum of plasma density and maximum of positive velocity gradients. It is shown that the stream structure of the solar wind is one of the principal factors in the formation of the extended transonic region.

INTRODUCTION

The suspected existence of radial plasma streams emerging from the Sun was verified in 1955 - 1960 /1-4/. These discoveries were accomplished using several independent experimental techniques. A widely accepted theoretical explanation of this solar wind phenomenon was given by Parker /5/. The solar wind starts as a subsonic flow at the coronal base, but very little is known about its transition to supersonic flow. It is shown here that this transition most likely occurs in the radial range from about 10 to 40 solar radii. This region has thus far been inaccessible to optical remote sensing and direct spacecraft observations. The only available technique at present is the radio occultation method.

The present state of solar wind theory reflects essentially the results of the experimental investigations. The Parker theory and most of the most recent theories produce simplified models of the radial, spherically symmetric outflow of the coronal matter. These theories do not predict the existence of peculiarities in the solar wind at its transition from subsonic to supersonic flow. The results presented here would appear to require modifications to the theory at solar distances between 10 and 40 solar radii.

RADIO OCCULTATION OBSERVATIONS

Recent occultation experiments, using both spacecraft and natural sources, provide a great amount of new data for the region R < 50 solar radii /6/. These measurements can be used to derive two physical parameters:

(1) the electron density, both the mean and its rms fluctuation,
(2) the solar wind velocity, again as a mean and its rms fluctuation.

The variation of the radio signal bandwidth with the radial distance of the line-of-sight, a result of spacecraft occultation experiments with Venera 10 /7/, shows a distinct step-like enhancement at elongations ε = 2.5° to 4.5° (radial distances of 10 to 18 solar radii).

Further evidence is provided by the radial dependence of the scintillation index m, defined by m = $<\delta I^2>^{1/2}/<I>$, where $<I>$ and $<\delta I^2>^{1/2}$ are the radio signal's intensity and its rms fluctuation, respectively. Enhanced radio wave scattering, reflected by unusually large values of m, is again revealed at radial distances of 10 to 25 solar radii /8,9/. These experiments /8/ were performed at 1.35 cm wavelength by the traditional interplanetary scintillation (IPS) method, but using water vapor masers instead of quasars.

A new series of coronal occultation experiments /10/ was performed in 1986-1987 at the wavelength λ = 2.9 m using quasars (strong scattering regime). The measurable quantity in this case is the apparent source size, also called the scattering angle θ. The radial dependence of θ reveals an anomalous enhancement of the scattering at radial distances from about 16 to 30 solar radii (Figure 1) in accordance with the above results.

Numerous radio occultation experiments using water vapor masers /8,11/, quasars /10,12-16/ and spacecraft /7-9,24/ reproduce the same features in independent occultation studies: the existence of an extended region of enhanced electron density fluctuations in the area between $\simeq$ 10-20 solar radii. The location and extent of this region and the level of turbulence change from one observation period to another, but the region's existence is a reproducible feature in all previous experiments.

Additional information can be derived from the spatial spectra of electron density fluctuations /17-20/ - the characteristic variation of the spectral index of the spatial turbulence with radial distance. The spectral data imply a smooth average change - an increase of the spectral index from 2.5 - 3.0 to 3.5 - 4.0 in the flow regime in the transonic region, in agreement with the observations of m and θ.

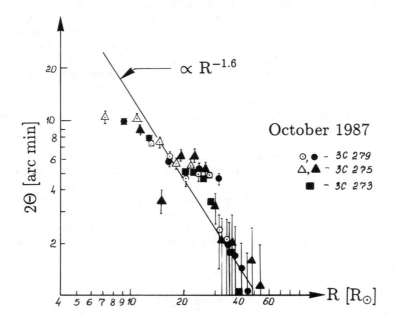

Fig. 1.   Scattering angle 2θ(R) for 3 quasars in October 1987

Still other data consistent with the above may be derived from dispersion measurement studies. The coronal electron content along the line of sight, derived from occultation observations of the Crab Nebula pulsar NPO531 /21/, indicates an electron density enhancement at radial distances ranging from 10 to 25 solar radii. Spacecraft radio group time delays /22/ also reveal these anomalies of the electron concentration.

Solar wind velocity measurements are essential for understanding the mechanism of supersonic solar wind formation. The mean line-of-sight solar wind velocity measurements from spacecraft experiments /7,18,23-26/ imply that the enhanced radio wave scattering at radial distances from 10 to 25 solar radii may be directly related to the transition from subsonic to supersonic solar wind flow. These observations indicate an abrupt rise of the mean velocity, coinciding in space with the region of enhanced radio wave scattering. Both theoretical predictions and the spacecraft experiments confirm the existence of a narrow local minimum in the solar wind velocity v(R) just before the main acceleration of the solar wind /26/. The position of this minimum is located at distances between $\simeq$ 10-20 solar radii and coincides with the sound velocity. The appreciable nonuniformity of the flow at radial distances ranging from 10 to 25 solar radii /24/ is of prime importance for understanding the physics of the transonic region. These velocity data provide compelling evidence of the simultaneous existence of both subsonic and supersonic streams along the line-of-sight. Moreover, they imply a correlation between the observed enhanced turbulence of the flow and the nonuniformity of the velocity field found in this mixed flow regime. The mixed flow regime may be characterized as a region in which subsonic and supersonic streams coexist and interact /6/.

Analyses of a numerous series of occultation experiments reveal in most cases the same features - a comparatively narrow region with diminished scattering preceding a broad zone with enhanced scattering. A comparison of radial velocity profiles from Venera-15 and -16 with the position of the diminished scattering region /18,27/ shows that diminished scattering corresponds to the largest positive gradients in the velocity (i.e. region of greatest solar wind acceleration). The scintillation index m(R) and scattering angle θ(R) have the same radial profiles as the mean electron density /21/. The position of the diminished scattering region and its overlying broad zone of enhanced scattering are subject to broad variations. A distinct correlation exists between the geometry of the transonic region and the solar wind velocity at large distances from the Sun /8/. The velocity data may be considered as a simple descriptive parameter of the solar activity state.

One immediate result of this study of the solar wind's formation is a determination of the radial distance and width of the transonic region. The data now available show that the locations of its inner and outer boundaries are influenced by changes in the solar activity /10/ and can vary considerably.

SUMMARY

1. Numerous radio occultation studies of the solar wind flow close to the Sun reveal an extended region of enhanced scattering, thereby implying an enhanced nonuniformity of the medium at radial distances $\simeq$ 10 - 30 solar radii from the Sun.

2. These complex investigations enable the association of the phenomenon of enhanced scattering with the mechanism of the sub-to-supersonic transition in the solar wind flow.

3. Detailed study of the radial dependence of scattering shows that the extended region of enhanced scattering is preceded by a comparatively

narrow region of diminished scattering that coincides with the largest positive velocity gradients /9,26,27/.

4. The spatial localization of the transonic region reveals both short-term and long-term variations in the radial distance of the inner and outer boundaries of ≃ 10-20 and 23-45 solar radii, respectively. The long-term changes show a distinct correlation with the general solar activity variations over the course of the 11-year solar cycle /8-16,24,26,27/.

5. The large-scale structure of the plasma flow in the transonic region is associated with the stream structure of the white-light solar corona observed close to the Sun /11/.

REFERENCES

1. S.K. Vsechsvytsky, G.M. Nikolsky et al., Astronomicheski J. (in Russian) 32, 165 (1955)
2. V.V. Vitkevich, Soviet Physics Doklady (in Russian) 101, 429 (1955)
3. K.I. Gringaus, V.V. Besrukich et al., Soviet Physics Doklady (in Russian) 131, 1301 (1960)
4. A. Hewish, Proc. Roy Soc. A228, 238 (1955)
5. E.N. Parker, Ap. J. 128, 664 (1958).
6. N.A. Lotova, Solar Phys. 117, 399 (1988)
7. M.A. Kolosov, O.I. Yakovlev, A.I. Efimov et al., Radio Sci. 17, 664 (1982)
8. N.A. Lotova, D.F. Blums, K.V. Vladimirskii, Astron. Astrophys. 150, 266 (1985)
9. S.N. Rubtsov, O.I. Yakovlev, A.I. Efimov, Kosmichsekie Issledovania (in Russian) 25, 620 (1987)
10. N.A. Lotova, Yu.I. Alekseev, Y.V. Nagelys, Kinematika i Physika Nebesnich Tel (in Russian) 3, 70 (1987)
11. N.A. Lotova, M.K. Bird et al., Astronomicheski. J. (in Russian), in press (1992)
12. N.A. Lotova and Y.V. Nagelys, Solar Phys. 117, 407 (1988)
13. N.A. Lotova, in Physics of the Outer Heliosphere, eds. S. Grzedzielski and D.E. Page, Pergamon Press, Oxford 1990, p. 397
14. N.A. Lotova, O.P. Medvedeva et al., Astronomicheski J. (in Russian) 66, 648 (1989)
15. N.A. Lotova, V.I. Kostromin et al., Kinematica i Physica Nebesnich Tel (in Russian) 5, 59 (1989)
16. N.A. Lotova, O.A. Korelov et al., Geomagnetizm i Aeronomy (in Russian) 31, 223 (1991)
17. O.I. Yakovlev, A.I. Efimov, V.M. Ruzmanov et al., Astronomicheski J. (in Russian) 57, 790 (1980)
18. O.I. Yakovlev, A.I. Efimov, S.N. Rubtsov, Kosmichsekie Issledovania (in Russian) 25, 251 (1987)
19. G.L. Tyler, J.F. Vesecky, M.A. Plume et al., Ap. J. 249, 318 (1981)
20. R. Woo, J.W. Armstrong, J. Geophys. Res. 84, 7288 (1979)
21. J.M. Weisberg, J.M. Rankin, R.R. Payne et al., Ap. J. 209, 252 (1976)
22. D.O. Muhleman and J.D. Anderson, Ap. J. 247, 1093 (1981)
23. N.A. Armand, A.I. Efimov, O.I. Yakovlev, Astron. Astrophys. 183, 135 (1987)
24. J.W. Armstrong and R. Woo, Astron. Astrophys. 103, 415 (1981)
25. S.L. Scott, W.A. Coles, G. Bourgois, Astron. Astrophys. 123, 207 (1983)
26. A.I. Efimov, I.V. Chashei, V.I. Shishov, O.I. Yakovlev, Kosmicheskie Issledovania (in Russian) 28, 581 (1990)
27. N.A. Savich, S.L. Azarch et al., Kosmicheskie Issledovania (in Russian) 25, 243 (1987)

THE LARGE-SCALE CONFIGURATION OF THE NEAR-SOLAR PLASMA

N. A. Lotova and O. A. Korelov

IZMIRAN, Troitsk, Moscow Region 142092, Russia

ABSTRACT

Results of occultation studies of the near Sun interplanetary medium are presented. The sub- to supersonic transition of the solar wind flow is studied by using several natural radio sounding sources simultaneously. The observations were carried out in June–July 1988 and in June–August 1989 at a wavelength λ = 2.9 m. Inner and outer boundaries of the sub- to supersonic transition region were determined from the radial dependence of the radio wave full scattering angle 2θ. Combining data from many sources at different heliolatitudes, a two-dimensional mapping of the solar wind transonic region was obtained.

INTRODUCTION

The main acceleration of the near-solar plasma proceeds in the transition region of the solar wind at radial distances of 10-30 solar radii. Almost radial supersonic flow is formed outside this region . This transition or transonic region is revealed in radio sounding experiments using both natural and man-made sources /1-4/. Due to the pronounced nonuniformity in the stream structure of the near-solar plasma, some specific mixed flow occurs there, which may be characterized as a coexistence and interaction of both sub- and supersonic streams. The intrinsic instability of the mixed flows causes significant peculiarities of the local solar wind parameters, in particular an excessive level of turbulence and enhanced radio wave scattering. This is accompanied by characteristic changes of turbulence spectra shape, velocity dispersion, and mean electron concentration. Parameters of the transonic region are quite variable with time. The rapid changes observed may be associated with large-scale processes on the Sun /5,6/. Results obtained over several years of observation demonstrate that pronounced slow systematic changes exist as well, which correlate with the slow systematic changes in the Sun over the course of the 11-year cycle /5-14/. Detailed investigation of these processes seems to be a vital part of the solar wind study.

OBSERVATIONS

The series of regular occultation observations reported here were started in 1985 /15/. Since IPS observations are in the strong scattering regime, the apparent source size was measured in this new series. New information was obtained about the transonic region geometry and its evolution in time. Until recently, only one or two sources were used in a given series of observations, thereby limiting the range of the transonic region geometry studies. It is not possible to differentiate between heliolatitude, radial and temporal changes from single source observations. Heliolatitude differences are of prime importance here. The existing data do not exclude a pronounced asymmetry of the transonic region, well correlated with that of the solar corona. Recognizing this, new experiments were performed, which better exploit the

advantages of the radioastronomical approach. Five sources that pass near the
Sun at almost the same time were used in 1988: the four quasars 3C138, 3C154,
3C165, 3C166, and the very strong compact source 3C144. The multi-source
observations may be used to perform a near real-time mapping of the circumsolar plasma. Even nine sources could be observed for the 1989 opportunity.

Using five sources it was possible to probe both the northern and southern
solar hemispheres over the one month time interval from 4 June until 2 July
1988. Nine sources were used in June - August 1989. The DKR-1000 radio telescope of the P.N. Lebedev Institute (Pushino) was used in a radio-
interferometer mode. Two halves of the EW-line were used as the interferometer antennas with an effective baseline of 476 m. For 3C144 the interferometer was formed by two sections of the same line with 910 and 497 m bases.
Daily measurements of the apparent source size, (full scattering angle 2θ)
were recorded. Both observational and processing methods are given in /5,15/.

The angular broadening data $2\theta(R)$ were taken using a one-dimensional array
with two baseline lengths. One must therefore be careful when intepreting
scattering variations by accounting for the anisotropic structure of inhomogeneities closer to the Sun /10,16-17/. The enhanced scattering is usually
preceded by a narrow region with diminished scattering /18/. This effect
arises from the solar wind transition process near the critical point /18/.
This process is accompanied by an enhancement of the mean electron density
/18/, which cannot be explained by scattering anisotropy. The broad solar
wind transonic region, which displays enhanced scattering properties and an
enhanced mean electron density, contains the critical point and can extend
typically from 10 out to 30 solar radii.

MAPPING THE TRANSONIC REGION

The radial dependence of the scattering angle $2\theta(R)$ was used to determine the
inner and outer boundaries of the enhanced scattering area, which may be
identified as inner R_{in} and outer R_{out} boundaries of the transonic region. An
asymptotic form of the radial dependence $2\theta \simeq R^{-1.6}$ /15/ was used in determining the boundary location, which is valid outside the transonic region
both in the subsonic flow area and in the area of well formed supersonic
solar wind /15/. Values of the R_{in} and R_{out}, as well as the heliolatitudes of
the sources fixed at the moment of the boundary passing, are presented in
Figure 1. Figure 1 shows two-dimensional mappings of the transonic region for
each year. The cross-hatched areas schematically illustrate the geometry of
the transonic region as a whole, as derived from our data. Figure 1 should
not be misinterpreted as a "snapshot" of the transonic region. The measurements lasted a full month in 1988 and three months in 1989. Solar rotation
and other time-dependent processes may considerably influence the results.
Nevertheless, the results imply an exceedingly strong asymmetry of the
transonic region structure, which cannot be masked by the temporal effects.
The most interesting results were obtained with 3C144 and 3C166 in 1988, for
which the trajectories pass very close to each another within a time interval
of two weeks. The 3C144 observations provided a reliable fix on the internal
boundary position very close to its absolute minimum. Two weeks later, the
3C166 observations revealed the same boundary at a distance twice as large.
This observation demonstrates that very powerful large-scale plasma streams
exist in the region of sub- to supersonic transition.

Another approach to characterizing the flow anisotropy is illustrated by
Figure 2, showing scattering level contours in the subsonic region for fixed
values of the scattering angle. It may be seen that the solar supercorona is
very asymmetric. Large-scale structure is apparent in the plasma flow. A
comparative analysis of both the transonic region and the that of supercorona

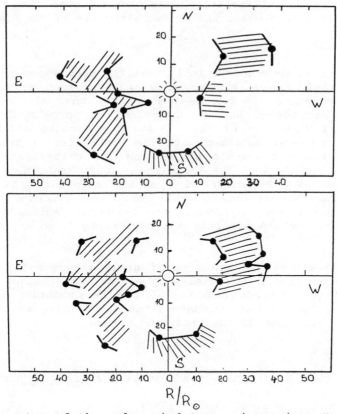

Fig. 1. The geometry of the solar wind transonic region. Upper panel: June-July 1988; Lower panel: June-August 1989.

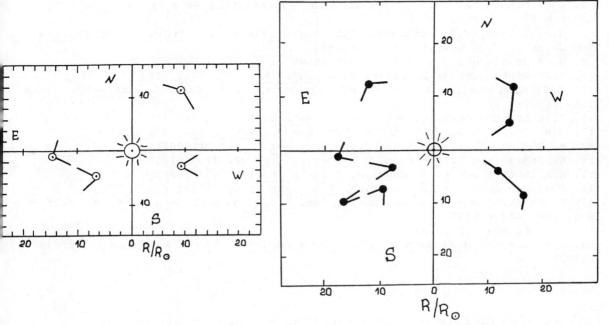

Fig. 2. Selected contours in the subsonic region for a fixed value of the scattering angle. Left panel: contour for θ = 5.0 arc min (1988); Right panel: contour for θ = 8.0 arc min (1989).

reveals their intimate interrelation, and demonstrates the existence of a common mechanism that determines the stream structure of the flow /19/.

CONCLUSIONS

The results summarized in Figure 1 imply that the pronounced asymmetry of the flow in the transonic region of the solar wind is most probably the rule, rather than an exception. It has been shown /6/ that the geometry of the solar wind transonic region and its location are correlated with velocity measured at radial distances of ca. 0.5 AU (i.e. region of the supersonic flow). This circumstance, together with the correlation between the large-scale structure of the transonic region and the white-light solar corona /19/, lead us to conclude that the large-scale configuration of the transonic region is governed by the solar wind stream structure. It may also be concluded that the observations provide immediate evidence of the stream structure of the solar wind flow within the transonic region, similar to that established earlier for the subsonic and supersonic flows on either side of this transition.

For future investigations, it would be of great interest to conduct radio-astronomical observations in parallel with optical observations of the solar corona. It would thus be possible to follow the evolution of large-scale events originating close to the Sun and later giving rise to the large-scale nonuniformities of the flow in the transonic region.

REFERENCES

1. N.A. Lotova, D.F. Blums, K.V. Vladimirskii, Astron. Astrophys. 150, 266 (1985)
2. N.A. Lotova, Solar Phys. 117, 399 (1988)
3. M.A. Kolosov, O.I. Yakovlev, A.I. Efimov et al., Radiotechnika i Electronika (in Russian), 23, 1829 (1978)
4. O.I. Yakovlev, A.I. Efimov, and S.N. Rubtsov, Astronomicheski J. (in Russian) 65, 1290 (1988)
5. N.A. Lotova, A.A. Rashkovetsky, P.B. Kazemirsky et al., Astronomicheksi J. (in Russian) 66, 114, (1989)
6. N.A. Lotova, K.V. Vladimirskii, A.A. Rashkovetsky et al., Geomagnetizm i Aeronomiiya (in Russian) 28, 722 (1988)
7. N.A. Lotova, V.I. Kostromin, P.B. Kazemirsky et al, Kinematika i Physika Nebesnich Tel (in Russian) 5, 59 (1989)
8. G. Bourgois and W.A. Coles, this issue
9. B.J. Rickett and W.A. Coles, J. Geophys. Res. 96, 1717 (1991)
10. J.W. Armstrong, W.A. Coles, M. Kojima, and B.J. Rickett, Astrophys. J. 358, 685 (1990)
11. J.W. Armstrong and R. Woo, Astron. Astrophys. 103, 415 (1981)
12. W.A. Coles, B.J. Rickett, and V.H. Rumsey, Nature 286, 239 (1980)
13. M. Kojima and T. Kakinuma, Space Sci. Rev. 53, 73 (1990)
14. M. Kojima and T. Kakinuma, J. Geophys. Res. 92, 7269 (1987)
15. N.A. Lotova and Y. V. Nagelys, Solar Phys. 117, 407 (1988)
16. P.A. Dennison and R.G. Blesing, Proc. ASA 2, 86 (1972)
17. R. Narayan, K.R. Anantharamaiah, and T.J. Cornwall, Mon. Not. R. Astr. Soc. 241, 403 (1989)
18. N.A. Lotova, this issue
19. N.A. Lotova, M.K. Bird et al., Astronomicheski J. (in Russian), in press (1992)

DISCONNECTION OF OPEN CORONAL MAGNETIC STRUCTURES

D. J. McComas,* J. L. Phillips,* A. J. Hundhausen** and J. T. Burkepile**

* Space Plasma Physics Group, Los Alamos National Laboratory, Los Alamos, NM 87545, U.S.A.
** High Altitude Observatory, Boulder, CO, U.S.A.

ABSTRACT

We have examined the Solar Maximum Mission coronagraph/polarimeter observations for evidence of magnetic disconnection of previously open magnetic structures and a number of likely examples have been found. Probable coronal disconnections typically appear as pinching off of helmet streamers followed by the release and outward acceleration of large U or V-shaped structures. The observed sequence of events is consistent with reconnection across the heliospheric current sheet between previously open magnetic field regions, and the creation of a detached magnetic structures which is open to interplanetary space at both ends. Sunward of the reconnection point, coronal disconnection events would return previously open magnetic flux to the Sun as closed field arches. Here we 1) describe one clear disconnection event on 1 June 1989; 2) examine the results of a limited survey of disconnection events; and 3) discuss the potential importance of coronal disconnections for maintaining open magnetic flux in interplanetary space.

INTRODUCTION

Coronal structures and transients are commonly observed in coronagraph observations of the Sun. Such observations measure the line of sight integrated Thomson scattered white light which is proportional to electron density. Owing to the high electrical conductivity of the coronal plasma, bright structures in coronagraph observations are generally interpreted as magnetic structures populated with higher density plasma. The typical coronal structure is characterized by a combination of "open" coronal holes which have roughly radial magnetic fields extending out into interplanetary space and low coronal densities and "closed" arch and loop structures which have higher coronal densities. Helmet streamers are regions where oppositely directed open fields are juxtaposed above closed field arches. These streamers map out into the heliospheric current sheet /1/. In addition to long lasting open and closed structures, the solar corona is routinely disturbed by coronal transient events. The most notable subset of coronal transients are called coronal mass ejections or CMEs /2,3/. Classical CMEs, which display loop-like structures rising out of the solar corona in sequences of coronagraph pictures, are generally interpreted as the eruption of previously closed field regions on the Sun. While there are other definitions of the term CME, we will use it to mean these eruptions of new flux and plasma in contrast to a broader range of coronal transients.

Most CMEs appear to leave long lasting "legs" which slowly fade into coronal holes. As such, CMEs appear to open previously magnetically closed regions of the Sun into long lasting open field regions. One long standing question of coronal physics is that if CMEs do open new magnetic flux into interplanetary space, why doesn't the amount of open flux, and hence interplanetary magnetic field (IMF) magnitude, grow without bound? The general question of the maintenance and evolution of open magnetic flux in interplanetary space has been discussed previously /4/ and is examined elsewhere in this proceedings /5/.

One important aspect of the maintenance of the solar magnetic flux in interplanetary space may be that at least some of the flux is returning to the Sun elsewhere in the corona; helmet streamers provide an ideal location for reconnection in the corona which would lead to the return of magnetic flux to the Sun /4,5/. Such reconnection would also lead to the release of "U" or "V" shaped structures, open to the outer heliosphere at both ends. One sequence of Solar Maximum Mission (SMM) coronagraph observations was recently published showing a U-shaped structure being detached and released from a helmet streamer on 27 June 1988 /4/. The primary purpose of this study is to extend the examination of the SMM coronagraph data for further evidence of such reconnection and the release of U-shaped magnetic structures.

OBSERVATIONS

Figure 1 displays a sequence of four white light chronograph pictures for a coronal transient event which occurred on June 1, 1989. The 6:32 UT panel shows a coronal streamer (straight up in this figure). By 11:10 UT a clear separation is forming at approximately two solar radii (arrow) between lower, sunward pointing arches and apparently disconnected U-shaped magnetic flux. This separation continues to develop (14:16 UT panel), and finally, by 17:05, the bulk of the outward pointing flux has risen, leaving a bundle of closed field arches tied to the Sun in its wake. It is worth noting that the sequence of images displayed here is only a small subset of those taken over this day; the other images show various intermediate configurations and a rapid stepping through of the images (as was shown at the Solar Wind 7 meeting) displays the disconnection event far better than the four panels of Figure 1.

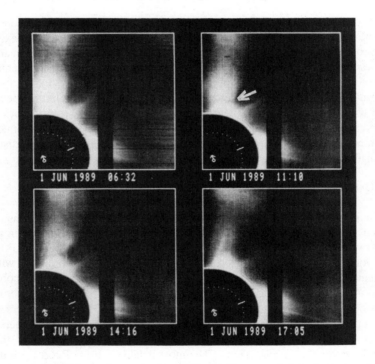

Fig. 1. Sequence of SMM coronagraph observations of a coronal detachment on 1 June 1989.

Differenced images of the central interval of this event have also been made using a pre-event image from 9:38 UT. Four of these differenced images are displayed in Figure 2. Lighter gray regions indicate an increase in electron density compared to the preevent image while darker gray indicates a deficit at later times. Clearly the region of detached plasma and flux grows and rises with time while an ever increasing deficit is observed along at least the left flank of this region. This configuration is qualitatively consistent with X-line type reconnection where plasma and open magnetic flux are brought into the reconnection region from the sides while reconnected flux is ejected sunward and antisunward.

Figure 3 schematically displays a time sequence for reconnection across a helmet streamer. The top panel shows a helmet streamer configuration; in the middle panel reconnection has begun. Such reconnection could begin on already open field lines or on the closed, innermost arches of the helmet structure and continue through to open fields. The closed loop in the center of the disconnected flux (second panel) schematically indicates that this disconnection started on closed field lines. In either case, however, once reconnection is occuring on open field lines, flux is returned to the Sun as long as the reconnection is sufficiently close to the Sun that newly created arches are bound and not freely expanding. In the bottom panel, reconnection has continued on the open fields across the helmet streamer, and a region of "U" or "V" shaped magnetic field has been released while newly closed arches have been returned to the Sun.

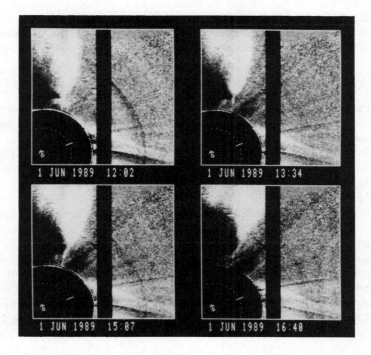

Fig. 2. Images of the coronal detachment differenced from a preevent image at 9:38 UT. Lighter (darker) gray indicates greater (lesser) intensity at later times compared to the preevent image.

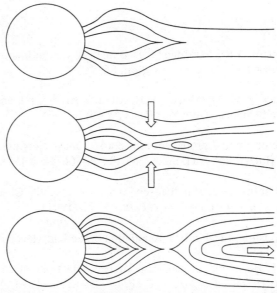

Fig. 3. Time sequence of schematic diagrams showing a coronal disconnection event. Once the initially open magnetic fields on either side of the helmet streamer are reconnecting, magnetic flux is returned to the Sun and a U-shaped magnetic field lines are released.

The general survey of the SMM coronagraph data /6/ only identified suggestions of disconnected structures in 3% of all coronal transients. However, this survey was not conducted with an emphasis on disconnection, and we were aware of at least several examples of apparent disconnections which were not identified in the survey. In order to asses the frequency of occurrence of coronal disconnection events, we reexamined an interval where no magnetic disconnections were noted in the initial survey. In particular, we reexamined the first 52 coronal transient events identified in 1988 (1/1/88 through 3/26/88). While there were no large, obvious events in this interval, we found six (11%) that displayed some reasonable evidence for reconnection in sequences of several frames. Another 13 events (23%) showed a "U" or "V" shaped structure in a single coronagraph frame, but since the appearance of such structures can also be caused by projection effects of bright streamers, single frame observations cannot be unambiguously interpreted.

DISCUSSION

Coronal disconnection events are not as obvious in sequences of coronagraph images as are large CMEs. However, there are a number of clear examples of coronal disconnections in the SMM coronagraph data set; in our reexamination of 52 coronal transient events we found reasonable evidence for disconnection in 6 events (>10%). In spite of the complications inherent in interpreting coronagraph observations, we conclude that disconnection events are not uncommon occurences in the solar corona. By and large the coronal disconnection events that we have examined have not appeared as the release of plasmoid-type CMEs with both the eruption of new flux from the Sun and its subsequent disconnection. Rather, we find coronal disconnection events generally occurring over helmet streamers on previously open magnetic fields.

We believe that the combination of CMEs, which remain at least partially attached to the Sun, opening new flux into interplanetary space /5/, and coronal disconnection events returning previously open flux to the Sun /4/ is a fundamental process in the evolution and maintenance of open magnetic flux in interplanetary space. Even though we have shown here that coronal disconnections occur more frequently than was previously appreciated, they are clearly less common that CMEs. Two possible explanations for this inequality are 1) CMEs having a "flux rope" type geometry in which only a fraction of the CME flux is newly opening /5/ and 2) magnetic reconnection routinely occurring across helmet streamers at a slow rate but only unusually large and impulsive disconnection events are observable remotely with coronagraphs.

Work at NCAR, sponsored by the NSF, was supported by NASA contracts S55989 and S55989A. Work at Los Alamos National Laboratory was conducted under the auspices of the US Department of Energy.

REFERENCES

1. J.T. Gosling, G. Borrini, J.R. Asbridge, S.J. Bame, W.C. Feldman, and R.T. Hansen, Coronal streamers in the solar wind at 1 AU, *J. Geophys. Res.*, 86, 5438 (1981)

2. J.T. Gosling, E. Hildner, R.M. MacQueen, R.H. Munro, A.I. Poland, and C.L. Ross, Mass ejections from the Sun: a view from Skylab, *J. Geophys. Res.*, 79, 4581 (1974)

3. A.J. Hundhausen, The origin and propagation of coronal mass ejections, *Proc.Sixth Int. Solar Wind Conf.*, ed. V.J. Pizzo, T. Holzer, and D.G. Sime, NCAR/TN-306+Proc 1988, p. 181.

4. D.J. McComas, J.L. Phillips, A.J. Hundhausen, and J.T. Burkepile, Observations of disconnection of open coronal magnetic structures, *Geophys. Res. Lett.*, 18, 73 (1991)

5. D.J. McComas, J.T. Gosling, J.L. Phillips, Regulation of the Interplanetary Magnetic Flux, this issue.

6. O.C. St. Cyr and J.T. Burkepile, A catalogue of mass ejections observed by the Solar Maximum Mission coronagraph, *NCAR Technical Note*, NCAR/TN-352+STR (1990)

7. O.C. St. Cyr and A.J. Hundhausen, On the interpretation of "halo" coronal mass ejections, *Proc.Sixth Int. Solar Wind Conf.*, ed. V.J. Pizzo, T. Holzer, and D.G. Sime, NCAR/TN-306+Proc 1988, p. 253.

CONCERNING SOLAR SOURCES FOR CYCLE 22 SOLAR WIND ACTIVITY IN THE HELIOSPHERE

J. D. Mihalov,* A. Barnes,* F. B. McDonald,** J. T. Burkepile*** and A. J. Hundhausen***

* NASA-Ames Research Center, Moffett Field, CA 94035-1000, U.S.A.
** Institute for Physical Science and Technology, University of Maryland, College Park, MD 20742, U.S.A.
*** High Altitude Observatory, National Center for Atmospheric Research, Boulder, CO 80307, U.S.A.

ABSTRACT

Beginning in 1989, the active phase of the present solar cycle became manifest in the outer heliosphere as large disturbances in solar wind velocity as observed by the Ames plasma analyzers aboard Pioneer 10 (46-50 AU heliocentric distance) and Pioneer 11 (~28 AU). During the previous several years plasma observations had shown little evidence for transient events in the outer heliosphere. Inner heliospheric baseline plasma observations from the Pioneer Venus Orbiter (0.7 AU) and IMP 8 (1 AU) are useful for attempts to correlate solar events with the outer heliospheric disturbances. With regard to the onset of activity at Pioneer 11, Pioneer Venus observations are pertinent, and some of these in turn correspond with CMEs (coronal mass ejections) observed in SMM coronagraph data. In particular, enhanced solar wind speeds observed at Pioneer Venus during December 1988 to February 1989 are associated with seven large solar wind shocks (or shock candidates); corresponding CMEs may be identified. Two of these seven shocks were identified as candidates for a precursor to the onset of the disturbances at Pioneer 11. At Pioneer 10 the disturbed period includes two large disturbances, associated with the passage of shocks. There are several candidate CMEs in the SMM observations, one of which may be associated with the second Pioneer 10 shock. In addition, large Hα solar flares can also be identified that may correspond with observed activity at Pioneer 10.

INTRODUCTION

During 1989, the increasing solar activity of the present solar cycle resulted in large disturbances of the solar wind in the outer heliosphere, as observed at Pioneers 10 and 11. By mid-1989, Pioneer 10 was at 46.8 AU heliocentric distance (4 deg N. latitude), and Pioneer 11 similarly was at 28.6 AU and 17 deg N. latitude.

ONSET OF SOLAR CYCLE ACTIVITY AT PIONEER 11

First, an onset of the current solar activity cycle, as observed at Pioneer 11, is considered. In Figure 1 the solar wind proton speed and density, and 0.8 - 1.6, and 3.2 - 5.2 MeV proton fluxes are shown, as observed at Pioneer 11 during the first half of 1989. Considering only the solar wind plasma data, the speed increase at day 64 was taken, somewhat arbitrarily, as the initial manifestation of Cycle 22 activity. Clearly, the proton flux data support this identification, and, furthermore, the approximately coincident flux increase can represent shock accelerated protons, and indicate the passage of an interplanetary shock. (The larger shock event at day 151 has been associated with the large, east quadrant solar flare of day 69, and has been discussed elsewhere /1/.)

An estimate of the time for the solar wind disturbance to propagate from the Sun to Pioneer 11 indicates that it originated at the Sun during November - December 1988. During this period the heliographic longitude of Pioneer 11 is similar to that of Venus, so that relevant inner heliospheric data may be available from the Pioneer Venus Orbiter. In addition, Venus was located in the vicinity of the Sun's west limb, as seen from Earth, so that Solar Maximum Mission (SMM) observations could be used to look for events that might be the source of the heliospheric disturbances. The Pioneer Venus solar wind speed averages during this period were unusually high. In fact, daily sample Pioneer Venus solar wind speeds averaged separately over Carrington solar rotations 1810 and 1811, in December 1988 and January 1989, each exceed 500 km/s, and therefore surpass all the other such averages, for both the preceding and following year, as shown in Figure 2, which also

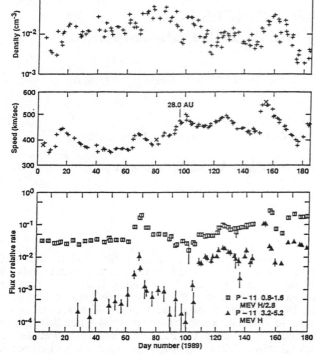

Fig. 1. Solar wind proton speed and number density, and 0.8 - 1.6, and 3.2 - 5.2 MeV proton data from Pioneer 11, for the first half of 1989. The solar wind densities are half-day averages. The solar wind speeds are also half-day averages, except for a few daily sample values, indicated by 'Xs', given when half-day averages are unavailable.

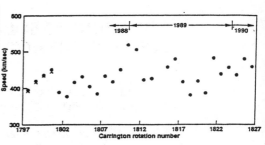

Fig. 2. Carrington rotation averages of Pioneer Venus Orbiter daily sample solar wind speeds.

seems to indicate more activity after Rotations 1810 and 1811, than before. More detailed examination of the available data indicates that the high speeds were associated with seven large shocks, or shock candidates. Furthermore, each shock (candidate) can be associated with a coronal mass ejection (CME) (from ref. /2/), except for two events for which there is no corresponding SMM coronagraph data. In Table 1 the Pioneer Venus shock candidates and corresponding CMEs are listed. In addition, the available speeds and accelerations of the CMEs, as observed at heliocentric distances inside of 6 solar radii by the coronagraph on SMM, match plausible acceleration profiles that include the post-shock solar wind speeds observed at Pioneer Venus (156 solar radii). These CME accelerations and/or speeds are available for the first, third and fifth entries of the Table. Of these three, the speeds of CME features are slower than the post-shock speeds for the first and third entries, and faster for the fifth entry.

TABLE 1 Correspondence of SMM CMEs with Pioneer Venus events possibly associated with outer heliospheric disturbances

Pioneer Venus Shock Candidate				Candidate SMM CME
DATE - TIME (UT)	ΔV (km/s)	V_{final} (km/s)	Proton Density Ratio	
12/6/88 1740 - 12/7/88 0530	125 ?	535 ?		12/5/88 1200 ff.
12/15/88 1507:00 ± 10s†	95	510	3.7	Data Gap 12/13.9 - 15.6/88
12/26/88 0310-0420	200	660	1.2	12/24/88 1217 ff.
1/1/89 0504:54 ± 13 s	215	640	~1.9	12/29/88 0622 ff.
1/11/89 1044:17 ± 1m 47s	205	560	3.4 •	1/9/89 1319 ff.
1/25/89 1914 - 1/26/89 0715	215 ?	660 ?		Data Gap 1/23.2 - 25.0/89
2/5/89 0210 - 0415	~110	730		2/3/89 <1514 ff.

• Has large overshoot
† Personal communication, R. Strangeway and C.T. Russell

If plausible assumptions are made for the propagation speeds to Pioneer 11, of the seven shock (candidates) of Table 1, one finds that the first and third events are those most likely to be associated with the day 64 onset of activity at Pioneer 11 (the other cases involve arrival times at Pioneer 11 much later than the observed disturbance). Because of the apparent weakness of the third shock, the first event seems most likely to account for the onset at Pioneer 11. A fairly prominent shock observed in the MIT solar wind plasma data (A. J. Lazarus, private communication) from Voyager 2 (located relatively close to Pioneer 11, but closer to the heliographic equator) probably represents another sample of the same shock front propagating through the outer heliosphere. At Voyager 2 the shock front was observed about 9 days earlier than at Pioneer 11, and with a speed increase almost 100 km/s greater. The earlier observation at the lower latitude of Voyager 2 can correspond to a greater propagation speed nearer the heliographic equator in this case. The mean propagation speed from Pioneer Venus to Pioneer 11 would be 527 km/s.

ONSET OF SOLAR CYCLE ACTIVITY AT PIONEER 10

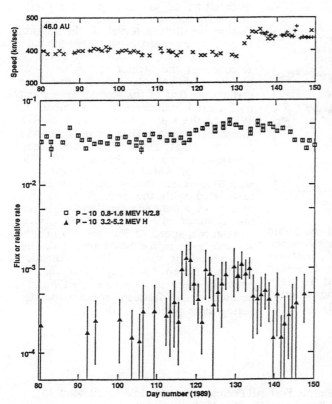

At Pioneer 10 an onset of the current solar activity cycle seems easier to identify than at Pioneer 11. In Figure 3 the solar wind proton speed, and energetic proton data as in Figure 1, are shown for a 70 day interval in 1989. No speed increase of comparable magnitude to the one that follows day 130 had been observed at this spacecraft for about the previous three years (speeds for 1985-7 are published in /3/, and for the beginning of 1988 in /4/). As with the event discussed above in connection with Figure 1, the energetic proton fluxes make it seem likely that this speed increase is an interplanetary shock. Furthermore, times in the inner solar system that would correspond with this onset, based on likely values of propagation speeds, are about the same as for the Figure 1 event. However, Pioneer 10 was located in the general direction opposite the Sun, as seen from Earth, or nearly 180 deg of solar longitude away from Pioneer 11.

Published solar wind data from IMP 8, for December 1988 /5/, show large disturbances (peak speeds of 700 - 800 km/s) that could correspond with this activity onset at Pioneer 10. The tabulation of Hα solar flares also contains large events that could correspond with this onset. One of those begins at 0826 UT, December 16, 1988, at E33 deg longitude relative to Earth (p. 15 of ref. /5/). The mean propagation speed to the Pioneer 10 onset from this flare is ~555 km/s.

Fig. 3. Solar wind proton speed, and 0.8 - 1.6, and 3.2 - 5.2 MeV proton data from Pioneer 10 during 1989. The solar wind proton parameters are half-day averages, except for daily sample speeds, indicated with 'Xs' and given when half-day averages are not available.

REMARKABLE LARGE SHOCKS OBSERVED NEAR 48 AU

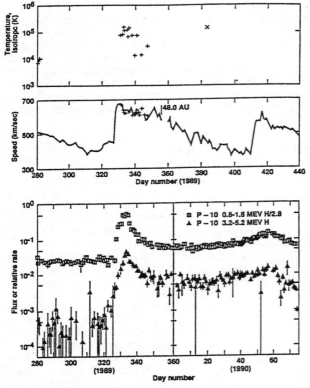

In Figure 4 are shown two remarkable large-amplitude solar wind speed increases observed near the end of 1989 at Pioneer 10, near 48 AU heliocentric distance. The associated energetic proton flux increases suggest that the speed increases actually represent the passage of interplanetary shocks. There appear to be sizable deflectons of the solar wind flow during this period of large-amplitude speed variations. In particular, while the second of the two large speed increases was occurring, the flow seemed to be deviated by up to ~18° from the Earth-oriented spacecraft spin axis, at one point. During the time interval of the Figure, Pioneer 10 emerged from behind the west limb of the Sun, as seen from Earth. A selection of the largest Hα solar flares observed during an appropriate time earlier than that of the second large event on Figure 4 includes one on October 24, 1989, in the west quadrant, at 1738 UT (Solar Geophysical Data - comprehensive reports, April 1990; 705 km/s mean propagation speed to Pioneer 10). The tabulation of SMM CMEs /2/, includes west limb events that, similarly, might be associated with the second event. An example is a complex ejection that began on October 19 (~735 km/s mean propagation speed to Pioneer 10). There are data gaps in the SMM record during this period. It seems wise to consider cautions raised by Neugebauer /6/, in pursuing these associations, particularly because of the frequent solar outbursts, during these periods.

Fig. 4. Solar wind proton speed and isotropic temperature, and 0.8 - 1.6, and 3.2 - 5.2 MeV proton data from Pioneer 10, near the end of 1989. Half-day averages of the solar wind parameters are indicated with '+' signs. Daily sample values of speed are given by a line. A solar wind proton temperature value indicated with an 'X' was computed manually.

REFERENCES

1. J. A. Van Allen and J. D. Mihalov, Forbush decreases and particle acceleration in the outer heliosphere, Geophys. Res. Lett., 17, 761, 1990.

2. O. C. St. Cyr and J. T. Burkepile, A catalogue of mass ejections observed by the Solar Maximum Mission coronagraph, NCAR Technical Note NCAR/TN-352+STR, 1990.

3. J. D. Mihalov, A. Barnes, A. J. Hundhausen, and E. J. Smith, Solar wind and coronal structure near sunspot minimum: Pioneer and SMM observations from 1985 - 1987, J. Geophys. Res., 95, 8231, 1990.

4. A. Barnes, Distant solar wind plasma - view from the Pioneers, in: Physics of the Outer Heliosphere, ed. S. Grzedzielski and D. E. Page, Pergamon Press, Oxford 1990, p. 235.

5. Solar Geophysical Data - comprehensive reports, NOAA, Boulder, Col., June 1989, p. 98.

6. M. Neugebauer, The problem of associating solar and interplanetary events, in: Proceedings of the Sixth International Solar Wind Conference, NCAR Technical Note NCAR/TN-306+Proc, ed. V. J. Pizzo, T. E. Holzer, and D. G. Sime, 1988, p. 243.

RECURRENCE: IMPLICATIONS FOR HELIOSPHERIC COSMIC RAY TRANSPORT

G. D. Parker*

Norwich University, Northfield, VT 05663, U.S.A.

ABSTRACT

An analysis procedure is developed to quantify recurrence tendencies in lists of events. I apply this technique to 27-day recurrent variations in the intensity of the galactic cosmic radiation.

Changes in the ground-level nucleonic intensity are significantly more recurrent than are intensity extrema. The recurrence periods of cosmic ray decreases do not respond to changes in the polarity of the heliospheric magnetic field. Corotation of quasi-static structures is a less important cause of recurrence than are the interplanetary disturbances responsible for abrupt cosmic ray changes.

TECHNIQUE

I offer a strategy for extracting the properties of cyclic or recurrent geophysical phenomena from lists of event times. For a list of N events, there exist $N(N-1)/2$ event pairs. If a recurrence pattern exists among the events, a histogram of these pairs as a function of event separation will show more pairs than expected from chance at preferred event separations.

In a time series of randomly occurring events, the probability per unit time that the nth subsequent event happens at time t is given by the exponential distribution

$$Q_n(t) = \frac{e^{-\lambda t} \lambda^n t^{n-1}}{(n-1)!}$$

(1)

where $n = 1,2,3,...$ and λ^{-1} is the average interval between adjacent events. Suppose we wish to observe a total number N of event occurrences. Define an "adjoining interval" as the variable interval between consecutive events. Among N events, the number of event pairs which are separated by n adjoining intervals is N-n. The expected total number of event pairs separated by an interevent time of t is

$$EP(t) = \sum_{n=1}^{N-1} (N-n) Q_n(t)$$

(2)

This function is compared with the observed number of event pairs at time separation t. In graphs which follow, "Excess Pairs" equals (the observed number of events separated by t) minus EP(t).

COSMIC RAY RECURRENCE PERIOD

Several mechanisms have been advanced for the origin of 27-day variations in the galactic cosmic ray intensity measured at Earth. These include enhanced convection and adiabatic cooling in recurrent fast solar wind streams /1/, transient solar wind disturbances from long-lived solar "hot-spots" whose geoeffectiveness varies with solar rotation /2/, corotating regions of magnetic turbulence which impedes

*The author acknowledges NSF support under grant ATM-9002543.

cosmic ray diffusion /3/, and a corotating heliospheric neutral sheet along which gradient drift transports cosmic rays /4/. The recurrence period imprinted on the cosmic ray record may be a property of the heliospheric regions traversed by cosmic rays reaching the inner solar system; because of solar differential rotation, changes in this period could result from global drifts. This study seeks observational measures of the relative importance of possible recurrence mechanisms.

What is the recurring feature in the ground level nucleonic intensity? What aspect of the time series is repeating? The definition of "event" may be varied to examine its effect on the strength and period of recurrence. The cosmic ray data used here are daily averaged, pressure corrected Deep River neutrons over 1964-1989.

Cosmic ray changes (Fig. 1) show more recurrence than do extrema (Fig. 2). Frequency components in the time series unrelated to solar rotation must affect the timing of extrema and reduce their recurrence. Both Figures show no evidence of the recurrence near 13 days which one would expect if rotation of a warped heliospheric current sheet were a major cause of recurrence.

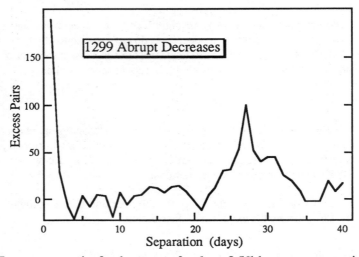

Fig. 1. Excess event pairs for decreases of at least 0.5% between consecutive days.

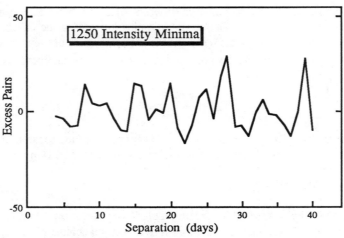

Fig. 2. Excess pairs for days of depressed intensity less than at least two preceding
           and two following days.

I have repeated the analysis of abrupt decreases for three distinct heliospheric magnetic polarity intervals. In each case there is a significant clustering of excess pairs near the solar rotation period (as in Fig. 1), and I determine a centroid period from values of Excess Pairs above the 2 σ level. With A<0 during 1964.0-1969.7, the centroid is 27.8 days. With A > 0 during 1971.1-1980.4, the centroid is 27.4 days. With A < 0 during 1980.7-1990.0, the centroid is 27.0 days.

The cosmic ray recurrence period does not change in response to a change in global polarity. Global drift inward over the poles (A>0) does not lengthen significantly the recurrence period. Either global drift is of minor importance to global transport or corotation is of minor importance to cosmic ray recurrence.

RECURRENCE AMPLITUDE HISTORY

Figure 3 employs two methods for determining recurrence amplitude -- (1) solid line, left ordinate axis: number of excess pairs due to recurrence per 1000 days; (2) dotted line, right ordinate axis: area under the 27-day peak of a maximum entropy power spectral density function. The amplitude of recurrence given by the dotted line peaks around the maxima of sunspot cycles 20 and 21, and does not change systematically with the sign of the heliospheric field.

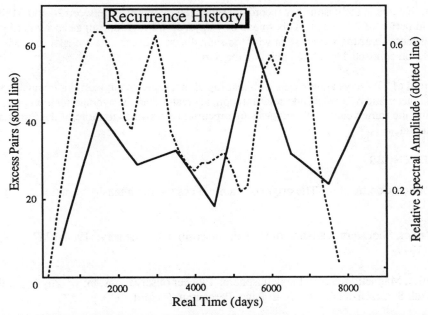

Fig. 3. The strength of recurrent cosmic ray variations over sunspot cycles 20 and 21.
Day 1 = 1 Jan 1964.

Around days 6500-7000 (1981-1982) there is a relatively large amplitude recurrence (as measured by spectral methods) but only an average number of recurrent events (as measured by the number of excess pairs). Therefore, the recurrent events present at this time are expected to be unusually large. I have examined this interval in detail by preparing Bartels plots of the cosmic ray intensity. During Bartels rotations 2021 through 2040, one finds 16 instances of cosmic ray decreases and 4 of non-decreases near Bartels day 24. In this epoch of strong recurrent features, large amplitude Forbush decreases such as that of 14 July 1982 are major contributors to recurrence. The instances of non-decreases suggest that the recurrence mechanism is intermittent. Events in this strong recurrence sequence are likely related to a very long-lived longitudinal asymmetry in solar activity which included the second most active flare producing region during 1972-1989 /5/.

Geomagnetic storm sudden commencements are often taken as proxy data for solar wind shocks which originate in transient solar disturbances. An investigation of 119 years of ssc's using the analysis techniques above /6/ demonstrates that ssc's recur at multiples of the solar rotation period. The Earth is recurrently susceptible to shocks which make Forbush decreases a prominent part of cosmic ray recurrence.

CONCLUSIONS

(1) The analysis of separation intervals in a list of solar-terrestrial events is a useful addition to the practical tools of time series analysis. Recurrence interval analysis can provide information on both the period of recurrence and the amplitude history. Diverse definitions of events may reveal differences in frequency of occurrence. Rigorous procedures for estimating uncertainty /6/ make quantifiable conclusions possible.

(2) Classical Forbush decreases, historically attributed to flares, are a prominent part of recurrence episodes. Operational distinctions between "flare associated" Forbush decreases and so-called recurrent Forbush decreases are problematic. Transient activity at long-lived solar active centers produces recurrent interplanetary disturbances which contribute substantially to cosmic ray recurrence.

(3) Recurrent features in the cosmic ray record are separated by intervals between 24 and 31 days (2σ), with a centroid period of 27.7 days. This spread of 7 days, or 1/4 rotation, suggests a volatility in the timing of recurrent events more consistent with transient disturbances from a hot spot than with a spiral modulation region connected to a fixed point on the Sun.

(4) The centroid of 27.7 days agrees well with that found in a separate autocorrelation analysis /7/. This period, not characteristic of low latitude solar rotation, is greater than the average rotation period of sunspots during the same years. This suggests that recurrence responds to more of the Sun than to just the sunspot latitudes.

REFERENCES

1. G. D. Parker, Comment on 'The origin of transient cosmic ray intensity variations', J. Geophys. Res. 83, 2711 (1978)

2. G. D. Parker, Short-term variation of the galactic cosmic ray intensity: 1964-1967, Planet. Space Sci. 25, 681 (1977)

3. E. J. Smith, Magnetic fields in the heliosphere: Pioneer observations, in: Physics of the Outer Heliosphere, ed. S. Grzedzielski and D. E. Page, Pergamon, Oxford 1990, p. 253.

4. G. Newkirk, Jr. and L. A. Fisk, Variation of cosmic rays and solar wind properties with respect to the heliospheric current sheet. 1. Five Gev protons and solar wind speed, J. Geophys. Res. 90, 3391 (1985)

5. P. McIntosh, The classification of sunspot groups, Solar Phys. 125, 251 (1990)

6. G. D. Parker, Detection of recurrence patterns in impulsive geomagnetic activity, J. Geophys. Res. submitted (1991)

7. G. D. Parker, Properties of 27-day recurrence of ground level cosmic rays, EOS 69, 1361 (1988)

CORONAL SOUNDING WITH ULYSSES: PRELIMINARY RESULTS FROM THE FIRST SOLAR CONJUNCTION

M. Pätzold,* M. K. Bird,** H. Volland,** P. Edenhofer,*** S. W. Asmar†
and J. P. Brenkle†

* DLR-Oberpfaffenhofen, NE-HF, 8031 Oberpfaffenhofen, Germany
** Radioastronomisches Inst., Universität Bonn, 5300 Bonn, Germany
*** Inst. für Hochfrequenztechnik, Universität Bochum, 4630 Bochum, Germany
† Jet Propulsion Laboratory, California Inst. of Technology, Pasadena,
CA 91009, U.S.A.

ABSTRACT

Radio-sounding observations of the solar corona between 4 $R_\odot$ and 115 $R_\odot$ were performed during the first superior solar conjunction phase of the Ulysses spacecraft in August/September 1991. As a first result of this Solar Corona Experiment (SCE), the total electron content inferred from dual-frequency ranging observations is presented here as a function of solar distance.

INTRODUCTION

The Ulysses spacecraft attained its first solar superior conjunction on 21 Aug 1991. Dual-frequency ranging and Doppler measurements at signal ray path offsets inside 40 solar radii ($R_\odot$) were performed nearly continously from 5 Aug to 5 Sep 1991. The closest approach of the signal ray path was 4.3 $R_\odot$ above the north pole of the Sun. With the exception of five days around closest approach the signal passed through regions of the solar corona at heliolatitudes less than 30 degrees. Emphasis is placed here on the dependence of the ranging observations with solar distance. Latitudinal effects and a detailed comparison with previous experiments will be treated in a subsequent paper. The spacecraft was configured to receive a S-band uplink (2112 MHz) and transmit S- and X-band downlinks (2293 MHz and 8408 MHz, respectively), both coherent with the uplink. The polarization of both downlinks was designed to be right-hand circularly polarized (RCP). A significant left-hand circularly polarized (LCP) component was detected, however, in both downlinks.

The principal ground stations used for SCE were the 34-m and 70-m antennas of the NASA Deep Space Network (DSN), located in Goldstone (California), Canberra (Australia) and Madrid (Spain). Dual-frequency ranging measurements could be performed only at these ground stations. Thirteen additional tracking passes in a "listen-only" mode were executed at the 30-m DLR/GSOC Deep Space Station in Weilheim, Germany. Six tracking passes were performed at the 32-m VLBI station at Medicina, Italy. Simultaneous dual-frequency Doppler observations of Ulysses were taken at the three European ground stations on three days.

SCIENTIFIC OBJECTIVES

The main scientific objective of SCE is to derive the large- and small-scale structural distribution of the solar corona. The large-scale structure is reflected in the radial and latitudinal distribution of the total coronal electron content, which is derived from dual-frequency ranging measurements. The delay of the S-band ranging code with respect to the X-band ranging code is proportional to the total electron content along the downlink ray path. The nominal ranging sample rate was one point every ten minutes.

Dual-frequency Doppler measurements provide only relative changes in electron content. Doppler observations are used to investigate the small scale structure of the corona, recording fluctuations and perhaps even coronal mass ejections. The measurements are nominally sampled with a rate of one point per second. Although the Doppler measurements cannot provide an absolute value in electron content, they are about 100 times more sensitive than the ranging observations /1/. They can also be used to derive the coronal turbulence spectrum.

Solar wind velocities can be inferred using multiple station observations by cross correlating the Doppler scintillations across the separate coronal ray paths /1/.

The elliptical polarization of both downlinks can be used to measure the Faraday Rotation along the coronal ray path. Combining these with the absolute electron content observations from dual-frequency ranging, the coronal magnetic field can be directly determined.

A more detailed description of SCE can be found in /2/. This research note presents only the first results from the dual-frequency ranging observations.

CORONAL ELECTRON CONTENT

The dual-frequency ranging measurements are merely the differential delay time of the S-band to the X-band ranging code, a quantity directly proportional to the total electron content. Figure 1 shows the total electron content inferred from the Ulysses measurements during solar conjunction as a function of solar distance. Each point represents an average over one tracking pass.

The data displayed in Figure 1 begin on 10 Jul 1991 at a solar distance of 104 R$\odot$. Due to power problems on board Ulysses, the project decided to switch off the S-band downlink from 25 Jul to 5 Aug 1991 (69.1 R$\odot$ to 41.8 R$\odot$) resulting in a data gap of about two weeks. Starting from 40 R$\odot$ and moving down to the closest approach point at 4.3 R$\odot$ (ingress phase), the electron content follows a power law dependence with an exponent $\alpha = -1.55$, as obtained in a linear regression analysis. The corresponding exponent during the egress phase was $\alpha = -1.21$. In order to save power for the on-board tape recorders, it became necessary to switch off the S-band again on 5 Sep.

FIRST INTERPRETATIONS

A modified Allen-Baumbach formula is commonly used to describe the variation of the coronal electron density as a function of distance from the Sun /3/:

$$N(R) = \frac{N_A}{R^6} + \frac{N_B}{R^{2+\gamma}} \quad el \cdot m^{-3} \tag{1}$$

with R in solar radii. The first term is significant only for R < 4 R$\odot$, a region not sounded by Ulysses. The total electron content is thus adequately represented by the integral of only the second term in (1) and will display a power law radial dependence such that $\alpha = -\gamma-1$, where α is the radial falloff exponent of the electron content measurements. Well out in interplanetary space, where the solar wind reaches its final asymptotic speed, the electron density falls off as the inverse square of the solar distance, with $\gamma = 0$. Moving down into the solar corona it is obvious that $\gamma \neq 0$ and that, by conservation of mass, the solar wind must be accelerating /4/. The Ulysses data in Figure 1 indicate that the acceleration region of the solar wind extends at least out to 40 R$\odot$. The dashed line denoted "$\gamma = 0$"

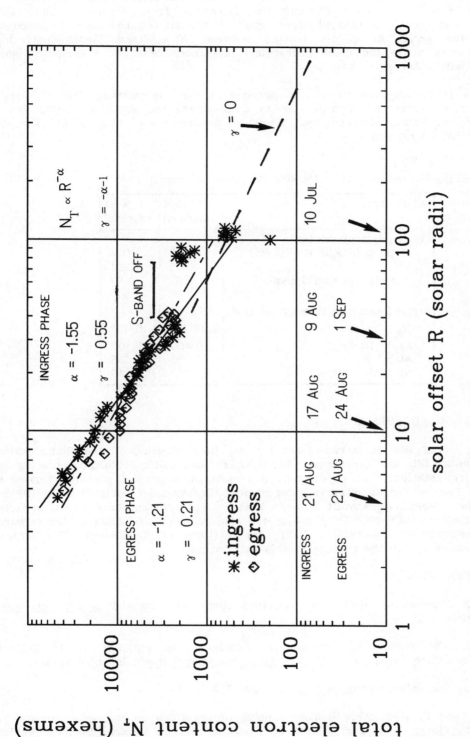

Fig. 1. Total electron content N_T in hexems (10^{16} el·m^{-2}) versus solar distance R in solar radii inferred from Ulysses dual-frequency ranging observations during thr first solar conjunction. Points between 4.3 R⊙ and 40 R⊙ were used for a linear regression analysis. The individual range points and their corresponding linear fits are given by the asterisks and solid line (ingress), and by diamonds and dot-dashed line (egress), respectively. The dashed line marked with $\gamma = 0$ indicates a falloff in electron content according to 1/R.

reflects what would be expected from an electron density distribution proportional to an inverse square law. At what radial distance the electron content begins to follow the "$\gamma = 0$" curve is difficult to determine from Figure 1, particularly because of the above mentioned data gap. Additional studies are required to understand the series of points plotted between 70 - 90 $R_\odot$. Anomalously high electron contents above the quiet background have been observed at previous solar occultations with other spacecraft /4/.

Table 1 lists preliminary values of the parameters for the coronal electron density model as derived from the Ulysses ranging data during the superior conjunction in August/September 1991. Extrapolated electron densities are also given for the distances 10 $R_\odot$, 60 $R_\odot$ and 1 AU.

TABLE 1 Coronal Electron Density Parameters: August 1991

solar limb	east (ingress)	west (egress)
fit to electron content:		
slope α (4 $R_\odot$ R 40 $R_\odot$)	-1.55	-1.21
γ	0.55	0.21
correlation coefficient	0.99	0.97
derived electron density (el/cm^3)		
N_B	$3.34 \cdot 10^6$	$1.15 \cdot 10^6$
at 10 $R_\odot$	9,400	7,100
at 60 $R_\odot$	98	135
at 1 AU	7	10

ACKNOWLEDGEMENTS

The authors would like to acknowledge the excellent support of the Radio-Science Support Team at JPL and the German Space Operations Center (GSOC), operating the Weilheim ground station. This paper present results of a project partially funded by the Deutsche Agentur für Raumfahrtangelegenheiten (DARA) under contract 500N9104 and by Deutsche Forschungsanstalt für Luft- und Raumfahrt, Institut für Hochfrequenztechnik (DLR-NE-HF). Funding was also provided by NASA for the research at the Jet Propulsion Laboratory, California Institute of Technology. The authors assume responsibility for the content of this publication.

REFERENCES

1. M.K. Bird, Coronal sounding with occulted spacecraft signals, Space Sci. Rev., 33, 99-126 (1982)

2. M.K. Bird, S.W. Asmar, J.P. Brenkle, P. Edenhofer, M. Pätzold, and H. Volland, The Coronal Sounding Experiment, Astron. Astrophys. Suppl. Ser., in press (1992)

3. D.B. Beard, The solar corona, Ap. J. 234, 696-706 (1979)

4. M.K. Bird and P. Edenhofer, Remote sensing observations of the solar corona, in: Physics of the Inner Heliosphere, eds. R. Schwenn and E. Marsch, Springer-Verlag, Heidelberg 1990, pp. 13-97.

ON SPECTRAL STRUCTURE EVOLUTION OF INTERPLANETARY MEDIUM PARAMETERS

A. Prigancová

Geophysical Institute of the Slovak Academy of Sciences, 842 28 Bratislava, Czechoslovakia

ABSTRACT

Based on NSSDC Composite omnitape data, the time series of a number of interplanetary medium parameters are examined as dynamical systems on a time scale of daily values. The interval of space data 1973-1982 gives only weak retrospective testing of unstable features in the development of spectra. The analysis of a broader interval for inferred data 1932-1988 reveals the regularity and stochastic nature of spectral features. The differences between individual spectral patterns are reported and discussed.

INTRODUCTION

The long-term and large-scale structure evolution of the solar wind (SW) and interplanetary magnetic field (IMF) features is of continuing concern /1, 2, 3/. The main attributes of the long-term dynamics of these parameters are mainly the solar cycle modulation, annual variations, and recurrent tendency /2,4/. Modulation within a solar cycle was reported for both the SW velocity and IMF sector structure /5,6,7/. Systematic changes in the SW velocity recurrent variation pattern in the course of the solar activity cycle are well known. The quite stable source structure for the geomagnetic disturbances generated by corotating structure in the SW flows was pointed out in /5/. Concerning the IMF sector structure evolution, alternating two- and four-sector types have been mentioned as the dominant structures /6/. However, more complex structures with high degree of variability can occur /6,7/. Recently, it was indicated that solar and geomagnetic data are consistent with a chaotic system behavior manifested on various time scales /8,9,10/.

To define regular and stochastic features in the SW and IMF dynamics within the recurrent variation frequency band, the spectral structure evolution of selected parameters is traced for an extended time interval.

DATA

To trace the recurrent features in the SW and IMF spectral structure pattern the NSSDC Composite omnitape data on the solar wind velocity (SWV), density (SWD), and IMF azimuthal components (Bx, By) were used. The time series of these interplanetary medium parameters were examined as dynamical systems on a time scale of daily values. Since the time series contains numerous data gaps, the interval of sufficient data coverage 1973-1982 was chosen. The occasional shorter segments of missing values within this interval were bridged by smoothing /4/.In order to broaden the data set range for the analysis of the inferred IMF sector structure (IMF SS), data were used here from the interval 1932-1988 /11,12/.

CALCULATION AND DISCUSSION

An FFT algorithm was applied to calculate the spectra. Their temporal evolution was followed by consideration of consecutive subintervals ($N = 10^9$), each successively shifted by one Bartels rotation throughout the whole interval. The sequence of individual spectra manifests the dynamics of the spectral power distribution pattern scanned rotation by rotation.

The IMF SS spectral structure evolution was analyzed using inferred and observed data. The observed IMF SS was represented by the spiral angle Θ calculated using the IMF Bx and By azimuthal components and then coded in a ± 1 manner for $45° < \Theta \leq 225°$ and $225° < \Theta \leq 45°$, respectively. The IMF Bx radial component as an indicator of IMF SS polarity demonstrated the identical spectral structure evolution.

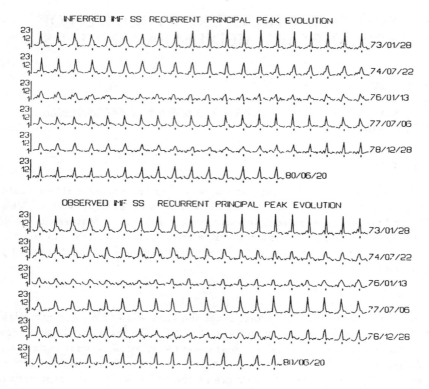

Fig. 1. The successive fragments of the spectral power profile (relative units used) for the recurrent principal period $T0$ frequency band. Spectra were obtained for input time series of about 1.5-y long shifted rotation by rotation within the interval 1973-1982. The date column (right) specifies the corresponding beginning of subintervals to which the first spectrum fragment in each row is related. The marks at the bottom of each row indicate the $T0 \approx 27d$ position.

In Figure 1 the time-sequence spectral fragments in the recurrent principal period $T0$ frequency band are displayed for observed data (bottom panel) with those of inferred IMF sector polarity (upper panel). The comparison of spectral patterns reveals a high degree of similarity, thereby strengthening the credibility of the inferred IMF SS data. Accordingly, this justifies their use for analysis within a broader data set range.

The long-term time profiles of the recurrent principal period $T0$ power, first harmonic $T1$ power and their power ratio (abbreviated below as $T0$ power, $T1$ power and $T0/T1$ power ratio) were obtained using inferred IMF SS data throughout solar cycles 17–21. The profiles displayed in Figure 2 show the IMF SS recurrent pattern tendency of being organized within a solar cycle.

This can be most properly seen in the $T0/T1$ power redistribution profile. The logarithmic scale has been used as in /6/, which allows one to follow easily the alternate predominance of $T0$ and $T1$ power. Due to one Bartels rotation shifting throughout the whole interval, the power redistribution pattern was obtained in detail. It is fully consistent with conclusions by Gonzalez and Gonzalez /6/ and Behannon et al. /7/ and confirms that the predominance of the $T0$ power is generally associated with two-sector IMF structure around solar maxima, while four-sector IMF structure stimulates $T1$ power prevailing around minima.

The most regular relationship is evident for minima in 1933, 1964, 1976 and related maxima epochs in 1937, 1968, 1979 during solar cycles 17, 20, 21. For extreme epochs of the remaining cycles, shifting in phase and quite unstable variability are evident. This is usually related to the significant suppression of the $T0$ and $T1$ power arising from deviation of the power maxima position in the frequency band of $T0 \approx 27d$ and $T1 \approx 13.5d$ as seen in Figure 2. The complexity of spectral structure evolution is generally reflective of its unpredictability, i.e. of its stochastic nature. This can be ex-

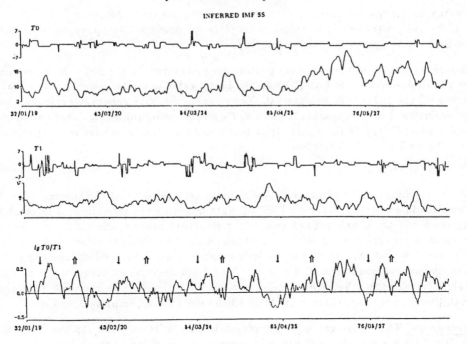

Fig. 2. The long-term time profiles of the $T0$ power (upper panel), $T1$ power (middle panel), and $T0/T1$ power ratio (bottom panel) as derived for inferred IMF SS data within the interval 1932-1988. The corresponding peak deviation profiles are also shown, using the frequency index, its zero value being adopted for $T0 \approx 27d$ and $T1 \approx 13.5d$, respectively. The dates along the horizontal axis specify beginnings of Bartels rotations being 150 rotations apart. The arrows indicate solar cycle minima ($\downarrow$) and maxima ($\Uparrow$) epochs.

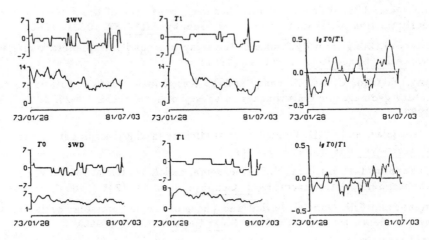

Fig. 3. The time profiles of the $T0$ (left) and $T1$ (mid) power (relative units used) with corresponding peak deviation profiles as derived for daily SWV (upper panel) and SWD (bottom panel) data within the interval 1973-1982. For details, see Figure 2. The $T0/T1$ power ratio for SWV and SWD data (right panels) are also shown. The dates of the beginning of the first and last Bartels rotation (115 rotations apart) are indicated on the horizontal axis.

plained by fluctuations in the location, heliospheric extension and configuration of the heliospheric neutral current sheet /3,6,7/. Obviously, the stochastic evolution of solar cycle development might also play a role /9/.

However, there are indications for periodicities in the IMF structure. The periods of about 2.5-

and 3.5-y can be recognized directly from the profiles presented. Moreover, the 1.5-y periodicity is likely to be real, although it is rather masked in all profiles. Its appearance is unlikely to be produced by beat frequency contamination /6/.

The long-term evolution of the recurrent pattern reconstructed for the SWV data reveals a variable character of the power distribution (Figure 3). The regular annual enhancements of the $T0$ power are apparent for the 1973–1975 subinterval. The quasi-periodicity generally evident in both $T0$ and $T1$ power evolution is quite apparent in the $T0/T1$ power ratio profile (e.g., there are indications for the 2.5-3.5-y periodicity). This regularity is much more subtle in the recurrent pattern evolution for SWD data, as Figure 3 illustrates.

SUMMARY

The main conclusions can be summarized as follows: (i) similarity of the spectral structure evolution pattern for inferred and observed IMF SS data confirms the reliability of inferred values; (ii) prevailing tendency for a well-defined two-sector structure around solar activity maximum phase and four-sector structure around minimum phase persists throughout solar cycles 17-21; (iii) suppression of the SW and IMF recurrent variation power is associated with deviations of the power maxima position; (iv) quasi-periodicity in the long-term spectral structure dynamics is distorted by stochastic features; (v) the IMF SS 1.5-y periodicity is likely to be real; (vi) recurrent SW and IMF variability requires the features of chaotic behaviour for a complete description.

Acknowledgments. The appreciation in expressed to the WDC-A R&S for the NSSDC omnitape and to L. Bittó for his valuable assistance in computer processing of these data.

REFERENCES

1. J. H. King, Solar wind parameters and coupling studies, in: Solar Wind–Magnetosphere Coupling, eds. Y. Kamide and J. A. Slavin, Terrapub, Tokyo 1986, p. 163

2. J. W. Freeman, and R. E. Lopez, Solar cycle variations in the solar wind, in: Solar Wind– Magnetosphere Coupling, eds. Y. Kamide and J. A. Slavin, Terrapub, Tokyo 1986, p. 179

3. J.T. Hoeksema, J.M. Wilcox, and P.H. Scherrer, Structure of the heliospheric current sheet in the early portion of sunspot cycle 21, J. Geophys. Res. 87, 10331 (1982)

4. A. Prigancová, Long-term dynamics of the solar wind and interplanetary magnetic field parameters, Adv. Space Res. 11, #1, 41 (1991)

5. G. Zhang, Y. Gao, C. Lu, Z. Feng, and Y. Xu, Quasi-stready corotating structure of interplanetary geomagnetic disturbances: a survey of solar cycles 13–20, J. Geophys. Res. 94, 1235 (1989)

6. A.L.C. Gonzalez, and W.D. Gonzalez, Periodicities in the interplanetary magnetic field polarity, J. Geophys. Res. 92, 4357 (1987)

7. K.W. Behannon, L.F. Burlaga, Y.T. Hoeksema, and L.W. Klein Spatial variation and evolution of heliospheric sector structure, J. Geophys. Res. 94, 1245 (1989)

8. J. Feymann, and S.B. Gabriel, Period and phase of the 88-year solar cycle and the Maunder minimum: evidence for a chaotic Sun, Solar Phys. 127, 393 (1990)

9. M.D. Mundt, W.B. Maguire II, and R.R.P. Chase, Chaos in the sunspot cycle: analysis and prediction, J. Geophys. Res. 96, 1705 (1990)

10. D.N. Baker, A.J. Klimas, R.I. McPherron, and J. Büchner, The evolution from weak to strong geomagnetic activity: an interpretation in terms of deterministic chaos, Geophys. Res. Lett. 17, 1841 (1990)

11. L. Svalgaard, Interplanetary magnetic sector structure 1926-1971, Danish Met. Inst., Charlottenlund 1972

12. H.E. Coffey (ed), Solar Geophysical Data, e.g. No. 469, NOAA, Boulder 1983, p. 34.

ON THE DETECTION OF HELIOSPHERIC INTERFACE PROPERTIES FROM INTERSTELLAR/INTERPLANETARY LYMAN α PROFILES: EXAMPLE OF BARANOV'S TWO-SHOCK MODEL

E. Quémerais,* Y. G. Malama,** R. Lallement* and J. L. Bertaux*

*Service d'Aeronomie du CNRS, BP No. 3, F91371 Verrières-le-Buisson, France
** Institute for Problems in Mechanics, Russian Academy of Sciences, Moscow, Russia

ABSTRACT

Interstellar/interplanetary neutral H atoms which are observed towards the upwind (UW) and downwind (DW) directions do not originate from the same region of the heliospheric interface . If the neutral/plasma coupling is not negligible , the variations of the plasma properties with the distance to the stagnation line are imprinted in the neutral atoms and in the relationship between the UW and DW Lyman α spectral profiles , whatever the distance to the heliospheric shock and to the heliopause .

This effect is checked both on the Prognoz 5/6 backscattered Lyman α data and in the case of the Baranov type interface.

Hydrogen absorption cell data interpretation is much improved when inhomogeneities of the interstellar wind are taken into account. Lyman α profiles derived from the Montecarlo neutral flow model of Malama (1990) show the predicted macroscopic effects.

These results tend to show that information on the type of the interface at the heliopause can be derived from precise Lyman α line shape measurements and from comparison with line shapes computed from a homogeneous interstellar flow model.

INTRODUCTION

The heliopause is the boundary which separates the region which contains the solar wind from the surrounding interstellar medium. Whereas the plasma component of the very local interstellar medium (VLISM) cannot penetrate this boundary , the neutral atoms can flow across the heliopause and fill the heliosphere . This creates a wind , mainly consisting of neutral H and He atoms , that can be observed in the inner part of the heliosphere through UV resonance scattering of solar photons at H Lyman α (121.6 nm) and He (58.4 nm) (Bertaux and Blamont , 1971). Since the interstellar medium is partly ionized , the flow of interstellar plasma interacts with the solar wind plasma flow and must compress the heliopause , giving the heliosphere an elongated shape along the downwind direction. In first approximation , the problem has a symmetry of revolution around the stagnation line , which is defined as the line parallel to the interstellar wind velocity vector and passing through the center of the sun.

Various models attempt to describe the interaction between the interstellar matter and the heliopause (Grzedzielski and Ratkiewicz-Landowska 1975 , Fahr et al 1986, Suess et al 1987, Holzer 1989, Baranov et al 1991). Some predict large changes of density and velocity distributions of the neutrals which have penetrated inside the heliosphere , those changes due to H-H$^+$ charge exchange coupling with the decelerating interstellar plasma protons , and inside the heliosphere with the solar protons. Assuming symmetry of revolution , these effects are expected to be dependent on the distance to the stagnation line. The Lyman α backscattered line profile measured from the inner part of the solar system (at about 1 AU) is very sensitive to the distribution parameters of the incoming H atoms , through both the line width and the line shift in the solar frame (Dopplershift).

HYDROGEN CELL OBSERVATIONS

Inside the heliosphere , H neutral atoms follow hyperbolic trajectories under the action of solar gravitation and solar radiation pressure at Lyman α . When approaching the sun , these neutrals are ionized by EUV photoionization , charge exchange with solar wind protons and collisions with solar wind electrons . In 1976 and 1977 two H absorption cells were flown aboard Pronoz 5 and Prognoz 6 spacecrafts. The cell acts as a negative filter , by absorbing photons whose wavelength is within 0.001 nm of the Lyman α wavelength in its frame of reference. The reduction factor R is defined by the ratio of the intensity measured when the cell is active and partly absorbs the interstellar emission to the intensity measured when the cell is inactive. It is obvious that R depends on the Lyman α line profile , on the line of sight and on the distribution parameters of the interstellar wind. Bertaux et al (1985) have used the Prognoz data to estimate the temperature and the bulk velocity of the incoming H atoms assuming that the neutral Hydrogen flow was homogeneous far away from the sun , as in the case of the hot model (Thomas 1978).

The geometry of scan provided us with data both for the upwind and the downwind directions at a given position of observation. Yet , the data in the upwind direction gave an estimate of 19 km/s $\pm$ 1 km/s for the bulk velocity whereas the downwind data were fitted with 25 km/s $\pm$ 1 km/s. The discrepancy between these two values can be understood by considering the origin of the scattering atoms. Indeed , it has been shown by Lallement and Bertaux (1991) that the atoms which contribute to the upwind emission come from the nose of the heliopause whereas those which contribute to the downwind emission originate from the sides. This shows that the flow must not be assumed homogeneous. Due to charge exchange processes , the atoms which cross the heliopause near the stagnation line are more decelerated than those that cross far from the nose (see next paragraph). An effect is to be expected too in density and temperature since deceleration increases the density and heats up the neutrals. It should be noted that the likeliness of such an effect depends on the rate of H-H$^+$ charge exchange which enables the coupling between the plasma and the neutrals.

We then created a model assuming that at a given radius from the sun (i.e. 100 AU) the distributions of velocity , density and temperature of the H atoms were some functions of the angle between the position vector of the point (The sun being the origin) and the interstellar velocity vector. This takes the symmetry of revolution around the wind axis into account. We then derived the distributions inside the heliosphere and computed the line shape and the reduction factor. This model is not very sensitive to density since we have neglected absorption along the photon path and assumed that we had only single scattering. By choosing step functions and linear functions , we have been able to improve the agreement between the model and the data (figure 1) , but we had to assume great differences in the values at the nose and on the side , e.g. near the wind axis we found a bulk velocity of 16 km/s $\pm$ 1 km/s and a temperature of 7000 K $\pm$ 500 K whereas for the points at an angle equal to or greater than 90 ° with the wind axis (the 'sides' of the heliopause) we found 36 km/s $\pm$ 1 km/s and 5000 K $\pm$ 500 K . Those parameters were derived with a density of neutral Hydrogen at 100 AU of 0.065 cm^{-3} , a ratio of radiative pressure to solar gravitation μ equal to 0.55 and a lifetime at 1 AU for the H neutrals of 1.2 10^6 s.

The results obtained here show that the Prognoz data suggest that the heliopause has really an effect on the neutral H atoms. Indeed , the data agree with a greater bulk velocity and a lower temperature for the atoms far from the stagnation line as expected from the plasma models. Yet , the absolute values given here are the results of a model which does not take multiple scattering , absorption along the light path and anisotropy of the solar wind into account. We have also assumed that at 100 AU the distributions functions of velocity are gaussian , yet Baranov's model for instance shows this is not true. More precise results will require to model those effects.

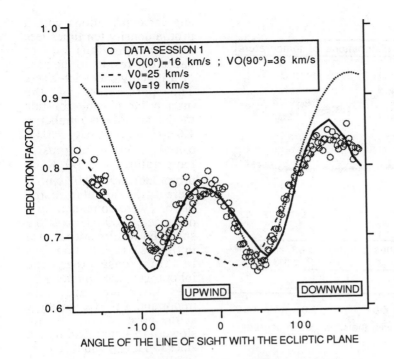

FIGURE 1
This graph shows the comparison between the data of Session 1 (see Lallement et al 1985) and the model in three cases. First an homo- geneous model with parameters 19 km/s and 6500 K for a lifetime at 1 AU of neutral H of 1.2 10^6 s and μ=0.65 .This model agrees well with upwind data.The second homogeneous model used the same values except for 25 km/s.In this case we have good agreement with downwind data. The best global fit is obtained with the inhomogeneous model with 16 km/s and 7000 K at the nose , 36 km/s and 5000 K on the sides , a lifetime of 1.2 10^6 s and μ=0.55.

BARANOV'S TWO SHOCK MODEL : MONTECARLO MODEL OF NEUTRAL FLOW

Plasma models expect a strong deceleration of the interstellar plasma near the stagnation line with a strong increase of both temperature and density and less important effects for the plasma more distant from the stagnation point , which means , on the sides , a plasma less decelerated and less hot. Here we present results obtained with a Montecarlo model of the neutral flow through a Baranov type two-shock interface (Malama 1990). This two-shock model (Baranov at al 1981) is a hydrodynamic model for the plasma interface between the solar wind and the interstellar wind. The solar wind plasma , with a speed ranging from 400 km/s to 800 km/s and a temperature of about 10^6 K , is supersonic and is decelerated through the inner shock before reaching the interface. It is still not sure whether the interstellar plasma flow is supersonic or not and whether there is a second shock (the bow shock) or not. The hydrodynamic model used here assumes that the Mach number of the interstellar wind is larger than 1.

The plasma model has been used with a Montecarlo simulation of the charge exchange processes between H and H^+ (Malama 1990) , which has enabled us to compute the distributions of the neutrals inside the solar system. It must be noted that the effect of the neutrals on the source functions of the plasma model has not been taken into account , which is necessary to solve the problem in a self consistent way. Yet , the results show the possible effects of the heliopause on the neutral H distributions.

In figures 2 to 4 , we present the distributions computed with two different plasma models. The values chosen for the interstellar plasma are Mach number = 2 ; V0 = 29 km/s (bulk velocity at 'infinity') ; H atom density = 0.04 cm^{-3} ; temperature = 7600 K ; proton density = 0.2 cm^{-3} . The distributions were computed at 80 AU from the sun , inside the inner shock which is at about 90 AU in the upwind direction for this model. The large proton density has been chosen to enhance the effects on H atoms, even if this value seems to be too large. We see from figure 2 that the atoms are globally heated up from 7600 K . Yet the temperature is larger at the nose where it reaches 12000 K than at the sides where it is about 9000 K. The difference is about 30 % . At the same time along the wind axis , the neutrals are slowed down from 29 km/s to 16 km/s at the nose and 21 km/s on the sides (figure 3). Again the difference between the sides and the nose is about 30 % and the more heated up neutrals are the more slowed down . The density angular distribution (figure 4) shows that there are fewer atoms on the sides than at the nose ; in fact the density at the nose is twice the value of density on the sides. We have also computed the

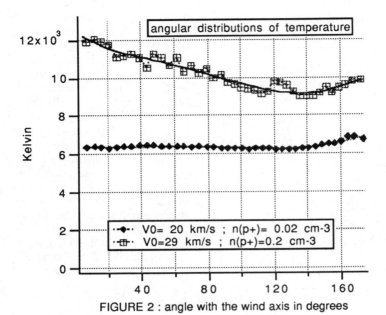

FIGURE 2 : angle with the wind axis in degrees

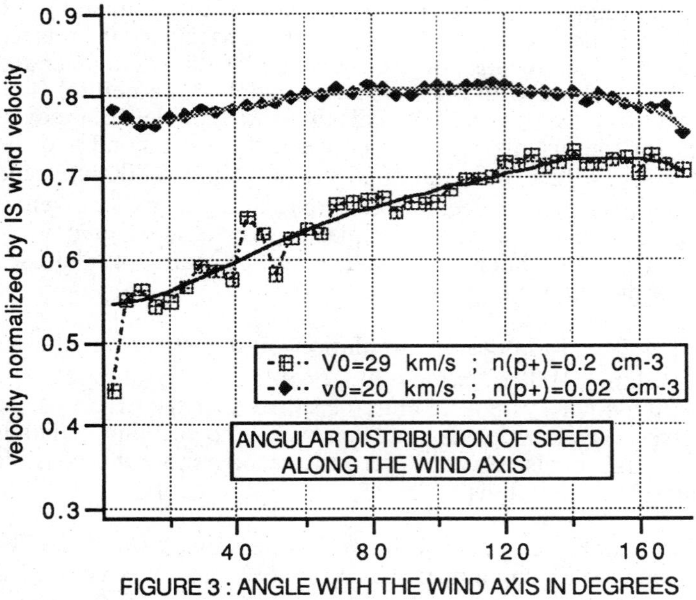

FIGURE 3 : ANGLE WITH THE WIND AXIS IN DEGREES

angular distributions with a proton density ten times less than the previous one to check our results in the case of a much weaker interface. The parameters of the interstellar plasma for this model are Mach number = 1.6 ; H atom density = 0.07 cm^{-3} ; V0 = 20 km/s ; temperature = 5500 K ; proton density = 0.02 cm^{-3}. Here , the angular distributions were computed for a distance of 100 AU from the sun. The results appear also in figures 2 to 4. It is obvious in this case that the inhomogeneities induced by the interface are less important. Yet , we see that , at the nose , the atoms are slower (15 km/s) than on the sides (16 km/s) and that there is still a greater density near the stagnation point , with about 25 % less on the sides of the heliosphere. The effects on temperature are not strong enough to be of interest.

From all this we can deduce that at the interface the neutral H atoms are slowed down and heated up , but that the effect is more important for those who cross near the stagnation point where the plasma is hotter , denser and more decelerated. So, through modeling of H-H$^+$ charge exchange processes with a two-shock plasma model of the heliopause , we see that the expected effects are qualitatively the same as those derived from the study of Prognoz 5/6 H cell data. An improvement of both models should provide better constraints on the interstellar wind parameters and on the heliopause.

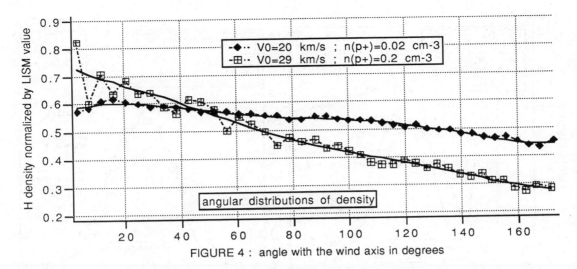

FIGURE 4 : angle with the wind axis in degrees

LYMAN α LINE PROFILE

A good diagnostic of the effect of the heliopause on the neutral Hydrogen can be given by the comparison of the downwind and upwind line profiles. Indeed, the atoms which contribute to the upwind lines of sight come mainly from the nose of the heliopause, whereas those which contribute to the downwind lines come from the sides. Then for a given position of observation, we can compare the upwind and downwind Lyman α profiles. By fitting the upwind profile with our homogeneous model , we find the values of the parameters of the flow near the stagnation line. We can then compare the downwind line profile computed with these values with the measured downwind profile. Discrepancies between the two profiles would be the proof of inhomogeneities of the interstellar flow created by the effects of the heliopause on the interstellar neutral Hydrogen distributions.

We have tried this scheme on the line profiles computed by the Montecarlo model using the two sets of parameters presented in the previous section. In the case of a proton density of 0.2 cm^{-3} , the upwind line was fitted with 17 km/s and 17000 K. The three downwind profiles displayed in figure 5 are very different. We see from the line shape that the Montecarlo profile is cooler and more Dopplershifted than the profile of the homogeneous model , which means that the atoms of the Montecarlo model were globally faster and cooler . And , since this is a downwind profile , these scattering atoms originate from the sides of the heliopause. The third curve shows the profile expected if there is no interaction between the interstellar flow and the heliopause , there is no deceleration and no warming up. These features appear also in figure 6 , in the case of a proton density of 0.02 cm^{-3} . The upwind homogeneous fit of the upwind Montecarlo line yielded 15 km/s and 8700 K whereas the external parameters before interaction were 20 km/s and 7200K. But , here too , the Montecarlo downwind profile is cooler and more Dopplershifted which means that the atoms on the sides are colder and move faster than those near the stagnation line.

This example has shown that we could see the possible effects of the heliopause on interstellar Hydrogen from Lyman α profile even if this effect is not large . In fact , for the second set of parameters , the difference of bulk velocity and temperature between the nose and the side is 1 km/s and 600 K respectively . So , whatever the real situation , if the heliopause imprints at least a variation of 1 km/s and 600 K in the neutral hydrogen distribution , we are able to detect its effects by comparing upwind and downwind Lyman α line profiles. Besides , inside the heliosphere , H atoms are mainly ionized by EUV photoionization or by charge exchange with solar protons. The distribution in the inner part of the heliosphere keeps track of what it was after crossing the heliopause and , since the backscattered intensity is integrated all along the line of sight , the discrepancies induced by charge exchange coupling can be seen whatever the real position of the heliopause from the sun.

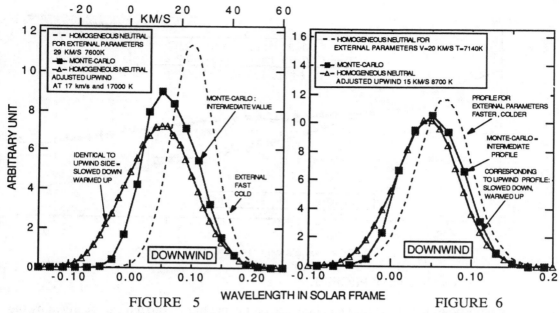

FIGURE 5   WAVELENGTH IN SOLAR FRAME   FIGURE 6

Both figures show the downwind Lyman α profile computed with three different models.From left to right for each figure, we have first, the downwind profile of the homogeneous model with the values adjusted with the upwind Montecarlo profile, then the downwind Montecarlo profile, and then the downwind profile of the homogeneous model with the values of the interstellar wind before interaction. Figure 5 : proton density 0.2 cm^{-3} ; Figure 6 : 0.02 cm^{-3} . The ordinate is in arbitrary unit since for each line profile the surface area has been normalized to 1.

CONCLUSION

As we have seen earlier , the Prognoz 5/6 data provide an interesting clue on the nature of the heliopause which qualitatively is consistent with results obtained from Baranov's two-shock model when charge exchange processes are taken into account. Yet , both models require improvements if we wish to get better constraints on the interstellar medium and the heliopause. We have seen also that , in principle , we are able to detect inhomogeneities imprinted to neutral H distributions of velocity and temperature by studying upwind and downwind line profiles.

Though precise measurements of Lyman α line profile have not yet been made , the use of Hydrogen absorption cells is a good way to get interesting data on this subject and other related aspects as anisotropies of the solar wind or possible departure from axisymmetry of the heliosphere caused by magnetic fields , for instance. Such data should be acquired by the SWAN experiment on the SOHO space mission scheduled around 1995.

REFERENCES

Baranov VB , Ermakov MK & Lebedev MG , **Sov. Astron. Lett. 7** , 206(1981)
Baranov VB , Lebedev MG & Malama YG ,   **Astrophys. J. 375** , 347 (1991)
Bertaux JL & Blamont JE , **Astr. Astrophys. 11** , 200 (1971)
Bertaux JL , Lallement R , Kurt VG & Mironova EN, **Astr. Astrophys. 150,** 1(1985)
Fahr HJ , **Adv. Space Res. 6** , 13 (1986)
Fahr HJ , Neutsch W , Grzedzielski S , Macek W , Ratkiewicz-Landowska R , **Space science reviews 43** , 329 (1986)
Fahr HJ , Ratkiewicz-Landowska R & Grzedzielski S , **Adv Space Res 6** , 389 (1986)
Holzer TE **Ann. rev. Astron. Astrophys.** (1989)
Lallement R , Bertaux JL , Dalaudier F , **Astron. Astrophys. 150** , 21 , (1985)
Lallement R & Bertaux JL , **Astron. Astrophys 231** , L3 (1991)
Malama YG , **Astrophys. and space science** vol. 176 , 21 (1991)
Suess ST , the heliopause , **Review of Geophysics** (1990)
Thomas GE , **Ann. Rev. Earth Planetary Sci. 6** , 173 (1978)

MHD MODELLING OF THE HELIOSPHERIC INTERFACE

R. Ratkiewicz

*Max-Planck-Institut für Aeronomie, Postfach 20, D-3411 Katlenburg-Lindau,
Germany (and Space Research Center, Bartycka 18a, 00-716 Warsaw, Poland)*

ABSTRACT

On the basis of the thin layer approximation the effect of the interstellar magnetic field on the heliospheric interface is studied. The results obtained are in a good agreement with previous predictions concerning the shape and distance of the solar wind termination shock.

The thin layer approximation (TLA) approach /3/ already applied to model the heliospheric boundary in a pure hydrodynamic case by Ratkiewicz /7/ can be generalized to a case of a homogeneous interstellar magnetic field of arbitrary magnitude and orientation. In the case of two hypersonically counterflowing gases besides the heliopause as a separatrix involving two self consistent flows, two shocks must also appear. The subsonic region between them is restricted to a thin layer with a constant velocity throughout it (Fig. 1). According to this assumption the heliopause is represented by a curve $r = r(\theta)$ derived from a third order, nonlinear differential equation (arising from the momentum equations as described by Ratkiewicz /7/) of the form:

$$rr''' = 2rr' + 3r'r'' - F_3[F_2'/F_2 - F_3F_1/F_2] \qquad (1)$$

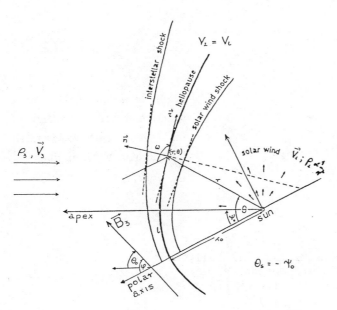

Fig.1 Assumed geometry:
θ_s-deviation angle of a symmetry axis;
θ_o-magnetic field inclination angle;
ω-tangent - symmetry axis angle;
ℓ-distance measured from the stagnation point

where,

$$F_1 = 2\pi r \sin\theta(r^2 + r'^2)^{1/2}\{(\rho_3 V_3^2 \cos 2\theta_s - B_3^2/4\pi \cos 2(\theta_o - \theta_s))\times$$

$$(r\cos\theta + r'\sin\theta)(r\sin\theta - r'\cos\theta)/(r^2 + r'^2) + \rho_1 V_1^2 rr'/(r^2 + r'^2)+$$

$$[\rho_3 V_3^2 \sin 2\theta_s - B_3^2/4\pi \sin 2(\theta_o - \theta_s)](r\cos\theta + r'\sin\theta)^2/(r^2 + r'^2)-$$

$$1/2[\rho_3 V_3^2 \sin 2\theta_s - B_3^2/4\pi \sin 2(\theta_o - \theta_s)]\}$$

$$F_2 = 2\pi r \sin\theta(r^2 + r'^2)^{3/2}\{(\rho_3 V_3^2 \cos 2\theta_s - B_3^2/4\pi \cos 2(\theta_o - \theta_s))\times$$

$$(r\cos\theta + r'\sin\theta)^2/(r^2 + r'^2) - \rho_1 V_1^2 r^2/(r^2 + r'^2)-$$

$$[\rho_3 V_3^2 \sin 2\theta_s - B_3^2/4\pi \sin 2(\theta_o - \theta_s)](r\cos\theta + r'\sin\theta)(r\sin\theta - r'\cos\theta)/(r^2 + r'^2)+$$

$$\rho_3 V_3^2 \sin^2\theta_s + B_3^2/8\pi \cos 2(\theta_o - \theta_s)\}$$

$$F_3 = r^2 + 2r'^2 - rr'' \tag{2}$$

where $\rho_1 = \rho_E r_E^2/r^2 \cdot \rho_E$ and V_1 are the solar wind density at the Earth's orbit and unperturbed solar wind velocity, respectively. $r_E = 1$ a.u., and r is the distance from the sun. ρ_3, V_3, B_3 are the density, velocity and magnetic field of the interstellar plasma, respectively.
The initial conditions are:

$$r(0) = r_0, \quad r'(0) = 0, \quad r''(0) = \left[-1/2(5M_a^2 - 9/2) + \sqrt{1/4(5M_a^2 - 9/2)^2 - 2(1 - M_a^2)}\right]r_o, \tag{3}$$

where r_o is the minimum heliocentric distance to the heliopause given by

$$r_o^2 = r_E^2 \rho_E V_1^2/(\rho_3 V_3^2 \cos^2\theta_s - B_3^2/8\pi \cos 2(\theta_o - \theta_s)), \quad M_a = V_3/V_a, \quad V_a^2 = B_3^2/4\pi\rho_3 \tag{4}$$

For an arbitrary magnetic field inclination angle θ_o, a deviation angle θ_s of a symmetry axis /4,5/ can be calculated from the formula

$$\rho_3 V_3^2 \sin 2\theta_s = B_3^2/4\pi \sin(2\theta_s - 2\theta_o) \quad -\pi/2 \leq \theta_s \leq \pi/2 \tag{5}$$

One might wish to determine the heliopause interface in the MHD case taking into account a specific value and direction of an arbitrary interstellar magnetic field. In order to do this the deviation angle θ_s should be found (Fig.2).

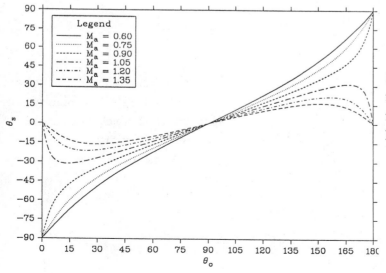

Fig.2 Deviation angle θ_s as a function of inclination angle θ_o for various values of interstellar Alfvénic Mach number M_a

Then the curve $r = r(\theta)$ representing the solution of equ.(1) for given M_a should be selected as the heliopause (Fig.3). To obtain the inner shock the procedure described by Fahr et al. /5/ can be applied. Finally, it is necessary to calculate the heliocentric distances to the heliopause and the termination shock.

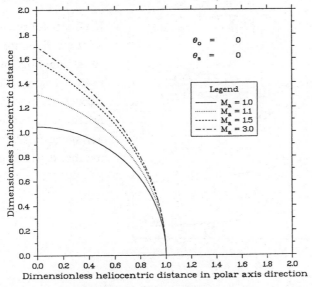

Fig.3 Heliopauses in mhd thin layer approximation

It is worthwhile to outline the difference in a resulting deviation angle for a weak $(M_a > 1)$ and strong $(M_a < 1)$ magnetic field, when it is parallel to the interstellar velocity (Fig.2). The minimum distance to the heliopause in both cases is shown schematically in Fig.4.

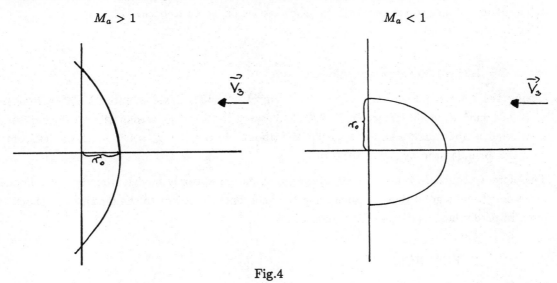

Fig.4

As shown in Fig.5 for $M_a > 1$ (weak magnetic field) the minimum distance to the heliopause does not depend strongly on the magnetic field inclination angle θ_o. On the contrary, for $M_a \approx 1$, this distance is extremely sensitive to variations of θ_o and for $\theta_o = 0°$ is more than 1.5 times larger than at $\theta_o = 90°$ for standard solar wind and interstellar plasmas conditions.

On the other hand the thin layer approximation approach can be used to discuss some properties of the solar wind termination shock (for $M_a \geq 1$) since by definition both curves are very close to each other.

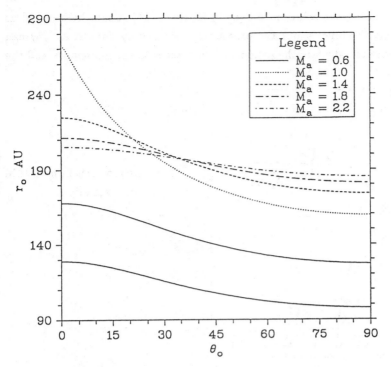

Fig.5 Heliocentric distance to the heliopause (5 curves) as a function of an inclination angle. The lowest curve corresponds to the heliocentric distance to the inner shock for Alfvénic Mach number equal to 0.6

In particular, if $M_a \rightarrow 1$, the heliospheric cavity becomes approximately spherical (compare Fig.3) as predicted by Parker /6/. The outer Mach number does not affect very much the position of the discontinuity surfaces in a pure hydrodynamic model /2/. If it is still valid in MHD case the heliocentric distance to the termination shock can be expected to occur at

$$R_o = r_o/C$$

where $C = 1.25 - 1.35$ (see Fig. 5 - lowest curve)

Taking the solar wind parameters, $n_E = 5/\text{cm}^3, V_1 = 400\text{km/sec}$ and the LISM parameters, $n_3 = 0.04/\text{cm}^3, V_3 = 23\text{km/sec}, M_a = 0.6$ the distance to the shock termination of the supersonic solar wind is about 100 a.u. for $\theta_o = 90°$ and about 130 a.u. for $\theta_o = 0°$, which one can obtain with the formula used by Axford and Ip /1/.

Therefore we conclude that the thin layer approximation not only models directly the heliopause but seems to be a good tool for calculating the heliocentric distance to the termination shock and often in predicting the shape of the inner shock.

REFERENCES

1. Axford, W.I., Ip, W.-H., *Adv. Space Res.* **6**, No 2, 27 (1986)

2. Baranov, V.B., **Pr-1126**, *Space Research Institute, AN SSSR* (1986)

3. Baranov, V.B., Krasnobayev, K.V., *Gidrodinamicheskaja teoria kosmicheskoj plasmy*, Nauka, Moscow, 1977

4. Fahr, H.J., Ratkiewicz, R., Grzedzielski, S., *Adv. Space Res.* **6**, No 1, 389 (1986)

5. Fahr, H.J., Grzedzielski, S., Ratkiewicz, R., *Annales Geophysicae* **6**, (4), 337 (1988)

6. Parker, E.N., *Interplanetary Dynamical Processes,* Interscience, New York, 1963

7. Ratkiewicz, R., *Astronomy and Astrophysics* **255**, No 1/2, 383 (1992)

IPS OBSERVATIONS OF THE SOLAR WIND VELOCITY AND MICROSCALE DENSITY IRREGULARITIES IN THE INNER SOLAR WIND

B. Rickett

ECE Department, University of California, San Diego, CA 92093-0407, U.S.A.

ABSTRACT

Preliminary results are presented from a recent interplanetary scintillation (IPS) experiment using the MERLIN radio-telescope array in England. The mean and rms velocity are studied versus distance from 10-100 R_o and latitudes 80 deg S to 50 deg N, at a time of high solar activity. Also estimated over this range is the axial ratio of the microscale density structures responsible for the scintillation. Whereas the mean velocity shows little systematic variation with either distance or latitude, the rms velocity and axial ratio increase toward the sun. The microscale density spectrum forms a continuum with the spectrum of compressive fluctuations reported from Helios measurements over 0.3-1.0 AU.

INTERPLANETARY SCINTILLATION (IPS) MEASUREMENTS

Radio propagation measurements have long shown the existence of a distribution of microscale density irregularities in the solar wind. As shown by Coles and Harmon /1/, these have small scales (10's to 100's Km) and can be characterized by a wavenumber spectrum that is power law in form, with a cut-off at small scales as in a turbulence spectrum. This paper presents preliminary results on the velocity and spatial anisotropy from a series of radio measurements using the IPS method.

The Multiple Element Radio-Linked INterferometer (MERLIN) array of 6 antennas operated from Jodrell Bank, England was used for a series of IPS observations from 28 April - 15 May, 1989. At each of the 6 antennas the fluctuating intensities (scintillations) were recorded at 1658 MHz from a set of small diameter radio sources as they passed close to the sun. Solar distances covered 10-100 R_o and solar latitudes -80 to +50 deg. Cross-correlation functions of intensity were computed for each of the 15 antenna pairs. The basic idea is that the motion of the diffraction pattern can be measured by fitting a model to the intensity cross-correlations across each baseline (5-120 Km). Hence the motion of the solar wind at the point of closest approach of the line-of-sight to the sun can be estimated /2-4/.

The model assumed for the correlation functions uses the linear superposition of signal components scattered along the line of sight. Each has the same spatial parameters, but there is a range of velocities (Ekers and Little /2/). The velocities are assumed radial with a mean (V_r) and an rms range (σ_r). This model is appropriate for weak scintillation conditions, which apply for distances beyond about 18 R_o. For data closer to the sun only the average velocity V_r will be properly estimated; a "strong scintillation" model will be needed for full analysis. The spatial parameters of the pattern are described by characteristic scales in directions along and perpendicular to the mean spiral field. Their ratio is used to determine an axial ratio (AR), which is usually greater than one; it is less than one if the apparent scale is smaller along the spiral than transverse to it. Preliminary results on the above-mentioned quantities have been derived from about 100 observations of 3 radio sources. An analysis, with a distribution of velocities in the perpendicular as well as radial directions, improves the fits to the correlations, but it is not yet complete.

RESULTS

The time series of the estimated parameters were examined for each source. Several high velocities were seen during 2-6 and on 11 May, following 9 flares of importance 2 or more in these intervals. Since such transients confuse the interpretation of any variations versus solar distance and latitude, they have been excluded from what follows. The effective solar distance and solar latitude of the scattering region change in a coupled fashion, as the line of sight moves with respect to the sun. Thus latitude and distance dependences must both be considered. Figure 1a shows V_r from the 3 radio sources versus latitude. There are no signs of fast polar streams, consistent with results from 0.5-1 AU at times of maximum solar activity /3,4/. We thus concentrate on distance dependence. Figure 1b shows V_r versus distance. The results show substantial variability over 10 to 50 R_o, but no clear signature of acceleration. In contrast Coles et al. /5/ were able to measure V_r versus distance along particular stream lines and often found an acceleration of about 100 km/s between 20 and 100 R_o.

Figure 1c shows the rms radial velocity (σ_r) versus distance, with values noticeably enhanced in the inner solar wind (10-40 R_o). σ_r has a contribution due to the spatial averaging of signals scattered from points along the line of sight with differing projected velocities. For a uniform spherical flow this contributes about 60 Km/s (and an 18% reduction in the mean radial velocity); however, if there is a range of radial flow velocities in the central part of the line of sight, σ_r represents the range of velocities present. However, such a line of sight contribution should not depend on distance. Earlier measurements showed an increase in σ_r inwards of about 20 R_o (/6,2/), which thus might be due to waves in the inner solar wind. Note, however, that the present data exhibit such an enhancement out to 40 R_o. Figure 1d shows the axial ratio of the scintillation pattern versus distance, and provides a lower limit to the axial ratio for the density microstructure (on scales 10-150 Km) that causes the IPS. Again this is enhanced inwards of about 40 R_o. These results agree with the measurements by Armstrong et al. /7/, who found the axial ratio to increase from near 4 at 10 R_o to over 10 at 2 R_o.

DISCUSSION AND CONCLUSIONS

The microscale density irregularities, studied by these and other radio probing methods /1,8/, fit well with the density fluctuations in the solar wind, investigated recently from Helios measurements /9,10/. The Helios plasma data from 0.3-1.0 AU have been analyzed as compressible turbulence, showing increased density fluctuations at high wavenumbers, particularly in the body of high speed streams where outward travelling Alfvenic turbulence dominates. The highest wavenumbers sampled were 1-2 orders of magnitude below the wavenumbers responsible for IPS, nevertheless the Helios spectra combined with the radio-probing results give strong evidence for a turbulent density spectrum. It is of particular interest that there is an enhancement in the density spectrum at high wavenumbers in the inner solar wind, seen as a flattening in the spectrum before a cut-off as in turbulent dissipation /1/. It seems possible that this enhancement near the sun is associated with the increased axial ratios and rms velocities observed out to 20-40 R_o and is related to an extended deposition of alfvenic energy. The present IPS data support the earlier result /7/ that the smallest density structures are elongated, but shows that their axial ratio slowly decreases out to about 40 R_o. Various theoretical analyses have addressed the question of densities in large amplitude Alfvenic turbulence; for example, Harmon /11/ emphasized the increasing compressibility with increasing wavenumber, particularly for obliquely propagating waves in low β plasma. Kuo, Whang & Lee /12/ discuss a "filamentation instability" in which Alfven waves interact with magnetoacoustic waves to produce filamentary density structures, particularly in the low β of the inner solar wind. Montgomery et al. /13/ analyzed density perturbations by assuming an equation of state for the plasma.

The IPS measurements relate to the local plasma parameters at the point where the line of sight passes closest to the sun. The IPS velocity (multiplied by 1.18 to correct for velocity projection effects) measures the apparent radial velocity of elongated density irregularities on very small scales. These irregularities are presumably aligned with the local magnetic field, and may then be subject to transverse motion due to longer wavelength Alfven waves. As discussed by Coles /14/, this can cause several effects: an upwards bias to the IPS radial velocity; a reduction of the apparent axial ratio; contributions to radial and tangential rms velocities. These ideas are particularly important given the dominance of outward travelling Alfvenic turbulence at 0.3 AU. There is a clear need reanalyze these IPS data with a model that includes these effects.

Fig. 1. MERLIN IPS observations from the radio sources indicated at the top. Note that the plots only show the nominal fitting error bars, which would be increased with the inclusion of systematic errors. (a) V_r mean velocity versus solar latitude. (b) V_r mean velocity versus solar distance. (c) σ_r RMS radial velocity versus distance. (d) Axial-ratio versus distance. Data closer to the sun than the dotted lines are nominally in strong scattering, and are thus less reliable.

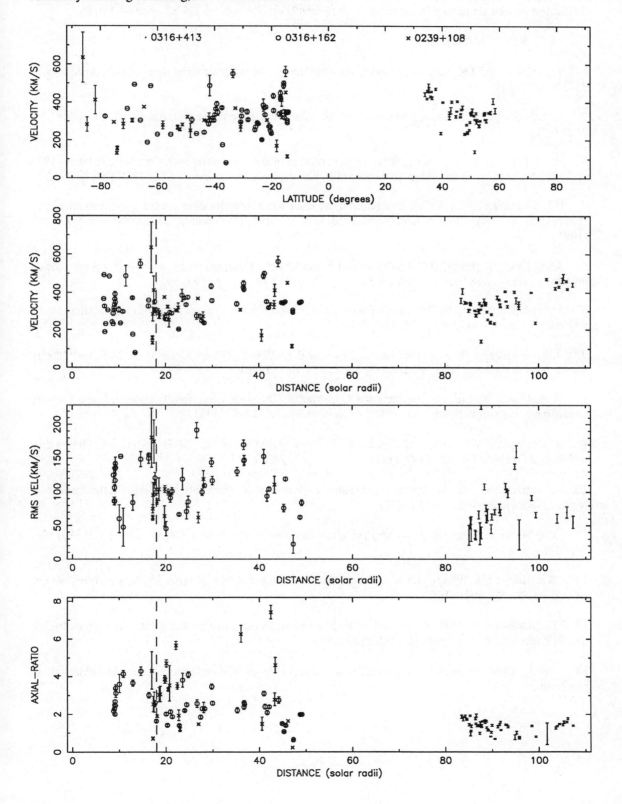

Conclusions. (1) The present data successfully use a new antenna array to probe the polar regions of the inner solar wind, however they show no sign of a fast polar stream in 1989 (high solar activity). (2) No substantial acceleration is seen in the average IPS speed between 10-100 R_o. (3) The flare activity in early May 1989 caused a sequence of transient velocity increases detected at distances of 20-100 R_o. (4) An increase in rms radial velocity is observed inwards of about 40 Ro. (5) Axial ratios measured for the diffraction pattern are mostly in the range 2-6 between 10-40 Ro and 1.0-2.5 between 80-110 R_o.

REFERENCES

1.   W.A. Coles and J.K. Harmon, Propagation observations of the solar wind near the sun, *Astrophys.J.*, 337, 1023 (1989)

2.   R.D. Ekers and L.T. Little, The motion of the solar wind close to the sun, *Astron. & Astrophys.*, 10, 310 (1971)

3.   M. Kojima and T. Kakinuma, Solar cycle evolution of the solar wind speed structure between 1973 and 1985 observed with the interplanetary scintillation method, *J.Geophys.Res.* 92, 7269 (1987)

4.   B.J. Rickett and W.A. Coles, Evolution of the solar wind structure over a solar cycle: interplanetary scintillation velocity measurements compared with coronal observations, *J.Geophys.Res.* 96, 1717 (1991)

5.   W.A. Coles, R. Esser, U-P. Lovhaug, and J. Markkanen, Comparison of solar wind velocity measurements with a theoretical acceleration model, *J.Geophys.Res.* 96, 13,849 (1991)

6.   J.W. Armstrong and R. Woo, Solear wind motion within 30 R_o: spacecraft radio scintillation observations, *Astron. & Astrophys.*, 103, 415 (1981)

7.   J.W. Armstrong, W.A. Coles, M. Kojima and B.J. Rickett, Observations of field-aligned density fluctuations in the inner solar wind, *Astrophys.J.*, 358, 685 (1990)

8.   W.A. Coles, W. Liu, J.K. Harmon and C.L. Martin, The solar wind density spectrum near the sun: results from voyager radio measurements, *J.Geophys.Res.* 96, 1745 (1991)

9.   R. Grappin, A. Mangeney and E. Marsch, On the origin of solar wind MHD turbulence: Helios data revisited, *J.Geophys.Res.* 95, 8197, (1990)

10.   E. Marsch and C-Y. Tu, Spectral and spatial evolution of compressible turbulence in the inner solar wind, *J.Geophys.Res.* 95, 11945 (1990)

11.   J.K. Harmon, Compressibility and cyclotron damping in the oblique alfven wave, *J.Geophys.Res.* 94, 15399 (1989)

12.   S.P. Kuo, M.H. Whang and M.C. Lee, Filamentation instability of large amplitude alfven waves, *J.Geophys.Res.* 93, 9621 (1988)

13.   D. Montgomery, M.R. Brown and W.H. Matthaeus, Density fluctuation spectra in magnetohydrodynamic turbulence, *J.Geophys.Res.* 92, 282 (1987)

14.   W.A. Coles, Interplanetary scintillation, alfven waves and anisotropic microstructure, these proceedings.

INDICATIONS FOR AXIAL ASYMMETRIES IN THE INTERPLANETARY HYDROGEN DISTRIBUTION DERIVED FROM PIONEER-10 LYMAN-ALPHA DATA

K. Scherer* and D. L. Judge**

*Institut für Astrophysik und Extraterrestrische Forschung, Universität Bonn, Bonn, Germany
** Department of Physics, University of Southern California, CA, U.S.A.

ABSTRACT

For the first time highly spatially resolved Pioneer 10 Lyman-α data are presented.The Lyman-α backscatterd resonance glow shows a clear and persistant asymmetry when resolved in space.

INTRODUCTION

In order to understand the data to be presented we first recall some facts about the Pioneer 10 Lyman-α photometer as mounted on the spacecraft (for more details see, e. g. /1/, and a theoretical approach may be found in /2,3/, which in the future should be compared with the data).

The Pioneer 10 probe is moving in the downwind-direction 3° above the ecliptic plane with its spin axis orientated toward the Earth. The optical axis of the photometer is tilted at an angle of 20° with respect to the space-craft rotation axis. Thus, the field of view (1°×10°) of the instrument rotates and sweeps out a conical viewing shell with a full cone angle of 40°. After the encounter with Jupiter the spacecraft moves more or less along a straight line in the downwind direction relative to the inflowing interstellar gas. The orbit geometry and the changes of the field of view are shown in Figure 1.

The clock angle of the instrument, φ, is measured counterclockwise from a plane parallel to the plane of the ecliptic as looking toward the detector from Earth. The angular resolution Δφ depends on the bitrate and ranges from 3° to 180° for each data sample. The actual resolution can be improved, however, because the data sampling period is not commensurable with the spacecraft spin period. This gives rise to a phase shift of the origin of the clock angle φ, thereby providing differential data. The differential data presented here covering the interval ranging from January 1976 to June 1978 have a mean resolution of about 1°.

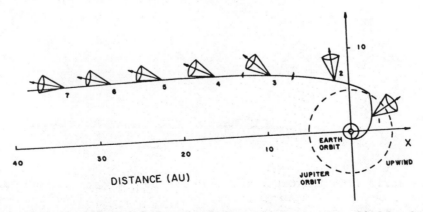

Fig. 1. Pioneer 10 spacecraft trajectory and photometer field of view. The bars indicate the position of the spacecraft during the observation period. (Taken from /4/)

REDUCTION AND PRESENTATION OF THE DATA

In Figure 2 we present one typical example of uncalibrated intensity of the data versus the clock angle φ. The linear plot shows an interesting feature,

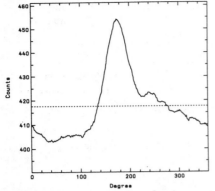

a **sinusoidal variation** of the glow dependence on the clock angle φ and therefore on the direction of view. This variation must be associated with either a nonuniform density distribution of atomic hydrogen or with a nonuniform illumination or ionisation of the local interstellar gas. The peak at $\varphi \simeq 180°$ in Figure 2 is a typical the contamination of the data due to a star.

Fig. 2. The uncalibrated Lyman-α glow intensity versus clock angle. The daily average of 25 Dec 1976 is shown.

While it often happens that a star contributes to the Lyman-α data most of the stars are quite bright compared to the Lyman-α glow. They are easily recognised by their characteristic Gaussian distribution signature and can thus be easily extracted from the data. To suitably handle the complete set of observational data, the Lyman-α glow intensity is developed into a Fourier series up to the fourth harmonic, while the stars are fitted by a Gaussian distribution, i. e.,the total observed glow is given by

$$I = a_0 + \sum_{n=1}^{4} \left[a_n \cos(n \cdot \varphi) + b_n \sin(n \cdot \varphi) \right] + \sum_{k=1}^{1,2} c_{3k-2} \exp\left(- \frac{\varphi - c_{3k-1}}{c_{3k}} \right) \qquad (1)$$

The last sum normally runs only up to 2 because more than two stars at different positions are usually not in the field of view. If there are more than two or if the stars are too close together, they must be handled much more carefully because the star signals overlap. In these few cases the stars are fitted first and then subtracted from the data after the Fourier decomposion is developed.

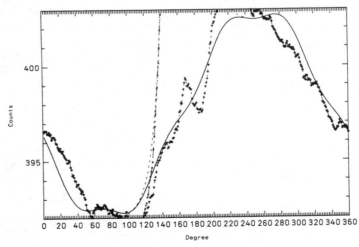

Fig. 3. The solid line indicates the best fit to the data without stars, the dashed line shows the fit with stars, and the dashed line connecting the crosses follows the orginal data. The plus signs indicate the data obtained after subtracting the contaminating stars.

In Figure 3 we give an example of a fit using equation (1) for the case of a high background stellar source. The dashed line starting at about 110° gives an impression of how such a star would contibute to the background glow. The fitting procedure in this case works as follows: First a least squares fit to the parameters of formula (1) is found. The contribution of the star is then subtracted from the original data, resulting in the curve indicated by the crosses in Figure 3. The "cleaned" data are again fit by equation (1), but without the last sum (so-lid line).

In the Fourier analysis of the glow of interest we found that the magnitude of the parameters a_n and b_n decreases by a factor of 10 for each consecutive n (e. g. $a_n/a_{n+1} \simeq b_n/b_{n+1} \simeq 10$). We therefore neglected all higher coefficients of the Fourier decomposition except n = 1. In the following a_0 is the **mean value** of the glow and the wave described by a_1,b_1 is the **first-order wave** or **first harmonic**. As expected, the difference in the standard deviation between the fits with and without stars is negligible.

In Figures 4 and 5 the Fourier coefficients are calibrated and normalised to a constant solar Lyman-α flux using the full disk solar 10 830 Å helium linewidth as a proxy for the solar variability, i. e. changes of the helium linewidth relative to the first day of observation are taken into account.

The mean value of the heliospheric Lyman-α glow obeys approximately a (1/r)-law, which is nicely inferred from the data shown in figure 4, as was also observed by Wu et al./4/. But there are more pronounced deviations from the 1/r law in the data presented here than were presented in /4/.

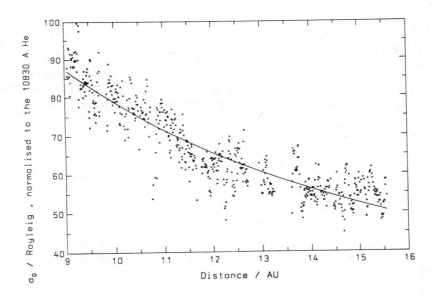

Fig. 4. The mean value of the Lyman-α glow versus distance. The 1/r dependence of the mean value is shown.

In Figure. 5 the origin of the clock angle φ and the position of the Lyman-α glow maximum are given in heliocentric ecliptic coordinates (β = ecliptic latitude, λ = ecliptic longitude). The dashed line in Figure 5 indicates the $\varphi = 0^\circ$ clock angle, while the solid line describes motion of the maximum of the first harmonic during the time interval of observation. The motion of the Earth is clearly seen in the variation of the 0° clock angle.

The first harmonic amplitude ranges from about 5% to 10% of the mean value
and is observed continously for the entire period of this analysis (ca. two
years). The maximum of the first harmonic amplitude is observed when viewing
(more or less) in the southward direction. Interpretation of the observed
features, which requires a modeling beyond the scope of this paper, will be
presented in a forthcoming article.

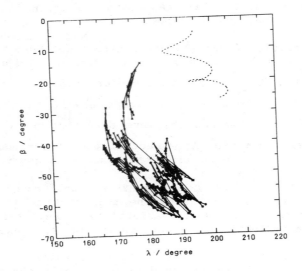

Fig. 5. Position of the maximum of the first harmonic in heliospheric ecliptic
coordinates (solid line). The dashed line gives the 0° clock angle.

ACKNOWLEDGEMENT

We would like our thanks to the Deutsche Forschungsgesellschaft (DFG) for
financial support within the framework of the project 'Heliowasserstoff'.
Further we acknowledge the NASA grant number NAG 2-146.

REFERENCES

1. F. M. Wu., K. Suzuki, R. W. Carlson, and D. L. Judge, Pioneer 10 Ultraviolet photometer observations of the interplanetary glow at heliocentric distances from 2 to 4 AU, Astrophys. J. 245, 1145 (1981)

2. H. J. Fahr and K. Scherer,   Heliospheric   neutral   gases   in   three-
dimensional   and   time-dependent   solar   force   and   ionization   fields,
Astron. Astrophys. 232, 556, (1990)

3. K. Scherer and H. J.Fahr, Fluctuations in the heliospheric hydrogen distribution induced by generalised and time-dependent interstellar boundary
conditions, Planet. Space Sci. 38, 1487 (1991)

4.) F. M. Wu., P. Gangopadhyay, H. S. Ogawa and D. L. Judge,   The   hydrogen
density of the local interstellar medium and an upper limit to the galactic
glow determined from Pioneer 10 ultraviolet photometer obervations, Astrophys. J. 331, 1004, (1988)

CORONAL HOLES AND SOLAR WIND STREAMS DURING THE SUNSPOT CYCLE

N. R. Sheeley Jr

E. O. Hulburt Center for Space Research, Naval Research Laboratory, Washington, DC 20375-5000, U.S.A.

ABSTRACT

Complementary synoptic observations of the Sun and interplanetary space have been obtained nearly continuously for more than two sunspot cycles and have led to new ideas about the origin of the solar wind. These observations show an inverse correlation between wind speed at Earth and magnetic flux tube expansion in the corona, with fast wind originating from slowly diverging tubes and vice versa. Although this result is consistent with the Skylab-era concept that fast wind originates from the center of a large isolated coronal hole, it implies that the wind may be even faster at the facing edges of like-polarity holes where the flux-tubes converge as they begin their outward extension. Thus, very fast wind ought to originate from the high-latitude edges of the circumpolar holes soon after sunspot maximum and from the mid-latitude necks of the polar-hole lobes during the declining phase of the cycle. The observed inverse correlation may be understood physically in terms of a model in which Alfven waves boost the wind to high speed provided that the wave energy flux is distributed approximately uniformly at the coronal base.

Introduction

The subject of coronal holes and high-speed solar wind streams commanded much excitement during the early 1970's when new instruments began to make routine observations of holes, and the wind obligingly became dominated by recurrent high-speed streams /1/. The complementary solar and interplanetary observations have continued almost continuously since that time, leading us to the corresponding phase of sunspot cycle 22. Thus, in preparation for the imminent declining phase of the cycle, we shall briefly review the observations that have accumulated since the Skylab era and have been summarized in previously published Bartels displays /2,3,4,5,6/.

The well correlated and simple recurrence patterns that existed during 1973-1976 disappeared at sunspot minimum and were never again quite as strong, even during 1982-1986 in the declining phase of the next cycle when high-speed streams returned. Sometimes, as in 1986, recurrent patterns of high speed occurred at Earth without coronal holes in the ±40° range of solar latitudes, suggesting perhaps that this fast wind originated from the polar holes. Sometimes, as in 1979-1981, low-latitude holes occurred without high speed streams, suggesting perhaps that the holes were "too small" or "too far toward the edge of the ±40° latitude range" to affect Earth. Sometimes, as in 1977 and 1985, fast wind was associated with "weak holes" whose contrast on X-ray or helium images was unusually low.

Although the central-meridian-passage times of the holes did not remain well correlated with wind speed, the polarity of those holes did remain well correlated with the polarity of the interplanetary magnetic field. This agreement persisted even through the years of sunspot maximum, degrading only near the 1986 sunspot minimum when the holes disappeared and the mean field strength became too weak to measure. At this point the interplanetary field lost its sectorial character, remaining positive throughout each rotation during spring when Earth lay at southern heliocentric latitudes where the Sun's field was positive, and negative in the fall when Earth lay at northern latitudes. This annual variation reinforces the speculation that the corresponding wind streams without low-latitude holes may have originated in the Sun's polar holes.

Attempts to Obtain an Empirical Understanding

The excellent hole/stream correlations that were obtained during the Skylab era suggested that fast wind ought to originate from the center of a large hole and slow wind ought to originate from its edge or from a small hole /7,8/. This conclusion also seemed to imply that the wind speed at Earth ought to be correlated with distance from the streamer belt in the corona, as measured by heliomagnetic latitude or source-surface field strength, for example. Such an idea was consistent with radio interplanetary scintillation (IPS) observations near sunspot minimum, which indicated that the wind was generally faster at high latitudes than in the ecliptic plane /9,10/. Thus, it was generally expected that the Earth would encounter fast wind if it lay on a field line originating sufficiently far inside the boundary of a large coronal hole, whether it be a low-latitude hole during the declining phase of the cycle or a polar hole at sunspot minimum.

The main difficulty with this Skylab-era concept was that the wind/latitude correlation varied during the sunspot cycle. Seasonal effects, associated with the 7.25° tilt of the Sun's axis, were very important near sunspot minimum, but not near maximum. During the Skylab mission, polar-hole lobes sometimes extended to only 40° latitude, but were nevertheless well correlated with high-speed wind at Earth. On the other hand, during the rising phase of the cycle, some holes came closer to the equator, but were not accompanied by fast wind at Earth /5,6/. In a quantitative study of this effect, Fry and Akasofu /11/ found a wind-speed/magnetic-latitude relation which was an order of magnitude steeper at sunspot minimum than at maximum. Gazis et al. /12/ reported similar results from their comparison of the Pioneer 11 and Voyager 2 spacecraft: Separated by only 16° in latitude, these two spacecraft gave comparable speeds prior to mid-1985, but much slower equatorial wind during the next two years.

A way out of this dilemma had already been provided by Levine et al. /13/. Using observations obtained during two months of the Skylab mission, they found a one-to-one correspondence between the occurrence of high-speed streams at Earth and the central-meridian-passage times of low-divergence regions in the Sun's corona. However, their result was not fully appreciated, partly due to unexplained discrepancies between their current-free extensions of the photospheric field and some observations: The extrapolated interplanetary field strengths were too small by a factor of about four, and the derived open-field regions were systematically smaller than the corresponding coronal holes /7/. These discrepancies are now thought to be due to an underestimate of photospheric field strength /14,15,16/, and to a previously inconsistent procedure for deriving the radial field from the observed line-of-sight field /17/.

The skepticism increased when Munro and Jackson /18/ reported that the north polar hole increased its angular size by a factor of seven as it extended outward from the Sun. Ignoring the difference between the expansion of an infinitesimal flux tube within a coronal hole and the total expansion of the entire hole, many people concluded that the high-speed wind must be a consequence of the large non-radial expansion, and the Levine et al. /13/ inverse correlation fell into obscurity.

The Levine et al. /13/ work was revived recently using observations over the entire interval 1967-1988 /19/. The inverse correlation which Levine et al. /13/ found during two months of 1973 occurs throughout the 22-year span. Unlike the coronal hole comparisons described above, the inferred speed waxes and wanes during the sunspot cycle without the need for any ad hoc adjustments. The better correlation during the declining phase of the recent cycle reflects the improvements in photospheric measurements which show less noise after 1981. This suggests that much of the detailed discrepancy between derived and observed wind speed may be due to limitations in the quality of the magnetic field measurements, rather than to a fundamental limitation of the correlation.

Within this data set, individual high-speed streams often were associated with open-field regions with low coronal expansion factors, but with little or no evidence of coronal holes in X-ray or helium images /5,20/. Figures 1

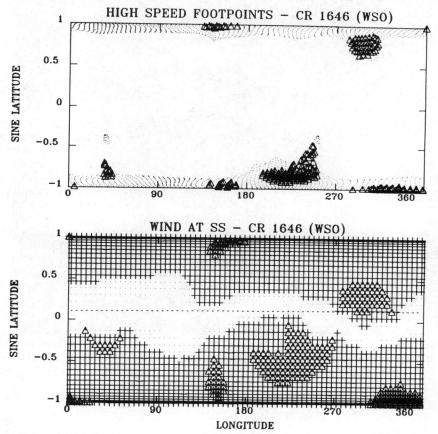

FIG. 1. Carrington maps of the photospheric distribution of open magnetic field lines (top) and of the source-surface distribution of flux-tube expansion factors (bottom), for the rotation that begins (at the right edge) on September 13, 1976. Open field lines, marked by dots or triangles in the upper map, spread out to fill the entire source surface, where they are indicated by corresponding triangles (expansion factors $f \leq 3.5$ and wind speeds at Earth $v \geq 650$ km s^{-1}), or by plusses ($3.5 < f \leq 9$; $650 > v \geq 550$ km s^{-1}), dots ($9 < f \leq 18$; $550 > v \geq 450$ km s^{-1}), or white areas ($f > 18$; $v < 450$ km s^{-1}). The dashed line indicates the sub-Earth track, which intersects "high-speed wind" at 300° longitude (September 18), three days prior to its observed arrival at Earth.

and 2 illustrate the case which Nolte et al. /20/ discussed. In Figure 1
the source-surface distribution of low-expansion flux tubes (bottom panel)
shows fast equatorial wind at 300° longitude (September 18, 1976), well
correlated with the observed 700 km s^{-1} recurrent stream at Earth on
September 20-21. The corresponding open-field region in the photosphere (top
panel) is located where Nolte et al. /20/ identified a very faint coronal
hole, barely visible in the northern hemisphere just east of disk center on
their September 16, 1976 X-ray image (Figure 2). Thus, the flux-tube
calculations show a clear origin of the high-speed stream while the X-ray
image shows only marginal evidence of an associated coronal hole.

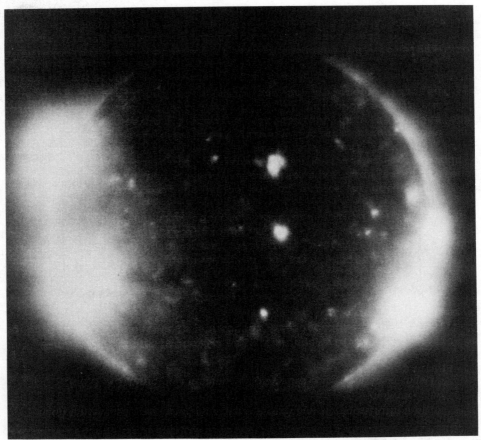

FIG. 2. A soft X-ray image of the Sun on September 16, 1976, showing only a
faint trace of a coronal hole in the northeast quadrant (upper left) where
fast footpoints were clearly visible in Figure 1. (AS&E photograph, courtesy
of J. M. Davis).

Implications of a Wind-Speed/Expansion-Factor Inverse Relation

It is instructive to consider the implications of a hypothetical inverse
relation between wind speed at 1 AU and flux-tube expansion in the corona
/19,21,22/. If each solar hemisphere were dominated by a single large polar
coronal hole whose field is strongest at its center, then the flux tubes at
the pole would diverge the least and give the fastest wind speed at 1 AU,
just as the Skylab-era concept predicted /7/.

However, this result stems from the fact that the field in a polar hole is
concentrated toward its center. If the field at the center of the hole were
to become weaker relative to that at the edge, then the expansion factor
would decrease, reaching unity (corresponding to a radial expansion) for a

flat profile and finally zero for a vanishing central field strength. In this latter case, the surrounding field lines would actually converge as they began their outward extension. If the central field reversed sign, then a gap would develop within the hole, and the fastest wind would come from the edges that faced each other across the gap. More generally, one would expect fast wind to originate from the facing edges of two like-polarity coronal holes. If the facing edges had considerably different field strengths, then the expansion factor would be discontinuous across the coronal interface, with the smaller value corresponding to the weaker photospheric field.

Thus, on the real Sun the fastest high-latitude wind should occur just after sunspot maximum when the weak polar field is surrounded by strong field migrating poleward from the sunspot belts. In contrast, the simple axisymmetric field expected at sunspot minimum would give speeds that decrease smoothly from moderately high values near the pole to very low values near the equator. A localized region of very fast wind would occur only when this symmetry is broken by the eruption of a bipolar region in the sunspot belts, and in this case the fast wind would originate from the facing edges of the polar hole and its low-latitude companion.

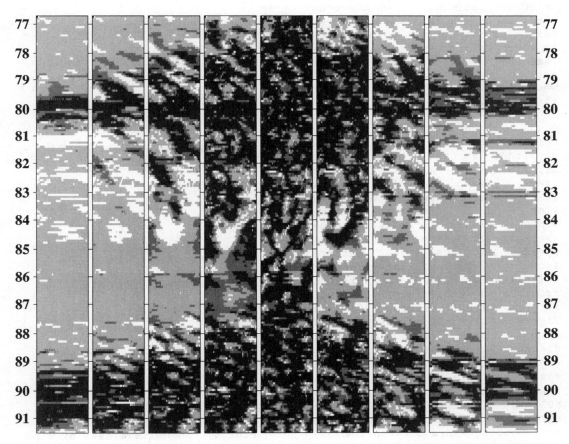

80S 60S 40S 20S   0   20N 40N 60N 80N

FIG. 3. A latitude-time display of coronal flux-tube expansion factor derived from Wilcox Solar Observatory measurements during 1976-1991, with small expansion (high speed) indicated by light shading and large expansion (low speed) by dark shading. Each panel is like a Bartels display at the indicated latitude, except that the 27-day Bartels period is replaced by the 27.3-day Carrington period.

Figure 3 summarizes the distribution of source-surface expansion factors as a function of both latitude and time during the sunspot cycle /23/. Each panel is like a Bartels display at that latitude, except that the 27-day Bartels format is replaced by the 27.3-day Carrington format. Consistent with the examples above, the very fast (white) wind forms above the poles just after sunspot maximum and gradually progresses toward the equator as the cycle advances. Apparently, the fastest wind does not quite reach the equator because many individual patches seem to have faster counterparts on one side of the equator or the other. A notable example is the recurrent equatorial pattern during 1990-1991, which is stronger in the southern hemisphere at 20-40° latitude. On the other hand, at higher latitudes, the very fast wind has not yet formed in the southern hemisphere (perhaps, like Penelope, waiting for Ulysses arrival a few years from now), whereas some large patches are already visible in the north.

Each latitude shows a different temporal variation, being of small amplitude at the equator, large amplitude at midlatitude, and nearly flat at high latitude (except near sunspot maximum, where slow wind is followed by fast wind). Figure 4 compares annual averages of these derived "wind speeds" and the speeds inferred from radio interplanetary scintillation (IPS) measurements from the University of California at San Diego (before 1982) and Nagoya University (after 1983) /24/. Except for the in-ecliptic disagreement during 1984-1989 (which also occurs when the spacecraft data are used), and the faster rise of the derived speed at 60° N in 1980, the overall agreement is about as good as the agreement obtained between the derived speed and the in situ speed in the ecliptic plane /19/. The wind speed difference between Pioneer 11 (16°N) and Voyager 2 (ecliptic) during 1984-1988, which Gazis _et al_. /12/ pointed out at the previous

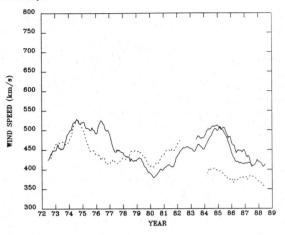

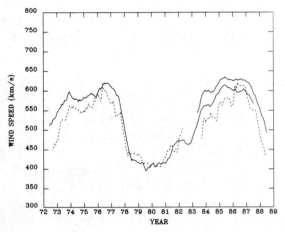

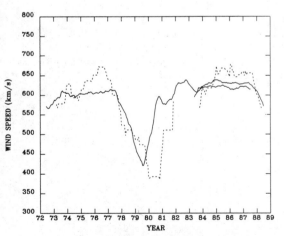

FIG. 4. Annual averages of wind speed derived from photospheric field measurements (solid curves) and from IPS measurements (dashed curves) in the ecliptic (top), at 30° N latitude (middle), and at 60° N latitude (bottom).

Solar Wind Conference, is also reproduced by the speeds derived from the WSO measurements /24/.

A Physical Basis for the Inverse Relation

At this point one might ask whether an inverse correlation between wind speed at 1 AU and flux-tube divergence in the corona is not contrary to our current understanding. After all, the Skylab observations indicated that fast wind accompanied big coronal holes with large non-radial expansions /8,18/, and elementary gas dynamic considerations indicated that enhanced acceleration ought to occur in coronal regions of large flux-tube divergence /25/.

As mentioned in the Introduction, part of the problem stems from confusion between the total expansion of an entire coronal hole and the expansion of an infinitesimal flux tube within the hole. The sevenfold expansion of the north polar hole during the Skylab mission does not seem very large now when compared to the infinite expansion of the individual flux tubes at the edge of the hole where the slow wind is supposed to originate. Also, part of the problem stems from confusion between the wind speed in the corona and at 1 AU. Holzer and Leer /26/ clarified this situation by showing for thermal models that whatever effect a rapid divergence may have on the wind speed in the corona, it will have little effect on the speed at 1 AU.

However, the problem at that time was not just the direction of the correlation, but also the speed of the wind at 1 AU. The thermal models gave values on the order of 300 km s^{-1}, which fell far short of the observed 700 km s^{-1} speeds. In a paper devoted to this problem, Leer et al. /27/ showed that realistic speeds could be reproduced if Alfven waves were included in the model, but that even these speeds would have little dependence on the coronal flux-tube divergence. However, the latter conclusion followed from the assumption that the Alfven wave amplitude is distributed uniformly at the coronal base. If this uniform distribution of wave amplitude is replaced with a uniform distribution of wave energy flux, then the calculated speeds at 1 AU are not only realistically fast, but also are inversely dependent on the coronal flux-tube divergence /28,29/.

The physical explanation may be understood as follows: As Leer et al. /27/ emphasized, the solar wind energy at Earth consists almost entirely of bulk-flow kinetic energy. This means that the wind speed at Earth is accurately determined by the ratio of energy flux to mass flux. Because both of these quantities are conserved along a flux tube, the wind speed at 1 AU may be determined from their ratio at any convenient location along the tube such as the coronal base or the solar wind critical point.

At first, it seemed necessary to explain how a magnetic flux tube could take a uniform energy flux at the coronal base and focus it at the critical point without also focussing an assumed uniform mass flux and thus maintaining a constant ratio. However, the essence of the argument is to recognize, as Parker /30/ did many years ago, that the mass flux is approximately uniform at the critical point and therefore not uniform at the coronal base. The constancy of the mass flux at the critical point follows from the fact that the gas density is roughly that of a hydrostatic atmosphere (which depends on gravity and temperature but not the expansion factor) and that the wind speed is approximately equal to the sound speed. Thus, if the wave energy flux were distributed uniformly at the coronal base, then the ratio of wave energy flux to mass flux would depend inversely on the coronal expansion factor, as would the wave-enhanced part of the wind speed at Earth. On the other hand, the thermally generated part would remain largely independent of the expansion factor, as Holzer and Leer /26/ found previously.

Leer et al.'s /27/ assumption of a uniform wave amplitude at the coronal base is roughly equivalent to a uniform wave energy flux at the solar wind critical point, which explains why their ratio of wave energy flux to mass flux was largely independent of flux-tube expansion. In contrast, a uniform distribution of wave energy flux at the coronal base implies that the wave amplitude ought to vary inversely as the square root of the magnetic field strength there -- a prediction that might be tested in the future by correlating field strengths with widths of coronal emission lines.

Summary

The essential points of this review are the following:

a. Empirical evidence accumulated over two sunspot cycles suggests an inverse relation between wind speed at Earth and magnetic flux-tube divergence in the corona.

b. Such a relation is consistent with Skylab-era ideas about isolated large polar coronal holes, but has additional consequences for more complicated configurations at other phases of the sunspot cycle. In particular, the fastest wind ought to come from weak holes or the facing edges of like-polarity holes where the large-scale field has a local minimum.

c. This relation is consistent with a mechanism in which Alfven waves boost the wind to high speed, provided that the wave flux at the coronal base is approximately uniform independent of the expansion factor. This latter condition has the testable implication that wave amplitude (and thus coronal line widths) be inversely correlated with magnetic field strength at the coronal base.

Acknowledgements

I am grateful to Y.-M. Wang (NRL) for useful conversations and to both him and A.G. Nash (ARC) for creating the multiple-latitude Bartels display shown in Figure 3. The long-term synoptic observations described in this paper were generously provided by the following people: (solar magnetograms and helium images) - J.E. Boyden and R.K. Ulrich (MWO/UCLA), J.W. Harvey (NSO/KP), J.T. Hoeksema and P.H. Scherrer (WSO/Stanford); (solar plasma and magnetic field data) - P.R. Gazis (Mycol and NASA/Ames), J.T. Gosling (LANL), A.J. Lazarus (MIT), E.J. Smith (JPL), R.D. Zwickl (NOAA); (radio IPS data) - B.J. Rickett (UCSD), M. Kojima (Nagoya). Financial support was provided by the Solar Physics Branch of the NASA Space Physics Division (DPR W-14,429) and by the Office of Naval Research.

References

1. Zirker, J.B. (ed.): 1977, Coronal Holes and High Speed Wind Streams, Colorado Assoc. Univ. Press, Boulder.

2. Sheeley, Jr., N.R., Harvey, J.W., and Feldman, W.C.: 1976, Solar Phys. 49, 271.

3. Sheeley, Jr., N.R., Asbridge, J.R., Bame, S.J., and Harvey, J.W.: 1977, Solar Phys. 52, 485.

4. Sheeley, Jr., N.R., DeVore, C.R., and Boris, J.P.: 1985, Solar Phys. 98, 219.

5. Sheeley, Jr., N.R. and Harvey, J.W.: 1978, Solar Phys. 59, 159.

6. Sheeley, Jr., N.R. and Harvey, J.W.: 1981, Solar Phys. 70, 237.

7. Hundhausen, A.J.: 1977, in Coronal Holes and High Speed Wind Streams, J.B. Zirker (ed.), Colorado Assoc. Univ. Press, Boulder, p. 225.

8. Nolte, J.T., Krieger, A.S., Timothy, A.F., Gold, R.E., Roelof, E.C., Vaiana, G., Lazarus, A.J., Sullivan, J.D., and McIntosh, P.S.: 1976, Solar Phys. 46, 303.

9. Coles, W.A. and Rickett, B.J.: 1976, J. Geophys. Res. 81, 4797.

10. Sime, D.G. and Rickett, B.J.: 1978, J. Geophys. Res. 83, 5757.

11. Fry, C.D. and Akasofu, S.-I.: 1987, Planet. Space Sci. 35, 913.

12. Gazis, P.R., Barnes, A., and Lazarus, A.J.: 1988, in Proc. 6th Int. Solar Wind Conf., V.J. Pizzo, T.E. Holzer, and D.G. Sime (eds.), NCAR Tech. Note 306, Boulder CO, p.563.

13. Levine, R.H., Altschuler, M.D., and Harvey, J.W.: 1977, J. Geophys. Res. 82, 1061.

14. Svalgaard, L., Duvall, T.L., and Scherrer, P.H.: 1978, Solar Phys. 58, 225.

15. Ulrich, R.K.: 1991, private communication.

16. Wang, Y.-M. and Sheeley, Jr., N.R.: 1988, J. Geophys. Res. 93, 11227.

17. Wang, Y.-M. and Sheeley, Jr., N.R.: 1992a, Astrophys. J. (in press).

18. Munro, R.H. and Jackson, B.V.: 1977, Astrophys. J. 213, 874.

19. Wang, Y.-M. and Sheeley, Jr., N.R.: 1990a, Astrophys. J. 355, 726.

20. Nolte, J.T., Davis, J.M., Gerassimenko, M., Lazarus, A.J., and Sullivan, J.D.: 1977, Geophys. Res. Letters 4, 291.

21. Sheeley, Jr., N.R. and Wang, Y.-M.: 1991, Solar Phys. 131, 165.

22. Wang, Y.-M. and Sheeley, Jr., N.R.: 1990b, Astrophys. J. 365, 372.

23. Wang, Y.-M., Sheeley, Jr., and Nash, A.G.: 1990, Nature 347, 439.

24. Sheeley, Jr., N.R., Swanson, E.T., and Wang, Y.-M.: 1991, J. Geophys. Res. 96, 13861.

25. Foukal, P.V.: 1990, Solar Astrophysics, John Wiley and Sons, Inc., New York, p.427.

26. Holzer, T.E. and Leer, E.: 1980, J. Geophys. Res. 85, 4665.

27. Leer, E., Holzer, T.E., and Fla, T.: 1982, Space Sci. Rev. 33, 161.

28. Wang, Y.-M. and Sheeley, Jr., N.R.: 1991, Astrophys. J. (Letters) 372, L45.

29. Wang, Y.-M. and Sheeley, Jr., N.R.: 1992b, Proc. Solar Wind 7 (these proceedings).

30. Parker, E.N.: 1965, Space. Sci. Rev. 4, 666.

OBSERVATIONAL STUDY OF THE IMF SPIRAL NORTH AND SOUTH OF THE CURRENT SHEET

C. W. Smith and J. W. Bieber

Bartol Research Institute, University of Delaware, Newark, DE 19716, U.S.A.

Abstract

We have analyzed 23 years of spacecraft observations spanning 27 AU. Our analysis reveals both an overwinding of the interplanetary magnetic field (IMF) and a sustained asymmetry between the northern and southern hemispheres of the heliosphere. Nonzero azimuthal field components at the source boundary may account for the observed overwinding. The north-south asymmetry, whereby the IMF spiral north of the current sheet is more tightly wound than the IMF spiral south of the current sheet, persists due to unknown sources. It is also shown that there exist significant, correlated departures from the Parker theory in the azimuthal component of the field.

Introduction

The Parker theory /1,2/ for the winding of the IMF argues that acceleration of the plasma near the Alfven radius leads to a nearly radial magnetic field geometry a few solar radii above the photosphere. Radio scintillation observations set the Alfven radius at 10 to 30 solar radii /3/. As the plasma expands outward, solar rotation causes the large-scale IMF to take the form of an Archimedean spiral /4,5,6,7,8,9/. The magnetic field lines near the solar equatorial plane are organized into "toward" and "away" sectors /10/. Within each sector the large-scale field is oriented either sunward or anti-sunward along the nominal spiral direction. These sectors are a manifestation of the bipolar structure of the heliosphere where toward and away fields occupy opposing hemispheres separated by a wavy current sheet that corotates with the sun /11,12,13,14/.

Overwinding of the IMF

The Parker prediction /1,2/ for the IMF winding angle, ψ^P, is:

$$\psi^P = \arctan\left[\frac{2\pi r \sin(\Theta)}{T V_{sw}}\left(1 - \frac{b}{r}\right)\right] \tag{1}$$

where r, V_{sw}, and Θ are the heliocentric distance, solar wind speed, and colatitude, respectively, at the point in question; b is the heliocentric distance of the source boundary; and T is the sidereal solar rotation period at the source latitude which we take to be 25.4 days /15/.

Figure 1 compares the orientation of the measured IMF with the predictions of equation (1) using a value for the source boundary radius, $b = 20R_\odot$ /3/. Figure 1 shows histograms of the orientation of the IMF projected onto the solar ecliptic plane for hourly measurements recorded on the NSSDC omnitape of near-Earth spacecraft /16/ and by the PVO spacecraft at 0.7 AU. The orientation is defined to be 0° in the anti-sunward direction and increases clockwise as viewed from the north. The two more narrowly peaked distributions shown in each plot are the theoretical predictions derived from equation (1) using the observed wind speed and the above parameters. In addition to the disparity between the widths of the theoretical and observed distributions, the theoretical result is shifted to smaller angles relative to the observed distribution. This represents an "overwinding" of the field relative to the predictions of equation (1).

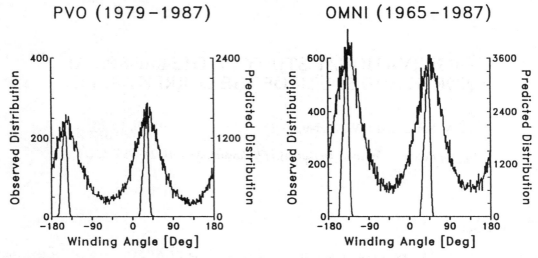

Fig. 1: Distribution of observed and predicted IMF orientations for 0.7 and 1 AU.

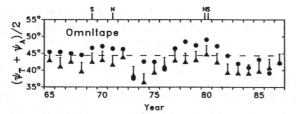

Fig. 2: Average winding angle of IMF at 1 AU (circles) and predicted value derived from equation (1). Arrows above panel denote years when the Sun's north (N) and south (S) polar magnetic fields reversed direction.

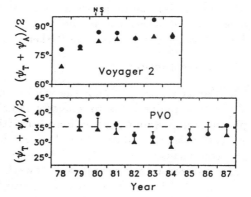

Fig. 3: Average winding angle measured by Voyager 2 (top) and PVO (bottom).

Circles in Figure 2 show the mean winding angle for each year of the omnitape data set of 1 AU measurements. The average winding angle is computed using a unit vector definition /17/ that separately treats toward and away sector measurements. A strong solar cycle dependence is evident with the largest mean winding angles seen during the years surrounding solar maximum. This is consistent with the lower mean wind speeds during this time /17/. Triangles in Figure 2 represent the expected winding angle derived from equation (1), the assumption $b = 20 R_\odot$, and the observed annual mean wind speed. The bar over each triangle depicts the predicted winding angle in the limit $b \rightarrow 0$. During the years surrounding solar maximum (1968-71 and 1977-83) the observed mean winding angle exceeds the upper limit theoretical prediction based on $b = 0$ and the lower limit solar rotation period, T = 25.4 days. The mean winding angle observed at 1 AU for these 11 years and for 4 other years exceeds the absolute upper limit set by the Parker theory /1,2/.

Figure 3 shows the mean winding angle as measured at 0.7 AU by PVO (bottom) and beyond 1 AU by Voyager 2 (top). The apparent absence of the $b \rightarrow 0$ limit from the Voyager results is due to the decreasing significance of b in the outer heliosphere. The PVO results again demonstrate that the mean winding angle exceeds the "best guess" value determined by $b = 20 R_\odot$ in 8 out of 9 years and exceeds the upper limit values given by $b = 0$ in 2 of the 9 years. Voyager 2 demonstrates a significant overwinding in the first 4 years following launch in late 1979. The observed wind speed is again insufficient to account for the observed mean winding angle.

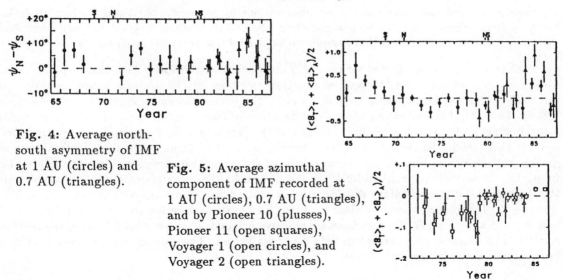

Fig. 4: Average north-south asymmetry of IMF at 1 AU (circles) and 0.7 AU (triangles).

Fig. 5: Average azimuthal component of IMF recorded at 1 AU (circles), 0.7 AU (triangles), and by Pioneer 10 (plusses), Pioneer 11 (open squares), Voyager 1 (open circles), and Voyager 2 (open triangles).

Source Field Azimuthal Components

A generalization of equation (1) to allow for azimuthal fields at the inner boundary leads to /17/:

$$\tan\left[\psi(r)\right] = \frac{2\pi r \sin(\Theta)}{TV_R}\left(1 - \frac{b}{r}\right) - \frac{B_T(b)}{B_R(b)}\frac{V_R(b)}{V_R(r)}\frac{r}{b}. \tag{2}$$

The second term on the right-hand side of equation (2) has the same dependence on heliocentric distance as does the standard Parker expression, but has a different dependence on latitude. If we take $b = 20R_\odot$ and adopt plausible values for the other parameters, a 3° augmentation of the IMF winding angle at 1 AU requires $\frac{B_T(b)}{B_R(b)} \simeq -0.02$. This represents a deviation of the IMF from the radial direction of only 1° at the inner boundary. Small amounts of the azimuthal fields from below the chromosphere may reach the corona through the activities of various solar processes, thereby providing an azimuthal field component with the correct sign to explain the observed overwinding /17/. Other processes are also being considered.

This extension of the Parker theory offers interesting implications for high latitude magnetic fields. Equation (1) implies that polar fields should be relatively straight and unaffected by solar rotation. Several authors /18,19,20/ have argued that the motion of supergranules may lead to large azimuthal field components over the solar poles, but these fluctuations will have random signs and average to the Parker prediction. Equation (2) suggests that the polar fields are wound in a consistent, long term manner. Under the above conditions, the polar field at 1 AU (10 AU) should display a winding angle $\simeq 5°$ ($\simeq 45°$). This result will not average to the Parker prediction and implies that the termination shock over the poles has a perpendicular shock structure.

North-South Asymmetry of the IMF

Prior to the solar magnetic reversal of 1969-71 and following the reversal of 1980, toward sectors correspond to northern heliospheric measurements and away sectors to southern heliospheric measurements. Between 1971 and 1980, the association is reversed. The difference between the mean winding angles of the toward and away sector populations thereby represent the difference between the winding angles of the two hemispheric fields /21,22,23/.

Figure 4 shows the yearly average of the computed winding angle asymmetry for the PVO (triangles) and omnitape (circles) data sets. Errors are determined in a manner prescribed by Smith and Bieber /23/. Results for the years 1969-71 and 1980 are not shown due to the changing state of the solar magnetic field. The average winding angle of the southern hemispheric spiral exceeds the winding angle of the northern spiral in only 9 of the 27 measurements, and only 1 of these 9 differs from zero by more than 1σ. The remaining 18 years have positive asymmetries. The average north-south asymmetry at 1 AU is $2.5° \pm 0.8°$ while at 0.7 AU it is $1.6° \pm 1.0°$ /23/.

We have performed the same analysis on observations by Voyager 1 & 2 and Pioneer 10 & 11, but have found these results to be unsatisfactory /23/. Nevertheless, a significant and correlated asymmetry is seen in the azimuthal component of the IMF as recorded in all six data sets.

Figure 5 shows the mean azimuthal (or "T") component of the IMF. This mean applies equal weight to both sector types, thereby eliminating any bias due to unequal sector coverage. If the IMF of the two hemispheres were symmetric, the means shown in Figure 5 would be zero. Pioneer 10 & 11 (bottom panel) observed a negative DC azimuthal component of the magnetic field starting in 1973 and extending through 1979. Voyager 1 & 2 observations (same panel) confirm this trend. Following the solar magnetic reversal of 1980, when the spacecraft are at greater heliocentric distances, the average value of the azimuthal component is indistinguishable from zero. Observations recorded by PVO and on the NSSDC omnitape (top panel) corroborate the Voyager and Pioneer findings with negative azimuthal field components during the period 1973 through 1979. However, there is a clear tendency for positive azimuthal components prior to 1970 and there is a suggestion of a trend toward positive azimuthal components following the reversal of 1980. The source of this anomalous field is not yet known. The observations are consistent with both solar and interplanetary sources.

Acknowledgements: The authors are grateful to the P.I.'s who provided data used in this analysis: A. Barnes, J. W. Belcher, E. J. Smith, N. F. Ness, and C. T. Russell. We thank J. King and the NSSDC for providing the omnitape data set of 1 AU measurements. The authors acknowledge helpful discussions with L. E. Burlaga, P. A. Evenson, N. F. Ness, C. T. Russell, and E. J. Smith. This work was supported by Jet Propulsion Laboratory contract 957921, NASA grant NAG 2- 553, and NSF grant ATM-9014806 to the Bartol Research Institute.

References

1. Parker, E. N., 1958, Astrophys. J., **128**, 664.
2. Parker, E. N., 1963, "Interplanetary Dynamical Processes", Interscience Publishers, New York
3. Lotova, N. A., Blums, D. F., and Vladimirskii, K. V., 1985, Astron. Astrophys., **150**, 226.
4. Rosenberg, R. L., 1970, J. Geophys. Res., **75**, 5310.
5. Neugebauer, M., 1976, J. Geophys. Res., **81**, 4664.
6. Behannon, K. W., 1978, Rev. Geophys., **16**, 125.
7. Thomas, B. T., and Smith, E. J., 1980, J. Geophys. Res., **85**, 6861.
8. Burlaga, L. F., Lepping, R. P., Behannon, K. W., Klein, L. W., and Neubauer, F. M., 1982, J. Geophys. Res., **87**, 4345.
9. Klein, L. W., Burlaga, L. F., and Ness, N. F., 1987, J. Geophys. Res., **92**, 9885.
10. Ness, N. F., and Wilcox, J. M., 1965, Science, **148**, 1592.
11. Rosenberg, R. L., and Coleman, P. J., Jr., 1969, J. Geophys. Res., **74**, 5611.
12. Wilcox, J. M., and Scherrer, P. H., 1972, J. Geophys. Res., **77**, 5385.
13. Smith, E. J., Tsurutani, B. T., and Rosenberg, R. L., 1978, J. Geophys. Res., **83**, 717.
14. Thomas, B. T., and Smith, E. J., 1981, J. Geophys. Res., **86**, 11105.
15. Beckers, J. M., 1981, in "The Sun as a Star", edited by S. Jordan, NASA Publication SP-450 Washington, D.C., pp. 11-64.
16. Couzens, D. A., and King, J. H., 1986, "Interplanetary Medium Data Book - Supplement 3, 1977-1986", Rep. NSSDC/WDC-A-R&S 86-04, NASA, Greenbelt, MD.
17. Smith, C. W., and Bieber, J. W., 1991, Astrophys. J., **370**, 435.
18. Jokipii, J. R., and Kota, J., 1989, Geophys. Res. Lett., **16**, 1.
19. Jokipii, J. R., and Kota, J., 1990, Proc. XXI ICRC (Adelaide), **6**, 104.
20. Hollweg, J. V., and Lee, M. A., 1989, Geophys. Res. Lett., **16**, 919.
21. Bieber, J. W., 1988, J. Geophys. Res., **93**, 5903.
22. Smith, C. W., and Bieber, J. W., 1990, Proc. XXI ICRC (Adelaide), **5**, 304.
23. Smith, C. W., and Bieber, J. W., 1991, J. Geophys. Res., submitted.

GLOBAL STRUCTURE OF THE ONCOMING LOCAL INTERSTELLAR PLASMA FLOW ABOUT A SIMPLIFIED HELIOPAUSE INTERFACE

S. S. Stahara,* R. Ratkiewicz** and J. R. Spreiter***

RMA Aerospace Inc., Mountain View, CA 94043, U.S.A.
*** Space Research Center, Bartycka 18a, 00-716 Warsaw, Poland*
*** Stanford University, Stanford, CA 94305, U.S.A.*

ABSTRACT

Global plasma flow results obtained from a gasdynamic convected-field MHD computational model /1/ are presented for the outer flow structure representing the interaction of the oncoming LISM plasma with a simplified heliopause interface shape. The heliopause shape employed is based on an analytic solution /2/ of a Newtonian thin layer theoretical model of the heliospheric interface. The plasma flow results presented demonstrate the sensitivity of the global outer flow on oncoming interstellar Mach number.

INTRODUCTION

The hypersonic spherically-symmetric solar wind blowing from the Sun interacts with the oncoming local interstellar gas in such a fashion that the dynamic interaction region between the two plasmas consists of four distinct zones. As illustrated in Fig.1, they are: (a) the hypersonic unperturbed solar wind; (b) a region of shocked solar wind plasma between the heliocentric termination shock and the heliopause interface surface; (c) an adjacent region of shocked interstellar plasma (if the oncoming LISM plasma velocity is supersonic); and (d) the undisturbed local interstellar plasma. The interaction between the two plasmas in these four zones can be solved in one of two ways, i.e. (1) by simultaneously solving the coupled outer (interstellar) and inner (solar wind) plasma flow problems, the resulting solution then providing the location of the heliopause interface surface

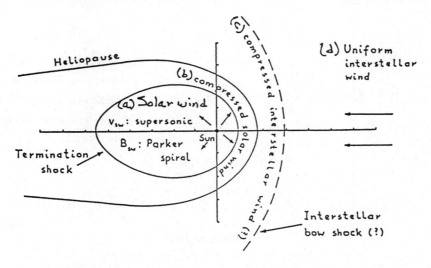

Fig. 1 Schematic of various flow zones of heliosphere configuration

that is everywhere in local pressure balance between the two plasmas, or (2) by an iterative approach in which the outer interstellar and inner solar wind plasma flows are solved separately and then consistently joined across a common heliopause interface boundary surface whose position is successively relocated via a local pressure balance between the two plasma flow regions after each completed outer/inner solution cycle.

METHODOLOGY

For the results reported here we have adopted the iterative outer/inner solution approach, and present global plasma flow results for the first outer problem solution cycle representing the interaction of an oncoming supersonic LISM plasma with a specified heliopause interface. For the initial shape of the heliopause interface surface required to initiate the outer flow calculation, we have employed the shape described by the analytical solution obtained by Ratkiewicz /2/.

The global LISM plasma flow solution about the specified heliopause interface proceeds as follows. With the heliopause interface surface known, the computational procedures developed by Stahara et al. /1/ require only a specification of the oncoming interstellar Mach number M_∞ and ratio of plasma specific heats γ to determine the entire global solution about the heliopause. The computational methods employed in the present application do not solve the full MHD equations but a subset of them, known as the gasdynamic convected-field equations, which are valid in the limit of weak interstellar magnetic field i.e., high interstellar Alfvenic Mach number. A highly accurate computational model based on these equations has been developed /1/, and the procedures have been successfully employed in detailed studies of the global interaction of the solar wind with virtually all the planets in the solar system. See /3/ for a recent review. Details of the algorithms employed and the accuracy of the methods are provided in /1/ and /3/. In brief summary, the computational model employs two separate plasma flow solvers to determine the global, steady, supersonic plasma flow past an arbitrary axisymmetric magnetoionopause obstacle. The first solver determines the solution from the subsolar nose region to a downstream location that is often but not always taken to be the terminator location. This solution algorithm advances the unsteady gasdynamic conservation equations forward in time until the steady state is attained. The global solution at the terminator or most downstream plane is then provided to a steady state, spatially

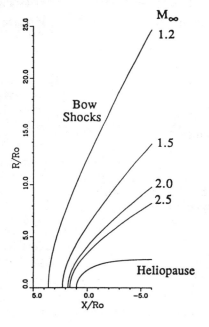

Fig. 2. Comparison of predicted interstellar bow shock shapes for various oncoming LISM Mach numbers past the heliopause shape given by Ratkiewicz /2/

marching solver that carries the solution tailward from that starting location downstream to any arbitrary location desired. Fig. 2 provides summary results from a series of such model calculations. Shown in that figure is a comparison of interstellar bow shock shapes and locations scaled by the heliopause nose radius R_0 for various oncoming LISM Mach numbers from $M_\infty = 1.2$ to 2.5 and $\gamma = 5/3$ for flow past the heliopause shape given by Ratkiewicz /2/. Of particular interest is the large sensitivity of the interstellar bow shock location to the oncoming LISM Mach number for these low Mach number flows. Fig. 3 provides contours of the density, temperature, velocity, and Mach number for the $M_\infty = 2.0$ outer flow solution.

In order to check the accuracy of the Newtonian pressure distribution approximation on which the determination of the initial heliopause interface shape given in Figs.2 and 3 was based, we provide results in the left plot of Fig.4 which displays the pressures on the heliopause surface as determined from the gasdynamic computational model for the four Mach numbers for which the bow shock locations were provided in Fig.2. These pressures are meant to be compared with the corresponding Newtonian surface pressure distributions shown in the right plot of Fig.4 at the same Mach numbers. The Newtonian pressure distribution depends on both M_∞ and γ and is

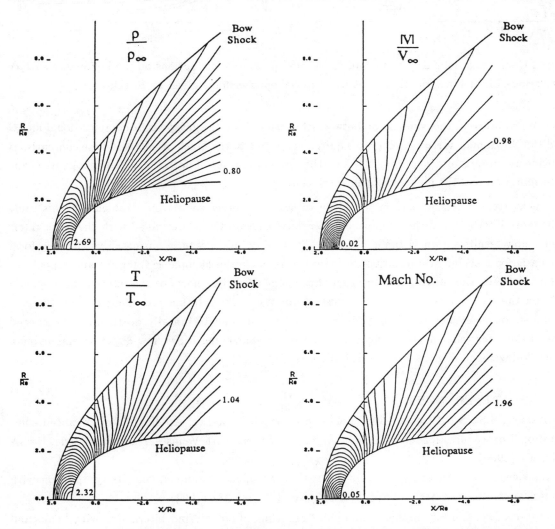

Fig. 3 Contours of density, velocity, temperature, and Mach number for $M_\infty = 2.0$ and $\gamma = 5/3$ LISM flow past the heliopause shape given by Ratkiewicz /2/

given by /3/,

$$\frac{p}{p_{stag}} = \frac{1}{K'} \left(\frac{1}{\gamma M_\infty^2} + (K' - \frac{1}{\gamma M_\infty^2}) \cos^2 \psi \right) \tag{1}$$

where

$$K' = \frac{1}{\gamma} \left(\frac{\gamma + 1}{2} \right)^{(\gamma+1)/(\gamma-1)} \left(\gamma - \frac{\gamma - 1}{2M_\infty^2} \right)^{-1/(\gamma-1)} \tag{2}$$

and ψ is the angle between the local normal at the heliopause surface and the oncoming LISM flow direction.

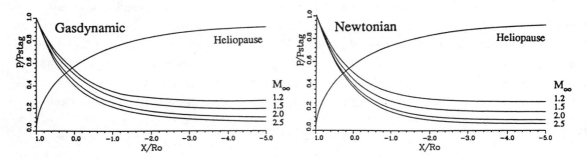

Fig. 4 Comparison of gasdynamic and Newtonian pressure distributions on the heliopause surface for various LISM Mach number flows past the heliopause shape given by Ratkiewicz /2/

As can be observed, the comparison between the results from the gasdynamic computational model and the Newtonian surface pressures is quite good for all of the Mach numbers. These comparisons provide an important consistency check on the overall accuracy of both the initial heliopause shape determination and the global outer flow solution.

The next step in this procedure would be to solve the inner solar wind problem for this same heliopause interface and then compare the surface pressure thus determined with the outer interstellar flow results shown in the left plot of Fig. 4. Any local pressure mismatch would then form the basis for a systematic iterative procedure to relocate the heliopause interface and repeat the outer/inner solution procedure. In order to account for finite effects of the interstellar magnetic field on the outer plasma flow, it is necessary to extend the solution procedure beyond the gasdynamic convected field level to a full MHD level. We plan to initially do that for the special case of the interstellar magnetic field aligned with the oncoming LISM flow direction employing a recently-developed aligned flow full MHD model /3/.

REFERENCES

1. Stahara, S.S., S.D. Klenke, B.C. Trudinger, and J.R. Spreiter, Application of advanced computational procedures for modeling solar-wind interactions with Venus – Theory and code, NASA CR 3267 (1980)

2. Ratkiewicz, R., An analytical solution for the heliopause boundary and its comparison with numerical solutions, Astron. Astrophys. 255, No. 1/2, 383-387 (1992)

3. Spreiter, J.R. and S.S. Stahara, Computer Modeling of Solar Wind Interaction with Venus and Mars, in: The Comparative Study of Venus and Mars; Atmospheres, Ionospheres and Solar Wind Interactions, AGU Monograph, in press

THE MAGNETIC FIELD IN THE HELIOSHEATH

S. T. Suess* and S. Nerney**

*Space Science Lab/ES52, NASA Marshall Space Flight Center, Huntsville,
AL 35812, U.S.A.
** San Juan College, Farmington, NM 87401, U.S.A.

ABSTRACT

The interplanetary magnetic field (IMF) behaves in a reasonably well-understood manner between the Sun and the heliospheric termination shock. At the shock, the azimuthal field is amplified by a factor of four (for a strong shock) and undergoes secular amplification in the heliosheath until the flow is fully turned into the downstream direction and has reached its asymptotic state in the distant heliotail. This amplification may lead to important MHD effects that can cause the shock to be closer to the Sun than otherwise expected. Here we further examine whether there are important MHD effects in the heliosheath. We do this by calculating the kinematic compression of the magnetic field in the heliosheath using an analytic incompressible flow model of the dynamics downstream of the shock. We conclude that it is likely that MHD effects are important in the heliosheath in a narrow cone about the upstream direction.

1. INTRODUCTION

The magnetic field in the heliosheath can be compressed until the local plasma β (ratio of internal to magnetic field energy densities) and β_v (the ratio of the kinetic to magnetic field energy densities) are smaller than unity. When this happens, MHD effects may dominate the flow, invalidating hydrodynamic (HD) model predictions for the distance to the termination shock. Here, we review the spherical kinematic magnetic field calculation /6/ and the spherical MHD problem /1,2,3,4/. Then, we show what heliosheath magnetic field will result from kinematic distortion in our HD global model of heliosheath flow /8,9/.

2. SPHERICAL FLOW: KINEMATIC MAGNETIC FIELD

The kinematic field calculation /6/, shows how the IMF develops into an Archimedian spiral as it is passively advected by the solar wind. In it, the flow field is assumed given by $\vec{v} = \hat{e}_r v_o$, where v_o is assumed constant, (e.g. 400 km/s). The magnetic field is found from $\nabla \times (\vec{v} \times \vec{B}) = 0$ and $\nabla \cdot \vec{B} = 0$, giving

$$B_r(r) = B_o(r_o/r)^2 \qquad\qquad B_\phi = -B_o \frac{\Omega r_o^2}{r v_o} sin\theta \qquad (1)$$

B_ϕ is zero at $r = 0$, θ is the polar angle with respect to the Sun's rotation axis, r_o is a reference radius, and Ω is the Sun's (constant) angular velocity.

The transverse field is amplified at the shock in proportion to the density. The transverse field then undergoes further amplification because the flow speed is slowing. If we continue the assumption of spherical, but now incompressible flow, and assume a strong shock, then the downstream flow is $\vec{v} = \hat{e}_r (v_o/4)(R_s/r)^2$ - the speed falls off as r^{-2} (R_s is the shock radius). The radial magnetic field is unaffected, but the transverse field is given by (1b) with v_o replaced by $(v_o/4)(R_s/r)^2$ so it *increases* with radius.

Fig. 1 shows fieldlines for a 50 AU termination shock with $v_o = 400$ km/s and illustrates several important points. First, while equatorial fieldlines undergo several full spiral turns inside the shock, near polar fieldlines undergo less than two turns. This results in easier cosmic ray access to the inner solar system near the poles and is one of the main reasons for the Ulysses mission. Second, the fourfold compression of the magnetic field at the shock has a large (visible) effect at all latitudes. Third, the secular increase in the azimuthal field beyond the shock is illustrated by the narrowing gap between successive spiral turns beyond the shock. Eventually the field amplification must lead to dynamic effects - there are magnetic pressure gradients in both the radial and meridional directions.

Fig. 1 Magnetic fieldlines upstream and downstream of the termination shock for spherically symmetric flow and a passive field. Fieldlines are plotted at the rotational equator (assuming the field is not identically zero there, contrary to the case for most ideal field geometries) and at polar angles of 3 through 10°. The shock is at 50 AU.

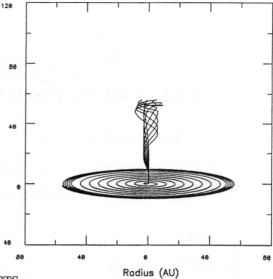

Radius (AU)

3. SPHERICAL FLOW: MHD EFFECTS

For spherical flow with a purely azimuthal magnetic field, the solutions are found in closed, analytic form /1,2,5/, shown in Fig. 2a for $\beta_v = 500$ (evaluated on the upstream side of the termination shock), where $\bar{\rho}, \bar{v}, \bar{p}$, and $\bar{b}$ are scaled density, flow speed, pressure, and transverse magnetic field. We also show the behavior with changing β_v (Fig. 2b) and with latitude (Fig. 2c). The main point is that $\bar{p}_T$, the total pressure, is reduced relative to the field-free pressure, which means that for a given interstellar pressure the terminal shock can lie closer to the Sun than predicted by HD theory - we call this the *Cranfill effect*. Fig. 2b shows that $\bar{p}_T$ decreases as β_v decreases. Fig. 2c shows that the MHD effect disappears at the poles, where the transverse field is zero. Two things can be deduced from this figure: First that the distance between the shock and the heliopause should be several times the shock radius for MHD effect to be important. Second, that the shock will be more distant over the poles than at the equator if MHD effects are important.

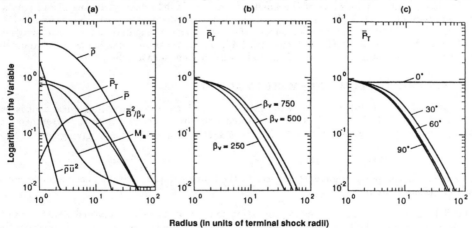

Radius (in units of terminal shock radii)

Fig. 2 (a) The Cranfill effect with $\beta_v = 500$. The radius has been scaled to an arbitrary shock radius, M_a is the Alfvén Mach number, and $\bar{p}_T$ is the total (internal + dynamic + magnetic) pressure. (b) $\bar{p}_T$ for $\beta_v = 250, 500, 750$. (c) $\bar{p}_T$ for $\beta_v = 500$ at polar angles of 0, 30, 60, and 90°.

3. NON-SPHERICAL FLOW: KINEMATIC MAGNETIC FIELD

We know that the heliosphere is moving with respect to the local interstellar medium at ~ 25 km/s /7/, which will lead to asymmetric flow. We therefore apply an extended version /8,9/ of Parker's /6/ model of flow outside the termination shock termination shock to compute the kinematic advection of the IMF in the heliosheath. This flow model assumes incompressible, irrotational flow with the velocity components given by

$$v_\varpi = \frac{v_s R_s^2 \varpi}{r^3}\left[1 - \frac{3R_s \varepsilon^{2/3} z}{2r^2}\right] \qquad v_z = v_s\left[\frac{R_s^2}{r^3} - \varepsilon^{2/3}\left(R_s^3\frac{(z^2 - \varpi^2/2)}{r^5} - 1\right)\right] \qquad (2)$$

where

$$r = (\varpi^2 + z^2)^{1/2} \qquad\qquad \varepsilon = \left(\frac{v_{i\infty}}{v_s}\right)^{\frac{3}{2}} \qquad (3)$$

R_s is the shock radius, ϕ is now the azimuthal angle *around the heliotail*, ϖ is the cylindrical radius away from the tail axis, and z is the distance down the heliotail. $v_{i\infty}$ is the flow speed of the solar system with respect to the interstellar plasma, ~ 25 km/s. v_s is the flow speed on the downstream side of the termination shock, ~ 100 km/s. Therefore, $\varepsilon = 0.125$. The magnetic field is calculated writing it in streamline coordinate form:

$$\vec{B} = B_s \hat{e}_s + B_t \hat{e}_t + B_\phi \hat{e}_\phi; \qquad \hat{e}_s = \vec{v}/\mid\vec{v}\mid; \qquad \hat{e}_t = \hat{e}_\phi \times \vec{v}/\mid\vec{v}\mid \qquad (4)$$

where B_s, B_t, and B_ϕ are the scalar components of $\vec{B}$ and $\vec{v}$ is given by (2). The transverse magnetic field components are found in closed form to be

$$\frac{B_\phi}{\varpi} = f(\psi, \phi); \qquad B_t \varpi \mid \vec{v} \mid = F(\psi, \phi); \qquad \nabla\psi = \varpi \hat{e}_\phi \times \vec{v} \qquad (5)$$

where $v = \mid \vec{v} \mid$ and $\vec{v} \cdot \nabla\psi = 0$ so ψ is a streamline constant, as is ϕ. Therefore, F and f in (8) are also streamline constants. The component of the magnetic field along the streamline is given by the ordinary differential equation

$$\vec{v} \cdot \nabla\left(\frac{B_s}{v}\right) - 2\frac{B_t}{v^3}\left[v_\varpi v_z\left(2\frac{\partial v_\varpi}{\partial \varpi} + \frac{v_\varpi}{\varpi}\right) + (v_z^2 - v_\varpi^2)\frac{\partial v_z}{\partial \varpi}\right] = 0 \qquad (6)$$

We solve this equation numerically.

Given $\vec{B}$ on the termination shock, the magnetic field in the heliosheath can now be computed. We use the kinematic model IMF from section 2, with the azimuthal component amplified by four under the assumption of a strong shock. The results are plotted as a function of distance along streamlines. In Fig. 3 we show the streamlines for $\varepsilon = 0.125$, with six of them labelled '1' through '6' to indicate those for which results will be shown.

Fig. 3 Streamlines used in computing the kinematic magnetic field in the heliosheath. The field will be plotted as a function of distance along the streamlines labelled '1' to '6' in this figure.

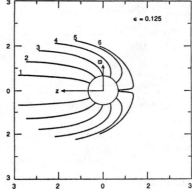

Fig. 4 shows the flow speed and magnetic field magnitude in the heliographic equator. The flow speed always starts at 100 km/s at the shock and falls off smoothly in the tailward direction while falling precipitously on the upstream side towards the stagnation point, going to zero at the stagnation point.

Fig. 4 Total velocity and magnetic field downstream of the termination shock in the solar equatorial plane and meridian defined by solar rotation. The distance along streamlines is in units of distance to the stagnation point in the upstream direction in the limit $\varepsilon \to 0$ /8/.

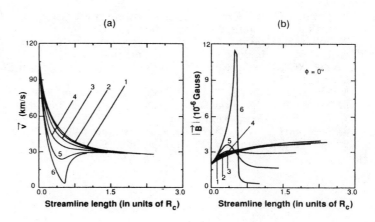

The heliosheath magnetic field in the heliographic equator (Fig. 4b) grows slowly in the tailward direction, but undergoes large amplification near the stagnation point - becoming infinite at the stagnation point. The results demonstrate in a simple way that the motion of the heliosphere through the interstellar medium causes a substantial difference from the spherical models described above.

To help understand whether upstream amplification is physically significant, we show the local plasma β and β_v (eqn (3)) along each of the six streamlines in Fig. 5. The β-values are scaled by their values at the termination shock and the logarithms are plotted. Therefore, each curve, corresponding to each of the streamlines in Fig. 3, starts at zero at the base of the streamline. Results are shown for the same plane shown in Fig. 4b.

Generally, the behavior of the plasma β can be divided into downstream and upstream behavior. Downstream, along curves 1, 2, 3, and 4, β and β_v generally decrease by an order of magnitude in a distance of two shock radii along the streamlines. Upstream, there is a sharp minimum near the heliopause. To appreciate the significance of this, we note that on the downstream side of the termination shock, $\beta \sim O[10]$ and $\beta_v \sim O[20-40]$. This means that in the downstream direction the magnetic field never dominates the dynamics. However, the magnetic field dominates the dynamics in the neighborhood of the stagnation point, being important in a $30°$ (half-angle) cone about the upstream direction. However, this volume is not large enough to cause a reduction in the distance to the termination shock of the magnitude hypothesized by Lee /4/ - a factor of two or more, although it could easily be much more than the 10% effect first suggested by Axford /1/. We conclude, then, by asserting that within a $30°$ cone around the upstream direction, MHD effects may reduce the distance to the termination shock by between 10% and 50%, but that outside this cone the effect will be negligible. A more quantitative evaluation will have to await numerical MHD solutions.

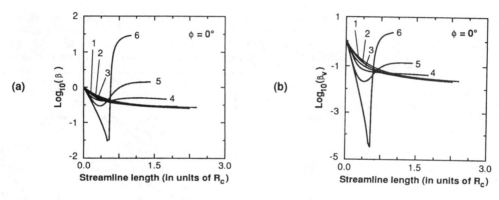

Fig. 5 Plasma β and β_v downstream of the termination shock in the solar equatorial plane. The distance along streamlines is in the same units as in Fig. 4 and the β-values are scaled to their values at the shock, in the solar equatorial plane.

Acknowledgement: This research has been supported by a grant from the Cosmic and Heliospheric Physics Branch of the Office of Space Science and Applications, U.S. National Aeronautics and Space Administration.

REFERENCES
1. Axford, W. I., in *Solar Wind* eds. C. P. Sonnett, P. J. Coleman, Jr., and J. W. Wilcox, **NASA SP-308**, 609-658 (1972).
2. Cranfill, C. W., Flow problems in astrophysical systems, Ph.D. dissertation, Univ. of Calif. at San Diego, Univ. Microfilms, Ann Arbor, Michigan (1974).
3. Holzer, T. E., *Ann. Rev., Astron. Astrophys.*, **27**, 199-234 (1989).
4. Lee, M. A., in *Proc. of the Sixth Internatl. Solar Wind Conf.*, eds. V. J. Pizzo, T. E. Holzer, and D. Sime, NCAR Tech. Note **NCAR/TN-306+Proc**, 635-650, Natl. Center For Atmos. Res., Boulder, Colorado (1988).
5. Nerney, S., Suess, S. T., and Schmahl, E. J., *Astron. Astrophys.*, in press (1991).
6. Parker, E. N., *Interplanetary Dynamical Processes*, Interscience, New York (1963).
7. Suess, S. T., *Rev. Geophys.*, **28**, 97-115 (1990).
8. Suess, S. T., Nerney, S., *J. Geophys. Res.*, **95**, 6403-6412 (1990).
9. Suess, S. T., Nerney, S., *J. Geophys. Res.*, **96**, 1883 (1991).

THE LATITUDINAL DEPENDENCE OF THE SOLAR IONIZATION RATE AS DEDUCED FROM THE PROGNOZ-6 LYMAN-ALPHA MEASUREMENTS

T. Summanen,* E. Kyrölä,* R. Lallement** and J. L. Bertaux**

* Finnish Meteorological Institute, PO Box 503, SF-00101 Helsinki, Finland
** Service d'Aéronomie du CNRS, BP No. 3, 91371 Verrières le Buisson, France

ABSTRACT

The latitudinal dependence of the solar ionization is studied using the Lyman-α measurements by the Prognoz-6 spacecraft during the solar minimum 1976-77. Applying a hot model for the interplanetary H-gas we have searched for an optimal ionization function to comply with the measurements. Using the optimal ionization function we have studied the latitudinal variation of the solar wind mass flux and the proton density.

INTRODUCTION

While waiting for the unprecedented Ulysses in situ measurements of the solar wind it is useful to remind that the probing of the three-dimensional structure and the temporal variation of the solar wind is possible only by using remote sensing methods. The principal methods which have been employed are the measurement of interplanetary scintillations (IPS) and the measurement of the interplanetary Lyman-α radiation. While the scintillation method is discussed elsewhere in this volume we concentrate here on the Lyman-α method and specifically how it can be used to study solar wind induced ionization.

The measurement of the Lyman-α radiation emanating from interplanetary neutral H-atoms can be modelled through the following formula:

$$I_m(\vec{r}_m, \vec{\Omega}, t) = \int_0^\infty F(\vec{r}(s), t)\sigma(\theta(s))n(\vec{r}(s), t)ds \qquad (1)$$

Here $\vec{r}_m$ and $\vec{\Omega}$ are the measurement coordinate and the viewing direction, respectively. Solar flux is denoted by F, σ is the scattering cross section depending on the angle of the scattering, and n is the interplanetary hydrogen density. For simplicity we have ignored the wavelength dependence of various quantities. We take the interplanetary material optically thin for the Lyman-α radiation i.e., we neglect multiple scatterings and absorption. We also neglect other sources of Lyman-α like galactic radiation.

The measurements of I_m in Eq. (1) can be used to investigate the solar wind if we are able to model its influence on the neutral hydrogen density n. We have employed the following modelling (see e.g., [1]):

$$n(\vec{r}, t) = \int d\vec{v}_0 W(\vec{v}_B + \vec{v}_0; T) \sum n_{cold}(\vec{r}, \vec{v}_0, t) e^{-\int_{-\infty}^{\vec{r}} \frac{\beta(\lambda(s), r(s), t)}{v(s)} ds} \qquad (2)$$

Here $W(\vec{v}_B + \vec{v}_0; T)$ is the velocity distribution of the neutrals far away from the Sun. We assume that it is a Maxwellian distribution. This may be too severe a simplification because neutrals must cross the shocked boundary region between the solar wind plasma and the interstellar plasma. The density of neutrals with the initial velocity $\vec{v}_0$ is denoted by n_{cold}. The summation symbolizes the presence of two trajectories for every space-point. For a constant solar radiation pressure we can solve the cold density easily but the more realistic case where solar cycle variations are included makes necessary a solution based on numerical simulation. The last factor in Eq. (2) is the total ionization experienced by a particle arrived at $\vec{r}$. The ionization rate β may depend on the heliographic latitude λ. The ionization is mainly due to the charge exchange reactions with the solar wind protons, only one fifth is due to the EUV-radiation from the Sun. The role of collisional ionization by electrons is estimated to be of a minor significance.

The formulas (1) and (2) introduce several quantities connected to the physics of the Sun. The solar radiation field enters as the original source of the Lyman-α radiation, the counterforce of the gravitation guiding the trajectories of H-atoms in the heliosphere and finally as an ionization field destroying the neutral H population. The solar wind itself enters only in the ionization factor. The remaining parameters are connected to the initial bulk velocity of the neutral wind and its initial temperature. The density of the interstellar flow is introduced as a scaling parameter.

It seems impossible to derive the unknown parameters of this problem by means of any straightforward inversion method. Therefore, in order to infer the parameters we must rely on the fitting of the experimental results with the model calculations. In this work we concentrate on studying the properties of the solar wind ionization and take the other intervening parameters as they have been estimated in earlier investigations. While this approach is not ideal and does not yield a unique estimation in the parameter space it seems to be the only method which is feasible to carry through. The parameters taken as fixed are listed in Table 1.

Table 1. Parameter values for the interstellar wind used in this work (See Ref. /2/). The fitting has indicated that the ecliptic latitude direction of the interstellar H-flow is rather 6° (same as for He-flow) than 7.5° used earlier.

$T = 8000 K$
$\mu = 0.75$
$Longitude = 254.5°$
$Latitude = 6°$
$v_0 = 20 km/s$

LATITUDINAL VARIATION OF THE IONIZATION RATE

The obvious first choice for the ionization rate is a spherically symmetric form. As this turned out to be inadequate to explain the measurements by Prognoz (see /3/ and references therein), Bertaux et al. made use of the following harmonic model originally introduced in /4/:

$$\beta(\lambda, r) = \frac{\beta_e r_e^2}{r^2}(1 - A\sin^2 \lambda) \tag{3}$$

The best fits for the Prognoz measurements were obtained with the value of $A = 0.4$.

We have generalized the ionization rate model to include arbitrary variations of the latitudinal dependence (for a full account of the work, see /5/). The model is defined by fixing the ionization rate at discrete heliographic latitudes:

$$\beta(\lambda_i, r) = \frac{\beta_e^i r_e^2}{r^2}, \quad i = 1..., M \tag{4}$$

Values of the ionization rate at other latitudes are calculated by a linear interpolation. We have taken

$M = 19$ but in the work reported here we have moreover assumed symmetry with respect to the solar equator i.e., actually $M = 10$. Making adjustments for coefficients β_e^i we have assumed β to decrease from the equator to poles. An example of the solar ionization rates used in this work has been depicted in Fig. 1.

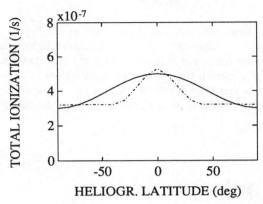

Fig. 1. Ionization models. Solid curve: harmonic model Eq. (3); dashed curve: Eq. (4).

To investigate different ionization models we have used Eqs. (1)-(2) to calculate the predicted Lyman-α intensity and compared these values to measurements by Prognoz 6-satellite /3/. To estimate the quality of a fit we have employed the standard deviation as

$$s = \sqrt{\frac{1}{N} \sum \left(\frac{I_i^{data} - I_i^{model}}{I_i^{data}}\right)^2} \tag{5}$$

where N is the number of data points selected for comparison (typically $N = 30$). An example of this fitting is shown in Fig. 2.

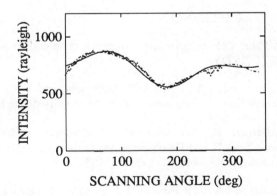

Fig. 2. Measured and predicted Lyman-α intensities. Solid curve: harmonic model Eq. (3); dashed curve: Eq. (4); dots: measured values.

SOLAR WIND VELOCITY AND DENSITY VARIATIONS

The ionization field determined by fitting can further be used to infer more specific solar wind characteristics. As mentioned in Introduction the charge exchange processes due to solar wind protons and the EUV ionization are the main causes of the ionization. We can therefore write

$$\beta(\lambda, r) = \frac{r_e^2}{r^2}[n_{pr}(\lambda)|\vec{v}_{rel}(\lambda)|\sigma_{exc}(|\vec{v}_{rel}(\lambda)|) + \beta_{ph}(\lambda)] \tag{6}$$

Here n_{pr} is the solar proton density, $\vec{v}_{rel}$ the relative velocity between the protons and the neutral H-atoms and σ_{exc} the charge exchange cross section. Due to the large relative velocity difference between the solar wind and neutral H-atoms we approximate the relative velocity by the velocity of the solar wind protons. We have modelled the solar wind velocity according to the models reviewed by Kojima and Kakinuma /6/. The figure 3 gives one example of the solar wind velocity models and the corresponding result for the proton mass flux.

CONCLUSIONS

In this work we have investigated the possibility to deduce solar wind latitudinal characteristics from the Lyman-α measurements. As emphasized in Introduction the path between the measurements and the solar wind properties is paved with many uncertainties. Many groups including ours are now working in order to diminish these uncertain assumptions. Studies on heliospheric interface, multiple scattering of Lyman-α and time-dependent effects are in good progress. The more refined modelling will challenge the future Lyman-α measurements on board the SOHO satellite /7/.

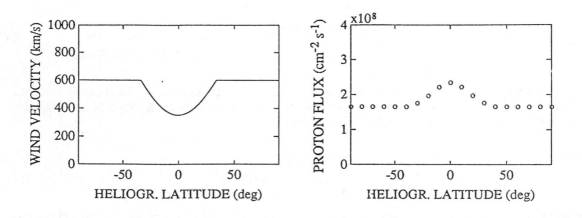

Fig. 3. The solar wind velocity model 4 from /6/ and the corresponding solar wind proton mass flux.

REFERENCES

1. H. J. Fahr, The interplanetary hydrogen cone and its solar cycle variations, **Astron. Astrophys.** 14, 263-274 (1971).

2. E. Chassefière, J. L. Bertaux, R. Lallement, and V. G. Kurt, Atomic hydrogen and helium densities of the interstellar medium measured in the vicinity of the Sun, **Astron. Astrophys.** 160, 229-242 (1986).

3. J. L. Bertaux, R. Lallement, and E. Chassefiere, Interstellar gas parameters and solar wind anisotropies deduced from H and He observations in the solar system, in **The Sun and the Heliosphere in Three Dimensions**, 435-440, D. Reidel (1986).

4. S. Kumar and A. L. B Broadfoot, Signatures of solar wind latitudinal structure in interplanetary Lyman-α emissions: Mariner 10 observations, **Astrophys. J.** 228, 302-311 (1979).

5. T. Summanen, Latitudinal distribution of solar wind as deduced from Lyman alpha measurements, Licentiate thesis, **Geophysical publications 28**, Finnish Meteorological Institute (1992).

6. M. Kojima and T. Kakinuma, Solar cycle evolution of solar wind speed structure between 1973 and 1985 observed with the interplanetary scintillation method, **J. Geophys. Res.** 92, 7269-7279 (1987).

7. J.L. Bertaux, R. Pellinen, E. Chassefière, E. Dimarellis, F. Goutail, T.E. Holzer, V. Kelhä, S. Korpela, E. Kyrölä, R. Lallement, K. Leppälä, G. Leppelmeier, I. Liede, H. Rautonen, J. Torsti, 'SWAN'- A Study of Solar Wind Anisotropies, in **The SOHO Misssion**, ESA SP-1104 (1988).

OBSERVATIONS OF THE SOLAR WIND NEAR
THE SUN FROM MICROWAVE IPS PHENOMENA

M. Tokumaru,* H. Mori,** T. Tanaka,** T. Kondo,*** H. Takaba*** and
Y. Koyama***

* Wakkanai Radio Observatory, Wakkanai, Hokkaido 097, Japan
** Radio Science Division, CRL, Koganei, Tokyo 184, Japan
*** Kashima Space Research Center, Kashima, Ibaraki 314, Japan

ABSTRACT

The solar wind plasma near the sun has been investigated using the power
spectra of interplanetary scintillations (IPS) observed at 2, 8 and 22 GHz in
1990. The results of a spectral fitting analysis indicate that the solar wind
is accelerated significantly from 5 to 30 Rs(solar radii). It is also found
that the spatial spectra of plasma turbulence flatten at smaller radial
distances.

INTRODUCTION

The shape of the power spectrum of interplanetary scintillations (IPS)
reflects some properties of the solar wind plasma intervening between radio
source and an observer on the earth. Several authors have discussed the motion
and the turbulence spectrum of the solar wind from observed IPS spectra
/1,2,3,4/. In the present paper, we briefly summarize the results of a power
spectrum analysis for IPS data obtained near the sun from the Kashima 34m
antenna observations.

IPS studies have been carried out IPS studies at 2, 8 and 22 GHz using the
Kashima 34m antenna /5/. The radio sources employed in our studies are two
intense quasars; 3C273B and 3C279 (at 2 and 8 GHz), and also the H_2O maser of
IRC20431 (at 22 GHz). The IPS observations for the H_2O maser allow us to probe
the solar wind down to 5Rs. The observations were performed during the periods
Sep.20–Oct.20 and Dec.20–26, when quasars and maser, respectively, pass
through solar conjunction. Here, we calculate IPS power spectra from IPS data
obtained in 1990, and analyze them by matching the theoretical model given in
the next section.

MODEL OF IPS POWER SPECTRUM

The temporal spectrum of IPS produced by a slab of thickness dz at distance z
is given by

$$P'(k_x,z)dz = T (2\pi\lambda r_e)^2/V(z)$$

$$\cdot dz \int_{-\infty}^{\infty} (k_x^2+(k_y/AR)^2)^{-\alpha/2}\exp(-k^2/k_{eff}^2)F_{diff}dk_y \qquad (1)$$

where F_{diff} is the Fresnel filter term,

$$F_{diff}(k_x, k_y) = 4\sin^2(k^2\lambda z/4\pi) \tag{2}$$

Here, $k=(k_x, k_y, k_z=0)$ is the three-dimensional wavenumber, λ the wavelength of observation, $V(z)$ the solar wind velocity projected onto the plane perpendicular to the direction z, r_e classical electron radius ($=2.8\times10^{-13}$cm), α the spectral index of density fluctuations, AR the axial ratio of irregularities, k_{eff} the wavenumber of the high frequency cutoff, and T the scintillation amplitude.

Assuming that the radio wave is weakly scattered in interplanetary space, the observed IPS spectrum is regarded as the sum of contributions from all such thin layers perpendicular to the line-of-sight. Since the power of density fluctuations shows a steep radial gradient, most of the contributions to the IPS spectrum come from the region of minimum distance from the sun along the ray path; i.e. at $z=Z=\cos\varepsilon$ where ε is the solar elongation angle. Therefore, the observed IPS spectrum $P(f)$ can be written by

$$P(f) = \int_0^\infty \delta(z-Z)P'(k_x=2\pi f/V(z),z)dz \tag{3}$$

where f is the temporal frequency ($=k_xV/2\pi$).

In this model, we neglect the effect of the random velocity, since the spectral shapes produced by the random velocity and the axial ratio are so similar that we cannot distinguish them in fitting the model. Therefore, the axial ratio of AR determined by spectral fitting analysis includes the contribution of random velocity as well as that of the intrinsic anisotropy.

SPECTRAL FITTING ANALYSIS

The typical IPS spectrum exhibits the Fresnel knee at f_F. Assuming $f_F = V/(\pi\lambda z)^{1/2}$, the solar wind velocity V at $z=Z$, can be derived from the break frequency in the spectrum. The slope of the asymptotic fall above the break frequency corresponds to the spectral index of density fluctuations α, and also the rounded shape of the Fresnel knee indicates the anisotropy of the solar wind AR. Thus, the solar wind parameters such as V, AR, α can be discriminated from each other if an IPS spectral curve is given.

In order to determine the solar wind parameters quantitatively, we have fitted the spectral model (3) to the observed IPS spectra. The free parameters determined from the fitting procedure include the wavenumber of high frequency cutoff k_{eff} and the scintillation amplitude T as well as V, AR, α. To satisfy the weak scattering condition, we have removed IPS data, which show a saturation in the diagram of scintillation index versus radial distance. We have employed the least square method iteratively to find the best fit, and finally obtained 245 sets of the optimal solution from the IPS data collected by Kashima 34m observations in 1990. The detailed description on the analysis will be given in a separated paper.

RESULTS AND DISCUSSION

Figure 1 shows results for V, AR, and α plotted as a function of the radial distance from the sun. The error estimated from the residuals does not exceed

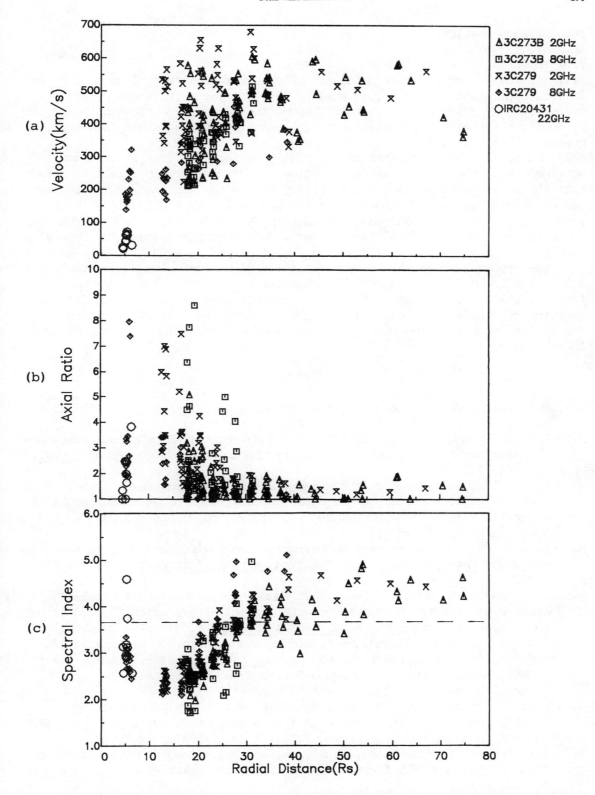

Fig. 1. (a) Solar wind velocity V, (b) axial ratio AR (which includes the effect of the random velocity), and (c) spectral index of density fluctuations α, plotted versus the radial distance from the sun. The dashed line in the panel (c) corresponds to Kolomogorov value of 11/3.

20% for each results.

From the figure, we conclude the following:

(1) The solar wind is accelerated significantly in the range from 5 to 30 Rs. This is consistent with previous results /3,4,6/. The multifrequency cross-spectral study /7,8/, although it is beyond the scope of the present paper, is a promising technique for calibrating the velocity estimates obtained from quasar observations.

(2) The axial ratio increases significantly inside of 30 Rs, showing good agreement with the results of Armstrong et al./9/ However, it should be noted that the increase of axial ratio is partly due to the increase of the random velocity near the sun. To distinguish the intrinsic axial ratio from the random velocity, we would need multi-station IPS observations using large antennas (e.g. Usuda 64m and Kashima 34m). This is being planned for future studies.

(3) The spectral index varies systematically with solar elongation from about 4.0 at > 30Rs to 2.5 around 10-20 Rs. Similar results have been presented from radio scattering observations of spacecraft beacons /3,10/. The flatter spectrum of density fluctuations near the sun might be related to energy deposition in driving the solar wind. An extended discussion must be deferred to subsequent IPS studies.

REFERENCES

1. W.A. Coles, Interplanetary scintillation, Space Sci. Rev. 21, 411-425 (1978)
2. W.A. Coles and J.K. Harmon, Interplanetary scintillation measurements of the electron density power spectrum in the solar wind, J. Geophys. Res. 83, 1413-1420 (1978)
3. G.L. Tyler, J.F. Vesecky, M.A. Plume, and H.T. Howard, Radio wave scattering observations of the solar corona: First-order measurements of expansion velocity and turbulence spectrum using Viking and Mariner 10 spacecraft, Astrophys. J. 249, 318-332 (1981)
4. S.L. Scott, W.A. Coles, and G. Bourgois, Solar wind observations near the sun using interplanetary scintillation, Astron. Astrophys. 123, 207-215 (1983)
5. M. Tokumaru, H. Mori, T. Tanaka, T. Kondo, H. Takaba, and Y. Koyama, Solar wind near the sun observed with interplanetary scintillation using three microwave frequencies, J. Geomag. Geoelectr. 43, 619-630 (1991)
6. J.W. Armstrong and R. Woo, Solar wind motion within 30 $R_\odot$: Spacecraft radio scintillation observations, Astron. Astrophys. 103, 415-421 (1981)
7. R. Woo, Multifrequency techniques for studying interplanetary scintillations, Astrophys. J. 201, 238-248 (1975)
8. S.L. Scott, B.J. Rickett, and J.W. Armstrong, The velocity and the density spectrum of the solar wind from simultaneous three-frequency IPS observations, Astron. Astrophys. 123, 191-206 (1983)
9. J.W. Armstrong, W.A. Coles, M. Kojima, and B.J. Rickett, Observations of field-aligned density fluctuations in the inner solar wind, Astrophys. J. 358, 685-692 (1990)
10. R. Woo and J.W. Armstrong, Spacecraft radio scattering observations of the power spectrum of electron density fluctuations in the solar wind, J.Geophys. Res. 84, 7288-7296 (1979)

ON THE POSSIBILITY OF REGRESSIVE DIAGNOSTICS OF THE NEUTRAL HELIOSPHERIC CURRENT SHEET SHAPE

L. Třísková and P. Šroubek

Geophysical Institute, Czechoslovak Academy of Sciences, 141 31 Praha 4, Czechoslovakia

ABSTRACT

The indices aa and Dst were analyzed for intervals with a stable polarity of the main solar dipole field. Applying a frequency filtration technique, preliminary results show that at least rough features of the IMF sector structure near the Earth's orbit and of the neutral line at the solar source surface can be obtained from the geomagnetic activity variation on the scale of Carrington rotations.

INITIAL CONSIDERATIONS

If the southward component of the interplanetary magnetic field (IMF) influences geomagnetic activity /1/, and if the probability of the $-B_z$ occurrence depends on the configuration of the heliospheric neutral current sheet, then the geomagnetic activity variation on the scale of one Carrington rotation contains information on both these factors. If the changes of geomagnetic activity due to the $-B_z$ component can be separated, their pattern in one Carrington rotation should yield information about the polarity of the IMF observed on the Earth, as well as the approximate shape of the neutral line on the source surface at times when the polarity of the main solar dipole field is stable (see Figure 1).

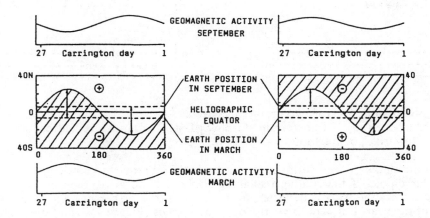

Fig. 1. Influence of the of the neutral heliospheric current sheet configuration and the polarity of the main solar dipole field on the variation of geomagnetic activity. Schematic illu stration, symbolic Carrington rotation, two-sector structure resulting from a tilted dipole axis. No asymmetry of solar hemispheres.

The occurrence probability of the southward IMF component is greater in the first half of the year (March) in the negative and in the second half of the year (September) in the positive sector. In these sectors, the Earth is situated at a larger distance from the neutral current sheet, i.e. in the region of greater solar wind velocities, when the southern solar hemisphere is negative.

DATA AND DATA PROCESSING

Daily values of the indices aa and Dst, which were available from 1868 and 1958 /2,3/ respectively, were reduced to Carrington days for each Carrington rotation investigated (a Carrington day is 1/27 of the synodic Carrington rotation, lasting 27.27 calendar days):

$$I_i^c = I_x \ (1 - y/100) + I_{x+1} \ (y + 1)/100 \qquad\qquad (1)$$

I_i^c is the mean value of index I for the i-th Carrington day which begins at $y/100$ of the x-th calendar day and ends at $(y+1)/100$ of the (x+1)st calendar day, I_x is the daily value for the x-th calendar day, and y is the percentage part of the date of the beginning of the i-th Carrington day.

The variation due to the IMF structure was separated by frequency filtration, and the sum of the first and second harmonic (periods of 27 and 13.5 Carrington days, respectively) S_i (I) was taken as the characterising parameter.

The positive value of $S_i(aa)$, and/or the negative value of $S_i(Dst)$ was considered to be evidence of the existence of the $-B_z$ component on a particular Carrington day. The Carrington days with positive $S_i(aa)$ and/or negative $S_i(Dst)$ were assigned negative IMF polarity in the first and positive polarity in the second half of the year, and for Carrington days with negative $S_i(aa)$ and/or positive $S_i(Dst)$ the opposite was supposed. An example of the polarity of the IMF derived there from as compared to the observed IMF polarity /4-6/ is shown in Figure 2.

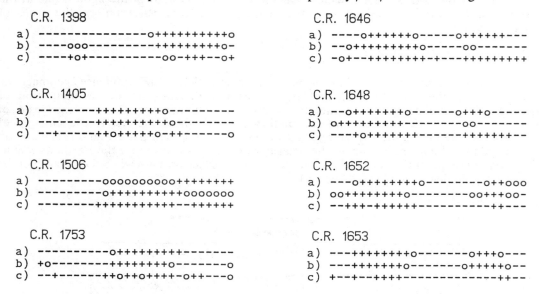

Fig. 2. IMF polarity in the Carrington rotations given:
a) determined from $S_i(aa)$,   b) determined from $S_i(Dst)$,   c) measured, after /4-6/.
  + away, - towards, o indefinite

Determining the shape of the neutral current sheet (NCS) requires knowledge of the polarity of the main solar dipole. If the southern solar hemisphere is negative, then on days of enhanced geomagnetic activity the NCS is deflected to the north in the first half, and to the south in the second half of the year; if it is positive, the opposite applies (Figure 1). The time required for the solar wind to transfer the configuration from the Sun to the Earth was considered to be constant, 4 days.

Comparison with the results obtained with the method of the potential field and maximum brightness contours /7/ indicates that the shapes of the curves correspond best if $+1^o$ of heliographic latitude is assigned to a positive unit of index aa and negative unit of index Dst, respectively. The results are shown in Figure 3.

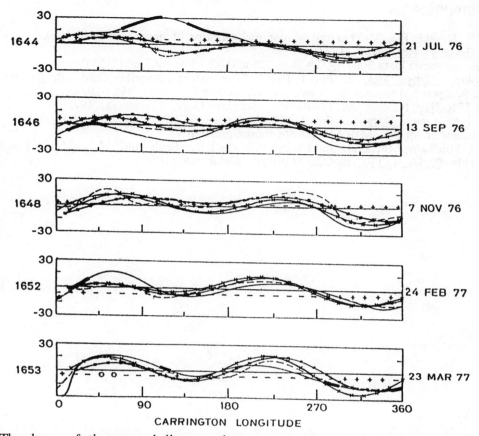

Fig. 3. The shape of the neutral line on the source surface obtained from potential field calculations and the maximum brightness contour method, adopted from /7/, as well as the results derived here from $S_i(aa)$ and $S_i(Dst)$.

- - - - - - maximum brightness contour ——————— potential field calculation
————■———— filtered aa estimation ————✗———— filtered Dst estimation

SUMMARY OF RESULTS

As indicated by Figures 2 and 3, reasonably reliable estimate of the polarity of the IMF at the Earth, as well as the shape of the neutral line on the source surface, can be obtained from the quasiperiodic changes of geomagnetic activity at times of stable polarity of the main solar dipole field.

The first harmonic provides basic information about the inclination of the NCS, the position of its maximum agreeing with the position of the geoactive sector (negative in the first half of the year, positive in the second half). In most cases the second harmonic specifies this information. Cases did occur, even for a well-defined two-sector structure, for which the second harmonic displayed a non-negligible amplitude. This greatly undermines there liability of this information, indicating that factors other than the -B_z component (e.g. solar wind velocity) may be acting. These discrepancies occurred mostly at times near the minima of the odd solar cycles.

The findings reported in this paper are consistent with the hypothesis of Russell and McPherron /1/. The method has been shown to be viable for approximately determining the IMF polarity at times when it was not regularly observed. The shape of the neutral line on the source surface was determined satisfactorily in the five cases shown. However, definitive judgment as to the usability of the procedure requires further comparisons with the results of the maximum brightness contour method and the potential field calculations. It is obvious that the method is not applicable to solstices.

REFERENCES

1. C.T. Russell, R.L. McPherron: Semiannual variation of geomagnetic Activity. J. Geophys. Res. 78, 92, (1973)
2. Geomagnetic Data 1980, IAGA Bull. 32f-k, IUGG Pub.-office (1983)
3. Solar Geophys. Data, 522, Part I., Dept of Commerce (Boulder Colo. USA), (1988)
4. L. Svalgaard: An Atlas of Interplanetary Sector Structure 1957-1974, SUIPR. Rep. No 629, Inst. Plasma Res., Stanford Univ, California, (1975)
5. Solar Geophys. Data, 423, Part I., (1979)
6. Measurement of IZMIRAN, private communication
7. J.T. Hoeksema: Structure and Evolution of the Large Scale and Heliospheric Magnetic Fields. CSSA ASTRO-84-OZ, Stanford, California, (1984)

ON THE LARGE-SCALE MAGNETIC FIELD STRUCTURE IN THE OUTER HELIOSPHERE

L. Třísková* and I. S. Veselovsky**

* Geophysical Institute, Czechoslovak Academy of Sciences, 141 31 Praha 4, Czechoslovakia
** Nuclear Physics Institute, Moscow State University, 119899 Moscow, Russia

ABSTRACT

In a vacuum reconnection model, the magnetic field in the outer heliosphere can be approximated by the sum of fields of solar and interstellar origin. The resulting heliospheric magnetic field depends on space-time coordinates and is determined by the mutual orientation of the main solar dipole, the interstellar field, and the solar rotation axis. A parallel and/or antiparallel component of the interstellar field results in an open and/or closed field line configuration, respectively.

INTRODUCTION

In the kinematic MHD approximation with a given velocity field, the interplanetary magnetic field can be characterized by simple linear equations. These equations may be solved explicitly under quasistationary boundary conditions in the case of a radially expanding solar wind. In the subsequent qualitative analysis we neglect the inhomogeneity of the outer magnetic field and analyze the field structure assuming a vacuum reconnection pattern (see also /1/).

MAGNETIC FIELD MODELS

Characteristics of the magnetic field in the outer heliosphere and in the local interstellar medium are not well known yet, e.g. /2, 3/. The model magnetic field in the outer heliosphere may be approximated by

$$\mathbf{B} = \mathbf{B_S} + \mathbf{B_O} \tag{1}$$

where $\mathbf{B_S}$ is the field of solar and $\mathbf{B_O}$ of interstellar origin.

It is assumed that, on the solar wind boundary surface r_O, $\mathbf{B_S}$ has a radial component only, $|\mathbf{B_S}| = |\mathbf{B_{rs}}| = $ const, the sense of this component being opposite in both hemispheres and determined by the "main solar dipole" μ_S orientation. This model represents a dipole-like structure of the coronal magnetic field with the thin current sheet at the magnetic equator. In the case of a "tilted rotating dipole" one obtains Parker-type formulae for the $\mathbf{B_S}$ components in interplanetary space,

$$B_{rs} = B_S \left(r_O / r \right)^2 \text{sign} (\cos \gamma)$$

$$B_{\Theta s} = 0 \tag{2}$$

$$B_{\varphi s} = - B_S \left(r_O / r \right)^2 \left(\Omega r / v \right) \sin \Theta \text{ sign} (\cos \gamma)$$

where spherical coordinates (r, Θ, φ) are used, Ω is the angular velocity of the rotation of the Sun, v is the solar wind velocity, assumed to be radial and constant, direction $\Theta = 0$ coincides with the rotation axis of the Sun, and

$$\cos \gamma = \cos \Theta \cos \alpha + \sin \Theta \sin \alpha \cos (\varphi_\mu - \varphi), \quad \varphi_\mu = \varphi_0 + \Omega (t - t_0),$$

where α is the angle between μ_s and Θ, and φ_0 is the initial phase. A homogeneous interstellar magnetic field $\mathbf{B}_0$ is assumed.

FIELD LINE STRUCTURE IN THE OUTER HELIOSPHERE

Magnetic field of interstellar origin is subjected to strong shielding when penetrating into the heliosphere /4/, reconnection of interplanetary and interstellar magnetic fields can take place only in the outer heliosphere. The resulting heliospheric magnetic field (1) depends on space-time coordinates $(\mathbf{r}, t)$ and, for a given time t, is determined by the mutual orientation of three vectors: μ, $\mathbf{B}_0$, Ω.

Some special cases are shown in Figures 1-3, which are calculated for $B_s = 5$ nT, $B_0 = 0.5$ nT, $r_0 = 1$ A.U., $v = 400$ km/s, $L = 3.16$ A.U., $g = 3.44$, and Ω is given by the sidereal rotation of the Sun.

Figure 1 represents open and closed magnetic configurations when $\Omega = 0$ for $\mathbf{B}_0$ and μ_s parallel, anti-parallel, and perpendicular. Field line equations may be expressed in dimensionless cylindrical (z, ρ) or Cartesian (z, x) coordinates using the scale length $L = r_0 (B_s / B_0)^{1/2}$.

a) $d\rho/dz = \rho [z + (\rho^2 + z^2)^{3/2} \operatorname{sign} z]^{-1}$,

b) $d\rho/dz = \rho [z - (\rho^2 + z^2)^{3/2} \operatorname{sign} z]^{-1}$,   (3)

c) $dx/dz = x [z + (x^2 + z^2)^{3/2} \operatorname{sign} x]^{-1}$.

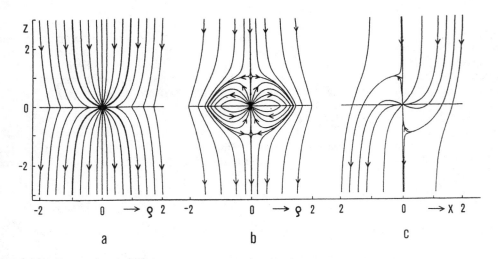

Fig. 1. Open and closed magnetic field configurations in the case of a "nonrotaing Sun", $\Omega = 0$:
a) $\mu_s \uparrow\uparrow \mathbf{B}_0$, b) $\mu_s \uparrow\downarrow \mathbf{B}_0$, c) $\mu_s \perp \mathbf{B}_0$. Z marks two neutral points.

The analytical solution of equation (3) yields equations of field lines in the form

$$x^2 + y^2 = 2x (x^2 + y^2)^{1/2} K^{-1} y^{-2} + C (x^2 + y^2) y^{-2}$$ (4)

which, transformed into polar coordinates may be written as

$$r^2 = (2K^{-1} \cos \varphi + C) \sin^{-2} \varphi$$ (5)

where $K = L^{-2}$ and C is the integration constant.

Figure 2 shows the configurations of the heliosphere when both vectors $\mathbf{B_0}$ and μ_s are perpendicular to the Sun's rotation axis. In the equatorial cross section ($z=0$), the field line equation in dimensionless variables with scale length L becomes

$$dy/dx = [y\text{-}gx(x^2+y^2)^{1/2}] \, \text{sign}(\cos\gamma) \, \{[x+gy(x^2+y^2)^{1/2}] \, \text{sign}(\cos\gamma) + (x^2+y^2)^{3/2}\}^{-1} \qquad (6)$$

Right-handed Cartesian coordinates are used with the z axis along Ω, and the x axis along $\mathbf{B_0}$,

$$g = \Omega L v^{-1}, \cos\gamma = \cos(\varphi_\mu - \varphi).$$

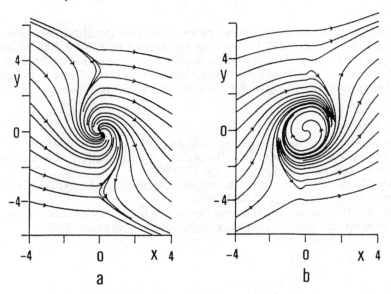

Fig. 2. Qualitative instantaneous patterns of the heliospheric magnetic field in the equatorial plane when both μ_s, $\mathbf{B_0}$ vectors are perpendicular to the Sun's rotation axis. The times are shown, when a) $\mu_s \uparrow\uparrow \mathbf{B_0}$, b) $\mu_s \uparrow\downarrow \mathbf{B_0}$

Figure 3 shows the magnetic field structure when $\Omega \quad \mu_s$ and $\mathbf{B_0} \quad \Omega$. The field line equation can be expressed in dimensionless coordinates as

$$dx/dy = gy \, [\text{-}gx + (x^2 + y^2 + z^2) \, \text{sign} \, z]^{-1} \qquad (7)$$

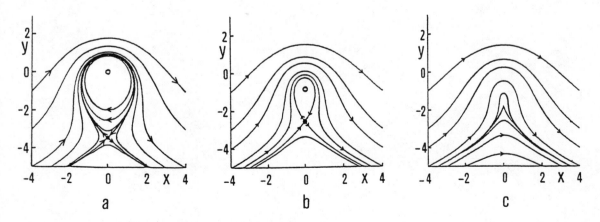

Fig. 3. Magnetic field lines in the outer heliosphere when $\mu_s \parallel \Omega$, $\mathbf{B_0} \perp \Omega$. Cross section are shown in planes parallel to the helioequator: a) $z = 0$, b) $z = 1.5$, c) $z = 1.72$.
● X-type neutral points, o O-type neutral points.

DISCUSSION AND CONCLUSION

A simple vacuum reconnection model was used with homogeneous magnetic field B_0 representing the interstellar magnetic field.

$$B_0 = B_{ext} \exp[(a-r)/\lambda], \quad r \leq a,$$

where B_{ext} is the (uniform) interstellar magnetic field, a is the radius of the heliosphere and λ is the skin lenght /4/. Our approximation holds for $(a-r)/\lambda << 1$, i.e. in the outer part of the heliosphere.

To represent adequately a real situation, more elaborate quantitative magnetic field models are needed taking into account MHD effects for an external flow around heloisphere. However, our main qualitative results, i. e. the occurrence of closed and open heliospheric configurations in dependence on the solar magnetic field orientation, may be valid. These heliospheric structures may be reflected in asymmetric filling of the heliosphere by galactic cosmic rays /5/.

REFERENCES
1. W. Macek, Reconnection at the heliopause. Adv. Space Res. 9, 257 (1980)
2. P.C.Frisch, The physical properties of the "local fluff". Adv.Space Res. 6, 345 (1986)
3. S.T. Suess, The heliopause. Rev. Geophys. 28, 97 (1990)
4. I.S. Veselovsky, On the screening of the interstellar electric and magnetic fields by the solar wind. Astrophys. and Space Sci. 86, 209 (1982)
5. I.S. Veselovsky, L.Třísková, On the filling of the heliosphere by galactic cosmic rays under various conditions on the Sun. Travaux géophysiques 38, in press (1992)

TRAVELLING INTERPLANETARY DISTURBANCES FROM RADIOASTRONOMICAL DATA

V. I. Vlasov

Lebedev Physical Institute, Leninsky Prospect 53, 117924 Moscow, Russia

ABSTRACT

Analysis of the all-sky scintillation index maps shows that several types of disturbances may exist in the interplanetary plasma. The average scale of travelling disturbances are equal 120 and 90 degrees in heliolongitude and latitude sections, respectively, and their average velocities are about $500 \div 700$ km/s. It also follows from radioastronomical data that the shock wave of flare origin are decelerated by travelling through the interplanetary medium and the shock velocity depends on the value of initial shock velocity. The radial dependence of shock wave speed may be represented in the power law form, $V \sim r^{-\alpha}$, by the index between $0.25 \leqslant \alpha \leqslant 1$. The reason of the difference of the index from $\alpha = 0.5$ value which is expected in the case of simple plasma sweeping are at present unknown. In particular for $\alpha > 0.5$ damping of shock wave energy must take place. The shock wave angular sizes agree with that of the interplanetary travelling disturbances average scales and differ from sizes of standart model of the interplanetary shock waves. The strong correlation between scintillation index variations and geomagnetic A_p-index take place. It allows using of the radioastronomical data for prediction of geomagnetic activity.

INTRODUCTION

Radioastronomy allows us to observe a large number of scintillating radiosources during a short time interval for mapping an all-sky scintillation index. The scintillation index m is a relative r.m.s. flux fluctuation. It characterizes the dispersion of electron dencity fluctuations. Scintillation index

mapping is now realised in USSR /1,2/ and England /3/. The
mapping made over a long period of time show that the
scintillation characteristics undergo strong time variations.
The scintillation index variations (and other parameters) are
caused by two main types of plasma perturbations. The first tipe
is the corotating disturbances of solar wind high-speed streams.
The second type of disturbances is a travelling events. These
disturbances are generated by short time events on the Sun. We
have many various data of 1975 to 1989 years scintillation index
mapping. Here we shall consider the travelling disturbances.

THE AVERAGE SCALE IN HELIOLONGITUDE AND LATITUDE CROSS-SECTIONS

All-sky scintillation indices as a function of the date are
shown in Figure 1. Figure 2 shows the maximum value of R_m of the

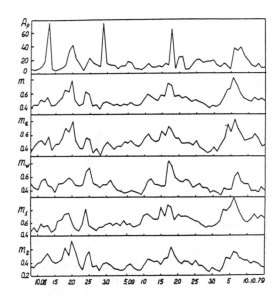

Fig.1. Scintillation index as a function of the date. m is
all-sky scintillation index. m_E is East-sky index, m_W is
West-sky index, m_1 - scintillation index on 0.5÷0.7 A.U.
heliocentric distances, m_2 - on distances >0.7 A.U. A_p -
geomagnetic index.

cross-correlation function $R(\Delta\theta, t)$ depending on
heliolongitudional distance $\Delta\theta$ between two interplanetary plasma
regions with coordinates $r_1=r_2=const$, $\varphi_1 \approx \varphi_2$ and $\theta_1 \neq \theta_2$. His
seen that the average scale of travelling disturbances of
interplanetary plasma is about $120°$ in the longitudinal section.
The same dependence of R_m but on heliolatitudional distance $\Delta\varphi$
in Figure 2 (curve 2) shows that the angular scale in the

heliolatitudional section is equal to approximately $75 \div 90^{\circ}$. Curve 3 in this figure represents R_m as a function of angular distance $\Delta\Psi$ between interplanetary plasma regions in oblique sections ($\Delta\Psi=\sqrt{\Delta\varphi^2+\Delta\theta^2}$).

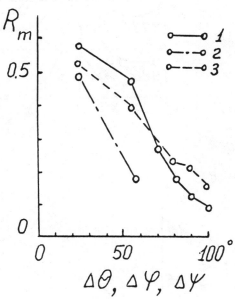

Fig.2.   Cross-correlation coefficients of scintillation index variations for two equidistant from the Sun interplanetary regions as a function of heliocentrical angles.

EAST-WEST ASYMMETRY OF TRAVELLING DISTURBANCES

An example of the cross-correlation functions of the scintillation index variation for two equidistant from the Sun and spaced on heliolongitude angle regions in Figure 3(a) show that the time delay of the correlation maximum $\tau \neq 0$. The sign of τ corresponds to the delay of the scintillation index variations of the west interplanetary regions relative to the east regions. The dependence of $\tau_{1,2}$ on heliolongitudional distance $\Delta\theta$ is shown in Figure 3(b). Figure 3(c) shows the time delay $\tau_{Ap,m}$ between the variations of the Ap-index and scintillation index as a function of heliolongitude $\Delta\theta$. We attribute the correlation delay between the scintillation index variations of the west interplanetary plasma regions from east regions to the east-west asymmetry of the travelling disturbances. Assuming inferred values of $\tau_{1,2}$ and average propagating disturbances velocity $v=500$ km·s^{-1}, the spatial structure of the disturbance can be represented schematically as shown in Figure 3(d). The asymmetry is about 20 to 30° in size

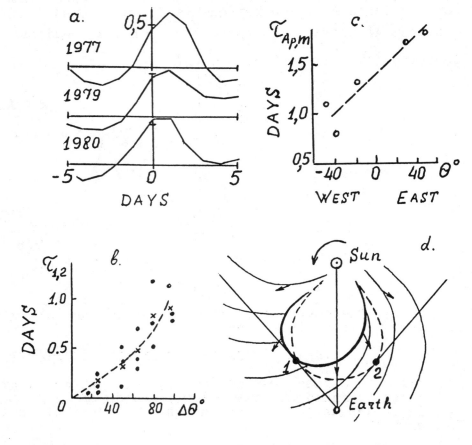

Fig.3 **a.** Examples of cross-correlation functions. **b,c** – the dependence of cross-correlation delay $\tau_{1,2}$ and $\tau_{Ap,m}$ on heliolongitudional distance $\Delta\theta$. **d.** – scheme of the travelling interplanetary disturbance East-West asymmetry.

at the heliocentric distance of ~0.8 A.U. The main reason for such asymmetry is perhaps that travelling disturbance interacts a spiral interplanetary magnetic field.

THE AVERAGE VELOCITY OF TRAVELLING DISTURBANCES

The average velocity of travelling disturbances producing the scintillation index variations can be estimated from the correlation time delay τ and corresponding heliocentric distance Δr between interplanetary plasma regions considered. Figure 4 illustrates the average dependence of $\tau(\Delta r)$. The sign of the delay, τ, corresponds to the direction away from the Sun. From this dependence it is seen that the average velocity of the interplanetary disturbance $V=\Delta r/\tau$ is about $500\div700$ km·s^{-1}. This estimate corresponds to the solar wind disturbance velocity determined using spacecraft /6/ and threepoint scintillation observations /7,8/.

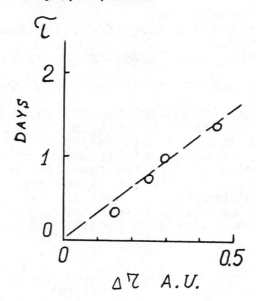

Fig.4. The correlation time delay τ as a function of heliocentric distace Δr.

THE EXPOSURE OF THE INTERPLANETARY SHOCK WAVES

The concrete distubances of the interplanetary plasma may be selected from number of solar wind disturbances by means of Δm_i maps /9,10/:

$$\Delta m_i = \frac{m_{i,j+1} - m_{i,j}}{m_{i,j+1} + m_{i,j}}$$

Where m is the scintillation index, i – scintillating sources number, j – observation date.
The Δm_i values may be selected by three parts:

$\Delta m_i = 0$ means that interplanetary plasma state do not change,

$\Delta m_i > 0$ – disturbance front cross the zonding region

$\Delta m_i < 0$ – disturbance front leaved interplanetary zonding region.

The maps of spatial distribution of $\Delta m_i > 0$ values are that of interplanetary disturbances. Continual row of these maps makes it possible to estimate the form, velocity and direction of disturbance propagation, the place of the disturbance source on the Sun et al.

INTERPLANETARY SHOCK GEOMETRY

The radio astronomical observations allows us to derive plane two-dementional image of the interplanetary plasma. But it is

possible to suppose that:
    1) Interplanetary shock front is expanded three-dimentional
       envelope.
    2) Shock has axial symmetry in heliolatitud and longitude
directions.
    3) Interplanetary shock generating center is plased near the
       Sun.
The examples of the interplanetary shock maps in heliocentrycal
(θ,φ), (r,φ) and (r,θ) coordinats are presented in Figure 5 and
6. Here the points indicate the position of the radiosources
that showed increasing scintillation index relative to the value
in the preceding day. We see the form, position and motion of
the shock wave front clearly. Symmetric picture of the shock in
Figure 5 is connected with that location near the solar central
meridian and travelling almost forward to the Earth. The shock
in Figure 6 is observed in profile and moved in West direction
from the Sun.

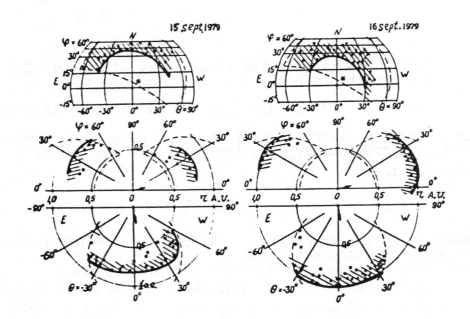

    Fig.5. The interolanetary shock maps in heliocentrical
    coordinats. Points indicate the positions of radiosources,
    that showed on increase of scintillation index relative to
    the value in preciding day. The shock wave propagaite from
    Sun to Earth.

The interplanetary shock angular sizes in heliolatitud and
heliolongitude cross-sectios on the average are $\langle l_\varphi \rangle \approx 70 \div 75^{\circ}$
and $\langle l_\theta \rangle > 115^{\circ}$. They agree with average parameters of the
interplanetary travelling disturbance model and differ from size
of standart model of the interplanetary shock wave.

THE VELOCITY OF INTERPLANETARY SHOCK WAVES
AT DIFFERENT HELIOCENTRIC DISTANCES

Interplanetary shock waves generated by solar flares are the most important type of travelling events. The velocity of shock fronts may be estimated from flare time and from the positions of shock front for some events:

$$V = \Delta r / \Delta t = (r_{j+1} - r_j)/(t_{j+1} - t_j)$$

Where r_j is interplanetary shock front heliocentric distant in the its main propagation direction, t_j – record tame of the shock front and the flare, j – date of observations.

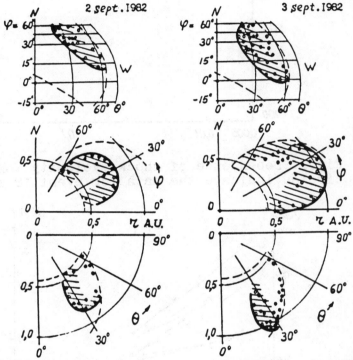

Fig.6. As in figure 5. The shock wave propagate in West direction from the Sun

The velocity data are shown on Figure 7 /10/. Unigue data were received for the event 18 august 1979. The interplanetary plasma parameters were measured by the spacecrafts "Voyager-1" /11-13/ and "Pioneer-11" and ground based radiotelescope /10,14/. The velocity data for this event are indicated by open circles on Figure 7. The observable data lead to the conclusion that the velocity decreases with increasing distance from the Sun. The radial dependence of shock wave speed may be represented in the power law form, $V \propto r^{-\alpha}$, by the index between $-0.25 \leqslant \alpha \leqslant 1$. The rate of decrease depends on the initial value of velocity V_o (Figure 8). The data correspond to the relation

$$\frac{dV}{dr} \propto V_o^2$$

These observations data do not conform with theoretical models of shock wave propagation /15/.

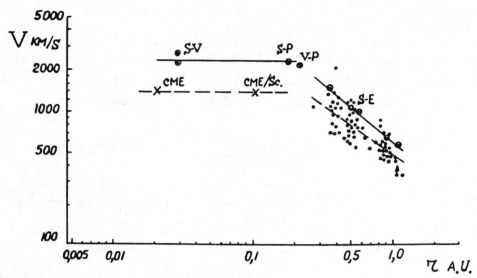

Fig.7. Dependence of shock velocity on distance **r**. Open circles indicate the data for event 18 august 1979.

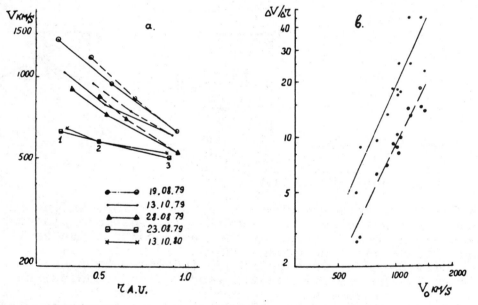

Fig.8. a. Examples of the dependence V(r) of high speed and low speed shock waves. b - Dependence of the decrease rate of shock velocity on that initial velocity V_o.

In conclusion we discuss the problem of using of radioastronomical observations of the interplanetary plasma for prediction of the geomagnetic disturbances. The comparison of the scintillation index data all-sky average and A_p-index of geomagnetic activity data shows sufficiently strong correlation for the periods of our observations. The correlation coefficient is equal 0.6 to 0.8 /16/. The forestall time of radioastronomical data relative to geomagnetic data is about 1 to 2 days. There is a favorable present for using the radioastronomical observations for the prediction of geomagnetic disturbances.

REFERENCES

1. V.I.Vlasov, V.I.Shishov, T.D.Shishova, On the large-scale structure of the interplanetary plasma, Letters to Astron.J. (Sov.) 2, 248 (1976)

2. V.I.Vlasov, Radio images of the interplanetary turbulent plasma, Astron.J. (Sov.) 56, 96 (1979)

3. G.R.Gupper, A.Hewish, A.Purvis, P.J.Duffett-Smith, Observing interplanetary disturbances from the ground, Nature 296, 635 (1982)

4. V.I.Vlasov, V.I.Shishov, T.D.Shishova, The travelling interplanetary disturbances structure, Geomagnetizm i aeronomiya 24, 541 (1984)

5. V.I.Vlasov, East-west asymmetry of travelling interplanetary disturbances, Geomagnetizm i aeronomiya 27, 657 (1987)

6. A.Hundhansen, Coronal Expansion and Solar Wind, "Mir", Moscow (1976)

7. W.A.Coles, B.J.Rickett, Solar Geoph. Data, Colorado, (1977-1981)

8. T.Kakinuma, Solar Wind Speed from IPS, Nagoya University, Toyokava, Japan (1977-1980)

9. V.I.Vlasov, Interplanetary shock waves on the IPS observations, Geomagnetizm i aeronomiya 21, 927 (1981)

10. V.I.Vlasov, Interplanetary shock waves velocity on the radioastronomical data, Geomagnetizm i aeronomiya 28, 1 (1988)

11. R.Woo, J.W.Armstrong, Measurements of a solar flare-generated shock wave at 13.1 Rs, Nature 292, 608 (1981)

12. A.Maxwell, M.Dryer, Measurements on a shock wave generated a solar flare, Nature 300, 237 (1982)

13. H.V.Cane, R.G.Stone, R.Woo, Velosity of the shock generated

by large east limb flare on August 18, 1979, <u>Geoph.Res.Lett.</u> 9, 897 (1982)

14. V.I.Vlasov, Radioastronomical observations of the shock wave from solar flare on August 18, 1979, <u>Geomagnetizm i aeronomiya</u> 26, 182 (1986)

15. M.Dryer, S.T.Wu, G.Gislason, S.M.Han, Z.K.Smith, J.F.Wang, D.F.Smart, M.A.Shea, <u>Astroph. Space Sci.</u> 105, 187 (1984)

16. V.I.Vlasov, V.I.Shishov, T.D.Shishova, The connection between variations of geomagnetic activity index and IPS parameters, <u>Geomagnetizm i aeronomiya</u> 25, 254 (1985)

A TWO-DIMENSIONAL MHD GLOBAL CORONAL MODEL: STEADY-STATE STREAMERS

A. -H. Wang,* S. T. Wu,* S. T.Suess** and G. Poletto***

* Center for Space Plasma and Aeronomic Research and Department of
Mechanical Engineering,The University of Alabama in Huntsville, Huntsville,
AL 35899, U.S.A.
** NASA Marshall Space Flight Center, Space Science Lab/ES52, Huntsville,
AL 35812, U.S.A.
*** Osservatorio di Arcetri, Firenze, Italy

ABSTRACT

A two-dimensional, time-dependent, numerical, MHD model for the simulation of coronal streamers from the solar surface ($r = 1R_\odot$) to $15R_\odot$ is presented. Three examples are given; for dipole, quadrupole and hexapole (Legendre polynomials P_1, P_2, and P_3) initial field topologies. The computed properties are density, temperature, velocity, and magnetic field. The calculation is set up as an initial-boundary value problem wherein a relaxation in time produces the steady state solution. In addition to the properties of the solutions, their accuracy is discussed. Besides solutions for dipole, quadrupole, and hexapole geometries, the model permits use of realistic values for the density and Alfvén speed while still meeting the requirement that the flow speed be super-Alfvénic at the outer boundary by extending the outer boundary to $15R_\odot$.

1. INTRODUCTION

We present results from a recently-developed numerical model of coronal structure. The reasons for a new model are to extend the outer boundary farther from the Sun and to gain the experience necessary for development of a three-dimensional model. In addition, an immediate application will be to the simulation of streamers in support of the Ultraviolet Coronagraph and Spectroheliograph (UVCS) and the Large Angle Spectrometric Coronagraph (LASCO) on the Solar Heliospheric Observatory (SoHO). These instruments will be able to measure the temperature, density, and flow vector in the corona so, with model calculations, it will be possible to estimate the magnetic field vector.

2. THE PHYSICAL AND NUMERICAL MODEL

We assume axisymmetric, single fluid, polytropic, time-dependent ideal magnetohydrodynamic flow and calculate the flow in a meridional plane defined by the axis of the magnetic field. The coordinates are (r, θ, ϕ) with ϕ being the ignorable coordinate. For the magnetic field boundary condition, we take the variation of the radial component at the lower boundary to be given by Legendre polynomials, so that the flow has reflective symmetry across the equator and the calculation need be done in only one quadrant. For P_1, the radial field thus has a dipole variation. The equations describing such flow can be found in many places (e.g. /3/).

The equations are solved in a computational domain extending from the Sun ($1R_\odot$) to $15R_\odot$, from the pole to the equator. It is assumed that meridional flow is zero at the pole and equator. The grid is divided so that there are 37 gridpoints in the radial direction and 22 gridpoints in the meridional direction, with the radial grid size slowly increasing with radius. The algorithm adopted here is the Full-Implicit Continuous Eulerian (FICE) scheme described by Hu and Wu /1/; for time stepping a second-order accurate forward differencing scheme is used and the step size is of the same order as given by the Courant condition. Smoothing is inserted when gradients become too large - i.e. at shocks (which do not occur here). At the inner boundary, the flow is subsonic and sub-Alfvénic so that some variables are calculated using compatibility relations /1/. We choose to specify the radial magnetic field, pressure (or temperature), and density. The meridional field, radial and meridional flow speed, and pressure are computed from the compatibility relations. At the outer boundary, the flow is restricted to being both supersonic and super-Alfvénic. In this case, all variables at the boundary can be calculated by simple linear extrapolation from the first (or first two) grid points inside the boundary.

We start with an essentially arbitrary initial state and allow the flow to relax in time while holding the boundary values constant. In the present case the initial flow field is a polytropic, hydrodynamic solution to the steady state radial flow equation of motion (e.g. /2/) superimposed on a potential magnetic field. That this is neither a self-consistent nor stable solution to the steady state MHD equations is irrelevant since the flow is allowed to evolve in time under the control of the equations of motion. The main concerns are that the numerical solution be stable and of sufficient accuracy to define the physically interesting aspects of the solution, and that the relaxation proceed long enough that an acceptably close approximation to the steady state has been reached. We address these issues briefly in section 4.

3. THE CALCULATIONS

As stated, we present results from three simulations; for a dipole ($B_r(R_\odot) \propto P_1(\cos\theta)$), a quadrupole ($B_r(R_\odot) \propto P_2(\cos\theta)$), and a hexapole ($B_r(R_\odot) \propto P_3(\cos\theta)$). At the lower boundary, the conditions are that $n = 2.25 \times 10^8 cm^{-3}$ and $T = 1.80 \times 10^6 K$. The polytropic index is chosen to be $\gamma = 1.05$. The magnetic field strength at the equator ($B_r(\theta = 90°)$) is 1.67 gauss so that $\beta = 0.5$ in all three cases (where β is the ratio of the internal pressure to the magnetic pressure at the lower boundary, at the equator). The final steady state magnetic field geometry for the three cases is shown in Fig. 1.

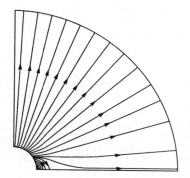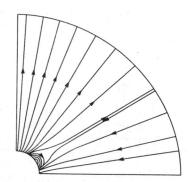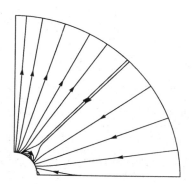

Fig. 1 The steady state magnetic field for the three cases. The left panel is for an initial dipole ($P_1(\cos\theta)$), the middle panel for an initial quadrupole ($P_2(\cos\theta)$), and the right panel for an initial hexapole ($P_3(\cos\theta)$). In all cases, $\beta = 0.5$ at the base, at the equator, in the initial state. At the same location, the total magnetic field strength is 1.67 Gauss in all three examples. The times allowed for the relaxation are: dipole - 22.22 hours, quadrupole - 16.67 hours, hexapole - 18.06 hours.

This figure shows the well-known property that the flow is essentially radial beyond $3 - 4R_\odot$. Having begun with large closed field volumes, only small magnetically closed volumes remain, underlying the coronal streamers. Flow is field aligned everywhere. Fieldlines which cross the outer boundary reach to ∞. The fieldlines are seen to diverge most rapidly on the edges of the close field regions and apparently most slowly near the center of open regions.

The radial velocity is shown in Fig. 2, at the pole and equator for the dipole; at the pole and centered over the mid-latitude streamer for the quadrupole; and at pole, over the mid-latitude streamer, and over the equatorial streamer for the hexapole. As is generally the case in this type of model, the flow speed in the center of the open regions is similar to the undisturbed initial flow speed - because the flow direction is approximately radial above a few solar radii. In the streamer, the flow speed is essentially zero on closed magnetic field lines and is greatly reduced on the open lines - the field has undergone rapid overexpansion on the flanks of the streamer. The density and temperature for the three examples are shown in Fig. 1, in the directions specified in Fig. 2. Most obviously, the density is enhanced in the closed field regions. There is, of course, also some depletion along rapidly diverging fieldlines.

Several physical aspects of such models as these need to be emphasized. First, the temperature that has been calculated is an "effective temperature." This is because a polytropic energy equation is assumed - with a polytropic index of 1.05, which is equivalent to a large amount of energy being added to the flow. Nowhere is the form of this energy specified, nor what the conversion and dissipation mechanisms are. However, it has been shown that a polytropic index on the order 1.05 is required to reproduce observations of coronal densities /5/.

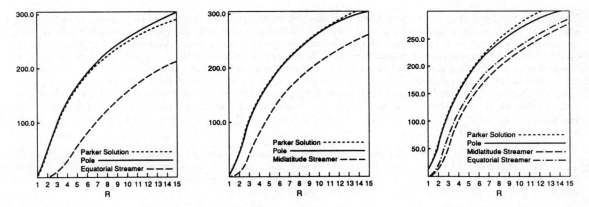

Fig. 2 The radial velocity for the three cases shown in Fig. 1. Left: Dipole field, showing the radial flow speed along a polar radius and an equatorial radius. Middle: Quadrupole field, showing the speed along a radius over the pole and over the mid-latitude streamer. Right: Hexapole field, showing the speed along a radius over the pole, over the mid-latitude streamer, and at the equator.

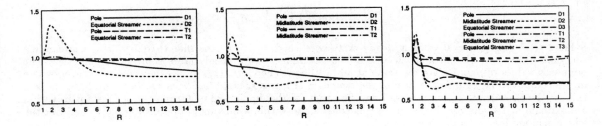

Fig. 3 The density and temperature for the three examples shown in Fig. 1. Left: Dipole field, showing the density and temperature, scaled to their starting values, along a polar radius and an equatorial radius. Middle: Quadrupole field, showing the temperature and density along a radius over the pole and over the mid-latitude streamer. Right: Hexapole field, showing the temperature and density along a radius over the pole, the mid-latitude streamer, and at the equator.

Second, the magnetically open regions, although equivalent to coronal hole flows, do not simulate coronal holes because the flow speeds are far too small. To obtain reasonable flow speeds in this model it would be necessary to have the temperature vary across the base of the open region - which is well within the capability of the model. Such a variation has been shown to reproduce all the known properties of coronal hole flow and lead to accurate simulations of the geometry, with the effective temperature being larger in the center of the hole than at the edge /5/.

In contrast to the open regions, the densities in the closed regions are similar to observed streamer densities and we feel this model is therefore a good approximation to streamer geometry. The temperature must still be qualified as an effective temperature, but can be used for diagnostic purposes in combination with planned observations on SoHO/UVCS.

4. ACCURACY AND STABILITY OF CALCULATIONS

This model has been found to be weakly subject to the Courant condition on size of time step. Therefore, the size of the time step decreases as the largest values of the temperature and magnetic field increase - along with the maximum sound and Alfvén speeds anywhere in the grid. Counteracting this, the higher characteristic speeds lead to a somewhat faster relaxation time. However, generally more time steps are required for smaller β calculations. The flow speed also plays an important role in determining the relaxation time to a steady state - the initial state is a disequilibrium configuration. This imbalance must have time to be advected from the base through the outer boundary. The physical time this takes can be estimated by taking a typical (but small) value for the flow speed and calculating how long it would take the plasma to flow at this speed from the base to the outer boundary. For example, at 150 km/s, to $15R_\odot$, this takes 18 hours (relaxation times we have used are given in Fig. 1).

A second consideration is gridpoint resolution. The grid used in these examples is about $4.5°$ in latitude and $0.24R_\odot$ in radius near the base - increasing slowly with radius. This is sufficient to adequately resolve the geometry and flow on the scale shown in Fig. 1. However, if finer scale information is required in, for example, the core of the streamers, a denser grid would be required.

Always a serious consideration in these time-dependent, non-cartesian MHD calculations is the conservation of magnetic flux - that $\nabla \cdot \vec{B} = 0$ is maintained at all times. The condition is maintained here through accurate differencing rather than a self-correcting scheme, but we are able to conserve magnetic flux divergence to within one part in 10^5. The numerical scheme is pressure-based so it is limited by stability to large and moderate β values (e.g. $\beta \geq 0.1$) - which turns out to be the same restriction for maintaining $\nabla \cdot \vec{B} = 0$ to the required degree.

Finally, the energy equation:

$$\left(\frac{\partial}{\partial t} + \vec{v} \cdot \nabla \right) \left(\frac{p}{\rho^\gamma} \right) = 0$$

reduces to $\vec{v} \cdot \nabla(p/\rho^\gamma) = 0$ when a steady state is reached, which means that (p/ρ^γ) is then a streamline constant. This becomes an analytic test of the achievement of a steady state solution in our case. The boundary values of p and ρ are the same at all latitudes. Therefore, (p/ρ^γ) has the same value everywhere in the computation regime as it has on the boundary if a steady state has been reached. We have checked this for the three cases shown in Fig. 1 and find that for the dipole and quadrupole it is constant to within a maximum of 1% and for the hexapole it is constant to within a maximum of 4% (average values over the whole grid are less than 1% in all cases).

5. NEW RESULTS

The utility of this model is that the outer boundary has been extended to $15R_\odot$. Although this is not a big advance conceptually, this and the stability and ruggedness of the code make it useful for simulating realistic coronal conditions. We present results for quadrupole and hexapole fields, with their accompanying midlatitude streamers and open magnetic field regions. The Alfvén speed ranges between 800 km/s and a few tens of km/s. This is lower than is believed appropriate for the corona /2/, but we expect our model will now enable simulations with higher Alfvén speeds.

Acknowledgement: AHW and STW are supported by NASA Grant NAGW-9. STS has been supported by a grant from the Cosmic and Heliospheric Physics Branch of NASA. GP acknowledges support from ASI (Italian Space Agency) and the University of Alabama in Huntsville.

REFERENCES

1. Hu, Y. Q., and Wu, S. T., *J. Comput. Phys.*, **55(1)**, 33 (1984).
2. Parker, E. N., *Interplanetary Dynamical Processes*, Interscience, New York (1963).
3. Steinolfson, R. S., Suess, S. T., and Wu, S. T., *Astrophys. J.*, **255**, 730-742 (1982).
4. Suess, S. T., *Solar and Stellar Coronal Structure and Dynamics*, R. C. Altrock, ed., Proc. of the 9th Sacramento Pk. Summer Symp., Sunspot, New Mexico, 130-139 (1988).
5. Suess, S. T., Richter, A. K., Winge, C. R., Jr., and Nerney, S., *Astrophys. J.*, **217**, 296-305 (1977).

A SIMULATION STUDY OF THE OUTER HELIOSPHERE INCLUDING THE SOLAR ROTATION EFFECT

H. Washimi and S. Nozawa

Solar-Terrestrial Environment Laboratory, Nagoya University, Toyokawa 442, Japan

ABSTRACT

An axisymmetric structure of the outer heliosphere is studied using the MHD computer simulation method. The solar rotation effect which results in the formation of the toroidal magnetic field, is taken into account in our system. It is shown that a magnetic neutral sheet is formed along the heliopause. The equatorial poloidal current which flows out of the sun along the equatorial neutral sheet in the solar wind plasma, flows beyond the terminal shock and is blown downward under the effect of the surrounding downward interstellar gas flow. In the heliosheath (between the terminal shock and the heliopause), this downward poloidal current is found to return step-by-step towards the sun crossing through the terminal shock at high latitudes in both upper and lower hemispheres. It is also found that the toroidal magnetic field increases in the heliosheath, and a self-pinch effect due to the magnetic pressure force begins to work there. By this effect, the outer heliosphere is found to expand in the axial direction while the terminal shock is contracted. Due to the self-pinch effect, a collimated channel of the subsonic outward plasma flow, in which the kinetic pressure is relatively high, while the magnetic intensity is low, is formed along the rotation axis in the heliosheath.

INTRODUCTION

It was about twenty years ago when the effect of the toroidal magnetic field, B_ϕ, was first discussed in the study of the outer heliosphere (Cranfill, 1971; Axford, 1972). In the approximation of the radially expanding flow, B_ϕ is obeyed by the relation, $B_\phi v R = const$, where v is the flow velocity, and R the distance from the sun. B_ϕ decreases with R in interplanetary space because v is constant in this space, and increases by the factor $(\gamma + 1)/(\gamma - 1)$ (=4 when the polytrope index γ is 5/3) at the terminal shock at R_s . The interesting point which was pointed out by Cranfill is that B_ϕ subsequently increases with R beyond the shock, i.e., $B_\phi \propto R$ because the plasma in the heliosheath is considered to have an incompressible flow while v changes with R^{-2}. Thus the magnetic pressure due to B_ϕ is expected to affect the global structure of the heliosphere. This problem was rediscussed recently by Lee(1988) and Suess(1990). But, because the magnetic pressure is small enough in comparison with the flow energy in the region up to the terminal shock, the effect of B_ϕ has not been considered to be dominant unless the distance of the heliopause from the sun, R_p, is much greater than R_s.

In this paper global structures of the the outer heliosphere under the interaction of the solar wind with the interstellar medium are studied by means of MHD simulation in the axisymmetric system. For the discussion of the above B_ϕ effect, our analysis includes the effect of solar rotation. The structures of the polidal magnetic field and the global current circuit are also discussed.

METHOD OF SIMULATION

We assume that both the solar wind plasma and the interstellar medium obey by the coupled

MHD and Maxwell equations. Cylindrical coordinates (r, ϕ, z) are used, and axial symmetry is assumed, so that the interstellar medium flow and the magnetic field are parallel to the solar rotation axis in our system. We have obtained the solar wind solution near the sun by including the solar rotation effect for the case of the magnetic dipole configuration on the photosphere (Washimi, 1990). The more distant solar wind solution can be obtained by using the solution near the sun as the inner condition in the larger simulation box with a relatively rough grid size. By using this procedure, we obtain the distant solar wind solution . It is worth noting that this expanding technique can be used only when all the information is carried out from inner to outer regions such as in the supersonic and super-Alfvenic solar wind plasma.

For simplicity the interstellar medium is assumed to be composed of protons and electrons only, and the effects of the neutral and high energy particles are neglected. The interstellar gas flow speed is assumed to be $-30km/sec$ (downward parallel to the axis). To confine the structure of the heliosphere inside the $300AU$ simulation box, the density and the pressure of the gas and the interstellar magnetic intensity in our model are rather high, i.e., $0.22/cc$, $2.4 \times 10^{-12}dyne \cdot cm^{-2}$ and $10^{-5}Gauss$, respectively. The solar wind plasma density and the pressure just inside of the initially prescribed terminal shock at $R_s = 48AU$ from the sun, are $1.3 \times 10^{-3}/cc$ and $1.4 \times 10^{-13}dyne \cdot cm^{-2}$, respectively, and the solar wind velocity v is $550km/sec$. The position of the heliopause at the initial time is set to be $R_p = 86AU$ from the sun in the upper plane ($z \geq 0$) and the same distance ($86AU$) from the rotation axis in the lower plane ($z < 0$). The inner boundary is located at $30AU$. There are 160 x 320 grid points for the physical domain 300AU x 600AU. The computation is performed using these initial conditions and boundary conditions to obtain a quasi-steady state by using the two-step Lax-Wendroff scheme. Because the position of the inner boundary is far enough from the initially prescribed terminal shock, the structure of the outer heliosphere is dependent on neither the initial nor the inner boundary conditions.

RESULTS AND DISCUSSION

In the region of the supersonic solar wind (i.e., inside of the terminal shock), the poloidal magnetic field $\mathbf{B}_p$ is oriented away from the sun in the upper hemisphere in our system. It is found in our quasi-steady solution that the field, which is carried by the plasma flow, changes its direction downward together with the plasma flow in the heliosheath. Thus $\mathbf{B}_p$ just inside the heliopause is oriented downward along the heliopause while the magnetic field in the interstellar medium remains upwards. Thus the formation of the neutral sheet in the upper hemisphere is found as depicted in Fig.1(a). In the lower hemisphere the neutral sheet is not clear due to noise. Due to the sharp return of the magnetic field from upward to downward in the high and middle latitudes of the upper hemisphere the formation of the additional neutral sheet is found in the high and middle latitudes inside of the heliosheath.

The equi-contour lines of the pressure in Fig.1(b) show that the structure of the terminal shock is kept sharp in our solution. It was discussed in the simulation study near the sun (Washimi, 1990) that the poloidal current goes out along the equatorial neutral sheet from the sun and comes in from the outer region towards the sun in the open magnetic field region of midle and high latitudes. It is found in Fig.1(a and c) that the equatorial neutral sheet of the region of supersonic solar wind extends beyond the terminal shock, and is blown downward in the heliosheath under the effect of the surrounding downward interstellar gas flow. Though the current structure is somewhat noisy, it is found that the poloidal current is also blown downward along the neutral sheet. It is interesting that the downward current returns back upward step-by-step. One of the return currents goes upwards along the heliopause from lower to upper heliospheres, and in the upper heliosphere this current parts from the heliopause and goes back to the sun, crossing the terminal shock at middle and high latitudes. The other return current goes back to the sun along the axis in the lower hemisphere. Thus both of the return currents are expected to be inward currents in the open region near the sun.

The Cranfill's argument was confirmed in our simulation in Fig.2: The amplidude of the B_ϕ incerases about 4 times at the terminal shock and it increases with $R(= (r^2+z^2)^{1/2})$ in the heliosheath. The contraction of scale of the terminal shock was also found. For the intensity of B_ϕ just inside of the initially prescribed terminal shock $1.2 \times 10^{-7}Gauss$, the scale of final pattern is $43AU$, and for $6 \times 10^{-7}Gauss$ and $1.2 \times 10^{-6}Gauss$, the scales are 41 and $33AU$, respectively. This contraction is, roughly speaking, consistent with the theoretical estimation given by Axford (1972). The

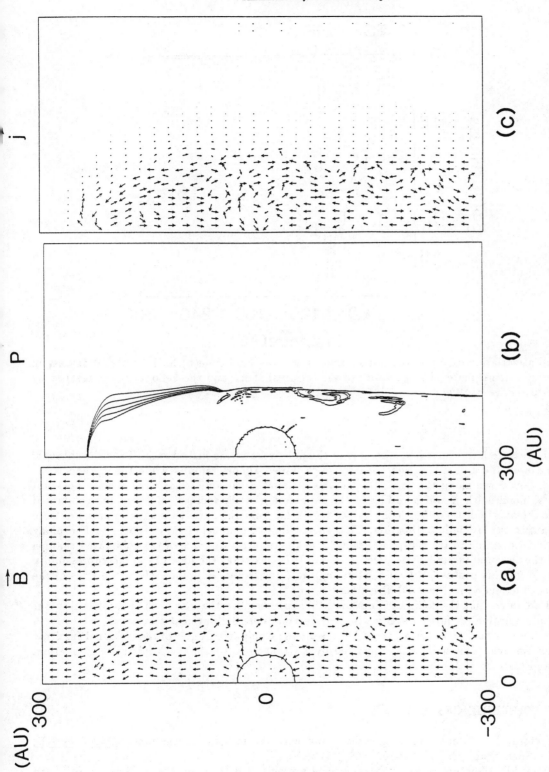

Fig.1 Global patterns of the poloidal magnetic field vectors(a),equi-pressure contours(b) and the poloidal current vectors(c).

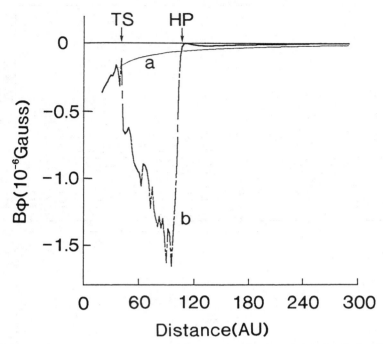

Fig.2 Radial variation of B_ϕ at the initial time(a) and the final time(b) for latitude 30 degrees in the upper hemisphere. The positions of the terminal shock and the heliopause are marked by TS and HP, respectively.

nonlinear pressure force due to B_ϕ also acts to the pole-ward in the heliosheath. Thus the subsonic solar wind near the axis in the heliosheath, which flowed almost radially when the B_ϕ effect was neglected, now flows along the axis. Due to this collimated subsonic solar wind flow, v does not decrease so much with R which results in the suppression of the increase of B_ϕ in the collimated flow near the axis. On the other hand, in the region far from the axis the magnetic pressure force is dominant in the latitudinal direction. Thus the collimated channel of the subsonic outflow along the axis is formed due to the self-pinch effect. The kinetic pressure force of the collimated channel balances the surrounding magnetic pressure force. It is clear that the idea of the incompressibility cannot hold for this process. It is also intersting that, due to this self-pinch effect, the scale of the heliopause is extended in the axial direction.

It should be noted that the collimated outflow still expands along the axis in the heliosheath in our quasi-steady solution. A further study will provide more detailed physical processes.

Computation was performed with Facom VP-200 at the Computer Center of the National Institute for Fusion Science. This work was supported by a Grant-in-Aid from the Ministry of Education, Science and Culture of Japan.

REFERENCES

1. W.I.Axford, The interaction of the solar wind with the interstellar medium, SOLAR WIND, NASA Spec. Publ., NASA SP-308, 609 (1972).
2. C.W.Cranfill, Flow problems in astrophysical systems, PH.D. dissertation, Univ. Calif. San Diego (1971).
3. M.A.Lee, The solar wind terminal shock and the heliosphere beyond, Proc. XI international Solar Wind Conference, NCAR, Boulder, Colo. (1988).
4. S.T.Suess, The heliopause, Rev. Geophys., 28, 97 (1990).
5. H.Washimi, Structure of the interplanetary magnetic field near the sun, Geophys. Res. Letters, 17, 33 (1990).

OBSERVATIONS OF LARGE-SCALE STRUCTURE IN THE INNER HELIOSPHERE WITH DOPPLER SCINTILLATION MEASUREMENTS

R. Woo and J. W. Armstrong

Jet Propulsion Laboratory, California Institute of Technology, Pasadena, CA 91109, U.S.A.

ABSTRACT

A study of large-scale structure in the inner heliosphere based on unique and extensive Pioneer Venus Doppler scintillation data has been initiated. Preliminary results include the first observation of corotating stream structure inside 35 R_o. Results on variations of the Doppler scintillations with solar cycle and heliocentric distance, and their implications in terms of solar wind structure and mass flux density, are discussed.

INTRODUCTION

Measurements of Doppler scintillations during the superior conjunctions of planetary spacecraft represent an important tool for remotely probing the highly-variable near-Sun solar wind that has not yet been observed in situ /1–3/. Between 1979 and 1990, the Pioneer Venus Orbiter (PVO) spacecraft underwent seven superior conjunctions spanning a full solar cycle (Figure 1). For periods of ±4 months surrounding each conjunction, the PVO radio path probed the solor wind within about 0.43 AU of the Sun. Moreover, because PVO returned almost continuous Venus data, tracking of the spacecraft by the NASA Deep Space Network, and hence collection of Doppler scintillation data, was extensive. The Pioneer Venus mission is, thus, well-suited for conducting the first investigation based on Doppler scintillations of large-scale and long-term solar wind structure in the inner heliosphere and over a full solar cycle. The purpose of this paper is to present results obtained to date.

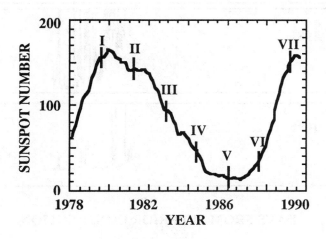

Fig. 1. Sunspot number versus time

DOPPLER SCINTILLATION MEASUREMENTS AND RESULTS

Estimates of rms Doppler scintillations σ_D are obtained as described in earlier studies /2–3/. They are based on 10-s Doppler measurements and computed over 3-min periods, thus sensitive to scale sizes less than about 5×10^4 km. For the current study, we additionally average the 3-min results over 1 hr. The use of 1-hr averages results in a more manageable-sized data set, and does not lead to significant loss in detail of the solar wind structure of interest.

So far, we have processed < 10% of the entire PVO data set. Shown in Figure 2 are 1-hr averages of σ_D over a period of ±120 days surrounding Superior Conjunction V (January 19, 1986), which occurred near solar minimum (Figure 1). Corresponding typical heliocentric distances are indicated at the top of Figure 2. Also included in Figure 2 are similar results obtained during Superior Conjunction II (7 April 1981) near solar maximum. For convenience of comparison, we have displayed the results for each conjunction twice, once with a maximum scale of 200 mHz and the other 2000 mHz.

Although the 1981 data are not yet fully analyzed, the difference between the conjunctions is striking. The scintillations are dominated by large-amplitude, short-lived structure in 1981, but show recurrent small-amplitude, longer-lived structure (marked by crosses about every 27 days in Figure 2) in 1985–1986. A recent comparison of the same 1981 scintillation results with Helios plasma measurements has shown that all of the observed large-amplitude short-lived structure were interplanetary shocks rather than corotating high speed streams /3/. On the other hand, the structures observed in 1985–1986 are probably associated with corotating high speed streams, which typically predominate during solar minimum conditions. Further insight will be provided by comparisons of these results with plasma measurements at the orbits of Venus (PVO) and Earth (IMP 7/8), studies which are already underway. In the meantime, it should be noted that the apparent absence of evidence for high speed streams in the scintillation data during a prolonged period in 1986 is consistent with the absence of high-speed streams observed in situ in the vicinity of the solar

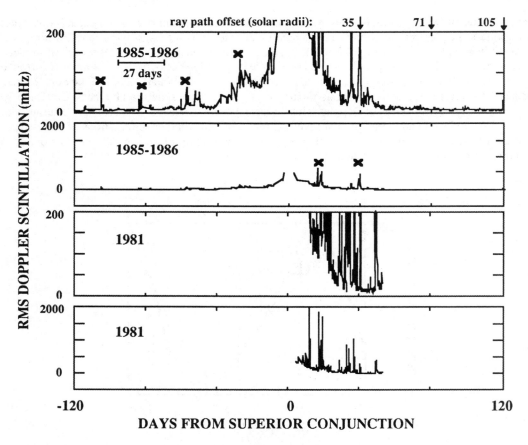

Fig. 2. Doppler scintillation as a function of time from superior conjunction

equator by PVO and IMP 7/8 in 1986 /4/. The exciting result of the 1985/1986 scintillation measurements is the evidence for substantial stream structure close to the Sun ($<35\,R_o$). This first detection of stream structure in the near-Sun solar wind shows that Doppler scintillation measurements can play an important role in elucidating the origin and evolution of stream structure close to the Sun and its relationship to coronal holes.

The PVO radio path probes the solar wind off the west and east limbs of the Sun prior to and after superior conjunction, respectively. Comparison of the west and east side PVO measurements in 1985/1986 reveals a distinct difference in the magnitudes of the scintillation structure corresponding to high-speed streams. This observed east-west asymmetry may be explained by the geometry of the measurements (Figure 3). Doppler scintillations are path-integrated measurements of electron density fluctuations and the component of solar wind velocity orthogonal to the radio path /1/. Because of the rapid fall-off of electron density fluctuations with heliocentric distance, the scintillation measurements essentially probe the region of solar wind in the vicinity of the closest approach point. From Figure 3, the higher scintillation amplitudes due to the streams observed following superior conjunction is not surprising, since the radio paths are better aligned with the streams closer to the Sun on the east side.

Shown in Figure 4 are the Doppler scintillation results of Figure 2 displayed as a function of heliocentric distance R. A line, representing a power-law fall-off with radial distance in the range of 20-100 R_o, has been drawn on each plot. Additional data pertaining to the radial variation of σ_D will be analyzed, but the results in Figure 2 suggest that during solar maximum the fall-off of the background Doppler scintillations may be steeper than during solar minimum. Previous studies have shown that for electron density fluctuations described by a Kolmogorov spatial wavenumber spectrum, $\sigma_D \propto \sigma_{ne} v^{5/6}$, where σ_{ne} is the rms electron density fluctuations and v is the solar wind speed transverse to the radio path /1/. With the assumption that $\sigma_{ne} \propto n_e$ and approximation that $v^{5/6} \approx v$, where n_e is electron density, then σ_D characterizes mass flux density. Furthermore, the radial variation of mass flux density is given by multiplying the radial variation of σ_D by $R^{-1/2}$, where R is heliocentric distance /1/. Thus, the results in Figure 4 imply that mass flux density varies as R^{-2} and $R^{-1.7}$ during solar maximum and solar minimum, respectively. In other words, the Doppler scintillation measurements suggest that the solar wind is approximately spherically symmetric during solar maximum but slightly converging in the ecliptic plane during solar minimum. This latter result is consistent with 1975 measurements during the

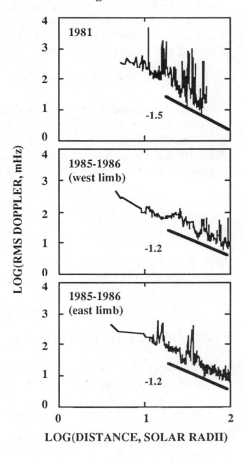

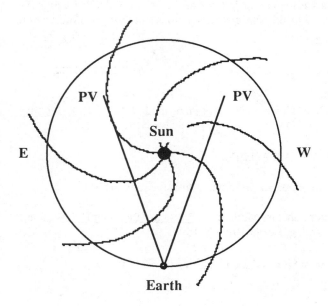

Fig. 3. East-west geometry

Fig. 4. Radial dependence of σ_D

previous solar minimum /1/, and may be related to the formation of polar coronal holes during solar minimum.

The results in Figure 4 show that the background Doppler scintillations (lower envelopes of the data plotted in Figure 4) are stronger during solar minimum than during solar maximum beyond 20 R_o. If the assumptions discussed above are independent of solar cycle, this implies that mass flux density in the ecliptic plane is higher during solar minimum than during solar maximum. More detailed analyses will be carried out in the future as more data are processed, but this result is consistent with the solar cycle variation of mass flux density observed by Helios and IMP 7/8 plasma measurements near 1 AU /5-6/.

CONCLUSIONS

Although the important region of the solar wind inside 0.3 AU has not yet been explored by in situ spacecraft, it can be studied with radio scintillation observations. Doppler (or phase) scintillations using spacecraft radio signals have so far yielded information on the spatial wavenumber spectrum of electron density fluctuations /7/ and on the evolution and propagation of interplanetary shocks /2/. The Doppler scintillation data analyzed in this paper show that corotating high speed streams apparently have formed inside 35 R_o. The value of Doppler scintillation measurements for studying stream structure would be significantly enhanced if scintillation and in situ measurements of the same streams are compared. Such comparisons are currently underway, which, together with a complete analysis of the unique PVO scintillation data, promise to provide further details of stream structure near the Sun.

Further processing and analysis of the extensive PVO scintillation data set will also be useful for studying the variation of mass flux density with heliocentric distance and solar cycle. Preliminary results indicate mass flux densities in the ecliptic plane that are higher at solar minimum than at solar maximum, a result consistent with that obtained by situ plasma measurements near 1 AU. There also appears to be a solar cycle variation in the radial dependence of mass flux density. This variation suggests a spherically symmetric solar wind during solar maximum, but a slightly converging solar wind during solar minimum in the ecliptic plane and in the heliocentric distance range of 20–100 R_o.

ACKNOWLEDGMENTS

It is a pleasure to thank C. Chang for her assistance in computing. We also wish to acknowledge the support received from the NASA Deep Space Network and the Pioneer Venus Project, especially L. Colin, R. Fimmel, G. Goltz, R. Jackson, D. Lozier, G. Roldan and R. Ryan. Discussions with P. Gazis are also gratefully acknowledged. This paper describes research carried out at the Jet Propulsion Laboratory, California Institute of Technology, under contract with the National Aeronautics and Space Administration.

REFERENCES

1. R. Woo, Astrophys. J., 219, 727 (1978).

2. R. Woo, J.W. Armstrong, N.R. Sheeley, Jr., R.A. Howard, M.J. Koomen, and D.J. Michels, J. Geophys. Res., 90, 154 (1985).

3. R. Woo and R. Schwenn, J. Geophys. Res., 96, 21227 (1991).

4. P. Gazis, A. Barnes, J. Mihalov, and A. Lazarus, paper presented at this meeting (1991).

5. R. Schwenn, Large-Scale Structure in the Interplanetary Medium, in Physics of the Inner Heliosphere I, eds. R. Schwenn and E. Marsch, Springer-Verlag, Berlin 1990, p. 159.

6. J.M. Ajello, A.I. Stewart, G.E. Thomas, and A. Graps, Astrophys. J., 317, 964 (1987).

7. R. Woo and J.W. Armstrong, J. Geophys. Res., 84, 7288 (1979).

Session 3

Minor Ions, Neutrals and Cosmic Rays in the Heliosphere

Conveners: Alfred Bürgi
Fred Ipavich
Eberhard Möbius

MINOR IONS – TRACERS FOR PHYSICAL PROCESSES IN THE HELIOSPHERE

P. Bochsler

Physikalisches Institut, University of Bern, Sidlerstrasse 5, CH - 3012 Bern, Switzerland

ABSTRACT

Minor ions can be used as tracers for physical processes in the heliosphere by monitoring their abundances, their charge states, and their dynamic properties in the solar wind. Two applications are discussed in some detail, more applications can be found in other contributions of this session. Here, we discuss the case of ^{3}He in the solar wind which is a particularly useful tracer for mixing processes within the radiative interior of the sun. The second example deals with the charge states of solar wind ions as temperature diagnostics for the inner corona. Weakly ionized species are sensitive indicators for various processes occurring in the inner heliosphere such as evaporation of cometary debris and the ionization and pick-up of infiltrated interstellar neutral gas. We also present results of a model calculation which demonstrates the usefulness of weakly ionized species as indicators for the formation and propagation of plasmoids within the solar corona.

INTRODUCTION

Minor ions have been successfully used as tracers for physical processes in the heliosphere in a wide spectrum of applications almost immediately after the first results of in situ measurements of minor ions in the solar wind became available. For instance, Hundhausen et al. /1/ and Bame et al. /2/ used the ionization states of oxygen, silicon and iron to determine coronal temperatures and temperature gradients. Later, Owocki and Scudder/3/ and Bürgi /4/ have shown in more details how solar wind ionization states can be applied as powerful diagnostics for processes occurring in the inner solar corona.

The reliable determination of the abundance of ^{3}He in the solar wind by means of the Apollo Foils /5/ has introduced a variety of important astrophysical applications including the determination of the cosmological D/H-abundance ratio in the protosun /6/. The precise measurement of the light noble gas fluxes, and the exact determination of the isotopic composition of these elements in the solar wind with the foil experiment provided a firm reference point for studies of light noble gases in the lunar soil and made it possible to use noble gases in lunar regolith samples as tracers for geological processes (e.g./7/). Conversely, the possibility to use ions of lunar origin in the solar wind to investigate large-scale sputtering processes on the lunar surface has been announced and demonstrated at this conference /8/. Solar wind minor ions have also been applied for some time as tracers for the transport of solar wind gases across the terrestrial bow shock into the magnetosphere /9,10,11/. Another example of an application of minor ions in heliospheric processes is the determination of the helium abundance in the

local interstellar medium by means of He^+-observations /12/. Also at this
conference, the absence of low ionization states of heavy elements in normal
solar wind has been confirmed and extended to significantly lower limits
based on ULYSSES-SWICS results, thus introducing another strong argument
against the possibility of the solar corona being contaminated with meteoritic and cometary debris /13/.

The above examples are tracer applications of minor ions in the classical
sense as widely practiced in geochemistry, hydrology, medicine etc.: In these
fields, the concentration of some minor constituent in a physical system is
used to study the relevant transport processes within the system assuming
that this constituent undergoes the same processes as the bulk material. The
notion of 'minor' is crucial: For instance, $^4He^{++}$ cannot be considered as a
minor component in the solar wind, since it carries 20% of the solar wind
momentum[1].

The concept 'tracer' has also been applied to problems where the concentration of an ion is not only considered in coordinate space but also in momentum space, opening a wide variety of additional applications. These applications include the study of the dynamical properties of minor ions in relation
to physical processes such as wave-particle interactions, shock-layer
crossings, or particle-particle interactions in the inner corona. Clearly,
such applications are possible because the solar wind plasma is not collision-dominated, hence, minor ions not only keep their chemical and charge
state identity over large distances in the interplanetary space but to some
extent they can also preserve their kinematic properties.

In this review, we will concentrate on two examples of classical tracer-
investigations by means of minor ions. The first case is related to the
'solar connection' of the solar wind; we discuss the significance of the
^3He-abundance in the solar wind and in the solar photosphere for transport
processes within the solar interior. The second example deals with the case
of ionization states as indicators for the existence of the frequently
discussed small plasmoids in the solar wind.

^3He IN THE SOLAR WIND – A TRACER FOR TRANSPORT PROCESSES IN THE INTERIOR OF THE SUN

The noble gases are the only elements from which we have direct information
on the isotopic composition in the solar atmosphere. The case of ^3He, an ion
which has been repeatedly measured in the solar wind /5,14,15,16/, is particularly interesting.

In the process of the conversion of hydrogen into helium in the solar proton-
proton cycle, ^3He is an intermediate product. It can be produced at comparatively moderate temperatures of $5 \cdot 10^6$ K since the Coulomb barrier between
hydrogen nuclei is relatively low. However, further processing is extremely
slow due to the efficient repulsion between He nuclei. At temperatures above
10^7 K the Coulomb barrier can no longer inhibit further reactions, and ^3He is
converted into ^4He, mainly by reactions with other ^3He nuclei. Thus, in
standard solar models, with no mixing involved, after $4.6 \cdot 10^9$ years three
shells with distinct isotopic compositions of helium are found. In the core,
the relative abundance of ^3He is depleted relative to the initial main-
sequence concentration (after deuterium burning) by an order of magnitude
(cf. Figure 1). In the outermost shell, containing about 10 % of the solar
mass, the helium isotopic composition remains unchanged. In the intermediate
layer, a ^3He-rich zone is gradually built up during the lifetime of the sun,

[1]However, a good case of 'minor' elements has impressively been presented in
the course of this conference at the visit of the mines of Lautenthal, where
concentrations of 2.8 g silver and 0.05 g gold per ton of rock are found.

reaching a maximum enrichment over the surface concentration by approximately
a factor of 30 at the solar age of 4.6 billion years.

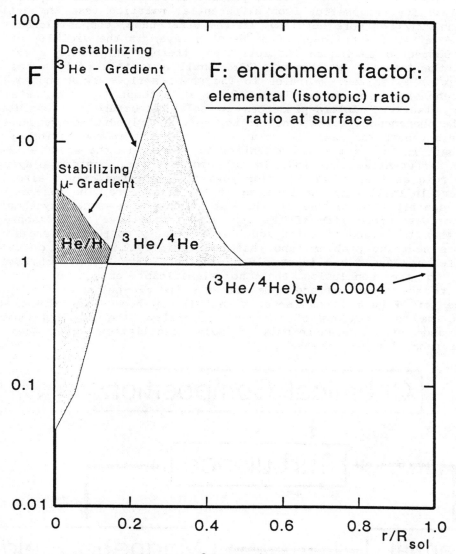

Fig. 1. Concentrations of ^{3}He and ^{4}He in the solar interior
(adapted from /17/).

The surface content of ^{3}He is an extremely sensitive and reliable (conservative) indicator for mixing processes between the radiative interior and the
outer convective zone. The lack of a significant contribution of ^{3}He from
solar nucleosynthesis at the solar surface is not only a clue for the absence
of transport processes working in the outer radiative zone at present but
also for the entire main sequence life of the sun in the past.

Turbulent mixing between different layers of the sun has been extensively
studied in connection with the solar neutrino problem /18,19,20/. The main
driver of turbulent mixing is considered to be radial differential rotation
leading to hydrodynamical instabilities, notably shear instability, which
produce small-scale mixing thereby exchanging rotational momentum and matter
between different layers (e.g. /21/). The role of the ^{3}He-abundance in this
context is not only passive, i.e. to indicate by its surface concentration
whether mixing is occurring or not. Radial gradients in chemical composition,
as illustrated in Figure 1, can trigger instabilities or produce stabilizing

forces, for instance, via the so-called μ-gradient mechanism: The layering of ^{4}He in the radiative core - in the sense of heavier species having increasing concentrations at deeper layers - could prevent turbulent mixing despite some generic instability originating from differential rotation (see the article by J.-P. Zahn /21/ for more details). On the other hand, compositional gradients - i.e. after the build-up of a ^{3}He-rich layer in the vicinity of the core - are suspected to induce instabilities, thereby triggering a secular turnover of the interior of the sun /22/. Until now, this problem has not been satisfactorily solved. We refer to the short article by Roxburgh /23/ for a comprehensive discussion. Figure 2 is an attempt to illustrate the complicated network of interdependences in solar structure and evolution. Note that the phenomenon 'solar wind' is not only a peripheral consequence of magnetic field generation near the solar surface, but that the action of the wind - at least in the past - was essential to decelerate the solar spin and to generate differential rotation. In this context ^{3}He, besides playing a significant role as tracer for interior physical processes, might also take an active part in initiating the turnover in the solar core. The absence of a significant contribution of ^{3}He in the outer convective zone, synthesized during the main-sequence life of the sun, might thus have two reasons: It could mean that the standard solar model is basically correct (disregarding the notorious neutrino problem) and that no significant exchange of matter between different layers has taken place. Another, less conventional view, could lead to the conclusion that no significant amount of ^{3}He was transported to the solar surface, just because the ^{3}He-bulge in the intermediate zone (cf. Figure 1) has never been built-up because of the secular turnover in the core regions of the sun. Hopefully, the ongoing neutrino experiments and refined measurements of solar oscillations will shed new light on the remaining grey areas.

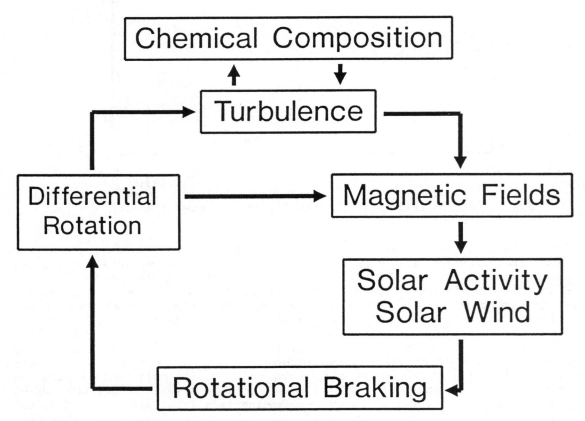

Fig. 2. The solar connection: Network of physical processes in relation to solar structure and evolution.

CHARGE STATE OF MINOR IONS IN PLASMOIDS

Steady state models of the solar wind have been considerably improved in recent years. One crucial step which has been achieved was to include $^4He^{++}$
ions as separate fluid into these models, thereby taking into account that
helium carries a significant fraction of momentum /24,25,26/. Although the
detailed mechanism which heats the solar corona is still not unambiguously
identified, modern models reproduce velocity and temperature profiles and the
ionization states of various elements for different types of solar wind
regimes satisfactorily just by assuming some heating term at the coronal base
as a free parameter in the energy equation /27/. From observations it is
clear that at least at the boundary between chromosphere and corona, the picture of a corona continuously fed by matter flowing across the transition
zone is not entirely correct. Brueckner and Bartoe /28/ observed small explosive events above the quiet sun with the HRTS-instrument (High Resolution
Telescope and Spectrograph). These events were so frequent that they could
convert sufficient energy to heat the corona. They were observed on small
spatial scales of typically 2 arc sec and lasted for a few minutes. The most
conspicuous features were strongly Doppler shifted C IV lines, frequently
blue and red shift appearing simultaneously and reaching values above 150
km/s. Although it is no longer believed that these explosive events could be
responsible for the coronal-heating, it is quite obvious that they are
produced by small-scale field annihilation /29/.

Related to the HRTS-observations, Pneuman /30/ has developed a solar wind
model invoking small-scale magnetic reconnection. Reviving the so-called
'melon-seed' mechanism of Schlüter /31/, he investigated the propulsion of
diamagnetic structures in the solar corona. In a subsequent paper, Cargill
and Pneuman /32/ studied the acceleration mechanism and the thermal evolution
of such structures ('plasmoids') in more details. A summary of the potential
importance of plasmoids in the solar wind has recently been presented by
Mullan /33/.

If the mass-, momentum- and energy-flow related to these plasmoids would make
a significant contribution to the overall flow of the solar wind, is it
possible to explain the observed charge state distributions at the orbit of
the earth with a solar wind fed from plasmoids? Or, how far does a solar wind
originating from fine structures remember its grainy source? In the following, we make an attempt to address some of these questions in a preliminary
manner, using a model of magnetically dominated plasmoids, as described by
Cargill and Pneuman /32/. We assume that a plasmoid is driven by the diverging magnetic field in a 'background-corona',and we establish the basic
equations ruling the motion and energy balance of a closed plasmoid.
Throughout the entire existence of a plasmoid, we assume that a pressure
balance holds across its boundaries:

$$p_i + \frac{B_i^2}{2\mu_0} = p_e + \frac{B_e^2}{2\mu_0} , \qquad (1)$$

where subscripts i denote quantities inside the boundary and e refers to
quantities outside.

It is assumed that a spherical plasmoid expands in a self-similar manner,
conserving its shape, mass, and magnetic flux, i.e.

$$B_i R^2 = \text{const.} \qquad (2)$$

where R denotes the radius of the plasmoid (typically 10^3 to 10^4 km). The
momentum balance (3) yields an equation describing the motion of the plasmoid
as it propagates along a radial coordinate r

$$\frac{d^2 r}{dt^2} = - \frac{GM_0}{r^2} - \frac{V}{M} \frac{d}{dr} \left[P_e + \frac{B_e^2}{2\mu_0} \right] .$$ (3)

The terms on the right hand side describe the effects of gravity (M_0 is the solar mass) and of magnetic buoyancy (V and M denote volume and mass of the plasmoid).

The balance of internal energy of the plasmoid is given by expression (4) (cf. Cargill and Pneuman /32/ their equation 2.18):

$$\frac{dT}{dt} = \frac{3(\gamma - 1) \; T}{4} \frac{dr}{dt} \frac{d}{dr} \ln \left[P_e + \frac{B_e^2}{2\mu_0} \right] + \frac{(H - R_L)(\gamma - 1)}{2 \, n_{i0} \cdot k} \frac{V}{V_0} .$$ (4)

Here, T is the internal gas temperature, H and R_L denote the volumetric heating and radiation loss rates (in W m^{-3}), V/V_0 describes the expansion of the plasmoid in relation to its initial volume. n_{i0} is the initial number density of electrons inside and outside the plasmoid, which is assumed to be 10^{16} m^{-3}.

Following Cargill and Pneuman, we adopt the cooling rates published by Hildner /34/, and we assume a divergence of the external magnetic field $B_e = B_{e0}(r_0/r)^2$. We do not further discuss the underlying assumptions and simplifications of the above equations in the limited frame of this paper. For the investigation of the ionization balance inside the plasmoid, we use the rates of Arnaud and Rothenflug /35/ not taking into account any effects of photoionization.

Equations (3) and (4) are now integrated numerically together with the equations describing ionization and recombination in an electron gas at temperature T and density n. For all runs, the initial temperature was T_0 = 20'000 K, and the initial velocity was assumed to be 10 km/s. The initial magnetic field strength B_{e0} was taken to be 20 gauss. The heating rate H was assumed to depend on the particle density: H = $H_0 \cdot n$, i.e. it was assumed that throughout the development of the plasmoid, the amount of energy introduced per unit time and particle was constant, of the order of a fraction of an eV per second and per particle. We do not specify the nature of the heating mechanism operating on the plasmoids: For example, an obvious assumption could be that the heating is related to the dissipation of current sheets and ongoing reconnection with the ambient external magnetic field during the motion of the plasmoid in the corona. This would then lead to the conclusion that the lifetime of plasmoids is limited - another boundary condition which has not been considered at this stage of the investigation.

In each case, the numerical integration was stopped after the temperature of the plasmoid had increased to more than 10^6 K. The results of the numerical integration are summarized in Figure 3. Figure 3a) shows the evolution of the radial distance of a plasmoid measured from the solar center. Figures 3b), 3c) and 3d) illustrate the evolution of velocities, temperatures, and the fraction of neutral hydrogen for different heating rates (0.25, 0.30, 0.40, and 0.50 eV/s per hydrogen nucleus). Figure 3e) describes the evolution of the number densities of hydrogen ions and atoms, and Figure 3f) shows the contributions of different oxygen ions within the plasmoid during its travel from the solar surface to five solar radii from the center. Although these results can only be a crude approximation to a real plasmoid, some interesting features emerge: First, as has already been noted by Cargill and Pneuman, the dependence of the radiation loss function from the temperature plays an essential role in the thermal evolution of the plasmoid. Evidence for this is most obvious in Figures 3c) and 3d). A 'knee' appears in all calculated

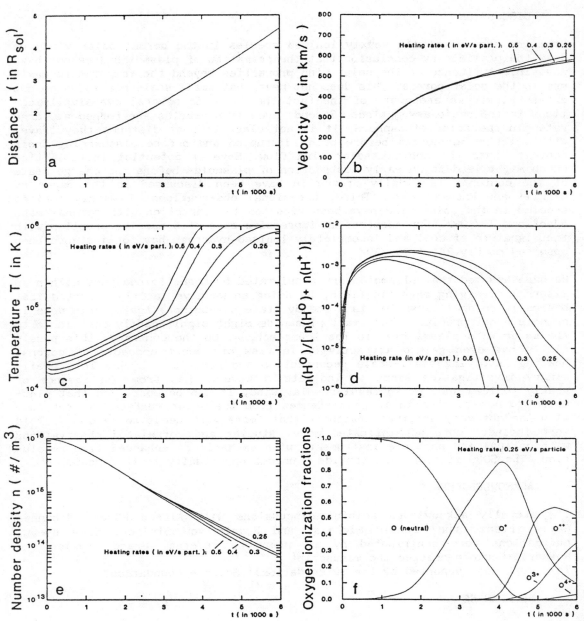

Fig. 3. Evolution of radial distance (a), velocity (b), temperature (c), neutral fraction of hydrogen (d), number density (e), and oxygen ionization fraction (f) with time.

curves at T = 80'000 K, i.e. at the temperature where according to Hildner /34/ the radiation losses drop drastically with temperature. Second, despite the rather large energy input into the plasmoid, it appears that, e.g., oxygen (Figure 3f) in this model remains in a rather low ionization state, contrary to what is actually observed in the solar wind /2,16,36/. This is because in this model a significant fraction of the energy input goes into work to expand the plasmoid against the external pressure. For instance, Figure 3f indicates that at a distance of 5 solar radii, most of the oxygen is in charge states 2+, 3+, and 4+, and from Figure 3d we find that a relatively large amount of hydrogen is not ionized within 1 R_{sol} from the solar surface.

DISCUSSION

From the absence of the weakly ionized species in the normal solar wind, we
draw the preliminary conclusion that the formation of plasmoids does not have
a visible influence on the solar wind properties beyond the temperature maximum in the solar corona. This does not mean that small-scale plasmoids do not
exist: First, we are aware of the fact that we made several oversimplifica-
tions in the basic assumptions. Second, even if plasmoids are formed at large
rates in the solar atmosphere, it is not clear to what distance they travel
within the solar corona before being disrupted and before discharging their
contents into the ambient solar wind flow. Several potential instabilities
which might lead to an early destruction of plasmoids before the charge state
of the contents is finally frozen in have been discussed in the paper by
Cargill and Pneuman /32/. Third, occasional observations of weakly ionized
species in the solar wind have been reported in connection with coronal mass
ejections /37,38/. Furthermore, an impressive case of a massive eruption with
the signature of cool and incompletely ionized hydrogen has for instance been
observed on May 24, 1979 /39/.

We have assumed that plasmoids are accelerated by magnetic buoyancy within an
external diverging magnetic field. By doing so we have tacitly accepted that
basically coronal expansion is a steady state phenomenon. Again, this does in
principle not preclude that small plasmoids might significantly contribute to
the solar wind plasma flow in the lower corona. On the contrary, this possibility has even gained attractivity in view of recent concepts of coronal
heating involving small-scale reconnection and nanoflares /40/. This preliminary investigation seems to indicate, however, that from the absence of
weakly ionized species in regular solar wind, it can be concluded that plasmoids do not survive to large distances from the solar surface. We conclude
with the not very original statement that 'more work needs to be done'. This
work includes more sophisticated model studies and correlated observations
with optical and particle instruments with emphasis on enhanced time resolution. The SOHO-mission will provide a unique opportunity to this goal.

ACKNOWLEDGEMENTS

I gratefully acknowledge helpful discussions with Alfred Bürgi, Johannes
Geiss, Franz Grün, Donald Michels, and Rudolf von Steiger. Some useful
suggestions were contributed by an anomynous referee. Grace Troxler and
Rosmarie Neukomm checked the manuscript.
This work was supported by the Swiss National Science Foundation.

REFERENCES

1. A.J. Hundhausen, H.E. Gilbert, and S.J. Bame, Ionization state of the
interplanetary plasma, J. Geophys. Res. 73, 5485-5493 (1968)

2. S.J. Bame, J.R. Asbridge, W.C. Feldman, and P.D. Kearney, The quiet
corona: Temperature and temperature gradient, Solar Physics 35, 137-153
(1974)

3. S.P. Owocki and J.D. Scudder, The effect of a non-Maxwellian electron
distribution on oxygen and iron ionization balances in the solar corona,
Astrophys. J. 270, 758-768 (1983)

4. A. Bürgi, Effects of non-Maxwellian electron velocity distribution
functions and nonspherical geometry on minor ions in the solar wind,
J. Geophys. Res. 92, 1057-1066 (1987)

5. J. Geiss, F. Bühler, H. Cerutti, P. Eberhardt, and Ch. Filleux, Solar wind
composition experiment, Apollo 16 Preliminary Sci.Report, NASA SP-315 (1972)

6. J. Geiss and H. Reeves, Cosmic and solar system abundances of deuterium and helium-3, Astron. Astrophys. 18, 126-132 (1972)

7. O. Eugster, N. Grögler, P. Eberhardt, and J. Geiss, Double drive tube 74001/2: Composition of noble gases trapped 3.7 AE ago, Proc. Lunar Planet. Sci. Conf. 11th (Pergamon Press), 1565-1592 (1980).

8. M. Hilchenbach, D. Hovestadt, B. Klecker, and E. Möbius, Detection of singly ionized energetic pick-up ions upstream of the earth's bow shock, this issue.

9. D.T. Young, H. Balsiger, and J. Geiss, Correlations of magnetospheric ion composition with geomagnetic and solar activity, J. Geophys. Res. 87, 9077-9096 (1982).

10. G. Gloeckler and D.C. Hamilton, AMPTE ion composition results, Physica Scripta T18, 73-84 (1987)

11. S.P. Christon, G. Gloeckler, D.C. Hamilton, and F.M. Ipavich, High-charge state solar wind carbon and oxygen ions in the magnetosphere, EOS 72, 237 (1991)

12. E. Möbius, D. Hovestadt, B. Klecker, M. Scholer, G. Gloeckler, and F.M. Ipavich, Direct observation of He^+ pick-up ions of interstellar origin in the solar wind, Nature 318, 426-429 (1985)

13. J. Geiss, ULYSSES: New results related to the FIP effect, this issue.

14. K.W. Ogilvie, M.A. Coplan, P. Bochsler, and J. Geiss, Abundance ratios of $^4He^{++}/^3He^{++}$ in the solar wind, J. Geophys. Res. 85, 6021-6024 (1980)

15. M.A. Coplan, K.W. Ogilvie, P. Bochsler, and J. Geiss, Interpretation of ^3He abundance variations in the solar wind, Solar Physics 93, 415-434 (1984)

16. P. Bochsler, Helium and oxygen in the solar wind: Dynamic properties and abundances of elements and helium isotopes as observed with the ISEE-3 Plasma Composition Experiment, Habilitationsschrift, University of Bern (1984)

17. P. Bochsler, J. Geiss, and A. Maeder, The abundance of ^3He in the solar wind - a constraint for models of solar evolution, Solar Physics 128, 203-215, (1990)

18. E. Schatzman, Turbulent transport, solar lithium and solar neutrinos, Astrophys. Letters 3, 139-140 (1969)

19. E. Schatzman and A. Maeder, Stellar evolution with turbulent mixing III. The solar model and the neutrino problem, Astron. Astrophys. 96, 1-16 (1981)

20. Y. Lebreton and A. Maeder, Stellar evolution with turbulent diffusion mixing VI. The solar model, surface ^7Li and ^3He abundances, solar neutrinos and oscillations, Astron. Astrophys. 175, 99-112 (1987)

21. J.-P. Zahn, Turbulent transport in stellar radiation zones: Causes and effects, in: Rotation and Mixing in Stellar Interiors, eds. M.-J. Goupil and J.-P. Zahn, Springer-Verlag, 1990, p.141-150.

22. F.W.W. Dilke and D.O. Gough, The solar spoon, Nature 240, 262 (1972)

23. I.W. Roxburgh, Problems of the solar interior, in: The Internal Solar Angular Velocity, eds. B.R. Durney and S. Sofia, D. Reidel, Dordrecht, 1987, pp. 1-5.

24. A. Bürgi and J. Geiss, Helium and minor ions in the corona and solar wind: Dynamics and charge states, Solar Physics 103, 347-383 (1986)

25. A. Bürgi, Proton and alpha particle fluxes in the solar wind: Results of a three-fluid model, J. Geophys. Res. in press (1992)

26. A. Bürgi, Dynamics of alpha particles in coronal streamer type geometries, this issue.

27. G.L. Withbroe, The temperature structure, mass, and energy flow in the corona and inner solar wind, Astrophys. J. 325, 442-467 (1988)

28. G.E. Brueckner and J.-D. F. Bartoe, Observations of high energy jets in the corona above the quiet sun, the heating of the corona and the acceleration of the solar wind, Astrophys. J. 272, 329-348 (1983)

29. K.P. Dere, J.-D. F. Bartoe, G.E. Brueckner, J. Ewing, P. Lund, Explosive events and magnetic reconnection in the solar atmosphere, J. Geophys. Res. 96, 9399-9407 (1991)

30. G.W. Pneuman, Ejection of magnetic fields from the sun: Acceleration of a solar wind containing diamagnetic plasmoids, Astrophys. J. 265, 468-482 (1983)

31. A. Schlüter, Solar radio emission and the acceleration of magnetic storm particles, in: IAU Symposium 4, Radio Astronomy, ed. H.C. van de Hulst, Cambridge University Press, Cambridge, 1957, p. 356.

32. P.J. Cargill and G.W. Pneuman, Diamagnetic propulsion and energy balance of magnetic elements in the solar chromosphere and transition region, Astrophys. J. 276, 369-378 (1984)

33. D.J. Mullan, Sources of the solar wind: What are the smallest-scale structures? Astron. Astrophys. 232, 520-535 (1990)

34. E. Hildner, The formation of solar quiescent prominences by condensation, Solar Physics 35, 123-136 (1974)

35. M. Arnaud and R. Rothenflug, An updated evaluation of recombination and ionization rates, Astron. Astrophys. Suppl. Ser. 60, 425-457 (1985)

36. K.W. Ogilvie, Analysis of O^{7+}/O^{6+} observations in the solar wind, J. Geophys. Res. 90, 9881-9884 (1985)

37. R. Schwenn, H. Rosenbauer, and K.H. Mühlhäuser, Singly-ionized helium in the driver gas of an interplanetary shock wave, Geophys. Res. Lett. 7, 201-204 (1980)

38. R.D. Zwickl, J.R. Asbridge, S.J. Bame, W.C. Feldman, and J.T. Gosling, He^+ and other unusual ions in the solar wind: A systematic search covering 1972-1980, J. Geophys. Res. 87, 7379-7388 (1988)

39. D.J. Michels, M.J. Koomen, R.A. Howard, N.R. Sheeley Jr., Cool material in the corona at 10 solar radii, EOS 62, 376 (1981)

40. E.N. Parker, Nanoflares and the solar X-ray corona, Astrophys. J. 330, 474-479 (1988)

DYNAMICS OF ALPHA PARTICLES IN CORONAL STREAMER TYPE GEOMETRIES

A. Bürgi

Physikalisches Institut, University of Bern, Sidlerstrasse 5, CH-3012 Bern, Switzerland

Abstract

The alpha to proton flux ratio in the solar wind is studied in a three-fluid solar wind model for an idealized coronal magnetic field geometry approximating a coronal streamer. It is found that flow tube geometries in coronal streamers can produce strong fractionation, and reduce the alpha to proton flux ratio in the solar wind to very low values, in agreement with observations in the solar wind near the heliospheric current sheet.

Introduction

The average alpha to proton flux ratio in the solar wind is $\langle F_\alpha/F_p \rangle \equiv \langle (n_\alpha u_\alpha)/(n_p u_p) \rangle \approx 4\%$. There are relatively small variations in the high speed solar wind, where $F_\alpha/F_p \approx 0.048$, while in the low speed solar wind the ratio is somewhat smaller — $F_\alpha/F_p \approx 0.038$ — and much more variable, cf. /1/. The helium abundance has a sharply localized minimum in the solar wind near the heliospheric current sheet at sector boundaries of the interplanetary magnetic field /2,3/. In the simplest coronal configuration, these current sheets are the extensions of coronal streamers into the solar wind.

The question of the low helium abundance in the solar wind near sector boundaries is studied with a theoretical model assuming an idealized magnetic field geometry near the cusp of a coronal streamer.

The Model

The model used is a steady state, three-fluid model. It is described in detail in /4/, and only a short characterization is given here. The model uses the continuity, momentum and energy equations for the three particle species p, α, e. The acceleration is given by the pressure gradients, electric field, gravity, and collisional friction. Charge neutrality and zero current are assumed. The energy equations for p, α, e contain terms describing adiabatic cooling, heat transfer by collisions, an electron heat flux, and a coronal heat source (but no radiative cooling). The heat source is assumed to heat the electrons only, and to decrease exponentially with scaleheight λ_H. The collisional heat flux law for electrons given by /5/ is used, and ion heat fluxes are neglected.

The flow is assumed to be in the radial direction, but with a flow tube cross section $a(r)$ varying nonradially as

$$a(r) = \left(\frac{r}{R_\odot} \right)^2 f_a(r) \tag{1}$$

where

$$f_a(r) = \frac{f_{m1} \exp\left(\frac{r-R_1}{\sigma_1}\right) + f_1}{\exp\left(\frac{r-R_1}{\sigma_1}\right) + 1} + \frac{f_{m2}}{1 + \left(\frac{r-R_2}{\sigma_2}\right)^2} \tag{2}$$

The first term on the right-hand side is parametrized as in /6/. f_{m1} describes the overall superradial divergence of the flux tube. The second term describes the local inflation of the flux tube near the cusp at the top of a coronal streamer, which is controlled by the parameter f_{m2}. R_1 and R_2 are the locations, σ_1 and σ_2 are the scale heights over which the respective expansion occurs.

Figure 1 (from /7/) shows an idealized two-dimensional coronal magnetic field geometry. The nonradial expansion factor for a flow tube close to a coronal streamer is shown in Figure 2, as well as the contributions of the two terms in equation 2. The parameters corresponding to these geometries are summarized in Table I. The contribution of the "hole"-type becomes dominant as the pole is approached, while the contribution of the "cusp"-type becomes dominant as the equator is approached, but only very close to it (cf. Figure 4b of /7/).

The free model parameters and boundary conditions are choosen as follows: $n_e = 3 \times 10^8 \mathrm{cm}^{-3}$, $n_\alpha/n_p = 0.05$ and $T_p = T_\alpha = T_e = 10^6 \mathrm{K}$ at $r = R_\odot$. The total heat input is taken to be $2.5 \times 10^5 \mathrm{erg/cm^2/s}$ and the heating scaleheight $\lambda_H = 0.3 R_\odot$. With these parameters, the resulting model solutions have maximum electron temperatures $\approx 2 \times 10^6 \mathrm{K}$ in the low corona.

Results

The influence of the "cusp"-type geometry on the flow is studied, varying f_{m2} from $f_{m2} = 0$ (radial) to $f_{m2} = 10$. The overall superradial divergence, which tends to be less important near the cusp of the streamer, is neglected in this parameter study ($f_{m1} = 1$). Figure 3 compares the three-fluid model solutions with $f_{m2} = 0$ (radial, dashed thin lines) and $f_{m2} = 8.7$ ("cusp", solid, thick lines).

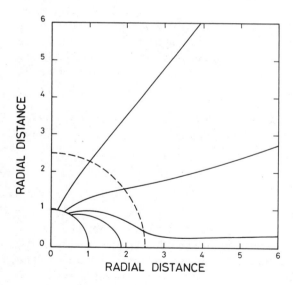

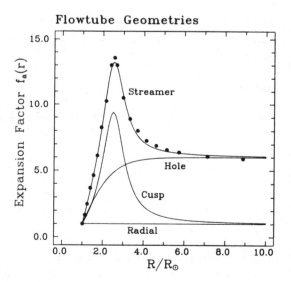

Figure 1: Geometry of magnetic field in a model by Wang and Sheeley /7/. The magnetic field is assumed to be current free everywhere except on a current sheet located in the equatorial plane outside $r = 2.5 R_\odot$. The field lines drawn have footpoints at latitudes $\lambda = 80°$, $70°$, $63°$, and $59°$.

Figure 2: Flux tube expansion factors $f_a(r)$. Dots: values from /7/ for the field line originating at heliolatitude $\lambda = 63°$ and approaching $\lambda \approx 3.5°$ as $r \to \infty$ in Figure 1. "Streamer": Analytic approximation using (2); "hole" and "cusp" are the two individual terms in (2) whose sum gives "streamer".

Table I:

Parameters for the areal expansion factors in Figure 2.					
	f_{m1}	R_1 $(R_\odot)$	σ_1 $(R_\odot)$	f_{m2}	R_2 $(R_\odot)$	σ_2 $(R_\odot)$
Radial:	1	...	...	0	...	...
Hole:	6	1	0.75	0	...	...
Streamer:	6	1	0.75	8.7	2.5	0.65
Cusp:	1	...	...	8.7	2.5	0.65

The dynamics of alphas in the corona is controlled to a large extent by frictional drag from Coulomb collisions with protons. The drag is proportional to

$$\frac{n_p}{T^{3/2}}\ (u_p - u_\alpha) = \frac{n_p u_p}{T^{3/2}}\left(1 - \frac{u_\alpha}{u_p}\right) = \frac{F_p}{T^{3/2}a(r)}\left(1 - \frac{u_\alpha}{u_p}\right) \tag{3}$$

where $F_p \equiv n_p u_p a(r)$ is a stream-tube constant and T the appropriate average ion temperature /4/. In the radial case the frictional drag is sufficient to give $u_\alpha \lesssim u_p$. In the cusp geometry, friction is reduced as the flux tube is inflated — increasing $a(r)$ decreases the proton flux $n_p u_p$ — and therefore $u_\alpha \ll u_p$ (Figure 3, bottom): Alpha particles become quasistatic in the corona, with a density scaleheight much shorter than the protons (Figure 3, top), and the alpha particle flux is greatly reduced. In the transition from the radial to the cusp geometry, the ion temperatures are also raised somewhat, thereby further weakening friction [cf. eq. 3]. However, the main reason for the poor frictional drag on alphas is the dilution of the proton flux $n_p u_p$ in the widely expanded part of the stream tube.

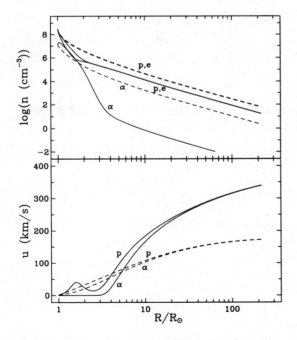

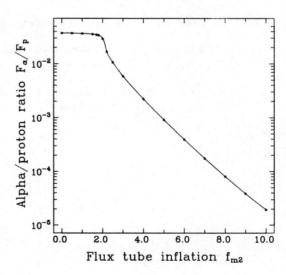

Figure 3: Comparison of densities (upper panel) and velocities (lower panel) for the radial (dashed, thin lines) and cusp geometry (solid, thick lines)

Figure 4: Alpha to proton flux ratio F_α/F_p as a function of the flux tube inflation factor f_{m2}.

The effect of the flux tube inflation in the cusp geometry on the alpha to proton flux ratio is shown in Figure 4. For the radial case we find $F_\alpha/F_p = 0.037$, very close to the value in the low speed solar wind. Inflating the flux tube by increasing f_{m2} decreases the flux ratio to arbitrarily small values.

Conclusions

Flux tubes which at large distances are located close to the heliospheric current sheet have a large local expansion (minimum of $|\mathbf{B}|$) near the neutral point at the top of a streamer. As a consequence, collisional friction is insufficient to drag helium into the solar wind along such a field line, alpha particles become quasistatic in the corona, and the alpha particle flux in the solar wind can be depressed to arbitrarily low values. This is in very good agreement with the observed minimum of F_α/F_p at the heliospheric current sheet above coronal streamers.

Acknowlegments

The author would like to thank P. Bochsler, J. Geiss, and Y.-M. Wang for helpful comments. This work was supported in part by the Swiss National Science Foundation.

References

1. S. J. Bame, J. R. Asbridge, W. C. Feldman, and J. T. Gosling, Evidence for a structure-free state at high solar wind speeds, *J. Geophys. Res.*, 82, 1487-1492 (1977).

2. G. Borrini, J. T. Gosling, S. J. Bame, W. C. Feldman, and J. M. Wilcox, Solar wind helium and hydrogen structure near the heliospheric current sheet: a signal of coronal streamers at 1 AU, *J. Geophys. Res.*, 86, 4565-4573 (1981).

3. J. T. Gosling, G. Borrini, J. R. Asbridge, S. J. Bame, W. C. Feldman, and R. T. Hansen, Coronal streamers in the solar wind at 1 AU, *J. Geophys. Res.*, 86, 5438-5448 (1981).

4. A. Bürgi, Proton and alpha particle fluxes in the solar wind: Results of a three-fluid model, *J. Geophys. Res.*, in press (1992).

5. J. M. Burgers, *Flow equations for composite gases*, Academic Press, New York, 1969,

6. R. A. Kopp, and T. E. Holzer, Dynamics of coronal hole regions: I. Steady polytropic flow with multiple critical points, *Sol. Phys.*, 49, 43-56 (1976).

7. Y.-M. Wang, and N. R. Sheeley Jr., Solar wind speed and coronal flux-tube expansion, *Astrophys. J.*, 355, 726- 732 (1990).

SILICON AND OXYGEN CHARGE STATE DISTRIBUTIONS AND RELATIVE ABUNDANCES IN THE SOLAR WIND MEASURED BY SWICS ON ULYSSES

A. B. Galvin,* F. M. Ipavich,* G. Gloeckler,* R. von Steiger** and B. Wilken†

*Department of Physics, University of Maryland, College Park, MD 20742, U.S.A.
** Physikalisches Institut, University of Bern, Sidlerstrasse 5, CH-3012 Bern, Switzerland
† Max Planck Institut für Aeronomie, D-3411 Katlenburg-Lindau, Germany

ABSTRACT

We report an initial survey of solar wind silicon and oxygen using data obtained with the Ulysses Solar Wind Ion Composition Spectrometer (SWICS). In this study, the O^{+7}/O^{+6} ratio is used to group silicon counts accumulated over a two month period. Results on Si charge state distributions, relative Si/O abundances, and associated proton kinetic temperature and speed distributions are presented.

INTRODUCTION

Measurements of ionization states and relative abundances of solar wind minor ions provide important information on solar source conditions and acceleration processes in the corona. Composition parameters vary among different types of solar wind [1] and typically provide a good indicator of the source region. Previous studies of solar wind silicon cover isolated cases of interstream, transient-related, and noncompressive density enhancement solar wind flow types [1, references therein]. There has been one long-term study of silicon abundances and velocities [2]. There have been no long-term studies of silicon charge state distributions nor of abundances as a function of solar wind flow type.

INSTRUMENTATION AND METHODOLOGY

The SWICS "main channel" [3] measures the time-of-flight (energy/mass), electrostatic deflection (energy/charge), and the residual energy deposited in a solid state detector to determine the mass, mass/charge, incident energy, bulk speed, and kinetic temperature of ions in the energy/charge range ~0.6 to 60 keV/e. The instrument cycle time is ~13 minutes, during which the deflection system steps through 64 logarithmically-spaced voltages. For ions in which both mass and mass/charge are determined, the background contribution under all solar wind conditions is essentially zero.

"Matrix elements", which are used extensively in this study, are 490 rate bins defined in terms of mass, m, and mass per charge, m/q, with typical full bin widths of ~3% in m/q and ~20% in m (some bins have less resolution). These rates are accumulated over an entire instrument cycle; energy/charge information is not retained. For ions with m and m/q values in the vicinity of silicon, the inherent resolution (FWHM) is 25-35% in m and 3% in m/q. In this study, corrections to silicon matrix elements due to count contributions by adjacent species (magnesium and sulfur) are made by assuming nominal solar wind (if available) or solar energetic particle (SEP) relative abundances [2,4] and charge state ratios from Arnaud and Rothenflug (1985) [5] (hereafter abbreviated A&R). Similarly, the matrix elements representing oxygen have been corrected for "spillover" contributions by carbon and nitrogen.

Using the tables of A&R and the O^{+7}/O^{+6} ratio calculated from the corresponding oxygen matrix elements, an oxygen "coronal ionization temperature", $T_C(O)$, is routinely derived for each 13-minute instrument cycle. The A&R tables represent ionization equilibria obtained by collisional processes only (low density case). Under these conditions the derived T_C corresponds to the "freezing-in" temperature for solar wind ions. In cases where non-Maxwellian conditions or non-collisional processes are prevalent at the freezing-in site, the derived T_C is still a useful tool for comparison purposes, although it may not in such cases accurately reflect the local electron temperature /6/.

Counting statistics for the less abundant silicon have been enhanced in this study by using long accumulation times. Two months of silicon data have been summed together, organized by the $T_C(O)$ derived for each 13-minute instrument cycle. The data obtained during a given cycle is placed into one of seven $T_C(O)$ groups (given in Table 1). The grouping by oxygen ionization temperature is an attempt to organize the data by solar wind flow type (see the Discussion section).

Out of a total of 6881 instrument cycles available in the selected survey time period, 14% were excluded from analysis due to data gaps existing within the given cycle. Another 12% of the data were not used because a given cycle had insufficient oxygen counts to determine a meaningful O^{+7}/O^{+6} ratio. The remaining cycles represent 1100 hours of solar wind data.

SURVEY RESULTS

An overview of solar wind conditions during the survey time period (16 September - 17 November 1991) is shown in Figure 1. During this time the Ulysses spacecraft traveled from 4.2 to 4.7 AU (heliocentric distance). According to the *Preliminary Solar Geophysical Data Reports*, this period encompasses solar activity of low, moderate, and high levels, including minor and major flares, disappearing filaments, and several favorably located coronal holes (6 equatorial and 4 near-equatorial). It is therefore anticipated that this data base contains a variety of solar wind flow types. The solar wind proton parameters given in the figure are measured by the SWICS "auxiliary channel" /3/. The proton densities have not been corrected for the instrument viewing angle (a function of the spacecraft trajectory) and should be considered as relative densities (good to ~25% of the absolute density values during this time period).

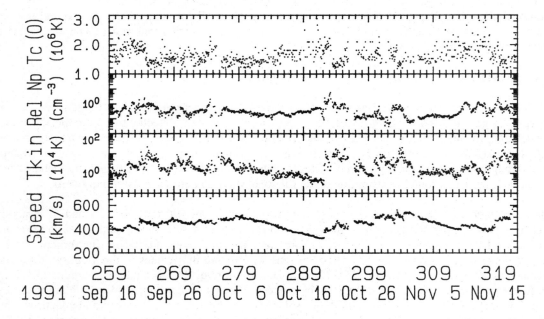

Fig. 1. Overview of the solar wind parameters measured by SWICS on Ulysses. From top to bottom: $T_C(O)$ from O^{+7}/O^{+6}; proton relative density; proton kinetic temperature; proton bulk speed.

TABLE 1 Survey Summary

$T_c(O)$ Group (a)	Nr of cases	$T_c(O)$ +7/+6	$T_c(Si)$, derived from adjacent charge state abundance ratios:					Si/O	Proton Tkin $(10^4 K)$	Proton Speed (k/s)
			+8/+7	+9/+8	+10/+9	+11/+10	+12/+11			
<1.2	837	1.1 ±0.1	1.1 ±0.1	1.4 ±0.1	1.6 ±0.1	1.6 ±0.1	-	0.12 ±.01	4.5 ±4.8(b)	467 ±36(b)
1.2-1.4	1610	1.3 ±0.1	1.1 ±0.1	1.4 ±0.1	1.5 ±0.1	1.6 ±0.1	-	0.12 ±.01	3.6 ±4.3	459 ±42
1.4-1.6	1251	1.5 ±0.1	1.1 ±0.1	1.3 ±0.1	1.5 ±0.1	1.6 ±0.1	-	0.13 ±.01	2.8 ±3.9	448 ±46
1.6-1.8	685	1.7 ±0.1	1.0 ±0.1	1.4 ±0.1	1.6 ±0.1	1.8 ±0.1	1.7 ±0.2	0.16 ±.01	2.4 ±2.3	435 ±49
1.8-2.0	414	1.9 ±0.1	1.2 ±0.1	1.3 ±0.1	1.6 ±0.1	1.8 ±0.1	1.7 ±0.1	0.16 ±.01	2.4 ±2.0	427 ±44
2.0-2.2	214	2.1 ±0.1	1.1 ±0.1	1.4 ±0.1	1.6 ±0.1	1.9 ±0.1	1.8 ±0.1	0.17 ±.02	2.2 ±1.9	419 ±49
>2.2	101	2.4 ±0.1	1.3 ±0.2	1.1 ±0.2	1.8 ±0.1	1.9 ±0.1	1.7 ±0.1	0.15 ±.02	1.7 ±1.4	432 ±76

(a) Coronal ionization temperatures (T_c) in units of $10^6 K$.
(b) Sigma of the distribution.

Figure 2 (Table 1) gives the A&R ionization temperatures derived from the ratios of adjacent silicon charge states as a function of the average O^{+7}/O^{+6} ionization temperature found within each $T_c(O)$ category. In determining the Si/O abundance ratios listed in Table 1, the contribution to oxygen from charge states other than +6, +7 has been estimated from the derived O^{+7}/O^{+6} ionization temperature and A&R. In a like manner the contribution to silicon from charge states +5, +6 has been estimated from A&R based on the Si^{+8}/Si^{+7} ionization temperature. Most of the uncertainty given in the table is systematic and the result of assumed detection efficiencies and uncertainties in spillover correction factors. Also given are the average proton kinetic temperatures and proton bulk speeds for each category. (Radial dependence has not been factored out of the kinetic temperature.)

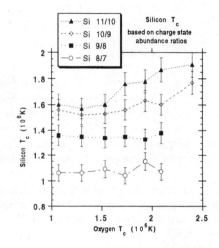

Fig. 2. Silicon ionization temperatures as a function of oxygen ionization temperature.

DISCUSSION

During the survey time period, the average $T_c(O)$ is ~1.5x10^6K, with the majority of cases at ~1.3x10^6K. The proton kinetic temperature shifts from a mean value ~ 5x10^4K (with a relatively broad distribution) for low $T_c(O)$ values, to a mean value < 2x10^4K (with a narrow distribution) for high $T_c(O)$. The observed trend of the kinetic temperature with $T_c(O)$ suggests that the $T_c(O)$ groups are successful (at least to some degree) in representing different types of solar wind flows. For example, high kinetic temperatures and low ionization states suggest coronal-hole associated flows, while low kinetic temperatures and high ionization states are indicative of transient-related flows /1/.

The ionization temperatures derived from the Si^{+8}/Si^{+7} and Si^{+9}/Si^{+8} ratios are insensitve to the oxygen ionization temperature. There is a general trend in all $T_c(O)$ groups for $T_c(Si)$ to increase with higher charge states, making it readily apparent that an isothermal equilbrium model for the silicon charge state distributions would be too simplistic. This result has been observed in previous studies and is at least partially explainable in terms of a multistage ionization freezing-in process in which higher charge states, with lower exchange rates, freeze in at the denser (hotter) lower coronal altitudes /1, references therein/. Bürgi /6/ has modelled the expected charge state ratios for silicon under the assumption of (1) Maxwellian velocity distributions for the electrons at the coronal freezing-in site and a rapid initial expansion of the solar wind, yielding low collisional coupling in the region of maximum coronal temperature (in this "hot" model the derived freezing-in temperatures reflect actual coronal temperatures), and (2) a more gradual expansion of the magnetic field with non-thermal tails in the local electron distribution, with the resultant derived temperatures overestimated (the "cool" model). The models are for low-speed solar wind forced to fit an oxygen charge state ratio $O^{+6}/O^{+7} \approx 3$ (A&R ~ 1.7 x 10^6K). A comparison of our results against the Bürgi models is given in Figure 3. The silicon charge state fraction (the relative abundance of a given ionization stage to the total species number) is plotted against charge state. The observed data have been broken up into two groups, $T_C(O) < 1.7 \times 10^6$K and $T_C(O) \geq 1.7 \times 10^6$K, which appear to exhibit fundamentally different trends in the charge state distributions. The first data group is not well-fit by either model. The second data group qualitatively follows the hot (Maxwellian) model.

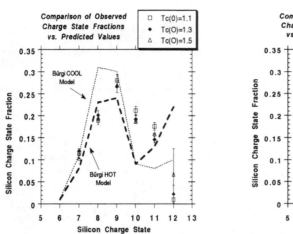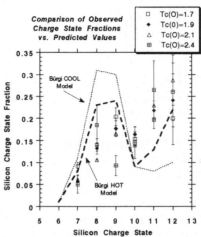

Fig. 3. The observed Si charge state fraction plotted against charge state for $T_C(O) < 1.7 \times 10^6$K and $T_C(O) \geq 1.7 \times 10^6$K data groups. The Bürgi /6/ predictions are shown for comparison.

The Si/O abundance ratios found in this survey are somewhat lower than the average value of 0.19 ± 0.04 previously reported for solar wind /2/ or the value 0.176 reported for SEPs /4/. As noted in these earlier measurements, the Si/O ratios indicate an enhancement relative to the solar system value. Such enhancements have been observed for elements in the solar wind and SEPs that are characterized by low first ionization potentials (FIP < 10 eV). The present study indicates that the degree of enhancement for silicon is reduced when $T_C(O) < 1.7 \times 10^6$K. This may indicate a difference between coronal hole associated flows and transient-related flows.

ACKNOWLEDGEMENTS

The authors thank C. Shafer for her assistance. This work was supported under grant JPL 955460.

REFERENCES

1. S.J. Bame, in: *Solar Wind Five*, ed. M. Neugebauer, Nasa Conference Publ. 2280, 1983, p. 573.

2. P. Bochsler, *J. Geophys. Res.*, **94**, 2365 (1989).

3. G. Gloeckler, J. Geiss, H. Balsiger, P. Bedini, J.C. Cain, J. Fischer, L.A. Fisk, A.B. Galvin, F. Gliem, D.C. Hamilton, J.V. Hollweg, F.M. Ipavich, R. Joos, S. Livi, R. Lundgren, U. Mall, J.F. McKenzie, K.W. Ogilvie, F. Ottens, W. Rieck, E.O. Tums, R. von Steiger, W. Weiss, and B. Wilken, *Astron. Astrophys. Suppl. Ser.*, **92**, 267 (1992).

4. H.H. Brenemann and E.C. Stone, *Astrophys. J.*, **299**, L57 (1985).

5. M. Arnaud and R. Rothenflug, *Astron. Astrophys. Suppl. Ser.*, **60**, 425 (1985).

6. A. Bürgi, *J. Geophys. Res.*, **92**, 1057 (1987).

IONS WITH LOW CHARGES IN THE SOLAR WIND AS MEASURED BY SWICS ON BOARD ULYSSES

J. Geiss,* K. W. Ogilvie,* R. von Steiger,** U. Mall,** G. Gloeckler,***
A. B. Galvin,*** F. Ipavich,*** B. Wilken† and F. Gliem‡

* NASA/Goddard Space Flight Center, Code 692, Greenbelt, MD 20770, U.S.A.
** University of Bern, Physikalisches Institut, Sidlerstrasse 5, 3000 Bern,
Switzerland
*** University of Maryland, Space Physics, College Park, MD 20742, U.S.A.
† Max-Planck-Institut für Aeronomie, 3411 Katlenburg-Lindau 3, Germany
‡ Technische Universität Braunschweig, Institut für Datenverarbeitungsanlagen,
Braunschweig, Germany

ABSTRACT

We present new data on rare ions in the solar wind. Using the Ulysses-SWICS
instrument with its very low background we have searched for low-charge ions
during a 6-day period of low-speed solar wind and established sensitive upper
limits for many species. In the solar wind, we found $He^{1+}/He^{2+} < 5 \cdot 1o^{-4}$. This
result and the charge state distributions of heavier elements indicate that
all components of the investigated ion population went through a regular coronal expansion and experienced the typical electron temperatures of 1 to 2
million Kelvin. We argue that the virtual absence of low-charge ions demonstrates a very low level of non-solar contamination in the source region of
the solar wind sample we studied. Since this sample showed the FIP effect
typical for low-speed solar wind, i.e. an enhancement in the abundances of
elements with low first ionisation potential, we conclude that this enhancement was caused by an ion-atom separation mechanism operating near the solar
surface and not by foreign material in the corona.

I INTRODUCTION

SWICS, the Solar Wind Ion Composition Spectrometer /1/ on board the ESA-NASA
spacecraft Ulysses is particularly suited to investigate rare ion species,
because (1) it simultaneously determines energy/charge (E/Q), mass/charge
(M/Q) and mass (M) of an ion, from a combination of E/Q, time-of-flight (TOF)
and total energy measurements; (2) it covers a wide range of energies, masses and mass/charge ratios; (3) it has an extremely low background, due to
the coincidence technique employed, i.e. two TOF pulses and one solid state
detector pulse.

In the heliosphere, ions with low charge could have a variety of origins.
Depending on location and mode of instrument operation, we could primarily
detect ions from four sources: atmospheres/magnetospheres of planets, ionized
interstellar gas atoms, ions resulting from disintegration or sputtering of
solid bodies (asteroids, comets, planets, satellites), and low-charge solar

ions. We present in this paper new evidence on the abundances of low-charge
ions and discuss the results in relation to the latter two sources:

He⁺ of Solar Origin

Parcels of solar wind gas containing significant amounts of He+ are occasionally observed /2,3/. They stem from a mass ejection or some similar event in
which solar gas passes the region of the corona without its electrons being
heated to the usual corona temperature. SWICS allows to study He+ at very
low abundance levels and to search for other low-charge ions which are expected in the solar wind at the time of He⁺-events.

Cometary and Asteroidal Matter in the Corona

When approching the sun, cometary and asteroidal matter will gradually evaporate and ionize. Such non-solar material in the coronal gas could be searched
for by solar wind ion mass spectrometry, cf./4/. Recently, Lemaire /5/ estimated the quantities of non-solar material entering the corona, and he concluded that they were sufficient to cause the FIP-effect, i.e. the overabundance of elements with low First Ionisation Potential which is observed in
the corona (cf. /6/), in flare particle populations (cf. /7,8,9/), and in the
solar wind (cf. /10/). Ions produced from solids in the vicinity of the sun
should have a broad distribution of charge states, reaching down to charges
much below those of the proper solar wind ions.

TABLE 1   Solar Wind Conditions and Experimental Parameters

SPACECRAFT AND INSTRUMENT

Period Investigated:	8. to 13. December, 1991.
Sun-Spacecraft Distance:	1.3 to 1.4 AU
Sun-Spacecraft-Earth Angle:	22 to 16 Degrees.
Arrival Directions Covered:	Up to 80 Degrees from the Sun-Spacecraft Line.

Velocity Ranges Covered:

He⁺ (Solar Wind):	320 to 420 km/s
Other Ions: Upper Limit:	3420 $(Q/M)^{1/2}$ km/s
Lower Limit:	320 km/s (Column 3, Table 2)
Lower Limit:	250 km/s (Column 2, Table 2)
Mass Limit (from TOF):	$M/Q < 17.47/[1-(V/500)^2]$

OBSERVED SOLAR WIND CONDITIONS

Velocity Range, Whole Period:	300 to 520 km/s
72% of the Time:	320 to 420 km/s
Freezing-in Temperatures (10^6K):	T $[O^{6+}/O^{7+}]$ = 1.69
	T $[C^{5+}/C^{6+}]$ = 1.42
	T $[C^{4+}/C^{5+}]$ = 1.32
Abundances Relative to Solar Abundances	C/O = 1.45
	Mg/O = 3.4

We have searched for such low-charge ions in the solar wind during a 6-day
period when the typical FIP pattern in elemental abundances was present in
the solar wind gas. We did not find any appreciable quantity of such ions
(Section II) and discuss this finding in Section III.

II EXPERIMENTAL RESULTS

All complete energy spectra obtained during the 6-day period given in table 1 were used in this investigation. SWICS receives ions from a cone with a half-width of 60 degrees around the earth-spacecraft line, enabling us to include ions as far away from the solar direction as 80 degrees. We note, however, that the duty cycle decreases with increasing angular distance from the sun.

Velocity and M/Q limits are detailed in table 1. Solar wind protons and, to a lesser extent, alphas cause accidental double coincidences in significant numbers. Accidental triple coincidences, mainly caused by alphas, are much rarer. By imposing a lower velocity limit in our composition analysis, we virtually eliminate the background due to these accidental coincidences for ions with high M/Q: with velocities above this limit, these ions register in energy/charge channels which are essentially free of protons and alphas. Two values were used as velocity limit: the 320km/s limit covers virtually the entire solar wind flux, the 250km/s limit was chosen to search for low-speed, "stray" ions.

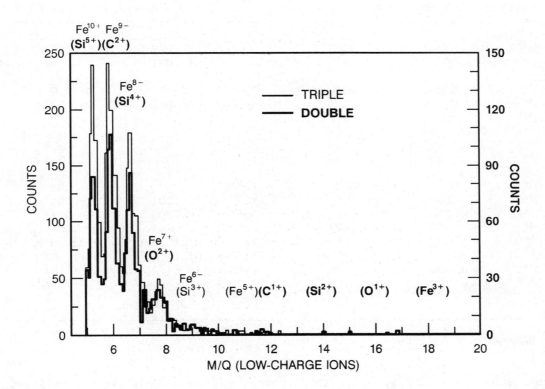

Figure 1. Mass/Charge Spectrum of the ions with M/Q $\geq$ 5 registered during the entire 6-day period. Whereas the multiply charged solar wind ions give mostly triple coincidences, low-charge ions would give mostly double coincidences. The figure demonstrates that low-charge ions like C^+, O^+, Si^{2+}, $Fe^{3+/4+/5+}$ are much rarer than the major Fe-ions which themselves have individual abundances of only $2 \cdot 10^{-4}$ relative to H^+.

In figure 1 all ions with M/Q $\geq$ 5 and with velocities larger than 320km/s registered during the 6-day period are shown. The M/Q scale given is valid

for the low-charge ions. The scale is slightly shifted for ions with high
charges, so that for instance Fe^{9+} appears very close to the position of the
C^{2+} ion. About 75% of the high-charge ions such as Fe^{7+} to Fe^{10+} give triple
coincidences and 25% give double coincidences. Low-charge ions such as C^+ or
C^{2+} give virtually only double coincidences. This difference helps in distinguishing ions with equal or similar M/Q ratios.

The spectra of total counts given in figure 1 were corrected with the instrument response function derived from laboratory calibrations which included
the appropriate energy range of the low charge ions we searched for. The resulting ion fluxes are presented in table 2, where abundances of individual
ion species are given relative to the sum of all ions of the respective ele-

TABLE 2   Abundances of Ions, Relative to all Ions of the Element.

Ion	Ion Abundance		Ion	Ion Abundance	
	V > 250 km/s	V > 320 km/s		V > 250 km/s	V > 320 km/s
He^{1+}	–	< .0005[a]	Ne^{1+}	< .022	< .015
He^{2+}	1.00	1.00	Ne^{2+}	< .021	< .018
C^{1+}	< .002	< .002	Mg^{1+}	–	< .010
C^{2+}	< .03	< .03	Mg^{2+}	< .013	< .010
C^{3+}	< .007	< .007			
C^{4+}	.06 ± .01	.06 ± .01	Si^{1+}	–	< .017
C^{5+}	.31 ± .05	.31 ± .05	Si^{2+}	< .010	< .008
C^{6+}	.63 ± .05	.63 ± .05	Si^{3+}	< .026	< .024
O^{1+}	< .002	< .002	Fe^{2+}	–	< .004
O^{2+}	< .009	< .008	Fe^{3+}	< .004	< .004
O^{3+}	< .015	< .015	Fe^{4+}	< .005	< .005
O^{4+}	< .010	< .010	Fe^{5+}	< .014	< .008
O^{5+}	< .015	< .015	Fe^{6+}	.012± .008	.012± .008
O^{6+}	.67 ± .04	.67 ± .04	Fe^{7+}	.07 ± .02	.07 ± .02
O^{7+}	.24 ± .04	.24 ± .04	Fe^{8+}	.19 ± .04	.19 ± .04
O^{8+}	.09 ± .02	.09 ± .02	Fe^{9+}	.21 ± .04	.21 ± .04
			Fe^{10+}	.18 ± .04	.18 ± .04

a)   320km/s < V < 420km/s

ment. Ions shown in figure 1 without parenthesis were positively identified;
for ions in parenthesis only upper limits could be derived. The reason is
that above M/Q = 10 there are very few counts in either the double or the
triple coincidence spectrum. We have more counts for ions with lower M/Q ratios, but here high-charge ions often mask ions with lower charges. The ratio
of double to triple coincidence counts is used in these cases to arrive at an
upper limit for the low-charge ion. The upper limits are conservative, they
include a 2σ statistical error and a factor of 1.5 for uncertainties in the
instrument function. The small differences between the upper limits in columns 2 and 3 of table 2 are not significant. The higher values in the 250km/s
column are due to an increase in accidental coincidences.

He^+ is a special case. We find significant fluxes of this ion at elevated
speed, up to twice the solar wind velocity. This is characteristic for He^+
produced from the neutral interstellar gas by solar EUV and picked up by the

magnetic field carried in the solar wind /11/. Therefore, in order to give a sensitive estimate for He^+ of solar origin we determined the He^+/He^{2+} ratio in the velocity range 320 to 420 km/s, which covers 72% of the period investigated (cf. Table 1). Only a minority of the interstellar He^+ ions is present in this velocity range, enabling us to derive the sensitive upper limit for solar He^+ given in Table 2.

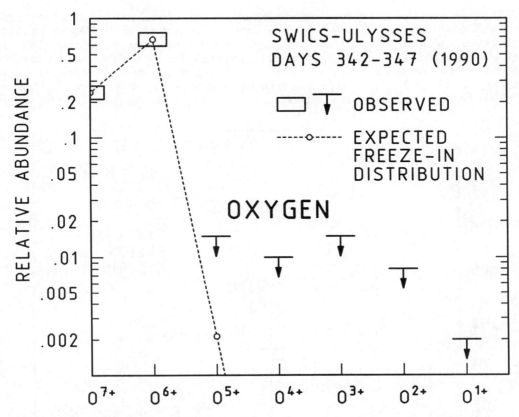

Figure 2. Charge state distribution of oxygen ions during the 6-day period investigated for V>320 km/s. Due to the low background of the instrument a very low abundance limit could be derived for O^+. The limits for O^{2+} to O^{5+} are higher because there are high-charge ions with equal or similar M/Q ratios (e.g. O^{2+} and Fe^{7+}).

The relative abundances of oxygen and iron ions for V>320km/s are plotted in figures 2 and 3. For reference purposes, abundances expected to arise from the coronal freezing-in process are also shown. Observations /12,13,14,15/ and dynamical freeze-in calculations /16/ show that (a) iron and carbon charge states freeze in at lower temperatures than oxygen charge states and, (b) for the same element, low charge ions freeze in later than high-charge ions (cf. our results in table 1). Therefore, we chose the following freezing-in temperatures (after /17/): 1.6 million Kelvin for O^{5+}/O^{6+} and a linear decrease in temperature from 1.3 to 1.0 million Kelvin for Fe^{10+} to Fe^{5+}

III DISCUSSION

The ions of C, O and Fe with the lowest charges we could identify are exactly those which ought to emerge from the coronal freezing-in process in detectable quantities (cf. figures 2 and 3). Thus we find C^{4+}, Fe^{7+} and Fe^{6+} with roughly the expected abundances, whereas C^{3+}, O^{5+}, or Fe^{5+} are undetectable.

We find no traces of ions that would not survive the passage through the dense inner part of the regular hot corona.

The absence of heavier low-charge ions of solar origin is consistent with the very low level of solar wind He$^+$ during the period considered here (table 2).

As pointed out in the introduction we expect solar wind ions originating from solid non-solar matter to have a wide charge spectrum, including ions with low charges. In this short publication we cannot discuss this assertion in detail, and summarise here only the main points. Solid bodies consisting of metal silicates and oxides will begin to evaporate and disintegrate at solar distances of several R$\odot$, bodies containing organics or ices even farther away. Evaporated material will quickly become ionized, mainly by EUV /18/.

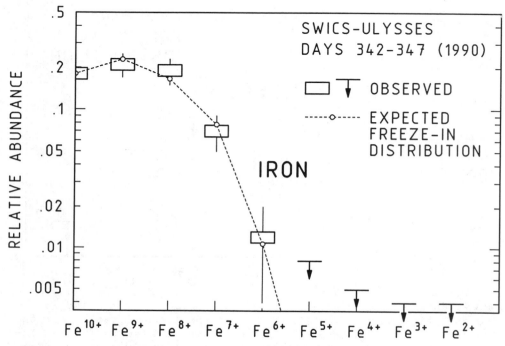

Figure 3. Charge state distribution of iron ions during the 6-day period investigated for V>320 km/s. For typical solar wind velocities Fe$^+$ is outside the range of the instrument (cf. table 1). For the other low-charge ions we find very low abundance limits because the background is low and interferences from ions of other elements are relatively unimportant.

Ionisation will not proceed to high charges, because there are no suitable photons and - beyond about 2R$\odot$ - electron densities are too low. The solid bodies will impart their velocities to these ions, i.e. the ions will start with velocities near the local escape speed and with significant non-radial velocity components. In the outer part of the evaporation zone, disintegration will be incomplete, particularly for larger bodies, and much of the mass loss will occur at perihelion and afterwards. i.e. the debris receives a vanishing or positive radial velocity component. Since the solar magnetic field is radially decreasing, even many ions starting with a velocity component towards the sun will be turned around before they get into the dense part of the corona and they will escape as low-charge ions. This mirror effect is quite significant. For instance if the local solar field is radially directed and decreases with r^{-n} (r = distance from the solar centre), the mirror point r$_m$ for an ion with zero total energy is related to the initial point r$_0$ and pitch angle φ_0 by

$$[r_m/r_0]^{n-1} = \sin^2\varphi_0$$

The mirror distance r_m is equal to the Keplerian perihelion distance for n=2 and larger for n>2. The average B-field decreases rapidly near the sun and even in the solar wind it decreases with r^{-2}. Therefore, mirroring should help in keeping these ions away from the dense region of the corona. Since at several solar radii the solar wind and escape speeds are similar, the low-charge ions should have no difficulty to be picked-up by the solar wind. Even if they do not become fully accomodated to their regular solar wind temperature, we would still detect a large portion of them (cf. the upper velocity limit and the acceptance angle given in table 1).

When a larger body disintegrates, the plasma cloud that is created could have enough energy density to push the coronal B-field away. Sufficiently massive clouds created near perihelion would emerge from near-solar space with their low-charge ion population. Even closer to the sun, the electrons in the cloud would not readily accomodate to the ambient coronal temperature, as in the case of coronal mass ejections. If there were plasma processes to heat the electrons it is most unlikely that such heating would create an electron energy distribution producing uniformly the same type of charge state distribution as an orderly corona expansion does. In other words we should have a different type of charge state distribution for low- and high-FIP elements, and this we do not observe.

We conclude that solar wind ions created from solid bodies with their wide ranges in size and perihelion distance should have charge state spectra different from those we actually observe. We argue therefore, that - at least during the period we have investigated - the observed FIP-effect was not primarily caused by foreign material in the corona, but by an ion-atom separation mechanism operating on the solar gas upon entering the corona.

ACKNOWLEDGEMENTS: The authors wish to thank Hans Balsiger and Peter Bochsler for valuable discussions, and Frank W. Ottens, Gary Burgess and Robert Wimmer for their help in the data reduction. J.G. was supported at GSFC by an Associateship of the US National Academy of Sciences. Development of SWICS and the work reported here was supported by the US National Aeronautics and Space Administration, the Swiss National Science Foundation, and the Bundesminister für Wissenschaft und Technologie of Germany.

REFERENCES

1. G. Gloeckler, J. Geiss, H. Balsiger, P. Bedini, J.C. Cain, J. Fischer, L.A. Fisk, A.B. Galvin, F. Gliem, D.C. Hamilton, J.V. Hollweg, F.M. Ipavich, R. Joos, S. Livi, R. Lundgren, U. Mall, J.F. McKenzie, K.W. Ogilvie, F.W. Ottens, W. Rieck, E.O. Tums, R. von Steiger, W. Weiss, and B. Wilken, The solar wind ion composition spectrometer, Astron. Astrophys. Suppl. 92, 267, 1992.

2. R. Schwenn, H. Rosenbauer and K.H. Mühlhäuser, Singly-ionized helium in the driver gas of an interplanetary shock wave, Geophys. Res. Lett. 7, 201, 1980.

3. J.T. Gosling J.R. Asbridge, S.J. Bame, W.C. Feldman and K.D. Zwickl, Observations of large fluxes of He^+ in the solar wind following interplanetary shock waves, Evidence for magnetic bottles, J. Geophys. Res. 85, 3431, 1980.

4. J. Geiss and P. Bochsler, Solar wind composition and what we expect to learn from out-of-ecliptic measurements, The Sun and the Heliosphere in three Dimensions, R. G. Marsden, Ed.,Reidel Publishing Co. 1986, p. 173.

5. J. Lemaire, Meteoritic ions in the corona and solar wind, Astrophys. J. 360, 288, 1990.

6. U. Feldman and K.G. Widing, Elemental abundances in the upper solar atmosphere, this issue, 1992.

7. H.H. Breneman and E.C. Stone, solar coronal and photospheric abundances from solar energetic particle measurements, Astrophys. J. 299, L57, 1985.

8. J.P. Meyer, The baseline composition of solar energetic particles, Astrophys. J. Suppl. Ser. 57, 151, 1985.

9. D.V. Reames, I.G. Richardson and L.M. Barbier, On the differences in elementabundances og energetic ions from corotating regions and from large solar events, Astrophys. J. Letters, 382, L43, 1991.

10. P. Bochsler and J. Geiss, Composition of the solar wind, in Solar System Plasma Physics, Geophysical Monograph 54, J. H. Waite, J.L. Burch and R.L. Moore, p. 133-141, 1989.

11. E. Möbius, The interaction of interstellar pick-up ions with the solar wind - probing the interstellar medium by in-situ measurements, in Physics of the Outer Heliossphere, S. Grzedzielski and D. E. Page, Ed., Pergamon Press, p. 345, 1991.

12. G.N. Zastenker, Yu.I. Yermolaev, Observations of solar wind stream with high abundance of heavy ions and relationwith coronal conditions, Planet. Space Sci 29, 1235-1240, 1981.

13. F.M. Ipavich, A.B. Galvin, J. Geiss, K.W. Ogilvie and F. Gliem, Solar wind iron and oxygen charge states and relative abundances measured by SWICS on Ulysses, Solar Wind Seven, Abstract 3.8, 1991.

14. R. von Steiger, S.P. Christon, G. Gloeckler, F.M. Ipavich, Variable carbon and oxygen abundances in the solar wind as observed in the Earth's magnetosheath by AMPTE/CCE, Astrophys. J., in press, 1991.

15. R. von Steiger, J. Geiss, H. Balsiger, U. Mall, G. Gloeckler, and A.B. Galvin, Magnesium, carbon and oxygen abundances in the solar wind as measured by SWICS on board Ulysses, this issue, 1992.

16. F. Bürgi and J. Geiss, Helium and minor ions in the corona and solar wind: dynamics and charge states, Solar Phys. 103, 347, 1986.

17. M. Arnaud and R. Rothenflug, An updated evaluation of recombination and ionisation rates, Astron. Astrophys. Suppl. Ser. 60, 425 - 457, 1985.

18. J. Geiss and P. Bochsler, Ion composition in the solar wind in relation to solar abundances, Rapports Isotopiques Dans Le Systeme Solaire, Cepadues-Editions, p. 213, 1985.

DETECTION OF SINGLY IONIZED ENERGETIC LUNAR PICK-UP IONS UPSTREAM OF EARTH'S BOW SHOCK

M. Hilchenbach,* D. Hovestadt,* B. Klecker* and E. Möbius**

* Max-Planck-Institut für Extraterrestrische Physik, Garching, Germany
** Space Science Center and Department of Physics, University of New Hampshire, Durham, NH, U.S.A.

ABSTRACT

Singly ionized suprathermal ions upstream of the earth's bow shock have been detected by using the time-of-flight spectrometer SULEICA on the AMPTE/IRM satellite. The data were collected between August and December 1985. The flux of the ions in the mass range between 23 and 37 amu is highly anisotropic towards the earth. The ions are observed with a period of about 29 days around new moon (± 3 days). The correlation of the energy of the ions with the solar wind speed and the interplanetary magnetic field orientation indicates the relation to the pick-up process. We conclude that the source of these pick-up ions is the moon. We argue that due to the impinging solar wind, atoms are sputtered off the lunar surface, ionised in the sputtering process or by ensuing photoionization and picked up by the solar wind.

I. INTRODUCTION

Lately the study of pick-up ions in the solar wind has become a valuable tool in investigations of neutral gas distributions in interplanetary space /1/. Neutral particles in interplanetary space do not interact with the magnetic field embedded in the solar wind. However, once they are ionized by solar UV radiation, charge exchange with solar wind ions or electron collisions, these ions are subject to the combined forces of the $-v_{sw}$ x B electric field and the interplanetary magnetic field B. In the inertial system pick-up ions initially move on cycloidal trajectories perpendicular to the local magnetic field. Their velocity varies between zero (neglecting their initial speed compared to the solar wind velocity) and a maximum value of $2 \cdot v_{sw} \cdot \sin \alpha$ which is determined by the solar wind velocity and the angle a between its flow direction and the local interplanetary magnetic field. On their journey away from the source - by means of pitch-angle scattering at magnetic fluctuations and adiabatic cooling in the expanding solar wind - the pick-up ions fill a sphere in velocity space centered around the solar wind with a radius equal to its velocity. In a variety of cases these characteristics allow the identification of neutral atom sources and their spatial distribution.

Positive identification of the interstellar He^+ pick-up ions from the neutral interstellar helium was reported by Möbius et al /2/. Pick-up ions of cometary origin were first detected by the ICE satellite as it moved through the tail of the comet Giacobini-Zinner /3, 4/. The atmospheres of unmagnetized planets such as Venus also produce a pick-up ion population (e.g., /5/). In addition to these gas distributions the surface of solid bodies like cosmic dust grains, asteroids, and moons are sources of pick-up ions. Atoms such as O, Mg, Al, Si, Ca and Fe can be released by sputtering, micro meteorite impact and/or vapourisation. The expected density distribution of pick-up ions from cosmic dust in the solar system has been discussed by Fahr and Ripken /6/.

In the following we will present evidence for the detection of singly ionized heavy pick-up ions in the solar wind upstream of the earth's bow shock. We will show that these ions are of lunar origin by making use of their correlation with the lunar phase and their distinct directional distribution.

II. SPACECRAFT AND INSTRUMENTATION

The observations reported here were made with the time-of-flight spectrometer SULEICA (SUprathermaL Energy Ionic Charge Analyzer) onboard the AMPTE/IRM (Active Magnetospheric Particle Tracer Explorer) satellite which was launched on August 16, 1984, into a highly elliptical orbit with an apogee of 18.7 RE. Between August and December it spent a large fraction of each orbit in the solar wind upstream of the earth's bow shock.

SULEICA combines the selection of incoming ions in the energy range 5 - 226 keV/Q according to their E/Q by electrostatic deflection with a subsequent time-of-flight (TOF) analysis and the final measurement of the residual ion energy in a silicon surface barrier detector (SSD). This combination allows to determine independently the energy (E), mass (M) and ionic charge (Q) of an incoming ion. The background is extremely low due to the applied triple coincidence technique, i.e., two TOF pulses (start and stop) and one SSD detector pulse. Directional information in the ecliptic plane is provided by a sectoring scheme using the satellite spin with 8 sectors for heavy ions. More detailed information may be found elsewhere /7/. The solar wind velocity was derived from the data of the 3-D plasma instrument /8/ and the magnetic field was determined with a fluxgate magnetometer /9/.

A survey of the pulse height analysis (PHA) data from the SULEICA instrument was made for the period August and December 1985 whenever the satellite was upstream of the bow shock. The time resolution of the PHA data is one complete energy stepping cycle which was 80 sec during the times of interest.

III. EXPERIMENTAL RESULTS

We observed periodic flux increases of suprathermal heavy ions. In Figure 1 we present the M/Q distribution of these ions flowing in the anti-sunward direction for the energy range 80 - 226 keV/Q. Two populations can be clearly distinguished: 14 to 20 and 23 to 37 amu. The width of these distributions in M/Q reflects the (E/Q)-resolution of the SULEICA instrument of $d(E/Q)/E/Q \approx 0.1$ and the additional effects of energy and angular straggling in the C-foil. The ratio $Q^* = E_{SSD} / (E/Q)$ has been derived from the solid state detector signal E_{SSD} and the energy/charge E/Q of the corresponding deflection voltage step. The observed Q^* is smaller than 1 and therefore charge state 1 can be inferred for these ions from the calibration data of the instrument. The first population most likely consists of O^+ ions. The second population may include ions such as Al^+, Si^+ or S^+. Molecular ions as O_2^+ or NO^+ can be excluded. On entering the TOF section of the sensor through the thin carbon foil, these ions would very probably break up into their constituents. Due to angular scattering, for the large majority of the cases, only one fragment reaches the solid state detector and the measured total energy is reduced accordingly /10/. The signature of fragmentation (Q^* peaked at values < 0.5) was not found in the Q^* distribution of this data set (not shown). A few ions are also seen in the mass range 40 and 60 amu which would be consistent with Ar^+, Ca^+ or Fe^+ ions, however, the statistics are marginal at this point.

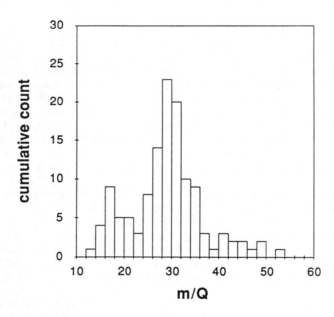

Fig. 1. Sample M/Q distribution of periodic suprathermal ions upstream of the earth's bow shock flowing in the anti-sunward direction for the E/Q range of 80 - 226 keV/Q. The data was collected between 21st August and 20th December 1985.

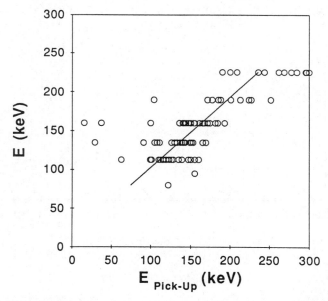

Fig. 2. Kinetic energy of ions in the mass range from 23 to 37 amu flowing in the anti-sunward direction. The kinetic energy is correlated with the maximum pick-up energy perpendicular to the IMF (R = 0.74).

In Figure 2 the energy of the incoming ions is plotted versus the maximum pick-up energy $E_{max} = 4 \cdot m/2 \cdot v_{sw}^2 \cdot \sin^2 \alpha$ perpendicular to the local interplanetary magnetic field. The correlation coefficient is about 0.74. Taking into account the finite resolution in E/Q of 0.1, the kinetic energy of the pick-up ions is well correlated with the maximum pick-up energy. The S/C spin axis was perpendicular to the ecliptic during the period of the observations and the distribution of the IMF directions generally has a maximum in the ecliptic plane. Therefore, preferably those ions on the cycloidal trajectory can be observed which move in the ecliptic plane and thus have the maximum pick-up energy. This means that the ion distribution still reflects the original cycloidal trajectories in the local interplanetary magnetic field and the ion distribution still contains the directional information of the local interplanetary magnetic field. Pitch-angle scattering, as observed for the interstellar He^+ pick-up ions /2/, has not yet become eminent for the ions. Therefore, it is concluded that the ions originate from a region within considerably less than one scattering mean free path length (< 0.2 AU, e.g. Mason et al /11/) upstream of the observer.

The directional distribution of the ions with mass 23 and 37 amu is highly anisotropic with a significant flow in the anti-sunward direction. The flux of these ions shows a distinct period of about 29 days and is peaked around new moon (Figure 3). The occurrence of the ions seems to be constrained to a configuration when the moon is located within ± 40° off the earth-sun line. For this time period the minimum average ion flux is ≈ 1 - 2 ions/cm$^2 \cdot$sr$\cdot$sec$\cdot$keV. No significant flow above the background level was observed in the sunward direction (Figure 3, lower panel). The flux of ions between 10 and 23 amu was analyzed in the same way. For these ions no obvious correlation with the lunar phase is observed, and a sunward flow prevails for several time periods (not shown here). Most probably energetic O^+ escaping from the earth's magnetosphere during magnetically disturbed times is a significant contributor to this distribution (see, e.g. Möbius et al /12/).

IV. DISCUSSION

In the solar wind upstream of the Earth's bow shock an ion population in the mass range 23 to 37 amu has been identified which shows:

• a flow in the anti-sunward direction;
• a correlation of the measured energy with the maximum pick-up energy perpendicular to the interplanetary magnetic field;
• a correlation of the ion flux with the lunar phase (observation around new moon only).

From the first observation the ions can be clearly identified as pick-up ions with their origin at a distance from the observer of considerably less than one mean free scattering length of the ions in the interplanetary magnetic field (< 0.2 AU). Their exclusive appearance at the lunar phase around new moon (± 3 days) leads to the conclusion that the moon itself is the source of these particles.

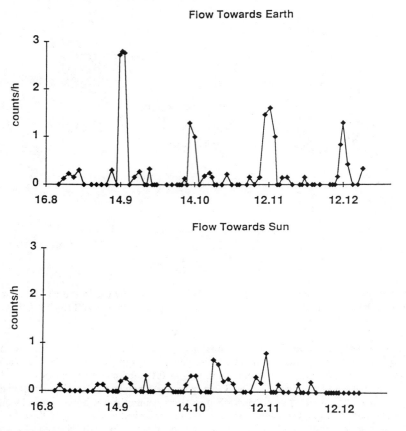

Fig. 3. Rate of suprathermal ions (23 to 37 amu, 80 to 226 keV/q) flowing earth- and sunward between August and December 1985. The dates refer to new moon in 1985.

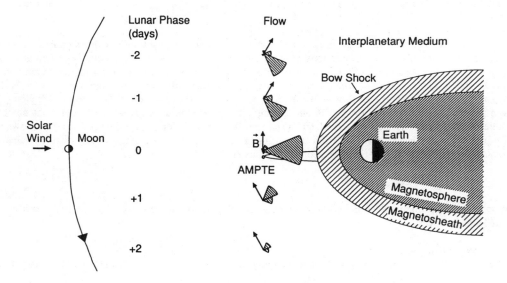

Fig. 4. Flow anisotropy (number of events in the respective sector, max. = 24 cts) of the pick-up ions in relation to the relative position of moon and satellite (23 to 37 amu, 80 to 226 keV/q). The schematic position of the earth and the orbits of the moon and satellite are plotted in the ecliptic plane.

The above conclusion can be further substantiated by a detailed analysis of the ion flow with respect to the actual location of the moon and the direction of the IMF during the observations. If the ions are of lunar origin and picked up by the solar wind, their trajectories are confined within a plane oriented perpendicular to the local interplanetary field (IMF) which includes the moon and the satellite. Therefore, the magnetic field vector should be preferentially perpendicular to the moon-satellite line during the pick-up ion observations. To illustrate the observed ion behaviour in this framework, the measured flow anisotropy has been plotted in Figure 4 in relation to the lunar phase (position of moon, earth and satellite are shown schematically in the ecliptic plane). As predicted above, the pick-up ions move perpendicular to the IMF. For new moon, the ions are observed just in the sun sector (sector 4) of the instrument. Prior to new moon they are detected in the sun sector and the adjacent sector 5, post new moon in the sun sector and sector 3. In addition, the IMF is oriented perpendicualar to the moon-satellite connection. This is reemphasized in Figure 5 where the azimuthal angle ϑ of the IMF is plotted against the lunar phase. In the days before new moon, the IMF direction is either $\vartheta = -130°$ to $-90°$ or after new moon $\vartheta = -90°$ to $-50°$. At new moon, the IMF is centered perpendicular to the earth-sun line. As the observed periodicity of 29 days is close to that of the solar rotation as seen from the earth (27 days), almost all observations were made in the same solar magnetic sector. The data set contains a few cases with the antiparallel magnetic field orientation. As the field turns around by $180°$, the ions should still move perpendicular to it, they just gyrate in the opposite direction (filled circles in Figure 5).

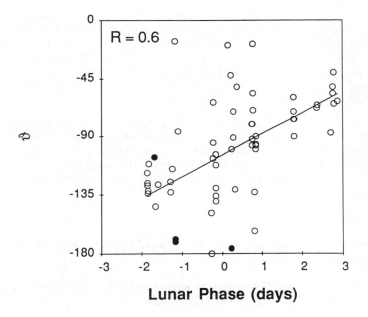

Fig. 5. Compilation of the azimuthal angle ϑ of the interplanetary magnetic field (IMF) in GSE coordinates during the observation of individual pick-up ions for the days around new moon. The magnetic field data samples are averaged over 60 sec intervals. The open circles refer to parallel, the filled circles to antiparallel IMF directions.

Therefore, not only the correct lunar phase but also the direction of the IMF is a necessary observation condition; i.e. the ions reach the satellite indeed on a pick-up trajectory perpendicular to the magnetic field which connects to the moon, and thus the moon can be uniquely identified as their source.

What are the possible mechanisms to inject the ions from the moon into the interplanetary space? Since the energy distribution of the ions is so close to the generic pick-up energy the ions have to be injected into the IMF with very low velocities relative to their source. The likely sources on the moon are a thin layer of lunar atmosphere or erosion processes on the lunar surface.

The lunar atmosphere consists besides other volatiles mainly of argon as a radioactive decay product of ^{40}K, which emanates from the interior of the moon /13, 14, 15/. However, only few of the observed ions are found in this mass regime. Since the kinetic energy of the pick-up ions is directly proportional to their mass, heavy ions such as calcium, argon and perhaps iron may well exceed the maximum E/Q step of the SULEICA instrument for solar wind velocities higher than average (e.g. Ar at $v_{sw} > 550$ km/sec) while others like silicon and aluminium fall clearly within the E/Q stepping range. The latter species fit well into the mass distribution from 23 to 37 amu.

Possible erosion processes are the sputtering mechanism by impinging solar wind ions or vapourisation due to micrometeorite impact. Sputtering by solar wind ions is a process which results in a relatively high yield of particles with sufficiently high energies to escape from the moon. The average kinetic energy of the sputtered particles spans from 5 and 10 eV /16/, which is well above the lunar escape velocity (about 1 eV for Al or Si). Therefore, nearly all sputtered atoms will move radially away from the lunar surface and will not contribute to the lunar atmosphere. About 1 particle per 100 impinging solar wind protons is produced by sputtering. Of these, at the most a few % escape as ions, the rest as neutrals (estimate from laboratory measurements, summarized in /16/). From recent laboratory measurements, Elphic et al. /17/ quote a sputtering yield of 10^{-5} ions per 1 keV proton from a surface with an elemental composition similar to the lunar surface.

These processes should reflect in some way the relative elemental abundance of the lunar surface. The lunar surface consists of oxygen ($\approx 60\%$), magnesium (4-5%), silicon (16-17%), aluminium (6-10%), calcium (4-6%) and iron (2-4%) (atomic %, /18/). This pattern of major elements on the lunar surface corresponds well to our observed ion mass distribution, with the exception of O^+ ions which seem to be underabundant in the observed ion distribution. Thus additional fractionation may play a role in the injection process /17/.

During their escape from the moon atoms are ionized by interaction with the solar wind and photoionization by solar UV radiation. The photoionization rate of silicon is about 50 times and the rate of aluminium is even about 10^4 times the rate of oxygen /19, 20/. This extreme difference in the ionisation rates may give rise to the observed filter effect, i.e., more Al or Si than O pick-up ions would be observed. A detailed discussion of these processes goes beyond the scope of this paper and will be resumed within a separate investigation.

ACKNOWLEDGEMENTS

We thank the many individuals at the Max-Planck-Institut for Extraterrestrial Physics, the Technical University of Braunschweig and the University of Maryland who contributed to the success of the SULEICA Instrument and the AMPTE/IRM satellite. The authors wish to thank Rainer Behrisch for valuable discussions and Miss Bernie Tucker for her technical assistance during the screening of the SULEICA PHA data. The work was supported in part by the German Bundesministerium für Forschung und Technologie and for one of the authors (E.M.) in part by NASA grant NAGW 2579.

REFERENCES

1. Möbius E, B.Klecker, D.Hovestadt, M.Scholer, Interaction of interstellar Pick-Up ions with the solar wind, Astrophysics and Space Science, 144, p.487, 1988

2. Möbius E., D.Hovestadt, B.Klecker, M.Scholer, G.Gloeckler and F.M.Ipavich, Direct observation of He^+ pick-up ions of interstellar origin in the solar wind, Nature, Vol 318, p.426, 1985

3. Gloeckler G., D.Hovestadt D., F.M. Ipavich, M.Scholer, B.Klecker and A.B.Galvin, Cometary pick-up ions observed near Giacobini-Zinner, Geophys. Res. Lett., Vol 13, NO. 3, p 251, 1986

4. Ipavich F.M., A.B. Galvin, G.Gloeckler, D.Hovestadt, B.Klecker and M.Scholer, Comet Giacobini-Zinner: In situ observations of energetic heavy ions, Science 232, p.366, 1986

5. Mikhalov, J.D ., and A. Barnes, Evidence for the acceleration of ionospheric O^+ in the magnetosheath of Venus, Geophys. Res. Lett., 8, 1277, 1981.

6. Fahr H.J. und H.W. Ripken, Dust-Plasma-Gas Interactions in the Helioshere, in: Properties and Interactions of Interplanetary Dust, R.H. Giese and P.Lamy, Ed., Reidel Publishing Co., p. 305, 1984

7. Möbius E., G.Gloeckler, D.Hovestadt, F.M.Ipavich, B.Klecker, M.Scholer, H.Arbinger, H.Höfner, E.Künneth, P.Laverenz, A.Luhn, E.O.Tums and H.Waldleben, The time-of-flight spectrometer SULEICA for ions of the energy range 5-270 keV/charge on the AMPTE/IRM, IEEE Trans.Geos. and Remote Sensing, GE-23, p. 274, 1985

8. Paschmann G., H.Loidl, P.Obermayer, M.Ertl, R.Laborenz, N.Scopke, W.Baumjohann, C.W.Carlson and D.W. Curtis, The plasma instrument for AMPTE/IRM, IEEE Trans.Geos. and Remote Sensing, GE-23, p. 262, 1985

9. Lühr H., N.Klöcker, W.Oelschlägel, B.Häusler and M.Acuna, The IRM fluxgate magnetometer, IEEE Trans.Geos. and Remote Sensing, GE-23, p. 259, 1985

10. Klecker B., E.Möbius, D.Hovestadt and M.Scholer, G.Gloeckler and F.M. Ipavich, Discovery of energetic molecular ions (NO^+ and O_2^+) in the storm time ring current, Geophys. Res. Lett., Vol 13, NO 7, p.632, 1986

11. Mason G.M., G. Gloeckler and D. Hovestadt, Temporal variations of nucleonic abundances in solar flare energetic particle events. I. Well connected events, Astrophys. J. 267, p.844, 1983

12. Möbius E., D.Hovestadt, B.Klecker, M.Scholer, F.M. Ipavich, C.W.Carlson, R.P.Lin, A burst of energetic O^+ ions during an upstream particle event, Geophys. Res. Lett., Vol. 13, No. 13, p. 1372, 1986

13. Siscoe G.L. and N.R. Mukherjee, Upperlimits on the lunar atmosphere determined from solar wind measurements, J.Geophys. Res., Vol 77, NO 31, p.6042, 1972

14. Manka R.H. and F.C.Michel, Lunar ion flux and energy, in: Photon and particle interactions with surfaces in space, R.J.L.Grard, Ed., Reidel Publishing Co, p. 429, 1973

15. Hodges R.R., J.H.Hoffman and F.S.Johnson, The lunar atmosphere, Icarus 21, p. 415, 1974

16. Behrisch R., Ed., Sputtering by Particle Bombardment I, Topics in Apllied Physics, Vol. 47, Springer-Verlag Berlin Heidelberg New York, 1981

17. Elphic R.C., H.O. Funsten, B.L. Barraclough, D.J. McComas, M.T. Paffett, D.T. Vaniman and G. Heiken, Lunar surface composition by solar wind-induced secondary ion mass spectrometry, submitted to Geophys. Res. Lett., Vol 18, p. 2165, 1991

18. Landolt-Börnstein, Astronomie and Astrophysik, Bd. 2, Springer-Verlag Berlin Heidelberg New York, p.258, 1981

19. von Steiger R. and J. Geiss, Supply of fractionated gases to the corona, Astron. Astrophys. 225, p. 222, 1989

20. Kohl J.L. and W.H. Parkinson, Measurement of the neutral-aluminum photoionization cross-section and parameters of the $3p^2P^0$-$3s3p^2$ $^2S_{1/2}$ autoionization doublet, Astrohys. J. 184, p.641, 1973

TECHNIQUES FOR THE REMOTE SENSING OF SPACE PLASMA IN THE HELIOSPHERE VIA ENERGETIC NEUTRAL ATOMS: A REVIEW

K. C. Hsieh,* C. C. Curtis,* C. Y. Fan* and M. A. Gruntman**

* *Department of Physics, University of Arizona, Tucson, AZ 85721, U.S.A.*
** *Space Sciences Center, University of Southern California, Los Angeles, CA 90089, U.S.A.*

ABSTRACT

Remote sensing of space plasma via its neutral component was ushered in by ground observations in 1950. Direct detection of energetic neutral atoms (ENA) at rocket altitudes was first reported in 1969. It is now recognized by the community that charge-exchange between energetic ions of space plasma and ambient neutral atoms is an important process and remote sensing space plasma via ENA should be implemented. In the past decade a host of techniques for direct ENA detection has been proposed and developed, but none of which is presently operating in space. We review some of the techniques for direct detection of ENA between 100 eV and 300 keV for diverse regions ranging from the heliospheric boundary to auroral zones, where the solar wind plays a crucial role.

INTRODUCTION

The presence of energetic protons in solar-terrestrial phenomena was first seen in an aurora in 1950. Figure 1 is the Doppler-shifted Hα (6563Å) emission spectrum /1/, from which Meinel deduced that protons of energies up to 57 keV were entering the upper atmosphere. Subsequent observations and theoretical works led Fan and Schulte /2/ to estimate the incident proton flux of energies > 100 keV. Bernstein et al. /3/ made the first in situ measurement of the neutral H flux (Figure 2) during a breakup aurora at ~250 km altitude in 1968. Using an identical instrument, but at ~800 km altitude and lower magnetic latitude, Wax et al. /4/ reported a high flux of neutral H at energies

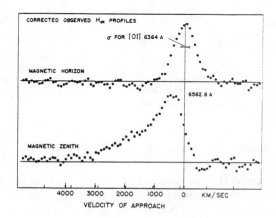

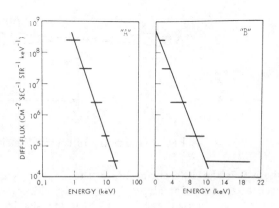

Fig. 1 First evidence of ENA in solar-terrestrial phenomena: Doppler-shifted Hα in magnetic zenith (lower) compared to the Hα line profile in the magnetic horizon (upper) /1/.

Fig. 2 First measured ENA spectrum at ~250km altitude. Data fitted to a power law in energy $\propto E^{-3.2}$ (left) and to an exponential law in energy with an e^- folding of ~ 1 keV (right) /2/.

> 1 keV. Such a flux at low latitude was not well understood. In 1972, Moritz /5/ reported the detection of protons of energies > 250 keV at equatorial low altitudes (< 600 km) and suggested that they were protons from the outer radiation belt that had become neutralized by charge exchange with exospheric H atoms, thereby crossing the magnetic field lines to reach the low altitudes. In 1959, Dessler and Parker /6/ had already suggested charge exchange between ring current ions and geocoronal H atoms as the mechanism for dissipating ring current energy. Subsequent particle detections on satellites /7, 8/ and low-latitude aurora observations from the ground /9-12/ lent support to the role of energetic neutral atoms (ENA) in the dynamics of solar-terrestrial phenomena. While US proposals for direct detection of ENA as a space plasma diagnostic were rejected, ENA instrumentation went ahead in the USSR in early 1980's /13-16/.

The interest in ENA was boosted when uncharged energetic particles emanating from Jupiter /17/ and Saturn /18/ were detected by Voyager 1 and convincingly analyzed in terms of ENA /19/. In 1985, the first study of the ring current using ENA was reported /20/. Using data from an ion spectrometer on ISEE-1 located outside of the ring current, where no ion flux was expected, Roelof /21/ constructed the first image of the ring current in ENA. More recently the possibility of using ENA to study the dynamics and structures of the heliosphere, such as transient shocks, corotating interaction regions, the termination shock, and anomalous cosmic rays, has also been explored with modeling /22-25/.

We review some of the techniques for the direct detection of ENA that have been developed and may be used for future missions. We shall complement the review by McEntire and Mitchell /26/.

PRINCIPLE OF ENA REMOTE SENSING AND DESIGN CONSIDERATIONS

Singly-charged ions of an energetic ion population that co-exists with a neutral gas, may be neutralized by charge exchange with the neutral atoms and become ENA. Freed from the magnetic fields, they travel in ballistic trajectories to remote observers wishing to sample the parent ion population in often inaccessible regions. Figure 3 illustrates the geometry of remote sensing an ion population via ENA. For a pixel of area A, the counting rate of ENA of the ith species of energy E coming from direction $\vec{s}$, along the line of sight, is

$$C_i(\vec{s}, E) = \int \int r^{-2} j_i(\vec{s}, \vec{E}, \vec{x}) \sum_k [\sigma_{ik}(E) n_k(\vec{x})] \exp[-D(s', E)] d\tau \, dA \qquad (1)$$

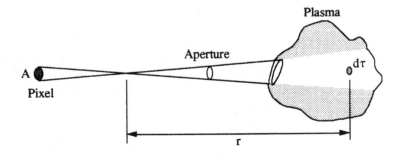

Fig. 3 Geometry of remote sensing an ion population via ENA.

where j_i is the differential flux of the singly charged ions of the ith species having the appropriate pitch angle to point along $\vec{s}$, σ_{ik} is the charge-exchange cross-section for ions of the ith species incident on a neutral gas target of the kth species, and n_k is the number density of the kth species of the neutral gas. The exponential term represents the extinction of ENA due to re-ionization by electrons, ions and photons along the way /21/. To obtain j_i via the measured set $\{C_i\}$, we must know $\{\sigma_{ik}\}$ and $\{n_k\}$. The cross sections are readily available, but $\{n_k\}$ and D must be determined by other means, e.g. in situ direct measurements or remotely by optical measurements. We concentrate on the techniques for obtaining $\{C_i\}$ from the ENA fluxes.

Three general features of ENA fluxes set the requirements for all ENA instruments. 1) ENA flux is low relative to j_i due to the magnitudes of σ_{ik}. Examples of estimated ENA fluxes are listed in Table 1. 2) ENA fluxes decrease rapidly with increasing energy due to the typical energy dependence of j_i and σ_{ik}. The precipitous drop in σ for both H^+ on H and O^+ on H at $E \gtrsim 300$ keV greatly diminishes the ENA flux at $E > 300$ keV. And 3) ENA fluxes are accompanied by intense H Lyα fluxes due to the resonant scattering of the bright solar radiation by the dominant member of $\{n_k\}$. Consequently, all ENA instruments must be designed with 1) large geometrical factor, 2) high ENA detection efficiency, 3) low energy threshold, 4) efficient UV and charged particle rejections, and 5) adequate resolutions for energy, mass and direction.

Table 1 Different Fluxes of Energetic Neutral Atoms

Source Region	Source Ions	Neutral Target	Estimated Flux*	
Heliosheath	Re-ordered solar wind	LISM***	10^2 at 300 eV	/22/
Heliosheath	Anomalous cosmic rays	LISM	$10^{-2}(>10$ keV$)$	/23/
Interplanetary	Energetic solar particles	Penetrating LISM	$10^0(>10$ keV$)$	/23/
Interplanetary	Co-rotating shock ions	Penetrating LISM	$10^{-2}(>10$ keV$)$	/23/
Interplanetary	Solar wind	Penetrating LISM	10^7 at 1 keV	/28/
Magnetosphere	Ring current	Exospheric atoms	$> 10^2$ at 10 keV	
Thermosphere	Precipitating ions	Thermospheric atoms	10^4 at 10 keV**	

*Differential flux in $(\text{cm}^2\text{sr s keV})^{-1}$ and integral flux in $(\text{cm}^2\text{sr s})^{-1}$.

**Based on /11/ for Earth.   ***Local intersteller medium.

In principle, one could construct an ENA instrument by adding a charged-particle repeller and an UV attenuator to an existing charge-particle analyzer designed for ions of similar masses and energies. No particle analyzer, however, can efficiently cover the entire energy interval from ~0.1 to 300 keV. We set the upper limit at 300 keV, because the ENA fluxes at energies > 300 keV are minimal and it is more difficult to repel ions of higher energies. The typical threshold of solid-state detectors divides this energy window into two regimes: ~0.1 to ~10 keV and ~10 to 300 keV.

ENA between ~10 - 300 keV can be detected directly. An appropriate direct-detection analyzer is the combination of time-of-flight (TOF) and energy (E) analyzers /e.g. 27/. For an ENA of $E \lesssim 20$ keV, only a TOF is registered to give the speed of the particle. For an ENA of higher energy, its E is also measured by a solid-state detector (efficiency ~1) to give its mass and energy. The sensitivity for both cases is $\sim 2 \times 10^{-2}$ G, where G is the geometrical factor. A thin foil attenuates UV and generates signal electrons during ENA passage. ENA between ~0.1 - ~10 keV suffer significant losses in foils thick enough to attenuate UV effectively. These ENA are ionized, selected electrostatically or magnetically, post-accelerated, and then detected. The most efficient means of ionization is electron-stripping by a thin foil (efficiency $\sim 3\%$). Bending the trajectories of the ionized ENA in an electrostatic or magnetic analyzer shields the detector from incident photons, thus lowering the threshold by using a thinner foil.

While ENA spectrometry allows us to sample the mass and energy distributions of a distant plasma, ENA imaging gives us a global view of the structures and dynamics of an extended plasma. Imaging can be done by a 1-dimensional imager on a spinning spacecraft or a 2-dimensional imager on a pointing platform. To view an object spanning more than 2π steradian, e.g., mapping a planetary magnetosphere or the heliosphere from within, a spinner is preferred. To view a more compact object, e.g., mapping a planetary magnetosphere from outside, a pointer is more efficient.

SAMPLE DESIGNS

We should first mention the instruments for in situ measurement of the interstellar gas (~ 2 eV), which is not ENA, but highly relevant to the production of ENA in the heliosphere. An instrument flown to 300 km altitude /28/ used a high-speed (up to 4.5×10^4 rpm) slotted-disk velocity selector which is highly efficient in photon rejection (10^{-5}), but mechanical and geometrical factor constraints do not favor its further use in space. A more recent attempt, carried out by the GAS instrument on ULYSSES /29/, detects the neutrals by the secondary electrons or ions emitted upon particle impact on a freshly deposited LiF layer. This technique lacks mass and energy identification. LiF effectively suppresses the noise due to H Lyα, but its response to particles depends on its purity, hence difficult to calibrate. Improvement of this technique for one-dimensional imaging has been demonstrated /30/ and for mass identification has been proposed /31/.

The first ENA instrument in space (Figure 4) was designed for H of energies $1 - 20$ keV /3/. Its stepping electrostatic analyzer provided no mass identification. Recent improvements in the time, directional, energy, and mass resolutions in space plasma analyzers /32/ should directly benefit the designs for ENA instruments in this energy window. Figure 5 shows such a development by McComas et al. /33/. The combination of electrostatic and TOF analyzers gives both mass and energy identification. The stripping foil is $\sim 1\mu g/cm^2$ of carbon.

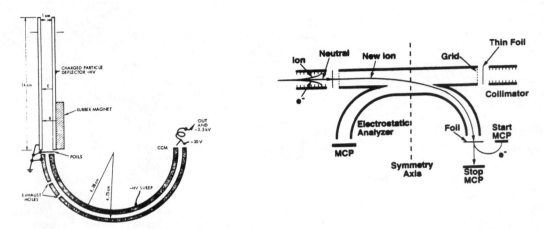

Fig. 4 The first space-borne ENA instrument /2/. Stepping electrostatic analyzer gave no mass identification of the ENA.

Fig. 5 Modern version of an ENA instrument using an electrostatic analyzer in conjunction with a TOF analyzer /33/.

Such a thin foil is also used as the "start" signal generator in the Neutral Solar Wind instrument (Figure 6) to be launched on RELIKT-2 /15/. The UV attenuation is handled by a "nuclear-track filter" /34/ having submicron-diameter channels, too small for efficient UV propagation, but large enough to pass ENA incident within a 5° cone to the TOF analyzer. Its small ENA throughput (10^{-2}) is compensated by its high Lyα attenuation (10^{-8}). The signal-to-noise ratio is further enhanced by the tripple-coincidence requirement ($D_1 D_2 D_3$). The ENA threshold is 0.5 keV.

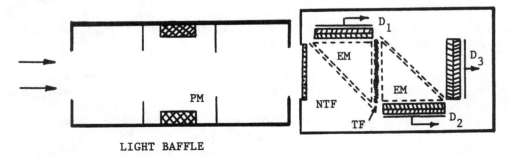

LIGHT BAFFLE

Fig. 6 An ENA instrument with a threshold of 500 eV without requiring ionization of the ENA /15/. The B field of the magnets (PM) and the E field between the baffles sweep away incident ions. A nuclear-track filter (NTF) dims the UV. Electrons from the thin foil (TF) give the start signals. Signals from D_1 (or D_2) and D_3 define the TOF.

For ENA of $E > 10$ keV there are two representative imaging spectrometer designs: a neutral atom camera INCA (Ion Neutral Camera) /26/, and an imaging spectrometer for ENA (ISENA) /35,36/. Both designs measure TOF/E. INCA (Figure 7) evolved from the ion analyzer MEPA on AMPTE CCE and will fly on CASSINI. ISENA (Figure 8), still under study, incorporates three different ideas: 1) a tandem TOF analyzer to provide particle identification at energies below the threshold of the solid-state detector, 2) 1-dimensional coded apertures in conjunction with a fan-shaped collimator to construct 2-dimensional images, and 3) one position-sensing detector for image construction to avoid effects of ENA scattering in the foil. INCA is a spinner and ISENA a pointer. INCA and ISENA combine ENA spectrometry with ENA imaging. The extended TOF flight paths (~10-cm) for mass resolution reduce the geometrical factor of the instruments, and, for INCA, worsen the angular resolution due to scattering in the foil, since INCA relies on the locations of ENA strikes on both the first foil and the end detector to do a two-point trajectory extrapolation. Simplicity is gained by having an ENA imager without mass resolution and an ENA spectrometer with only coarse directional resolution. McEntire and Mitchell /26/ discussed two such imagers, one using a pinhole (their Figure 5) and the other a slit (their Figure 6). The first still suffers from ambiguities in angular resolution due to scattering in the foil. The second does not take advantage of the more abundant ENA fluxes at energies < 20 keV. Alternative designs are needed.

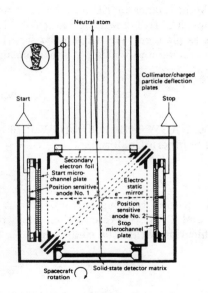

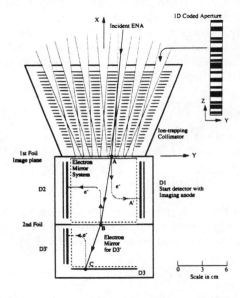

Fig. 7 Schematic of INCA (Ion Neutral Camera), Figure 7 of /26/. The FOV in the plane normal to the page is 120°.

Fig. 8 Schematic of ISENA (Imaging Spectrometer for ENA), Figure 2 of /36/. The FOV is 45° × 45°

FURTHER DEVELOPMENT

The ENA instrument designs reviewed above share many components, most of which, such as parallel-plate electrostatic collimator, carbon foils, TOF/E analyzers, electrostatic analyzers, and position-sensing detectors, have excellent flight performances as elements in charged-particle analyzers on space missions. Their application to ENA instruments requires caution, since ENA fluxes are so much lower (see Table 1) than the ion fluxes in the same mass and energy windows. Problems that may be ignored in the case of charged-particles cannot be overlooked. For example, noise contributions from scattering in the collimator and production of ENA from charged-particles going through any significant lengths of gas surrounding the spacecraft and along the collimator would be of concern to ENA experiments, as is the ubiquitous intense UV radiation. Testing better designs and materials is very much needed and has begun /e.g. 37/. Two examples are given.

Foils: Thin foils used as both UV filter and TOF signal generator face the conflicting needs for high UV attenuation and low particle stopping power. Recent tests found that layered foils such as

Si/C are more effective than pure C foils of the same areal density (in $\mu g/cm^2$) in attenuating Lyα photons /38/. To compare the effectiveness of foils for high UV attenuation, low stopping power and high secondary electron yield, more testing must be done systematically, as in /39,40/.

Coded aperture (CA): Substitution of a coded (patterned) array of small apertures for a single opening can increase the throughput of a pin-hole camera, but the resulting overlapping images must be deconvolved to reconstruct the object. Two-dimensional CA have been used in hard X-ray and γ-ray astronomy. Using CA for ENA imagers was proposed more recently /e.g. 16/. Given the need of charged-particle repelling collimators, geometries using 1-dimensional codes seem preferable /35/. Some codes are more sensitive to detector noise and counting statistics, and some codes are more suitable for viewing extended objects than for cluster of point sources. Optimization of codes for particular applications is underway.

CONCLUSION

Our brief review is more like an introduction, since much of the subject is yet to be realized. There is still need for refinement of designs and testing of concepts. There is the chance that INCA will be on CASSINI to Saturn, STOF (Superthermal TOF) on SOHO will be modified to explore the ENA flux from solar events, and IMI (Inner Magnetospheric Imager in EUV and ENA) will be in the queue for a NASA new start before 1998. We hope RELIKT-2 will be launched with its neutral solar wind and magnetospheric ENA instruments as planned. We will not be surprised if even more innovative techniques emerge as the momentum for remote sensing of space plasma via ENA continues to grow.

ACKNOWLEDGEMENT

This work is in part supported by NASA grants NAG5-260, NAGW 1419, and NAGW 1676 to the University of Arizona and NAG2-146 to the University of Southern California.

ADDENDUM

Recently we found that another group has been remotely sensing the ring current in ENA with passively cooled solid-state detectors on satellites SEEP and CRRES. We cite their most recent reference /41/ and hope this review will open up communication within the community.

REFERENCES

1. A. B. Meinel, Doppler-shifted auroral hydrogen emission, <u>Astrophys. J.</u> 113, 50 (1951)

2. C. Y. Fan and D. H. Schulte, Variations in the auroral spectrum, <u>Astrophys. J.</u> 120, 563 (1954)

3. W. Bernstein, G. T. Inouye, N. L. Sanders and R. L. Wax, Measurements of precipitated 1-20 keV protons and electrons during a breakup aurora, <u>J. Geophys. Res.</u> 74, 3601 (1969)

4. R. L. Wax, W. R. Simpson and W. Bernstein, Large fluxes of 1-keV atomic hydrogen at 800 km, <u>J. Geophys. Res.</u> 75, 6390 (1970)

5. J. Moritz, Energetic protons at low equatorial altitudes: a newly discovered radiation belt phenomenon and its explanation, <u>Z. Geophys.</u> 38, 701 (1972)

6. A. J. Dessler and E. N. Parker, Hydromagnetic theory of geomagnetic storm, <u>J. Geophys. Res.</u> 64, 2239 (1959)

7. D. Hovestadt, B. Häusler and M. Scholer, Observation of energetic particles at very low altitudes near the geomagnetic equator, <u>Phys. Rev. Lett.</u> 28, 1340 (1972)

8. P. F. Mizera and J. B. Black, Observation of ring current protons at low altitudes, J. Geophys. Res. 78, 1058 (1973)

9. B. A. Tinsley, Neutral atom precipitation - a review, J. Atmos. Terr. Phys. 43, 617 (1981)

10. B. A. Tinsley, Y. Sahai, M. A. Biondi and J. W. Meriwether, Energetic particle precipitation during magnetic storms and relationship to equatorial thermospheric heating, J. Geophys. Res. 93, #A1, 270 (1988)

11. M. Ishimoto and M. R. Torr, The role of He$^+$ precipitation in a mid-latitude aurora, J. Geophys. Res. 92, 3284 (1987)

12. M. Ishimoto, M. R. Torr, P. G. Richards and D. G. Torr, The role of energetic O$^+$ precipitation in a mid-latitude aurora, J. Geophys. Res. 91, 5793 (1986)

13. M. A. Gruntman and V. A. Morozov, H atom detection and energy analysis by use of thin foils and TOF technique, J. Phys. E: Sci. Instrum., 15, 1356 (1982)

14. M. A. Gruntman and V. B. Leonas, Neutral solar wind: possibility of experimental study, Preprint 825, Space Research Institute (IKI), Academy of Sciences, Moscow, 1983.

15. M. A. Gruntman, S. Grzedzielski and V. B. Leonas, Neutral solar wind experiment, in: Physics of the Outer Heliosphere, ed. S. Grzedzielski and D. E. Page, Pergamon Press, Warsaw, Poland, 1989, p.355-358.

16. M. A. Gruntman and V. B. Leonas, Experimental opportunity of planetary magnetosphere imaging in energetic neutral particles, Preprint 1181, Space Research Institute (IKI), Academy of Sciences, Moscow, 1986.

17. E. Kirsch, S. M. Krimigis, J. W. Kohl and E. P. Keath, Upper limits for x-ray and energetic neutral particle emission from Jupiter: Voyager 1 results, Phys. Rev. Lett. 8, 169 (1981a)

18. E. Kirsch, S. M. Krimigis, W.-H. Ip and G. Gloeckler, X-ray and energetic neutral particle emission from Saturn's magnetosphere, Nature 292, 718 (1981a)

19. E. C. Roelof, D. G. Mitchell and D. J. Williams, Energetic neutral atoms (E $\sim$ 50 keV) from the ring current: IMP 7/8 and ISEE-1, J. Geophys. Res. 90, 10991 (1985)

20. A. F. Cheng, Energetic neutral particles from Jupiter and Saturn, J. Geophys. Res. 91, 4524 (1986)

21. E. C. Roelof, Energetic neutral atom image of a storm-time ring current, Geophys. Res. Lett. 14, 652 (1987)

22. M. A. Gruntman, Anisotropy of the Energetic Neutral Atom Flux in the Heliosphere, Plan. Space Sci., in press (1992)

23. K. C. Hsieh, K. L. Shih, J. R. Jokipii, and S. Grzedzielski, Probing the heliosphere with energetic hydrogen atoms, Astrophys. J., in press (1992)

24. E. C. Roelof, Imaging heliospheric shocks using energetic neutral atoms, this issue

25. K. C. Hsieh, K. L. Shih, J. R. Jokipii and M. A. Gruntman, Sensing the solar wind termination shock from earth's orbit, this issue

26. R. W. McEntire and D. G. Mitchell, Instrumentation for global magnetospheric imaging via energetic neutral atoms, in: Solar System Plasma Physics, ed. J. H. Waite Jr., J. L. Burch and R. L. Moore, AGU, Washington DC 1989, p. 69-80.

27. B. Wilken and W. Stüdemann, A compact time-of-flight mass spectrometer with electrostatic mirrors, Nucl. Instrum. Meth. 222, 587 (1984)

28. J. Moore, Jr. and C. B. Opal, A slotted disk velocity selector for the detection of energetic atoms above the atmosphere, Space Sci. Instrum. 1, 377 (1975)

29. M. Witte, H. Rosenbauer, E. Keppler, H. Fahr, P. Hemmerich, H. Lauche, A. Loidl and R. Zwick, The interstellar neutral-gas experiment on ULYSSES, Astron. Astrophys., in press (1991)

30. M. A. Gruntman, MASTIF: Mass analysis of secondaries by time-of-flight technique: new approach to secondary ion mass spectrometry, Rev. Sci. Instrum., 60, 3188 (1989)

31. M. A. Gruntman, In-situ measurement of the composition (hydrogen, deuterium, and oxygen atoms) of interstellar gas, Report #102M, Space Sciences Center, University of Southern California, February 1991.

32. D. T. Young, Space plasma mass spectroscopy below 60 keV, in: Solar System Plasma Physics, ed. J. H. Waite Jr., Burch, J. L., and Moore, R. L., AGU, Washington, DC 1989, p. 143-157.

33. D. J. McComas, B. L. Barraclough, R. C. Elphic, H. O. Funsten, III and M. F. Thomsen, Magnetospheric imaging with low-energy neutral atoms, Proc. Nat. Acad. Sci. U.S.A. 88, 9598 (1991)

34. M. A. Gruntman, Submicron structures: promising filters in EUV - a review, Proceedings SPIE Internl. Symp., 1549, 385 (1991).

35. C. C. Curtis and K. C. Hsieh, Remote sensing of planetary magnetospheres: imaging via energetic neutral atoms, in: Solar System Plasma Physics, ed. J. H. Waite Jr., J. L. Burch and R. L. Moore, AGU, Washington DC 1989, p. 247-251.

36. K. C. Hsieh and C. C. Curtis, Remote sensing planetary magnetospheres: mass and energy analysis of energetic neutral atoms, in: Solar System Plasma Physics, ed. J. H. Waite Jr., J. L. Burch and R. L. Moore, AGU, Washington DC 1989, p. 159-164.

37. E. P. Keath, G. B. Andrews, A. F. Cheng, S. M. Krimigis, B. H. Mauk, D. G. Mitchell and D. J. Williams, Instrumentation for energetic neutral atom imaging of magnetosphere, in: Solar System Plasma Physics, ed. J. H. Waite Jr., J. L. Burch and R. L. Moore, AGU, Washington DC 1989, p. 165-170.

38. K. C. Hsieh, B. R. Sandel, V. A. Drake and R. S. King, H Lyman a transmittance of thin C and Si/C foils for keV particle detectors, Nucl. Instrum. Meth. B61, 187 (1991)

39. M. A. Gruntman, A. A. Kozochkina, and V. B. Leonas, Multielectron secondary emission from thin foils bombarded by accelerated beams of atoms, JETP Lett., 51, 22 (1990)

40. M. Rubel, B. Emmoth, H. Bergsaker, M. A. Gruntman, and V. Kh. Liechtenstein, Ion beam analysis of thin carbon foils for particle diagnostics in the interplanetary space, Nucl. Instrum. Meth., B47, 202 (1990)

41. H. D. Voss, E. Herzberg, A. G. Ghielmetti, S. J. Battel, K. L. Appert, B. R. Higgins, D. O. Murray, and R. R. Vondrak, The medium energy ion mass and neutral atom spectrometer (ONR-307-8-3), J. Spacecraft & Rockets, in press (1992)

SENSING THE SOLAR-WIND TERMINATION SHOCK FROM EARTH'S ORBIT

K. C. Hsieh,* K. L. Shih,* J. R. Jokipii** and M. A. Gruntman***

* *Department of Physics, University of Arizona, Tucson, AZ 85721, U.S.A.*
** *Department of Planetary Science, University of Arizona, Tucson, AZ 85721, U.S.A.*
*** *Space Sciences Center, University of Southern California, Los Angeles, CA 90089, U.S.A.*

ABSTRACT

The solar-wind termination shock is inaccessible for repeated in situ investigation. We examine, therefore, the possibility of remote sensing the entire heliopause from Earth's orbit using the energetic neutral atoms (ENA) produced by charge exchange between energetic ions and the neutral atoms of the interstellar medium at and beyond the termination shock. We estimate the ENA fluxes at Earth's orbit coming from the thermalized solar-wind ions and the shock-accelerated anomalous cosmic rays (ACR) at the heliospheric boundary.

INTRODUCTION

Whether the crossing of the heliosphere boundary is discernable is an urgent question for Voyagers 1 and 2 and Pioneers 10 and 11. Even if these spacecraft detected the interface, they would provide only four single point measurements of a vast structure. Independent of the wide range of estimates on the location of the interface /1-5/, it is clear that the boundary of the heliosphere is inaccessible for repeated in situ investigations. We seek, therefore, ways to study that remote region from a more accessible location, i.e. Earth's orbit.

To sense the heliosphere remotely, the observer must rely on emissaries with impeccable credentials, i.e. they must originate specifically from the boundary and can reach the observer with least diversions along the way. Such requirements quickly rule out all charged particles and any resonant scattering of solar radiations. Even the 2-3 kHz radiation coming from plasma oscillations at the termination shock are likely to be damped or deflected by disturbances in the solar wind. This leaves the energetic neutral atoms (ENA) produced at and just beyond the interface the sole emissary /6-8/.

At and beyond the termination shock, solar-wind ions, including pick-up ions originating from penetrating interstellar neutrals, are thermalized and accelerated by the shock. Solar modulation prevents any of these ions of energies < 10 MeV from re-entering the heliosphere. When these ions exchange charges with the neutral atoms of the local interstellar medium (LISM), ENA are produced in that distinct region defined by the termination shock. Because charge exchange has negligible momentum and energy transfer, the resulting ENA spectra reflect the spectra of the shock-processed ions, modified only by the energy-dependence of the well known charge-exchange cross-sections /9/. Because the shock-processed ions are randomized, there is a sun-ward component of the ENA flux. Unaffected by the magnetic field along their paths, these ENA serve as emissaries from the termination shock to the inner heliosphere, including Earth's orbit.

RESULTS FROM MODELING

In the region between the termination shock and the heliopause, the interaction between the thermalized solar-wind ions and the neutral atoms of LISM produces a low-energy component of ENA. Curve A in Figure 1 is the spectrum of such a component for a model having a temperature of 10^6 K, an ion density 4 times that of the solar wind upstream from the shock (5×10^{-3} cm^{-3}) and a thickness of 25 AU. Some of this ion population may be accelerated by the termination shock to become the unmodulated anomalous cosmic rays (UACR) /3/. The interaction between UACR and the LISM neutrals produces a more energetic ENA flux. A spectrum using the ACR model of Jokipii /3/, where 90° is solar equator and 0° is solar north pole, is shown as curve B in Figure 1.

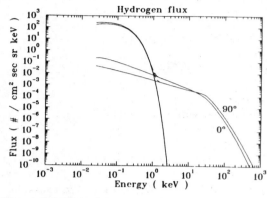

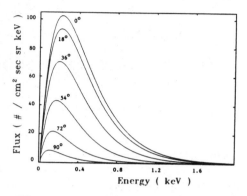

Fig. 1 Spectra of neutral H atoms of termination-shock origin. A: lower energy component is from thermalized solar wind; B: the higher energy one is from UACR /3/.

Fig. 2 Angular distribution of H spectra from thermalized solar wind based on the shock model of Baranov/2/. Angles are measured from the apex.

For a more detailed look at the production of ENA of the lower energy group, we take Baranov's two-shock model of the solar-wind/LISM interaction /2/. The ENA flux coming from the thermalized ions is highly anisotropic in both flux level and spectral shape. An example of this anisotropy is shown in Figure 2, where energy spectra from 6 different look directions, with respect to the apex of the heliosphere, are labeled /7/. For the higher energy ENA coming from UACR, the rotational-symmetric ACR model /3/ produces a rotational-symmetric ENA distribution shown in (> 10keV) Figure 3, relatively flat and little azimuthel variation. The maximum flux of 8×10^{-3} (cm^2 sr s)$^{-1}$ occurs near $\theta = 30°$ and 150° and the step size is 5×10^{-5} (cm^2 sr s)$^{-1}$.

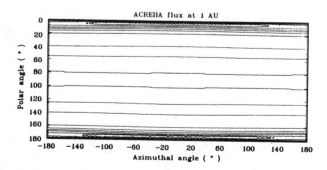

Fig. 3 Isoflux contours of H (>10 keV) coming from UACR seen at orbit of Earth.

The above figures show that ENA production at and beyond the termination shock, are model-dependent; especially at low energies, it is highly sensitive to the interstellar-plasma and solar-wind parameters. Measurements of ENA at Earth's orbit should give us information concerning the outer most part of the heliosphere. The task of detecting these ENA may be complicated by the much more intense fluxes of background particles and radiation, i.e. energetic interplanetary ions, cosmic rays, UV photons, and other interplanetary ENA fluxes. Leaving the problems of ENA detection to reference 9, we compare below the other expected interplanetary ENA fluxes of energies >10 keV at Earth's orbit with that of UACR origin /8/. The highest interplanetary ENA fluxes should come from the occasional energetic solar particle (ESP) events. The maximum ENA flux coming from a corotating interaction region (CIR) appear comparable to those from the termination shock (Figure 1). ENA flux coming from quiet-time interplanetary ions is negligible.

We examine the spatial distributions of the aforementioned interplanetary ENA fluxes (> 10 keV). Figure 4 shows the simulated CIR seen in energetic hydrogen atoms (EHA) (>10 keV) from a vantage point at 1 AU. The $0°$ azimuth in each case is anti-solar and the sun sits at $\alpha = \pm 180°$ on the equator. The two panels are separated by 6.2 days. The combined effect of the r-dependence of the CIR ion flux, the column length of the CIR along each line-of-sight, and the distributions of the interplanetary neutral H and He determine the EHA isoflux contours. Although the CIR expands latitudinally with increasing r, the latitudinal extend of the outer portions of the CIR viewed from 1 AU actually diminishes with r. Therefore, the CIR appears in EHA as a band within $\pm 40°$ about the equatorial plane, as we see between $\alpha = -180°$ and $105°$ in the upper panel. The crescent-shaped structure between $\alpha = 105°$ and $150°$ is the projection of the cross-section of the CIR near 1 AU. (A local maximum flux of 1.2×10^{-2} (cm^2 sr s)$^{-1}$ is found at $\alpha = 130°$ and the contour step size is 2×10^{-3} (cm^2 sr s)$^{-1}$.) The lager span in polar angle, $> 40°$ on either side of the equator, is due to the closeness of the vantage point to the CIR. Viewing the CIR from the convex side (leading edge) limits the span in polar angle, while viewing it from the concave side (trailing edge), especially from a close distance, widens the latitudinal span to the two poles. The latter is seen in the lower panel, 6.2 days later, when the observer is engulfed by the CIR. Comparing the two panels, we see that the structure of a CIR can be discerned in its EHA emission over a period of time. Since the CIR is a transient event and since even during the passage of a CIR there are portions of the sky unaffected by the CIR, ENA from the termination shock can still reach the observer unhampered.

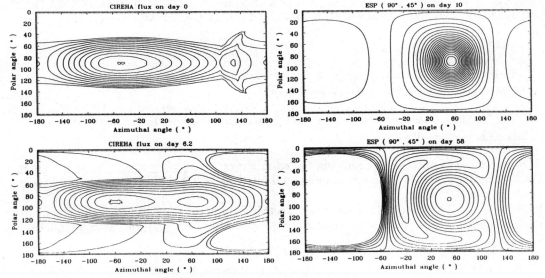

Fig. 4 Isoflux contours of EHA of CIR origin. The two panels are separated by 6.2 days. (See text.)

Fig. 5 Isoflux contours of EHA of ESP origin. The two panels are separated by 48 days. (See text.)

Less long enduring than a CIR is the transient ESP event. Figure 5 shows two stages of the simulated ESP plume seen in EHA (> 10 keV) from a fixed point at 1 AU. The upper panel is of the ESP plume just passed 1 AU with its leading edge approaching 4 AU and the lower panel is of the same plume at a later date with its leading edge approaching 20 AU. The combined effect of the r-dependence of the ESP flux, the column length of the plume along each line-of-sight, and the distributions of the interplanetary neutral H and He determine isoflux contours. For the upper panel, plume is contained within 4 AU, so that the ESP flux is highest at the portion that is farthest from the sun ($\alpha = 64°$) and the longest column length intercepting the plume is also approximately in that direction. The interplanetary H and He, however, are strongly peaked at $\alpha = 0°$, according to our model. The result is, therefore, a single maximum in EHA flux shifted to $\alpha = 56°$ in the equator, enclosed by the contour of maximum EHA flux of 8.8×10^{-1} $(cm^2 \text{ sr s})^{-1}$. (The contour step size is 4×10^{-2} $(cm^2 \text{ sr s})^{-1}$.) The contour of minimum EHA flux of 2×10^{-2} $(cm^2 \text{ sr s})^{-1}$ on the left ($\alpha < -75°$) and that on the right ($\alpha > 140°$) terminate the plume; the field-of-view bounded by this contour is void of EHA of ESP origin. The azimuthal asymmetry is due to the fact that the observe is not on the axis of the plume. A greater asymmetry is seen in the lower panel, where the plume presents a local minimum of $\sim 6 \times 10^{-2}$ $(cm^2 \text{ sr s})^{-1}$ at $\alpha = 49°$, a slower rise to a ridge of $\sim 2.6 \times 10^{-1}$ $(cm^2 \text{ sr s})^{-1}$ at $\alpha \sim 110°$ and drops to the terminal contour of 2×10^{-2} $(cm^2 \text{ sr s})^{-1}$ at $\alpha > 145°$ on the right, and a steep rise to a summit of $\sim 5 \times 10^{-1}$ $(cm^2 \text{ sr s})^{-1}$ at $\alpha \sim -24°$ and a precipitous drop to the terminal contour ($\alpha < -57°$) on the left. The field-of-view void of EHA flux of ESP origin is noticeably larger than that in the upper panel. This is due to the larger distance between the observer and the plume. Again, even in the presence of an ESP, there are portions of the sky clear for ENA from the termination shock to reach the observer at 1 AU.

CONCLUSION

Based on simple, but realistic models /7,8/, we have shown the ENA flux coming from the solar-wind termination shock and those generated in the occasional passage of a CIR or an ESP event. The flux level and spatial distribution of the ENA flux coming from the termination shock appear different from the others. Detecting this distinct ENA population is our only means to study the dynamics and structure of the solar-wind termination shock and to sample the spectra of the unmodulated low energy anomalous cosmic rays at Earth's orbit.

ACKNOWLEDGEMENT

This work is in part supported by NASA grants NAG5-260, NAGW 1419, NAGW 1676 (KCH), NAGW-1931 and NSF grant ATM-8922151 (JRJ), and NASA grant NAG2-146 (MAG).

REFERENCES

1. R. L. McNutt Jr., Remote sensing of the termination of the solar wind via in situ plasma measurements, Adv. Space Res. 9, #4, 235-238 (1989)

2. V. B. Baranov, Gasdynamics of the solar wind interaction with the interstellar medium, Space Sci. Rev. 52, 89-120 (1990)

3. J. R. Jokipii, The anomalous component of cosmic rays, in: Physics of the Outer Heliosphere, ed. S. Grzedzielski and D. E. Page, Pergamon, Oxford 1990, p. 169-178.

4. W. R. Webber, The interstellar cosmic ray spectrum and energy density: interplanetary cosmic ray gradients and new estimate of the boundary of the heliosphere, Astron. Astrophys. 179, 277 (1987)

5. W. S. Kurth, D. A. Gurnett, F. L. Scarf and R. L. Poynter, Detection of a radio emission at 3 kHz in the outer heliosphere, Nature 312, 27 (1987)

6. K. C. Hsieh, K. L. Shih, J. R. Jokipii and M. A. Gruntman, Imaging the Heliosphere from Earth's orbit, EOS, Transaction, AGU 71, #43, 1520 (1990)

7. M. A. Gruntman, Anisotropy of the energetic neutral atom flux in the heliosphere, Planet. Space Sci., in press (1992)

8. K. C. Hsieh, K. L. Shih, J. R. Jokipii and S. Grzedzielski, Probing the heliosphere with energetic hydrogen atoms, Astrophys. J., in press (1992)

9. K. C. Hsieh, C. C. Curtis, C. Y. Fan and M. A. Gruntman, Techniques for the remote sensing of space plasma in the heliosphere via energetic neutral atoms: a review, this issue.

SOLAR WIND IRON AND OXYGEN CHARGE STATES AND RELATIVE ABUNDANCES MEASURED BY SWICS ON ULYSSES

F. M. Ipavich,* A. B. Galvin,* J. Geiss,** K. W. Ogilvie*** and F. Gliem†

* Department of Physics, University of Maryland, College Park, MD 20742, U.S.A.
** Physikalisches Institut, University of Bern, Sidlerstrasse 5, CH-3012 Bern, Switzerland
*** NASA/GSFC, Code 692, Greenbelt, MD 20771, U.S.A.
† Tech. University Braunschweig, D-3300 Braunschweig, Germany

We present the results of a survey of iron and oxygen charge state distributions as detected by the Solar Wind Ion Composition Spectrometer (SWICS) on the Ulysses spacecraft. The results are categorized in terms of the ionization temperature determined every ~ 13 minute instrument cycle from the O^{+7} to O^{+6} ratio. Our preliminary Fe/O abundance ratio is 0.12 averaged over this 3 month survey. The Fe charge state distributions accumulated during times of high oxygen ionization temperatures are clearly non-isothermal.

INTRODUCTION

At low coronal altitudes, where the solar wind originates, the state of ionization for a given element is a function of the local electron temperature. Beyond some small heliocentric distance, different for each ion but typically ~ few solar radii, the ionization state of the expanding wind is "frozen in" and remains essentially unchanged during its passage through the upper solar atmosphere and interplanetary space. Solar wind charge state measurements therefore reflect the temperature profile of the solar corona. Solar wind abundance ratios, compared with those in the photosphere, can illuminate the selection effects that operate during the formation and acceleration of the solar wind.

The SWICS experiment /1/ relies on simultaneous measurements of energy per charge, time of flight, and total energy to determine the mass, mass per charge, and incident energy of solar wind ions in the energy-per-charge range of 0.6 to 60 keV/e. The total energy measurement is made by solid state detectors after the ions are accelerated (by about 22 kV in the data reported here) and allows the separation of ions with the same mass per charge but different mass (e.g., Si^{+8} and Fe^{+16}).

In this paper we present an analysis of oxygen and iron ions over the time interval: 1990 day 341 to 1991 day 67. The heliocentric distance of the Ulysses spacecraft varied from 1.3 to 2.3 AU during this interval. For each complete SWICS voltage cycle (64 logarithmic E/Q steps with a cycle time of about 13 minutes) in this ~ 3 month interval, the oxygen ionization temperature was estimated from the ratio O^{+7}/O^{+6} using the tables in /2/ and placed into one of 4 categories: > 2.0, between 1.6 and 2.0, between 1.3 and 1.6, and < 1.3 x 10^6 K. Only the SWICS "Matrix Elements" are utilized. Every incident ion is classified real-time by the on-board data processing unit (DPU) into one of these ~ 500 elements in mass vs. mass/charge space. The elements have a bin width of 20% in mass and 3% in mass/charge; the instrumental resolution (FWHM) for typical solar wind iron ions is ~ 35% in mass and ~ 3% in mass/charge. A total of 4 such matrices, one for each oxygen temperature category, was accumulated for the entire time interval.

Analysis based on Matrix Elements has the advantage of not being limited by the low spacecraft telemetry and not being influenced by various instrument priorities. Disadvantages are that there is no energy information available (since the Matrix Elements are summed over an entire voltage cycle by the DPU), and that the resolution is not as good as is available using individual pulse-height events. For the time interval in this paper the SWICS voltage cycle began at the highest E/Q value and then decreased until a certain counting rate was measured (usually corresponding to the He^{++} solar wind peak), after which the voltage steps increased. Because of this "turn-around" feature, certain E/Q steps could be covered twice, once or not at all. At this point in our analysis we estimate that this effect, which we corrected for on a statistical basis, leads to an uncertainty of about 20% in the Fe/O abundance results presented below.

We emphasize that our survey represents an unbiased sample of the solar wind during this time period. The only conditions imposed on each instrument cycle (~13 min) were that there be no data gap and that at least 4 oxygen counts be recorded (98% of the instrument cycles met this count criterion).

OBSERVATIONS

Figure 1 displays the distribution of oxygen ionization temperatures during our survey interval. The median temperature is about 1.7×10^6 K. These data are uncorrected for instrument efficiency effects; we estimate that inclusion of these efficiencies would decrease the indicated temperatures by 0.1×10^6 K.

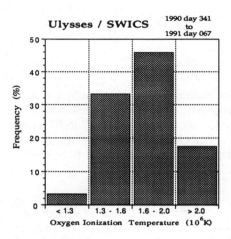

Fig 1. *Frequency distribution of the oxygen ionization temperature during the survey period.*

The oxygen ionization temperature does not correlate strongly with the solar wind speed. There is, however, a correlation with the hydrogen kinetic temperature as measured with the SWICS auxiliary channel (consisting of a solid state detector but no time-of-flight telescope). This correlation is demonstrated in Figure 2, which indicates a noticeably higher hydrogen kinetic temperature in the lowest oxygen ionization category ($< 1.3 \times 10^6$ K) than in the highest category ($> 2.0 \times 10^6$ K).

The mass and mass/charge resolutions of the SWICS matrix elements are demonstrated by the contour plot in Figure 3, corresponding to a section of the entire matrix accumulated during those cycles with an oxygen temperature between 1.6 and 2.0×10^6 K. The indicated bin numbers are those assigned by the DPU. Fe, S and Si are predicted to correspond to mass bins ~24, 22 and 21, respectively. M/Q bin 72 is the predicted position of Fe^{+8}. All Fe charge states from 8 to 16 are seen to be resolved, with the exception of Fe^{+15}, which appears as a

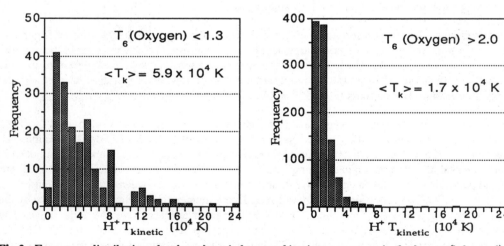

Fig 2. *Frequency distributions for the solar wind proton kinetic temperature in the lowest (left panel) and highest (right panel) oxygen ionization temperature categories.*

shoulder on the Fe^{+16} peak. The 3 peaks visible in the mass bin range 20 to 22 represent (from left to right): ~90% Si^{+9} and 10% S^{+10}; ~70% Si^{+8} and 30% S^{+9}; and ~30% Si^{+7} and 70% S^{+8}. The vertical extension of the Fe^{+12} peak at M/Q bin 58 corresponds to the presence of S^{+7}. A subtle effect in the SWICS instrument is that ions with different M but identical M/Q have slightly different times of flight (caused by different energy losses in the carbon foil) and hence a slightly different computed M/Q. For example, computer simulations of the SWICS response predict that Fe^{+16} will be centered at M/Q bin location 48.1, while for Si^{+8} the bin location should be 48.8; careful inspection of Figure 3 shows that this effect is indeed present.

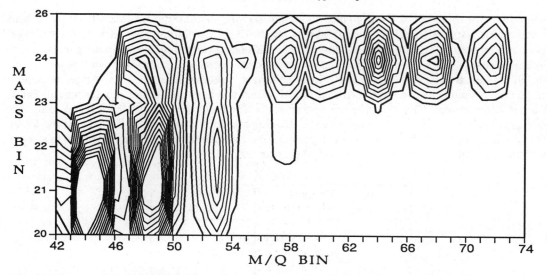

Fig. 3. *Contour plot of the Mass and Mass/Charge matrix bins in the mass range of silicon to iron. The contours are linearly spaced by 150 counts from 450 to 2400 counts.*

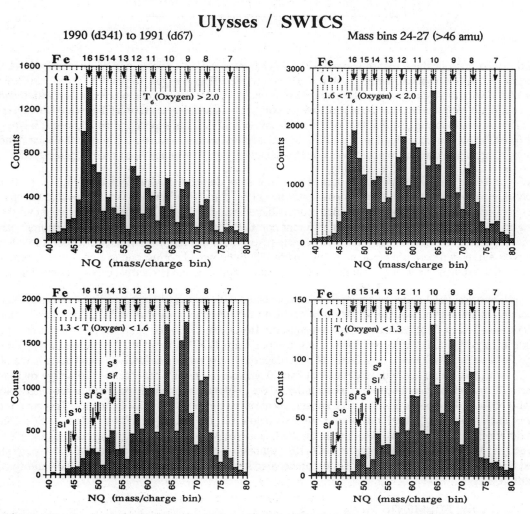

Fig. 4. *Charge state distributions for Fe ions in each of the four oxygen temperature categories.*

By summing over mass bins 24 and greater we obtain an M/Q distribution for predominantly Fe ions, uncorrected for instrument efficiencies. Figure 4 shows these distributions for each of the 4 oxygen temperature categories. The arrows show the M/Q bin positions predicted by the computer simulations for the indicated charge states. The predicted and observed peaks generally agree to within a fraction of a bin width. In the lowest two categories Si and S ions "spill over" into the chosen Fe mass range. In order to calculate the Fe abundance we have corrected for this effect and have also taken into account all known instrument efficiencies. The resulting abundance ratios for Fe/O for each of the oxygen ionization categories are given in Table 1. The error due to counting statistics is very small (± 0.004 for the lowest temperature category, ± 0.001 for the other categories). We estimate the systematic error (due to stepping reversal and uncertainty in instrument efficiencies) is about ± 0.03.

TABLE 1

Oxygen Ionization Temperature Range (10^6 K)	Fe/O
> 2.0	0.13
1.6-2.0	0.12
1.3-1.6	0.12
< 1.3	0.11

DISCUSSION

We have found that our three month survey period is characterized by a wide range of oxygen ionization temperatures, with a median value of about 1.6×10^6 K. The oxygen ionization temperature is anticorrelated with the local hydrogen kinetic temperature. A similar anticorrelation of iron ionization and hydrogen kinetic temperatures has been reported for coronal hole and driver plasma solar wind flow types /3/. The observations suggest that the solar wind from coronal holes is exposed to an additional heating source (presumably wave dissipation), resulting in a nearly isothermal expansion, whereas the solar wind from high-temperature coronal regions cools nearly adiabatically and hence has a large temperature gradient.

This suggestion is further supported by the Fe charge state distributions presented in Figure 4. For example Figure 4a indicates the simultaneous presence of Fe^{+8} and Fe^{+16}, which is not expected for any single temperature /2/. Although no detailed models have been developed, it appears reasonable that the freezing-in of different Fe charge states at different coronal altitudes (hence different temperatures) can explain our results. In this scenario, the charge state distribution in Figure 4a would correspond to a hot corona with a large temperature gradient, while that in Figure 4d would correspond to a cool corona with a relatively small gradient producing a charge state distribution similar to the isothermal cases presented in /2/.

Our overall average Fe/O abundance ratio is 0.12, consistent with previously reported Fe/O values of 0.10 for interstream solar wind /4/, 0.124 for magnetosheath measurements /5/ during "high pressure" solar wind (combination of flare-associated and coronal-hole flows), and 0.14 for plasma sheet measurements /5/. Our average value is, however, lower than the long-term Fe/O value of 0.19 reported in /6/. Our average Fe/O value is enhanced by a factor of ~3 compared to the meteoritic-derived Fe/O abundance of 0.038 /7/ (which is consistent with the ratio obtained from soft X-ray flux observations of active regions and flares /8/ and which may be more representative of the photospheric abundance than the spectroscopic value of 0.55 -- see discussion in /8/). This enhancement relative to photospheric abundances is typical for elements, such as iron, with first ionization potentials below ~10 eV (the so-called FIP effect, /5,8/). Our results do not indicate a dramatic decrease in the Fe/O abundance with decreasing temperature, an effect observed for Mg/O /9/, and possibly for Si/O /10/ (although it should be noted that all of these surveys covered different time intervals). A more comprehensive survey is underway to determine the validity of these differences.

ACKNOWLEDGMENTS

The authors thank C. Shafer for assisting with the data analysis. This work was supported under JPL Contract No. 955460.

REFERENCES

1. G. Gloeckler, J. Geiss, H. Balsiger, P. Bedini, J.C. Cain, J. Fischer, L.A. Fisk, A.B. Galvin, F. Gliem, D.C. Hamilton, J.V. Hollweg, F.M. Ipavich, R. Joos, S. Livi, R. Lundgren, U. Mall, J.F. McKenzie, K.W. Ogilvie, F. Ottens, W. Rieck, E.O. Tums, R. von Steiger, W. Weiss, and B. Wilken, The solar wind ion composition spectrometer, *Astron. Astrophys. Suppl. Ser.*, **92**, 267, (1992).

2. M. Arnaud and R. Rothenflug, An updated evaluation of recombination and ionization rates, *Astron. Astrophys. Suppl. Ser.*, **60**, 425 (1985).

3. F.M. Ipavich, A.B. Galvin, G. Gloeckler, D. Hovestadt, S.J. Bame, B. Klecker, M. Scholer, L.A. Fisk, and C.Y. Fan, Solar wind Fe and CNO measurements in high-speed flows, *J. Geophys. Res.*, **91**, 4133 (1986).

4. S.J. Bame, J.R. Asbridge, W.C. Feldman, and P.D. Kearny, The quiet corona: temperature and temperature gradient, *Solar Physics*, **35**, 137 (1975).

5. G. Gloeckler and J. Geiss, The abundance of elements and isotopes in the solar wind, *Cosmic Abundances of Matter,* ed. C.J. Waddington, AIP Conference Proceedings, **183**, 1989, p. 49.

6. P. Bochsler, Velocity and abundance of silicon ions in the solar wind, *J. Geophys. Res.*, **94**, 2365 (1989).

7. E. Anders and N. Grevesse, *Geochim. Cosmochim. Acta,* **53**, 197, 1989.

8. K.G. Widing and U. Feldman, Elemental abundances and their variations in the upper solar atmosphere, this issue.

9. R. von Steiger, J. Geiss, G. Gloeckler, H. Balsiger, A.B. Galvin, U. Mall, and B. Wilken, Magnesium, carbon, and oxygen abundances in different solar wind types, as measured by SWICS on Ulysses, this issue.

10. A.B. Galvin, F.M. Ipavich, G. Gloeckler, R. von Steiger, and B. Wilken, Silicon and oxygen charge state distributions and relative abundances in the solar wind measured by SWICS on Ulysses, this issue.

COMPOSITION SIGNATURES IN SEP EVENTS: INITIAL RESULTS FROM THE COSPIN LET EXPERIMENT ON ULYSSES

R. G. Marsden, T. R. Sanderson, A. M. Heras and K. -P. Wenzel

Space Science Department of ESA, ESTEC, PO Box 299, 2200 AG Noordwijk, The Netherlands

ABSTRACT

We present new observations of the elemental composition in solar energetic particles (SEP) events in interplanetary space as measured by the COSPIN Low Energy Telescope (LET) instrument on the Ulysses spacecraft during the first months of the in-ecliptic phase of the mission. We compare these measurements with previous SEP abundance determinations made using data from Phobos 2 and examine the extent to which the separation in longitude of the flare site and the nominal magnetic connection point of the spacecraft at the Sun affects the measured composition. We use these results to infer the probable source of the particle populations observed at Ulysses in association with major solar activity in March, June and July 1991.

INTRODUCTION

Recent studies of solar energetic particle abundances by Cane et al. /1, 2/ have shown that the composition measured in large SEP events in the energy range $\sim 1 - 10$ MeV/nucleon is a function of the longitude of the source region with respect to the observer. Those events occurring in regions for which magnetic connection to the observer is possible (i.e., near W 60° for an observer at the Earth) tend to show an enhancement of heavy elements compared with an event-averaged or "baseline" composition, whereas heavy elements tend to be depleted in poorly connected events. These findings have been interpreted in terms of two sources for the observed particles: flare-heated material in the case of well connected events, and corona and/or solar wind material accelerated by coronal and interplanetary shocks in the case of poorly connected events. We have examined the composition in five SEP events that occurred during the first months of the Ulysses mission, together with data obtained from the SSD/MPAe/IKI LET instrument on Phobos 2, in the light of the above model and find that our results are consistent with this picture. The Ulysses events include large increases on 23/24 March 1991 and 6–10 June 1991 associated with major flaring activity from Regions 6555 and 6659.

INSTRUMENTATION AND ANALYSIS

The energetic particle observations reported here were made with the LET instruments carried on board Ulysses and Phobos 2 as part of the COSPIN and ESTER experiment packages, respectively /3, 4/. The LET sensor is a four-element solid-state detector telescope operating in the single- and double dE/dX vs. E mode, allowing the measurement of protons, alpha particles and heavy elements up to and including iron to be made with good elemental resolution. On-board particle identification of groups of species is used to enhance the sample of rarer particles selected for pulse-height analysis (PHA). Normalization of the PHA sample is achieved by means of counting rate information which is available for all particles stopping in the detector telescope. In the case of the Ulysses data, we have calculated heavy element abundances with respect to O in the energy range 5.5–7.5 MeV/nucleon (4.25–6.75 MeV/nucleon for He, C and N). The Phobos observations are in the energy range 4.25–8.5 MeV/nucleon. Further details concerning the instrument can be found in /5/ and a discussion of the analysis technique is given in /6/.

OBSERVATIONS

Ulysses

The data set from which we have selected the events for the present study covers the period from switch-on (23 October 1990) up to mid-July 1991. It should be noted, however, that the coverage provided by the experimenter data tapes for this period is not yet complete. Three of the events (25 January, 24 February and 23 March 1991, referred to hereafter as Events U1, U2 and U3) have been discussed in /7/ and a detailed description of LET observations of Event U3 is given in /8/. Of the remaining two events, the first (U4) was associated with the highly active Region 6659 that produced five X12+ flares in the first half of June 1991. As seen at Ulysses this was a compound event, comprising at least three intensity enhancements in a 7-day period beginning on 6 June. An interplanetary shock was observed at the spacecraft on 7 June at approximately 1540 UT (S. Hoang, private communication). At the time of this activity, Ulysses was 120° east of the Sun-Earth line at a radial distance of 3.2 AU. The fifth event (U5) occurred at the end of June, one solar rotation after the event earlier in that month, and shows similarities with this latter increase. Returning Region 6659 (assigned as Region 6703) produced a number of flares, including long duration M6 and X1/2B events on 28 June and 7 July, respectively. Starting on 29 June, particle intensities at Ulysses (located 135° east at 3.5 AU) were enhanced for more than 10 days as a result of this activity and, as in the case of event U4, the profile was complex.

It should be noted that, even though the spacecraft was at different heliolongitudes and radial distances at the times of these events, it remained close to the *same* nominal spiral field line, connecting back to the Sun within 45° of the western limb as seen from the Earth.

Phobos 2

The Phobos 2 data set covers the interval July 1988 to March 1989. For the purpose of this work we have expanded the set of SEP events used in our previous studies (e.g. /6/) to include all events with measurable fluxes of O and Fe nuclei.

RESULTS AND DISCUSSION

Our SEP event-averaged abundance measurements (relative to O) of He, C, N, Ne, Mg, Si and Fe for each of the Ulysses events are given in Table 1. The values are typical for large SEP events and show no significant enhancements in the heavy elements. Figure 1 shows the Fe/O ratios for all events in the combined Ulysses-Phobos data set plotted as a function of the heliolongitudinal separation between the source flare and the magnetic connection point. For the Ulysses events, we have used preliminary solar wind velocity data (S. Bame, private communication) to compute the connection longitudes. In the case of the Phobos 2 events, for which no solar wind data are available, we have assumed a constant speed of 400 km s^{-1}. As can be seen, there is a tendency for those events for which the longitudinal separation angle is large and negative (corresponding to sources east of the connection point) to have low Fe/O ratios (~ 0.1 or less), whereas events which are well-connected (separation $\sim \pm 45°$) tend to have enhanced ratios. The 24 February event (U2) does not follow this trend. However, the source location is uncertain and local shock acceleration, possibly associated with a high speed stream, may be responsible for the heavy element depletion. In general, then, our results are in agreement with the findings of Cane et al. /1, 2/.

A feature of the events observed at Ulysses at solar maximum is that they are generally Fe-poor, whereas the Phobos 2 data set, acquired during the rising phase of cycle 22, contains a mixture of Fe-rich and Fe-poor events. This is possibly explained by the high frequency of transient events at solar maximum that perturb the interplanetary medium to such an extent that efficient magnetic connection between the source at the Sun and an observer beyond 1 AU rarely occurs.

Fig. 1. Fe/O ratios as a function of the separation between magnetic connection and source longitudes for the combined Ulysses (filled circles) and Phobos 2 (inverted triangles) SEP event set. Positive angles denote sources west of connection point.

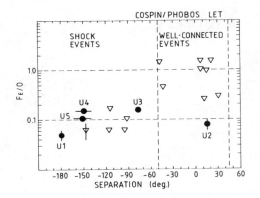

TABLE 1 SEP Abundances Measured by the LET on Ulysses (5.5 – 7.5 MeV/n)

	Event U1	Event U2	Event U3	Event U4			Event U5
Element	25 Jan 91	24 Feb 91	23 Mar 91	06 Jun 91	08 Jun 91	10 Jun 91	29 Jun 91
He/O†	75±4	68±2	62±1	69±3	69±2	71±2	66±2
C/O†	0.55±0.05	0.46±0.03	0.43±0.01	0.53±0.04	0.38±0.02	0.49±0.03	0.54±0.03
N/O†	0.11±0.02	0.13±0.01	0.12±0.01	0.12±0.02	0.12±0.01	0.15±0.01	0.13±0.01
Ne/O	0.16±0.04	0.13±0.01	0.14±0.01	0.19±0.03	0.15±0.02	0.15±0.02	0.15±0.02
Mg/O	0.13±0.03	0.22±0.03	0.25±0.01	0.20±0.03	0.19±0.02	0.16±0.02	0.22±0.02
Si/O	0.11±0.03	0.15±0.02	0.19±0.01	0.09±0.02	0.11±0.01	0.15±0.02	0.11±0.02
Fe/O	0.05±0.02	0.08±0.02	0.15±0.01	0.18±0.03	0.10±0.01	0.18±0.02	0.11±0.02

† 4.25–6.75 MeV/n

In Figures 2, 3 and 4 we plot the time-intensity profiles of C,N,O nuclei (3.2–7.5 MeV/n) and $Z \geq 10$ nuclei (3.9–9.5 MeV/n) for the periods corresponding to events U3, U4 and U5, together with the Fe/O ratio in the energy interval 5.5–7.5 MeV/n. Also shown in Figures 2 and 3 is the Fe/O ratio measured at higher energy (13–30 MeV/n). The ratios in each energy interval remain approximately constant throughout each of the three events and, based on the above discussion, the generally low values of Fe/O suggest that the bulk of the heavy ions were shock-accelerated rather than being samples of flare-heated material. In the case of the two events for which the ratio could be determined in both energy intervals there is a clear energy dependence, the ratio being a factor ~ 2 lower at higher energies. This effect is the result of the difference in the energy

Fig. 2. a) Time-intensity profiles for C,N,O and $Z \geq 10$ nuclei for the interval 23–28 March 1991.
b) Fe/O ratio in the energy ranges 5.5–7.5 MeV/n (filled circles) and 13–30 MeV/n (open circles).

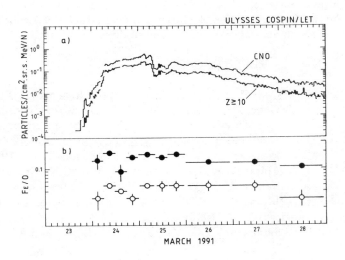

Fig. 3.   As for Figure 2, but for the interval 5–12 June 1991.

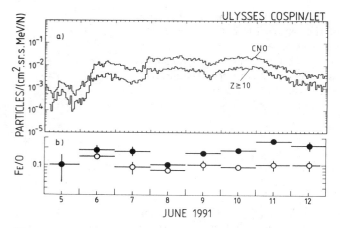

Fig. 4.   a) As for Figure 2, but for the interval 30 June–11 July 1991. b) Fe/O ratio in the energy range 5.5–7.5 MeV/n.

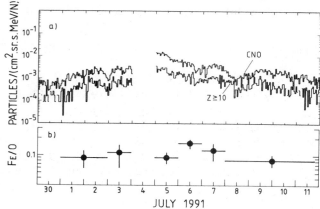

spectra of the two species.

CONCLUSIONS

We find that the SEP events studied tend to group into two populations when organised according to the separation in heliolongitude of the flare site and the magnetic connection point of the spacecraft. Well-connected events show enhanced Fe/O ratios, whereas poorly connected events tend to have low Fe/O ratios. Based on the model of Cane *et al.* /1, 2/, our observations suggest that the particle populations detected by the LET at the location of Ulysses in association with the major flaring activity of March, June and July 1991, which are characterised by low Fe/O ratios, are shock-accelerated rather than of direct flare origin.

REFERENCES

1. H.V. Cane, D.V. Reames, and T.T. von Rosenvinge, Proc. 21st ICRC, Adelaide, 5, 370 (1990)
2. H.V. Cane, D.V. Reames, and T.T. von Rosenvinge, Ap.J. 373, 675 (1991)
3. J.A. Simpson et al., Astron.&Astrophy. Suppl., in press (1991)
4. V.V. Afonin et al., Nucl. Instr. Meth. A290, 223 (1990)
5. R.G. Marsden et al., Nucl. Instr. Meth. A290, 211 (1990)
6. R.G. Marsden, K.-P. Wenzel, V.V. Afonin, K. Gringauz, M. Witte, A.K. Richter, G. Erdős, A. Somogyi, A. Varga, and L. Varhalmi, Planet. Space Sci. 39, 57 (1991)
7. R.G. Marsden, T.R. Sanderson, A.M. Heras, and K.-P. Wenzel, Proc. 22nd ICRC, Dublin, paper SH 5.1.6 (1991)
8. T.R. Sanderson, R.G. Marsden, A.M. Heras, and K.-P. Wenzel, this issue.

SOLAR WIND COMPOSITION FROM SECTOR BOUNDARY CROSSINGS AND CORONAL MASS EJECTIONS

K. W. Ogilvie,* M. A. Coplan** and J. Geiss***

*NASA Goddard Space Flight Center, Code 692, Greenbelt, MD 20771, U.S.A.
** Institute for Physical Science and Technology, University of Maryland, College Park, MD 20742, U.S.A.
*** Physikalisches Institut, University of Bern, Bern 3012, Switzerland

ABSTRACT

Using the Ion Composition Instrument (ICI) on board the ISEE-3/lCE spacecraft, average abundances of ^{4}He, ^{3}He, O, Ne, Si and Fe have been determined over extended periods. In this paper the abundances of ^{4}He, O, Ne, Si and Mg obtained by the ICI in the region of sector boundary crossings (SBCs), magnetic clouds and bi-directional streaming events (BDSs) are compared with the average abundances. Both magnetic clouds and BDSs are associated with coronal mass ejections (CMEs). No variation of abundance is seen to occur at SBCs except for helium, as has already been observed. In CME-related material, the abundance of neon appears to be high and variable, in agreement with recent analysis of spectroscopic observations of active regions. We find that our observations can be correlated with the magnetic topology in the corona.

INTRODUCTION

As a result of observations made by the Ion Composition Instrument (ICI) on the ISEE-3/ICE spacecraft, average abundances have been obtained in the solar wind over the period from August 1978 to December 1982 /1/. Abundances of the ions of ^{4}He, ^{3}He, O, Ne, Si, and Fe are available and were compared with abundances obtained from solar energetic particle (SEP) observations /2/ with which they agree well. Comparisons between both the ICI and SEP abundances and spectroscopically determined photospheric abundances show discrepancies, some of which have been attributed to a first ionization potential (FIP) effect /3/, whereby elements with FIPs above 10-12 eV are reduced in abundance in the solar wind (corona) with respect to the photosphere.

The ICI acquires mass per charge (M/Q) spectra, with a resolution $M/\Delta M \approx 30$, FWHM, when the solar wind speed is below 600 km sec^{-1}, which is approximately 95% of the time /4/. The abundance of ^{4}He, which is approximately two orders of magnitude greater than the next most abundant element, oxygen, is deduced separately from all the other minor ions by an individual fitting scheme. The abundances of the other minor ions are then obtained together by a method described below.

ANALYSIS

Recent analysis of remote spectroscopic observations has shown that there are anomalous local elemental abundances near the sites of solar activity /5/, /6/, /7/. We have made abundance observations in the solar wind at times when the plasma sampled by the ICI can be associated with well-defined solar structures or events. For this study we analyzed the elements in Table 1 which have first ionization potentials that span a wide range and are therefore well suited to an investigation of the FIP effect.

TABLE 1 Elements Analyzed by the ICI

Element	First Ionization Potential (eV)	Dominant M/Q Values
He	24.6	2.0
Ne	21.6	2.5
O	13.6	2.28, 2.67
Si	8.1	2.33, 2.56
Mg	7.6	2.4, 3.11

The density, temperature and bulk speed of helium are first obtained by fitting a one dimensional convected Maxwellian velocity distribution function to the observations. Individual spectra, or small groups of spectra, are then analyzed over the range $2.2 < M/Q < 5.6$ using a constrained linear least squares routine with a single freezing-in temperature, T_c, as a parameter. Two slightly different least-squares fitting schemes, developed by Bochsler, Geiss and Kunz /8/ and by D'Annunzio, Ogilvie and Coplan /9/, respectively, have been used to derive abundances of the minor ions from the data. They give essentially identical results. In each scheme a model solar wind is convoluted with measured instrument functions and the ion abundances adjusted to minimize our fitting parameter θ, defined as the square of the difference between calculated and observed counts divided by the calculated counts at each setting of the instrument and summed over the range $2.2 < M/Q < 5.6$. The results include an estimate of the "freezing-in" temperature, the temperature which characterizes the charge state distribution of the ions of each element the temperature which characterizes the charge state distribution of the ions of each element. From the count rates at 30 different M/Q settings, we deduce densities for 16 different minor ion charge states. Mathematically, our analysis is over determined; however, our criterion of a minimum in θ is a necessary but not sufficient criterion that we have found the most physically meaningful set of elemental abundances. We have explored large regions of parameter space to ensure that our procedure leads to a global rather than a local minimum, and the overall consistency of the results provides good evidence of this. Charge state tables by Shull and Von Steenberg are used in this work /10/.

Our fitting routine uses constraints, and the results are sensitive to the nature of the constraints and to the weights that are applied to them. Densities are constrained to be non-negative and the N/O ratio has been set at 0.125 because the overlap of oxygen and nitrogen charge states in our spectra does not permit us to determine the abundance of nitrogen independently. Furthermore, since a large proportion of the carbon in the solar wind is in the C^{+6} state, it cannot be distinguished by the ICI from much more abundant $^4He^{2+}$. We therefore constrain C/O to be 0.45, a value determined by Gloeckler et al. /11/ in the solar wind by a method involving ion time-of-flight. The actual values of the constraints and constraint weights were based on several hundred runs made on spectra picked at random, and using them produces good fits. An example of a typical, but not exceptional, analysis is shown in Figure 1; the data are for a period unrelated to any known solar event, and the abundances obtained agree well with our long-term average values.

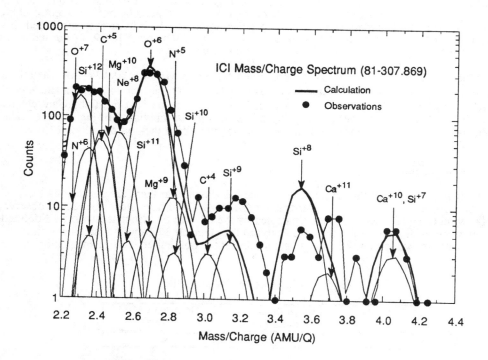

Fig. 1. Comparison of an ICI mass per charge spectrum and the calculated fit. The contributions from each of the ion species are shown and their sum compared with the experimental observations. The accumulation time was 30 minutes.

For each period of interest, the spectra were individually examined and those acquired over periods during which solar wind conditions changed appreciably were eliminated. The remaining spectra were then analyzed to obtain the freezing-in temperatures and the number densities of the minor ions. Provided that the θ was sufficiently low, the abundances in the plasma local to the spacecraft were calculated from the number densities. When conditions were very stable, data from successive spectra accumulated over 30 minutes could be added; this helped to reduce statistical uncertainties. In these cases θ was never greater than 5.

We have searched for systematic errors introduced by our assumptions of a fixed C/O ratio and a single coronal temperature. Both the C/O ratio and coronal temperature can be changed over substantial ranges without appreciably affecting the Ne/O ratio. Typically, a 20% change in the C/O ratio, changes the derived Ne/O ratio by 8%, and a 20% change in T_C changes the Ne/O ratio by 13%. The assumption that the distribution of charge states can be described accurately by a single freezing-in temperature, T_C may lead to systematic errors, however relaxation of this assumption would introduce considerable complication into the analysis which we feel is not justified at this time.

Observations at Sector Boundary Crossings

Magnetic sector boundary crossings are detected when the magnetometer on a spacecraft is carried through the current sheet separating the sides of a helmet streamer, Figure 2. Previous work on the solar wind signatures of sector boundary crossings /3/ has shown that the He/H ratio decreases during such a crossing; we were interested in changes which might occur in the abundances of other ions. We measured abundances within and on both sides of three extremely well-defined crossings which were well separated from compressions due to high speed streams. In all, 35 spectra were analyzed. The observations are summarized in Table 3; they show the abundance of He (referred to oxygen) to decrease slightly with respect to the average solar wind value, in agreement with /3/. For neon, however, which has approximately the same FIP as helium, there is no decrease; this is also the case for the other minor ions.

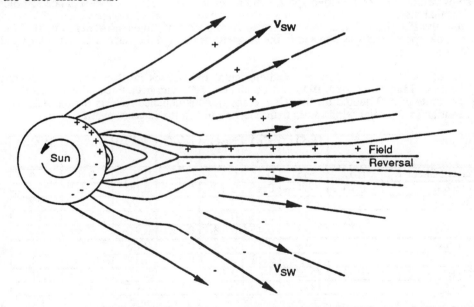

Fig. 2. Schematic diagram of the magnetic field configuration at a sector boundary.

Observations of Coronal Mass Ejecta

Coronal mass ejections provide another opportunity to study material from a known and active region of the corona as it passes over the spacecraft. The passage of the ejecta can be determined from a number of well established signatures. Bi-directional streaming /14/ of both electrons and ions is one signature of the passage of CME ejecta. Another is the passage of low density, low temperature solar wind and a large steady directional change in the interplanetary magnetic field, due to the presence of a so called magnetic cloud /15/. Figure 3 from Lepping et al. /16/ shows the large-scale topology postulated for a magnetic cloud and the twisted magnetic field lines that cause the angular change and the bi-directional streaming of charged particles.

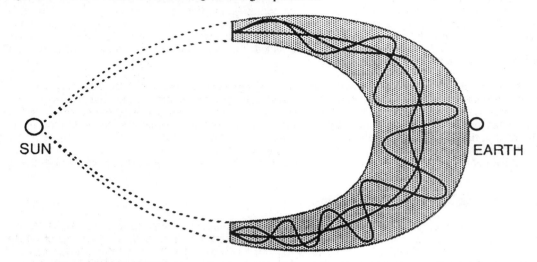

Fig. 3. Schematic diagram of a magnetic cloud showing the twisted magnetic field lines.

For the present work, we have selected and analyzed 21 spectra from magnetic clouds and 22 spectra from bi-directional streaming events. None of these events show the transient high helium abundance sometimes observed. The abundances reported here are believed to be of the bulk CME material. In Table 3 we show results combining cloud spectra and bi-directional spectra as two separate samples of plasma from coronal mass ejecta. Also shown are the abundances from SEP observations /2/, average abundances from the ICI /1/, and abundances derived from a time-of-flight instrument on AMPTE /11/, made when the space craft was situated in the magnetosheath, and the earth was in a solar wind flow from a coronal hole.

TABLE 3 Coronal temperatures and elemental abundances for different solar structures. The SBC, Magnetic Cloud, BDS and Combined CME data have uncertainties of 25%. The limits of ±20 quoted for the ICI average He/O abundance reflects the natural variation in the solar wind rather than uncertainties in the measurement.

Structure	T_c $(10^6$ K)	He/O	C/O	Ne/O	Mg/O	Si/O	Ne/Mg	Spectra
SBC Current Sheet	1.67	65	0.45	0.17	0.22	0.09	0.8	14
SBC Adjacent	1.76	81	0.45	0.19	0.13	0.09	1.5	25
Magnetic Cloud	2.1	42	0.45	0.24	0.12	0.09	2.0	22
BDS	2.4	58	0.45	0.30	0.12	0.10	2.5	22
Combined CME	2.2	52	0.45	0.28	0.12	0.10	2.3	44
Coronal Hole	-	48	0.55	0.14	0.08	0.05	1.75	-
SEP	-	-	0.41	0.14	0.19	0.18	0.74	-
ICI Average	-	75±20	-	0.17	-	0.19	1.1	-
Photosphere	-	115±9	0.39	0.17	0.045	0.042	3.2	-

In the second line of the table, we show our results obtained in the sector boundary current sheet. The measurement in the current sheet was obtained over a period of 2-6 hours. Partial densities of the ions have been summed and the sums divided by the summed densities of oxygen. We see that one difference between these abundances and the abundances near the current sheet boundary (SBC Adjacent in Table 3) is that the He/O ratio is markedly reduced in the current sheet. The reduction is less than that obtained by the superposed epoch analysis made by Borini et al. /13/, but this can be explained by the difference in time resolution between the measurements. The abundance of neon, the other element with high FIP, appears to be constant across the boundary. The magnesium abundance appears to be higher in the current sheet, but the difference is at the limit of significance.

The abundances obtained from the CME material occupy lines 3 to 5 in the table. Those from the magnetic clouds and from the BDSs are consistent with being the same, providing some evidence that both of these measurements come from the same class. The He/O ratio is significantly low in each case.. T_c, the coronal temperature is significantly higher than normal, the magnesium and silicon abundances are within the normal range, but the neon abundance is high and variable. These latter observations fall within the range of recent spectroscopic analyses of Skylab data by Feldman and Widing /5/, /6/, /7/. They show that the value of Ne/Mg varies from 1.1 to 4.6 near active regions.

DISCUSSION OF RESULTS

An interpretation of our measurement is illustrated in Table 4 where the measured helium and neon abundances are characterized along with the magnetic field configurations in the region of origin

TABLE 4   Magnetic Field Geometries and Helium and Neon Abundances

	CME	SBC Current Sheet	SBC Adjacent	Coronal Hole	High Speed Wind
Field Geometry	Closed	Open	Uncertain	Open	Open
He/O	Normal	Normal	Normal/High	Low	Low
Ne/O	High	Normal	Normal	Normal/Low	Normal/Low
Reference	This Work	This Work	This Work	/11/	/17/

The data in Table 4 suggest a connection between low helium abundance and open magnetic field structures, and normal or high helium abundance with closed fields. This in agreement with views expressed by Bame et al. /17/. Similarly, for neon the "normal" solar wind abundance, 0.15 would appear in open field structures, and a higher abundance, characteristic of active regions, in closed field structures. Widing and Feldman /15/ have suggested that the composition of material can resemble that of the photosphere or that of the corona, depending upon whether the topology of the local magnetic field is open or closed. This idea is consistent with our observations, if the photospheric abundance of neon (not directly determined - see Doshek and Bhatia /18/) is appreciably higher than the solar wind value, as suggested by Feldman and Widing /6/.

CONCLUSIONS

From the comparison of abundance measurements at CMEs and sector boundary crossings in the solar wind with the long-term average abundances in the solar wind and with spectroscopic observations of the sun, we conclude:

1. The abundance of neon in SBC current sheets, is not reduced compared to normal solar wind values. This agrees with AMPTE measurements taken in coronal hole-related streams that are also open magnetic structures.

2. Our samples of magnetic clouds and BDSs have the same composition, supporting the contention that both are CMEs.

3. The abundance of helium, oxygen, neon and magnesium in CMEs is the same as the average solar wind, except that the proportion of Ne is increased and seems variable.

4. The magnetic topology in the corona appears to be correlated with minor ion abundances; helium and neon are particularly interesting in this regard.

REFERENCES

1. M. A. Coplan, K. W. Ogilvie, P. Bochsler, and J. Geiss, Space-based measurements of elemental abundances and their relation to solar abundances, Solar Phys., 128, 195, (1990).

2. H. H. Breneman, and E. C. Stone, Solar coronal and photospheric abundances from solar energetic particle measurements, Ap. J., 299, L57, (1985).

3. D. Hovestadt, Nuclear composition of cosmic rays, in: Solar Wind 3, ed. C. T. Russell, UCLA, Inst.itute of Geophysics and Planetary Physics, 1974, p2.

4. M. A. Coplan, K. W. Ogilvie, P. Bochsler, and J. Geiss, Ion composition experiment. IEEE Trans. Geosci. Electronics, **GE-16**, 185, (1978).

5. K. G. Widing and U. Feldman, Abundance variations in the outer solar atmosphere observed in Skylab spectroheliograms, Ap. J., **344**, 1046, (1989).

6. U. Feldman and K. G. Widing, Photospheric abundances of oxygen, neon and argon derived from the XUV spectrum of an impulsive flare, Ap. J., **363**, 292, (1990).

7. U. Feldman, K. G. Widing, and P.A. Lund, On the anomalous abundances of the 2×10^4 to 2×10^5 K solar atmosphere above a sun spot, Ap. J. Lett., in press, (1991).

8. P. Bochsler, J. Geiss and S. Kunz, Abundances of carbon, oxygen, and neon in the solar wind during the period from August 1978 to June 1982, Solar Phys., **103**, 177, (1986).

9. C. M. D'Annunzio, K. W. Ogilvie and M. A. Coplan, Ion composition at a sector boundary, EOS, **67**, 1142, (1986).

10. J. M. Shull and M. Von Steenberg, The ionization equilibrium of astrophysically abundant elements, Ap. J. Suppl., **48**, 95, (1982).

11. G. Gloeckler and J. Geiss, The abundances of elements and isotopes in the solar wind, in AIP Conf. Proc., **183**, ed C. J. Waddington, Am. Inst. of Physics, (1989).

12. J. T. Gosling,, G. Borini, J. R. Asbridge, S. J. Bame, W. C. Feldman and R. T. Hanson, Coronal streamers in the solar wind, J. Geophys. Res. **86**, 5438, (1981).

13. G. Borini, J. T. Gosling, S. J. Bame and W. C. Feldman, Solar wind helium and hydrogen structure near the heliospheric current sheet, J. Geophys. Res., **86**, 4565, (1981).

14. J. T. Gosling, D. N. Baker, S. J. Bame, W. C. Feldman, R. D. Zwickl, Bi-directional solar wind electron heat flux events, J. Geophys. Res., **92**, 8519, (1987).

15. L. F. Burlaga, L. Klein, N. R. Sheeley, Jr., D. J. Michels, R. A. Howard, M. J. Koomen, R. Schwenn, and H. Rosenbauer, A magnetic cloud and a coronal mass ejection, Geophys. Res. Lett., **9**, 1317, (1982).

16. R. P. Lepping, J. A. Jones, L. F. Burlaga, Magnetic field structure of interplanetary magnetic clouds at 1 AU, J. Geophys. Res., **95**, 11957, (1990).

17. S. J. Bame, J. R. Asbridge and J. T. Gosling, Evidence for structure-free state at high solar wind speeds, J. Geophys. Res., **82**, 1487, (1977).

18. G. A. Doschek and A. K. Bhatia, Solar coronal Ar/Fe and Ne/Mg abundance ratios derived from ultraviolet forbidden line spectra, Ap. J., **358**, 338, (1990).

IMAGING HELIOSPHERIC SHOCKS USING ENERGETIC NEUTRAL ATOMS

E. C. Roelof

Johns Hopkins University/Applied Physics Laboratory, Johns Hopkins Road, Laurel. MD 20723-6099, U.S.A.

ABSTRACT

In order to explore the feasiblity of energetic neutral atom (ENA) imaging of shock-associated energetic proton populations in the heliosphere, computer-simulated ENA images have been generated based on Voyager 1/2 energetic ion measurements. One favorable vantage point for ENA shock imaging is from the Cassini spacecraft's orbit around Saturn at 10 AU. These images, calibrated relative to the measured shock-associated proton fluxes, yield an absolute estimate of ENA fluxes which indicates that useful heliospheric ENA imaging can be accomplished with present technology.

INTRODUCTION TO ENA IMAGING OF HELIOSPHERIC SHOCK STRUCTURE

Energetic neutral atoms (ENA) are produced whenever energetic singly-charged ions undergo charge-exchange (electron capture) collisions with a background population of neutral atoms. If the neutral atom population is much colder than that of the ions, then the ENA essentially preserve the velocity that the ion had just before the collision. Consequently, the energetic ion population will "glow" in ENA in all directions and an ENA image of the ion distribution can be recorded by a properly designed ENA "camera". Throughout the heliosphere, there is an energetic ion population : shock-associated ions (SAI), mainly protons, with energies extending well above 100 keV. There is also a relatively cold neutral atom population: interstellar atoms, mainly hydrogen, with densities ~0.1 cm^{-3} (except within a few AU of the sun, where the effects of ionization, radiation pressure, and gravitational focusing modify the interstellar atom distribution). Roelof [1] first described the scientific potential for ENA imaging of shock-associated ion structures, and Krimigis and Roelof [2] proposed that such imaging could be carried out by the ENA imaging camera (INCA) on the Cassini mission to Saturn whenever the spacecraft was near apoapsis in its orbit about the giant planet. Hsieh et al. [3,4] considered the same problem from a viewpoint at 1 AU with regard to whether or not the ENA produced by SAI will preclude the possibility of measuring ENA produced by unmodulated anomolous cosmic rays with energies above 10 keV. We will return to this latter topic at the end of the paper.

The general relationship between the ENA unidirectional intensity (differential in energy), and the ion flux that produces the ENA is given (for situations wherein the probablility of re-ionization of the ENA is ignorable), by the line integral

$$j_{ena} = \sigma \int ds\, n\, j_{ion} \tag{1}$$

where n is the cold neutral atom density and σ is the (energy-dependent) charge-exchange cross-section for the conversion of the singly-charged ion into an ENA. When there are several species of cold atoms present, the product σn is summed over all the species. The unidirectional ENA flux in any direction is calculated by taking the line integral (1) along the corresponding line of sight from the position where j_{ena} is evaluated outward to a distance beyond which j_{ion} is zero. Only the unidirectional ion flux that is directed toward the observer contributes to the line integral. Equation (1) can be scaled conveniently by introducing the characteristic distance a = 1 AU = 1.5 x 10^{13} cm, the cross-section σ_{16} (expressed in units of 10^{-16} cm^2), the interstellar neutral density n_0 (cm^{-3}) just inside the heliopause, and a characteristic ion flux j_0 which we identify with the peak ion flux at the helioradius of the observation point:

$$j_{ena}/j_0 = \sigma_{16}\, n_0\, I \tag{2}$$

$$I = 0.0015\ (cm) \int (ds/a)\ (n/n_0)\ (j_{ion}/j_0) \tag{3}$$

The hydrogen density essentially equals n_0 everywhere beyond a few AU helioradius, but nearer the sun, it decreases in a complicated manner that is solar-cycle dependent. We follow Holzer [5] and take, as a first approximation, $n = n_0 \exp(-r_0/r)$, with $r_0 = 3.3$ AU. The quantity I depends upon ion energy only through the ratio j_{ion}/j_0, so that if the spatial distribution function is only a weak function of energy over the energy range of interest, the morphology of the images can be deduced just from I alone. Estimates of integral ENA fluxes over a range of energies $\int dE$ j_{ena} can then be obtained from the integral $\int dE$ σ_{16} j_0. We now proceed to make more precise estimates by defining models for energetic proton distributions associated both with corotating interaction regions and driven shocks. We shall focus our analysis on ENA observations from 10 AU so that they are relevant to the upcoming NASA/ESA Cassini/Huyghens mission to Saturn.

QUANTITATIVE MODELS FOR SHOCK-ASSOCIATED ION DISTRIBUTIONS

The spatial and spectral characteristics of SAI above 35 keV have been measured out to 25 AU and up to 30° heliolatitude by the Low Energy Charged Particle (LECP) Experiment on the Voyager 1 and 2 spacecraft and analyzed by Gold et al. /6/. During 1984, Voyager 2 observed corotating SAI events while it was moving outward from 13 AU in the heliographic equatorial plane. Voyager 1 was moving outward from 18 AU and climbing above 22° heliolatitude. Gold et al. derived an average spectrum for the 8 largest events of this period. Because SAI fluxes decrease with increasing radial distance (see discussion of this point below), we adopt that spectrum as a conservative model for major corotating SAI events at 10 AU. The Voyager 2 spectrum has been tabulated in the TABLE presented below. It is assumed that the ions are predominantly protons. In order to model ENA production down to 10 keV, the Voyager spectrum, which is measured only down to 35 keV, has been extrapolated by eye to lower energies. Also included are the values of the charge-exchange cross-section ($\sigma_{16} = \sigma \times 10^{16}$) /7/ which allows us to calculate the integral $\int dE$ σ_{16} j_{ion} and thereby evaluate the integral ENA flux using (2) and (3) once we have computed I from the geometry of the models.

TABLE: Integral ENA Energy Coefficient Derived from Voyager Energetic Ions*							
E(keV)	10	20	40	70	100	200	400
σ_{16} (cm^2)	7.75	4.45	1.67	0.377	0.101	6.09(-3)	1.76(-4)
j_{ion} (cm^{2}sr s keV)$^{-1}$	[0.70]	[0.55]	0.35	0.20	0.125	0.036	0.010
$\int dE$ σ_{16} j_{ion} (sr s)$^{-1}$	[71.6]	[34.6]	8.81	1.36	0.31	9(-3)	1(-4)
*Ion fluxes in [brackets] are extrapolated downward in energy from 35 keV							

By the very nature of their production and propagation, SAI populations are bounded by interplanetary shocks. This means that an ENA image of the SAI will capture the global 3-dimensional structure of whatever shocks are propagating in the heliosphere at the time when the image is formed. For example, during the declining phase of the solar cycle, coronal holes are the dominant sources of 27-day recurrent solar wind streams that form quasi-spiral forward-reverse shock pairs beyond 1 AU as a consequence of the fast-slow interactions between the streams. During the declining phase of solar activity, there are usually two streams per rotation in the inner heliosphere. However, between 5 and 10 AU, there is a transition from two corotating interaction regions per solar rotation to a single merged interaction region (MIR) near 10 AU /8/. ENA images of these banner-shaped structures will immediately reveal information on the latitude dependence of the solar wind velocity. On the other hand, the large driven shocks that sweep outwards through the heliosphere during the years of solar maximum begin as radial shells that occupy several steradians of solid angle (as viewed from the sun), and confine intense energetic storm particle (ESP) events behind them. These ESP events "coast" outward and are usually accompanied by Forbush decreases in the cosmic ray intensity. The shock-on-shock interactions cause multiple ESP events injected from the continuous activity of major active regions to become entrained into a single compound event that can take several months to pass over a spacecraft in the outer heliosphere (r > 20 AU).

The starting point for the models is the undisturbed solar wind. There is a well-measured tendency for the solar wind velocity to increase away from the heliosperic equator, so we write $V = V_0 + \Delta V |\sin \Lambda|$, where Λ is the heliolatitude. Typical values in the outer heliosphere for moderate-sized recurrent solar wind streams might be $V_0 = 450$ km s^{-1} and $\Delta V = 150$ km s^{-1}. The ion fluxes themselves may be expected first to begin to increase with latitude away from the helio-equator (due to increasing shock strengths), but then corotating fluxes are likely to decrease again towards the poles as the stream-stream interactions weaken. The comparsion of Voyager 1 and 2 fluxes in 1984 /6/ showed an average negative latitude gradient of 3%/deg when Voyager 1 was above 20^0. Consequently we give these fluxes a latitude dependence of $(V/V_0) \cos \Lambda$. With the solar wind parameters we have chosen, this would maximize the ion flux at $\Lambda = 16^0$ (at a fixed helioradius). We then construct an MIR population centered on a single Parker spiral with a Gaussian radial dependence whose width (or s.d.) ℓ is constant in the outer helisophere, but narrows at smaller radii, i.e., $\ell = \ell_0 \exp(-\ell_0/r)$. This gives a semi-quantitative representation of the development of MIR-associated ion fluxes, neglecting the weaker one of the usual pair of streams inside a few AU. We choose $\ell_0 = 2$ AU because major 25-day recurrent ion events near 10 AU can last up to 6 days. It remains only to model the r-dependence of these fluxes. As we can see from the TABLE, the Voyager 2 LECP fluxes of 35 keV ions averaged about 0.4 (cm^2 sr s keV)$^{-1}$ for SAI during 1984 near 13 AU whereas in 1978 when both Voyagers were near 1 AU, a representative flux was about 10 (cm^2 sr s keV)$^{-1}$ [R. B. Decker, private communication]. Although the epochs are different, the class of events is similar, so if we take these two fluxes as representative, their ratio gives a rough radial flux dependence of $r^{-1.5}$.

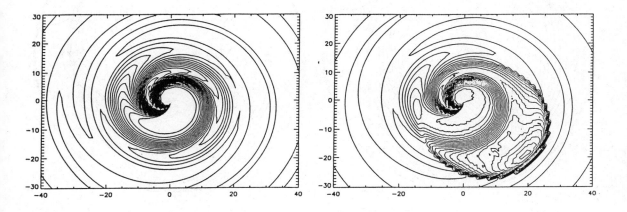

Fig. 1 Hypothetical ENA images of (a) MIR-associated energetic ions (left); and (b) with the addition of ESP fluxes associated with a driven shock (right). The plots are dimensioned in degrees, as viewed from 20 AU due north of the sun, so 3^0 corresponds to about 1 AU in this plane. Consequently the viewing point from 10 AU for the subsequent figures is at angular coordinates (0^0,-26.6^0).

Because we do not wish the model to be restricted purely to corotating SAI events, we also formulate a 3-dimensional energetic storm particle (ESP) flux enhancement for a large driven shock. Realistically, such a SAI event occurs in the presence of pre-existing MIR and other shock events. We therefore add the transient shock event on top of the above-described MIR model fluxes. Actually, a large driven shock would distort the MIR structure, but superposition seems to be a satisfactory zero-order approximation to a much more complex situation. Superposition yields a lower bound on the ENA fluxes, because the interaction of the driven shock would actually accelerate the pre-existing MIR ion population as it sweeps them up. The leading edge of the shock is taken to be an ellipsoid of revolution with its focus at the sun. For the present calculation, the axis of revolution is tilted 20^0 north of the helioequator some 45^0 west of the sun-Saturn line. The shock cuts this axis at a maximum radius $R_{max} = 10$ AU (on the Saturn side) and a minimum radius $R_{min} = 4$ AU (on the opposite side of the sun). The flux along the surface $R = R_S$ falls off as $(10 \text{ AU}/R_S)^{-1.5}$ and SAI fluxes inside the shock are reduced by the factor $\exp(R/L - R_{max}/L)$. These two functions mimic the r-dependence deduced above for the corotating shocks, plus an attenuation as one moves away from the nose of the driven shock. These additional SAI fluxes are normalized such that when the nose of the shock crosses 10 AU (as it is just doing for our example), the fluxes there are 5 times the peak undisturbed MIR fluxes at 10 AU.

The global configurations of the models are revealed in Figure 1 which depicts hypothetical ENA images as they would appear from 20 AU due north of the sun. The quantity that is linearly contoured is I from Equation (3). Fig. 1a shows the ENA image of the SAI population for a single (dominant) MIR, while in Fig. 1b the ESP protons associated with a driven shock are superimposed upon the pre-existing MIR population. The discontinuities in the contours are caused by the finite angular separation of the lines of sight on a $1^O \times 1^O$ grid.

Now that we have an overview of the models, let us return to the vicinity of Saturn and view these same events in the sunward hemisphere in Figure 2. As in Fig. 1, the quantity I is linearly contoured in 25 levels for the MIR-associated event in Fig. 2a and the ESP event in Fig. 2b. The angular resolution of the line-of-sight grid is now $2^O \times 2^O$, so as to correspond more closely to the actual instrument resolution attainable for ENA hydrogen above 50 keV. To aid in quantifying these simulations, selected contours are extracted from these images and labeled in the corresponding Figs. 2c and 2d. The quantity contoured is $100\, I$.

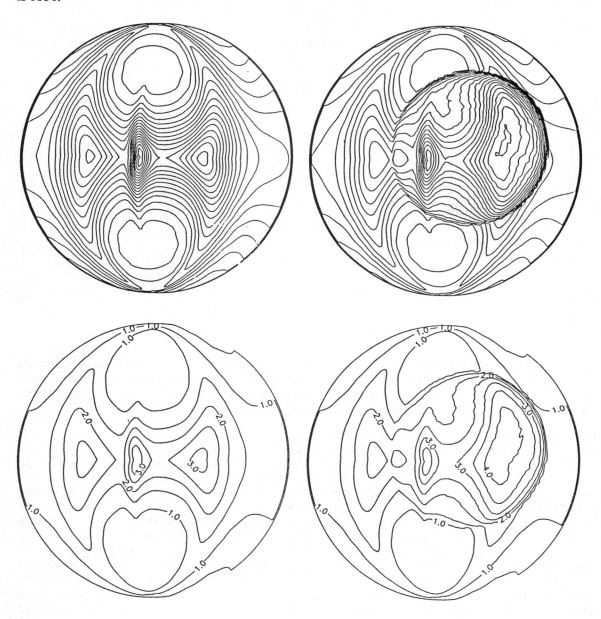

Fig. 2 Sunward hemisphere views from Saturn (10 AU) of simulated ENA images of (a) MIR-associated proton fluxes and (b) ESP event protons. Angular resolution of line-of-sight directions is $2^O \times 2^O$. Selected contours are labeled in (c) and (d) in units of $100\, I$,.the quantity defined in Equation (3).

DISCUSSION OF SIMULATED ENA IMAGES AND QUANTITATIVE INFERENCES

The first pattern that strikes the eye in Fig. 2a is the manner in which the latitude dependence of the solar wind velocity is revealed by the contours. The first winding of the MIR spiral is visible just to the left of the sun (which lies at the center of the image); compare with the view from north of the sun in Fig. 1a where 3° of angular displacement is equivalent to about 1 AU in radial distance. The outer ridge of this winding appears to be nearly normal to the equator because the model solar wind velocity is increasing almost linearly with latitude near the equator. This dependence is echoed in the next outward appearance of the MIR to the west of the sun, and again after a complete 360° winding (the furthest outward east of the sun). Such consistency in an observed image would indicate a quasi-stationary corotating MIR for the MIR, and conversely, any inconsistencies would imply temporal evolution. Of course, the entire pattern would rotate once every 25 days and thus provide more detailed checks on quasi-stationarity or evolution. Evolution would be strikingly apparent in the ENA images of an ESP event. Fig. 2b captures only one phase of the history of what must be a very complex and dynamic structure that can be followed from a few AU (once it emerges from the ionization-produced cavity in the interstellar H-atoms), until it passes over the observer and then appears mainly as a broad brightening of the antisunward hemisphere. These simulations show that ENA images can reveal the global structure of shock-associated ions, and hence of the topology of the shocks themselves.

We now take up the crucial question of whether or not such images can be obtained with current technology. The issue is one of both foreground and background counting rates for the ENA cameras. We can estimate the foreground rates from the contours of $100\,I$ in Figs. 2c and 2d. From the TABLE, we estimate the integral ENA hydrogen flux coefficient for $E > 10$ keV to be 71.6 $(\text{sr s})^{-1}$. We are taking $n_0 = 0.1$ cm^{-3}. Since $100\,I$ runs from about 1 to 3 or 4 in the sunward hemisphere, Equation (2) predicts integral ENA hydrogen fluxes of between 0.07 and 0.2 to 0.3 $(\text{cm}^2\text{ sr s})^{-1}$ above 10 keV. ENA cameras currently under development /9,10/ have effective geometric factors $g_{eff} = \varepsilon g$ of about 1 cm^2 sr (including the counting efficiency ε), so the counting rate of the instrument will be around 0.1 s^{-1} above 10 keV while viewing the sunward hemisphere. If triple coincidence techniques are utilized, the background rate due to interplanetary Lyman-α radiation for such an instrument can be kept down to 0.01 s^{-1} or less. Thus, the foreground/background ratio is acceptable.

However, the foreground rate itself must be considered. It must be high enough so that a useful image can be formed on the time scale of the temporal changes within the image itself. For heliospheric shock imaging from 10 AU, that means that an image must be obtained before the shock moves a substantial fraction of an AU. Therefore the camera must have an exposure time of no longer than a few days with an angular resolution of at least 1 AU/10 AU, or 6°. A typical ENA camera instantaneously views about 0.1 sr of the sky, and hence must scan in some manner so as to cover as much of the sky as possible. Scanning therefore introduces a "duty cycle" factor which equals the fraction of the total scan time that a particular line of sight is viewed by the detector. This fraction is design-dependent; for a fan-shaped collimator that is wide-angle in the plane containing the scan axis and narrow-angle ($\Delta\phi$) transverse to it, the duty cycle factor is $f = \Delta\phi/2\pi$. Let us choose 6° x 6° pixels and $\Delta\phi = 6°$, so that the instantaneous viewing solid angle of 0.1 sr is divided up into 10 pixels and $f = 1/60$. Then the foreground rate for a region of the sky 6° x 6° will be about $(0.1$ to $0.3\text{ s}^{-1})(1/10)(1/60) = 0.6$ to 1.8 counts/pixel-hour or 15 to 43 counts/pixel-day. The maximum allowable exposure time of 3 days will therefore give Poisson counting uncertainties ranging from 15% to 9%, respectively, from the dimmer and the brightest pixels. As pointed out above, background counts should not be high enough to contribute to the Poisson statistics. These statistics are adequate for interpreting the image in terms of global topology, particularly in the portions of the image where the contrast is strongest--at the edge of the shock structure (see Fig. 2).

There remains the question of how the ENA fluxes associated with shocks will affect the detectability of ENA produced in the outer heliosphere by unmodulated anomalous cosmic rays (ACR) charge-exchanging with the interstellar H-atoms. Hsieh et al. /4/ estimate that there is a rather isotropic flux of ENA hydrogen between 10 and 1000 keV of about 0.008 $(\text{cm}^2\text{sr s})^{-1}$ at 1 AU. According to the previous discussion, this is about an order of magnitude less than the lowest integral ENA fluxes in the sunward hemisphere. However, the models were constructed from the larger MIR and ESP events measured by the Voyager 2 LECP Experiment during 1984, and, in any case, one would not look sunward from 10 AU in order to measure ENA from the outer heliosphere. The model fluxes in the antisunward direction for the MIR (undisturbed by any ESP event) are shown in Fig. 3. The contours of $100\,I$ are now drawn every 0.25 units. The ENA flux is nearly constant at low latitudes because, as can be seen from Fig. 1a, the viewing point at 10 AU is immersed in the denser portion of the MIR-associated ion fluxes. The line-of-sight integrals are carried out to a helioradius of 50 AU, but because the model ion fluxes fall off as $r^{-1.5}$, the "local arm" of the MIR dominates the integral. The fluxes drop off significantly toward higher latitudes, as they also do in the sunward hemisphere. <u>Although the question deserves more thorough study, it appears that it may be possible to measure ENA from unmodulated ACR during periods of lower interplanetary activity, as suggested by Hsieh et al. /3,4/.</u>

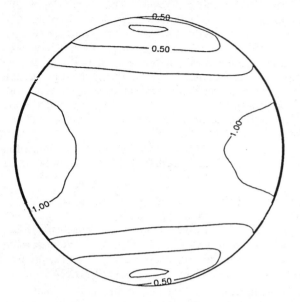

Fig. 3. Antisunward hemisphere view of MIR model values of the quantity 100 I.

Acknowledgments. I am grateful to K. C. Hsieh for providing me preprints of the referenced publications. This research was supported in part by NASA Grants NAG 2-642 and NAGW 2691 to The Johns Hopkins University/ Applied Physics Laboratory.

REFERENCES

1. E. C. Roelof, Imaging heliospheric shocks using energetic neutral atoms, <u>EOS 72</u>, #17, 226 (1991)

2. S. M. Krimigis and E. C. Roelof, Energetic neutral atom imaging of Saturn, Titan, and the outer heliosphere, <u>EOS 72</u>,#17, 236 (1991)

3. K. C. Hsieh, K. L. Shih, J. R. Jokipii, and M. A. Gruntman, Sensing the solar-wind termination shock from earth's orbit, this conference (1992)

4. K. C. Hsieh, K. L. Shih, J. R. Jokipii, and S. Grzedzielski, Probing the heliosphere with energetic hydrogen atoms, <u>Astrophys. J.</u>, in press (1992)

5. T. E. Holzer, Neutral hydrogen in interplanetary space, <u>Rev. Geophys. Space Phys. 15</u>, 467 (1977)

6. R. E. Gold, R. B. Decker, S. M. Krimigis, L. J. Lanzerotti,. and C. G. Maclennan, The latitude and radial dependence of shock acceleration in the heliosphere, <u>J. Geophys. Res. 93</u>, 991 (1988)

7. C. F. Barnett, H. T. Hunter, M. I. Kirkpatrick, I. Alvarez, C. Cisnernos, and R. A. Phaneuf, Atomic Data for Fusion: Collisions of H, H_2, He and Li atoms and ions with atoms and molecules, <u>Oak Ridge National Laboratory, ORNL-6086/V1</u> (1990)

8. Y. C. Whang and L. F. Burlaga, Simulation of period doubling of recurrent solar wind structures, <u>J. Geophys. Res. 95</u>, 20663 (1990)

9. R. W. McEntire and D. G. Mitchell, Instrumentation for the imaging of energetic neutral atoms, <u>Proc. Yosemite 1988 Outstanding Problems in Solar System Plasma Physics: Theory and Instrumentation</u>, eds. J. H. Waite, J. L. Burch, and R. L. Moore, AGU Geophys. Monograph <u>54</u>, 69 (1989)

10. K. C. Hsieh, C. C. Curtis, C. Y. Fan, and M. A. Gruntman, Techniques for the remote sensing of space plasma in the heliosphere via energetic neutral atoms: A review, this conference (1992)

PICK-UP IONS DEPENDENCE/INFLUENCE ON THE SOLAR WIND CHARACTERISTICS; POSSIBLE RELEVANCE TO THE DIAGNOSTICS OF HELIOSPHERIC PLASMA–VLISM INTERACTIONS

D. Rucinski

Space Research Centre of the Polish Academy of Sciences, Ordona 21, 01-237 Warsaw, Poland

ABSTRACT

Using the "hot" kinetic model of the distribution of interstellar neutral atoms in interplanetary space and including all important ionizing processes acting on neutral gas, the local production rates and expected total fluxes of various pick-up ions are calculated at different regions of the heliosphere. Basing on the comprehensive HELIOS data on the global parameters and their dependencies at different types (velocities) of the solar wind, the variability (weak for He^+, ~ factor of 2 for H^+) of the expected pick-up fluxes due to actually prevailing (slow/fast) solar wind type is discussed. The enrichment of the solar wind composition in singly ionized pick-up ions is calculated and it is shown that H^+ and He^+ pick-up ions may attain in the outer heliosphere ~ 10% of the primary proton and α-particle abundancies.

INTRODUCTION

According to the present understanding of gas-plasma interactions in inter-planetary space, "pick-up" ions are considered as the intermediate stage in the conversion of the neutral interstellar atoms to the anomalous cosmic rays (ACR). The existence of the neutral diffuse component in the he-liosphere was predicted theoretically by Fahr /1/ and Blum and Fahr /2/ and soon confirmed observationally by first measurements of solar backscattered radiation in Lyman-α /3,4/ and He-584$\overset{\circ}{A}$ /5,6/ lines. Following the general scenario of the interaction processes in interplanetary medium, the neutral gas penetrating into the heliosphere is subject to the various ionization reactions and becomes partly ionized in the vicinity of the Sun. The newly created ions (so-called "pick-up" ions) are picked-up by the frozen magne-tic field of the solar wind and are convected outwards with the solar wind flow. These ions undergo acceleration (most probably at the termination shock) and become the anomalous cosmic rays /7,8/, detected experimentally in early 1970's /9,10/. The pick-up ions considered to be a seed population for ACR were also identified experimentally in interplanetary space and the first measurements of He^+ fluxes were realized during AMPTE/IRM mission /11/. In recent years the advanced theoretical description of the evolution of pick-up ion distribution function due to pitch-angle scattering, adiaba-tic cooling, energy diffusion was developed /12,13,14/ and the effects of hydromagnetic wave excitation by these ions were studied /15/. It was also suggested that the analysis of the pick-up ion spectra may serve as a new (alternative to the backscattered glow data) source of information about the interstellar wind properties and conditions in VLISM /16/. In this paper we intend to analyze the dependencies of the pick-up ion fluxes on the cha-racteristics of the actually prevailing solar wind type and the influence of the pick-up ions on the evolution of the solar wind composition.

THEORETICAL APPROACH

In order to calculate the local production rates and relevant total fluxes of various pick-up ions at different solar conditions, we follow the approach of Vasyliunas and Siscoe /17/ modified however by the exact treatment of such reactions (like electron impact) which rates do not fall like $1/r^2$ with heliocentric distance. To express the distribution of the neutral interstellar "background" the "hot" model solutions /18,19/ are generally adopted, modified however for the exact treatment of the above-mentioned "non-standard" ionizing processes /20/. Assuming that the total destruction frequency of neutrals - $\beta_T(r)$ may be represented as a sum of all contributions due to photoionization, various charge-exchange reactions and electron impact ionization in the following form:

$$\beta_T(r) = \beta_{Ph}(r) + \beta_{Ex}(r) + \beta_{El}(r) \tag{1}$$

the resulting neutral gas density at local point $\vec{r} = (r,\Theta)$ according to the formula in /21/ (see there for explanations) is given by:

$$n(\vec{r})=\iiint_{d^3v} f_\infty(v_\infty(\vec{r},\vec{v})) \cdot \exp\left[-\frac{(\beta_{Ph}^0 + \beta_{Ex}^0)r_0^2\theta'(\vec{r},\vec{v})}{\Pi(\vec{r},\vec{v})v_\infty}\right] \cdot \exp\left[-\frac{\int_0^{\theta'}\beta_{El}' r'^2 d\theta^*}{\Pi(\vec{r},\vec{v})v_\infty}\right] d^3\vec{v} \tag{2}$$

The first exponential term in Eq.(2) /expressed analytically/ describes the ionization losses accumulated along atom's trajectory due to all considered ionizing processes which rates drop like $1/r^2$ with the solar distance. The second term (calculated numerically) contains similar losses, but related to the other (i.e."non-standard") ionizing reactions, which rates do not obey $1/r^2$ dependence. Using Eq.(2) one may straightforward express the local production rate $P(\vec{r})$ [in $cm^{-3}s^{-1}$] and resulting total flux $F(\vec{r})$ [in $cm^{-2}s^{-1}$] of newly created ions as:

$$P(\vec{r}) = n(\vec{r}) \cdot \beta_T^+(r) \tag{3}$$

$$F(\vec{r}) = \int_{r_s}^{r} n(\vec{r}') \cdot \beta_T^+(r') \cdot \left(\frac{r'}{r}\right)^2 dr' \tag{4}$$

It is worthwhile to note that the quantity β_T in Eq.(1) may in general differ (especially significantly in the case of He^{++} pick-up ions) from the quantity β_T^+ in Eqs.(3) and (4). One should not misinterpret this difference as due to the fact that the integration variables in Eq.(2) and Eq.(4) are different, i.e. that in the first case the integration is carried along neutral atom's trajectory, whereas in the second one along a solar wind parcel trajectory. The key reason of this difference is the fact that β_T corresponds to all processes responsible for the ionization (i.e destruction) of the considered neutral component, whereas β_T^+ reflects the sum of the rates of only these ionizing processes which are responsible for the production of the specific type of the pick-up ion (for detailed explanation see /22/).

DEPENDENCE OF HE^+ AND H^+ FLUXES ON SOLAR WIND CHARACTERISTICS

The comprehensive analysis of HELIOS data indicates that the high- and low speed solar wind types are significantly different in their plasma parameters (density, temperature-solar distance relations). The compiled characteristics (based on data reported by Schwenn /23/ and Marsch et al. /24/) of global plasma properties containing observed (at 0.3-1.0 AU) and inferred (0.014-0.3 AU) gradients in the solar wind temperature and data on electron (= proton) density was presented recently by Rucinski and Fahr /21/ (see Table 1 of their paper) for five categories of the solar wind of different velocities. It was shown that noticeable differences especially in solar wind densities (up to factor of 3 when comparing slow and fast types) should lead to the differences in ionization rates, which in turn affect the production of pick-up ions. Taking into account all important for H and He ionization processes (see compilation by Grzedzielski and Rucinski /25/) one may easily calculate the production rates and resulting fluxes of H^+ and He^+ ions as a function of the solar wind type (slow/fast). The detailed analysis of this problem was presented recently by Rucinski and Fahr /21/; below the main conclusions are summarized.

Helium component

Neutral interstellar helium, due to the high ionization potential is able to penetrate deep into the Solar System and the bulk of ionization losses occurs for this constituent within the region of ~ 1-2 AU around the Sun. The accelerated (by the solar gravitational potential) helium atom may co-ver this region in a few months, therefore thinking about global long-term (≥ 1 year) variations of the solar wind type it is justified to assume that helium flow is embedded (in the period of most efficient ionization and related with that creation of He^+ pick-up ions) in the same type of the solar wind. Since the cross-section for single charge-exchange with solar wind protons is very low for helium (~ $5 \cdot 10^{-18}$ cm^2) and double charge-exchange with solar α-particles does not contribute to the production of He^+ pick-up ions, therefore analyzing production processes in the inner helio-sphere (~ 1 AU) as a function of the solar wind type (assuming that pho-toionization rate is constant and independent of the solar wind type; β_{Ph}^{He} at 1 AU is assumed to be equal to $6.8 \cdot 10^{-8}$ s^{-1} for purposes of this paper) one may practically consider the changes of electron impact efficiencies at dif-ferent solar wind conditions. Applying the explicit formula for $\beta_{El}^{He}(r)$ from Rucinski and Fahr /21/ and inserting the resulting value for this quantity to Eq. (4) one may straightforward calculate expected He^+ fluxes for various solar wind types.

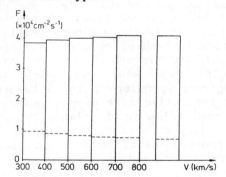

Fig. 1. Calculated total He^+ ion fluxes at 1 AU for different solar wind types (velocities). The upper (solid line) level corresponds to the downwind axis; the dashed one to the upwind axis.

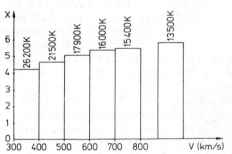

Fig. 2. Ratio of downwind to upwind He^+ fluxes at 1 AU. The temperature above each column corresponds to to the temperature which the in-terstellar gas should have (in case of negligence of electron impact in interpretational model) in order to lead to the specific ratios of the downwind to upwind fluxes. The separate column indi-cates the expected ratio (for the assumed T_∞^{He} = 13500 K) when electron impact effect is comple-tely neglected.

As it is shown in Fig. 1 the calculated fluxes at 1 AU are rather weakly de-pendent on solar wind velocity and expected discrepancies are within 15%, what remains in agreement with the conclusions extracted from AMPTE/IRM data /16/. It is due to the fact that electron impact is still a minor effect in comparison to photoionization and helium is weakly coupled to the solar wind plasma. However an interesting correlation may be found: on the upwind side slightly higher He^+ fluxes are connected with smaller velocities, whereas in the anti-apex direction the situation is just opposite. It is due to the fact that on the upwind side the small increase of β_{E1} in β_T^+ overcompensates the slight depletion of neutral density due to higher β_{E1} in β_T and on the downwind side the situation is opposite. As it results, more pronounced (ranging up to 25%) variations may be expected in the downwind to upwind ratios (see Fig.2). This ratio in case of He^+ pick-up ions is quite sensitive (at 1 AU) to changes of ionization rates values. As it is shown in Fig.2 the influence of the electron impact ionization may diminish this ratio by up to 25% in comparison to the theoretical case, when one neglects totally this effect. On the other hand it is well known (providing that the values of rates are properly controlled) that downwind to upwind ratio is a good indicator of interstellar wind temperature, which is still not known very accurately. If one therefore tries to derive helium temperature from downwind/upwind ratios using not enough refined interpretational model it may come out that the derived temperature has only "artificial" meaning and may disagree with the (unknown *a priori*) physical one by factor of ~ 2, since the value of the considered ratio may reflect not only the real temperature but also possible inaccuracies in the adopted efficiencies of ionization processes.

Hydrogen component

It is well known that interstellar hydrogen gas in contradistinction to helium component becomes relatively easily ionized in an extended region around the Sun. The patterns of the density distribution in the inner Solar System are shaped as the accumulated result of ionization processes acting on hydrogen during a few years of it's inflow towards the Sun. Since during this relatively long time different types of the solar wind may contribute to the formation of the hydrogen density pattern, it is thereby more realistic to use in Eq.(2) just the mean contribution instead of specific one related to the specific type of the solar wind. On the other hand in analyzing the production of pick-up ions in specific solar wind type one should insert into Eq.(4) the specific values of β_T corresponding to the actually considered type. The solar wind influence on ionization of hydrogen is twofold: by charge-exchange with solar wind protons, which is of primary importance and by electron impact which plays secondary role. Additionally, photoionization (treated here as constant and independent of solar wind type, with $\beta_{Ph}^H = 7.0 \cdot 10^{-8} s^{-1}$) contributes by $\approx$ 10-20% to the total rate.

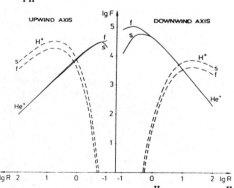

Fig.3. Expected H^+ (dashed curves) and He^+ (solid curves) total fluxes along upwind and downwind directions as a function of heliocentric distance R (in AU). The curves denoted by "s" correspond to the slow (300-400 km/s) solar wind and these by "f" to the fast (700-800 km/s) one.

Using the formulae for $\beta_{Ex}^H(r)$ and $\beta_{E1}^H(r)$ from Rucinski and Fahr /21/ and applying them for each of five types of the solar wind one may find that due to the significant differences in proton densities and cross-sections dependencies on energy the ionization rate for slow solar wind exceeds that for high speed solar wind by factor of ~ 2, and by the same factor H^+ fluxes in slow solar wind stream are expected to be higher than in the fast one (see Fig.3)

EXPECTED HEAVY PICK-UP ION FLUXES

It is well known from the astrophysical observations that interstellar medium contains also heavier elements like C, N, O, Ne and others with relatively low abundancies, which do not exceed 10^{-3} that of hydrogen /26/. Although up to now there carried out neither observations of the interplanetary glow backscattered by heavy components, nor direct measurements of heavier interplanetary pick-up ions, their existence in the heliosphere was confirmed by the registration of the relevant ACR spectra. In the present paper, we consider these heavier elements, which were observationally identified in ACR and which due to their relatively high abundancies in the interstellar medium seem to be most promising for the potential future detection in form of pick-up ions, namely: oxygen (O), neon (Ne), nitrogen (N) and carbon (C). For the computational purposes, we assume that interstellar gas containing heavy elements is inflowing into the Solar System with bulk velocity V = 20 km/s and it's thermal spread is characterized by $T = 10^4$K. The cosmic abundancies of these elements are taken from the compilation of Matteucci /27/, where the relative abundancies are as follows:

$$C/H = 3 \cdot 10^{-4}; \quad N/H = 4 \cdot 10^{-5}; \quad O/H = 5 \cdot 10^{-4}; \quad Ne/H = 7.9 \cdot 10^{-5}$$

The ionization rates corresponding to solar minimum and solar maximum conditions due to all important reactions for these elements (see Axford /28/) are taken from Siscoe and Mukherjee /29/ and Fahr /30/. For the present purposes, we assume that the heliospheric interface region is ideally transparent for all considered species (i.e. the transmissivity through the heliospheric boundary layer is equal to 100%) and we adopt Fahr's /30/ factors ξ describing (for each element) the degree of it's fractional ionization in VLISM; namely: $\xi(C)=0.99$; $\xi(N)=0.25$; $\xi(O)=0.32$ and $\xi(Ne)=0.0$. The validity of the last two assumption is disputable, since our knowledge on these problems is rather poor, but potential pick-up ion measurements may verify this doubts. As it results just from the relative abundancies and fractional ionization in VLISM and ionization rates in the heliosphere oxygen and neon are the most promising candidates among heavier elements for creation of noticeable fluxes of pick-up ions. Carbon, due to extremely high fractional ionization in VLISM and very high photoionization rates inside the heliosphere is practically absent in the inner part of the Solar System because of prior ionization. Nitrogen is less abundant that neon and it's fractional ionization in VLISM is higher. Considering then O and Ne as the main heavier constituents which potentially could be detected in pick-up measurements in the inner (R < 10 AU) heliosphere one should however be aware of quite important differences in their density distribution in the circumsolar area. Neon, due to the high ionization potential is able (similarly to helium) to penetrate to the close vicinities of the Sun before being ionized, and the strong gravitational focusing leads to the formation of the pronounced cone structure of the enhanced (by factor 3-8 at 1 AU depending on the solar cycle activity) density along downwind axis. The presence of the strong downwind/upwind asymmetry up to factor of 10 at 1 AU is expected (see Fig.4). Oxygen, in contrary to the hasty expectations, behaves in significantly different manner. Due to much higher ionization rates (ranging up to $10^{-6} s^{-1}$ at solar maximum) the oxygen density near the Earth's orbit is well below value of oxygen density in VLISM and practically no characteristic cone structure occurs in the inner part of the Solar System. Some slight enhancement of the downwind density (not exceeding 50%) may but not must occur along downwind axis, but only at larger (R > 10 AU) distances. The oxygen density may critically depend on the solar activity and during the extremely high activity the oxygen azimuthal density profiles may resemble those of hydrogen where stronger depletion is present in the downwind region. These density patterns of both considered elements affect of course the production rate and expected total fluxes of O$^+$ and Ne$^+$ pick-up ions. As it is shown in Fig. 4 the maximum of the Ne$^+$ pick-up ion flux is located inside or close to the Earth's orbit in the anti-apex direction and significant asymmetry (by factor of about 10) between upwind hemisphere and the cone region should occur. This provides congenial circumstances for potential Ne$^+$ detection from the Earth-bound satellite, since neon downwind fluxes at 1 AU may reach ≈ 60 cm^{-2}s^{-1} (at solar minimum) and exceed notice-ably fluxes of other heavier constituents. Oxygen fluxes should not expose such distinct downwind/upwind differences. The expected maximum values (at solar minimum) may attain 10-20 cm^{-2}s^{-1} in a quite extended region between 1 and 10 AU. The expected N$^+$ fluxes are fairly low (≤ 2 cm^{-2}s^{-1}).

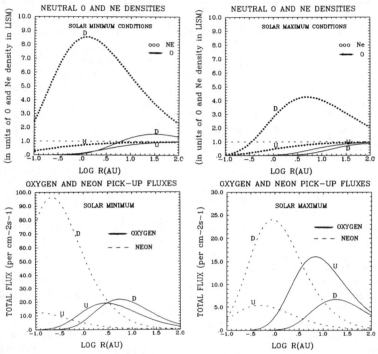

Fig. 4 Neutral O (solid line) and Ne (dashed line) densities /upper level/ and expected O$^+$ and Ne$^+$ pick-up ion fluxes /lower level/ along down-wind (D) and upwind (U) directions for solar minimum /left panel/ and solar maximum /right panel/ conditions.

As it results from the calculations the fluxes of N$^+$ and Ne$^+$ ions are (excluding narrow cone region) almost identical outside 5 AU. This re-mains in agreement with the N and Ne abundancies as derived from ACR data by Cummings and Stone /31/. It becomes also evident that the ratio of O$^+$/Ne$^+$ fluxes is slowly increasing with heliocentric distance. For the solar mini-mum conditions it raises in the upwind heliosphere from about 12 at 10 AU up to 18 at 50 AU. It is therefore by factor of 2-3 higher from the ratio of O/Ne abundancies as derived from ACR observations /31/. This may suggest that O$^+$ production calculated here is overestimated due to the idealistic assumption that no filtration effects occur at the heliospheric boundary. Theoretical estimates of Bleszynski /32/ indicate at ≈ 30% depletion of oxy-gen density after passage through the heliospheric boundary layer. The simi-lar analysis of Fahr (on the basis of Parker's model) leads to even lower transmissivity of oxygen through the interface region /33/. Therefore the potential future measurements of pick-up fluxes supported by the detailed comparison for different species may be extremely helpful in the estimation of at least two key factors known up to now very poorly: the fractional io-nization in VLISM and transparency of the heliospheric interface.

SUMMARY OF PICK-UP IONS INFLUENCE ON SOLAR WIND COMPOSITION

The creation of pick-up ions affects the chemical composition of the solar wind. The fluxes of newly-born ions at 1 AU are very low and do not exceed about 10^{-4} of the original proton flux, but the abundancies of the newly-created species increase systematically with the heliocentric distance. As one may notice from Fig.5 the H^+ pick-up ions may contribute by $\approx$ 10% at 50 AU and $\approx$ 20% at 80 AU to the total abundance of the proton population convected outwards with the solar wind. At the same distances, the abundance of newly-created He^+ ions attains 7 and 12% of the primary solar wind α-particles abundance, respectively. The corresponding fluxes of He^{++} ions are about 20-30 times lower than He^+ fluxes and this ratio reflects the differences in the values of the relevant β_T quantities which determine the creation rate of both considered species of helium ions /22/. The heavier ions become also more abundant in the outer heliosphere and for example the expected abundance of O^+ ions at 50 AU reaches about 10^{-5} of the of primary protons. The expected (at quiet Sun conditions) abundancies of different pick-up ions as a function of distance from the Sun are shown in Fig.6.

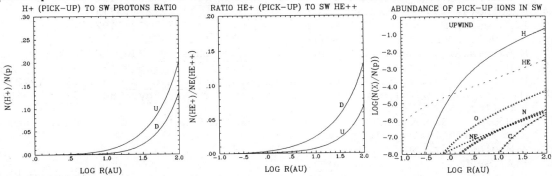

Fig.5. Ratios of H^+(pick-up) to solar wind protons /left panel/ and He^+(pick-up) to solar α-particles /right panel/ along downwind (D) and upwind (U) axis

Fig.6. Ratio of abundancies of various pick-ups to SW protons for low solar activity

It is also worth to point out that in the charge-exchange reactions besides the above-mentioned pick-up ions, also a new population of energetic neutral atoms (ENA) is simultaneously originated. According to Grzedzielski and Rucinski /25/ up to 6-10% of the total energy carried by the distant (at 50-80 AU) solar wind may be transported by the energetic neutrals.

PROSPECTS FOR THE FUTURE

The experimental confirmation of the presence of heavy constituents in the heliosphere (either in form of neutral atoms or pick-up ions) become an ambitious task for the future space experiments. Besides the planned ORFEUS experiment /34/ aiming at the first optical measurements of the solar EUV radiation resonantly backscattered by heavier elements in the interplanetary space, the attempts to detect heavy pick-up ions seem to be promising. Future space missions like SOHO and CASSINI equipped with sensitive pick-up ion detectors provide favourable opportunities for potential detection of heavier ions. Referring to the theoretical reconaissance discussed above one may explicitly conclude that the Earth-bound spacecraft (like SOHO) may be particularly predestinated to the detection and diagnostics of Ne^+ pick-up ions. Operating permanently at 1 AU such spacecraft may cross the cone region of especially enhanced Ne^+ fluxes and may also monitor the expected pronounced downwind/upwind asymmetry. Moreover, the possible simultaneous detection of He^+ fluxes may allow for the more accurate derivation of interstellar wind parameters. CASSINI mission directed towards Saturn may nicely complement the potential capabilities of SOHO and enlarge chances of detection of O^+ ions. CASSINI's expected trajectory /16/ provides particularly attractive conditions for the search of heavier pick-up species. In the initial phase, during circumterrestrial orbiting there is a chance to identify Ne^+ cone structure in the downwind region, whereas in the upwind hemisphere one may expect the potential detection of O^+ fluxes which for the low solar activity exceed by factor of 2 those corresponding to Ne^+. Later, during the cruise phase towards Saturn the potential continuous monitoring of O^+ and Ne^+ fluxes at the expected level of the order of 10-20 $cm^{-2}s^{-1}$ may allow for the comparative study of these constituents. It may lead to the verification of the amount of oxygen present in interplanetary space and to the indication of it's fractional ionization in VLISM and determination of the filtration effect at the heliospheric boundary layer.

REFERENCES

1. H.J.Fahr, Astrophys. Space Sci. 2, 474 (1968)
2. P.W.Blum and H.J.Fahr, Astron. Astrophys. 4, 280 (1970)
3. J.L.Bertaux and J.E.Blamont, Astron. Astrophys. 11, 200 (1971)
4. G.E.Thomas and R.F.Krassa, Astron. Astrophys. 11, 218 (1971)
5. F.Paresce, S.Bowyer and S.Kumar, Astrophys. J. 187, 633 (1974)
6. C.S.Weller and R.R.Meier, Astrophys.J. 193, 471 (1974)
7. M.E.Pesses, J.R.Jokipii and D.Eichler, Astrophys. J. 246, L85 (1981)
8. J.R.Jokipii, J. Geophys. Res. 91,2929 (1986)
9. M.Garcia-Munoz, G.M.Mason and J.A.Simpson, Astrophys. J. 182, L81 (1973)
10. D.Hovestadt, O.Vollmer, G.Gloeckler and C.Y.Fan, Phys. Rev. Lett. 31, 650 (1973)
11. E.Möbius, D.Hovestadt, B.Klecker and M.Scholer, G.Gloeckler and F.M.Ipavich, Nature 318, 426 (1985)
12. E.Möbius, Adv. Space Res. 6, # 1, 199 (1986)
13. P.A.Isenberg, J. Geophys. Res. 92, 1067 (1987)
14. E.Möbius, B.Klecker, D.Hovestadt and M.Scholer, Astrophys. Space Sci. 144, 487 (1988)
15. M.A.Lee and W.H.Ip, J. Geophys. Res. 92, 11041 (1987)
16. E.Möbius, in: Physics of the Outer Heliosphere, ed. S.Grzedzielski and D.E.Page, Pergamon Press, 1990 p.345
17. V.M.Vasyliunas, G.L.Siscoe, J. Geophys. Res. 81, 1247 (1976)
18. H.J.Fahr, Astron. Astrophys. 14, 263 (1971)
19. G.E.Thomas, Ann. Rev. Earth Planet Sci. 6, 173, 1978
20. D.Rucinski and H.J.Fahr, Astron. Astrophys. 224, 290 (1989)
21. D.Rucinski and H.J.Fahr, Ann. Geophys. 9, 102 (1991)
22. R.Ratkiewicz, D.Rucinski and W.H.Ip, Astron. Astrophys. 230, 227 (1990)
23. R.Schwenn, in: Solar Wind Five, NASA CP-2280, 1983, p.489
24. E.Marsch, W.G.Pillipp, K.M.Thieme, H.Rosenbauer, J. Geophys. Res. 94, 6893 (1989)
25. S.Grzedzielski and D.Rucinski, in: Physics of the Outer Heliosphere, ed. S.Grzedzielski and D.E.Page, Pergamon Press, 1990, p.367
26. A.G.W.Cameron, Space Sci. Rev. 15, 121 (1973)
27. F.Matteucci, MPA report 477 (1989)
28. W.I.Axford, in: Solar Wind, ed. C.P.Sonnett, P.J.Coleman,Jr. and J.M.Wilcox, NASA SP-308, 1972, p.609
29. G.L.Siscoe and N.R.Mukherjee, J. Geophys. Res. 77, 6042 (1972)
30. H.J.Fahr, in: Physics of the Outer Heliosphere, ed. S.Grzedzielski and D.E.Page, Pergamon Press, 1990, p.327
31. A.C.Cummings and E.C.Stone, in: Proceedings of the Sixth International Solar Wind Conference, ed. V.J.Pizzo, T.E.Holzer and D.G.Sime, NCAR/TN 306, Boulder, 1988, p.599
32. S.Bleszynski, Astron. Astrophys. 180, 201 (1987)
33. H.J.Fahr, Astron. Astrophys. 241, 251 (1991)
34. S.Bowyer and H.J.Fahr, in: Physics of the Outer Heliosphere, ed. S.Grzedzielski and D.E.Page, Pergamon Press, 1990, p.29

MAGNESIUM, CARBON, AND OXYGEN ABUNDANCES IN DIFFERENT SOLAR WIND FLOW TYPES, AS MEASURED BY SWICS ON ULYSSES

R. von Steiger,* J. Geiss,* G. Gloeckler,** H. Balsiger,* A. B. Galvin,**
U. Mall* and B. Wilken***

* Physikalisches Institut, University of Bern, Sidlerstrasse 5, CH-3012 Bern,
Switzerland
** University of Maryland, College Park, MD 20742, U.S.A.
*** MPI für Aeronomie, D-3411 Katlenburg-Lindau, Germany

ABSTRACT

The SWICS instrument on the Ulysses spacecraft has now gathered an almost continuous database of in–ecliptic solar wind composition data for many months. In this paper, we concentrate on the elements Mg, C, and O. We report relative abundances C/O and Mg/O and correlate their variations with different solar wind conditions and flow types, as inferred from the coronal freezing-in temperature. We find average abundance ratios of C/O = 0.62 and Mg/O = 0.13 over a period of almost 3 months. The daily Mg/O averages during this period are clearly correlated with the coronal temperature, while the same correlation is weaker for the C/O daily averages. This finding is briefly discussed in relation to the FIP fractionation effect.

INTRODUCTION

The SWICS instrument on the Ulysses spacecraft /1/ is a new kind of sensor, and this is the first time it has been flown in the solar wind. The sensor uses an energy per charge E/q measurement, followed by a time-of-flight mass per charge M/Q telescope, and a total energy E detector. Each incident ion is thereby classified by its mass, charge, and energy separately. However, due to the limited telemetry, the data cannot be transmitted for every ion, but only for a sample of 30 (8) ions per spin of 12 s at high (low) bit rate. The other ions are bundled into count rates at the cost of some resolution in M, M/Q, and/or E/q. In this work, we exclusively use the samples of direct events (called PHA data) and therefore can exploit the full resolution of the instrument, but at the cost of limited statistical precision. The resulting two-dimensional $M–M/Q$–spectra (see Fig. 1), produced with a time resolution of 13 minutes, are ideally suited to the study of solar wind composition and its variations over a wide range of time scales.

DATA SELECTION AND ANALYSIS

We have analyzed the PHA data of a continous period of almost 3 months, days 134–216, 1991. The procedure is as follows: The direct pulse height (PHA) events are collected on a daily basis. Each PHA event contains the energy and time-of-flight of an incident ion, which (using the E/q set at the deflection system) is then converted to M and M/Q of the ion. In the resulting 2–dimensional $M–M/Q$ spectra we define elliptical regions corresponding to the charge states investigated: C^{6+}, C^{5+}, C^{4+}, O^{8+}, O^{7+}, O^{6+}, and Mg^{10+}. Using pre-flight calibration data, the number of counts falling in each of these elliptical regions is converted to the differential flux of the corresponding charge state in the solar wind, and by integrating the differential flux over E/q the density is found. The density ratios C^{6+}/C^{5+} and O^{7+}/O^{6+} are then converted to the freezing-in temperatures T_{C65} and T_{O76} respectively, using the tables of Arnaud and Rothenflug (1985) /2/. These are the coronal temperatures near the freezing-in point of the two charge states. Finally, summing over all charge states of an element, the elemental abundances are obtained for C and O. In the case of Mg, the density of Mg^{10+} is multiplied by a temperature dependent factor (in the range between 1 and 1.5

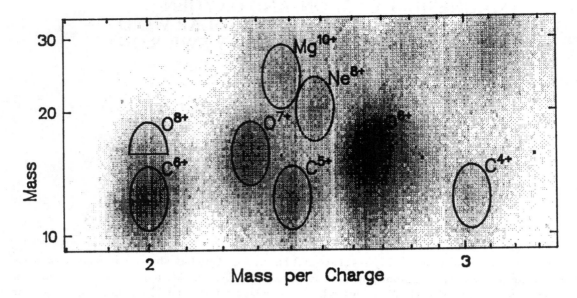

Fig. 1. $M - M/Q$ matrix (cut-out) of all PHA events of days 134–216, 1991. The seven charge states investigated here are indicated by their elliptical regions. The counts for O^{8+} are accumulated only in the upper half of its ellipse and then doubled, to prevent inerference with C^{6+}.

for the temperatures encountered here), which is also taken from Arnaud and Rothenflug (1985). In this way the contribution of charge states other than 10+ are taken into account. Additionally, the Mg density is corrected by a factor 1/0.79 to obtain the total Mg density, because the isotope ^{24}Mg is 79% of total magnesium.

The accuracy and absence of a bias of this method depends on the careful choice of the elliptical regions for each charge state. The ellipses in Fig. 1 are the ones actually used in this work. For checking their accuracy, we also have calculated charge state ratios using ellipses smaller by linear factors of 0.8 and 0.6 in both the M and the M/Q direction. Smaller ellipses produce a larger statistical uncertainty, while larger ellipses can be contaminated by neighboring charge state ions. For a typical period, we have found the results given in Table 1

TABLE 1 Error estimate of ellipse method as a function of ellipse size.

Ellipse size	C/O	Mg/O	T_{C65}	T_{O76}
as in Fig. 1	$\equiv 1$	$\equiv 1$	$\equiv 1$	$\equiv 1$
80%	1.00 ± 0.05	0.95 ± 0.07	1.01 ± 0.01	1.00 ± 0.01
60%	1.02 ± 0.07	0.91 ± 0.11	1.02 ± 0.02	0.99 ± 0.02

We observe that the C/O ratio and the freezing-in temperatures are practically unaffected by ellipse size, while there is a systematic shift in the Mg/O ratio. This shift is interpreted as due to the spilling over of Ne^{8+} into the larger Mg^{10+} boxes. We therefore take the box as in Fig. 1 (for better statistics), but correct the so obtained Mg/O ratio by a factor of 0.91 for Ne spillover. In total, we estimate that the systematic error introduced by our choice of the boxes is not larger than 10% for the abundance ratios, and it is negligible for the temperatures.

RESULTS

The results of our analysis are collected in Figures 2–5. In Fig. 2, 72 daily averages of C/O and Mg/O are given. It is obvious that the fluctuations in these ratios are significant, not statistical.

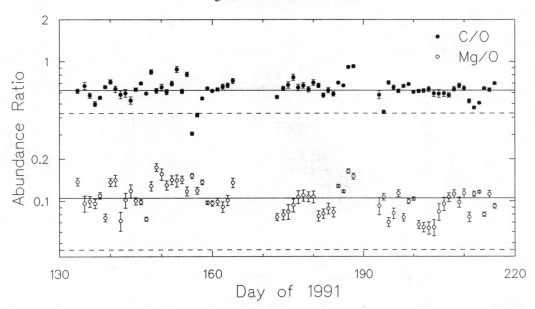

Fig. 2. Daily averages of the C/O and Mg/O abundance ratios. The deviations from the long-term average value (solid line) appear to be larger than the statistical uncertainty indicated by the error bars. Also, both ratios are clearly enriched relative to the solar ratio (dashed line). The two data gaps around days 170 and 190 respectively are due to satellite operations, not instrument performance.

However, there are only occasionally trends to be seen over periods of several days, which indicates that the typical time scale of these fluctuations is shorter than one day.

Still, we can find markedly strong correlations in this data set. In Fig. 3 the temperatures T_{O76} and T_{C65} are plotted, revealing the expected strong positive correlation. Also, we observe $T_{O76} > T_{C65}$ in all cases, which means that the O charge states freeze-in nearer to the coronal temperature maximum than the C charge states do (cf. /3/).

We then take the above temperatures as an indicator for the solar wind type and look for a correlation in the abundance ratios to them. The results are given in Figure 4 and 5. We find a strong correlation between Mg/O and coronal temperature T_{O76}, and a weaker correlation between C/O and T_{C65}. Also, the relative variations in Mg/O are clearly larger than in C/O.

Finally, when accumulating the data over the whole period considered, rather than dividing it into daily pieces, we obtain the average abundance ratios given in Table 2.

<u>TABLE 2</u> Average solar wind abundance ratios on days 134–216, 1991.

Element	SW Ratio	SW Ratio to solar ratio
C/O	0.66 ± 0.10	1.54
Mg/O	0.13 ± 0.02	2.82

CONCLUSIONS

We have found new long-term averages of the C/O and Mg/O solar wind abundance ratios in the period of days 134–195, 1991. When compared to the solar ratios of Anders and Grevesse (1989) /4/, we observe the well-known first ionization potential (FIP) fractionation effect of the solar wind: Carbon is only slightly enriched with respect to the high–FIP elements, while the enrichment of Magnesium is by a factor of ≈ 3.

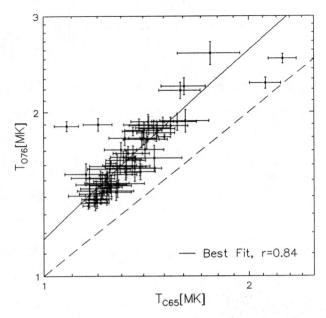

Fig. 3.   Correlogram of the inferred corona temperatures T_{C65} and T_{O76}, calculated from C^{6+}/C^{5+} and O^{7+}/O^{6+}. As expected, they are strongly correlated, but T_{O76} is found to be significantly higher than T_{C65}. The solid line is a best fit to the data, while on the dashed line $T_{C65} = T_{C65}$.

However, significant deviations from these long-term averages are found: The FIP fractionation appears not to produce the same enrichments of C/O and Mg/O in all solar wind types. We find that solar wind from hot coronal sources is enriched more strongly in C and Mg (and possibly in all low–FIP elements) relative to O than solar wind from cooler sources. Therefore, the FIP fractionation effect can be thought to be stronger in or below the hot parts of the corona, while it is weaker or even absent in or below the cooler parts.

This confirms and extends the findings of von Steiger *et al.* (1992) /5/, who have made similar observations in the case of C/O using magnetosheath plasma data of the CHEM instrument on AMPTE/CCE.

It must be said that the C/O *vs.* T_{C65} correlation is weaker in this data set than in the AMPTE data. This is probably due to the following reason: For simplicity, we have quite arbitrarily divided the data base into daily pieces, disregarding the fact that the solar wind type does not normally change at midnight UT. Therefore, different solar wind types can be mixed unless one particular type persists for the better part of a day. On the other hand, the AMPTE data necessarily only contain high-pressure solar wind data, and no such mixing did occur.

However, these new data are still far superior to the AMPTE solar wind data in at least three respects: (1) The free solar wind is observed, not the thermalized solar wind plasma behind the Earth's bow shock. (2) The full velocity distribution function is observed, while on AMPTE the lowest energy part is missing. (3) Due to the permanent exposure of SWICS to the solar wind, the statistics are better by orders of magnitude.

We conclude that the FIP–fractionation effect is significantly different in different solar wind flow types. In the future, we will extend this study to other elements, and we will replace the crude daily averaging by a method which breaks the dataset into time slices during which the solar wind type does not change. The data of the SWICS instrument are ideal for this purpose.

ACKNOWLEDGEMENTS

We wish to thank P. Bochsler, A. Bürgi, F. Ipavich, and R. Wimmer for many helpful comments and discussions. This work was supported in part by the Swiss National Science Foundation.

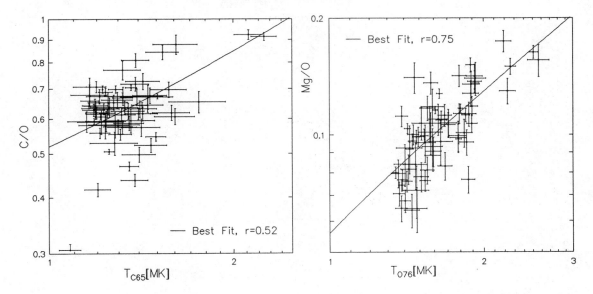

Fig. 4. *(Left)* C/O abundance ratio *vs.* coronal temperature T_{C65}. There is a (weak) positive correlation, indicating that solar wind from hotter coronal sources has a higher than average C/O ratio.

Fig. 5. *(Right)* Mg/O abundance ratio *vs.* coronal temperature T_{O76}. There is a clear positive correlation, indicating that solar wind from hotter coronal sources has a significantly higher than average Mg/O ratio.

REFERENCES

1. G. Gloeckler, J. Geiss, H. Balsiger, P. Bedini, J. C. Cain, J. Fischer, L. A. Fisk, A. B. Galvin, F. Gliem, D. C. Hamilton, J. V. Hollweg, F. M. Ipavich, R. Joos, S. Livi, R. Lundgren, U. Mall, J. F. McKenzie, K. W. Ogilvie, F. Ottens, W. Rieck, E. O. Tums, R. von Steiger, W. Weiss, and B. Wilken, The solar wind ion composition spectrometer, A&A Suppl. 92, 267 (1992)

2. M. Arnaud and R. Rothenflug, An updated evaluation of recombination and ionization rates, A&A Suppl. 60, 425 (1985)

3. S. P. Owocki, T. E. Holzer, and A. J. Hundhausen, The solar wind ionization state as a coronal temperature diagnostic, Ap. J. 275, 354 (1983)

4. E. Anders and N. Grevesse, Abundances of the elements: Meteoritic and solar, Geochim. Cosmochim. Acta 53, 197 (1988)

5. R. von Steiger, S. P. Christon, G. Gloeckler, and F. M. Ipavich, Variable carbon and oxygen abundances in the solar wind as observed in Earth's magnetosheath by AMPTE/CCE, Ap. J., in press (1992)

ELEMENTAL ABUNDANCES AND THEIR VARIATIONS IN THE UPPER SOLAR ATMOSPHERE

K. G. Widing and U. Feldman

E. O. Hulburt Center for Space Research, Naval Research Laboratory, Washington, DC 20375-5000, U.S.A.

ABSTRACT

In the standard FIP pattern of element abundances observed in the bulk corona and solar wind, elements with first ionization potential (FIP) below ~ 10 eV are enriched relative to elements with higher FIP /1/. We have surveyed element abundances in a variety of emitting structures on the sun using imaged spectra of the sun in the 200-600 Å range from SKYLAB and diagnostic flux ratios of soft X-ray lines between 10 and 20 Å. The resulting element abundances show the imprint of the FIP effect, but the characteristics of this pattern are variable and greatly enhanced in some features, and are correlated with the magnetic morphology. The enrichment factor reaches values of 10 to 15 in diffuse features with open magnetic fields. The O/Ne abundance ratio is constant in the SKYLAB sample (in agreement with the standard FIP pattern), but shows variations up to a factor of two in soft X-ray flares. Soft X-ray abundance ratios involving Fe suggest that the photospheric abundance of iron is equal to the meteoritic rather than the spectroscopic ratio.

INTRODUCTION

The anomalous composition of the solar wind and solar energetic particles (SEP) compared to the photosphere has long been known, as well as its probable origin in a FIP-dependent element separation at the base of the solar atmosphere /2/. Confirmation of the anomalous abundances by direct observation of the corona itself has been limited to the study of a few elements, with limited accuracy.

The main thrust of the observations reported here is that: (1) not only is the composition of the corona different from the photosphere, but it is highly variable, and (2) the composition of specific solar features depends on the magnetic topology.

The standard FIP pattern of anomalous abundances observed in the solar wind and solar energetic particles is schematically illustrated in Figure 1. As such, the coronal overabundance factor of 4 and other properties measured in the bulk solar wind and SEPs represent a global average of the solar plasma.

The approach adopted in this paper is to examine how the characteristics of the standard FIP-effect illustrated in Figure 1 change when we look at abundances in various localized emission regions on the sun. We will find, for example, that the overabundance factor, or "the FIP bias" is spatially variable, and can be much greater than a factor of 4. We can also examine whether the relative abundances of the elements on the high-FIP plateau (FIP > 11 eV) stay constant or not when the FIP bias changes by studying the O/Ne abundance ratio. There is also a question whether the position of the step between 9 and 11 eV is variable or not.

FIG. 1. The standard FIP-effect observed in coronal abundances. Elements with first ionization potentials (FIP) < 9 eV are 4 times more abundant in the corona than elements with FIP > 11 eV compared to the photosphere.

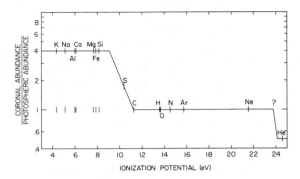

A frequently used observational approach is to compare ions of two different elements which are formed at approximately the same temperature in the solar atmosphere. The line intensity ratio then may have only a weak dependence on temperature and density, and it becomes an abundance diagnostic. The following analysis also assumes that the line emission is formed by electron collisional excitation in ionization equilibrium.

In the first part of this paper element abundances derived from the NRL spectroheliograph on SKYLAB are discussed. This instrument produced dispersed images of the whole sun between 170 Å and 600 Å (essentially slitless spectra of all emitting features on the sun). In the second part of the paper element abundances derived from soft X-ray spectra of active regions and flares in the 10-21 Å range are discussed.

ABUNDANCES FROM SKYLAB

One of the important abundance diagnostics in the SKYLAB spectroheliograms is the group of strong lines at 400 Å where a multiplet of Ne VI is overlapped by lines from a multiplet of Mg VI. These ions are formed at nearly the same temperature, so that the intensity ratio of their lines becomes a diagnostic of the Ne/Mg abundance ratio. In Figure 2 the dispersed images of an impulsive flare photographed on the disk of the sun (lower frame) are compared with the images of a polar plume (upper frame). The plume images are the short, bump-like features extending above the polar limb of the sun into the corona.

What this comparison shows is the effect of a 10 times lower abundance of neon relative to magnesium in the plume compared with the flare. We note, for example, that the three strongest lines of Ne VI in the flare - which has essentially photospheric abundances - can barely be seen in the plume. At the same time, lines of Mg VI, Na VIII, and Ca IX remain fairly strong; it is as though the plume was relatively enriched in low FIP element material by a factor of 10 compared to neon. Details of the abundance determination in the plume are given in Table 1. Here it will only be noted that the relative abundances of the three low-FIP elements Mg, Ca, and Na are hardly changed (within a factor of 2) between photosphere, SEPs, and polar plumes although the FIP bias (Ne/Mg abundance ratio) reaches an order of magnitude.

Because of their open magnetic fields, polar plumes have been suggested to be source regions of the solar wind. The present composition results suggest that plumes could well be one of the channels by which the corona acquires its anomalous composition. However, further tests of this idea depend on an accurate assessment of the mass flux into the corona from polar plumes, which is not easy to do.

Figure 3 shows the results of a survey of the Ne/Mg abundance ratio in various typical features observed on SKYLAB spectroheliograms. Since the neon abundance is used as a proxy for the high FIP elements and magnesium

FIG. 2. Dispersed images of an
impulsive flare in the 400 Å
region (below) compared with images
of a polar plume extending above
the solar limb (above). This shows
the effect of a 10 X lower abundance
of neon relative to magnesium in the
plume. Note that the two strongest
lines of Ne VI in the flare can
barely be seen in the plume.

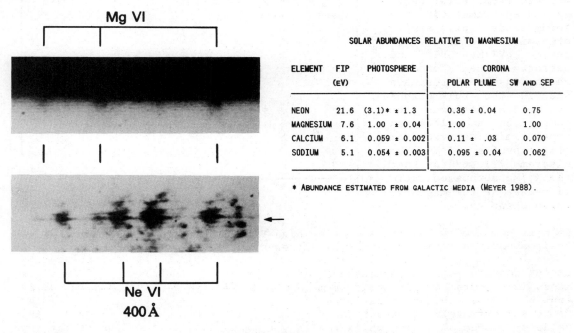

TABLE 1

SOLAR ABUNDANCES RELATIVE TO MAGNESIUM

ELEMENT	FIP (EV)	PHOTOSPHERE	CORONA POLAR PLUME	SW AND SEP
NEON	21.6	(3.1)* ± 1.3	0.36 ± 0.04	0.75
MAGNESIUM	7.6	1.00 ± 0.04	1.00	1.00
CALCIUM	6.1	0.059 ± 0.002	0.11 ± .03	0.070
SODIUM	5.1	0.054 ± 0.003	0.095 ± 0.04	0.062

* ABUNDANCE ESTIMATED FROM GALACTIC MEDIA (MEYER 1988).

for the low FIP group, this is also a survey of the FIP bias in the sense of
Figure 1 if we assume that the decrease in the observed Ne/Mg ratio is due
to an increasing overabundance of magnesium, and form the ratio: bias =
$(Ne/Mg)_{phot}/(Ne/Mg) = 3.6/(Ne/Mg)$.

What this diagram (Figure 3) shows is a sequence in which the observed Ne/Mg
abundance ratio decreases as the magnetic field opens up. The features with
photospheric-like abundance ratios are associated with compact, loop
configurations, whereas the features with low abundance ratios are
associated with open or diverging fields. Another characteristic of the
features with small Ne/Mg abundance ratios is that they have low surface
brightnesses and correspondingly low values of the electron density. We
also call attention to the Ne/Mg abundance ratio of ~ 1.3 labelled
"TRANSITION ZONE - INNER CORONA". The global emission of the solar
atmosphere in Ne VI and Mg VI (as distinct from the localized emission
sources) arises in a thin spherical shell which projects into a bright
emission ring surrounding the sun, and forming - as it were - a transition-
zone to the outer million degree corona. The Ne VI/Mg VI intensity ratios
of these rings therefore provide a bulk measurement of the abundance ratio
in the inner layers of the solar atmosphere. Another feature of the Ne/Mg
ratio in the transition zone is that it does not appear to change
significantly over polar coronal holes.

Finally, we note in Figure 3 that the Ne/Mg abundance ratio varies from a
value of 3.6 in impulsive flares to 0.24 in open, diverging fields compared
with a global average of 0.8 in the bulk corona. This suggests that a
number of channels with different magnetic field configurations (and not

just the open field ones) are active in supplying mass to the corona.

FIG. 3.   Survey of the
Ne/Mg abundance ratio on
SKYLAB spectroheliograms
(adapted from a figure in
/3/.   Filled circles:
detailed emission measure
analysis of particular
examples.   Open circles
with closed error bars:
estimated range in the
Ne/Mg ratio for a sample
of similar features based
on Ne VI/Mg VI line
ratios.   The abundance
ratios in the photosphere
and in the bulk corona
/4/   are   shown   for
comparison, as well as an
upper limit derived from
slit spectra /5/.   The
diagram illustrates that
the Ne/Mg abundance ratio
decreases as the magnetic
field opens up.

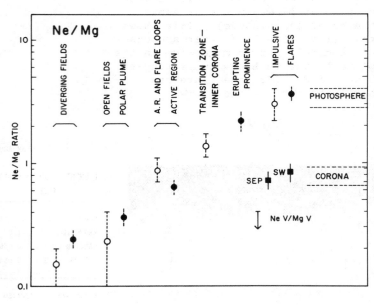

We next compare two elements on the high-FIP plateau, O with Ne, to see if
their abundance ratio remains constant or not (Fig. 4).

Except for the soft X-ray measurements in Figure 4, it would seem
that the O/Ne abundance ratio remains approximately constant at a
value of about 7, even while the
Ne/Mg abundance ratio and the FIP
bias are changing by a factor of 4
or 5.   The determination of the
O/Ne abundance ratios from solar
soft X-ray flux measurements is
presented in the next section.

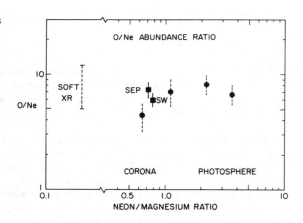

FIG. 4.   Filled circles: O/Ne
abundance ratio plotted against the
Ne/Mg abundance ratio for four
SKYLAB events.   Filled squares:
coronal abundance ratios from solar
wind /6/ and SEP measurements /4/.

ABUNDANCES FROM SOLAR SOFT X-RAYS

The spectral line diagnostics in this region are mainly the hydrogen-like
and helium-like ions of oxygen, neon, and magnesium which are located
between 10 and 20 Å.   There also are strong lines of Fe XVII and Fe XVIII in
this wavelength region to compare with.   As with the SKYLAB Ne VI/Mg VI
abundance diagnostic, line ratios of different ions which are not sensitive
to temperature are used as abundance diagnostics.   Because of the high
temperature and densities required, the results refer mainly to active
regions and flares; coronal densities in open-field regions are too low to
yield measurable X-ray spectra.   The abundance variations will be
illustrated with the recent, unpublished work of McKenzie and Feldman /7/.

McKenzie and Feldman first compared two low FIP elements: Fe with Mg. In Figure 5 they plotted for each flare event the observed flux ratio of Fe XVIII to Mg XI against a second flux ratio which gives the temperature.

FIG. 5. The photon flux ratio of Fe XVIII (14.21 Å) to Mg XI (9.17 Å) observed in solar flares plotted against the temperature-related ratio of Fe XVII (15.01 Å) to the Fe XVIII line.

FIG. 6. The flux ratio of Fe XVII to the Ne IX forbidden line at 13.70 Å plotted as in Fig. 5 for flare observations.

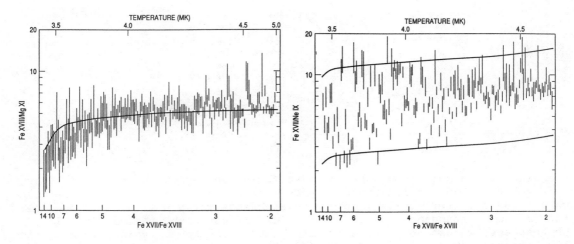

The curve drawn through the data is the theoretical flux ratio calculated with a best-fitting Fe/Mg abundance ratio of 0.8. The abundance ratio is the free parameter of the curve. An abundance ratio of 0.8 is essentially the same as the solar system abundance ratio based on C1 chondrites. So the Fe/Mg abundance ratio stays essentially constant at the photospheric value.

McKenzie and Feldman next compared low FIP iron with high FIP neon using the observed flux ratio of Fe XVII to Ne IX in flares (Fig. 6). We now see wide variations in the flux ratios, and changes in the Fe/Ne abundance ratio. It now takes two theoretical line ratio curves to bound the data, which differ by about a factor of 4.3 in the Fe/Ne abundance ratio. Note that the authors have put neon in the denominator of the abundance ratio (reversing the behavior of the Ne/Mg ratio in Fig. 3.). The authors further identify the lower bounding curve with the Fe/Ne abundance ratio in the photosphere. This can then be compared with other determinations in the photosphere.

In a previous study Strong et al. /8/ also used the Fe XVII/Ne IX flux ratio to study the Fe/Ne abundance ratio in 200 active regions observed on SMM. They concluded that there is a significant variation (factor of 7) in the relative abundance of Fe to Ne from one active region to another.

With the solar soft X-ray data we can again examine the O/Ne abundance ratio. In Figure 7 the flux ratio of O VIII to Ne IX for a sample of active regions is plotted. We see that two theoretical curves are required: the lower curve requires an O/Ne abundance ratio of 5, while the upper curve requires a ratio of 12. However, a majority of the events in the plot occupy a narrower band between these two extremes, in rough agreement with the average SKYLAB, solar wind, and SEP abundance ratio of 7 in Figure 4. McKenzie and Feldman interpreted the O/Ne abundance variation as a shift in the position of the step in Figure 1 toward higher ionization potentials so that oxygen in some flares behaves like an intermediate FIP element while neon remains high FIP.

In Table 2, McKenzie and Feldman summarized their results in terms of the minimum and maximum abundance ratios which bounded their data sets.

FIG. 7. The flux ratio of O VIII (18.97 Å) to Ne IX (13.45 Å) plotted as in Fig. 5 for active regions.

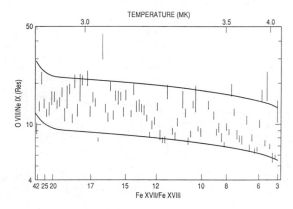

TABLE 2 ABUNDANCE RATIOS

RATIO	THIS PAPER*		VARIATION FACTOR	PHOTOSPHERE**	
	MINIMUM	MAXIMUM		SPECTROSCOPIC	METEORITIC
FE/MG (F)	0.80 ± 0.19		...	1.23 ± 0.17	0.85 ± 0.04
FE/NE (A)	0.27	1.12	4.1	0.38 ± 0.09	0.26 ± 0.06
FE/O (F)	0.034	0.104	3.1	0.055 ± 0.006	0.038 ± 0.003
FE/O (A)	0.041	0.098	2.4		
O/NE (A)	5.0	12.0	2.4	6.92 ± 1.69	6.92 ± 1.69

(A) = ACTIVE REGIONS, (F) = FLARES

* MCKENZIE AND FELDMAN /7/

** ANDERS AND GREVESSE /9/

Comparing these numbers we see no significant variation in the Fe/Mg ratio, a factor of 4.1 variation in the Fe/Ne ratio, a factor of 2-3 in the Fe/O ratio, and a factor of 2.4 in the O/Ne abundance ratio. If we remember that the minimum abundance ratios were identified with the ratios in the photosphere, we can compare them with the photospheric ratios given in the last two columns. J.-P. Meyer /3/ has pointed out that there is, in general, excellent agreement between the solar system abundances based on carbonaceus chondrites and the spectroscopic values, except in the case of Fe. For the ratios involving the Fe abundance in Table 2 we see that there is excellent agreement between the soft X-ray minimum ratios and the meteoritic abundance ratios. This suggests that the photospheric abundance of iron is equal to the meteoritic rather than the spectroscopic abundance.

In conclusion, it is interesting to think that the relative accuracy in the soft X-ray flux observations has reached the point that one can fine-tune the abundances of certain elements in the photosphere. We thank J.-P. Meyer for his illuminating comments on the manuscript.

REFERENCES

1. J. Geiss, this issue.

2. J.-P. Meyer, Astrophys. Jour. Supp. 57, 151 and 173, 1985.

3. J.-P. Meyer, Adv. Space Res. 11, No. 1, 269, 1991.

4. H.H. Breneman and E.C. Stone, Astrophys. Jour. 299, L57, 1985.

5. G.A. Doschek and A.K. Bhatia, Astrophys. Jour. 358, 338, 1990.

6. G. Gloeckler and J. Geiss, in: Cosmic Abundances of Matter, ed. C.J. Waddington (New York: AIP), 1989, p. 49.

7. D.L. McKenzie and U. Feldman, Astrophys. Jour. in press (1992).

8. K.T. Strong, J.R. Lemen, and G.A. Linford, Adv. Space Res. 11, (1) 151, 1991.

9. E. Anders and N. Grevesse, Geochim. Cosmochim. Acta 53, 197, 1989.

HELIUM ABUNDANCE, ACCELERATION, AND HEATING AND LARGE-SCALE STRUCTURE OF THE SOLAR WIND

Y. I. Yermolaev

Space Research Institute, Academy of Sciences, Profsoyuznaya 84/32, Moscow 117 810, Russia

ABSTRACT

This paper summarizes the latest results on the research of solar wind helium ions observed on broad the Prognoz 7 satellite. On the basis of solar wind parameter distributions on the "density-velocity plane" we selected five different types of the solar wind streams which may be connected with the well known solar corona structure and phenomena: (1) the heliospheric current sheet, (2) streams from closed coronal magnetic field regions (streamers), (3) streams from open magnetic field regions (coronal holes), (4) disturbed solar wind streams at large angles relative to the axis of non-stationary events in the corona, and (5) disturbed streams near the axis of non-stationary events including the matter of coronal mass ejections. In the streams from streamers the helium abundance decreases with increasing flux in agreement with hypothesis by Gosling et al.(1981) /24/ and in the streams from coronal holes it increases with increasing flux in agreement with model by Geiss et al.(1970) /12/. Preferential α-particle heating and acceleration are generally observed in streams from coronal holes. On the average, equalizations of α-particle and proton velocities and temperatures due to Coulomb collisions are expected to be located at heliocentric distances ~ 7 AU and ~ 20 AU, respectively.

INTRODUCTION

The direct measurements of different ions in solar wind provided important information about processes in the solar corona and in the interplanetary space. Parameters such as the relative abundance and the ionization state of different ion components do not change in the interplanetary space and may be used for diagnostics of the solar corona regions from which solar wind streams originate. On the other hand, the hydrodynamical parameters (density, bulk velocity and kinetic temperature) of different ion components change in the interplanetary space and may be used for study of solar wind dynamics /1-8/.

The main objectives of this paper are as follows: (1) classification of solar wind streams and investigation of helium parameter variations in different streams; (2) comparison of helium abundance with models of solar wind composition formation; (3) study of variations in temperature ratio and velocity difference of α-particles and protons; and (4) investigation of role of Coulomb collisions in dynamics of helium velocity and temperature relative to protons. For these purposes we use the proton and α-particle parameteres and magnetic field measured by means of SCS-04 energy-mass-spectrometer /9/ and magnetometer SG-70 /10/ on broad the Prognoz 7 satellite during November 1978 to June 1979.

HELIUM VARIATIONS IN DIFFERENT STREAMS OF SOLAR WIND

On the basis of measurements of proton and α-particle bulk parameters and magnetic field on board the Prognoz 7 satellite the structure of solar wind was studied and was compared with the known structure of solar corona /8,11/. The distributions of several solar wind parameters on the density-velocity plane, $\log n - V_p$, (i.e., the two-dimension dependences on the density, $n = \Sigma n_i \approx n_p$, and proton velocity, V_p) are presented in Fig 1. The thin lines in Fig. 1a and others show lines of constant level which separate the regions with different numbers of single measurements within an elemental bin with size $\Delta \log n = 0.1$ and $\Delta V_p = 10$ km/s: N≥3, >20 and >65. The solid lines which are the lines of constant level for the appropriate parameter were drawn by the linear interpolation using the mean values of parameters obtained within these bins.

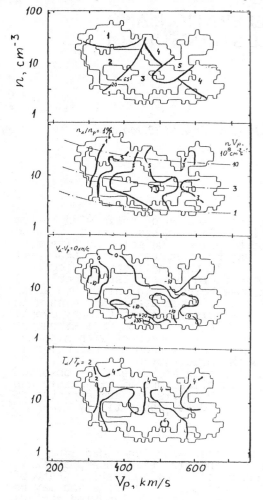

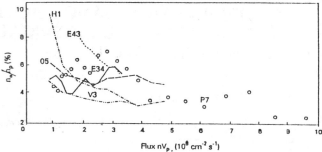

Fig.2. Average dependences of relative helium abundance on the solar wind flux as observed onboard Vela 3, Explorer 34 and 43, Heos 1, OGO 5 and Prognoz 7 (open circles)

Fig.1. Distributions on the "density – velocity plane" of number of measurements in the elementary bins (thin lines) and a) different types of solar wind streams (see Table and text), b) helium abundance (thick lines) and flux (dash-dotted lines) c) α-particle and proton velocity difference and d) their temperature ratio.

Table. Average parameters for five types of solar wind streams

SW type (part.%)	V_p, km/s	n, cm^{-3}	B, nT	T_p, 10^4 K	β	n_α/n_p, %
1 (8)	351±45	29.6±10.0	6.3±2.3	5.4± 4.4	2.0±1.7	1.7±1.5
2 (33)	359 33	9.6 4.2	7.2 3.0	5.4 5.1	0.5 0.5	4.4 4.0
3 (49)	449 52	6.1 3.4	6.9 2.7	9.8 8.5	0.6 0.6	6.6 6.0
4 (6)	573 79	8.3 7.9	9.5 3.5	14.6 12.5	0.6 0.5	3.5 3.5
5 (4)	515 44	8.6 5.5	11.4 8.0	18.8 14.1	1.0 0.8	4.8 4.5

The values and behaviour of several parameters on the density-velocity plane allowed us to identify five regions (the types of solar wind streams) which may be connected with the well known solar corona structure and events: (1) heliospheric current sheet, (2) streams from closed coronal magnetic field regions (streamers), (3) streams from open magnetic field regions (coronal holes), (4) disturbed solar wind streams at large angles relative to the axis of non-stationary events in the corona, and (5) disturbed streams near axis of non-stationary events including the matter of coronal mass ejections /8,11/.

The distribution of relative abundance of α-particles is illustrated by the line $n_\alpha/n_p = 1$, 3, 5, and 7 % in Fig.1b . This Figure also shows the lines of constant flux (dash-dotted line): $nV_p/(10^8 cm^{-2} s^{-1})$ = 1, 3, and 10. There are three different domains in the Figure: in regions 1,2 and 4,5 (see Fig.1b) the relative helium abundance decreases with increasing density, and in region 3 it increases. The differences in the three domains depending on velocity are as follows: (1) regions 1,2 cover the velocity range 270-450 km/s, region 3 covers the range 350-600 km/s, and regions 4,5 cover the range 450-700 km/s; (2) in regions 1,2 abundance is increasing, on the average, from ~ 1 to ~ 6 % with increasing velocity, in region 3 it remains at 5-6 %, and in regions 4,5 it decreases from ~ 5 to ~ 3%. The change in the solar wind flux is connected with variation in the relative contribution of each• of the five regions on the density-velocity plane (see Fig.1b): within $nV_p/(10^8 cm^{-2} s^{-1})$ = 1-3 the main contribution is from region 3, within 3-20 the main contribution is from region 2. Hence in these regions the mean dependences of helium abundance on the flux are similar to the dependences on the density: the relative abundance of α-particles increases from ~ 4 to ~ 7 % with increase in the flux from ~ 1 to ~ 3 and decreases from ~ 6 to ~ 2 % with increase in the flux from ~ 3 to ~20.

One of the effects possibly influencing the escape of heavy ions from the solar corona to the interplanetary space is their Coulomb friction with the major, i.e. proton, component of plasma. In particular, from the theoretical considerations /12/ it was assumed that there is a minimum plasma flux threshold above which ions, heavier than protons, effectively escape into the interplanetary space due to Coulomb friction. The dependence of the relative α-particle abundance on the solar wind ion flux as observed onboard Vela 3 /13/, Explorer 34 and 43 /14/, Heos 1 /15/, OGO 5 /1/ and Prognoz 7 (circles) /16/ is illustrated in Fig.2. In the flux range (3-5) the α-particle abundance behaves similarly to the results of earlier space experiments. In the flux range (1-3) the Prognoz 7 data qualitatively agree only with the experimental data from the Explorer 34 satellite. A cause of quantitative and qualitative differences with other experiments on board the Explorer 43, Heos 1, OGO 5 and Vela 3 may be connected with different measurement technique: mass-analysis methods were employed only in the Prognoz 7 and Explorer 34 measurements while all other data were derived using energy-analysis methods which may be a source of great uncertainty in the helium abundance estimations for weak solar wind fluxes mainly corresponding to the hot solar wind /8/. Thus, the data obtained speak in favour of the hypothesis about the Coulomb friction effect on the escape of heavy ions into the interplanetary space /12/ only within the range $nV_p = (1-3)*10^8 cm^{-2} s^{-1}$ (e.g. mainly in streams from coronal holes).

The identification of solar wind stream types on the density-velocity plane (see Figs.1a,c and d) allowed us to suggest that the comparative behaviour of α-particle and proton bulk velocities and kinetic temperatures, which reflects mainly the plasma dynamics in the interplanetary space, is associated with the types of solar wind. The processes of preferential acceleration and heating of the α-particles ($V_\alpha > V_p$ and $T_\alpha > 4*T_p$) are observed in the streams from coronal holes and at the boundary of the regions of heliospheric current sheet and streams from coronal streamers. In the streams from coronal holes the non-equilibrium state increases with decreasing density. Such anticorrelations with the density can be due to the fact that the decreasing density leads to a reduction in the number and efficiency of Coulomb collisions of ions /8,11/.

In the low velocity ($V_p < 330$ km/s) streams from streamers (region 2 on the density-velocity plane) as well as in the streams disturbed by non-stationary events on the Sun (regions 4 and 5), on the average the α-particles are moving slower than protons and the temperature ratio of α-particles and protons is lower than 4. Some causes of α-particle and proton velocity and temperature differences are discussed below. However, it should be noted that velocities and temperatures of ion components are not equal to each other in the slow, dense and cold solar wind streams from coronal streamers, where the number and efficiency of Coulomb collisions are sufficiently great.

DYNAMICS OF HELIUM VELOCITY AND TEMPERATURE

For an explanation of the observed bulk velocity difference of protons and heavier ions a mechanism was proposed which assumed mixing of different-velocity solar wind streams with various contents of ion components /17,18/. For example, if two streams with $V_{\alpha 1} = V_{p1} = V_1 \neq V_{\alpha 2} = V_{p2} = V_2$ and $(n_\alpha/n_p)_1 \neq (n_\alpha/n_p)_2$ are mixed, the mean velocity difference is

$$V_\alpha - V_p = \frac{n_{p1} n_{p2}}{(n_{\alpha 1} + n_{\alpha 2})(n_{p1} + n_{p2})} (V_1 - V_2)[(n_\alpha/n_p)_1 - (n_\alpha/n_p)_2]. \qquad (1)$$

On the basis of the Prognoz 7 observations we tried to check this hypothesis. Our data show that $(n_\alpha/n_p) \cdot vs. V_p$ and $(V_\alpha - V_p) \cdot vs. V_p$ dependences are similar to each other /7/. (The same results were observed with another spacecraft /1/.) This fact agrees with equation 1: for instance, the increase in relative abundance of α-particles [i.e. $(n_\alpha/n_p)_2 > (n_\alpha/n_p)_1$] with increasing velocity $(V_{p2} > V_{p1})$ must lead to positive velocity difference of α-particles and protons in accordance with eq.1. Thus the Prognoz 7 data speak in favour of hypothesis that mixing of different-velocity solar wind streams with different helium abundances may lead to different bulk velocities of α-particles and protons in the solar wind.

A second important conclusion was drawn from eq.1: if the change in relative abundance of different ions heavier than protons in the mixing streams is proportional [i.e. $(n_i/n_p)_2/(n_i/n_p)_1 = const$], the difference of their bulk velocities with respect to the proton velocity will be constant (i.e. $V_i - V_p$ = const for all the ions). The proportional change in helium, oxygen, silicon, and iron abundance in the low- and middle-velocity streams was observed with Prognoz 7 /7/. Moreover, the measurements on the ISEE 3 showed that heavy ion velocities are close to the α-particle velocity /19/.

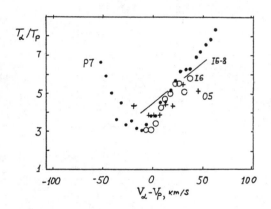

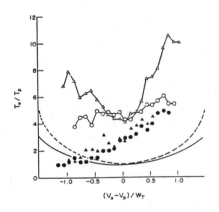

Fig 3. Average dependence of temperature ratio T_α/T_p on velocity difference $V_\alpha - V_p$ from the Prognoz 7 (dotted line), OGO 5 (crosses) and IMP 6-8 (straight line and open circles)

Fig. 4. Average dependence of T_α/T_p on $(V_\alpha - V_p)/W_t$ (W_t -average thermal velocity) for collisionless plasma ($\tau_e/\tau_c < 0.1$) - open symbols and collisional plasma ($\tau_e/\tau_c > 0.1$) - closed symbols; low helium abundance ($n_\alpha/n_p < 4\%$) - circles and high one ($n_\alpha/n_p < 5\%$) triangles. The same dependence calculated in /22/ for collisional plasma with n_α/n_p = 5% (solid line) and 2.5 % (dashed line)

An additional information about mechanisms of equilibrium distortion between protons and α-particles may be provided by analysis of relationship between the α-particle preferential acceleration and their preferential heating. Fig.3 presents the dependence of T_α/T_p on $V_\alpha - V_p$, derived from space experiments on board the OGO 5 and IMP 6-8 satellites /20,21/ and Prognoz 7 data. The Prognoz 7 data show the direct correlation of T_α/T_p and $V_\alpha - V_p$ parameters for the $V_\alpha - V_p$

range from 0 to ~ 70 km/s. Besides T_α/T_p increases with $V_\alpha-V_p$ decreasing within the velocity difference $\leq$ -10 km/s range whereas for other experiments there are no systematic data for the velocity difference $<$ -20 km/s range. It should note that in the OGO 5 data /20/ there is a certain indication that temperature ratio increases from ~ 3.9 at $V_\alpha-V_p \approx 0$ to ~ 4.4 at $V_\alpha-V_p$ = -20 km/s. On the whole a conclusion from the Prognoz 7 data can be made that the temperature ratio correlates with the absolute value of solar wind α-particle and proton velocity difference $|V_\alpha-V_p|$.

The source of free energy for heating of minor ions may be the difference of velocities of ions and protons in the solar wind. Hernandez and Marsch showed in the theoretical work /22/ that friction forces between two ion distributions, shifted in velocity space relatively to each other, heat both components, so that the α-particle and proton temperature ratio should change from 1 (for small velocity difference) to a value about $(m_p n_p)/(m_\alpha n_\alpha)$ (for large velocity difference). For the quantitative comparison of the Prognoz 7 results with the above hypothesis, Fig.4 shows the dependence of the α-particle and proton temperature ratio on the their velocity difference normalized to their mean thermal velocity $X = (V_\alpha-V_p)/W_T = (V_\alpha-V_p)/[2k(T_\alpha/m_\alpha + T_p/m_p)]^{1/2}$ for four different double intervals: for collisionless or collisional plasma and for plasma with low or high relative helium abundance. The results of model calculations /22/ are also showed.

Conclusions of model /22/ which only deals with collisional plasma (e.g. dense and slow streams from streamers and heliospheric current sheet) do not agree with Prognoz 7 results: (1) there is no temperature ratio dependence on helium abundance, and (2) the temperature ratio depends on the sign of value X. Thus, the dependence of α-particle and proton temperature ratio on the absolute value of relative difference of their velocities, $|X|$, is observed only in collisionless streams with high helium abundance. These streams are generally connected with coronal holes. Therefore, analysis of Prognoz 7 data contradicts the hypothesis by Hernandez and Marsch /22/ and points to a much more complicated character of the dependences considered. The fact that the temperature ratio measured on board Prognoz 7 is, on the average, much larger than those calculated according to the model /22/ implies that there are additional mechanisms of α-particles heating. In particular, the assumption can be made that the preferential heating of α-particles is the consequence of developing oscillations (instabilities) associated with the α-particle and proton velocity difference and differs for various types of solar wind streams /7/.

Total influence of the Coulomb collisions on α-particle and proton velocity difference and temperature ratio in the solar wind is shown in Figs.5 /7,16/. These figures present two-dimensional histograms of $V_\alpha-V_p$ dependence on the ratio of the stream expansion time to the time of momentum exchange due to Coulomb collisions τ_e/τ_s, and T_α/T_p dependence on the ratio of expansion time to the time of energy exchange due to Coulomb collisions τ_e/τ_c. The time ratios were calculated by means of the following formulae:

$$\tau_e/\tau_c = \text{const}\ \frac{r\ n\ \exp(-X^2)}{V_p\ T_{eff}^{3/2}},\quad X = \frac{|V_\alpha-V_p|}{[2k(T_p/m_p+T_\alpha/m_\alpha)]^{1/2}},\quad T_{eff} = \frac{m_p\ T_p + m_\alpha\ T_\alpha}{m_p + m_\alpha},\quad (2)$$

$$\tau_e/\tau_s = \text{const}\ \frac{r\ n\ G(X)}{V_p\ |V_\alpha-V_p|\ [2k(T_p/m_p + T_\alpha/m_\alpha)]},\qquad (3)$$

where r is a heliocentric distance, and G(X) is related to the error function /1,7,23/. On the two-dimensional histograms the data are presented as contour lines which denote the following: dotted line means a number of measurements in the bin >3, dashed line - >10, dash-dotted line - >20, solid line - >65.

Apart from systematic reduction of α-particle and proton velocity difference, if the time ratio grows, two facts attract attention: (1) in the collisional solar wind (time ratio >1) α-particles on the average move slower than protons; (2) as the time ratio grows, the observed deviations of experimental values $V_\alpha - V_p$ from their mean values, indicated as black circles in Figs.5, strongly decrease independently of the sign of this deviation (although $V_\alpha > V_p$ cases are more often observed than $V_\alpha < V_p$). As is seen from left figure, in general the Prognoz 7 data are in agreement with Explorer 43, OGO 5 and HEOS 1 data /1/; and they confirm the conclusion that Coulomb collisions influence the reduction of velocity differences of various ion components in the solar wind. Unlike α-particle and proton velocity difference the deviation of their temperature ratio is about constant for all the values of time ratio τ_e/τ_c, and average temperature ratio decreases as time ratio grows. As is seen from right figure, in general the Prognoz 7 data are in good agreement with OGO 5, IMP 6 /16/ and ISEE 3 /25/ data and demonstrate an important role of Coulomb collisions in equalization of temperatures of solar wind ion components in the region of flow parameters, where velocity difference of components is small enough.

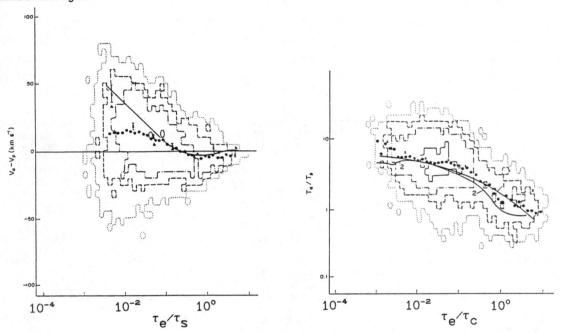

Fig.5. Histograms of dependence of a) velocity difference and b) temperature ratio of α-particles and protons on time ratios τ_e/τ_s and τ_e/τ_c. Points - Prognoz 7 averaged data, crosses - OGO 5, triangles - Explorer 43, thick lines - Heos 2 (a) or IMP 6 (b-1) or ISEE 3 (b-2).

The Prognoz 7 data allow one to assess approximately the heliocentric distances at which α-particle and proton velocities and temperatures are equalized due to Coulomb collisions. Figs.5 show that for solar wind at the distance $r = 1$ A.U. on the average $\tau_e/\tau_s \sim 0.03$ and $\tau_e/\tau_c \sim 0.05$; the conditions of $V_\alpha - V_p = 0$ and $T_\alpha/T_p = 1$ are achieved for $\tau_e/\tau_s \sim 0.2$ and $\tau_e/\tau_c \sim 1$, respectively. This means that to equalize velocities and temperatures of this components, a seven-time increase in relation to the average τ_e/τ_s and a 20-time increase in relation to the average τ_e/τ_c is required. If, with an increase of heliocentric distance $r > 1$ A.U. the density changes according to the mass conservation law $n \sim r^{-2}$, temperature changes according to adiabatic law $T_p \sim r^{-4/3}$ and velocity V_p is approximately constant, then the time ratio $\tau_e/\tau_c \cong r*const$. It may also be assumed that $\tau_e/\tau_s \cong r*const$. In such a case on the average the velocities of the components will be equalized at the heliocentric distance $r_V \sim 7$ A.U., and temperatures at $r \sim 20$ A.U. With this approach r_V and r_T estimations have an accuracy up to a factor of about 2 /7/.

DISCUSSION AND CONCLUSION

On the basis of the Prognoz 7 measurements we selected five types of the solar wind streams and suggested that different types of streams may be connected with the well known solar corona structure and phenomena: (1) the heliospheric current sheet, (2) streams from closed coronal magnetic field regions (streamers), (3) streams from open magnetic field regions (coronal holes), (4) disturbed solar wind streams at large angles relative to the axis of non-stationary events in the corona, and (5) disturbed streams near the axis of non-stationary events including the matter of coronal mass ejections. The events selected according to this classification (i.e., over density and velocity) are supposed to have more distinct dependence on hydrodynamical parameters of the solar wind since they are of the same origin. Such an approach allows one to study more reliably as processes of formation of solar wind streams as their dynamics in the interplanetary space.

The Prognoz 7 data agree with hypohtesis about the formation of the solar wind ion composition in the streamers by Gosling et al. /24/ and show that mechanism of its formation discussed by Geiss et al. /12/ may be implemented only in the streams from coronal holes.

The behaviour of bulk velocities and kinetic temperatures of α-particles relative to protons is studied for various types of the solar wind streams which are associated with solar wind and solar corona structures. Distributions of velocity difference and temperature ratio of α-particles and protons on the density-velocity plane show that preferential α-particle acceleration $(V_\alpha > V_p)$ and heating $(T_\alpha/T_p \geq 4)$ are observed on the whole in the streams from the coronal holes.

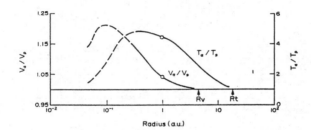

Fig. 6. Schematic view of dynamics with heliocentric distance ratios of bulk velocities V_α/V_p and kinetic temperatures T_α/T_p

The Prognoz 7 data allow us to propose (basing on α-particle and proton measurements) the following scenario of deviation from thermodynamic equilibrium of the solar wind ion components. At heliocentric distances from 10 to 25 solar radii, where strong inhomogeneous plasma parameters are observed (see, for instance, results of radio-occultation measurements /25/), different-velocity plasma streams with different minor ion abundances (usually their abunances are slightly higher in faster streams) are mixing; as a result flows with non-equilibrium bulk velocities $V_i \neq V_p$ (usually $V_i > V_p$) are formed. Besides, other known mechanisms of preferential heavy ion accelerations may take place, especially ion interaction with plasma waves. Then, due to energy contained in the ion bulk velocity difference, faster heating of heavy ions occurs. As a result the temperature ratio T_i/T_p increases and velocity difference V_i-V_p decreases while heliocentric distance increases up to ~ 0.3 A.U., and then the T_i/T_p begins to decrease (see e.g. /26/). The ion velocities are equalized, on the average, at the heliocentric distance ~ 7 A.U. and ion temperatures - at the distance ~ 20 A.U. due Coulomb collisions. Fig. 6 shows schematically the common dynamics of the average values V_α/V_p and T_α/T_p over a wide range of heliocentric distances. The Helios 1,2 observations at distances 0.3 - 1.0 AU /26/ and numerous measurements at 1 AU /1/ were taken into account.

REFERENCES

1. M. Neugebauer, Observation of solar wind helium, Fund. Cosmic Phys. 7, 131 (1981)

2. S. J. Bame, Solar wind minor ions - recent observations, in: Solar Wind Five, ed. M. Neugebauer, NASA, 1983, p. 573

3. I. S. Veselovsky, Space Plasma Physics, Science and Technology Results. Space Research 22, VINITI, Moscow, 1984 (in Russian)

4. J. Geiss, Diagnostics of corona by in-situ composition measurements at 1 AU. in: Future Mission in Solar, Heliospheric and Space Plasma Physics, Garmisch-Partenkirshen, Germany, ESA SP-235, 1985

5. P. Bochsler, Solar wind ion composition, Physica Scripta T18, 55 (1987)

6. G. N. Zastenker, Yu. I. Yermolaev, V. I. Zhuravlev, N. L. Borodkova, Z. Nemecek, Y. Safrankova, Large- and middlescale phenomena in the interplanetary medium: Prognoz 7, 8, 10 observations, Adv. Space Res. 9, #1, 117 (1989)

7. Yu. I. Yermolaev and V. V. Stupin, Some alpha-particle heating and acceleration mechanisms in the solar wind: Prognoz 7 measurements, Planet. Space Sci. 38, #10, 1305 (1990)

8. Yu. I. Yermolaev, Large-scale structure of solar wind and its relationship with solar corona: Prognoz 7 observations, Planet. Space Sci. 39, #10, 1351 (1991)

9. O. L. Vaisberg, L. S. Gorn, Yu. I. Yermolaev et al., Experiment on diagnostic of interplanetary and magnetospheric plasma onboard Venera 11, 12 spacecraft and Prognoz 7 satellite, Kosmicheskie Issledovaniya. 17, #5, 780, (1979) (in Russian)

10. E. A. Gavrilova, E. G. Eroshenko, V. A. Styazhkin, N. A. Eismont, K. Danov and P. Petrov, Visualization of magnetic field data on board the Prognoz 7 satellite. Preprint N 1064, Space Research Institute, USSR Academy of Sciences, Moscow (1986) (in Russian)

11. Yu. I. Yermolaev, Large-scale structure of the solar wind and the solar corona: Prognoz 7 observations, Adv. Space Res. 11, # 1, 75, 1991

12. J. Geiss, P. Hirt and H. Leutwyler, On acceleration and motion of ions in corona and solar wind, Solar Phys. 12, 458 (1970)

13. J. Hirshberg, J. R. Asbridge and D. E. Robbins, Velocity and flux dependence of the solar wind abundance, J. Geophys. Res. 77, 3583 (1972)

14. K. W. Ogilvie, Helium abundance variations, J. Geophys. Res. 77, 4227 (1972)

15. G. Moreno and F. Palmiotto. Variations of α-particle abundance in the solar wind, Solar Phys. 30, 207 (1973)

16. Yu. I. Yermolaev, G. N. Zastenker and V. V. Stupin, Relationships between bulk parameters of solar wind protons and alpha-particles: Prognoz 7 selective measurements, Preprint No 1575, Space Research Institute, Academy of Sciences of USSR, Moscow (1990)

17. M. Eyni and R. Steinitz, A model for the interaction of solar wind streams, in: Proc. COSPAR Symposium B, ed. Shea M. A., Smart D. F., Wu S. T., Tel Aviv, 1977, p. 101

18. I. S. Veselovsky, On evolution of strong unhomogeneties in solarwind plasma, Geomagnetism and Aeronomics 18, 3 (1978) (in Russian)

19. J. Schmid, P. Bochsler and J. Geiss, Velocity of iron ions in the solar wind, J. Geophys. Res. 92, #A9, 9901 (1987)

20. M. Neugebauer, Observations of solar wind helium. in: Solar Wind Four, ed. H. Rosenbauer, Report N MPAE-W-100-81-31, 1981, p. 425

21. M. Neugebauer and W. C. Feldman, Relation between superheating and superacceleration of helium in the solar wind, Solar Phys. 63, 201 (1979)

22. R. Hernandez and E. Marsch, Collisional time scales for temperature and velocity exchange between drifting Maxwellians, J. Geophys. Res. 90, # A11, 11062 (1985)

23. L. W. Klein, K. W. Ogilvie and L. F. Burlaga, Coulomb collisions in the solar wind, J. Geophys. Res. 90, # A8, 7389 (1985)

24. J. T. Gosling, G. Borrini, J. R. Asbridge, S. J. Bame, W. C. Feldman and R. T. Hansen, Coronal streamers in the solar wild at 1 AU, J. Geophys. Res. 86, 5438 (1981)

25. O. I. Yakovlev, A. I. Efimov and S. N. Rubtsov, Dynamics and turbulence of solar wind in region of its formation: radaosounding measurements onboard Venera 15 and Venera 16 space probes, Kossmicheskie Issledovaniya. 25, # A2, 251, (1987) (in Russian)

26. E. Marsch, K. -H. Muhlhauser, H. Rosenbauer, R. Schwenn and F. M. Neubauer, Solar wind helium ions: observation of the Helios solar probes between 0.3 and 1 AU, J. Geophys. Res. 87, #A1, 35 (1982)

Kinetic Physics, Waves and Turbulence

Conveners: Bruno Bavassano
Martin Lee
André Mangeney

BERNSTEIN WAVES IN THE SOLAR WIND

K. Baumgärtel and K. Sauer

Institut für Kosmosforschung, O-1199 Berlin, Germany

ABSTRACT

Pronounced emission peaks at a series of frequencies in the electron cyclotron harmonic (ECH) frequency range, observed at a lithium neutralgas release in the solar wind, are interpreted in terms of Bernstein waves (electrostatic waves in the ECH frequency range propagating perpendicular to the ambient magnetic field). The observed distinct frequencies are selected from a broad band spectrum, likely excited by photoionization of Li atoms, by the condition of zero group velocity in the satellite frame. This gives clear evidence for the solar wind to be capable of carrying Bernstein waves. The analysis is based on an previous paper of the authors /1/.

INTRODUCTION

Roeder et al. /2/ reported on an unpredicted observation of electrostatic emissions near the harmonics of the electron cyclotron frequency, following a lithium release in the solar wind upstream of the earth's bow shock. The release experiment was part of the AMPTE program /3/. The spacecraft diagnostics observed narrow-band emissions at 435 Hz, 732 Hz, 1014 Hz, 1292 Hz and 1566 Hz, each of them with an amplitude of approximately 10^{-5} V/m. These frequencies correspond, respectively, to 1.8, 3.0, 4.2, 5.3 and 6.4 times the local electron cyclotron frequency (see Fig. 1b). The harmonic emissions occurred at times from 50 s to 250 s after release, when the neutral gas has distributed over such a large volume due to radial expansion that the Li plasma produced from it by photoionization does no longer act as an obstacle to the solar wind, as it did immediately after the injection. The emissions were most pronounced at angles near perpendicular to the solar wind magnetic field, suggesting a predominantly electrostatic phenomenon. Attempts to explain the emissions in ref. /2/ as a result of an instability driven by the relative motion between Li ions and the solar wind failed which brought the authors to the conclusion that the observations are inconsistent with generation mechanisms based on both ion beam instability types and beam-cyclotron instabilities.

THE MODEL

In this paper a theory is presented that does not refer to an instability, but attributes the observed emission to the combined action of two effects in the equilibrium solar wind plasma: (1) The abrupt creation of ions and electrons by photoionization and their subsequent spatial separation through motion along

different trajectories comprise a time-dependent space charge
density that acts as source for broad-band excitation of
Bernstein waves. (2) The frequencies preferentially measured by
the spacecraft diagnostics are just those which make the group
velocity in the observer (i.e.,satellite) system vanish. In this
way, the observed frequencies can be predicted fairly accurately
and the order of magnitude of the measured electric field
strength can be roughly reproduced.

At a time of about 50 s after release, the plasma conditions
around the centre of injection have recovered to pre-release
values. Li ions and electrons, however, still continue to be
created by photoionization from the more and more rarefied
neutral gas, but they behave as test particles rather than as
plasma. Immediately after ionization, electrons and ions can be
considered at rest and thus start moving along cycloidal
trajectories in a plane perpendicular to the solar wind magnetic
field **B** with a guiding center velocity $\mathbf{v}_\perp$, the component of the
solar wind velocity $\mathbf{v}_{sw}$ perpendicular to **B**. Since both radius and
time of the ion gyration are much larger than the scale sizes of
interest, we treat the ions as unmagnetized, i.e., fix them at
the position of ionization. An electron moves along a trajectory
$\mathbf{r}_e(t)$ which can generally be written

$$\mathbf{r}_e(t) = \mathbf{r}_o + R[\cos(\omega_c t + \eta), \sin(\omega_c t + \eta), 0] + \mathbf{v}_\perp \qquad (1)$$

in a reference frame whose z-axis points in direction of **B** ($\mathbf{r}_o$
point of ionization, ω_c electron cyclotron frequency, $R = v_\perp \omega_c$
electron gyration radius, η arbitrary phase). In the following,
we shall use the test-particle approach /4,5/ to calculate the
quasistatic electric field set up in the solar wind plasma by a
single ionization process, i.e., by a charge density

$$\rho(\mathbf{r}, t) = e[\delta(\mathbf{r} - \mathbf{r}_o) - \delta(\mathbf{r} - \mathbf{r}_e(t)] \, \Theta(t) \qquad (2)$$

due to an ion at rest and an electron following the trajectory
$\mathbf{r}_e(t)$, both created at t=0.

ELECTRIC FIELD CALCULATION

Using Fourier/Laplace (F/L) transform technique in space/time the
eletric field at the satellite position **r**=0 can be written

$$E(t) = \frac{-i}{(2\pi)^4} \int d^3k \int d\omega \frac{\rho(\omega, \mathbf{k})}{k \epsilon_o \epsilon_L(\omega - \mathbf{k}\mathbf{v}_{sw}, \mathbf{k})} e^{-i\omega t} \qquad (3)$$

where

$$\rho(\omega, \mathbf{k}) = ie^{-i\mathbf{k}\mathbf{r}_o} \left[\frac{1}{\omega^2} - \sum_{n=-\infty}^{\infty} J_n(k_\perp R) \frac{e^{in(\eta \vartheta - \pi/2)}}{\omega + n\omega_c - k_\perp v_\perp} \right] \qquad (4)$$

is the F/L transform of the charge density of Eq. (2). ϵ_L denotes
the quasistatic dielectric function in the solar wind rest frame
(thus the Doppler-shifted frequency $\omega - \mathbf{k}\mathbf{v}_{sw}$ appears as argument
instead of ω) and $\mathbf{k} = (k_\perp \cos \vartheta, k_\perp \sin \vartheta, k_\parallel)$. In order to simplify
the analysis and to get rid of the arbitrary phase η we retain
only the leading term n=0 in Eq. (4). Physically, this means that
the gyrating electron is treated as time-independent charge

density ring with a smeared-out charge -e.

As a first step in evaluating E(t) we carry out the ω-integration in Eq. (3) with the help of the residue theorem. This yields polarization fields, induced by the two test particles in the solar wind plasma, and field contributions

$$E_s(t) = \frac{e}{\epsilon_o} \frac{i}{(2\pi)^3} \int d^3k \frac{e^{-i\mathbf{k}\mathbf{r}_o}}{k} \left[\frac{1}{\omega_s + \mathbf{k}\mathbf{v}_{sw}} - \frac{J_o(k_\perp R)}{\omega_s + k_\parallel v_\parallel} \right] \frac{e^{-i(\omega_s + \mathbf{k}\mathbf{v}_{sw})t}}{\epsilon_L'} \quad (5)$$

associated with the the zeros $\omega_s(\mathbf{k})$, s=1,2,... of $\epsilon_L(\omega,\mathbf{k})$. The latter represent electrostatic fields excited in response to the birth of the two particles. In order to accomplish an approximative analytical evaluation of the remaining k-integral $(d^3k = k_\perp dk_\perp d\vartheta dk_\parallel)$ we proceed along the following steps: (1) We employ that ϵ_L and consequently ω_s do not depend on ϑ. This allows the ϑ-integration to be carried out. (2) Quasistatic waves in a magnetized warm plasma can propagate without damping only perpendicular to the magnetic field and are highly damped in the remaining directions. Assuming that the angle of propagation is limited to less than δ from that perpendicular to **B**, i.e., $|k_\parallel/k_\perp| \leq \tan \delta \approx \delta$, the $k_\parallel$-integration can be simply approximated. This combines to reduce Eq. (5) to the form

$$E_s(t) \sim \sum_\pm \frac{e^{\pm i\pi/4}\delta}{(v_\perp t)^{1/2}} \int_0^\infty dk_\perp \frac{k_\perp^{1/2}}{\epsilon_L'} \left[\frac{1}{\omega_s \mp k_\perp v_\perp} - \frac{J_o(k_\perp R)}{\omega_s} \right] e^{-i(\omega_s \mp k_\perp v_\perp)t} \quad (6)$$

(for details see /1/). The integrand now exhibits a structure suitable to apply the Saddle-Point method. The condition for the existence of a saddle point k_s is

$$\frac{d}{dk_\perp} [\omega_s(k_\perp) \pm k_\perp v_\perp] = v_s^{group} \pm v_\perp = 0 \quad (7)$$

where v_s^{group} denotes the group velocity associated with the s-th branch of the dispersion curves. Eq. (7) expresses the fact that a saddle point exists if for a certain $k_\perp$ the group velocity coincides in magnitude with the flow velocity of the solar wind perpendicular to the magnetic field. When computing the dispersion curves for Bernstein waves /6/ for solar wind conditions during the release experiment ($\omega_p/\omega_c=91$, $T_e=18eV$, i.e., $v_{th}=2.5\times10^6$m/s, $v_\perp=4.2\times10^5$m/s) it is seen that all modes exhibit negative group velocity (ω decreases with growing $k_\perp$) and thus a saddle point can only be realized with the postitive sign in Eq. (7). Fig. 1a illustrates the conditions for which Eq. (7) is matched. All of the first five branches admit a saddle point, indicated by the region of vanishing derivative $d\omega_s/dk_\perp$. The corresponding frequencies are seen to agree excellently with those of the observed peaks (Fig.1b). The lack of well-pronounced zero derivative regions for higher harmonic numbers explains the high-frequency cutoff of the observed frequency chain.

We would like to emphasize that the frequency selection rule provided by the Saddle-Point method operates independently of the particular mechanism that is responsible for the presence of a broad-band spectrum of quasistatic eigenmodes. Seen in this view, Fig. 1 comprises the principal result of the paper. This gives clear evidence for the physical origin of the enhanced radiation, and, consequently, for the capability of the solar wind to support Bernstein waves.

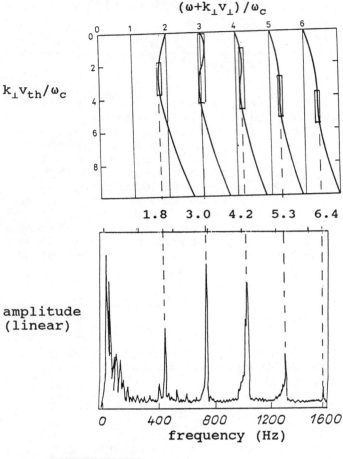

$(\omega+k_\perp v_\perp)/\omega_c$

$k_\perp v_{th}/\omega_c$

Fig.1a
Graphical scheme for finding frequencies and wavenumbers that make the group velocity of solar wind Bernstein waves vanish in the satellite system

Fig. 1b
Observed spectrum of the electric field amplitude (after *Roeder et al.* [1987])

amplitude
(linear)

frequency (Hz)

REFERENCES

/1/ Baumgärtel,K., and K.Sauer, Interpretation of electron cyclotron harmonic waves observed at an AMPTE lithium release in the solar wind, JGR 94, A9, 11983, 1989

/2/ Roeder,J.L.,H.C.Koons,R.H.Holzworth,R.R.Anderson,O.H.Bauer, D.A.Gurnett,G.Haerendel,B.Häusler,and R.Treumann, Electron cyclotron harmonic waves observed by the AMPTE-IRM plasma experiment following a lithium release in the solar wind, JGR 92, 5768, 1987

/3/ Bryant,D.A.,S.M.Krimigis, and G.Haerendel, Outline of the active magnetospheric particle tracer explorer (AMPTE) mission, IEEE Trans.Geosci.Remote Sensing, GE-23, 177, 1985

/4/ Krall,N.A.,and A.W.Trivelpiece, Principles of Plasma Physics, pp.556ff., McGraw-Hill, New York, 1973

/5/ Stone,P.M.,and P.L.Auer,Excitation of electrostatic waves near electron cyclotron harmonic frequencies, Phys.Rev.138, 1965

/6/ Crawford,F.W.,and J.A.Tataronis, Absolute instabilities of perpendicularly propagating cyclotron harmonic plasma waves, J.Appl.Phys. 36, 2930, 1965

INNER HELIOSPHERE OBSERVATIONS OF MHD TURBULENCE IN THE SOLAR WIND. CHALLENGES TO THEORY

R. Bruno

Istituto di Fisica dello Spazio Interplanetario, Consiglio Nazionale delle Ricerche, cp 27, 00044 Frascati, Italy

ABSTRACT

The interplanetary medium is permeated with fluctuations of magnetic field and plasma parameters over a wide range of frequencies. Most of these fluctuations although having a rather turbulent nature clearly show their Alfvénic character. Particularly interesting are observations of such a turbulence when performed in the inner heliosphere since it is between the Alfvénic point and 1 AU that MHD fluctuations experience most of their evolution towards a state of fully developed fluid turbulence. Mechanisms which govern this evolution are not fully understood yet but, the existence of phenomena of local generation of turbulence are strongly suggested by theory to be essential for this process. A comparison between Helios observations and some theoretical predictions about local generation and evolution of MHD turbulence are briefly reviewed.

Because of the limited space available for this review the author is forced to omit most of the references which will be found in three excellent and recent reviews by Marsch /3/, Schwenn /5/ and Roberts and Goldstein /4/.

WHY INNER HELIOSPHERE AND SOLAR CYCLE MINIMUM

It is relevant to the study of SW turbulence to distinguish between the inner and outer heliosphere. There are good physical reasons to do so and obviously the fortuitous location of the Earth at 1 AU doesn't have anything to do with this distinction. As a matter of fact, within 1 AU the interplanetary medium still conserves most of the structure characteristic of the low corona where the wind is born. Beyond 1 AU, the plasma is strongly reprocessed by dynamical stream-stream interaction due to an increasing bending of the magnetic field spiral and by the associated onset of shocks. Moreover, periods of time around minimum of solar activity cycle are particularly appealing since at that time the interplanetary medium appears, to an observer confined in the ecliptic plane, highly structured into recurrent high velocity streams coming from high latitude open magnetic field line regions and slow plasma originating from equatorial closed field line regions. Moreover, the stationarity of the wind structure at minimum gives the possibility to observe plasma, flowing from the same photospheric region, at different heliocentric distances. Around maximum of solar cycle the same observer would collect essentially slow wind since the source regions of fast wind shrink around the poles /see review 5/.

STREAM STRUCTURE AND LARGE SCALE ALFVÉNICITY

Considering averages lasting several hours, low density and high temperature characterize the fast plasma with respect to the slow one, while magnetic field intensity doesn't show particular differences. Rotations of the field vector from an inward to an outward configuration or vice versa are always located within slow plasma while fast plasma is generally unipolar. On shorter scales differences are even more evident. Fast plasma shows higher fluctuations in velocity, temperature and field orientation, while slow plasma has stronger fluctuations in density. The rather constant field intensity and number density suggest the possibility of finding Alfvénic fluctuations within the fast plasma. As a matter of fact, long period Alfvénic correlations are largely found within the trailing edge of high velocity streams at solar minimum /3,4/. We can be confident that this scenario is true also during solar maximum although it is quite difficult to observe from the ecliptic, plasma coming from long lasting high velocity streams. It is noteworthy to mention that observations performed at solar maximum showed the highest Alfvénic correlations being within those regions characterized by small velocity gradients and imbedded in the slow wind rather than fast wind /3,4/. This is not at odds with the above view since the observed very short-lasting fast wind regions certainly do not resemble the long-lasting high velocity streams seen at cycle minimum and any comparison is meningless. However, different degrees of Alfvénic correlations are ubiquitously found in the solar wind at least close to the Sun, showing that the production mechanism at the base of the corona is effective in both fast and slow wind.

Together with Alfvénic fluctuations, pressure balance structures (PBS's hereafter) dominate at hourly scales the solar wind /3,4/. Alfvénic fluctuations dominate at 0.3 AU in both fast and slow streams and by 1 AU there is a role exchange with PBS's in the slow wind. In the fast one they still dominate although their presence is strongly reduced /3/. During maximum, when the coronal configuration is such that high velocity streams are almost absent on the ecliptic, the radial evolution of fluctuations is similar to the one experienced by the slow wind during cycle minimum /3/. The fact that the solar cycle doesn't appreciably influence the behaviour of PBS's /3,4/ suggests the local generation of these structures, outside the source regions of the wind. For other types of fluctuations such as slow magnetosonic waves, the identification is uncertain since these modes are quickly damped, and a search for fast magnetosonic waves gave results not statistically relevant /see review 3/.

Fig. 1. Field and velocity total power spectra within slow (left panels) and fast (right panels) plasma, and at 0.3 AU (upper panels) and close to 1 AU (lower panels). Field intensity has been expressed in Alfvén units (km·sec^{-1}). Power density and wave numbers are reported in logarithmic scales, r_a in linear scale. The four smaller panels show the Alfvén ratio for the same intervals. The data set refers to the same plasma parcel as seen by Helios at two different heliocentric distances after two solar rotations. Solid diagonal lines indicate the expected slope for a fully developed turbulence. Plasma and magnetic field data shown in this figure and in the next one have been provided by H. Rosenbauer and F. Mariani, respectively.

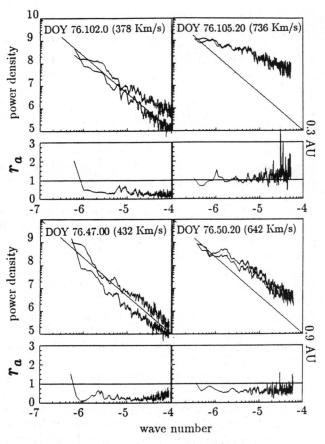

Spectral features of MHD turbulence related to solar wind stream structure is an important issue in order to understand the turbulent evolution and dynamics of the solar wind. Striking differences can be seen in spectra shown in Fig. 1. Fast wind at 0.3 AU at low wave number (k hereafter)

shows spectral indexes considerably smaller than those predicted by Kolmogorov being close to -1 and a higher power level with respect to slow plasma. This difference strongly increases at higher k's where the Alfvénic regime is stronger as also confirmed by the Alfvén ratio (r_a hereafter) very close to 1 between 10^{-6} and 10^{-5}km^{-1}. At 0.3 AU, the slow wind already shows a well developed turbulence with a spectral index close to $-\frac{5}{3}$ for both field and velocity fluctuations although, some tendency towards a smaller index is shown in the high k's of field fluctuations. The rather low values assumed by r_a around 0.3 is indicating the low Alfvénicity of the fluctuations. However, the radial evolution is such that by 1 AU spectra within fast streams tend to assume indexes close to $-\frac{5}{3}$ and fluctuations become smaller and much less Alfvénic /3,4/. In fact, the degree of correlation weakens with distance such that large amplitude Alfvén waves as long as 14 hours in the s/c frame have been found only at cycle minimum within fast streams and close to the Sun /3,4/. Almost absent is the radial evolution in the slow plasma except for the steepening towards $-\frac{5}{3}$ of the spectral index of the field fluctuations for the highest k's. The clear tendency of r_a in fast plasma towards values less than unity is a against dynamic alignment theory according to which solar wind Alfvénic turbulence would relax towards a state with complete absence of minority modes (z^-) and a consequent tendency towards a perfect equipartition of the energy ($r_a \rightarrow 1$). An excess of magnetic over kinetic energy has been obtained in numerical simulations for developed homogeneous MHD turbulence, and a tendency of r_a towards values less than unity within fast streams indicates a relaxation towards "standard" turbulence /2/. However, r_a close to the Sun shows that it is the kinetic energy that dominates the lowest and also the highest frequency range and that Alfvénic fluctuations are found in between, around 10^{-4} Hz. The dominance at low frequecy is due to kinetic energy in streams, less clear is the situation at the highest frequencies which is a new and unpredicted observational result.

The above observations show that fluctuations are in two different states depending on the stream structure. However, there isn't a clear cut difference. Large part of daily variations of the specific turbulent energy (kinetic + magnetic) seems to be due to only two large-scale parameters: the proton thermal speed c_s characterizing the internal state of the plasma and the normalized proton density fluctuations $\frac{\delta n}{n}$ describing the degree of compressibility of the turbulence /2 and references therein/. Both amplitude of the fluctuations and the spectral index within the inertial range depend on the proton temperature. Low temperature leads to a low level of turbulence with a steep index (-1.8) while high temperature leads to a high level of turbulence and a flatter spectral index (-1.2) /2/.

The changing of the spectrum slope with distance has been recognized as an evidence that non-linear interactions are at work to transfer energy from the largest to the smallest scales /3,4/. In order to have nonlinear interactions, in the incompressible limit, Alfvénic fluctuations with an inward sense of propagation have to be present. Dividing the inward from the outward Alfvénic contribution we discover that the power associated with outward Alfvénic modes (e$^+$ hereafter) correlates with proton thermal speed c_s while, for the inward modes (e$^-$ hereafter), power strongly depends on $\frac{\delta n}{n}$ /2/. In the slow wind inward and outward contributions have almost the same power level and a clear Kolmogorov-like index indicates that this turbulence is already fully developed by 0.3 AU. Different is the situation shown by fast streams /2,3/. Spectra of e$^+$ which are quite flat at 0.3 AU (index$\approx$ -1), strongly evolve with distance. The highest frequencies evolve faster than the lowest ones /3/, but the tendency is toward a fully developed turbulence. Completely new, unpredicted and not yet understood is the evolution of e$^-$ /3/. Its spectrum can be divided into two frequency bands around, $3 \cdot 10^{-4}$ Hz. While the lowest band has a typical turbulent index even close to the Sun, the highest band evolves from an almost flat spectrum at 0.3 AU to a fully developed turbulent spectrum at 1 AU /3/. On the contrary, within slow wind there isn't such a frequency dependence and e$^-$ has already at 0.3 AU a fully developed spectrum throughout the whole frequency band.

THE NATURE OF e⁻ MODES AND THEIR ROLE

The existence of this spectral break-point reveal the existence of different generation mechanisms depending on the frequency range. If we believe the high frequency range populated by real inward propagating waves then they must have a local origin in the interplanetary space beyond the Alfvénic critical point. The following generation mechanisms seem to be the likeliest candidate for this spectral feature /see review 3/ although, the fact that r_a is greater than unity in this frequency range as observed in Helios data close to the Sun (Fig. 1) /see also plots shown in review 3/, cast new constraints to be taken into account. One possible mechanism is the parametric decay of outward Alfvén waves e⁺ which generate one of the daughter waves backward propagating as e⁻ in the same high frequency range. Another possibility is that e⁻ modes in the high frequency range might be created by e⁺ modes at low frequencies.

At low frequency, the observation of the radial depletion of the normalized cross-helicity (σ_c hereafter) suggests the idea that more and more e⁻ modes are locally generated /4/. Velocity shear through Kelvin-Helmholtz instability seems to be the main mechanism. Its validity has been proved in simulations by a two-dimensional, incompressible MHD spectral code and it seems to work also in the solar wind as observed at the location of a large velocity shear at 0.3 AU /3,4/. A spectral analysis performed right after this shear revealed a strong presence of e⁻ modes which rapidly decreased moving away from the shear. However, the above mechanism doesn't seem to be very common in the solar wind and it has been shown /1,3/ that the radial decrease of σ_c is mostly due to a radial decrease of e⁺ rather than an increase of e⁻ . Several studies /3,4/ performed in the low frequency range have highlighted the opposite behavior of e⁻ and e⁺ as a function of magnetic field intensity and density fluctuations (B_{comp} and n_{comp} hereafter) /1,2,3/; while e⁻ increases with increasing B_{comp} and n_{comp}, e⁺ decreases. These and other considerations /3/ have cast doubts on the nature of e⁻ modes in the low frequency range such that they are largely considered as the spectral signature of convected structures probably related to flow tubes emanating from the supergranular photospheric network at the base of the corona.

Depletions of σ_c cannot always be ascribed to compressive phenomena. Regions where strong decreases of σ_c are simply due to sudden drops of e⁺ not caused by any compressive phenomenon are frequently found within those regions which are notoriously Alfvénic, namely the trailing edges. These cases show the existence of high velocity regions where Alfvénic fluctuations are completely absent either because they were never produced in those plasma regions or because they have already been absorbed to heat the plasma. Other cases of very low σ_c due to a complete absence of outward modes have been observed in low velocity regions and modeled /5/. These peculiar regions show fully developed turbulence, an extremely low Alfvén ratio (~ 0.2), and the very low σ_c doesn't seem to be influenced by compressible fluctuations. A possible explanation for these regions is based on the existence of magnetic field directional turnings (MFDT hereafter). These structures should have a constant magnetic field intensity showing spatial changes in direction, constant velocity vector, constant proton density and temperature and total pressure balance, similar to TD's. If we imagine Alfvén waves propagating in a medium filled with MFDT's then velocity fluctuations would result from Alfvénic fluctuations only, while magnetic fluctuations would have also the contribution of the MFDT's leading to very low values of r_a . However, the not complete agreement with observations shows that a satisfactory solution doesn't exist yet. A decrease of σ_c within an Alfvénic region can also be caused by decoupling between magnetic field and velocity fluctuations due to strong changes in velocity as shown in Fig. 2. There is a clear change in velocity within day 105. Within the same day there is a clear decrease of σ_c which has a different nature from the larger depletion observed one day before which is due to local generation phenomena /3,4/. In fact, there isn't any clear increase of e⁻ and both field and density compressions (b_{comp} and n_{comp}, respectively) are extremely low. In this case the reduction of σ_c is probably due to strong fluctuations in wind velocity. When this happens the velocity minimum variance direction deviates from the ambient magnetic field direction while the minimum variance direction of the magnetic field remains almost unaffected and aligned to the ambient field as shown in the last panel. This decoupling between field and velocity fluctuations causes a decrease of the Alfvénicity of the fluctuations (σ_c) which doesn't have anything to do with local generation of e⁻ modes. This phenomenon becomes less localized with increasing heliocentric distance. By the time the stream has reached 1 AU, the velocity minimum variance direction is at several degrees with respect to the ambient field while magnetic field fluctuations still keep their minimum variance rather close to it /4/. Moreover decoupling phenomena affect smaller scales and σ_C is clearly higher wherever field and velocity mantain the same minimum variance direction close to the ambient magnetic field (study by Bruno and Bavassano still in progress).

Fig. 2. Cross-helicity analysis performed for the stream of day 103:18 of 1976 at 0.29 AU. Wave number band is between 3 and $7 \cdot 10^{-7}$ Km^{-1}. and e^+ and e^- are in units of 10^9 $((km \cdot sec^{-1})^2/km^{-1})$. Results from minimum variance analysis performed for 6-hour running intervals are plotted in the bottom panel. The thick(thin) line shows the angle between velocity(field) minimum variance direction and the ambient magnetic field direction.

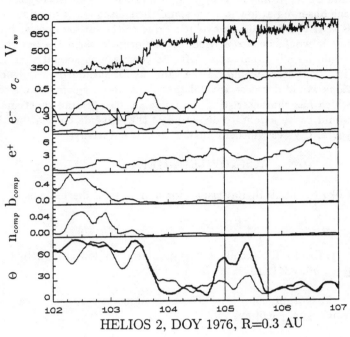

HELIOS 2, DOY 1976, R=0.3 AU

All this goes in the same direction of observations showing that fluctuations become isotropized with increasing distance /3,4/.

CONCLUSIONS AND OPEN PROBLEMS

Alfvénic fluctuations are ubiquitous in the solar wind. The fact that a perfect Alfvénic correlation is far from being found, not even in fast streams close to the Sun, suggests that pure Alfvén waves might be superimposed to an already well developed "standard" turbulence as the one found in the slow wind /2/. Close to the Sun, within fast streams, the relative importance of Alfvénic vs standard turbulence is bigger than in slow wind. At larger distances, with increasing stream-stream interaction and local generation of structures, the Alfvénic character of the fluctuations progressively fades away. Voyager's observations showed that plasma regions which have not experienced dynamical interactions with neighbouring plasma conserve the Alfvénic character of the fluctuations at distances as far as 8 AU /3/. However, the fact that by 1 AU the spectra of e^+ generally reach the level of e^- and that r_a is far from equipartition (~ 0.3) is an indication that the power which was associated to the e^+ modes by 0.3 AU has been converted to heat the plasma and that the Alfvénic character of the fluctuations has vanished.

Local generation of Alfvénic fluctuations by velocity shear is thought to be the main cause for the depletion of the Alfvénicity with distance /4/. However, this is not a common phenomenon /1/ and it is hard to believe that a few localized sources would be able to fill the whole interplanetary space with freshly generated e^- modes which would interact with e^+ modes of solar origin. On the other hand, since the radial evolution of Alfvénic fluctuations is active everywhere in the SW, also other mechanisms like coupling between outward modes and structures must play a significant role leading to a decrease of Alfvénicity and reducing the Alfvénic turbulence to standard turbulence. The destructive effects of compressive phenomena on e^+ /1/ strongly support this view.

Within fast streams and close to the ecliptic, Alfvénic turbulence evolves towards standard turbulence mainly because of the dynamical interaction between fast and slow streams. What would then the situation be around the poles where no such an interaction exists ? One possibility is that Alfvénic turbulence might not relax towards standard turbulence because there is no slower plasma to interact with /2/. The opposite possibility is that since the magnetic field would be smaller far from the ecliptic it would lead to an isotropization of the fluctuations faster than the one observed at low latitudes and the subsequent evolution would be faster /4/. The middle course is that there would be evolution due to the interaction with convected plasma and field structures but it would be slower than on the ecliptic since the power associated with Alfvénic fluctuations would largely dominate. Alfvénic correlations should then last longer than on the ecliptic plane. The final state would be the same as for standard turbulence since the amplitude of the fluctuations would decrease with distance and become comparable to the underlying solar wind structures which has characteristics of a universal background spectrum. At this point nonlinear interactions would cause a definite depletion of the Alfvénic correlation. Where should this happen ? In a few years, Ulysses will help answer to this question.

ACKNOWLEDGMENTS

I like to thank B. Bavassano for many useful discussions.

REFERENCES

1. Bruno, R., and B. Bavassano, Origin of low cross-helicity regions in the solar wind, J. Geophys. Res., 96, 7841, 1991.

2. Grappin, R., M. Velli, A. Mangeney, "Alfvénic" versus "standard" turbulence in the solar wind, Ann. Geophys., 9, 416, 1991.

3. Marsch, E., MHD turbulence in the solar wind, Physics of the Inner Heliosphere, edited by R. Schwenn and E. Marsch, Springer-Verlag, Berlin, 1991.

4. Roberts, D. A., M. L. Goldstein, Turbulence and waves in the solar wind, U.S. National Report 1987-1990, Rev. Geophys., suppl., p. 932, 1991.

5. Schwenn, R., Large-scale structure of interplanetary medium, Physics of the Inner Heliosphere edited by R. Schwenn and E. Marsch, Springer-Verlag, Berlin, 1991.

6. Tu, C.-Y., and E. Marsch, A case of very low cross-helicity fluctuations in the solar wind, Ann. Geophys., 9, 319, 1991.

MULTIFRACTALS IN THE SOLAR WIND

L. F. Burlaga

Laboratory for Extraterrestrial Physics, NASA/Goddard Space Flight Center, Code 692, Greenbelt, MD 20771, U.S.A.

ABSTRACT

Multifractals have been observed in the solar wind in several contexts. The velocity fluctuations observed by Voyager 2 near 8 AU have the structure of intermittent turbulence which has multifractal scaling symmetry. The velocity fluctuations in corotating streams at 1 AU and near 6 AU also have multifractal structure, and the structure evolves significantly between 1 AU and 6 AU. Multifractal scaling has also been observed in the magnetic field strength, density and temperature in recurrent streams at 1 AU and in large-scale fluctuations the magnetic field strength at 25 AU.

INTRODUCTION

Multifractals /1/ refer to a class of fluctuating fields with certain scaling properties. The concept arises in many physical processes, including the phenomenon of intermittent turbulence. Typically, turbulence is observed as a velocity profile $V(t)$ at a fixed point as a fluid containing the turbulence is convected past the observer. One can generally assume that there is little change in the turbulence during the observation (Taylor's hypothesis) so that one is effectively sampling the structure of the turbulence along some line through the fluid. It is customary to describe the turbulence in terms of a basic component called an "eddy", and it is assumed that there is a hierarchy of eddies of different sizes, the smaller eddies being most abundant. An observation made over a time interval τ_n picks out an eddy of size $l_n \sim V\tau_n$, and the characteristic velocity of an eddy of size l_n is $\langle\Delta V_n\rangle \sim \langle V(t + \tau_n) - V(t) \rangle$, where the average is over the measured time series. By considering the velocity on different scales, one obtains the characteristic velocity of eddies over a range of scales. The burstiness of the velocity profile, which is the hallmark of intermittency, is described by considering moments of ΔV, viz. $\langle\Delta V_n{}^p\rangle$, where p is a positive real number. Multifractal symmetry is present if $\langle\Delta V_n{}^p\rangle = \langle S_n{}^p\rangle \sim \tau_n{}^{s(p)}$, where $s(p)$ is a non-linear function of p. Thus, each moment scales as a power of τ_n, as in a self-affine fractal, but the exponent is a nonlinear function of p.

Kolmogorov's spectrum /2/ can be derived by postulating that a big eddy produces smaller eddies, each of which produces smaller eddies in the same ratio, etc. and which fill the available space /3/. This model gives the familiar 5/3 law for the power spectrum and it gives $s(p) = p/3$, but it does not describe intermittency. A similar model /3/ in which the eddies do not fill the available space gives the relation $s(p) = (D - 2) p/3 + (3 - D)$, where D is the fractal dimension of the space occupied by the turbulence. In this case, the turbulence is intermittent, and the burstiness arises from the mixture of turbulent and non-turbulent regions, which is described by the fractal dimension D. Kolmogorov /4/ introduced a statistical model of intermittent turbulent based on a lognormal distribution which gives $s(p) = p/3 + (\mu/18)(3 - p) p$, where the parameter μ describes the intermittency. A more general model proposed by Paladin and Vulpiani /5/ gives $s(p) = p/3 - \ln_2 <\beta^{(1-p/3)}>_{P(\beta)}$. The above models provide clear predictions which can be tested by examining observations of $<\Delta V_n^p>$.

INTERMITTENT TURBULENCE AND MULTIFRACTALS IN RECURRENT STREAMS

The interplanetary medium is often turbulent, especially beyond 1 AU /6/. Voyager 2 obtained a continuous series of high resolution measurements of the bulk speed of the solar wind from day 186 to day 191, 1981 while it was near 8.5 AU. The observations during this five-day interval represent a sample of a turbulent region whose radial dimension is about 1 AU. Moments of ΔV_n were computed for various time scales from the measured velocity profile. Figure 1 shows the existence of a linear relation between log $<\Delta V_n^p>$ and log τ_n over a wide range of p, and the slopes increase with p, consistent with the existence of multi-fractal symmetry. From the slopes of these lines and similar lines for other values of p obtained by a least squares

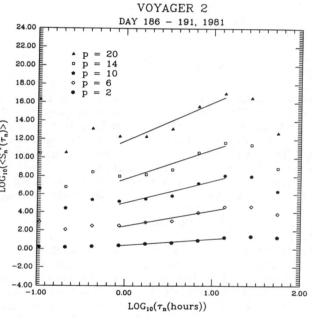

VOYAGER 2
DAY 186 - 191, 1981

Fig. 1

fit, one obtains s(p) as shown by the heavy dots in Figure 2. The error bars represent the uncertainty in s(p) derived from the least squares fit. It is evident that s(p) increases non-linearly with p. Theoretical values for s(p) are also shown in Figure 2. The non-intermittent Kolmogorov (K41) model /3/ predicts a linear relation with slope s = 1/3 is inconsistent with the observations for p > 6. The constant β model /3/ also predicts a straight

line for s(p) with a slope
that can be adjusted, but
it too is inconsistent with
the non-linear behavior
which is evident for large p.
The lognormal model /3,4,5/
shown by the curve marked
LN in Figure 2 agrees with
the observations within the
errors for all but the largest
p; it is marginally consistent
with the observations. The
random β model /5/ shown by
the solid curve in Figure 2
provides a good fit to the
data. This same curve
describes the observations of
intermittent turbulence made in
the laboratory /5/. Thus, we
have the remarkable result that
the compressible MHD turbulence
on a scale of 1 AU in the solar
wind near Saturn has the same
multifractal structure as the
gaseous turbulence observed in a
laboratory on Earth.

Multifractal structure has
also been observed in recurrent
streams at 1 AU on scales from
4 hr to 32 hr and in the
corresponding compound streams
near 6 AU on scales from 2 hr
to 16 hr /7/. Figure 3 shows
the curves s(p) vs p for both
the 1 AU data from IMP (solid
dots) and the 6 AU data from
Pioneer 11 (open squares).
For p > 5, the Pioneer 11 data
give s(p) consistent with the
random β model /5/ and the
log-normal model /3,4/ of
intermittent turbulence, and
the IMP data fall on a curve
with the same slope as that
derived from the K41 model of
turbulence /2,3/. However,
both the Pioneer 10 and the
IMP results depart significantly
from the models of intermittent

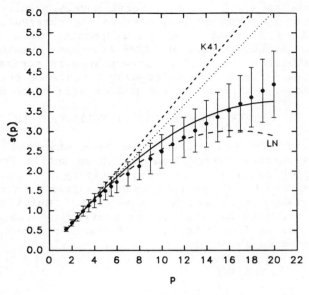

Fig. 2

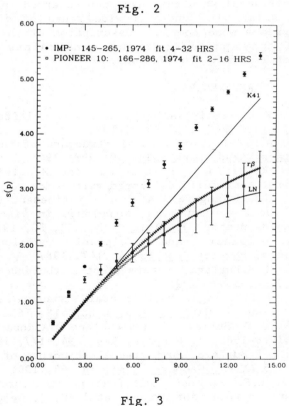

Fig. 3

turbulence. The conventional models predict s(p) = 2/3 (corresponding to a power spectrum of 5/3) whereas the observations show s(2) = 1.1 at 6 AU and s(2) = 1.18 at 1 AU (corresponding to a $f^{-2.1}$ and $f^{-2.2}$, respectively), consistent with the spectral exponents obtained in previous studies /8/. The turbulence observed in interplanetary streams on scales from 2 hours to 32 hours differs from Kolmogorov turbulence owing to the presence of jumps and discontinuities, which constrain s(2) to be near 1 rather than 2/3.

OTHER OBSERVATIONS OF MULTIFRACTALS

Multifractal structure has been observed in the magnetic field strength measured by Voyager near 25 AU on scales from 16 hours to 26 days during a period of increasing solar activity /9/ and in the magnetic field strength measured by IMP at 1 AU on scales from 2 hours to 32 hours during a period of decreasing solar activity /10/. Multifractal structure has also been observed in the temperature profile and the density profile measured at 1 AU on scales from 2 hours to 32 hours in association with the recurrent streams discussed above /10/.

SUMMARY

Multifractal structure has been observed in the interplanetary magnetic field and solar wind under a wide range of circumstances. These observations provide a more complete description of the fluctuations than power spectra, and they provide strong constraints on models of turbulence and interplanetary fluctuations.

REFERENCES

1. K.R.Sreenivasan, Fractals and multifractals in fluid turbulence, Ann. Rev. Fluid Mech., 23, 539, 1991.
2. A.N.Kolmogorov, Local structure of turbulence in incompressible fluid, Dokladay Akad. Nauk SSSR, 30, 299, 1941.
3. K.U.Frisch, P.-L.Sulem, and M.Nelkin, A simple dynamical model of intermittent fully developed turbulence, J. Fluid Mech., 87, 719, 1978.
4. A.N.Kolmogorov, A. N., A refinement of previous hypothesis concerning the local structure of turbulence in viscous incompressible fluid at high Reynolds number, J. Fluid Mech., 13, 82, 1962.
5. G.Paladin, and A.Vulpiani, Anomalous scaling laws in multifractal objects, Physics Reports, 4, 147, 1987.
6. L.F.Burlaga, Intermittent turbulence in the solar wind, J. Geophys. Res., 96, 5847, 1991.
7. L.F.Burlaga, Multifractal structure in recurrent streams at 1 AU and near 6 AU, J. Geophys. Res. Lett., 18, 1651, 1991,
8. L. F. Burlaga, W.H.Mish, and D.A.Roberts, Large-scale fluctuations at 1 AU: 1978-1982, J. Geophys. Res., 94, 177, 1989.
9. L.F.Burlaga, Multifractal structure of the interplanetary magnetic field near 25 AU, J. Geophys. Res., 18, 69, 1991.
10. L.F. Burlaga, Multifractal structure of the interplanetary field and plasma in recurrent streams at 1 AU, J. Geophys. Res., in press, 1991.

ANISOTROPY IN THE SPECTRA OF ALFVÉNIC MHD FLUCTUATIONS IN THE SOLAR WIND

V. Carbone, F. Malara and P. Veltri

Dipartimento di Fisica, Universitá della Calabria, 87030 Roges di Rende, Cosenza, Italy

ABSTRACT. A model of an anisotropic turbulence spectrum is fitted on the results of a minimum variance analysis of Solar Wind fluctuations /1/, obtaining information on the energy distribution in the k-space of both the alfvénic and magnetosonic polarizations at various distances from the Sun.

Minimum variance analyses /1-3/ have revealed the existence of an important anisotropy overall the entire spectrum of the magnetic field fluctuations δB in the Solar Wind. The minimum and maximum variance directions are parallel to B_0 and perpendicular to the plane containing B_0 and v_{SW}, respectively, B_0 being the mean magnetic field and and v_{SW} the Solar Wind velocity. This anisotropy tends to decrease in the outer heliosphere /3/ but is stronger during alfvénic periods /1/, indicating a possible link between anisotropy and δv-δB correlation. It would be useful to derive explicit information on the wave polarization and wavevectors distributions, in order to have some insight in the physical mechanisms, which are responsible for the production of the observed anisotropy. Matthaeus *et al*. /4/ have calculated, by a direct analysis of Solar Wind data, the two-points magnetic field correlation tensor and have found that two populations are present in t he Solar Wind fluctuations with wavevectors parallel and perpendicular to B_0, respectively. Their analysis was somewhat limited by the assumption of cylindrical symmetry around B_0 which follows from the lack of information in the direction out of the ecliptic plane; moreover no information on the wave polarization was obtained. Dobrowolny *et al*. /5/ assuming (i) statistically normal correlations, (ii) an unrealistic gaussian spectrum (iii) a magnetosonic polarization much smaller than the alfvénic one, were able to derive some information on the wavevector distribution; in the present paper we follow a similar idea but in a more realistic context: namely, we avoid the above hypotheses assuming only statistical homogeneity and power-law anisotropic spectra. This model is fitted on the results by Bavassano *et al*. /1/ who calculated the eigenvalues of the variance matrix evaluated over five time intervals { τ_n=168 s, 8 min, 22.5 min, 1 h, 3 h} for three observations at heliocentric distances R=0.29 AU, 0.65 AU, 0.87 AU.

The magnetic field B in the spacecraft frame of reference can be expanded in Fourier series under the statistical homogeneity hypothesis:

$$B(r, t) = \sum_{k, \omega} b(k, \omega - k \cdot v_{SW}) \exp \{i (k \cdot r - \omega t)\} \qquad (1)$$

where $b(k, \omega)$ is the Fourier amplitude of the magnetic field in the Solar Wind frame of reference and v_{SW} is the Solar Wind speed. Since $\nabla \cdot B = 0$ we can express $b(k, \omega)$ as a superposition of two polarizations both perpendicular to the wavevector k

$$b(k,\omega) = b^{[1]}(k,\omega) \, e^{[1]}(k) + b^{[2]}(k,\omega) \, e^{[2]}(k)$$

where

$$e^{[1]}(k) = i \, (k \times B_0) / |k \times B_0| \, , \qquad e^{[2]}(k) = i \, (k / |k|) \times e^{[1]}(k) \qquad (2)$$

are polarization vectors parallel to the alfvénic and magnetosonic waves polarization, respectively. The variance matrix, as calculated by Bavassano *et al*. /1/ for a given time interval of duration τ_n, is defined by $S_{ij}^{(\tau_n)} = <\delta B_i^{(\tau_n)} \delta B_j^{(\tau_n)}>_{\tau_n}$, where $\delta B^{(\tau_n)} = B - <B>_{\tau_n}$, and $< >_{\tau_n}$ indicates an average over the time basis τ_n. The form (1) is used to calculate $S_{ij}^{(\tau_n)}$ in the hypothesis of statistical stationary fluctuations, obtaining

$$S_{ij}^{(\tau_n)} \cong \sum_{\alpha=1}^{2} \int d^3k \; |G_{\tau_n}(k \cdot v)|^2 \; U^{(\alpha)}(k) \; e_i^{(\alpha)}(k) \; e_j^{(\alpha)}(-k) \tag{3}$$

where $U^{(\alpha)}(k)$ represents the spectral energy density for the two polarizations, $G_\tau(\omega) = g_{\Delta t}(\omega) - g_\tau(\omega)$,

$$g_t(\omega) = \frac{1 - \exp(-i\omega t)}{i\omega t}$$

and Δt is the data set time step ($\Delta t = 6$ s). In the expression (3) cross correlations between the polarizations $\alpha=1$ and $\alpha=2$ has been neglected with respect to self-correlation terms. This condition is exactly satisfied if the two polarizations can be considered as a superposition of linear modes:

$$b^{[1]}(k,\omega) = b^{[A]}(k) \, \delta_{\omega,\omega^{[A]}} \quad , \quad b^{[2]}(k,\omega) = b^{[F]}(k) \, \delta_{\omega,\omega^{[F]}} + b^{[S]}(k) \, \delta_{\omega,\omega^{[S]}}$$

where $\omega^{[A]}$, $\omega^{[F]}$ and $\omega^{[S]}$ are the dispersion relations of the Alfven, Fast and Slow modes respectively. The bandpass function $|G_\tau(\omega)|^2$ selects the frequency range $\tau^{-1} < \omega < \Delta t^{-1}$ in the expression (3).

The anisotropy of the fluctuations could be ascribed to two different effects: i) an anisotropic energy distribution in the wave vectors space, ii) differences between the energy content of the alfvénic polarization $\alpha = 1$ and the "compressible" polarization $\alpha = 2$. To evidentiate the contribution of each of these effects to the observed anisotropy, we assume the following phenomenological expression for the energy spectra in the inertial range

$$U^{(\alpha)}(k) = \frac{C^{(\alpha)}}{\{(\ell_x^{(\alpha)} k_x)^2 + (\ell_y^{(\alpha)} k_y)^2 + (\ell_z^{(\alpha)} k_z)^2\}^{1+\mu^{(\alpha)}/2}} \quad , \quad \alpha = 1, 2 \tag{4}$$

where $C^{(\alpha)}$, $\ell_i^{(\alpha)}$ and $\mu^{(\alpha)}$ are free parameters. Moreover, the x-axis is perpendicular to the B_0-v_{SW} plane and the z-axis is parallel to the mean magnetic field. The values of the free parameters in (4) are determined by fitting the eigenvalues $\lambda_i^{(\tau_n)}$ (i = 1,2,3) of the variance matrix (1) on the corresponding average eigenvalues $f_i^{(\tau_n)}$ calculated by Bavassano *et al.* /1/ using a standard χ^2 expression. The search of the minimum requires very long computational times since several numerical evaluations of triple integrals in the expression (3) are necessary. Replacing the smooth filter function $|G_{\tau_n}(k \cdot v)|^2$ by the following function:

$$F_{\tau_n}(k) = 1, \text{ if } k_{min}(\tau_n) \leq |k| \leq k_{max}$$
$$F_{\tau_n}(k) = 0, \text{ otherwise}$$

($k_{min}(\tau_n)$ corresponding to the time basis τ_n and k_{max} to the time step Δt) a good approximation for S_{ij} is obtained and the computational times are strongly reduced. The goodness of this approximation has been tested a posteriori. The values of the free parameters corresponding to the minimum χ^2 are shown in Tables 1 and 2, along with the energy content in each polarization: $E^{(\alpha)} = \int d^3k \, U^{(\alpha)}(k)$. The value of $\ell_z^{(\alpha)}$ has been set equal to 1 without any loss of generality. We have tested the values shown in the Tables by calculating the corresponding eigenvalues $\{\lambda_1, \lambda_2, \lambda_3\}$ of the expression (3) of the variance matrix using both $|G_{\tau_n}(k \cdot v)|^2$ and $F_{\tau_n}(k)$ as filter functions and comparing them with the data set. In Figure 1 the spectra of $\sigma = \lambda_1 + \lambda_2 + \lambda_3$ and of the ratios λ_2/λ_1 and λ_3/λ_1 calculated in both ways are shown along with the corresponding values and the error bars obtained by Bavassano *et al.* /1/.

The data set contains a detailed information on the energy distribution at different wavelengths. As a consequence, the fitting procedure is able to give a sufficiently accurate determination of the spectral indices $\mu^{(1)}$ and $\mu^{(2)}$. It turns out that the spectral index of the magnetosonic polarization is always larger by a factor 1.3-1.4 than that of the alfvénic polarization; this corresponds to a steeper spectrum for the magnetosonic polarization. Similarly, also the energy contents $E^{(1)}$ and $E^{(2)}$ are accurately determined. It is worth noting that the values of $E^{(1)}$ and $E^{(2)}$ are normalized to B_0^2; this quantity varies with the distance from the Sun. However, the ratio $E^{(1)}/E^{(2)}$ is independent of the normalization and represents a meaningful physical parameter. It shows that the alfvénic polarization is always more energetic than the magnetosonic polarization ($E^{(1)}/E^{(2)} = 2\text{-}3$).

TABLE 1: Alfvénic Polarization $\alpha = 1$

R (AU)	$C^{(1)}$	$\mu^{(1)}$	$\ell_x^{(1)}$	$\ell_y^{(1)}$	$\ell_z^{(1)}$	$E^{(1)}$
0.29	7.2	1.10	100	30	1.0	0.25
0.65	7.8	1.23	70	20	1.0	0.26
0.87	5.3	1.31	50	10	1.0	0.30

TABLE 2: Magnetosonic Polarization $\alpha = 2$

R (AU)	$C^{(2)}$	$\mu^{(2)}$	$\ell_x^{(2)}$	$\ell_y^{(2)}$	$\ell_z^{(2)}$	$E^{(2)}$
0.29	1.1	1.46	110	1.3	1.0	0.13
0.65	1.7	1.73	100	1.2	1.0	0.12
0.87	0.9	1.81	90	0.7	1.0	0.10

The parameters $\ell_x^{(\alpha)}$ and $\ell_y^{(\alpha)}$ determine the shape of the energy distributions of the two polarizations in the $\mathbf{k}$ space. The fitting procedure does not allow to determine such parameters in a manner so accurate as that of the spectral indices and of the energy contents, thus the values of $\ell_i^{(\alpha)}$ in the Tables should be considered as rough estimations. However, within the uncertainties we can say that: i) concerning the alfvénic polarization $\ell_x^{(1)} \gg \ell_y^{(1)} \gg 1$. This means that the main anisotropy in the alfvénic spectrum is a strong flattening onto the $\mathbf{B_0}\text{-}\mathbf{v}_{SW}$ plane (the yz-plane); moreover, within this plane fluctuations with wavevectors nearly parallel to $\mathbf{B_0}$ dominate with respect to those with $\mathbf{k}$ perpendicular to $\mathbf{B_0}$. Then, the Alfvén-like fluctuation propagate essentially parallel to the mean magnetic field. ii) Concerning the magnetosonic polarization $\ell_x^{(2)} \gg 1$, $\ell_y^{(2)} \approx 1$. Then, the magnetosonic spectrum is also strongly flattened on the $\mathbf{B_0}\text{-}\mathbf{v}_{SW}$ plane, but within this plane the energy distribution does not present an important anisotropy. In summary, the wavevectors of both polarization lay essentially parallel to the $\mathbf{B_0}\text{-}\mathbf{v}_{SW}$ plane; alfvénic fluctuations dominate in the direction parallel to $\mathbf{B_0}$, while wavevectors quasi-perpendicular to $\mathbf{B_0}$ essentially belong to the magnetosonic polarization. Since the wavevector distributions of both polarizations are flat on the $\mathbf{B_0}\text{-}\mathbf{v}_{SW}$ plane, from the equations (2) it can be deduced that the alfvénic polarization vectors $\mathbf{e}^{[1]}(\mathbf{k})$ are essentially perpendicular to the $\mathbf{B_0}\text{-}\mathbf{v}_{SW}$ plane, while the magnetosonic polarization vectors $\mathbf{e}^{[2]}(\mathbf{k})$ essentially lay on such a plane. Then, the most energetic component (the x-component) of the magnetic field fluctuations $\delta\mathbf{B}$ is essentially due to the alfvénic mode, while the others (the y- and z-components) are essentially due to the magnetosonic mode.

The above results indicates that the alfvénic and magnetosonic polarizations are both present in the Solar Wind fluctuations, but show quite different features. The higher spectral index and the lower energy content of the magnetosonic polarization indicate that a physical process which acts differently on the two polarizations is probably at work. It is tempting to suppose that these differences might be due to a kinetic dissipation mechanism because it is known that: i) the Landau damping is more effective on the compressible (magnetosonic) waves than on the Alfvén waves, thus explaining the difference in the energy content; ii) its damping rate, proportional to the wave frequency, is stronger at short wavelengths thus explaining the steeper spectrum of the magnetosonic polarization.

Concerning the energy distribution in the $\mathbf{k}$ space, this could result from two physical effects in competition: refraction due to the spherical symmetry of the Solar Wind, which tends to align the wavevectors of the fluctuations along the radial direction /6/; Alfvén effect in nonlinear interactions, which tends to align the wavevectors along the mean magnetic field as a result of the different rates of transfer towards small scales of parallel and perpendicular wavevectors /7,8/. This latter mechanism, which is due to the fact that the group velocity of the interacting waves is parallel to B_0, should be effective only for the alfvénic polarization. Actually, only this polarization shows dominant parallel wavevectors.

Let us consider now the variations with the distance from the Sun: i) the wavevector distributions are progressively less flat in the B_0-v_{SW} plane; ii) In that plane the wave vector distribution of the alfvénic polarization is always aligned with the mean magnetic field direction, but it become progressively broader; iii) the ratio of the energy content $E^{(1)}/E^{(2)}$ is progressively increased. These considerations seem to indicate that the physical effects responsible of the flattening in the B_0-v_{SW} plane and of the alignment with the mean magnetic field direction become progressively less important, while the effects responsible for the dissipation of the magnetosonic polarizations continues to be effective.

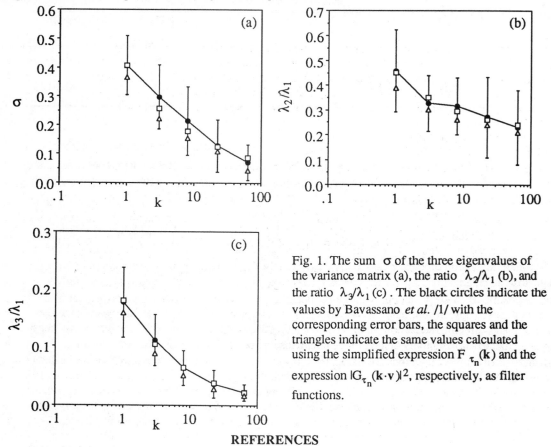

Fig. 1. The sum σ of the three eigenvalues of the variance matrix (a), the ratio λ_2/λ_1 (b), and the ratio λ_3/λ_1 (c) . The black circles indicate the values by Bavassano *et al.* /1/ with the corresponding error bars, the squares and the triangles indicate the same values calculated using the simplified expression $F_{\tau_n}(\mathbf{k})$ and the expression $|G_{\tau_n}(\mathbf{k}\cdot\mathbf{v})|^2$, respectively, as filter functions.

REFERENCES

1. B. Bavassano, M. Dobrowolny, G. Fanfoni, F. Mariani, and N. F. Ness, *Solar Phys.*, **78**, 373 (1982).
2. J. W. Belcher and L. Davis, *J. Geophys. Res.*, **76**, 3534 (1971).
3. L. Klein, D. A. Roberts, and M. L. Goldstein, *J. Geophys. Res.*, **96**, 3779 (1991).
4. M. Dobrowolny, A. Mangeney, and P. Veltri, *Astron. Astrophys.*, **83**, 26 (1982).
5. W. H. Matthaeus, M. L. Goldstein, and D. A. Roberts, *J. Geophys. Res.*, **95**, 20,673 (1990).
6. A. Barnes, in *Solar System Plasma Physics* , vol. 1, ed. E. N. Parker, C. F. Kennel, and L. J. Lanzerotti, p. 249, North-Holland, Amsterdam (1979).
7. J.V. Shebalin, W.H. Matthaeus, and D. Montgomery, *J.Plasma Phys.*, **29**, 525 (1983).
8. V. Carbone, and P. Veltri, *Geophys. Astrophys. Fluid Dynamics*, **52**, 153 (1990)

INTERPLANETARY SCINTILLATION, ALFVÉN WAVES AND ANISOTROPIC MICROSTRUCTURE

W. A. Coles

ECE Department, University of California, San Diego, CA 92093-0407, U.S.A.

ABSTRACT

Observations of interplanetary scintillation show that the spatial spectrum of electron density in the inner solar wind follows a Kolmogorov power-law at large scales. However at scales of 10's to 100's of km there is a well-defined excess. Angular scattering observations show this excess region to be highly anisotropic, at least inside 10 R_s. Here we explore the effect of Alfvén waves on scattering observations, the relationship of the waves to the density fluctuations, and ways of estimating the wave amplitudes.

INTRODUCTION

A variety of scattering phenomena can be used to probe the solar wind. These include: intensity scintillation, angular scattering, spectral broadening, pulse broadening, and phase scintillation. Most of these phenomena are caused by density fluctuations on very small scales (10 to 300 km). Only phase scintillations can probe scales larger than 1000 km. Such observations, discussed recently by Coles et al. /1/ and Harmon and Coles /2/ show an overall Kolmogorov power-law as one would expect of a turbulent process, but they also show a distinctly flatter portion of the spatial spectrum in the scale range of 3 to 300 km. It is convenient both experimentally and theoretically to study the small-scale fluctuations of electron density in terms of a structure function rather than a power spectrum. Some examples of observed structure functions of the path-integrated electron density from Coles et al. /1/ are shown in figure 1. The exponent of these structure functions is 5/3 (the Kolmogorov value) at scales greater than 1000 km. One could regard the flatter region as caused by an additional variance above the background Kolmogorov spectrum. It is interesting to note that, when a transient enhancement in the density fluctuations occurs, the Kolmogorov portion of the spectrum increases but the flatter portion does not. Thus the spectrum appears to steepen during a transient /2,3/. The cause of the flatter or enhanced portion of the spectrum is not known although various mechanisms have been proposed /1,2/.

Density fluctuations at scales of 1 to 30 km have recently been studied by measuring angular scattering with a radio interferometer by Armstrong et al. /5/. They showed that, inside of 10 R_s, the structure can be highly anisotropic. The sense of this anisotropy is that the irregularities are usually extended radially but, perhaps 5% of the time, their orientation rotates by about 90°. One presumes that the irregularities are, in fact, field-aligned. Armstrong et al. /5/ used a model with ellipsoidal symmetry, i.e. a structure function of the form $D(\vec{s}) = fn[(s_{par}/s_1)^2 + (s_{perp}/s_2)^2]$, to interpret their observations. The axial ratios s_1/s_2 of their measurements are reproduced in figure 2. This anisotropy may be useful in trying to understand the origin of the flatter or enhanced portion of the spectrum discussed above. It could be, for example, an indicator of the filamentation instability /6/ or simply an extension of the fine structure observed in coronal plumes out into the solar wind. It would be useful to know if the flatter region arises from a different physical mechanism than does the large scale Kolmogorov portion of the spectrum. It might be possible to determine this, for example, by measuring the anisotropy of the flatter region and the larger scales simultaneously. Such a measurement is possible, using the VLA and the VLBA, but has not yet been attempted.

The existence of Alfvén waves in the inner solar wind is an important factor. Such waves have been used to solve the "problem" that MHD models of the solar wind flow which are based entirely on thermal conduction do not give high enough velocities at 1 AU. Alfvén waves can transfer energy to the flow directly /7,8/. In addition large amplitude waves can damp by various mechanisms both heating the plasma and causing compressive turbulence /6,9/. This heating will also increase the flow velocity and the compressive turbulence could be related to the observed flatter region in the spatial spectrum of electron density. Wave amplitudes have not been directly measured at the distances of interest (2 to 20 R_s). However one can make an estimate of the wave amplitudes using a recent comparison of velocity measurements near the sun with theoretical acceleration models /7/. This comparison showed good agreement with an Alfvén wave assisted model about half the time. Thus we can use the wave flux of this model to estimate the effects of Alfvén waves on radio scattering observations. Here we have two objectives: to avoid errors in interpretation of the observations caused by failure to properly account for the presence of waves; and, if possible, to use the observations to estimate the wave amplitudes.

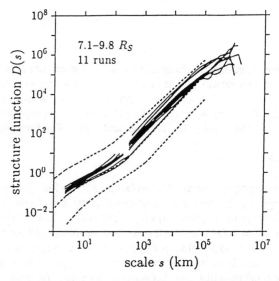

Figure 1. Plot showing structure functions derived from Voyager superior conjunction data between 7.1 and 9.8 R_s. The dashed lines represent model structure functions derived from a variety of observations by Coles and Harmon /4/ for distances of 5, 10 and 20 R_s (top to bottom).

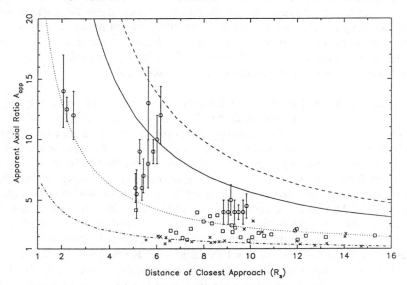

Figure 2. Axial ratio of angular scattering observations. Symbols with error bars are from the VLA, others are from Culgoora /4/. Theoretical curves are drawn in solid lines for various models /7/. The higher theoretical curves are from models with lower wave amplitudes and consequently the lower velocities.

EFFECT OF ALFVEN WAVES ON INTERFEROMETRY

A radio interferometer is an imaging instrument, however it actually measures the electric field correlation function $\Gamma(s)=<E(r)E^*(r+s)>/<|E|^2>$ which is the fourier transform of the brightness distribution (the image). Although we have no direct evidence we will assume that the structure is aligned with the magnetic field (because its mean orientation is radial close to the sun). Thus if an Alfvén wave is present in the scattering volume the direction of the anisotropy will fluctuate by some angle $\delta\theta=\tan^{-1}(\delta B/B)$. The scattering volume in the solar wind is much longer than the correlation length for Alfvén waves, therefore we must expect to see an incoherent superposition of many such waves. Clearly this will have the effect of "isotropizing" the apparent brightness distribution.

A calculation showing the apparent axial ratio as a function of wave amplitude, for various values of the intrinsic axial ratio, is shown in figure 3. One can see that the waves are very effective at reducing the apparent axial ratio. Measurements of the axial ratio thus set an upper bound on the wave amplitude than can be present in the scattering volume. For example if one were to observe an axial ratio of 6.5 it would imply that the wave amplitude $\delta B/B$ must be less than 0.2 because even if the instrinsic structure were completely two dimensional it would appear to have an axial ratio of 6.5.

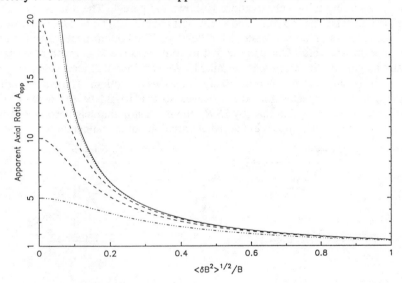

Figure 3. Calculations of apparent axial ratio versus wave amplitude, for various intrinsic axial ratios. The intrinsic axial ratios are (top to bottom) 100, 50, 20, 10 and 5.

To compare with observations one must actually integrate the effect of waves from all locations on the line-of-sight. To do this one needs a model of the change of wave amplitude with distance. The results of a family of such calculations are overplotted on figure 2, based on the model fitting of Coles et al /7/. The assumption in this calculation is that the intrinsic axial ratio is infinite, thus the calculated axial ratio is the largest that is allowed for that Alfvén wave amplitude. One can see that the range of theoretical models is quite consistent with the observed anisotropies. Furthermore the actual anisotropy could be much larger than observed; that is one must consider the possibility that the microstructure at scales in the vicinity of 10 km could be essentially two dimensional.

EFFECT OF ALFVEN WAVES ON INTENSITY SCINTILLATIONS

The effect of Alfvén waves on intensity scintillation is only beginning to be studied. Preliminary indications are that anisotropy is a very important factor. If the microstructure is highly anisotropic then Alfvén waves appear to convect the pattern in the direction of propagation at the Alfvén velocity. However if the structure is relatively isotropic (say axial ratio of 2 or less) then the waves appear to convect the scintillation pattern perpendicular to the direction of propagation. Unfortunately, as we have just discussed, we cannot tell if the microstructure outside of 15 R_s is really anisotropic or not because the presense of waves will make it appear isotropic to angular scattering observations. In fact it is even cannot measure the spatial anisotropy directly. However the anisotropy of vector velocity and magnetic field observations should show some indication if this were true.

We can model the effect of velocity fluctuations in both the radial and tangential directions. However we cannot distinguish between radial velocity variations due to waves and radial velocity variations due to the presence of multiple streams with different velocities in the scattering volume. In weak scintillation, outside of 25 R_s at EISCAT for example, one can calculate the cross correlation of intensity as a superposition of independent cross correlation functions each with a different velocity /10/. We have modelled the velocity as a two dimensional distribution for which the mean is radial (assumed to be the flow velocity) and with an rms velocity both radial and tangential. A spread in radial velocity is expected even in a constant speed spherical diverging flow, because the pattern only sees the velocity component perpendicular to the line of sight. This would reduce the apparent mean velocity to 85% of the mean flow speed and also causes an apparent rms velocity in the radial direction of 15%. If the scattering volume were partially filled by a high speed stream this apparent rms velocity in the radial direction would increas substantially. However neither spherical divergence or high speed streams would cause an apparent rms velocity ut in the tangential direction.

The result of such a model fitting to two observations is shown in figure 4. The results for a given observation are shown in three panels. The top panel shows the best overall fit. The middle panel shows a fit in which the tangential velocity variation was assumed to be zero. The bottom panel shows a fit in which the radial velocity variation was zero. One can see that in both cases the fit is much better when variation in both radial and tangential directions was permitted. We conclude that an rms velocity tangential to the mean flow is required and that this is most likely due to wave motion. The wave amplitudes are consistent with what would be expected for Alfvén waves so that would be the default assumption. However this also requires the assumption that, by 25 R_s, the microstructure has become essentially isotropic. If it were not the waves would appear in the radial variation not the tangential variation.

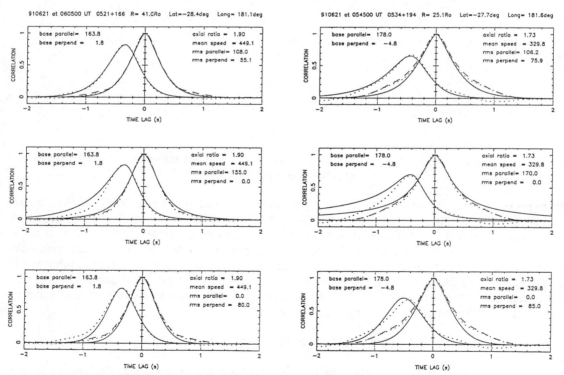

Figure 4. Observed cross correlations of intensity from EISCAT compared with a velocity distribution model. The observations are dashed and dot-dashed lines, the models are solid lines. The left panels display data from 6:05 UT June 21, 1991 at a distance of 41 R_s; the right panels from 5:45 UT on the same day at 25.1 R_s. The latitude and longitude of the scattering volume mapped back to the solar surface were almost the same, about -28° and 181° respectively.

The observations in figure 4 were selected to be at almost the same latitude and longitude but at different distances, so one can estimate the variation of tangential velocity with distance uncomplicated by potential variation with latitude or longitude. In both cases the formal errors are about 10 km/s at a two sigma level. Apparently in this case the wave amplitude decreased from 76 km/s to 55 km/s between 25 and 41 R_s. It should be evident that this discussion is in a very preliminary state, but it is a promising approach. This analysis is being pursued with a great deal of data collected at the EISCAT observatory over the last two years and also with a set of data taken at the MERLIN observatory in the spring of 1989 /12/.

SUMMARY

To obtain a clear view of the waves and turbulence in the solar wind one needs direct observations of the velocity and magnetic field vectors as these variables carry the energy. However near the sun, where wave dissipation is a particularly interesting process, we will be restricted to indirect measurements of density fluctuations for the forseeable future. Thus a reasonable approach is to compare density, velocity and magnetic field observations further from the sun with each other, as done by Marsch and Tu /11/ for example. With the understanding so gained we can hope to understand what density spectra observed nearer the sun imply about waves and turbulent dissipation.

In this review I have pointed out some interesting characteristics of the density spectra observed near the sun; that is the excess energy in a microscale regime, and the very high anisotropy of that regime. I have also tried to indicate the importance of waves in accelerating the flow, in the generation of density fluctuations, and in their effect on scintillation observations. From a variety of sources we know that Alfvén waves are present in the solar wind and we can develop models of the wave amplitude versus distance which are compatible with velocity measurements. Thus we can study the effect of these waves on observations of angular scattering and intensity scintillation.

At present it appears that the observations are often consistent with a model in which Alfvén waves increase in amplitude as they propagate outwards from the sun until they become nonlinear, after which they begin to damp. Such waves would help accelerate the wind and reduce the apparent anisotropy of the density microstructure which may thus be even more filamentary than it first appears. However if the density structure remained as anisotropic as this at larger distances the IPS correlation functions would not show a tangential velocity variation. Furthermore one would expect to see some signature of anisotropy in the spacecraft velocity and magnetic field observations. Thus we speculate that at some distance between 10 and 25 R_s, nonlinear processes stir the microstructure so as to truely "isotropize" it.

ACKNOWLEGEMENTS

The author would like to acknowledge very helpful discussions with Ruth Esser and Barney Rickett. The work was partially supported by the NSF under grant ATM-90 08608 and partially by the Norwegian Research Council (NAVF).

REFERENCES

1. W. A. Coles, W. Liu, J. K. Harmon and C. L. Martin, The Solar Wind Density Spectrum Near the Sun: Results from Voyager Radio Measurements, *J. Geophys. Res.* **96**, 1745-1755, 1991.

2. J. K. Harmon and W. A. Coles, The Density Fluctuation Spectrum of the Inner Solar Wind, *this issue*.

3. R. Woo and J. W. Armstrong, Observations of the Electron Density Spectrum Near the Sun During Interplanetary Shocks, *this issue*.

4. W. A. Coles and J. K. Harmon, Propagation Observations of the Solar Wind Near the Sun, *Astrophys. J.* **337**, 1023, 1989.

5. J. W. Armstrong, W. A. Coles, M. Kojima and B. J. Rickett, Observations of Field Aligned Density Fluctuations in the Inner Solar Wind, *Astrophys. J.* **358**, 685-692, 1990.

6. S. P. Kuo, M. H. Whang and M. C. Lee, Filamentation Instability of Large-amplitude Alfvén Waves, *J. Geophys. Res.* **93**, 9621, 1988.

7.   W. A. Coles, R. Esser, U-P Lovhaug and Jussi Markkanen, Comparison of Solar Wind Velocity Measurements with a Theoretical Acceleration Model, *J. Geophys. Res.* **96**, 13849-13859, 1991.

8.   Hollweg, J. V., Alfvén Waves in a Two-fluid Model of the Solar Wind, *J. Geophys. Res.* **181**, 547, 1973.

9.   Leer, E. and T. E. Holzer, Energy Addition in the Solar Wind, *J. Geophys. Res.* **85**, 4681, 1980.

10. R. D. Ekers and L. T. Little, The Motion of the Solar Wind Close to the Sun, *Astron. Astrophys.* **10**, 310, 1971.

11. E. Marsch and C.-Y. Tu, Spectral and Spatial Evolution of Compressive Turbulence in the Inner Solar Wind, *J. Geophys. Res.* **95**, 11945, 1990.

12. Rickett, B. J., The Large Scale Variation of Small-Scale Properties of the Solar Wind Derived from Radiowave Propagation Measurements, *this issue*.

EVOLUTION OF TRAPPED RADIATION IN A 3-DIMENSIONAL HELIOSPHERE: A COMPUTER SIMULATION

A. Czechowski and S. Grzedzielski

Space Research Centre, Polish Academy of Sciences, Ordona 21, 01-237 Warsaw, Poland

ABSTRACT

Using a numerical ray-tracing technique, the time evolution of the ~3 kHz radiation trapped in a nonspherical heliospheric cavity and interacting with corotating interaction regions (CIR) of different latitudinal extent is studied. Upward frequency drift (~1 kHz/yr for a small heliosphere /7/) is confirmed. Prolonged interaction with CIRs can significantly affect the spectrum of radiation.

INTRODUCTION

We present preliminary results of a numerical study of propagation of VLF radiation trapped in a 3-D heliosphere. The general aim is to understand the spectral and temporal characteristics of the 2-3 kHz band observed in the distant heliosphere by Voyager 1 and Voyager 2 /1,2,3/ in terms of Fermi-type interactions of radio quanta with the dynamical features of the solar wind such as CIRs, shocks etc. (for discussion of the possible origin of the signal see /4,5,6/). In our previous papers /7,8/ based on a simplified, spherically symmetric model of the heliosphere, we have shown that the positive frequency drift seen in the Voyager data may be caused by this type of interaction provided that the radiation is trapped in the heliospheric cavity. We address here the question whether similar conclusions can be made in a more realistic model, which includes corotating structures, latitudinal variations and a nonspherical heliopause.

THE MODEL

In the present study we make the following assumptions:
-Heliopause: ellipsoid, with the Sun offset from the centre;
-Termination shock: sphere;
-Solar wind: (plasma density)$\cdot r^2$=constant along Parker's spirals (the constant is modulated in longitude and latitude). The density profile is given by an analytical Ansatz (1);
-Plasma density outside the shock: constant within the heliopause;
-Initial frequency: 3 kHz.
We assume the radiation to be trapped inside the heliosphere with the external boundary determined by the heliopause. Containment of frequencies up to 3.5 kHz requires external plasma densities ≈ 0.15 cm^{-3}. The question whether and why such densities could appear is not addressed in this paper (see ref. /9/).

Ansatz for plasma density: To approximately describe the density variation due to CIRs we take the plasma density inside the shock in the form corresponding to rigidly corotating spiral structure:

$$n(\mathbf{r},t)=\text{const} \ \frac{1+a\cdot|\cos\theta|^{K}\cdot F[\cos\ n\cdot(\phi+\Omega\cdot(t-r/v))]}{r^2} \qquad (1)$$

Here (r,θ,ϕ) are polar coordinates centred at the Sun ($\theta=90°$ at the pole), $\Omega=\Omega_{\odot}$ is the Sun rotation rate and v the solar wind speed. The $|\cos\theta|^{K}$ factor ($K=0,1,2..$) was introduced to describe the expected waning of the density contrasts in the CIRs with increasing heliolatitude. $F[x]$ is the profile function with maximum value=1 and the average $\int dx F[x]=0$, a is the amplitude parameter for density variations and n the number of spiral arms. The profile function $F[x]$ was taken in the form

$$F[x]=\frac{\alpha}{1+\delta-x} - \Delta \qquad (2)$$

where α,δ and Δ are constants.

Heliosphere: The following parameters are used:
ax, ay, az: semiaxes of the heliopause ellipsoid (A.U.);
rsh: radius of the termination shock (A.U.);
dstgn: distance from shock to the stagnation point (A.U.).
The data quoted in the Figs. 1 and 2 refer to two cases: "large" heliosphere with rsh=80, ax=200, ay=az=120, dstgn=30 and "small" heliosphere with rsh=50, ax=100, ay=az=70, dstgn=20.

METHOD

The ray tracing equations of geometrical optics are solved numerically for a variety of initial conditions (typically, we vary at random the direction of the wave vector with the initial frequency and the starting point fixed). The frequency variation with time is monitored. Results are presented as frequency/drift rate histograms for a given moment after injection.

The heliopause is regarded as a reflecting surface. To mimic the effect of surface irregularities, at each reflection event the direction of the normal vector to the surface is randomized within a cone of opening angle 2χ (for data presented here $\cos\chi=0.9$).

RESULTS

-An initially monochromatic (3 kHz) radiation drifts and spreads in frequency. The drift rate depends sensitively on the dimensions of the heliosphere , the latitudinal extent of the corotating regions ,solar wind speed and the amplitude of the density variations. To generate a drift rate of up to 1 kHz/yr the heliosphere has to be small (average radius about 100 A.U.).
-The intuitively expected decrease of the drift rate when the latitudinal extent of the CIRs becomes more narrow is confirmed. Fig.1 compares the cases K=0 (no latitudinal damping) and K=2 for a large heliosphere. Although for K=2 the CIRs still extend to ~45° the drift rate is halved. Fig.2 gives the results for a small heliosphere.

-The frequency distribution has a two-component structure. It consists of a peak with positive frequency drift and a low frequency tail that results from negative drifts of those photons that were temporarily trapped inside the density troughs formed by the CIRs. The drift rate of the peak region agrees reasonably well with the predictions of our previous, spherically symmetric model ("approximate formula" in Figs.1,2). The peak is eroded in time as more and more photons join the tail.

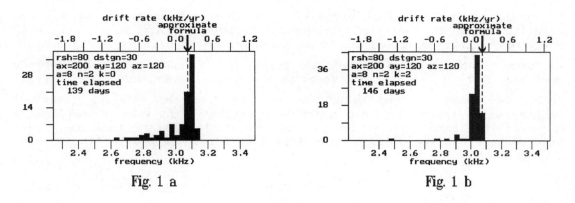

Fig. 1 a Fig. 1 b

Fig.1 Frequency/time averaged drift rate plots in a "large" heliosphere for a) K=0 (no latitudinal modulation of CIRs) and b) K=2 (damping by $|\cos\theta|^2$).

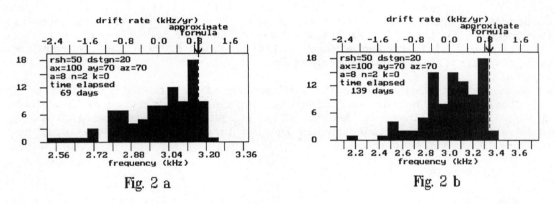

Fig. 2 a Fig. 2 b

Fig.2 Frequency/time averaged drift rate plots in a "small" heliosphere for a) 69 days and b) 139 days after photons injection.

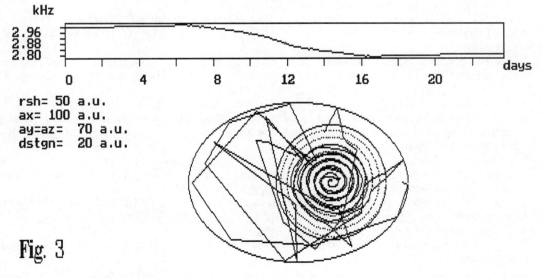

Fig. 3

Fig.3 Example of a ray path temporarily trapped in a CIR (projection onto equator plane). Upper panel shows frequency versus time. Frequency slowly increases except for large downward shift during the period of trapping.

Ray trapping in CIRs: Each event of this type is accompanied by a large negative shift in radiation frequency. A rough estimation of the total frequency change is given by $\Delta\omega/\omega = -2 \cdot R \cdot \Omega_\odot/c$ where R is the heliocentric distance at which the ray enters and leaves the CIR through its low-density open end. $R \cdot \Omega_\odot$ is the apparent speed of the channel's entrance. Due to rigid corotation this open end is always receding from the entering (leaving) photon, which translates into negative frequency shift when viewed from a non-rotating frame. The effect is enhanced by the highly idealized model assumed here for the CIRs. A typical path of a ray trapped in CIR is shown in Fig. 3.

DISCUSSION

The results support our previous conclusion that the radiation trapped in the heliosphere exhibits positive frequency drift. Also the magnitude of the drift and its dependence on the heliospheric parameters agree in essence with /7/. The new element is that this drift applies only to the peak component of the frequency spectrum. A new feature is also the phenomenon of ray trapping by the CIRs. We think, however, that to determine its effect on the frequency distribution a more realistic model of the CIRs is required. Such investigation is now in progress.

We also note that the reflection of photons from the CIRs restricted in heliolatitude may not be sufficient to explain high drift rates (~1 kHz/yr up to ~1.8 kHz/yr in recent events /3/). A possible implication is that the solar wind at high latitudes should involve large density fluctuations, even if the CIRs are absent there.

REFERENCES

1. W.S. Kurth, D.A. Gurnett, F.L. Scarf and R.L. Poynter, A new radio emission at 3 kHz in the outer heliosphere, Nature 312, 27 (1984)
2. W.S. Kurth, Radio noise in the heliospheric cavity, in: Physics of the Outer Heliosphere, COSPAR Colloquia, vol. 1, ed. S. Grzedzielski and D.E. Page, Pergamon, Oxford 1990, p. 267–275.
3. W.S. Kurth and D.A. Gurnett, New observations of the low frequency interplanetary radio emissions, Geoph. Res. Lett. (submitted for publication)
4. W.M. Macek, I.H. Cairns, W.S. Kurth and D.A. Gurnett, Low-frequency radio emissions in the outer heliosphere: constraints on emission processes, Journ. Geoph. Res. 96, A3, 3801–3806 (1991)
5. R.L. McNutt, Geoph. Res. Lett. 15, 1307–1310 (1988)
6. S. Grzedzielski and A.J. Lazarus, 2-3 kHz continuum emissions as possible indications of global heliosphere "breathing", Journ. Geoph. Res. (submitted for publication)
7. A. Czechowski and S. Grzedzielski, Frequency drift of 3 kHz interplanetary radio emissions: evidence of Fermi accelerated trapped radiation in a small heliosphere?, Nature 344, 640 (1990)
8. A. Czechowski and S. Grzedzielski, Trapped radiation in the outer heliosphere, in: Physics of the Outer Heliosphere, COSPAR Colloquia, vol. 1, ed. S. Grzedzielski and D.E. Page, Pergamon, Oxford 1990, p. 281–284.
9. S. Grzedzielski, The large-scale structure in the outer solar system, this issue.

ACCELERATION OF ALPHA PARTICLES BY ION-CYCLOTRON WAVES AND PROTON THERMAL ANISOTROPY

L. Gomberoff* and R. Hernández**

* Departamento de Física, Facultad de Ciencias, Universidad de Chile,
Casilla 653, Santiago, Chile
** Departamento de Física, Facultad de Ciencias, Universidad de Concepción,
Casilla 3-C, Concepión, Chile

ABSTRACT

It was shown recently /1/ that an initial alpha–proton drift velocity can lead to the generation of left hand polarized electromagnetic ion–cyclotron waves which can accelerate alpha particles to velocities well in excess of the proton bulk velocity. In the aforementioned reference, the solar wind protons were modeled by a biMaxwellian distribution function. However, the occurrence of a secondary peak is a persistent feature of the observations. Here, a drifting Maxwellian of secondary protons is added to the model. It is shown that the beam does not affect the dispersion relation of the ion–cyclotron waves. A simple model of the magnetic field close to the source of the fast solar wind shows that a proton–thermal anisotropy can generate the required initial alpha–proton drift velocity.

INTRODUCTION

It is well known that quasi-linear resonant cyclotron interaction of solar wind alpha particles with parallel propagating left-hand polarized ion–cyclotron waves can accelerate alpha particles to velocities in excess of the proton bulk velocity /2,3/. This mechanism has been thought to be at least a first step mechanism leading to the observed differential speeds.

However, it was pointed out in /2/ that in all treatments of the resonant interaction, alpha particles were either neglected or at best treated as test particles in the dispersion relation of the ion–cyclotron waves. If the dispersive effects of the alpha particles are taken into account, the stop band around the alpha particle gyrofrequency produces a gap in the resonant alpha particle velocity rendering this resonant acceleration mechanism impossible /2/.

It has been recently shown /1/ that when the alpha particle drift velocity exceeds the proton bulk velocity, the stop band around the alpha particle gyrofrequency disappears. Under these conditions, alpha particles can be accelerated to velocities well in excess of the proton bulk velocity.

In /1/ the solar wind was modeled by biMaxwellian protons and the alpha particles by a drifting Maxwellian. However, observations performed on board of Helios 1 and 2 indicate that the presence of a secondary beam of hot protons is a persistent feature of the high speed streams /4/. Therefore, we shall assume a more realistic model for the solar wind by considering a double–humped proton distribution consisting of an anisotropic core and a hot beam. It will be shown that the new term does not affect the results found in /1/.

The stop band around the alpha-particle gyrofrequency disappears only when the alpha particle drift velocity exceeds the proton velocity. The way in which this happens has remained unexplained.

It will be shown here that if at the origin of the fast solar wind the magnetic field is convergent, a proton thermal anisotropy can produce the small alpha-proton drift velocity that is required in order to trigger the acceleration mechanism.

THE DISPERSION RELATION

The solar wind is modeled by a proton distribution function consisting of an anisotropic core and a beam, described by a biMaxwellian and a Maxwellian distribution function respectively, and alpha particles described by a drifting Maxwellian distribution function. Assuming $\omega = \omega_r + i\omega_i$ and $\beta_{\|l} \leq 1$, from the real part of the dispersion relation of left-hand polarized ion-cyclotron waves /1/ we obtain:

$$y^2 = \frac{\delta_1 x^2}{1-x} + \frac{\delta_2(x-yU_b)^2}{1-x+yU_b} + \frac{4\eta(x-yU_\alpha)^2}{1-2x+2yU_\alpha}, \tag{1}$$

On the other hand, the imaginary part of the dispersion relation gives:

$$\gamma = (\frac{\pi}{\beta_{\|pc}})^{1/2}\frac{\delta_1[A_p(1-x)-x]}{F(y,x)}e^{-\frac{(1-x)^2}{y^2\beta_{\|pc}}} - (\frac{\pi}{\beta_{\|pb}})^{1/2}\frac{\delta_2(x-yU_b)}{F(y,x)}e^{-\frac{(x-yU_p-1)^2}{y^2\beta_{\|pb}}}$$

$$-(\frac{\pi}{\beta_{\|\alpha}})^{1/2}\frac{\eta(x-yU_\alpha)}{F(y,x)}e^{-\frac{(2x-2yU_\alpha-1)^2}{4y^2\beta_{\|\alpha}}}, \tag{2}$$

where

$$F(y,x) = y[\frac{\delta_1 x(2-x)}{(1-x)^2} + \frac{\delta_2(x-yU_b)(2-x+yU_b)}{(1-x+yU_b)^2} + \frac{8\eta(x-yU_\alpha)(1-x+yU_\alpha)}{(1-2x+2yU_\alpha)^2}]. \tag{3}$$

In equations (1) and (2) $y = k_\| V_A/\Omega_p$, $x = \omega/\Omega_p$, $A_p = (T_\perp/T_\| - 1)$, $\beta_{\|l} = 2kT_{\|l}/m_l V_A^2$, $\delta_1 = n_{pc}/n_p$, $\delta_2 = n_{pb}/n_p$, $n_p = n_{pc} + n_{pb}$, $\eta = n_\alpha/n_p$, $U_b = V_b/V_A$, $U_\alpha = V_\alpha/V_A$, and $V_A = B_0/(4\pi n_p m_p)^{1/2}$ is the Alfvén velocity.

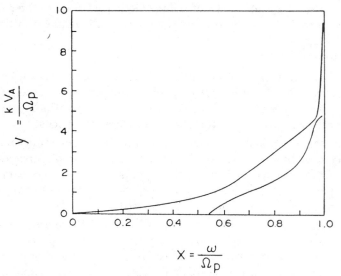

Fig. 1. Wavenumber $y = k_\| V_A/\Omega_p$ vs frequency $x = \omega/\Omega_p$ for $\delta_c = 0.9$, $\delta_b = 0.1$, $\eta = 0.04$, $U_b = 1.2$, and $U_\alpha = 0.1$.

In Figure 1, the dispersion relation given by equation (1) is illustrated. We can see that there is no difference with the dispersion relation obtained by taking $\delta_1 = 1$ and $\delta_2 = 0$, (see Figure 1b of /1/). On the other hand, the last two terms in equation (2) are thermal corrections to the proton–cyclotron instability. These corrections are expected to have stabilizing effects on the proton-cyclotron instability /5/.

The growth/damping rate given by equation (2) is illustrated in Figure 2 for plasma parameters consistent with solar wind observations /4/ We can see that there is practically no difference between our Figure 2 and Figure 3d of /1/ in the region $\Omega_\alpha < \omega < \Omega_p$. The region $0 < \omega < \Omega_\alpha$, which is weakly unstable when thermal effects of the proton beam and the alpha–particle beam are not taken into account, is now completely stabilized. This means that thermal effects due to the beams on the proton-cyclotron instability are negligible for the parameters indicated in Figure 2.

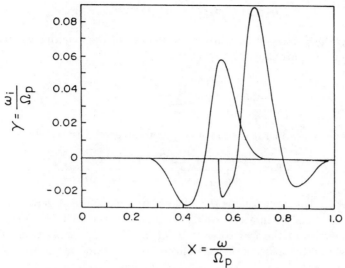

Fig. 2. Growth/damding rate $\gamma = \omega_i/\Omega_p$ vs frequency $x = \omega/\Omega_p$, for $A_p = 4$, $\beta_{\|p} = 0.1$, $T_{pb}/T_{pc} = 2$ $(\beta_{\|b} = 0.2)$, $T_\alpha/T_p = 4$ $(\beta_{\|\alpha} = 0.1)$. The other parameters are like in Figure 1.

The initial alpha-proton drift velocity required to trigger the acceleration mechanism cannot be explained by this model. To this end, in the next section we present some simple ideas concerning the initial drift velocity.

THE INITIAL ALPHA–PROTON DRIFT VELOCITY

Let us write the equation of motion for particles of type l in the following form:

$$\frac{D^{(l)}\vec{V}_l}{Dt} + \frac{1}{m_l n_l}\vec{\nabla}\cdot\bar{P}_l - \frac{e_l}{m_l}\{\vec{E} + \frac{1}{c}\vec{V}_l\times\vec{B}\} + \frac{GM_\odot}{r^2}\hat{r} = \{\frac{\partial\vec{V}_l}{\partial t}\}_{c,w}, \tag{4}$$

where $D^{(l)}/Dt = \frac{\partial}{\partial t} + \vec{U}_l\cdot\vec{\nabla}$ is the convective derivative, $\bar{P}$ is the pressure tensor, $\vec{U}_l$ is the fluid velocity, $\vec{E} \simeq -\vec{\nabla}\cdot\bar{P}_e/en_e$, $\vec{B}$ is the external magnetic field, $GM_\odot/r^2$ is the Sun gravitational acceleration, and the last term contains all other interactions including the collisions between the particles and wave-particle interactions.

Let us now introduce the magnitudes $\vec{V}_d = \vec{V}_\alpha - \vec{V}_p$, $\vec{V}_{sw} = \hat{\rho}_\alpha\vec{V}_\alpha + \hat{\rho}_p\vec{V}_p$, where $\hat{\rho}_{\alpha,p} = \rho_{\alpha,p}/(\rho_\alpha+\rho_p)$, and $\rho_{\alpha,p} = m_{\alpha,p}n_{\alpha,p}$, and make the following assumptions:(1) when $\vec{U}_d$ is zero or very small, the collision and the wave-particle interaction term can be neglected, (2) the distances involved are sufficiently short so that $\nabla\hat{\rho}_{\alpha,p} \simeq 0$, (3) the magnetic field is weakly inhomogeneous, and (4), the pressure tensor can be written in the form $\bar{P} = p_\|\hat{b}\hat{b} + p_\perp\{\bar{1} - \hat{b}\hat{b}\}$, where $p_{\|(\perp)}$ is the pressure parallel (perpendicular) to the external magnetic field whose direction is $\hat{b}$. Introducing $p = \frac{1}{3}(p_\| + 2p_\perp) = nT$, where T is the kinetic temperature of the corresponding species, and noticing that $\vec{V}_{sw} \parallel \vec{V}_d \parallel \vec{B}$, equation (4) reduces to

$$\frac{D\vec{V}_d}{Dt} = -\vec{v}_d\cdot\vec{\nabla}\vec{V}_{sw} - \{\frac{T_\alpha}{m_\alpha}\frac{3A_\alpha}{2A_\alpha + 3} - \frac{T_p}{m_p}\frac{3A_p}{2A_P + 3} - \frac{T_e}{m_p}\frac{2A_e}{2A_e + 3}\}\{(\hat{b}\cdot\vec{\nabla}LnB)\hat{b} - \hat{b}\cdot\vec{\nabla}\hat{b}\}. \tag{5}$$

In order to evaluate the last term in equation (5), we consider a simple model of the magnetic field which agrees with the observed topology in the vicinity of the solar corona /6/.

Thus, let us assume cylindrical symmetry and let r be the axis of symmetry, $\vec{B}(r,\rho) = B_r\hat{r} + B_\rho\hat{\rho}$, where $\hat{\rho}$ is in the radial direction, $B_r = B_0\{r_0/r\}^\gamma$, where B_0, r_0, and γ are free parameters /6/. Assuming that $\partial B_r/\partial r$ does not depend on ρ, from $\vec{\nabla} \cdot \vec{B} = 0$ it follows that:

$$\vec{B}(r,\rho) \simeq B_r\{\hat{r} + \frac{1}{2}\gamma\frac{\rho}{r}\hat{\rho}\}. \tag{6}$$

Moreover, if $\rho \ll r$, i.e., the particle flux is around a vicinity of the symmetry axis, then the last term in equation (5) can be approximated by $-(\gamma/r)\hat{r}$.

Thus, equation (5) reduces to:

$$\frac{d(V_{sw}V_d)}{dr} = \frac{V_0}{r}, \tag{7}$$

where V_0 does not depend on r. The last equation can be integrated to obtain:

$$V_d = -\frac{3\gamma}{4}\frac{\alpha_p^2}{V_{sw}}\{\frac{A_p}{A_p + 3/2} - \frac{T}{M}\frac{A_\alpha}{A_\alpha + 3/2} + \frac{1}{2}\frac{T_e}{T_p}\frac{A_e}{A_e + 3/2}\}ln(r/r_0), \tag{8}$$

where r_0 is some reference distance where the plasma is injected with zero drift velocity in a magnetic field characterized by the index γ, which is negative if the field lines are convergent /6/.

If the observed properties of the fast solar wind at 0.3 AU bare some resemblance with the properties of the solar wind after crossing the small layer under consideration, then $A_p > 1$, $A_\alpha < 0$, $A_e \le 0$, $T = M$, $T_p \gg T_e$. For these values the sum of the first two terms in the curly brackets is positive and the last one is negative. However, the electron contribution is much smaller and can be neglected.

Thus, the difference of the pressure gradient forces can lead to a positiv alpha–proton drift velocity.

SUMMARY

We have improved the solar wind model of /1/ by considering a double–humped proton distribution. We have also included in the expression for the growth/damping rate thermal contributions due to the proton beam and the alpha-particle beam.

A comparison with the results of /1/ shows that thermal effects of the beams on the gowth/damping rate are negligible. Therefore, the simple model used in /1/ in order to study the left-hand polarized electromagnetic ion-cyclotron waves is not affected in any sensible way by the addition of a beam to the proton distribution function.

In /1/ it was shown that the proposed acceleration mechanism can take place only when the alpha particle drift velocity exceeds the proton velocity. The way in which the alpha particle velocity exceeds the proton velocity cannot be explained by the resonant interaction.

To this end, we showed that if at the origin of the fast solar wind the magnetic field is convergent /7/, a proton thermal anisotropy can generate the required initial alpha–proton drift velocity. The result given by equation (8) is a small quantity independently of the actual values assigned to the various parameters involved.

REFERENCES

1. L. Gomberoff, and R. Elgueta, J. Geophys.Res. 96, 9801, (1991)

2. P. A. Isenberg, J. Geophys. Res. 89, 2133, (1984)

3. P. A. Isenberg and J. V. Hollweg, J. Geophys. Res. 88, 3923, (1983)

4. E. Marsch, K. H. Muhlhauser, R. Schwenn, H. Rosenbauer, W. Pillip,
and F. M. Neubauer, J. Geophys. Res. 87, 52 (1982)

5. L. Gomberoff and P. Vega, J. Geophys. Res. 92, 7728, (1987)

6. Tyan Yeh, Solar Phys. 56, 439, (1977)

7. W. I. Axford and J. M. McKenzie, these Proceedings

MHD TURBULENCE: THEORY/SIMULATIONS

R. Grappin, A. Mangeney and M. Velli

Observatoire de Paris-Meudon, 92195 Meudon, France

ABSTRACT

We discuss some aspects of "Alfvénic" turbulence which can be found in numerical experiments. We present in particular preliminary results of direct numerical simulations taking into account one of the main feature of the solar wind, namely its spherical expansion.

INTRODUCTION

The supersonic solar wind convects fluctuations both in velocity and magnetic field with substantial relative energy: how fast do these fluctuations give their energy back to the wind, either accelerating or heating the wind via nonlinear interactions? One may distinguish the large regions with non zero average magnetic field ("magnetic sectors"), in which one finds Alfvén-like propagating fluctuations from the neutral sheet where turbulent reconnection is likely to occur. While the energy transfer is probably fastest near the neutral sheet /1/, near the ecliptic plane, the "Alfvénic turbulence" may dominate far from it /2/ and it is important to study its nonlinear evolution. One of the interesting problems is how the waves are able to deposit their energy and momentum sufficiently far from the sun, so as to produce the hot, high-speed streams: this seems to require a relatively slow nonlinear evolution. Our discussion will be illustrated by results from numerical simulations: we shall consider in turn some aspects of "Alfvén waves turbulence" in an incompressible medium, in a compressible one and finally in an expanding medium.

HOMOGENEOUS INCOMPRESSIBLE "ALFVEN WAVES TURBULENCE"

Consider a homogeneous incompressible medium with a uniform magnetic field $B°$. A linear analysis shows that the plasma admits only one kind of oscillations, Alfvén waves, with two eigenmodes propagating in opposite directions along the average magnetic field with the Alfvén speed $V_a = B°/\sqrt{(4\pi\rho)}$, ρ being the density. In general, nonlinear distortions appear in a time of the order of l/b_{rms}, where $b = \delta B/\sqrt{(4\pi\rho)}$; δB is the magnetic field fluctuation and l its characteristic size. Fig. 1 shows this in the particular case where the relative fluctuation is of order unity: $\delta B \approx B°$. The fields shown are the Elsässer fields $z^\pm = u \pm b$ (resp. right and left) which measure the amplitudes of each Alfvén mode; $B°$ is horizontal and rightward.

However, when the amplitude of the fluctuation is smaller than $B°$ (which usually occurs in the high wavenumber part of a turbulent spectrum), each individual "collision" as in fig.1 is too short for the energy exchange to be complete. Iroshnikov (1963) and Kraichnan (1965) /3/ (henceforth IK) proposed a model of weak turbulence for Alfvén waves, which amounts to assuming that the energy exchanges in successive collisions add up in an incoherent manner so that nonlinear transfer is delayed /4/. The numerical evidence for this delaying effect is only indirect, via tentative measurements of spectral slopes which the IK theory predicts differ from the Kolmogorov value /5/. Within the solar wind, since the first proposal /6/ to interpret the spectral range between one day and one minute with IK turbulence, detailed analyses have indicated that the spectral slope is in fact closer to the Kolmogorov value /7/.

A still stronger delaying effect is built in the nonlinear terms of the incompressible MHD equations: since the nonlinear kernel is of the form z^-z^+, it vanishes when the amplitude of one of the two Alfvén modes (z^+ or z^-) becomes very small (for a fixed value of $\delta B/B^\circ$), so that the effective interaction time becomes very large. This effect, first observed numerically by Pouquet and Patterson /8/, is certainly relevant for the solar wind, which is often depleted in one Alfvén mode, the dominant mode being the one propagating away from the sun /9/. Hence, the emergence of the well-developed spectrum observed in situ in the solar wind should be slow. Dobrowolny, Mangeney and Veltri /4/ argued that in fact nonlinear interactions delay even more, by amplifying any initial imbalance between the two fields $z^\pm$. This evolution is observed both in propagation experiments (fig.1) and in reconnection configurations, where it appears as an increasing alignment between velocity and magnetic field (fig. 2).

The solar wind turbulence actually evolves in the *reverse* way /10/: it starts near the sun with a high correlation and ends up with a low correlation. A way out of this dilemma /11/ possibly lies in the existence of large-scale shear flows /6/ which would be a reservoir of both Alfvén modes which could progressively replace the initial small-scale disequilibrium between the two modes. However, it is not clear whether this process is fast enough to do the job in the required time (say between 0.3 and 1 AU), and the large scale shear energy is not necessarily available everywhere in the heliospheric cavity /2/.

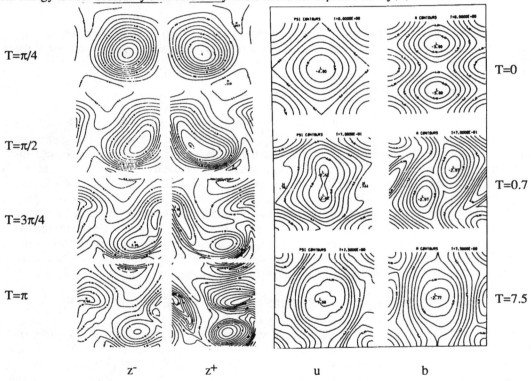

T=π/4

T=π/2

T=3π/4

T=π

z^- z^+ u   b

T=0

T=0.7

T=7.5

Fig. 1. Collision of two incompressible Alfvén vortices along a uniform horizontal magnetic field. Stream functions for z^- (left) and z^+ (right). (From /12/).

Fig. 2. Evolution of the incompressible Orszag-Tang vortex from a state with low velocity-magnetic field correlation (45%) towards a state with large correlation (85%) (From /13/).

TURBULENCE WITH u≈b

A maximal correlation stops nonlinear evolution only in the incompressible limit. Indeed, consider what would be in the incompressible case a pure Alfvén mode within a uniform field B° (i.e. assume $u=b$) but now with a finite sound velocity. If the (magnetic) pressure is not constant, our "pure" Alfvén mode appears to be actually made of a superposition of magnetosonic and Alfvén modes, which interact nonlinearly. Fig. 3a shows for an initial Mach number $M=\delta u/c_S=0.6$, where δu is the rms velocity fluctuation (with equal sound and Alfvén speeds) how a "pure" z^+ vortex is distorted, while some z^- fluctuations are generated.

It is interesting to ask about the propagation properties of such waves. While the propagation (from right to left) of the dominant z^+ vortex can be easily followed, no rightward propagating structure in the z^- mode is seen. In fact, the z^- component shows up largely in the form of compressive fluctuations: the stationary regime which sets in is characterized by an amplitude ratio z^-/z^+ of about 20% and also a comparable level of relative density fluctuation. The small scales features, which have smaller relative amplitudes, might show more clearly Alfvénic propagative properties. To emphasize the dynamics of the solenoidal small scales, we plot in fig. 3b the contours of the curl of both Elsässer fields. It is seen that the extrema of the + and - vorticities are located at about the same place in the box, corresponding to the folding of the z^+ eddy. Both form abrupt ridges along the folds of the eddies, which propagate *for both fields* in one and the same direction, instead of opposite directions as in a linear analysis.

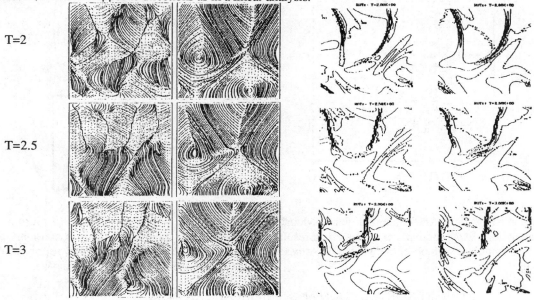

T=2

T=2.5

T=3

Fig. 3. Evolution of a single "Alfvén" mode along a uniform horizontal magnetic field, with Mach number M=0.6 and Alfvén Mach number unity (perfect gas, periodic boundary conditions). The z^- amplitude is initially zero everywhere. (a) left: Stream functions for z^- and z^+; (b) right: vorticity contours curl(z^-) and curl(z^+). Time increases from top (T=2) to bottom (T=3).

The phenomenon is less apparent at smaller Mach numbers: when the Mach number is 0.3 (instead of 0.6 in fig. 3), the coherent vorticity structures of the dominant field have a much shorter life-time. Surprisingly enough however, the phenomenon shows up again in the incompressible case: a collision of two Alfvén modes as in fig. 1 does generate one-way propagating vortical features, while now the large scale vortices propagate "normally" in opposite directions. The propagation direction of vortical features is determined by the dominant field (either z^+ or z^-). The fact that the high Mach number configuration is more similar to the incompressible one than the low Mach number one could be related with the two regimes of nearly incompressible turbulence proposed by /14/. Finally, we may remark that the amplitude of the fluctuation is large ($\delta B/B° = 1$): in that case one can hardly expect the (linear) dispersion relation to be very precise; however, the phenomenon persists when one considers a smaller ratio, as $\delta B/B° = 0.5$.

What about the level of density fluctuations? A small level of density fluctuations has long been a criterion of identification of Alfvén waves in the wind. Now, it is known /15/ that solenoidal flows generate some acoustic noise, even in the absence of boundaries. Kliatskin /16/ predicted that the rms level of density fluctuation should be simply related to the turbulent Mach number: $\delta n/n = M^2$. This amounts in fact to equating the internal and kinetic turbulent energy, while neglecting temperature fluctuations.

In the simple configurations considered above, namely starting with *constant* density and only one Elsässer component ($u=b$), one finds that after a time of the order of a nonlinear time the level of the density fluctuation settles around Kliatskin's relation in a diagram of M^2 versus $\delta n/n$, as is shown in fig. 4 which illustrates the case of two different values of the initial Mach number (black symbols). The same figure also shows observational data (crosses) at periods of one hour: the observational cloud is tangent to Kliatskin's diagonal and lies well above it, i.e. the observed density fluctuation is significantly lower than expected from the numerical results. As is discussed in /2/, even taking into account the slowing down effects mentioned above, there is plenty of time during advection by the solar wind for the density fluctuation to reach a quasi-stationary level. A possible clue to understand the discrepancy, which we shall explore now, lies in the spherical expansion of the wind.

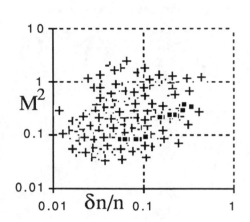

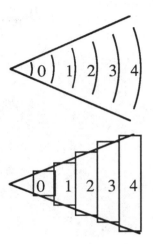

Fig. 4. Turbulent Mach number versus density fluctuations $\delta n/n$: comparison of numerical results (black symbols) with observational results (crosses, /17/).

Fig. 5. Sketch of the evolution of a plasma box advected by a spherical wind with constant speed (top) exact evolution (bottom) approximate evolution in the limit of small angular size.

NONLINEAR EVOLUTION AND SPHERICAL EXPANSION

The simplest way to obtain no evolution at all is to choose carefully the initial conditions as a (three-dimensional) initial fluctuation with zero z^- amplitude, i.e. $u=b$ as before, but also with constant magnetic and kinetic pressure. Such a wave will propagate without distortion, whatever the Mach number, in a homogeneous medium. This is however no longer true in the expanding solar wind.

Consider the simplified model of a solar wind expanding spherically at *constant speed*, and let us follow a cubic box of plasma, i.e., follow the wind along the radial direction passing through the center of the box. In this frame, (if the angular size of the box as seen from the sun is small enough), the box expands in the transverse directions, but its radial size remains constant as it recedes from the sun (see fig. 5), so that the average density $\rho°$ decreases as $1/R^2$, where R is the heliocentric distance. The result is an anisotropic damping: transverse components of the velocity decrease as $1/R$ while, because of the flux conservation, the normalized magnetic field's radial component ($b=B/\sqrt{(4\pi\rho°)}$) decreases as $1/R$. This effect dominates at low frequencies (periods larger than 2 days), and leads to an increasing departure from kinetic-magnetic equipartition, hence *diminishes* the imbalance between Elsässer fields $z^{\pm}$, which somewhat resembles the evolution observed in the solar wind /18/. At higher frequencies, the Alfvén coupling with the mean field will enforce a common decay rate for both velocity and magnetic fields (in velocity unit) at the same intermediate rate as $1/\sqrt{R}$: this is the WKB result. However, nonlinear couplings could deeply modify this picture by coupling the transverse and radial directions; reversely, nonlinear evolution may either be *halted* by the expansion, as happens in the one-dimensional case (Burgers equation) /19/, or *accelerated*, because the expansion perturbs otherwise stable states. We illustrate now these opposite effects of expansion with two different numerical examples: the Orszag-Tang (OT) vortex and an Alfvén wave.

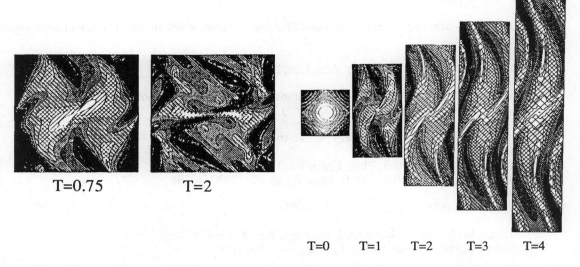

Fig. 6. Vorticity contours of the OT vortex (left) without expansion /20/ and (right) with expansion.

Fig. 6 shows the evolution of the OT vortex with M=0.6, both in the homogeneous case /20/ and in the expanding case. In the expanding case, the rms velocity fluctuation is equal to the transverse expansion speed; the boundary conditions are periodic in comobile coordinates. The expanding OT vortex evolution is clearly *frozen* at a stage comparable to that reached at T=0.75 by the standard vortex. Our second example is provided by the three-dimensional Alfvén wave defined at the top of this section. Its evolution depends on the angle between the wavevector and the radial direction. In the special case where the angle vanishes, one observes no nonlinear interactions, but solely the (linear) generation of non-zero z^- amplitude, since the two Alfvén modes propagating in opposite directions are no longer pure eigenmodes of the systems: however the density remains constant and the fluctuations solenoidal /21/. When the angle is finite, however, the transverse expansion forces the average field and the wavevector to depart from one another, so that nonlinear interactions set in:

(i) The average magnetic field turns toward the transverse direction because the radial component is damped (conservation of the magnetic flux): this reproduces the evolution of Parker's spiral field outside the corotation region. At the same time the wavevectors turn toward the radial direction. In fact, this turning corresponds to the wavevector direction being fixed in the comobile system of coordinates.

(ii) A one-dimensional nonlinear cascade occurs along this direction, leading to a significant production of small scales, largely compressible, with a density fluctuation $\delta n/n \approx M^2$.

CONCLUSION

We have tried to show that the nonlinear evolution of MHD fluctuations and in particular of Alfvén waves is a complex phenomenon: one observes the formation of small-scale coherent structures, as well as a stationary level of density fluctuations simply related to the Mach number. We presented results of direct numerical simulations taking the spherical expansion of the solar wind into account: these are the continuation of the models of Tu, Pu and Wei /22/ who studied (in an isotropic approximation) the inhibition effect of the expansion. The expansion in fact leads to very anisotropic effects in the plasma frame, and so the turbulent evolution should actually depend strongly on initial conditions, and also on nonlinear isotropization /23/. This non-universality of turbulent evolution could help to explain why the evolution apparently proceeds at different rates in the hot, high-speed streams and in cold, slow streams at solar minimum /2/.

ACKNOWLEDGMENTS

We thank the Conseil scientifique du CCVR for allowing computational time (project 2522). We thank J. Léorat for critical reading of the manuscript and U. Frisch for an interesting discussion.

REFERENCES

1.   Roberts, D. A., Observation and simulation of the radial evolution and stream structure of solar wind turbulence, this issue.

2.   Grappin R., M. Velli, A. Mangeney, <u>Ann. Geophys</u>. 9, 416 (1991).

3.   Iroshnikov, P. S., Astron. J. SSSR 40, 742, transl. in <u>Soviet Astron</u>. 7, 566, (1963); Kraichnan, R. H., <u>Phys.Fluids</u>, 8, 1385, (1965).

4.   Dobrowolny, M., A. Mangeney, and P.-L. Veltri, <u>Phys. Rev. Lett</u>., 45, 144, (1980).

5.   Biskamp, D., and H. Welther, <u>Phys. Fluids</u> B1, 1964, (1989); Politano H., A. Pouquet, P.L. Sulem, <u>Phys. Fluids</u> B1, 2330 (1989).

6.   Coleman, P. J., <u>Astrophys. J</u>., 153, 371, (1968).

7.   Matthaeus, W. H., M. L. Goldstein, <u>J. Geophys. Res</u>. 87 (1982); Bavassano B. and J. Smith, <u>J. Geophys. Res</u>. 91, 1706 (1986).

8.   Pouquet A. and S. Patterson, <u>J. Fluid Mech</u> 85, 305 (1978).

9.   Belcher, J. W., and L. Davis, <u>J. Geophys. Res</u>., 76, 3534, (1971).

10.   Roberts, D. A., M. L. Goldstein, L. W. Klein, and W. H. Matthaeus, <u>J. Geophys. Res</u>., 92, 12023 (1987).

11.   Roberts, D. A., and M. L. Goldstein, in <u>Proc. of the Third Int. Conference on Supercomputing</u>, ed. L. P. Kartashev and S. I. Kartashev, p. 370, International Supercomputing Institute, Boston, (1988).

12.   Grappin R. <u>Phys. Fluids</u> 29, 2433 (1986).

13.   <u>Images de la Physique</u>, CNRS, Paris, (1981); Pouquet A., U. Frisch, M. Meneguzzi <u>Phys. Rev</u>. 33A, 4266 (1986).

14.   Zank G.P., The theory of nearly incompressible turbulence, this issue, and personal communication.

15.   Lighthill M.J., <u>Proc. R. Soc. London</u> A 211, 564, (1952); Montgomery, D., M. R. Brown, and W. H. Matthaeus, <u>J. Geophys. Res</u>., 92, 282 (1987).

16.   Kliatskin, V. I., <u>Izv. Atmospheric and Oceanic Physics</u>, 2, 474, (1966).

17.   Grappin R., A. Mangeney and E. Marsch, Why does Alfvénic turbulence survive in the slow winds of solar maximum?, to be submitted (1991).

18.   Velli M., R. Grappin, A. Mangeney, Waves from the Sun?, Geophys. & Astroph. Fluid Dyn., in press (1991); Zhou Y. and W.H. Matthaeus, <u>J. Geophys. Res</u>. 95A, 10291 (1990).

19.   Velli M., R. Grappin, A. Mangeney, MHD turbulence in an expanding atmosphere, in OSL workshop on The electromechanical coupling of the solar atmosphere, to appear in Proc. of the American Institute of Physics (1991).

20.   Dahlburg R.B. and J.M. Picone 1989, <u>Phys. Fluids</u> B1, 2153 (1989).

21.   Velli, M., R., Grappin, A., Mangeney, <u>Phys. Rev. Lett</u>., 63, 1807 (1989).

22.   Tu C.-Y., Z. Y. Pu, F. S. Wei, <u>J. Geophys. Res</u>., 89, 9695 (1984). Tu C.-Y., <u>J. Geophys. Res</u>., 93, 7 (1988).

23.   Carbone V., P. Veltri, <u>Geophys. Astrophys. Fluid Dyn</u>. 52, 153 (1990).

PECULIARITIES OF THE MHD DISCONTINUITIES INTERACTIONS IN THE SOLAR WIND

S. A. Grib* and E. A. Pushkar**

*Academy of Sciences, Main Astronomical Observatory (Pulkovo),
St Petersburg 196140, Russia
**MASI, VTUZ-ZIL, Moscow 109068, Russia

ABSTRACT

Different types of the solar wind shock waves, rotational and contact discontinuities interactions are considered. It is shown that there are key parameters values that describe the catastrophic reconstruction of the flow structure. The double shock waves ensemble's generation and the problem of the magnetic field reconnection are discussed in the frame of the nonlinear mhd discontinuities interactions problem. The nonflare origin of some fast solar wind shock waves is indicated. The appearance of the slow shock wave in the result of nonshock discontinuities collission is underlined. The possibilities of the examination of the nonstationary solar wind phenomena with the help of the proposed method are shown.

INTRODUCTION

It is wellknown that the MHD discontinuities play significant role in the dynamics of the solar corona and of the solar wind. The coronal discontinuities with a large and defenite jump of an electron density across them are observed in the solar corona /1/. The interplanetary shock pair with the contact discontinuity between shocks appeared in the solar wind flow was discussed many times long since /2/. The problem of the double shock pair was also considered with the help of the numerical simulation /3/ without considering the nonlinear MHD solar wind descontinuities interaction. The slow MHD shock wave generation in the low β solar wind plasma was discussed recently/4/. By the way it was firstly proclaimed in /5/ that the slow shock waves may appear in the result of dual interactions between solar wind discontinuities. Then in /6/ the dual interaction between two forward solar wind shock waves in connection with the so calles R-event and /B/ – change was treated.
It is possible to suppose that interaction of the rotational(A) discontinuity with the contact (C) one may be significant in the solar corona and especially in the solar wind flow because of the presence of many plasma inhomogenieties and the relictive contact surfaces. The obliqueness of the solar wind discontinuities interaction may be also important because of the statement /7/ that from 647 prominent discontinuities of the ISEE3 data there were more rotational than tangential discontinuities in all types of the solar wind flow.
So this paper deals with the oblique solar wind MHD discontinuities interactions in the free flow in contrast to the well de-

fined normal one dimensional interaction of solar wind shock waves /8/

METHODOLOGY

Here we use the method of Pushkar's generalized polars /9/ and the numerical code with the iteration /IO/ which is different from that used in /6/ and /II/ for special values of the interaction parameters.

The oblique interaction between rotational (A) and contact discontinuity C: $\underset{\rightarrow}{A}C$ may be considered as the model for S+A(S+ forward fast shock wase), S+S+ and S+S+ interactions (S+ = reverse shock wave). For all types of discontinuities interactions (initial discontinuity) we'll have that the sum of the jumps for the physical value across the strong discontinuities must be equal to the jump on the initial discontinuity:

$$\sum (\Delta_f u_i, \Delta_s u_i, \delta_f u_i, \delta_s u_i, \Delta_A u_i, \Delta_c u_i) = \Delta_c u_i \qquad (I)$$

where $\Delta_c u_i = u_{io} - u_{io}$ is the initial discontinuity for the value of u_i which may correspond to entropy, density ρ, the velocity components and magnetic field components. The indexes f and s are used for fast and slow waves, A -for rotational or Alfven discontinuity, δ corresponds to the gradual cleange across the rarefaction wave.

Instead of usual MND Rankine-Hugoniot relations for oblique shock wave the generalized shock relation is used.

In the result of oblique MND discontinuities collision we'll have contact discontinuity C' on which boundary conditions are valid:

$$f_1 \equiv p^{(1)}(\psi_+^{(1)}, \psi_-^{(1)}) - p^{(2)}(\psi_+^{(2)}, \psi_-^{(2)}) = 0,$$
$$f_2 \equiv v_x^{(1)}(\psi_+^{(1)}, \psi_-^{(1)}) - v_x^{(2)}(\psi_+^{(2)}, \psi_-^{(2)}) = 0, \qquad (2)$$
$$f_3 \equiv v_y^{(1)}(\psi_+^{(1)}, \psi_-^{(1)}) - v_y^{(2)}(\psi_+^{(2)}, \psi_-^{(2)}) = 0,$$
$$f_4 \equiv B_x^{(1)}(\psi_+^{(1)}, \psi_-^{(1)}) - B_x^{(2)}(\psi_+^{(2)}, \psi_-^{(2)}) = 0.$$

Here the first relation corresponds to the conservation of pressure, the second and the third of flow velocity and the forth - of the intensity of the magnetic fiels - all across the contact surface (index (I) corresponds to the region above and (2) - below the contact surface), "+" relates to fast and "-" to slow waves.

The idea of method is: at first to describe qualitatively the result of the interaction, i.e. in the frame of the selfsimiliar formalism to define the type of waves which give the solution for (2) and secondly to determine the intensities of all reflected and refracted waves, and the parameters of contact surfase.

It must be kept in mind that the flow after the arriving discontinuity has to be hyperbolic.

MAIN RESULTS

For the interaction between rotational (A) and contact discontinuity (C) on which the proton concentration decreases we have for plasma parameter $\beta = 2\rho /B^2 < 1$

a) $AC \rightarrow R_+AS_-C'S_-AR_+$ if the angle $\psi = B_o\hat{}C = 30^\circ$ and the angle of the collision is 15°. b) $\underset{\rightarrow}{A} C \rightarrow S_+R_-C'R_-AS_+$ for $\psi = 45^\circ$, c) $\underset{\rightarrow}{A} C \rightarrow R_+\underset{\leftarrow}{S_-}C'\underset{\rightarrow}{S_-}AS_+$ for $\psi = I5^\circ$

In first case we have slow shock wave generation(reverse and forward), in second case there is double shock ware ensemble as in/3/ and in C)- the fast shock wave generation in the result of the refraction of the rotational discontinuity. It is possible to compare the solution in first case with the problem of the reconnection of the magnetic fields where also there are two slow shock waves and no fast shock waves. If we change the value from 15° to 5° we may have the catastrophic reconstruction of the solution C from R+ do S+). If we take $\psi = 60^{\circ}$ with the increase of the plasma concentration across C and for $\varphi = 15^{\circ}$, we have $AC-S+R\,C\,'R-AS+$ These forward wases may be interpreted like the waves for the magnetic clouds described in /12, 13/. For the collision between fast shock wave and the rotational discontinuity A we have

$S+A \to R+AS\,C\,S\_R+$ for Mach number of the shock M=2, $\beta=0,53, \psi = 25^{\circ}$ and $\varphi = 10^{\circ}$. So the fast shock wave may disappear in the result of the collision with the rotational discontinuity. For the shock wave going after another one the typical picture is $R+AS\,CR\_S+$ as in /6/. Now it is possible to conclude that the shock waves might be generated inside the solar wind flow without any flare and without the collision of two streams as it was indicated in /14/ from the Voyagers I and 2 data.

REFERENCES

I. S.L.Koutchmy, Small scale coronal structures, in: Proceed. of the 9 Sacramento Peak Summer workshop on Solar and Stellar Coronal Structure and Dynamics, ed. R.C.Altrock, New Mexico 1988, p.208.

2. J.S.Hirshherg, J.Barne, and D.E.Robbins, Solar Phys 23,467 (1975).

3. R.S.Steinolfson and M.Dryer, J.Geophys.Res.80, # 10,1223 (1975).

4. R.S.Steinolfson and A.J.Hundhausen, J.Geophys. Res.94,# A2, 1222(1989).

5. F.M.Neubauer.J.Geophys. Res. 81, # 13, 2248(1986).

6. S.A.Grib, in: Solar Maximum Year. Proceedings of International conference. 2.Ac.of Sci.USSR 1981, p.34.

7. M.Neugebauer, C.J.Alexander, J.Geophys. Res.96, # A6,9409 (1991).

8. S.A.Grib, B.E.Brunelli, M.Dryer, and W-W.Shen., J.Geophys. Pes. 84, 5907 (1979).

9. E.A.Pushkar, M. ZG 3, 103 (1977)/

10. A.A.Barmin, E.A.Pushkar, M ZG 1, 131 (1990)

11. S.A.Grib, Space Sci: Rev.32, 43 (1982)

12. L.F.Burlaga and E.Sittler, F.Mariani, R.Schwenn, J.Geophys. Res.86, # A 8, 6673 (1981).

13. L.F. Burlaga, L.Klein, R.Schweenn and H.Rosenbauer, Geophys. Res.Gett, 9, # 12, 1317 (1982).

14. E.T.Sarris, in: First SOLTIP Symposium, Liblice, Czechoslovakia, 1991, p.16.

THE DENSITY FLUCTUATION POWER
SPECTRUM OF THE INNER SOLAR WIND

J. K. Harmon* and W. A. Coles**

National Astronomy and Ionosphere Center, Arecibo, PR 00613, U.S.A.
*** Department of Electrical and Computer Engineering, University of California
at San Diego, La Jolla, CA 92093, U.S.A.*

ABSTRACT

Recent work has been done at Arecibo and UCSD to arrive at a self-consistent model of the
density fluctuation power spectrum inside 25 R_S based on a comparison of spectral broadening,
angular broadening, intensity scintillation, and phase scintillation measurements. It is found that
the density spectrum has a very distinctive shape which deviates from a simple Kolmogorov power
law at high wave number (small scales).

INTRODUCTION

We began making near-Sun spectral broadening measurements at Arecibo a decade ago /1/ and
our recent work on modelling the density spectrum developed largely from attempts to reconcile
the Arecibo data with earlier results. Our first detailed treatment of the near-Sun spectrum was
presented in Coles and Harmon /2/. In this paper we argued that various radio propagation
results reported over the years were mutually consistent provided the density spectrum conformed
to a distinctive shape consisting of a Kolmogorov power law at large scales (10^3–10^6 km), a local
flattening at small scales (10–100 km), and in inner scale of 10 km or less. As a partial test of
this model we recently analyzed Voyager radio data to obtain density spectra with large dynamic
range /3/. The Voyager results were the first to directly show the spectral break between the
Kolmogorov and flattened sections. They also provided a clearer picture of spectral shape changes
during transients.

THE SPECTRUM MODEL OF COLES AND HARMON (1989)

The spectrum model consisting of a Kolmogorov power law with high wave number flattening and
inner scale was proposed by us /2/ based on a comparison of Arecibo radar spectral broadening
data with numerous other data. A very useful plot which helped us to arrive at our model is shown
in Fig. 1. Here we plot examples of the structure function $D(s)$, a convenient quantity which gives
the distribution of fluctuation power with scale size rather than spatial wave number. The data in
Fig. 1 include: radar spectral broadening measurements at 5 R_S (*a, b*), 10 R_S (*c, d, e*), and 20
R_S (*f*); spacecraft spectral broadening measurements at 5 and 10 R_S (*squares*); angular broadening
measurements at 10 R_S (*dashed lines*); intensity scintillation measurements at 10, 20, and 50 R_S
(*triangles*); VLBI phase scintillation measurements at 22 R_S (*g*); extrapolated spacecraft phase
scintillation measurements at 5, 10, 20, and 50 R_S (*solid straight lines*).

The strongest evidence for an inner scale comes from angular broadening measurements /4,5,6/,
which consistently show the $D(s) \propto s^2$ indicative of a spectral cutoff. Some of the radar spectral
broadening $D(s)$ also show a break consistent with an inner scale. The flattening behind the inner
scale dominates the spectral broadening /1,2,7,8,9/ and intensity scintillation /10/ measurements
between 5 and 15 R_S. Phase scintillation measurements /7/ are sensitive to scales larger than 10^3

km; at these scales the spectrum is approximately Kolmogorov (except very near the Sun, where the power law is slightly shallower).

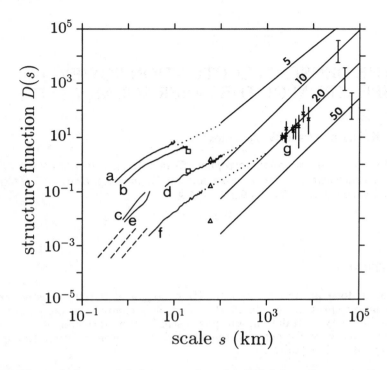

Fig. 1. Plots of the spatial structure function $D(s)$ based on a variety of data obtained at various radial distances. See text for explanation. Adapted from /2/.

RESULTS FROM VOYAGER RADIO MEASUREMENTS

Although our model suggested that the spectrum had to undergo a break between the local flattening and the larger scales, there was no direct evidence of such a break in any published data. We decided that a good partial test of the model could be made by reanalyzing spacecraft radio propagation data to obtain simultaneous measurements of the spectral broadening and phase scintillation structure functions. Using Voyager data obtained from the Radio Science Group at Stanford, we computed "composite" structure functions from a total of 28 observing runs. Each composite structure function was formed from a spectral broadening structure function covering scales smaller than 10^3 km and a phase scintillation structure function covering scales larger than 10^3 km. Both the spectral broadening and phase scintillation $D(s)$ were averaged over the full duration of an observation (typically 1–2 hours). One innovation was the use of the X-band signal to correct the S-band signal for spurious slow drifts; this permitted longer integration of the S-band RF spectrum before estimating the spectral broadening structure function, resulting in reduced estimation error.

In Fig. 2 we show 8 typical composite structure functions obtained between 7.1 and 20.5 R_S. Also shown are curves at 5, 10, and 20 R_S representing the Coles and Harmon model /2/ (cf. Fig. 1). It is clear that the Voyager structure functions are in excellent agreement with our model in terms of both shape and amplitude. The upward break between the flattened region and the Kolmogorov region can be seen directly at the large-s end of the spectral broadening structure function in most cases.

Several transients were seen in the Voyager data set in which the spectrum departed from its quiescent shape. In each case this took the form of an increase in the level of $D(s)$ at large scales along with a steepening of $D(s)$ in the normally flattened region, so as to make the overall spectrum appear more like a single power law. An example of a typical transient is shown in Fig. 3. Here the spectrum has the quiescent form on 14 September 1980, undergoes a transient on 15 September, and falls back to a quiescent state by 20 September. This sort of behavior is similar in character to transients reported earlier by Woo and Armstrong /11/ and by us /1,2/.

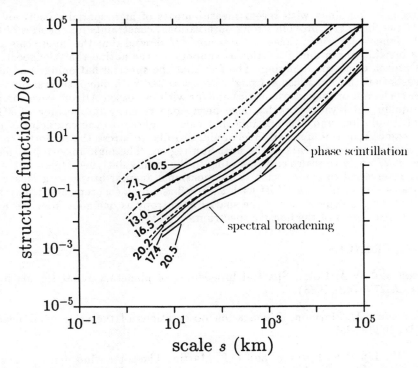

Fig. 2. Eight composite structure functions from the Voyager analysis. The numbers give the radial distances in solar radii. The Coles and Harmon model at 5, 10 and 20 R_S is shown for comparison (*dashed curves*). Adapted from /3/.

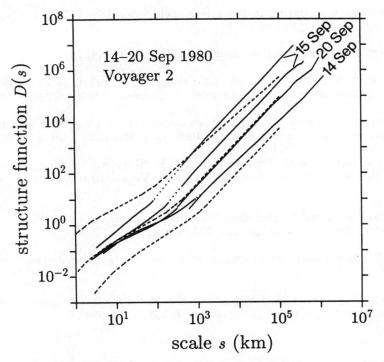

Fig. 3. Voyager 2 composite structure functions showing the transient of 15 September 1980 (scaled to 10 R_S). The Coles and Harmon model at 5, 10 and 20 R_S is shown for comparison (*dashed curves*). Adapted from /3/.

DISCUSSION

The foregoing results, combined with recent measurements of high spatial anisotropy in density fluctuations near the Sun /12/, provide strong observational constraints for theories of the plasma turbulence in the inner solar wind. Coles and Harmon /2/ showed that the inner scale is comparable with the ion inertial length and cited this as support for the notion that the spectral cutoff is associated with proton cyclotron damping. The fact that the spectral flattening is adjacent to the inner scale suggests that this enhancement might be associated with one of several possible plasma instabilities (ion cyclotron or magnetoacoustic) and/or with increased Alfvén wave compressibility at high wave number. Relevant work has also been done recently in the area of MHD turbulence theory and simulation, and it is possible that new insights into the character of the density turbulence may come from this quarter. We would also like to stress the importance of making additional propagation measurements in the near-Sun region. The high anisotropies measured by Armstrong *et al.* /12/ apply to scales corresponding to the locally flattened region of the spectrum. It would be very interesting to extend such two-dimensional angular broadening measurements to larger (VLBI) scales. By combining VLBI with shorter-baseline measurements it may be possible to measure the wavelength dependence of the spatial anisotropy as one goes from the Kolmogorov region to the flattened region of the density spectrum.

REFERENCES

1. J.K. Harmon and W.A. Coles, Spectral broadening of planetary radar signals by the solar wind, *Astrophys. J.* 270, 748 (1983).

2. W.A. Coles and J.K. Harmon, Propagation observations of the solar wind near the Sun, *Astrophys. J.* 337, 1023 (1989).

3. W.A. Coles, W. Liu, J.K. Harmon, and C.L. Martin, The solar wind density spectrum near the Sun: Results from Voyager radio measurements, *J. Geophys. Res.* 96, #A2, 1745 (1991).

4. R.G. Blessing and P.A. Dennison, Coronal broadening of the Crab Nebula 1969–1971, *Proc. ASA* 2, 84 (1972).

5. B.D. Ward, Radiowave scattering in the interplanetary medium (Ph.D. thesis), Adelaide University (1975).

6. W.A. Coles, Interplanetary scintillation, *Space Sci. Rev.* 21, 411 (1978).

7. R. Woo and J.W. Armstrong, Spacecraft radio scattering observations of the power spectrum of electron density fluctuations in the solar wind, *J. Geophys. Res.* 84, 7288 (1979).

8. J.M. Martin, Voyager microwave scintillation measurements of solar wind parameters, *Sci. Rep. D208-1985*, Cent. for Radar Astron., Stanford University, Stanford, Calif. (1985).

9. O.I. Yakovlev, A.I. Efimov, V.M. Razmanov, and V.K. Shtrykov, Inhomogeneous structure and velocity of the circumsolar plasma based on data of the Venera-10 station, *Sov. Astron.* 24, 454 (1980).

10. S.L. Scott, W.A. Coles, and G. Bourgois, Solar wind observations near the Sun using interplanetary scintillation, *Astron. Astrophys.* 123, 207 (1983).

11. R. Woo and J.W. Armstrong, Measurements of a solar flare-generated shock wave at 13.1 R_S, *Nature* 292, 608 (1981).

12. J.W. Armstrong, W.A. Coles, M. Kojima, and B.J. Rickett, Observations of field-aligned density fluctuations in the inner solar wind, *Astrophys. J.* 358, 685 (1990).

INTERPLANETARY FAST SHOCK DIAGNOSIS WITH THE RADIO RECEIVER ON ULYSSES

S. Hoang,* F.Pantellini,* C. C. Harvey,* C. Lacombe,* A. Mangeney,*
N. Meyer-Vernet,* C. Perche,* J.-L. Steinberg,* D. Lengyel-Frey,**
R. J. MacDowall,** R. G. Stone** and R. J. Forsyth***

* Observatoire de Paris-Meudon, Departement de Recherche Spatiale,
 URA 264 CNRS, F-92195 Meudon Cedex, France
** NASA Goddard Space Flight Center, Greenbelt, MD, U.S.A.
*** The Blackett Laboratory, Imperial College of Science and Technology,
London SW7 2BZ, U.K.

ABSTRACT

The Radio receiver on Ulysses records the quasi-thermal noise which allows a determination of the density and temperature of the cold (core) electrons of the solar wind. Seven interplanetary fast forward or reverse shocks are identified from the density and temperature profiles, together with the magnetic field profile from the Magnetometer experiment. Upstream of the three strongest shocks, bursts of non-thermal waves are observed at the electron plasma frequency f_{peu}. The more perpendicular the shock, the longer is the time interval during which these upstream bursts are observed. For one of the strongest shocks we also observe two kinds of upstream electromagnetic radiation : radiation at $2 f_{peu}$, and radiation at the downstream electron plasma frequency, which propagates into the less dense upstream regions.

INTRODUCTION

The Radio receiver from the Ulysses Unified Radio and Plasma Wave (URAP) Experiment in the solar wind continuously records the local quasi-thermal plasma noise (the so-called plasma line) on which are often superimposed various radio and plasma waves. The quasi-thermal noise (QTN) is observed near the electron plasma frequency f_{pe} /1,2/ when f_{pe} is sufficiently high, above approximately 6 to 10 kHz, for the Debye length to be short enough compared to the antenna length. Analysis using the complete QTN spectrum yields the density and temperature of both the cold and hot electron populations. The total electron density can however be deduced simply from the spectral cut-off at the plasma frequency, without the use of sophisticated model-dependent calculations. This method of plasma diagnosis, using a long, thin electrical antenna, was applied successfully to observations from ISEE-3/ICE to measure the electron densities and temperatures in the solar wind /3/, and in the tail of the comet Giacobini-Zinner /4/.

On radio spectrogrammes, the time evolution of the plasma line clearly shows that many Interplanetary (IP) large-scale structures are crossed by Ulysses, several of which can be identified as fast shocks. The companion paper by Lengyel-Frey et al. (this issue) discusses the low-frequency electric and magnetic turbulence in fast shocks, as observed by the URAP experiment. Here, we present the results of preliminary analysis of the density and higher frequency radio observations for several shocks observed by Ulysses (launched in October 1990) early in the mission, when it was still in the ecliptic plane.

OBSERVATIONS

We will use 7 examples of fast IP shocks, as listed in Table 1. To identify these shocks, we have used two data sets : the dynamic spectra from the Radio receiver /5/, and some of the 32-64 s average plots of the IP magnetic field from the Magnetometer experiment /6/. Shocks are identified by a rapid change of f_{pe} and a simultaneous rapid change, in the same sense, of the magnetic field strength, together with the occurrence of wave turbulence in the vicinity. Figure 1 shows examples of time-series of wave spectra observed during the crossing of two of the fast shocks thus selected.

Next, we analyse radio spectra acquired over consecutive 128-s intervals during which 64 frequencies are sampled sequentially. Figure 2 displays two sequences of such spectra, which show a variety of waves in addition to the QTN spectrum itself. From each sweep, the total electron density N_e and the temperature T_c of the cold (core) component can be determined by QTN spectroscopy, assuming the plasma has a bi-Maxwellian electron distribution function /1/. N_e and T_c can be determined to within 10% and 30% respectively. The density and temperature of the hot component cannot be determined with useful accuracy due to the frequent presence of non-thermal waves at frequencies above the plasma line. Therefore we use only N_e and T_c for the shock identification.

RESULTS

Shock Parameters

Hereafter, the indices u and d respectively denote the upstream and downstream values. Table 1 gives the ratio across the shock transition of the total density N_{ed}/N_{eu}, of the core temperature T_{cd}/T_{cu} and of the modulus of the magnetic field B_d/B_u for 7 fast shocks. The upstream beta factor β_u is estimated by assuming that the proton temperature is half the electron temperature. The angle θ_{Bn} between the upstream magnetic field and the shock normal, and the Alfvén Mach number M_A, are deduced from β_u, N_{ed}/N_{eu} and B_d/B_u using the Rankine-Hugoniot relations.

TABLE 1 List of Fast Shocks

Date	Time UT	Type	N_{eu} cm^{-3}	$\dfrac{N_{ed}}{N_{eu}}$	T_{cu} $10^5 K$	$\dfrac{T_{cd}}{T_{cu}}$	B_u nT	$\dfrac{B_d}{B_u}$	β_u	M_A	θ_{Bn}
09 Dec 90/343	19 16	F	2.8	1.5	1.6	1.4	3.9	1.4	1.5	1.9	$45° - 50°$
15 Dec 90/349	16 43	R	3.2	1.4	2.1	1.2	6.5	1.3	0.8	1.6	$50° - 55°$
16 Dec 90/350	07 52	R	2.0	1.4	2.1	1.0	3.7	1.4	1.6	1.9	$70° - 75°$
14 Jan 91/014	07 53	F	3.2	1.6	1.2	1.2	3.2	1.7	2.0	2.4	$> 70°$
07 Apr 91/097	04 45	F	1.7	2.6	0.7	1.6	2.1	2.4	1.5	3.6	$80° - 90°$
27 May 91/147	04 08	F	2.1	2.4	1.4	1.7	4.2	2.3	0.8	2.9	$65° - 70°$
27 May 91/147	19 04	R	1.5	1.7	1.6	1.1	3.5	1.5	1.0	1.8	$30° - 35°$

It can be seen that the density jump and the Mach number are generally stronger for forward (F) shocks than for reverse (R) shocks. The correlation between the jumps of T_c and N_e is similar to that found by Feldman et al. /7/.

Waves at f_{pe} and $2 f_{pe}$

Bursts of non-thermal waves at f_{peu} are observed superimposed on the QTN upstream of 3 shocks on our list : i) during several tens of minutes before the ramp of the 7 Apr 1991 shock (Figure 1a), ii) for 6 mn before the ramp of the shock on 27 May 1991 at 04 08 (Figure 1b), iii) for about 1 mn before the ramp of the 9 Dec 1990 shock (not shown). These 3 shocks are forward shocks, and have the largest T_c jumps shown in Table 1. The time interval during which these bursts of electron plasma waves are observed increases as θ_{Bn} increases. Such f_{peu} waves have been observed frequently upstream of interplanetary shocks /8/. Their impulsiveness seems to indicate that they are, at least partly, electrostatic Langmuir waves.

For the shock of Figure 1b (27 May 1991 at 04 08) we observe not only waves at $f_{peu} \simeq 13$ kHz but also at $2 f_{peu} \simeq 26$ kHz (Figure 2a) for about 25 mn upstream of the ramp. To our knowledge, this $2 f_{peu}$ radiation, which is very common upstream of the Earth's bow shock /9/, has never been reported for IP shocks, probably because of the high frequency resolution needed. However, one of us (R.J.M.) has found one case of such radiation for an IP shock observed by the ISEE-3 radio experiment.

Figure 1b also displays upstream waves at $f \simeq 21$ kHz (Figure 2a), which is close to f_{ped} (Figure 2b). We consider that these waves are mainly electromagnetic (e.m.) because they are not impulsive and are observed in a narrow band well above the local f_{peu}. They are detected during more than one day before the shock crossing, but with a frequency drift much smaller than that of standard type II solar radio bursts. We thus observe in the less dense region the propagation of e.m. waves at the plasma frequency of the denser downstream region. Burgess et al. /10/ observed such radiation in the Earth's electron foreshock close to a solar wind density jump which was not a shock. This upstream radiation has never before been documented, either for IP shocks or for the Earth's bow shock.

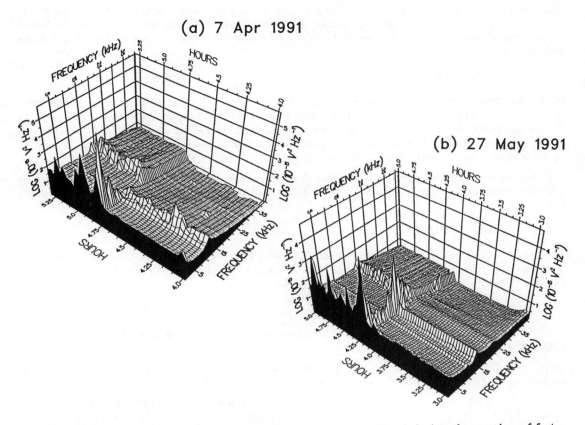

Fig. 1. Two examples of time-series of radio wave spectra observed during the crossing of fast shocks on 7 Apr 1991 (a) and 27 May 1991 (b). The peak at 9.5 kHz is due to a radio frequency interference on board.

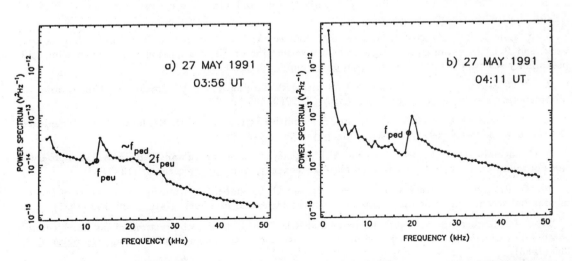

Fig. 2. Spectra observed upstream (a) and downstream (b) of the fast shock on 27 May 1991; in the upstream region, emissions at downstream f_{ped} and at twice upstream f_{peu} show up clearly on the thermal noise spectrum.

It has been shown by Steinberg *et al.* /11/ that if there is a density overshoot in the Earth's bow shock, the downstream f_{ped} radiation cannot propagate upstream. These observations could help to understand the generation and propagation of the radiation of type II bursts, which are known to be produced in the vicinity of IP shocks /12/.

CONCLUSION

In this preliminary study of interplanetary shocks, the density and temperature of the core component of the electron distribution are deduced from quasi-thermal electric noise measurements. The resulting density and temperature are then used together with the magnetic field to characterize the shocks. The observations of upstream Langmuir waves indicate that the dynamics of the fast electrons thought to generate them could be controlled by the shock normal angle and the shock strength.

The observation of electromagnetic radiation at the downstream electron plasma frequency in the region upstream from the shock may also give some indication of the internal structure of the interplanetary shocks.

AKNOWLEDGEMENTS

The URAP experiment is a joint project of NASA/GSFC, Observatoire de Paris, CRPE and University of Minnesota. The french contribution has been financed by Centre National d'Etudes Spatiales and partly by CNRS.

REFERENCES

1. N. Meyer-Vernet and C. Perche, Tool kit for antennae and thermal noise near the plasma frequency, J. Geophys. Res. 94, 2405 (1989)

2. Y.F. Chateau and N. Meyer-Vernet, Electrostatic noise in non-Maxwellian plasmas : generic properties and "Kappa" distributions, J. Geophys. Res. 96, 5825 (1991)

3. P. Couturier, S. Hoang, N. Meyer-Vernet and J.-L. Steinberg, Quasi-thermal noise in a stable plasma at rest : theory and observations from ISEE 3, J. Geophys. Res. 86, 11127 (1981)

4. N. Meyer-Vernet, P. Couturier, S. Hoang, C. Perche and J.-L. Steinberg, Physical parameters for hot and cold electron populations in comet Giacobini-Zinner with the ICE radio experiment, Geophys. Res. Lett. 13, 279 (1986)

5. R.G. Stone, J.L.B., J.C., P.C., Y.d.C., N.C.-W., M.D.D., J.F., K.G., M.L.G., C.C.H., S.H., R.H., M.L.K., P.J.K., B.K., R.K., A.L., D.L.-F., R.J.M., R.M., C.A.M., A.M., N.M., S.M., G.N., M.J.R., J.L.S., E.T., C.d.V., F.W. and P.Z., The unified radio and plasma wave investigation on Ulysses, Astron. Astrophys. Suppl., in press

6. A. Balogh, T.J. Beek, R.J. Forsyth, P.C. Hedgecock, R.J. Marquedant, E.J. Smith, D.J. Southwood and B.T. Tsurutani, The magnetic field investigation on Ulysses mission : instrumentation and preliminary scientific results, Astron. Astrophys. Suppl., in press

7. W.C. Feldman, J.R. Asbridge, S.J. Bame, J.T. Gosling and R.D. Zwickl, Electron heating at interplanetary shocks, Solar Wind Five, NASA CP-2280, 403 (1983)

8. C.F. Kennel, F.L. Scarf, F.V. Coroniti, E.J. Smith and D.A. Gurnett, Nonlocal plasma turbulence associated with interplanetary shocks, J. Geophys. Res. 87, 17 (1982)

9. C. Lacombe, C.C. Harvey, S. Hoang, A. Mangeney, J.-L. Steinberg and D. Burgess, ISEE observations of radiation at twice the solar wind plasma frequency, Ann. Geophysicae 6, 113 (1988)

10. D. Burgess, C.C. Harvey, J.-L. Steinberg and C. Lacombe, Simultaneous observation of fundamental and second harmonic radio emission from the terrestrial foreshock, Nature 330, 732 (1987)

11. J.-L. Steinberg, S. Hoang, C. Lacombe and R.D. Zwickl, ISEE-3 observations of the Earth's radio continuum through the bow shock and magnetosheath and in the magnetosphere, Ann. Geophysicae 6, 309 (1988)

12. D. Lengyel-Frey, Location of the radio emitting regions of interplanetary shocks, J. Geophys. Res., in press

PARTICLE ORBITS AND ACCELERATION IN A FILAMENTARY CURRENT SHEET MODEL

B. Kliem

Institute for Astrophysics, D-O-1591 Potsdam, Germany

ABSTRACT

Test particle orbits in the 2D Fadeev equilibrium with a perpendicular electric field added are analyzed to show that impulsive bursty reconnection, which has been proposed as a model for fragmentary energy release in solar flares, may account also for particle acceleration to (near) relativistic energies within a fraction of a second. The convective electric field connected with magnetic island dynamics can play an important role in the acceleration process.

INTRODUCTION

The acceleration of charged particles is a ubiquitous phenomenon in astrophysical plasmas. Also in the solar system there are many sources of energetic particles (e.g. /1/). Solar flares do not only produce the most dramatic particle events in interplanetary space, they challenge the theory of acceleration because the accelerated particles probably carry away a significant amount of the released energy (e.g. /2/). Electrons and ions may reach relativistic energies after acceleration times of the order of a second. Maximum energies are, for electrons, $W_{e,max} \sim 10^8 \ eV$, and for protons, $W_{i,max} \sim 10^{10} \ eV$ (although the main part of energy typically appears to go to electrons in the energy range $W_e \sim 10^4 - 10^5 eV$). Understanding particle acceleration in solar flares is important for at least two reasons. (1) Because of the fraction of energy contained and because of the association of the bulk of the energetic particles with the impulsive flare phase, which is also the phase of the highest *rate* of energy release, understanding the acceleration of particles holds a key to understanding the flare phenomenon itself. (2) If the original particle spectrum and its temporal development are known, measurements of the interplanetary particle flux can be more reliably used to infer properties of the intervening coronal and interplanetary medium. For solar flares it is not known whether particles are energized at shock waves, in turbulent wave fields, at potential jumps, or in reconnection regions, and by which mechanism. There is no doubt that the reconnection of magnetic field lines is a central element of the flare energy release and one is tempted to suppose that it directly produces also the bulk of the accelerated particles, but it is also possible that reconnection plays only a minor role in the acceleration process (producing, e.g., only the seed particles required for shock wave acceleration).

Magnetic field line reconnection implies the occurrence of parallel electric fields, of convective electric fields, $\mathbf{E}_{conv} = -\mathbf{u} \times \mathbf{B}/c$, and possibly (if the field develops sufficiently small scales $L_B \lesssim r_{ci}$) also the occurrence of polarization electric fields. All three field components contribute to particle acceleration. It is questionable whether polarization fields really occur in the low-β coronal plasma because the current density exceeds the threshold for onset of microinstabilities for $L_B \lesssim 4\beta^{-1}r_{ci}$. Further, at the high magnetic Reynolds numbers in the corona ($R_m \gtrsim 10^8$), the convective electric field exceeds the Dreicer field (which is the upper limit to the classical parallel electric field) by many orders of magnitude. Only in the presence of anomalous effects, such as localized anomalous resitivity caused by a current-driven microinstability, space charge electric fields ($\mathbf{E}_{sc}$) may build up also along the magnetic field to a magnitude comparable to, or possibly even higher as, E_{conv}, due to the large inductance of the whole coronal circuit. The two components, $\mathbf{E}_{conv}$ and $\mathbf{E}_{sc}$, may be considered as alternative accelerating agents (although it might turn out that they do not occur independently of each other). In this paper some aspects of acceleration by $\mathbf{E}_{conv}$ are considered. Perpendicular to $\mathbf{B}$, in which direction $\mathbf{E}_{conv}$ always points, particles are tied to the field lines and may in general not be accelerated — with two exceptions:

(1) at rotational discontinuities and inversion planes and at magnetic X- and O-lines particles perform a meander-like motion and decouple from the fluid,

(2) particles drift across the field due to inhomogeneities in $\mathbf{B}$; those particles that enter the region where an $\mathbf{E}_{conv}$ is induced in an inhomogeneous $\mathbf{B}$ from a region with a different local flow velocity $\mathbf{u}$ and are collisionless at the scales of the $\mathbf{u}$ and $\mathbf{B}$ variations experience energy gain or loss, respectively, depending on the sign of $\mathbf{E}_{conv}$.

TEST PARTICLE ORBITS IN A FILAMENTARY CURRENT SHEET

We consider a geometry containing multiple X- and O-lines — magnetic islands in a filamentary current sheet, which may have been produced by a tearing instability and are susceptible to dynamic motion and, possibly, coalescence. The radiation signatures of solar flare particles at radio and hard X-ray wavelengths are often resolved into single spikes with durations in the millisecond range; this indicates that the energy release and particle acceleration may occur in a fragmentary manner in a highly structured and dynamic source volume containing many reconnection sites (e.g., /3/). The impulsive bursty model of reconnection /4/, which supposes repeated island formation and coalescence, may account for fragmentary energy release /5/. Numerical simulations of coalescence revealed typical flow velocities $u \sim (0.1 - 1)\, V_A$, hence $\mathbf{E}_{conv} \sim 10^4 - 10^5\, V/m$ may be expected during this instability in the lower solar corona ($N_e = 10^{9.5}\, cm^{-3}$, $B = 200\, G$).

In order to simplify the consideration of some basic effects, a stationary analytic magnetic field model is employed. Particle trajectories are calculated in the nonrelativistic approximation (using a predictor-corrector scheme) in the Fadeev equilibrium

$$B_x = B_0 \frac{\sinh(z/L)}{\cosh(z/L) + \epsilon \cos(x/L)} \tag{1a}$$

$$B_z = B_0 \frac{\epsilon \sin(x/L)}{\cosh(z/L) + \epsilon \cos(x/L)} \tag{1b}$$

where ϵ ($= 0.3$ here) controls the current localization within the islands. The separation of the effects of $\mathbf{E}_{conv}$ from the effects of $\mathbf{E}_{\parallel}$ is most easily achieved in the idealized case $B_y \equiv 0$. An electric field into the invariant direction is formally added to simulate the occurrence of convective electric fields due to island dynamics in the coalescence process. The simplest choice, $E_y = const.$, implies a flow that is directed to or away from the centre of the islands at each point, which is not characteristic of coalescence. Instead, neighbouring islands approach each other as a whole, this is modeled by a periodic electric field $E_y(x)$ which changes sign at the position of the O-lines ($x/L\pi = \pm 1, 3, 5, \ldots$). With $E_y \propto \cos(x/2L)$, acceleration and deceleration of particles trapped in the configuration would exactly cancel. We take into account that the islands are deforming while approaching or repelling each other, with different magnitude of $|\mathbf{u}|$ and $|\mathbf{B}|$ on either side of the O-lines. This results in a convective electric field that is asymmetric about the O-lines and may be modeled as (cf. Figure 1)

$$E_y = E_{y0} \cos(\frac{x}{2L}) \cdot \cos^2(\frac{x}{4L}) \tag{2}$$

The test particle orbits have the following general characteristics (independent of the sign of charge). [1] Meander orbits exist at the O-lines, at the X-lines in the downstream region (e.g. in Fig. 1 at $x = 0$ in the x-y-plane), and at the X-lines in the upstream region (e.g. at $x = 0$ in the y-z-plane). $\mathbf{E}_{conv}$ accelerates(decelerates) a particle with meander orbit if the orbit is located in plasma which is streaming towards(away from) the O- or X-line. Particles which are accelerated along meander orbits at an O-line remain stably trapped at the O-line, hence unlimited acceleration is possible at O-lines. This can be understood as follows. During each approximately semi-circular orbit the particle gains energy, its Larmor radius is increasing, hence the particle *must* cross the O-line to repeat the process at the other side. This is even true for unequal field strengths on either side of the O-line. (The situation is similar to a magnetic inversion plane [$\epsilon \to 0$].) Meander orbits are not stable at X-lines due to the motion of the plasma across the separatrices connected with $E_y \neq 0$.

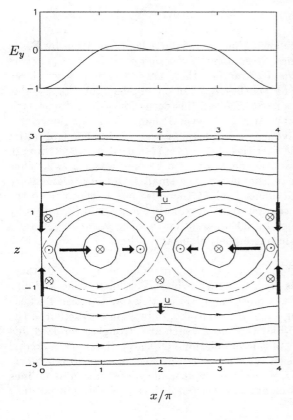

Fig. 1. The considered configuration $\mathbf{B}(x, z)$, $E_y(x)$. The implied flow $\mathbf{u}$ is indicated by heavy arrows. Arrows out of the plane give the direction of the ∇B - drift for positive charges.

[2] Particles drift in the $\pm y$ direction due to gradients and curvature in the magnetic field. The drift motion in the neighbourhood of an O- or X-line is *oppositely directed* to the meander motion about the O- or X-line. Hence, in order to gain energy due to the drift along an O- or X-line, the orbit must be located in plasma that is streaming away from the O- or X-line. Unless the particles are magnetically guided to return to the vicinity of an O- or X-line with appropriately directed $\mathbf{E}_{conv}$, there will in general be only one period of strong acceleration due to the drift. In the filamentary current sheet particles with orbits close to the separatrices are repeatedly guided into the vicinity of the X-lines and experience several acceleration periods (with, on average, progressively smaller energy gain), see Figure 2. Concerning particle acceleration, multiple X-line reconnection is thus far superior in comparison to single X-line reconnection.

[3] The analytic model (1), (2) implies a net $\mathbf{E} \times \mathbf{B}$ – drift of the particles across the field lines towards ($E_{y0} < 0$) or away from ($E_{y0} > 0$) the O-lines (for $B_0 < 0$), while in reality all particles move with the field lines for $R_m \gg 1$ (exept for the small reconnection regions around the X-lines). (To avoid unrealistically strong forcing of test particles toward the O-lines, $-E_{y0}$ must be chosen small enough so that the particles drift much less than an island half width during the time their trajectories are followed.)

Figure 2 shows the orbit of a particle that has gained energy during several periods of drift close to an X-line, gradually drifted to the neighbouring O-line and finally obtained high energy during meander motion at the O-line. This orbit was followed for the characteristic time of island approach, $\tau_{CI} \sim \pi L/u$ (i.e. $[\pi V_A/u]$ Alfvén times), equivalent to complete $\mathbf{E} \times \mathbf{B}$ – drift of the particle from the X-line to the O-line. It is seen, however, that the meander motion starts already at $\approx \tau_{CI}/3$ due to the large cyclotron radius acquired near the X-line.

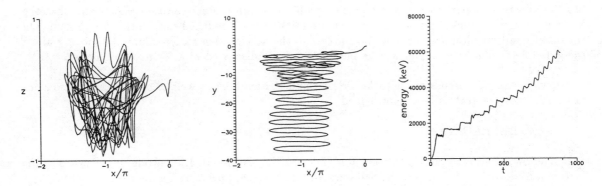

Fig. 2. Orbit of a proton with high final energy for $L = 4\beta^{-1}r_{ci} = 2\,10^4\,cm$ (for $N_e = 10^{9.5}\,cm^{-3}$, $-B_0 = 200\,G$, $T = 2.5\,10^6\,K$), $u = 0.32\,V_A = 1380\,km\,s^{-1}$, and $E_{y0} = uB_0/c$. The initial energy of $76.5\,keV$ corresponds to $r_{ci}(t = 0) = 0.01\,L$. Spatial coordinates are normalized to L and time is given in units of ω_{ci}^{-1}.

DISCUSSION AND CONCLUSIONS

The properties of the particle orbits lead to the following picture of particle acceleration by $\mathbf{E}_{conv}$ in a filamentary current sheet with island dynamics. *Regardless of the sign of* $\mathbf{u}$ *(i.e. of E_{yo}) there are always regions of strong acceleration due to the drift motion*, these are the regions where large velocity arrows point away from an X-line (at $x/L\pi = \pm 0, 4, 8, \ldots$ in our model); for $E_{y0} < 0$ this occurs inside the separatrices (as in Figures 1, 2) and for $E_{y0} > 0$ this occurs outside the separatrices. Particles initially close to the separatrices may "traverse" several islands and experience this type of acceleration. The drift acceleration effectively acts on particles over a large range of initial energies: for runs with 10^3 initially monoenergetic particles each and $r_c(t = 0)/L$ stepped through the range $10^{-4} - 10^{-2}$, the maximum final test particle energy differed by only a factor ≈ 2. Thus even thermal electrons in the corona (which are collisionless at the small scales of a filamentary current sheet) can be strongly energized. The mechanism might therefore be of relevance to the bulk of the flare particles. A good estimate of the characteristic maximum energy gain (acquired by at least one per cent of the test particles) is obtained by the assumption that the particles generally do not drift more than an island half length (πL in our model) along the X-lines.

For appropriate asymmetry of $\mathbf{E}_{conv}$ about the O-lines ($E_{y0} < 0$ in our model with $B_0 < 0$) there exists another population of accelerated particles with meander orbits at O-lines. The stability of the meander orbits leads to the production of the most energetic particles at O-lines. Because the majority of particles has small cyclotron radii, $r_{ce,i} \ll L$, initially, only a small fraction of the particles, those with start positions close to an O-line and those which are strongly energized due to a close encounter with an X-line, becomes part of this population. Of these, the more energetic ones are more strongly accelerated because particles with too small cyclotron radii do not "feel" much of the asymmetry of $\mathbf{E}_{conv}$ about the O-line. Consequently, meander-acceleration at O-lines may be important for the explanation of the most energetic particles, it does probably not produce the bulk of the accelerated particles.

In three dimensions, the coalescence process, started at a particular value of y, propagates in $\pm y$ direction with approximately Alfvén velocity. The acceleration region ($\mathbf{E}_{conv} \neq 0$) has therefore a finite extent $\triangle y \sim V_A \tau_{CI}$. The maximum achievable energy is thus $\triangle W_{max} = e(L_{CI}/2)(V_A B/c)$ where L_{CI} is the distance between islands and $L_{CI} > 2\pi \cdot 4\beta^{-1} r_{ci}$. The parameters of Figure 2 lead to $\triangle y > (V_A/u) \cdot 6\,10^4\ cm$ and $\triangle W_{max} > 10^8\ eV$. An upper limit on $\triangle y$ (and hence on L_{CI} and L) is set by the maximum size of elementary energy release regions in flares estimated from observations ($\sim 2\,10^7\ cm$). The characteristic magnitudes of L and L_{CI} have yet to be determined.

The acceleration by $\mathbf{E}_{conv}$ obviously rests on the differences in electric field strength on either sides of the trapping O-lines; in reality these differences are possibly much smaller than those in our model (2), but for $E_{conv} \gg E_{\parallel}$ (as one may generally expect for $R_m \gg 1$ in the absence of anomalous effects) they should still be larger than $E_{\parallel}$. A nonvanishing guide field component B_y has the effects that particles may be quickly guided through the acceleration region and that the ∇B - drift does no longer point into the $\pm y$ direction — with reduced maximum energy gain. On the other hand, much more particles have access to the acceleration region simply by moving along the guide field. The mechanism will be more extensively discussed in a forthcoming paper.

Acknowledgement. Discussions with M. Scholer and F. Jamitzky as well as permission to use computer routines are gratefully acknowledged.

REFERENCES

1. H. Kunow, G. Wibberenz, G. Green, R. Müller-Mellin and M.-B. Kallenrode, in: *Physics of the Inner Heliosphere*, Vol. 2, ed. R. Schwenn and E. Marsch, Springer, Berlin 1991, p. 243
2. D.B. Melrose, *Aust. J. Phys.* 43, 703 (1990)
3. M.J. Aschwanden and M. Güdel, The coevolution of decimetric millisecond spikes and hard X-ray emission during solar flares, *Ap. J.*, in press (1991)
4. E.R. Priest, *Rep. Prog. Phys.* 48, 955 (1985)
5. B. Kliem, *Astron. Nachr.* 311, 399 (1990)

PLASMA DENSITY FLUCTUATIONS AS INDICATORS OF WAVE PROCESSES IN SOLAR WIND MHD WAVES

J. A. Leckband and S. R. Spangler

Department of Physics and Astronomy, University of Iowa, Iowa City, IA 52242, U.S.A.

ABSTRACT

We have recently completed an analysis of 20 intervals of ISEE magnetic field and plasma density data from times when the spacecraft were in the ion foreshock. The purpose is to investigate the relationship between the wave magnetic field and plasma density. As shown in a previous investigation (Spangler et al. /1/) such correlations are helpful in revealing important wave processes. The intervals under present consideration ranged in duration from roughly half an hour to three hours. As found by Spangler et al. /1/ from a smaller data set, density fluctuations due to oblique wave propagation and ponderomotive effects are present. However, we have also obtained several new results. The first is a strong correlation between the fluctuating plasma density and the x component of the wave magnetic field, where the x direction is defined by the average solar wind magnetic field. In the ideal case of parallel-propagating plane MHD waves, there would be no x component of the wave magnetic field. This $\delta n - b_x$ correlation is found in all twenty intervals and must be considered a ubiquitous foreshock phenomenon. We have also seen evidence for two distinct populations of waves. The first (and dominant) population is the nearly parallel-propagating MHD waves in the spacecraft frequency range of $0.015 - 0.05$ Hz. The second population of waves are nearly perpendicular propagating waves with frequencies consistently at $0.3 - 0.4$ times the higher frequency waves. Our proposed explanation is that the second population of waves are magnetosonic waves produced by nonlinear parametric instabilities of the MHD waves.

I. DATA ANALYSIS

We have gathered and analyzed data from 20 episodes of MHD wave activity in the Earth's ion foreshock region. We used data from the Los Alamos plasma experiment of Bame et al. /2/ and the UCLA magnetometer experiment of Russell /3/ aboard ISEE 2 when the spacecraft was in the ion foreshock in 1977 and late 1978. The plasma data had resolution of 3 seconds while in the high data rate mode and 12 seconds while in the low data rate mode. The magnetometer gave a 12 second boxcar averaged vector magnetic field measurement every 4 seconds. The vector magnetic field time series were low pass filtered using a Gaussian filter of 500–1000 seconds. Euler angles were calculated from these smoothed fields and used to rotate the vector magnetic field into a frame in which the x direction constituted the mean field direction. Therefore, a parallel-propagating MHD wave will have fluctuations in the $y - z$ plane. The waves were then analyzed by constructing power spectra of the rotated vector magnetic field and density time series. Specific MHD modes can also be differentiated by performing cross correlations and cross spectra by the mean lagged product method between the various time series.

II. $\delta n - b_x$ CORRELATION

An unexpected finding is that there exists a large correlation between the density fluctuations (δn) and the component of the magnetic field parallel to the mean field (b_x). This correlation is present in all of our episodes with correlation coefficients ranging from 0.34 to 0.77. Figure 1 explicitly illustrates this correlation in the time series for a 30 minute period on Dec. 8, 1977. The correlation exists for a wide range of frequencies as shown by Figure 2. Generally for all the episodes this correlation can be attributed at the upper frequencies ($> 20\,\text{mHz}$) to oblique propagation of the

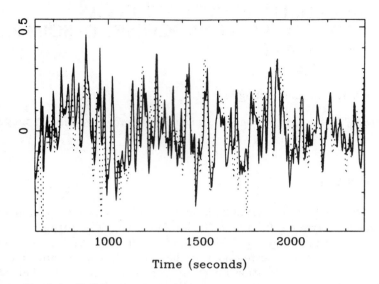

Fig. 1.   Normalized δn (full line) and b_x (dashed line) 13:10–13:40 UT Dec. 8, 1977. The x axis denotes seconds elapsed after 13:00 UT.

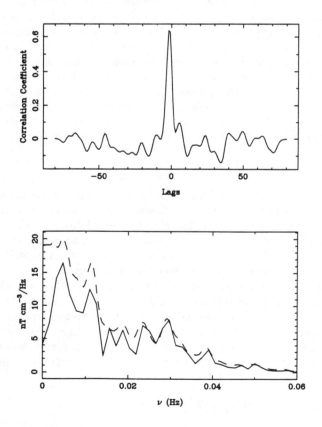

Fig. 2.   (Top) Correlation coefficient of δn and b_x as a function of lag. One lag is of 4 seconds duration for the Dec. 8, 1977 episode. (Bottom) Cross spectra of δn and b_x (full line). The dashed line is the product of the two time series power spectra, $\sqrt{P(b_x) \times P(\delta n)}$.

MHD waves by 10° to 20°. However, the persistence of the correlation at lower frequencies is not as easily explained. Our proposal is that the lower frequency $\delta n - b_x$ correlation is due to a second population of waves that are lower in amplitude and frequency than the well-known parallel-propagating MHD waves but are propagating nearly perpendicular to the magnetic field.

III. TWO-WAVE POPULATIONS

The existence of the two-wave populations is easily seen in Figure 3. The top panel shows the minimum variance direction as a function of frequency for an episode on Oct. 7, 1978 of 3 hours duration. The bottom panel shows the combined magnetic power spectra (full line) and the ambient solar wind power level (dotted line). The $3-9$ mHz waves are close to perpendicularly propagating

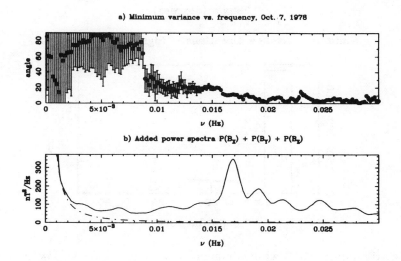

Fig. 3.   (Top) Angle the minimum variance direction makes with the mean magnetic field as a function of frequency for an episode on 1:50–4:50 UT Oct. 7, 1978. The error bars are calculated by assuming that the solar wind power level is a measure of the noise in the data. (Bottom) Combined magnetic power spectra for Oct. 7, 1978. The solar wind power estimate is denoted by the dotted line and is a power law of $-5/3$.

and the higher amplitude $15-20$ mHz waves are slightly obliquely propagating. This pattern is repeated in much the same fashion for all of our episodes. Figure 4 shows the distribution of minimum variance angles for the two populations for all the episodes. This graph was constructed by selecting a power threshold to count only frequencies with significant power for each episode and accordingly binning the minimum variance angles.

The most likely identification of the perpendicular waves are that they are magnetoacoustic waves since magnetoacoustic waves are known to be compressive and can propagate across field lines with fluctuations in the parallel component of the magnetic field. Their generating mechanism is unclear, however. It is unlikely that they are produced at the same location as the parallel-propagating waves since damping will occur as they convect downstream. The perpendicular waves may be produced locally by nonlinear instabilities acting on the large-amplitude parallel waves. In particular, filamentation instabilities (Viñas and Goldstein /4/) produce quasi-perpendicular magnetoacoustic waves and are characterized by a very broad growth rate spectrum. This is consistent with our data since the lower frequency waves generally do not have discrete spectral components but are broad band in nature. The daughter wave produced by this instability also has a low phase velocity which would place the convected fluctuations in a lower part of the frequency range than the pump waves.

Minimum variance directions – Low frequency waves

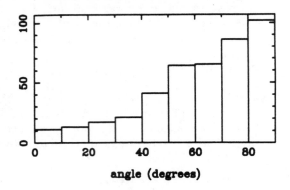

Minimum variance directions – High frequency waves

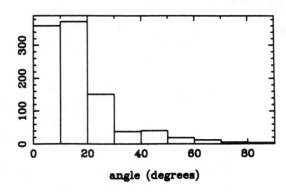

Fig. 4.   Distribution of MHD wave minimum variance directions for low-frequency (top) and high-frequency (bottom) waves.

IV. SUMMARY

Large correlations between the fluctuating density and the component of the magnetic field parallel to the mean field have been found for waves in the Earth's foreshock region. These correlations are ubiquitous and exist over a broad frequency range. We have also detected the existence of two-wave populations in the foreshock. The first population is nearly parallel-propagating MHD waves and the second is interpreted as lower frequency perpendicular propagating magnetosonic waves. The second population may be produced by parametric instabilities operating on the large amplitude MHD waves. Evidence that supports this view are the large minimum variance directions, the large $\delta n - b_x$ correlations, and the predicted wavenumbers for maximum growth of the magnetoacoustic instability are coincident with the frequencies of the magnetosonic waves.

REFERENCES

1. S.R. Spangler, S. Fuselier, A. Fey, and G. Anderson, <u>J. Geophys. Res.</u> 93, 845 (1988)

2. S.J. Bame, J.R. Asbridge, H.E. Felthauser, J.P. Glore, G. Paschmann, P. Hemmerich, K. Lehmann, and H. Rosenbauer, <u>IEEE Trans. Geosci. Electron.</u> GE-16, 216 (1978)

3. C.T. Russell, <u>IEEE Trans. Geosci. Electron.</u> GE-16, 239 (1978)

4. A.F. Viñas, and M.L. Goldstein, <u>J. Plasma Phys.</u> (1991)

PLASMA WAVE PHENOMENA AT INTERPLANETARY SHOCKS OBSERVED BY THE ULYSSES URAP EXPERIMENT

D. Lengyel-Frey,[a] R. J. MacDowall,[b] R. G. Stone,[b] S. Hoang,[c]
F. Pantellini,[c] C. Harvey,[c] A. Mangeney,[c] P. Kellogg,[d] J. Thiessen,[d]
P. Canu,[e] N. Cornilleau[e] and R. Forsyth[f]

[a] University of Maryland, College Park, MD 20742, U.S.A.
[b] Goddard Space Flight Center, Greenbelt, MD 20771, U.S.A.
[c] Observatoire de Paris, Meudon, France
[d] University of Minnesota, Minneapolis, MN, U.S.A.
[e] CRPE/CNET, 92131 Issy les Moulineaux, France
[f] Imperial College, London, U.K.

ABSTRACT

We present Ulysses URAP observations of plasma waves at seven
interplanetary shocks detected between approximately 1 and 3 AU. The URAP
data allows ready correlation of wave phenomena from .1 Hz to 1 MHz. Wave
phenomena observed in the shock vicinity include abrupt changes in the
quasi-thermal noise continuum, Langmuir wave activity, ion acoustic
noise, whistler waves and low frequency electrostatic waves. We focus on
the forward/reverse shock pair of May 27, 1991 to demonstrate the
characteristics of the URAP data.

INTRODUCTION

The Ulysses URAP (Unified Radio and Plasma Wave) experiment is designed
to detect plasma waves in the immediate vicinity of the spacecraft, as
well as to measure radio waves propagating to the spacecraft from remote
sources. The experiment provides a direct determination of electron
density at the spacecraft, and, as such, can be used to identify solar
wind density structures propagating past the spacecraft. Since its launch
in October 1990 the experiment has detected a large number of solar wind
density discontinuities, some of which have been identified as
interplanetary shocks through a comparison of URAP data with Ulysses
magnetometer data. We have done a preliminary investigation of seven of
these shocks, which have been observed within 5 degrees of the ecliptic
and at distances from 1 to 3 AU. The seven shocks are listed in Table 1
of Hoang et al. (this issue). In this study we categorize the various
wave phenomena observed in the vicinity of the shocks.

DATA

The Ulysses URAP experiment has been described in detail elsewhere /1,2/.
Briefly, it consists of a 72-m dipole electric field antenna in the spin
plane of the spacecraft, a 7.5-m monopole antenna along the spin axis,
and two orthogonal search coil magnetic antennas. The various receivers
of the URAP experiment cover a wavelength range of .1 Hz to 1 MHz. Only
electric field wave measurements are made above 500 Hz, whereas both E
and B field wave measurements are obtained from .1 to 500 Hz. The URAP
instruments used in this analysis are the Radio Astronomy Receiver (RAR)

covering a frequency range of approximately 1 to 1000 kHz, the Plasma
Frequency Receiver (PFR) with observing frequencies of .5 to 35 kHz, and
the Waveform Analyzer (WFA) ranging from .08 to 448 Hz. In addition, a
Fast Envelope Sampler (FES) can capture rapidly varying signals in one of
several frequency ranges with up to 1 msec time resolution.

ANALYSIS OF MAY 27, 1991 SHOCK

To illustrate the general features of URAP observations of plasma waves
at shocks, we present the example of a forward/reverse shock pair
observed on May 27, 1991. The forward shock occurred at about 04:08 UT,
and the reverse shock was observed at about 19:04 UT. An overview plot
which presents dynamic spectra of the wave data from three of the URAP
instruments is shown in Figure 1, a greyshaded plot of frequency versus
time, with the degree of darkness proportional to the wave intensity. The
topmost panel, ranging from 1 kHz to 1000 kHz, represents data from the
RAR. The RAR data is characterized by an enhanced continuum emission,
known as the thermal noise continuum, the low frequency edge of which is
known as the plasma line. The plasma line occurs at the local plasma
frequency at the spacecraft. An abrupt increase in the frequency of the
plasma line is observed at the forward shock, corresponding to the jump
in density at shock passage. An abrupt decrease in density is observed at
the reverse shock. Interplanetary type III radio bursts are evident at
the highest frequencies. A prominent type III burst which propagates to
lower frequencies is labeled. A detailed analysis of the RAR spectrum
yields electron temperature and density of the plasma at the spacecraft
(Hoang et al., this issue).

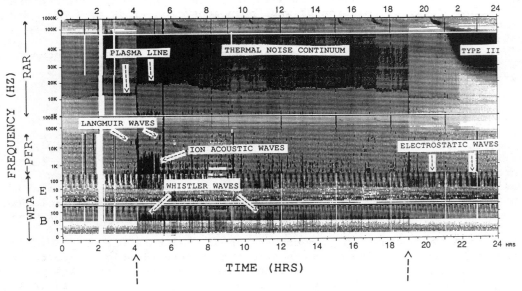

Fig. 1. Dynamic spectra of plasma wave observations of May 27, 1991 from
the Ulysses URAP experiment. The range of frequencies for each instrument
is shown at left. Times of shock passage are marked by dashed arrows at
bottom. Examples of various plasma wave phenomena are labeled. A vertical
interference pattern occurs at the highest frequencies of WFA E and B.

PFR observations are shown below the RAR panel, in the frequency range of
.5 kHz to 35 kHz. The ion acoustic noise observed in this frequency range
is clearly enhanced at shock passage and persists into the downstream
region of the forward shock. A brief burst of ion acoustic waves also is
seen at the reverse shock. In addition, brief bursts of Langmuir wave
activity are observed both before and after shock passage. Correlated
electric and magnetic waves are observed by the WFA instrument below 500
Hz. These waves begin abruptly at the forward shock and continue
sporadically for many hours downstream. Previous studies /3,4/ have
identified electromagnetic waves downstream of interplanetary shocks as
whistler waves. We have computed the index of refraction from the ratio
of magnetic and electric field energies and obtain typical values of
about 300 at 28 Hz. This is in reasonable agreement with an index of
refraction of 240 obtained using the whistler dispersion relation and the
observed magnetic field strength of 10 nT during this interval, thereby
providing evidence for the whistler mode interpretation.

After the reverse shock the magnetic wave activity quickly vanishes while
the electric field waves become significantly enhanced and remain strong
for many days thereafter. The identification of this low frequency
electrostatic wave remains unclear. The persistence of these waves long
after shock passage suggests that the electrostatic waves are not
directly related to the shock fronts or shocked plasma. This is not true
of the magnetic waves, which are clearly enhanced within the
forward/reverse shock boundaries.

GENERAL RESULTS

We summarize the general properties of plasma waves observed at shocks by
the Ulysses URAP experiment as follows:

Langmuir wave activity is seen at all shocks analyzed in this study.
These waves can be observed either upstream or downstream within several
hours of shock passage. Very often they are observed within minutes of
shock passage. Doppler-shifted ion acoustic waves are detected at all
shocks by the PFR. In all cases the downstream activity is more intense
and of longer duration than the upstream waves. Typically a burst of
exceptionally intense emission occurs within minutes of shock passage.

As found in previous studies /3,4/, magnetic waves begin, or are
significantly enhanced, at shock passage and persist for hours into the
enhanced density region downstream of the shock. In some cases magnetic
turbulence can last days. For instance, wave activity commences
approximately 1 day before the December 15 shock, at an abrupt density
increase occurring at approximately 14:30 on December 14. The wave
levels remain strongly elevated until the passage of the reverse shock of
December 16 at about 08:00. The nature of long-duration magnetic wave
activity is unclear. It may be related to the continuous background of
magnetic fluctuations found by the Helios spacecraft /5/ within 1 AU.
Alternatively, waves propagating from the vicinity of shocks and
discontinuities (or local instabilities due to propagating particles) may
be responsible.

Low frequency (less than about 100 Hz) electrostatic waves observed by
the WFA occur in conjunction with some of the shocks. In previous studies

/3/ these waves have been detected predominantly in the downstream
region. We observe these waves both upstream and downstream of shocks. In
the case of the April 7 shock, they begin about 10 hours before shock
passage, at the time of the appearance of higher frequency ion acoustic
waves. Low frequency electrostatic waves are observed to begin during
the interval between the May 27 shocks and persist for many days
thereafter. In this case they have no obvious association with the higher
frequency ion acoustic waves. The persistence of a low frequency
electrostatic component for many days in the solar wind has not, to our
knowledge, been previously reported.

SUMMARY

The combined data from the URAP instruments permit ready correlation of
wave phenomena over a wide frequency range, permitting rapid
identification of wave phenomena for subsequent quantitative study.
Plasma wave phenomena observed by the Ulysses URAP experiment in the
vicinity of IP shocks between 1 and 3 AU is similar to wave activity
reported in previous studies of shocks at 1 AU. A detailed comparison of
wave properties as a function of shock parameters is in progress. Such
studies will form a baseline for comparative studies of shocks at high
heliospheric latitude.

Acknowledgements: The URAP experiment is a cooperative effort of
NASA/GSFC, Observatoire de Paris, CRPE, and the University of Minnesota.
D.L-F. wishes to acknowledge the support of NASA grant NAG5-1134.

REFERENCES

1. R.G. Stone et al. (26 coauthors),The ISPM Unified Radio and Plasma
Wave Experiment, ESA Special Publication SP-1050(1983)

2. R.G. Stone et al. (31 coauthors),The Ulysses Unified Radio and
Plasma Wave Investigation,Astron. and Astrophys. Suppl.,in press(1991).

3. C.F. Kennel, F.L. Scarf, F.V. Coroniti, E.J. Smith, and D.A.
Gurnett, Nonlocal plasma turbulence associated with interplanetary
shocks, J.Geophys.Res. 87,17(1982)

4. D.A. Gurnett, F.M. Neubauer, and R. Schwenn, Plasma wave turbulence
associated with an interplanetary shock, J.Geophys.Res. 84,541(1979)

5. F.M. Neubauer, G. Musmann, and G. Dehmel, J.Geophys.Res.
82,3201(1977)

TWO DIMENSIONAL PIC SIMULATIONS OF PLASMA HEATING BY THE DISSIPATION OF ALFVÉN WAVES

P. C. Liewer,* T. J. Krücken,** R. D. Ferraro,* V. K. Decyk*** and B. E. Goldstein*

* Jet Propulsion Laboratory, California Institute of Technology, Pasadena, CA 91109, U.S.A.
** Applied Physics, California Institute of Technology, Pasadena, CA 91125, U.S.A.
*** Physics Department, University of California, Los Angeles, CA 90024, U.S.A.

ABSTRACT

Two dimensional plasma particle simulations of the evolution of large amplitude circularly polarized Alfvén waves propagating parallel to the magnetic field show that the waves decay via both one- and two- dimensional parametric decay instabilities. For parameters studied, one-dimensional processes dominate the simulations, but two dimensional decay processes, including the recently predicted filamentation instability are also observed. The daughter waves generated by the parametric decay are primarily damped by the ions, leading to ion heating. The parametric decay processes efficiently convert the ordered fluid ion motion in the Alfvén wave into ion thermal energy. These processes may be important for the dissipation of Alfvén waves in the solar wind, the corona and other space plasma environments. The computations were performed on the Intel Touchstone parallel supercomputer.

INTRODUCTION

Large amplitude Alfvén waves propagating parallel to the magnetic field are observed in the solar wind, upstream of Jupiter's bow shock and in other space plasma environments. Large amplitude Alfvén waves thought to be present close to the sun are often invoked as a source of solar wind and coronal heating and acceleration, but the mechanisms by which these waves are dissipated are unclear. Understanding the evolution of large amplitude Alfvén waves is important for understanding wave spectra and plasma heating in space plasma environments.

While large amplitude, circularly polarized Alfvén waves propagating along the magnetic field are not damped by the plasma, they are known to be subject to parametric decay instabilities, e.g., the Alfvén wave decays spontaneously into a daughter wave and one or more sideband waves/1-4/. In one-dimensional parametric instabilities, all daughter and side-band waves propagate along the field; in two-dimensional processes, obliquely and perpendicularly propagating waves are also present. Here we present results of two dimensional plasma particle-in-cell computer simulations of the decay of a large amplitude circularly polarized Alfvén wave propagating parallel to the field. We find that the Alfvén wave decays via simultaneous one- and two-dimensional parametric decay instabilities. For parameters studied, the familiar one-dimensional decay instability dominates, but two-dimensional decay processes, including the recently predicted filamentation instability/4/ are also observed. The computations are performed in parallel on concurrent supercomputers including the Intel Delta Touchstone computer, allowing kinetic treatment of both electrons and ions. The daughter waves generated by the parametric decay are primarily damped by the ions, leading to ion heating, as observed in the one-dimensional simulations of Terasawa et al./2/. The parametric decay processes efficiently convert the ordered kinetic energy of ions in the Alfvén wave into ion thermal energy. The electrons are also observed to heat, apparently due to both wave damping and equilibration with the ions. At the end of the simulations, longer wavelength amplitude Alfvén waves, generated by the decay instability, dominate the magnetic fluctuation spectrum.

BACKGROUND AND MODEL

In parametric Alfvén wave instabilities, the initial "pump" Alfvèn wave $(\omega_0, \vec{k}_0)$ with $\vec{k}_0 \| \vec{B}_0$ beats with a daughter wave $(\omega_d, \vec{k}_d)$ which is initially excited from the noise and forms a pair of sideband waves $(\omega_\pm, \vec{k}_\pm)$. Frequency and wavenumber matching requires that $(\omega_\pm = \omega_0 \pm \omega_d, \ \vec{k}_\pm = \vec{k}_0 \pm \vec{k}_d)$. Generally, if three of the four waves are eigenmodes, *i.e.*, the $\omega, \vec{k}$ are solutions of the dispersion relation, a parametric instability can occur.

In the past, one-dimensional parametric decay instabilities, the decay and modulation instabilities have been studied analytically/1/ and numerically via computer simulation/2/. In the first case the daughter wave is an ion sound wave with a shorter wavelength than the pump Alfvèn wave; the sidebands are Alfvén waves. The modulational instability is analogous with the daughter ion sound wave having a longer wavelength than the pump. Recently, Viñas and Goldstein /3,4/ have analyzed the linear stability of large amplitude parallel-propagating Alfvén waves to two dimensional parametric decay processes using a two-fluid model. In this study, they predict for some parameters a strong two-dimensional filamentation instability, also studied by Kuo *et al.*/5/ which creates field-aligned density filaments. The one dimensional decay processes can be studied in simpler 1D models. On the other hand, the filamentation instability can not be treated in a one dimensional model since the daughter "wave" is a nearly purely growing, magnetostatic perturbation, i.e. $Re(\omega) \simeq 0$, with $\vec{k} \perp \vec{k}_0$. The pump wave gets "ducted" in aligned density and magnetic field filaments due to this magnetostatic perturbation. Sidebands are obliquely propagating magneto-acoustic waves.

In the analytic fluid treatments, the linear growth rates of the parametric decay instabilities are found subject to the approximations that the amplitude of the pump wave is constant and kinetic effects are negligible. Computer simulations follow the evolution of the pump wave self-consistently with any growing decay instabilities and include kinetic effects on the waves.

The numerical model used in this work is a standard two-dimensional electromagnetic plasma particle-in-cell code: the orbits of the plasma electrons and ions are evolved self-consistently with Maxwell's equations for the fields. Maxwell's equations are solved using fast Fourier transform techniques. The code is described in Krücken *et al.*/6/. The parallel implementation uses the same algorithms as the 2D electrostatic code described by Ferraro *et al.*/7/ and Liewer and Decyk/8/.

SIMULATION RESULTS

To start the simulation, a right-hand circularly polarized Alfvén wave is excited with uniform amplitude across the simulation domain propagating parallel to the static magnetic field ($\hat{y}$ direction). The wave is then observed to decay in time via parametric instabilities. Once the wave launch is turned off, the code conserves energy as the parametric processes evolve.

In the simulations, the initial pump wave is observed to decay parametrically with several decay processes occurring simultaneously. Figure 1 shows the time history of the Alfvén pump wave with $\delta B/B = 0.6$ and $\lambda_0 = L_y/3$ for a simulation with a 256x256 grid. Other parameters are $\kappa = k_0 V_A/\omega_{ci} = 1.1$, $m_i/m_e = 100$, $\omega_{ce} = \omega_{pe}$, $T_e = T_i$, and $\beta_e = 0.22$. The figure shows the time history of the real and imaginary parts of the amplitude of the Fourier mode corresponding to the pump. The high frequency oscillation shows the real frequency of the pump and the envelope shows its depletion. To identify the various parametric decay processes, time histories of Fourier modes of the magnetic and density fluctuations are stored and the frequencies of the Fourier modes, as well as the polarization of observed daughter and sideband waves, determined from the stored information via a post-processor. The real frequency of the pump is $\omega_r \simeq 0.023$ and the observed decay rate for this mode is $\gamma/\omega_r \approx 1.4 \times 10^{-2}$.

Figure 2 shows the energy balance for this simulation. Plotted as a function of time are the total energy, ion energy, electron energy, and electromagnetic field energy, *e.g.*, the energy in the magnetic and electric field fluctuations. The particle energies are determined by summing the kinetic energy of all the particles and thus include thermal as well as ordered "fluid" motion such as that due to the plasma $\delta\mathbf{E} \times \mathbf{B}$ motion in the initial Alfvén pump wave. It can be seen that most of the energy initially in the electromagnetic fields of the pump wave is transferred to the ions and electrons. This energy transfer is mediated by the parametric instabilities: the ion acoustic and oblique magnetosonic waves created in the parametric processes are damped by the plasma, leading to ion and electron heating. The much slower decrease in ion energy and increase in electron energy after $t\omega_{pe} \simeq 4000$ is due to collisional electron-ion temperature equilibration, where the collisions are due to the rather small number of particle per cell used in these simulations (8 particles – electrons plus ions – per cell).

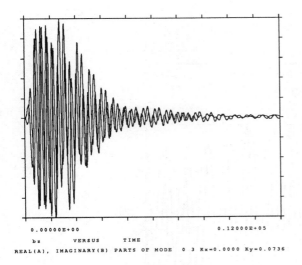

Fig. 1 – Time history of the Alfvén pump wave, $\delta B/B_z$ vs. t, for a simulation with $\lambda_0 = L_y/3$. The envelope shows the decay of the pump due to parametric instabilities; the oscillations show the real frequency.

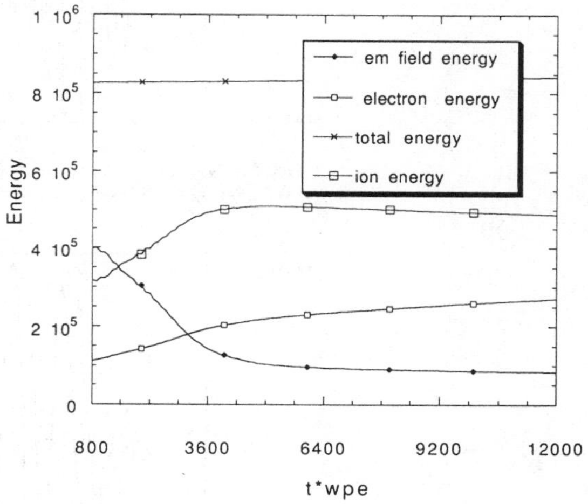

Fig. 2 – Time history of energy balance in the simulation, showing total, ion, electron and electromagnetic field energies. The pump wave energy is transfered to the plasma by damping of daughter and sideband waves.

In this simulation, many modes were excited as the pump wave decayed. Analysis of the Fourier modes of the magnetic and electric field and density fluctuations led to the identification of six parametric instabilities, all occurring simultaneously, but with different growth rates. Three of the processes were one-dimensional decay instabilities (Alfvén wave into ion acoustic daughter plus backward propagating Alfvén wave) with all $\vec{k}$'s aligned along the pump; two were two-dimensional extensions of the decay with slightly obliquely propagating daughter ion acoustic and magneto-acoustic side bands (predicted by Viñas and Goldstein/4/); and one was a filamentation instability/4,5/ e.g., an Alfvén wave decay into an oblique magneto-acoustic wave plus a perpendicular magnetostatic mode with $\omega_r \approx 0$. For example, labeling each mode by the Fourier modes numbers (m,n) where $\vec{k} = (2\pi/L_x)m\hat{x} + (2\pi/L_y)n\hat{y}$, the 1D decay of $(0,3)_{pump\ Alfven} \rightarrow (0,5)_{ion\ acoustic} + (0,2)_{backward\ Alfven}$ was observed with a growth rate $\gamma/\omega_0 \approx 0.1$ The frequencies satisfy $\omega_0 = \omega_d + \omega_s$ with $\omega_d = 0.017$ and $\omega_s = 0.006$. Figures 3a and b show the time history of the daughter and side band modes for this process. Shown in Fig. 3a and 3b, respectively, are the time histories of the density fluctuation of the $(0,5)$ ion acoustic mode and the $\hat{z}$ magnetic fluctuation of the $(0,2)$ circularly polarized backward propagating Alfvén wave. It can be seen that after the initial growth phase, the ion acoustic wave is heavily damped, presumably by the plasma ions since the phase velocity of this mode, and all the parametrically excited ion acoustic modes, falls well within the ion distribution function for this $T_i \approx T_e$ plasma. This appears to be the dominant processes for heating the ions. The electrons heat more due to damping oblique modes. The backward propagating Alfvén wave, however, is not damped by the plasma and, because of its lower amplitude ($\delta B/B \approx 0.2$), further parametric decays to the $(0,1)$ mode are apparently occurring on a much slower time scale. Except for this $(0,2)$ Alfvén wave and the $(0,1)$ Alfvén wave, all daughter and sideband waves are damped by the plasma so that the fi-

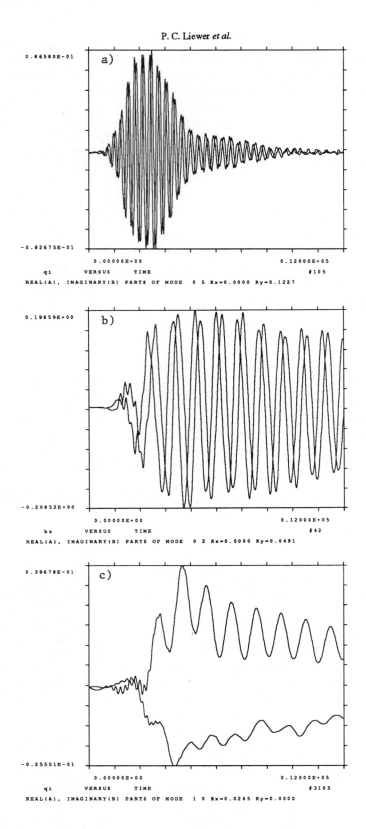

Fig. 3 – Time history of three Fourier Modes showing decay and a filamentation instabilities. (a) Density fluctuation δn vs. t for $(0, 5)$ ion acoustic daughter wave showing parametric growth and subsequent plasma damping; (b) Magnetic fluctuation of $(0, 2)$ backward Alfvén sideband to above ion acoustic wave (see text); (c) Density fluctuation of $(1, 0)$ perpendicular density fluctuation of filamentation instability showing $\gamma \gg \omega_r$.

nal magnetic fluctuation spectrum is dominated by these parallel propagating Alfvén waves. The oblique analogy of this process, $(0,3)_{pump\ Alfven} \rightarrow (1,5)_{ion\ acoustic} + (1,2)_{oblique\ magneto-acoustic}$, was also observed with a comparable growth rate $(\gamma/\omega_r \approx 0.09)$. The filamentation instability observed, was $(0,3)_{pump\ Alfven} \rightarrow (1,0)_{magnetostatic} + (1,3)_{oblique\ magnetosonic}$. The time history of the density fluctuation for this purely perpendicular mode is shown in Fig. 3c; the power spectrum shows that this mode satisfies $\omega_r \approx 0$. Note that the use of a discrete grid limits the available decay channels, e.g., the $(0,1)$ Alfvén wave is the longest wavelength mode allowed in the simulation.

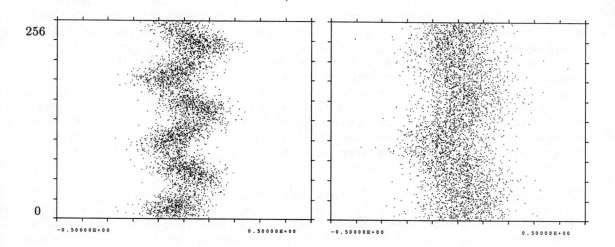

Fig. 4 – Ion phase space y versus v_x early $(t\omega_{pe} = 2000)$ and late $(t\omega_{pe} = 10,000)$ in the simulation. Much of the ordered ion energy in the pump Alfvén wave has been thermalized, leading to about a factor of 2 increase in the ion thermal velocity.

Figure 4 shows the ion phase space y versus v_x at two times in the same simulation as Figs. 1-3. In Figure 4 at $t\omega_{pe} = 2000$(left), the ion "fluid" motion in the pump Alfvén wave $(\lambda_0 = L_y/3)$ is visible. Ions in an Alfvén wave have a perpendicular fluid velocity $\delta v_\perp/v_A \approx \delta B/B$; here, $\delta v_\perp \approx 0.08$ and initially the ion thermal velocity is $v_{ti} = 0.05$. Figure 4 at $t\omega_{pe} = 10,000$ shows that the ions have been heated by damping on the parallel and oblique ion acoustic modes: the distribution is significantly broader with $v_{ti} \approx 0.09$ and a temperature increase by almost a factor of 4. The ion fluid motion in the dominant longer wavelength $(0,2)$ Alfvén wave (Fig. 3b) is visible in this plot. In all the simulations, we observe an increase in the ion thermal velocity by an amount of comparable to their fluid velocity in the pump wave.

DISCUSSION

These two-dimensional plasma particle-in-cell simulations have shown that right-hand circularly polarized Alfvén waves are rapidly dissipated when the amplitude is large $(\delta B/B > 0.1)$. The final state observed in the simulations consists of longer wavelength, lower amplitude parallel-propagating Alfvén waves. Any parametric decay of these lower amplitude Alfvén waves are apparently too slow to be observed. The time scale for a single wave decay process is typically $\tau \sim \gamma^{-1} \sim (10 - 10^2)/\omega_0$. To relate to solar wind time and space scales, we take the decay length of an Alfvén wave to be $\delta \sim (\omega_0/\gamma)\lambda$, where γ is the growth rate of the absolute instability. Thus a large nonlinear Alfvén wave will decay parametrically on a spatial scale of $\delta \sim (10-100)\lambda$. Since spacecraft measure $f = V_{sw}/\lambda$, where $V_{sw} \sim 400\ km/s$ is the solar wind speed, the decay length for an Alfvén wave of measured frequency f is $\delta \sim (10 - 100)V_{sw}/f \sim 4 \times (10^3 - 10^4)/f_{Hz}\ km$.

For example, a solitary large amplitude Alfvén wave of measured frequency $f \sim 10^{-3}$ Hz will decay parametrically in a distance $\delta \sim 4 \times (10^6 - 10^7) \, km \sim (0.03 - 0.3) \, AU$. Thus these processes are fast enough to influence the evolution of the spectrum of solar wind fluctuations and the plasma temperatures. Note that if the amplitude of an Alfvén wave are sufficiently low, however, the growth rates of parametric instabilities become zero and no damping is expected from these processes. The analysis presented here is for a single, large amplitude wave whereas the solar wind has a broad, turbulent spectrum of waves. It has been suggested that this may stabilize the parametric decay modes.

The dissipation of the Alfvén waves is very rich in the plasma processes involved: Both wave-wave (the parametric instabilities) and wave-particle (electron and ion landau and/or cyclotron damping) processes are involved in the conversion of the pump wave energy into plasma thermal energy. Both one- and two-dimensional decay instabilities, including the traditional decay and the newer filamentation instability have been shown to be robust in the presence of plasma kinetic effects. Many parametric instabilities were found to occur simultaneously in the simulations; which processes are observed to occur depends on the parameters of the simulation (mass ration, plasma β, pump wave number, grid resolution etc.). Viñas and Goldstein/3/ also predicted a one-dimensional electromagnetic modulation instability which has not been identified in our simulations. Because of computational limitations imposed by a full particle code, only a small region in parameter space as been investigated. Moreover, nonlinear Alfvén wave coupling processes treated in MHD turbulence calculations are not well treated in a particle code. These simulations have only scratched the surface of the wealth of physics involved in the evolution and dissipation of large amplitude Alfvén waves.

This work was supported in part by NASA/Cosmic and Heliospheric Physics and in part by the NSF under Cooperative Agreement No. CCR-880961. Computations were performed on the Caltech Concurrent Supercomputing Facilities. One of the authors (T.J.K.) was partially supported by the Max-Planck-Institut für Plasmaphysik (IPP) at Garching and the Max-Planck-Gesellschaft at München, Germany and partially supported by the Office of Fusion Energy/USDOE.

REFERENCES

1. H. K. Wong and M. L. Goldstein, *J. Geophys. Res., 91,* 5617 (1986).

2. T. Terasawa, M. Hoshino, J.-I. Sakai and T. Hada, *J. Geophys. Res. 91, A4,* 4171 (1986).

3. A. F. Viñas and M. L. Goldstein, Parametric Instabilities of Circularly Polarized, Large Amplitude, Dispersive Alfvèn Waves: Excitation of Parallel Propagating Electromagnetic Daughter Waves, *J. Plasma Phys., 46,* 107 (1991).

4. A. F. Viñas and M. L. Goldstein, Parametric Instabilities of Circularly Polarized, Large Amplitude, Dispersive Alfvèn Waves: Excitation of Obliquely Propagating Electromagnetic Daughter Waves, *J. Plasma Phys. , 46,* 129 (1991).

5. S. P. Kuo, H. H. Whang and M. C. Lee, *J. Geophys. Res. 93, A9,* 9621 (1988).

6. T. J. Krücken, P. C. Liewer, R. D. Ferraro, and V. K. Decyk, A 2D Electromagnetic PIC Code for Distributed Memory Parallel Computers, in *Proceedings of 6th Distributed Memory Computing Conference,* (IEEE Computer Society Press, Los Alamitos, CA) 452 (1991).

7. R. D. Ferraro, P. C. Liewer, and V. K. Decyk, A 2D Electrostatic PIC Code for the Mark III Hypercube, in *Proceedings of 5th Distributed Memory Computing Conference, Vol. I* (IEEE Computer Society Press, Los Alamitos, CA) 440 (1990).

8. P. C. Liewer and V. K. Decyk, A General Concurrent Algorithm for Plasma Particle-in-Cell Simulation Codes, *J. Computational Physics, 85,* 302 (1989).

ENERGY DISSIPATION BY ALFVÉN WAVES PROPAGATING IN AN INHOMOGENEOUS MAGNETIC FIELD

F. Malara,* P. Veltri,* C. Chiuderi** and G. Einaudi**

* Departimento di Fisica, Universitá della Calabria, 87030 Roges di Rende, Italy
** Dipartimento di Fisica e Scienza dello Spazio, Universitá di Firenze,
largo E. Fermi 5, 50125 Firenze, Italy

ABSTRACT. The evolution of incompressible MHD disturbance propagating in an inhomogeneous magnetic field is studied by means of numerical simulations. Small scales are generated in the perturbation which is dissipated within few Alfvén times. This phenomenon could be relevant in the problem of coronal heating.

The dissipation of propagating waves is one of the mechanism which could contribute to the non-radiative heating of astrophysical plasmas and, in particular, of the Solar Corona. Dissipation of hydromagnetic waves moving up from the photosphere can represent one of the possible heating mechanisms. Indeed, the Corona is threaded by magnetic fields and, on the other hand, motions are observed in the range of 10 - 30 km s^{-1} (rms) which could be interpreted as propagating (hydromagnetic) waves. The investigation of this mechanism must face two problems: i) is the amount of energy which is carried by the waves adequate to the observed heating? ii) are the waves able of depositing the right amount of energy in the right place with the appropriate rate? Concerning the first question, it has been argued /1/ that the observed motions may contain enough energy to heat the Corona. The present study gives some contribution to answer the question ii).We will show that, when the background magnetic fields is inhomogeneous, the dynamical evolution of an hydromagnetic wave leads to the formation of strong gradients and to an efficient dissipation of the wave energy into heat. Normal mode analysis applied to the propagation in an inhomogeneous magnetic field has shown the existence of fast-dissipating solutions: resonant modes /1-7/ in which strong gradients are present in a thin layer, and resistive modes /8/ in which small scale structures are present in a wider region, of the order of the inhomogeneity scale. In this paper we consider an initial-value approach which is able to give information on the time necessary for a wave to generate the small scales and how the available energy distributes on the set of the normal modes.

As a simplified model we consider the evolution of an incompressible MHD wave (Alfvén + slow magnetosonic) in a two-dimensional configuration, which is described by the following equations:

$$\frac{\partial Z^{\sigma}}{\partial t} + \left(Z^{-\sigma}\cdot\nabla\right)Z^{\sigma} + \nabla P = \frac{1}{S}\nabla^2 Z^{\sigma} \quad, \quad \sigma = \pm 1 \qquad ; \qquad \nabla^2 P = -\sum_{i,j=1}^{2}\frac{\partial Z_i^+}{\partial x_j}\frac{\partial Z_j^-}{\partial x_i} \qquad (1)$$

where $Z^{\sigma}=v + \sigma\, B/\sqrt{(4\pi\rho)}$, P, v, B, ρ and S are the total pressure, the velocity, the magnetic field, the density, and the Reynolds number, respectively. These equations have been solved using a numerical code /9/. We consider a static equilibrium structure with an inhomogeneous magnetic field:

$$v^{(0)} = 0 \qquad , \qquad B^{(0)} = B_0\{1 + \Delta[\text{tgh}(x/a) - b(x/a)]\}e_y \qquad (2)$$

where a represents the shear length and the constant b is such that $\partial B_y/\partial x=0$ at the boundaries $x=\pm\ell$. The total pressure $P^{(0)}$ is constant in order to ensure the equilibrium, so the kinetic pressure is given by $p^{(0)}(x)=P^{(0)} - [B^{(0)}(x)]^2/(8\pi)$. A smooth, backward propagating disturbance with a given wavelength $\delta Z^+(x,y,t=0)=\text{Re}\{\delta Z^+(x)\exp(iky)\}$ is superposed at t=0. The boundary conditions are: periodicity in the propagation (y) direction, and free-slip in the perpendicular direction; these boundary conditions are

Figure 1

run 1: kinetic energy at t=0

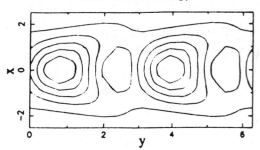

run 2: kinetic energy at t=0

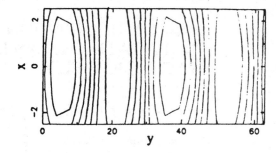

run 1: kinetic energy at t=10

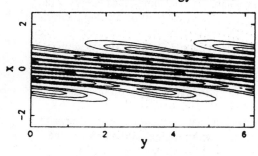

run 2: kinetic energy at t=80

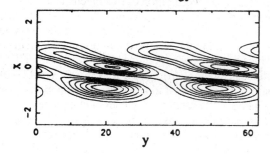

run 1: kinetic energy at t=20

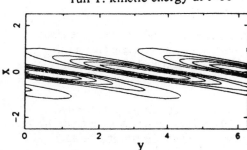

run 2: kinetic energy at t=120

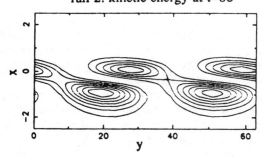

run 1: kinetic energy at t=45

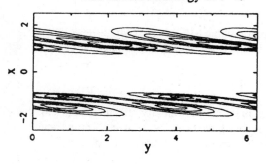

run 2: kinetic energy at t=250

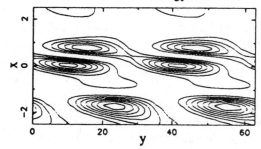

discussed in more detail elsewhere /9/. Five runs of the numerical code have been performed with different values for the wavenumber k, the Reynolds number S and the initial wave energy to the equilibrium structure energy ratio ε; these values are summarized in Table 1. The value of Δ is Δ=0.5

TABLE 1

	run 1	run 2	run 3	run 4	run 5
ka	1	0.1	1	1	0.1
S	10^3	10^3	10^4	10^3	10^3
ε	2.5×10^{-5}	2.5×10^{-5}	2.5×10^{-5}	0.01	0.01
τ_d	≈30	≈250	≈65	≈30	≈300
$\tau_{d,hom}$	10^3	10^5	10^4	10^3	10^5

for all the runs. In Figure 1 contour plots of the kinetic energy are shown for run 1 and 2. The results show that two fundamental dynamical mechanisms are at work during the time evolution: i) phase-mixing /10/: due to the inhomogeneity of the Alfvén velocity in the transverse direction, different points in the wavefront propagates with different velocities. The initial perturbation is elongated and folded in an oblique direction; the result is the formation of small scales along the transverse direction. ii) energy pinching: the fluctuations of Z_+ and Z_- are coupled each other and to the inhomogeneity of the equilibrium structure through the pressure gradient terms and source-like terms ($\delta Z^{-\sigma} \cdot \nabla)Z^\sigma_0$. These terms are nonvanishing only in the inhomogeneity region; their effect is a concentration of the energy of the perturbation at the center of the shear region. The wavy structures generated by the phase-mixing are similar to the resistive normal modes /8/, while the concentration of the energy at the center of the shear is reminiscent of a resonant mode. Both effects are in general present, even if their relative importance depends on the wavenumber ka.

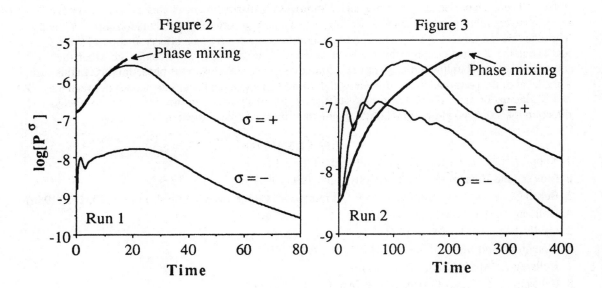

Figure 2 — Run 1
Figure 3 — Run 2

In runs 1 to 3 the parameter ε is very small so that nonlinear effects are negligible. In Figures 2 and 3 the power P^σ dissipated by the fluctuations of Z^+ and Z^- is shown vs. time for runs 1 and 2, respectively. Since the equilibrium is uniform in the y-direction, we define P^σ as:

$$P^\sigma(t) = \frac{1}{E^{(0)}} \sum_{m>0} \int_{-\ell}^{\ell} \left(\left| \frac{\partial Z^\sigma_{xm}}{\partial x} \right|^2 + m^2 \left| Z^\sigma_{xm} \right|^2 + \left| \frac{\partial Z^\sigma_{ym}}{\partial x} \right|^2 + m^2 \left| Z^\sigma_{ym} \right|^2 \right) dx$$

where $E^{(0)}$ is the energy of the equilibrium structure and the index m identifies the m-th coefficient of a Fourier expansion in the y-direction. In the first stage the dissipated power P^+ increases with the time following a quadratic law; this corresponds to the formation of small scales in the transverse direction. The effect of the phase-mixing can be easily modelized neglecting the pressure terms in equations (1); the resulting power P^+ is also plotted in Figures 2 and 3 (heavy line) for comparison. The first evolution stage is well reproduced by the phase-mixing model only for short wavelengths (k a=1). For longer wavelengths (ka=0.1) the dissipated power grows faster than the prediction of the phase-mixing model; then, in this case a mechanism faster than phase-mixing works to produce small scales. We identify such a mechanism as the energy pinching. Actually, the time scale of the source terms ($\delta Z^{-\sigma} \cdot \nabla) Z^\sigma_0$ is independent on k whereas the efficiency of the phase-mixing decreases with k. It follows that at long wavelengths the energy pinching effect is more efficient than the phase-mixing. Indeed, looking at Figure 1 it can be seen that when k a=0.1 the concentration of the fluctuating energy in the center of the shear is much more relevant than the stretching due to the phase-mixing.

The formation of small scales continues up to the maximum dissipated power. From this time on P exponentially decreases and more than one decay rate seems to be present. This behavior suggests that when P^+ reaches the maximum, the perturbation contains a superposition of normal modes, both resonant and resistive, which has been formed during the previous stage. The following stage then represents the exponential decay of such normal modes. An important information obtained from the simulations is the typical dissipation time τ_d defined as the time at which the perturbation energy is decreases by a factor 1/e with respect to the initial value. T his quantity increases with the wavelength and with the Reynolds number; when S increases from 10^3 to 10^4 τ_d is multiplied by a factor ≈ 2. However, τ_d it is much shorter than the corresponding time $\tau_{d,hom}$ necessary to dissipate the wave when it propagates in a homogeneous magnetic field. The values of τ_d and $\tau_{d,hom}$ are given for comparison in Table 1. Then, the presence of inhomogeneity improves the dissipation mechanisms. The inhomogeneity in the equilibrium structure gives also origin to a coupling between the fluctuations of Z^+ and Z^-. The latter are not present in the initial condition, but are generated by such effects. It appears (Figs. 2 and 3) that the Z^- fluctuations grow to a level which is higher at longer wavelengths, i.e. when the importance of the coupling terms (pressure + source) is greater relative to the phase-mixing. In this case, if the level of the perturbation is sufficiently high (run 5) effective nonlinear interactions between the Z^+ and Z^- fluctuations take place. The result is the generation of a power-law spectrum in the parallel y-direction; the corresponding formation of small scales increases the dissipation of the wave.

REFERENCES

1. Kappraff J.M. and Tataronis J.A.: J. Plasma Phys., 18, 209 (1977).

2. Mok Y. and Einaudi G.: J. Plasma Phys., 33, 199 (1985).

3. Bertin G., Einaudi G., and Pegoraro F.: Comments Plasma Physics Control. Fusion, 10, 173 (1986).

4. Hollweg J.: Ap. J., 312, 880 (1987).

5. Davila J.M.: Ap. J. , 317, 514 (1987).

6. Einaudi G. and Mok Y.: Ap. J. , 319, 520 (1987).

7. Hollweg J.: Ap. J. , 320, 875 (1987).

8. Califano F., Chiuderi C., Einaudi G.: Ap. J. , 365, 757 (1991).

9. Malara F., Veltri P. and Carbone V.:Phys. Fluids, submitted (1991).

10. Heyvaerts J. and Priest E.: Astron. Astrophys. , 127, 220 (1983).

CHARACTERISTICS OF POWER SPECTRA OF LOW FREQUENCY MAGNETIC FIELD FLUCTUATIONS IN THE TERRESTRIAL FORESHOCK

G. Mann* and H. Lühr**

* Zentralinstitut für Astrophysik, O-1501 Tremsdorf, Germany
** Institut für Geophysik der Technischen Universität, W-3300 Braunschweig, Germany

ABSTRACT

In the terrestrial foreshock high energetic ions appear in association with large amplitude low frequency magnetic field fluctuations. The characteristics of power spectra of those fluctuations are statistically investigated by inspection of 21 different high energetic ion events by means of the data of the AMPTE/IRM spacecraft. Many of the observed features of the power spectra are reflected by a recently presented strong Alfvénic turbulence model.

INTRODUCTION

Owing to the interaction of the solar wind with the bow shock of the Earth a small part of the protons are accelerated and reflected upstream at the Earth's bow shock. In the quasi-parallel region of the terrestrial foreshock they appear in the form of a diffuse nearly isotropic shell distribution in velocity space. These ions are accompanied by large amplitude magnetic field fluctuations representing an example of strong magneto-hydrodynamic turbulence in space plasmas /1, 2, 3, 4/. These fluctuations play an important role in the diffuse ion acceleration since they scatter the ions back to the bow shock /5/.

By inspection of 57 magnetic field power spectra during 21 different diffuse ion events observed by the magnetometer on board of the AMPTE/IRM satellite /6/ we found typical characteristic features of those spectra. A recently presented strong magneto-hydrodynamic turbulence model /7/ in terms of an ideal gas of Alfvén solitons is able to reproduce many of the before mentioned properties of the magnetic field power spectra especially in the low frequency region, i.e. up to one third of the proton gyrofrequency.

STATISTICAL ANALYSIS

Although each of the 21 events has its own individuality the power spectra during the event of September 13, 1985 exhibit all typical features. This is statistically confirmed by the other events. Figure 1a contains magnetic field power spectra obtained in the undisturbed solar wind. Here the wave activity is rather low. The transverse power spectrum shows a power law fall-off with a spectral index of 1.6 which is typical for the undisturbed solar wind /8/. During the diffuse ion event both the transverse and compressional wave energy are significantly enhanced over the whole frequency range as displayed in Figure 1b and 1c. In the early phase of the event the power spectrum (Fig. 1b) shows a dominant peak at $f_{max} = 40$ mHz. But it cannot be regarded as being well approximated by a power law decay. Note that only in 40% of all 57 power spectra the fall-off could reasonably be fitted by a power law with a mean spectral index of 2.66. At the frequency f_{max} of the dominant maximum the ratio between the compressional wave power and the transverse one has a mean value of 0.08. Furthermore, we derived the relationship between f_{max} and the magnitude of the magnetic field strength $|\bar{B}|$ averaged over the time interval of the spectrum, i.e. $f_{max}(mHz)/|\bar{B}|(nT)=$ 5.2 $\pm$ 1.3 (cf./9/). As in the power spectra of Figure 1b and 1c we found an enhanced wave activity at $f_w \sim 14\ f_{max} \sim 0.6$ Hz. It originates from the discrete whistler wave packets representing a common feature of upstream waves /2/.

A STRONG ALFVENIC TURBULENCE MODEL

The upstream waves around 40 mHz were identified as magnetosonic and Alfvénic waves /10/. Since their amplitudes are of the same order of magnitudes as the background magnetic field, a nonlinear treatment is necessary. Here we use the derivative nonlinear Schrödinger equation (DNLS) describing the evolution of quasi-parallel magnetohydrodynamic waves /11, 12/. Their soliton solutions are well known /11, 12, 13/.

In the recently presented strong Alfvénic turbulence model /7/ the turbulent state is assumed to be made up of an ensemble of non-overlapping circularly polarized Alfvén solitons governed by the DNLS. The resulting transverse and compressional power spectra have the form

$$P_{\perp} = \frac{N}{2\pi L}\ \frac{32\ \pi^2}{a^2 u_{max}^2}\ q^2 e^{-2q-1} \tag{1}$$

$$P_c = \frac{N}{2\pi L}\ \frac{16}{a^2 u_{max}^2}\ K_o(q)^2 \tag{2}$$

respectively. Here, K_o, u_{max} and q denote the modified Bessel function, the maximum amplitude of the soliton and a dimensionless frequency, respectively. a is given by $1/a = 4(1 - c_s^2 / v_A^2)$ with c_s as the ion sound speed and v_A as the Alfvén velocity.

DISCUSSION

The presented turbulence model is able to predict a number of observed features of upstream wave power spectra. The relation between q and the frequency f in the satellite reference frame is given by $f = q|a|u_{max}^2 \cdot v/4\pi d$ with v as the relative velocity between the wave frame and the satellite frame and d as the ion inertial length. The transverse wave power spectrum shows a broad local maximum at $q = 1$ and an exponential decay, and it exceeds the compressional one beyond 0.6 f_{max}. On the other hand the compressional power spectrum is monotonically decaying with an exponential law and it is the dominant component for $f < 0.6\ f_{max}$. The model agrees with the observations for frequencies below 0.15 Hz very well. Just that is the region of the validity of magneto-hydrodynamics, i.e. for frequencies well below the ion cyclotron frequency (near 0.4 Hz) (cf./10/).

The presented model is appropriate to describe strong magneto-hydrodynamic turbulence. But it is not able to reproduce the whole magnetic field power spectrum of upstream waves. A complete understanding of upstream wave turbulence requires the inclusion of the discrete whistler wave packets which play an important role in the energy transfer to higher frequencies.

Figure 1

The transverse (+) and compressional (x) magnetic field power
spectra in the undisturbed
solar wind (a) and during a
diffuse ion event (b and c)
on September 13, 1984. The
power spectra have been calculated over time intervals of
640 s using 0.3125 s averages.
Across the bottom the start
time of the intervals is given.
(a) The full line represents
    a power law fall off with
    a spectral index of 1.6.
    The dashed line indicates
    the spacecraft noise level.
(b) The dashed line represents
    a power law decay with
    a spectral index of 2.7.
(c) The full and dashed lines
    are fits of the theoretical-
    ly derived transverse and
    compressional wave power
    according to Eqs. (1) and
    (2).
The diffuse ion event started
at 13:03 UT and disappeared
at 13:23 UT on September 13,
1984.

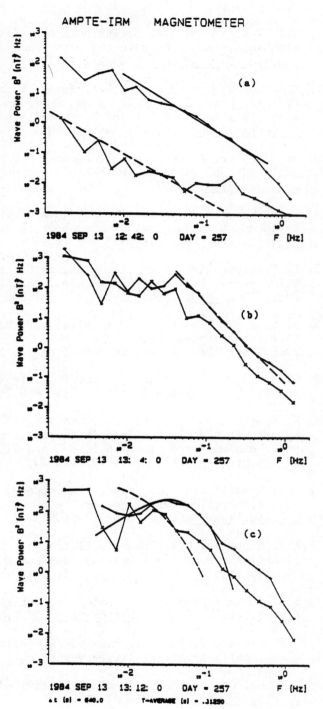

REFERENCES

/1/ G. Paschmann, N. Sckopke, S.J.Bame, J.R. Asbridge, J.T. Gosling,
D.T. Russell and E.W. Greenstadt, Association of low-frequency waves
with superthermal ions in the upstream solar wind,
Geophys. Res. Lett. 6, 209–212 (1979)

/2/ M.M. Hoppe, C.T. Russell, L.A. Frank, T.E. Eastman and E.W. Greenstadt,
Upstream bydromagnetic waves and their association with backstreaming
ion populations: ISEE 1 and 2 observations,
J. Geophys. Res. 86, 4471–4492 (1981)

/3/ J. Elaoufir, A. Mangeney, T. Passot, C.C. Harvey and C.T. Russell, Large
amplitude MHD waves in the Earth's proton foreshock.
Ann. Geophys. 8, 297–314 (1990)

/4/ G. Le and C.T. Russell, Observations of the magnetic field fluctuation
enhancement in the Earth's foreshock region,
Geophys. Res. Lett. 17, 905–908 (1990)

/5/ M. Scholer, Diffuse acceleration, in Collisionless Shocks in the
Heliosphere: Review of Current Research, B.T. Tsurutani and R.G. Stone,
eds., AGU, GM–34 Washingtion, pp. 287–302 (1985)

/6/ H. Lühr, N. Klöcker, W. Oelschlägel, B. Häusler and M. Acuna, The IRM
fluxgate magnetometer,
IEEE Trans. Geosci. Remote Sens., GE–23, 259–261 (1985)

/7/ G. Mann, On low frequency magnetic field turbulence upstream of Earth's
bow shock, Adv.Space Res. 11, (9)249–(9)252 (1991)

/8/ K. Denskat and F.M. Neubauer, Statistical properties of low frequency
magnetic field fluctuations in the solar wind from 0.29 to 1.0 AU during
solar minimum conditions: Helios I and II,
J. Geophys. Res. 87, 2215–2223 (1982)

/9/ C.T. Russell and M.M. Hoppe, The dependence of upstream wave periods
on the interplanetary magnetic field,
Geophys. Res. Lett. 8, 615–617 (1981)

/10/M.M. Hoppe and C.T. Russell, Plasma rest frame frequencies and
polarizations of low frequency upstream waves : ISEE 1 and 2
observations, J. Geophys. Res. 88, 2021–2027 (1983)

/11/S.R. Spangler and J.P. Sheerin, Properties of Alfvén solitons in a
finite beta plasma, J. Plasma Phys. 27, 193–188 (1982)

/12/C.F. Kennel, B. Buti, T. Hada and R. Pellat, Nonlinear despersive,
elliptically polarized Alfvén waves, Phys. Fluids 31, 1949–1961 (1988)

/13/G. Mann, On nonlinear circularly polarized Alfvén solitons,
J. Plasma Phys. 40, 281–287 (1988)

SURFACE AND BODY WAVES IN SOLAR WIND FLOW TUBES

G. Mann,* E. Marsch** and B. Roberts***

* Zentralinstitut für Astrophysik, O-1501 Tremsdorf, Germany
** Max-Planck-Institut für Aeronomie, W-3411 Katlenburg-Lindau, Germany
*** University of St Andrews, The Mathematical Institute, St Andrews, KY16 9SS, U.K.

ABSTRACT

In 1963 Parker already assumed that the solar wind might be fine-structured in form of flow tubes. Such spatial structures can give rise to surface and body waves with characteristic frequencies. These waves are studied here by means of the ideal magnetohydrodynamic equations. The resulting dispersion relations are discussed for typical parameters of solar wind flow tubes observed by the two HELIOS probes. These waves might be able to transport photospheric oscillations into the interplanetary space.

INTRODUCTION

Parker /1/ firstly pointed out that the plasma and the magnetic field of the solar wind might be fine-structured in the form of flow tubes, sometimes also called "spaghetti structures". They are considered to be anchored in the chromospheric network and to extend through the corona into the interplanetary space. Recently, Thieme et al. /2/ reported on some plasma signatures reminiscent of such flow tubes in high-speed solar wind streams by using data of the two HELIOS spacecrafts. Such spatial structures can support surface and body waves which could be excited by several mechanisms at the foot points of the flow tubes, for example by convective photospheric motions or perhaps by chromospheric solar oscillations. Thus, these waves could be able to transport solar atmospheric oscillations into interplanetary space.

The flow tubes are modelled as a finite–width magnetic slab with an enhanced internal flow directed along the magnetic field. The surface and body waves are treated by means of the linearized ideal magnetohydrodynamic equations. Adopting typical plasma parameters of solar wind flow tubes as observed by the two HELIOS spacecrafts /2/, the resulting dispersion relations provide typical periods of these waves in the range from 9 up to 450 minutes.

At solar wind tangential discontinuities /3/ as well as in connection with many solar phenomena (e.g. jets, helmet streamers, plasma sheets etc.) surface and body waves have already been investigated /4,5,6,7/.

DISPERSION RELATIONS OF SURFACE AND BODY WAVES

The flow tube is considered to be a magnetic slab with a finite width of 2a in x direction. The slab is infinitely extended in y and z direction. Both the background magnetic field and the flow velocity are directed along the axis of the slab (z axis). In order to derive the dispersion relations of surface and body waves in such flow tubes we employ the linearized ideal magnetohydrodynamic equations supplemented by an isotropic equation of state, and we require continuity of the normal component of the fluid velocity and of the total gas pressure across the slab boundaries. Furthermore, gravitational effects are ignored. Then, the linearized magnetohydrodynamic equations yield the differential equation

$$0 = \frac{d^2 v_{1x}}{dx^2} + m^2 v_{1x} \tag{1}$$

with $\tilde{v}_{1x} = v_{1x}(x) exp[i(kz - \omega t)]$. Here, $\tilde{v}_{1x}$ denotes the x-component of the disturbance of the fluid velocity. The ansatz

$$v_{1x}(x) = \begin{cases} e^{-n_e(x-a)} & x \geq a \\ A sinh(n_i x) & -a \leq x \leq +a \\ -e^{n_e(x+a)} & x \leq -a \end{cases} \tag{2}$$

for the symmetric (sausage) surface wave ($m_i^2 < 0$) and

$$v_{1x}(x) = \begin{cases} e^{-n_e(x-a)} & x \geq a \\ A sin(m_i x) & -a \leq x \leq +a \\ -e^{-n_e(x+a)} & x \leq -a \end{cases} \tag{3}$$

for the symmetric body wave ($m_i^2 > 0$) leads to the following dispersion relation

$$\frac{\rho_{oi}}{\rho_{oe}} \frac{n_e}{n_i} \frac{[(u - v_{oi})^2 - v_{Ai}^2]}{[(u - v_{oe})^2 - v_{Ae}^2]} \frac{(u - v_{oe})}{(u - v_{oi})} = \begin{cases} -tanh(n_i a) & \text{(sausage)} \\ -coth(n_i a) & \text{(kink)} \end{cases} \tag{4}$$

for the symmetric (sausage) and antisymmetric (kink) surface waves and

$$\frac{\rho_{oi}}{\rho_{oe}} \frac{n_e}{m_i} \frac{[(u - v_{oi})^2 - v_{Ai}^2]}{[(u - v_{oe})^2 - v_{Ae}^2]} \frac{(u - v_{oe})}{(u - v_{oi})} = \begin{cases} +tan(m_i a) & \text{(sausage)} \\ -cot(m_i a) & \text{(kink)} \end{cases} \tag{5}$$

for the corresponding body waves with

$$m_j^2 = -n_j^2 = k^2 \frac{[(u - v_{oj})^2 - v_{Aj}^2][(u - v_{oj})^2 - c_{sj}^2]}{(v_{Aj}^2 + c_{sj}^2)[(u - v_{oj})^2 - c_{Tj}^2]} \tag{6}$$

and $c_{Tj}^2 = c_{sj}^2 v_{Aj}^2 / (v_{Aj}^2 + c_{sj}^2)$. Here, c_{sj}, v_{Aj} and v_{oj} denote the external ($j = e$) and internal ($j = i$) sound speed, Alfvén velocity, and flow velocity, respectively. $u = \omega/k$ is the phase velocity with the circle frequency ω and the wave number k. The ansatz for the antisymmetric modes are chosen in an equivalent manner as in Eqs. (2) and (3). Here, the waves are divided in a symmetric and antisymmetric mode with respect to the symmetry properties of the density perturbations of the wave. Note that the body waves are connected with spatially oscillating disturbances within the slab in contrast to the surface waves. For vanishing flow velocities ($v_{oj} = 0$) the dispersion relations of Eqs. (1) and (2) take the well-known form given by Edwin and Roberts /8/.

DISCUSSION

In Figure 1 the dispersion relations of Eqs. (4) and (5) are illustrated for typical parameters of solar wind flow tubes as reported by Thieme et al. /2/. These parameters provide the following relation between the different speeds.

$$c_{Ti} < c_{Te} < c_{si} = v_{Ai} < c_{se} < v_{Ae} << v_{oe} < v_{oi}$$

The Figure 1 shows that the surface and body waves are existing as freely propagating modes in discrete frequency and wave number bands. The corresponding frequency f can be determined by $f = U \mid ka \mid v_{oe}/(2\pi a)$. ($U = u/v_{oe}$). Adopting a half width of the flow tube of $a = 1.6 \cdot 10^6$ km one finds the following characteristic frequency bands as summarized in the Table 1. Such a tube width corresponds to an angular width of $1°$ at 1 AU (cf. Thieme et al. /2/).

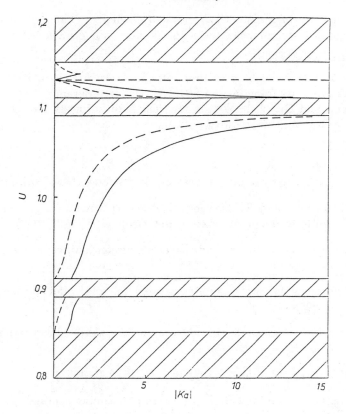

Fig. 1. Dispersion relation (phase speed $U = u/v_{oe}$ as a function of $|ka|$) of the surface and body waves in solar wind flow tubes according to the Eqs. (4) and (5) using the following parameters: $v_{oi} = 750$ km/s, $v_{oe} = 650$ km/s, $v_{Ai} = c_{si} = 65$ km/s, $v_{Ae} = 100$ km/s and $c_{se} = 70$ km/s /2/. Surface waves appear only in the interval $1.13 < U < 1.15$. All other modes are body waves. The full and dashed lines represent the symmetric (sausage) and antisymmetric (kink) modes, respectively.

TABLE 1   Frequency bands deduced from the results of Figure 1 with $a = 1.6 \cdot 10^6$ km, with sb, sausage body mode; kb, kink body mode; ss, sausage surface mode, ks, kink surface mode.

phase speed band			frequency band ($f[10^{-5}$ Hz])			mode
0.85	< U <	0.89	3.72	< f <	8.97	sb
			0	< f <	4.48	kb
0.91	< U <	1.09	4.97	< f <	190.0	sb
			0	< f <	127.0	kb
1.11	< U <	1.13	0	< f <	93.5	sb
			0	< f <	46.8	kb
1.13	< U <	1.15	0	< f <	8.74	ss
			0	< f <	9.77	ks

Thus, the discussed waves are accompanied with typical periods in the range between 9 and 450 minutes. Such periods are typical for convective photospheric motions and chromospheric oscillations. Just that is the region where the solar wind flow tubes are anchored. Therefore, the presented surface and body waves are able to transport such oscillations into the solar wind.

Of course, we have only discussed a simple model of solar wind flow tubes. We restricted ourselves to a slab geometry. But, Edwin and Roberts /9/ have shown that surface and body waves have similar quantitative properties in a cylindrical and slab geometry. Thus, we believe that the presented results are qualitatively independent on special flow tube geometries.

ACKNOWLEDGEMENT

One of the authors (G. Mann) is indebted to Prof. Dr. E. Priest for his invitation to visit the University of St. Andrews in the frame of the S.E.R.C. visitor program. He thanks Prof. Dr. E. Priest, Dr. B. Roberts and Dr. P. M. Edwin for their kind hospitaly during his stay in St. Andrews where a part of this work was carried out.

REFERENCES

1. E. N. Parker, Interplanetary dynamical processes, *Interscience*, New York (1963)

2. K. M. Thieme, E. Marsch, and R. Schwenn, Spatial structures in high-speed streams as signatures of fine structures in coronal holes, *Ann. Geophys.* 8, 713 (1990)

3. J. V. Hollweg, Surface waves on solar wind tangential discontinuities, *J. Geophys. Res.* 87, 8065 (1982)

4. J. F. McKenzie, Hydromagnetic wave coupling between the solar wind and the plasma sheet, *J. Geophys. Res.* 76, 2958 (1971)

5. A. S. Narayanan and K. Somasunderam, Alfvén surface waves along coronal streamers, *Solar Phys.* 109, 357 (1985)

6. B. Roberts, On MHD solitons in jets, *Astrophys. J.* 318, 590 (1987)

7. L. C. Lee, S. Wang, C. Q. Wei, and B. T. Tsurutani, Streaming sausage, kink and tearing instabilities in a current sheet with applications to the Earth's magnetotail, *J. Geophys. Res.* 93, 7354 (1988)

8. P. M. Edwin and B. Roberts, Wave propagation in a magnetically structured atmosphere, *Solar Phys.* 76, 239 (1982)

9. P. M. Edwin and B. Roberts, Wave propagation in a magnetic cylinder, *Solar Phys.* 88, 179 (1983)

INTRODUCTION TO KINETIC PHYSICS, WAVES AND TURBULENCE IN THE SOLAR WIND

E. Marsch

Max-Planck-Institut für Aeronomie, W-3411 Katlenburg-Lindau, Germany

ABSTRACT

This paper is meant to set the stage for some of the following presentations on kinetic physics, waves and turbulence in the solar wind. Firstly, a summary of some key observations is given, and the nature and possible origin of the fluctuations are discussed. Secondly, emphasis is placed on the kinetic aspects of the dissipation of turbulence and the related heating of the protons. Relevant features of the velocity distributions reflecting the wave-particle interactions are shortly discussed. Then some modern topics of turbulence such as intermittency, multifractals, selfsimilar scaling and observations of the structure function are dealt with. Finally, we address theoretical issues and problems associated with models based on two-scale energy transfer equations that have been proposed to describe the spatial and spectral evolution of MHD turbulence in the inhomogeneous solar wind.

INTRODUCTION

Kinetic physics, waves and turbulence in the solar wind is a vast subject that encompasses numerous original reasearch articles and reviews written during more than three decades of in-situ measurements in interplanetary space. Of course, we can not do justice even to the most recent developements in this reasearch field in such short an introduction, and thus at the outset we apologize for omissions and refer to the literature. This is, for the last few years, fully covered in the review by Roberts and Goldstein /1/. Another recent review by Marsch /2/ summarizes and evaluates the Helios results from the inner heliosphere and the theoretical attempts to describe them. Mangeney et al. /3/, Hollweg /4/ and Zhou and Matthaeus /5/ dwell on various crucial nonlinear and transport aspects of interplanetary fluctuations and other debated issues of the turbulence models developed by Marsch and Tu /7/, Tu and Marsch /8/, Zhou and Matthaeus /9,10/ and Velli et al./11,12/. Here we will not address these problems extensively, but instead concentrate on those observations providing evidence for turbulence dissipation and on some related theoretical aspects of kinetic plasma physics. Also, data analyses involving the structure function for a discussion of spatial intermittency of turbulence are briefly reviewed.

SUMMARY OF OBSERVATIONS

The solar wind is not the spherically symmetric, stationary expanding plasma of the theory; observations show a broad band of fluctuations in the plasma and electromagnetic properties of the wind, ranging from the large-scale (scales of days) stream structure, meso-scale (scales of hours to days) plasma flow tubes /13/ and convected magnetic structures /14/ to the small-scale (scales of hours and less) Alfvénic fluctuations. The most important conclusion of the many studies /1,2/ carried out in the eighties is that these fluctuations are not only remnants of coronal processes but do in fact dynamically and nonlinearly evolve while propagating in the expanding solar wind. The solar rotation plays an important role by coupling, at a sufficiently large distance from the sun, fluctuations at disparate scales. As a result, interplanetary velocity shears /16/, dilution and compression effects associated with the stream interactions in the vicinity of the ecliptic plane appear to affect strongly the MHD turbulence at smaller scales. Since the solar rotation is diminished at high latitudes, some recent papers /17-19/ also speculate about a different turbulent state of the solar wind over the suns polar regions.

The existence of meso-scale (lasting several hours and longer) magnetic field and plasma structures has recently been established convincingly by means of various data sets. Spatial structures in high-speed streams have been interpreted as signatures of coronal fine structures, such as plumes in polar coronal holes (Thieme et al. /13/). These flow tubes have typical angular diameters of 2-3 degrees, are fairly well pressured balanced and tend to be washed out near the earths orbit. Pressure balanced structures are visible by their distinct anti-correlation between the fluctuations in thermal and magnetic pressure and are ubiquitously observed in the outer heliosphere /20/. The nature of these fluctuations is still an issue and under investigation. Theoretical schemes of nearly incompressible magnetohydrodynamics have been advanced /21,22/, leading to an understanding of density fluctuations in terms of pseudosound and of correlated density-temperature fluctuations /23/. Power spectra of fluctuations of solar wind density and temperature are available for the Helios orbital range /24,25/. They generally indicate Kolmogorov-type slopes with flattening in the higher-frequency range. It is clear that much more data analyses and theoretical studies are needed to comprehend fully the nature of compressive fluctuations in the solar wind.

Distinctive planar magnetic structures have been detected /26/, in which the magnetic field changes in directions parallel to planes inclined to the ecliptic. Such events usually last for several hours and occur near the heliospheric sector boundaries or current sheet. The planes contain the spiral direction. They have been considered as belonging to magnetic tongues extending from coronal loops into interplanetary space and appear to be clearly different from magnetic clouds /27/. They more resemble the magnetic field directional turnings /14/, i.e. structures that have been identified by their comparatively low cross helicity and preponderance of magnetic over kinetic turbulent energy. These structures are convected in the wind and are pressure balanced and unaffected by compressive fluctuations. Their power spectra have a -5/3 slope, indicating nonlinear turbulent evolution. Structures mostly constitute the so-called background spectrum of the inward oriented Elsässer field at the meso-scales /28,16/, to which yet other compressive low-cross-helicity regions in the inner solar wind may also contribute /29/.

Depending on the coronal magnetic field topology (open or closed fields) and the related structure of solar wind streams, there occur, with varying intensity, high-frequency fluctuations mainly of solar origin which are superimposed on and intermingled with the convected structures. These Alfvénic fluctuations have extensively been studied in the past years, in particular by means of Elsässer variables that naturally decompose the fluctuations into their sunward and earthward propagating parts /6,15,24,30,31/. Here we will only shortly summarize some observational findings. The Fig. 1 provides an overview of the variations in the turbulence pattern with wind speed: Top panel, energy e^+ of outward mode, then sound speed c_s, energy of the inward mode and finally the relative density fluctuation $\delta n/n$, plotted as daily averages versus time for the Helios 1 primary mission (after Grappin et al. /15/). Nine frequency bands correspond to $f_n = \text{day}^{-1}$ 2^{-n}; e.g. $f_9 = 5.9 \times 10^{-3}$ Hz. Note the striking correlation between e^+ and c_s and between e^- and $\delta n/n$. Apparently, the variations are selfsimilar, although the turbulent energies may change by orders of magnitude as a function of time and stream structure. In short, the observed spatial and spectral features can be broken in two classes characterised by the coronal magnetic field being open or closed (coronal hole or streamer belt near activity minimum), or by the solar wind speed being high (or low): Alfvén waves, yes (no); turbulence level high (low); excess of magnetic over kinetic turbulent energy, no (yes); spectral slope of e^+ flat (steep), exponent equals -1 (-5/3); density fluctuations weak (intense); proton temperature high (low); inferred outer coronal heating strong (weak); interplanetary heating by a turbulent cascade strong (weak). A concise overview of the radial evolution of the turbulence spectra is given by Fig. 3 of Tu et al. /31/.

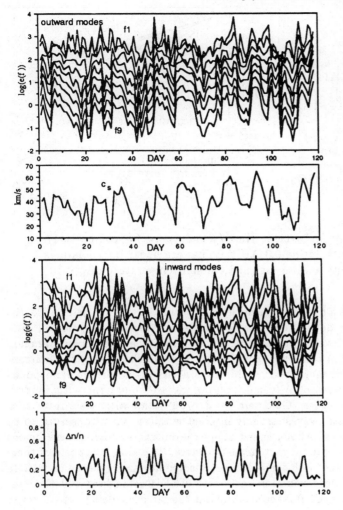

Fig. 1   Overview of the turbulence characteristics as observed by Helios 1 during the first 120 days of the mission. From top to bottom: Turbulent energy e^+ of the outward mode, sound speed given by the proton thermal speed, energy e^- of the inward mode, and finally rms value of the relative density fluctuations. Turbulent energies are given in nine octaves corresponding to the frequency bands f_n = day$^{-1}/2^n$, with n = 0,1,..,9. Individual points give daily averages plotted versus time in days. Note the selfsimilar variations of the turbulent intensities and the close correlations of e^+ with c_s and e^- with $\delta n/n$, particularly at high and low frequencies, respectively.

Spectral slopes and proton temperature are closely interrelated. The spectral index m^+ of e^+ in a simple power-law fit like $e^\pm \propto f^{-m^\pm}$ is given by $m^+ \approx 1.67 - \log(c_s/c_s^*)$ with reference speed $c_s^* = 21.5$ km/s (see Grappin et al. /15/). This relation well interpolates between the various slopes obtained in ref./24/ on a statistical basis of many individual spectra. Fig. 2 shows the daily variations /30/ of m^+ and m^- and c_s in the left panel (a) and right panel (b) the proton temperature gradients /33/ for different solar wind speeds ranging from 800 km/s ($m^+ = 1.2$, $c_s = 63$ km/s) down to 300 km/s ($m^+ = 1.6$, $c_s = 16$ km/s). To the right of 0.3 AU, the profiles are measured and represent the average proton temperature gradients. To the left, the lines are inward interpolations between the interplanetary temperatures at 0.3 AU and an assumed coronal temperature of two million degrees. This figure emphasizes the close connection between turbulence evolution and thermodynamics, as it was earlier pointed out in Tu's solar wind model of the damping of interplanetary Alfvénic fluctuations /32/.

DISSIPATION AND KINETIC HEATING

Observational evidence exists on the damping of the fluctuations that manifests itself in nonadibatic temperature profiles of the ions as shown in Fig. 2(b). Yet, the details of the underlying nonlinear microscopic physics have to be worked out, and the associated wave-particle interactions are not fully understood. Schwartz and Marsch /34/ have established, by exploiting the radial line-up configurations of the two Helios spacecraft, that the proton magnetic moment is not conserved but tends to increase radially. Tu /32/ has demonstrated that the measured trends can be reproduced in turbulence models if the turbulent dissipation rate, which is equated with

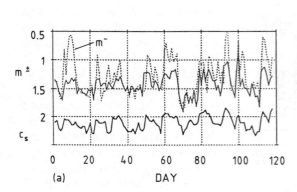

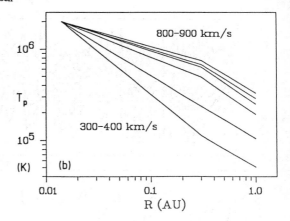

Fig. 2   (a) Spectral exponents $m^{\pm}$ versus time (sound speed for comparison);
(b) Proton temperature gradients; inside 0.3 AU the lines are linear interpolations
between the Helios measurements and an assumed coronal temperature of 2×10^6
K.

the nonlinear cascading rate, is used as a heating rate in the ion thermal energy equation. Clearly,
the precise heating rate will sensitively depend on the velocity distribution in the cyclotron resonance regime /35/ and the fluctuation level of the relevant kinetic waves near the ion gyrofrequencies. Sparse observations are available in this frequency domain /2/, which indicate the frequent
ocurrence of field-aligned cyclotron and magnetoacoustic waves with left- and right-hand polarization. Recent magnetic helicity spectra of Alfvénic fluctuations are consistent with random
circular polarization also at all MHD scales (Goldstein et al. /36/) and with scale invariance
from 10^{-5} to 0.1 km^{-1}. Magnetoacoustic waves strongly interact with ion beams observed to be
drifting at about 1.5 to 2 times the local Alfvén speed along the magnetic field. These beams
are marginally unstable and prone to excite magnetoacoustic waves. The kinetic aspects of these
phenomena and of instabilities associated with temperature anisotropies are reviewed in /2/. In
conclusion, we may say that the kinetic dissipation regime of MHD fluctuations and the subtle nonlinear effects involved certainly need further studies and are far from being satisfactorily
understood.

STRUCTURE FUNCTION AND SPATIAL INTERMITTENCY

Aside of spectral analysis there is an equivalent method for studying fluctuations, which is to
investigate directly the velocity differences of the solar wind between two points in space and
time. Burlaga /37/ pioneered the use of the structure function in order to seek for multifractals
and spatial intermittency of the speed fluctuations. This is demonstrated in Fig. 3 showing
Voyager 2 data obtained at 8.5 AU. The structure function reads:

$$S^p(\tau) = <\mid V(t+\tau) - V(t) \mid^p> \tag{1}$$

For turbulence with gaussian statistics a universal scaling law $S^p(\tau) \propto \tau^{s(p)}$ should hold with
$s(p) = p/3$ for a -5/3 Kolmogorov spectrum. In case of spatial intermittency each p-th order
structure function may have its own index and contains new information. For example, if the
turbulent eddies are not space-filling but at all scales only occupy a fraction of a volume element
given by the factor $2^{(D-3)}$, then $s(p) = 3-D+p(D-2)/3$ with the fractal dimension D. The Fig. 3
gives the logarithm of S^p versus time lag τ for the range from 0.85 to 13.6 hours. Note that a
scaling law applies but the exponent $s(p)$ is not a linear function of p but is less steeply increasing.
This departure was concluded to show that interplanetary turbulence is indeed intermittent. In
the observational section we have discussed different convected structures and coherent wave
features that do in fact not comply with the notion of developed gaussian turbulence. The
departure from Kolmogorov scaling becomes visibly evident from higher-order correlations but is

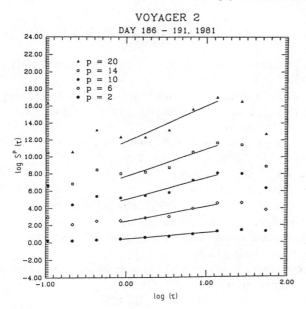

VOYAGER 2
DAY 186 – 191, 1981

Fig. 3 Logarithm of the p-th order structure function $S^p(\tau)$ of the solar wind speed versus time lag τ for selected values of the selfsimilarity scaling exponent p. Note the linear least-squares fits indicating that a power law describes the temporal dependence of $S^p(\tau)$ in the time range $0.85 \leq \tau \leq 13.6$ hours. The exponents are $s(p) = 1.68$, 2.52, 3.24 for $p = 6,10,14$, a result clearly showing a departure from the nonintermittent Kolmogorov result $s(p) = p/3$.

less clear from the autocorrelation function that is equivalent to the spectrum. The recent work of Burlaga /37/ opens in my mind a very promising avenue to analyse and understand solar wind turbulence from a new theoretical vantage point. Static structure functions have even been calculated analytically /38/ for Navier-Stokes flows. This approach may also be useful for MHD turbulence. Possible connections between intermittent turbulence and deterministic chaos have recently been investigated (see e.g. reference /39/). We are still waiting for applications of these modern concepts of chaos theory to solar wind MHD fluctuations.

SPECTRAL TRANSFER EQUATIONS

Considerable efforts have been made to model theoretically the observed turbulence /4,5,7-12,32/, following the early ideas of Coleman /40/. The various approaches have been hampered by difficulties in the physical interpretation of the cross-correlation between the two Elsässer fields or Alfvén waves that propagate oppositely in the inhomogeneous solar wind. Large-scale inhomogeneity leads to a partial reflection of waves coming from the corona. Hollweg /4/ contributed to a clarification of this issue by showing that the resulting corrections are merely of second order in the smallness parameter wavelength over density scale height; thus the WKB scheme for waves remains essentially unaffected. However, the structures discussed before and convected small-scale turbulence also have to be incorporated in the models. Unlike Alfvén waves the structures may have significant departures from equipartition of kinetic and magnetic energy /14/, and thus theoretically the residual energy can be a leading order term /5/, and consequently mixing terms become important. There remains substantial ambiguity in the current interpretation of the nature of the observed fluctuations in terms of waves and/or structures. Tu and Marsch /41/ offer in this volume a new interpretation and try to develop a unifying theoretical scheme for solar wind turbulence. Matthaeus et al. /42/ recently published observational evidence for the presence of quasi-two-dimensional nearly incompressible fluctuations that have wave vectors nearly transverse to the mean magnetic field. Much work, of course, remains to be done to achieve what could be considered a consistent model. In particular, many open problems remain as to the nature of the compressive fluctuations and to their role in the transport of the overall turbulent energy.

REFERENCES

1. Roberts, D.A., and M.L. Goldstein, *Reviews of Geophysics*, Vol. 96, Supplement, 932 (1991)

2. Marsch, E., in *Physics of the Inner Heliosphere*, Vol. 2, R. Schwenn and E. Marsch (Eds.), Springer-Verlag, Heidelberg, 159 (1990)

3. Mangeney, A., R. Grappin, and M. Velli, in *Advances in Solar System Magnetohydrodynamics*, E.R. Priest and A.W. Hood (Eds.), Cambridge University Press, p. 344 (1991)

4. Hollweg, J.V., *J. Geophys. Res.*, 95, 14873 (1990)

5. Zhou, Y., and W.H. Matthaeus, *J. Geophys. Res.*, 95, 14863 (1990)

6. Tu, C.-Y., E. Marsch, and K.M. Thieme, *J. Geophys. Res.*, 94, 739 (1989)

7. Marsch, E., and C.-Y. Tu, *J. Plasma Phys.*, 41, 479 (1989)

8. Tu, C.-Y., and E. Marsch, *J. Plasma Phys.*, 44, 103 (1990)

9. Zhou, Y., and W.H. Matthaeus, *Geophys. Res. Lett.*, 16, 755 (1989)

10. Zhou, Y., and W.H. Matthaeus, *J. Geophys. Res.*, 95, 14881 (1990)

11. Velli, M., R. Grappin, and A. Mangeney, in *Plasma Phenomena in the Solar Atmosphere*, M.A. Dubois (Ed.), Editions de Physique, Orsay (1989)

12. Velli, M., R. Grappin, and A. Mangeney, *Proc. NATO ARW "Numerical modeling of solar and stallar MHD"*, Computer Phys. Comm., in press (1991)

13. Thieme, K.M., R. Schwenn, and E. Marsch, *Ann. Geophys.*, 8, 713 (1990)

14. Tu, C.-Y., and E. Marsch, *Ann. Geophys.*, 9, 319 (1991)

15. Grappin, R., A. Mangeney, and E. Marsch, *J. Geophys. Res.*, 95, 8197 (1990)

16. Roberts, D.A., *this volume* (1991)

17. Roberts, D.A., *Geophys. Res. Lett.*, 17, 567 (1990)

18. Jokipii, J.R., and J. Kota, *Geophys. Res. Lett.*, 16, 1 (1989)

19. Roberts, D.A., L.W. Klein, M.L. Goldstein, and W.H. Matthaeus, *J. Geophys. Res.*, 92, 11021 (1987)

20. Roberts, D.A., M.L. Goldstein, L.W. Klein, and W.H. Matthaeus, *J. Geophys. Res.*, 92, 12023 (1987)

21. Montgomery, D., M. Brown, and W.H. Matthaeus, *J. Geophys. Res.*, 92, 282, (1987)

22. Matthaeus, W.H., L.W. Klein, S. Gosh, and M.R. Brown, *J. Geophys. Res.*, 96, 5421 (1991)

23. Zank, G.P., W.H. Matthaeus, and L.W. Klein, *Geophys. Res. Lett.*, 17, 1239 (1990)

24. Tu, C.-Y., E. Marsch, and H. Rosenbauer, *Ann. Geophys.*, 9, 748, (1991)

25. Marsch, E., and Tu, C.-Y., *J. Geophys. Res.*, 95, 11945 (1990)

26. Nakagawa, T., A. Nishida, and T. Saito, *J. Geophys. Res.*, 94, 11761 (1989)

27. Burlaga, L.F.E., in *Physics of the Inner Heliosphere*, Vol.2, R. Schwenn and E. Marsch (Eds.), Springer-Verlag, Heidelberg, 1 (1991)

28. Tu, C.-Y., and E. Marsch, *J. Geophys. Res.*, 95, 4337 (1990)

29. Bruno, R. and R. Bavassano, *J. Geophys. Res.*, 96, 7841 (1991)

30. Grappin, R., A. Mangeney, and M. Velli, *Ann. Geophys.*, 9, 416 (1991)

31. Tu, C.-Y., E. Marsch, and H. Rosenbauer, *Geophys. Res. Lett.*, 17, 283, (1990)

32. Tu, C.-Y., *J. Geophys. Res.*, 93, 7 (1988)

33. Thieme, K.M., E. Marsch, H. Rosenbauer, *J. Geophys. Res.*, 94, 2673 (1988)

34. Schwartz, S.J., and E. Marsch, *J. Geophys. Res.*, 88, 9919 (1983)

35. Marsch, E., C.K. Goertz, and K. Richter, *J. Geophys. Res.*, 87, 5030 (1982)

36. Goldstein, M.L., D.A. Roberts, and C.A. Fitch, *Geophys. Res. Lett.*, 18, 1505 (1991)

37. Burlaga, L.F., *Geophys. Res. Lett.*, 18, 1651 (1991); *J. Geophys. Res.*, 96, 5847 (1991)

38. Effinger, H., and S. Grossmann, *Z. Phys. B – Condensed Matter*, 66, 289 (1987)

39. Eggers, J., and S. Grossmann, *Phys. Fluids*, A3, 1958 (1991)

40. Coleman, P.J., *Astrophys. J.*, 153, 371 (1968)

41. Tu, C.-Y., and E. Marsch, *this volume*, (1991)

42. Matthaeus, H.W., M.L. Goldstein, and D.A. Roberts, *J. Geophys. Res.*, 95, 20673 (1991)

ELECTRIC FIELD FLUCTUATIONS AND POSSIBLE DYNAMO EFFECTS IN THE SOLAR WIND

E. Marsch* and C. Y. Tu*,**

*Max-Planck-Institut für Aeronomie, W-3411 Katlenburg-Lindau, Germany
**Department of Geophysics, Peking University, Beijing 100871, China

ABSTRACT

Magnetohydrodynamic fluctuations in different kinds of solar wind have been investigated. Electric field fluctuation spectra have been obtained from the observed velocity and magnetic field fluctuations. The mean electromotive force $\mathcal{E}$, generated by the turbulent motion of the solar wind plasma and field, turns out to depend upon the nature and Alfvénicity of the fluctuations. Dynamo theory predicts a linear relationship between $\mathcal{E}$ and the mean magnetic field $\mathbf{B}_o$. Correlation studies carried out with the intention to establish this so-called alpha effect have given negative results.

INTRODUCTION

Magnetohydrodynamic fluctuations in the solar wind have been investigated in much detail and in the recent past quite successfully described in the terms of MHD turbulence phenomenology and theory (for recent reviews see e.g. Marsch /1/, Mangeney et al. /2/, Roberts /3/ and further references therein). Whereas much effort has been spent on analysing and understanding the nature and origin of the fluctuations and their energy and helicity spectra, almost no attention has been payed to the fluctuation electric fields and their related power spectra, let alone to the question of a possible turbulent dynamo in the solar wind and interplanetary medium. The present paper will briefly address this problem. The effect that the Alfvén wave pressure has on the background solar wind flow has been thoroughly studied (see e.g. the review by Leer et al., /4/) and its prominent role in driving high-speed flows is generally accepted. Likewise, one may expect an electromotive force $\mathcal{E}$ to be associated with the fluctutions. This term would then show up in the induction equation in addition to the average convection electric field, a result that has been worked out in the framework of mean-field electrodynamics as described in the classical monographs by Moffatt /5/ and Krause and Rädler /6/ on this subject and on dynamo theory.

ELECTRIC FIELD FLUCTUATIONS AND THE ELECTROMOTIVE FORCE $\mathcal{E}$

Mean-field electrodynamics is based on the assumpion that a two-scale decomposition of the flow velocity and magnetic field is meaningful in such a way that the average magnetofluid can be described by mean fields varying on the large temporal and spatial scales T and L. Superposed on these are small-scale random field components that strongly vary on the small scales of time t and space ℓ (of the order of the correlation time and length of the fluctuations). Thus we decompose the fields of the magnetofluid such that

$$\mathbf{V} = \mathbf{V}_o + \delta\mathbf{V}, \qquad \mathbf{V}_o = <\mathbf{V}>, \qquad <\delta\mathbf{V}> = 0 \tag{1a}$$

$$\mathbf{B} = \mathbf{B}_o + \delta\mathbf{B}, \qquad \mathbf{B}_o = <\mathbf{B}>, \qquad <\delta\mathbf{B}> = 0 \tag{1b}$$

represent the full velocity and magnetic fields which are then inserted in the standard MHD equations. The mean fields are given by appropriate ensemble averages, which are indicated by brackets (for a discussion of this subject see the papers by Matthaeus et al. /7/ and Matthaeus and Goldstein /8,9/). In the ensemble averages of the momentum and induction equations only terms of second order in the fluctuations survive. These represent the turbulent stresses $\underline{\underline{T}}$ and electromotive force $\mathcal{E}$, given by the expressions:

$$\underline{\underline{T}} = \; < \rho_o \delta\mathbf{V}\delta\mathbf{V} - \frac{1}{4\pi}\delta\mathbf{B}\delta\mathbf{B} + \frac{1}{4\pi}\delta B^2 \underline{\underline{1}} > \tag{2}$$

$$\mathcal{E} = \; < \delta\mathbf{V} \times \delta\mathbf{B} > \tag{3}$$

For the sake of simplicity we assumed incompressible fluctuations with plasma density ρ_o in deriving (2). Note that in the case of Alfvénic fluctuations the well known correlation holds $\delta\mathbf{V} = \pm\delta\mathbf{B}/\sqrt{4\pi\rho_o}$. Consequently, the stress tensor reduces to the Alfvén wave pressure $\underline{\underline{P}}_A = \underline{\underline{1}}\frac{1}{8\pi} < \delta B^2 >$ and there is no turbulent electric field $\mathcal{E}$, since the fluctuations of velocity and magnetic field are aligned.

In general, the full stress tensor $\underline{\underline{T}}$ has to be added to the kinetic and magnetic stresses acting on the mean flow. Here we only quote Faradays law reading

$$\frac{\partial}{\partial t}\mathbf{B}_o = \boldsymbol{\nabla} \times (\mathbf{V}_o \times \mathbf{B}_o) + \lambda\boldsymbol{\nabla}^2\mathbf{B}_o + \boldsymbol{\nabla} \times \mathcal{E} \tag{4}$$

where the magnetic diffusivity is denoted by λ. Note that due to rare collisions the classical conductivity is numerically rather high and thus λ rather low in the solar wind, but its applicability is questionable (Montgomery /10/). For the relevant anomalous transport processes see the recent review by Marsch /11/. The electric field $\mathbf{E}'$ in the rest frame of the moving plasma is given by

$$\mathbf{E}' = \mathbf{E} + \frac{1}{c}(\mathbf{V} \times \mathbf{B}) \tag{5}$$

and vanishes for an ideal conductor. Inserting (1a,b) in (5) and ensemble averaging yields

$$\mathbf{E}_o = -\frac{1}{c}(\mathbf{V}_o \times \mathbf{B}_o) - \frac{1}{c}\mathcal{E} \tag{6}$$

Here the first term defines the normal convection electric field in the mean-flow velocity frame, in which now also a turbulent electromotive force $\mathcal{E}$ exists. This may lead to a "turbulent diffusion" of the mean magnetic flux such that the mean field lines are enabled to slip through the mean flow, and thus the frozen-in-field law does not hold any more in this frame.

The turbulent EMF $\mathcal{E}$ will also affect the spatial and temporal evolution of the spectral energy of the fluctuations. Theoretical models and spectral transfer equations have recently been proposed by various authors (Zhou and Matthaeus /12,13/; Marsch and Tu /14/; Tu and Marsch /15/; Velli et al. /16/) to describe the fluctuations in terms of the Elsässer fields

$$\mathbf{Z}^{\pm} = \mathbf{V}_o \pm \mathbf{V}_{Ao} \tag{7}$$

where $\mathbf{V}_{Ao} = \mathbf{B}/\sqrt{4\pi\rho_o}$ is the Alfvén velocity. The EMF in these variables is given by

$$\mathcal{E} = -\sqrt{\pi\rho_o} < \delta\mathbf{Z}^+ \times \delta\mathbf{Z}^- > \tag{8}$$

The electromotive force $\mathcal{E}$ is closely related to the general correlation tensor

$$\underline{\underline{Q}}^{\alpha\beta}(\mathbf{x},t;\boldsymbol{\xi},\tau) = <\delta\mathbf{Z}^\alpha(\mathbf{x}-\frac{1}{2}\boldsymbol{\xi},t-\frac{1}{2}\tau)\delta\mathbf{Z}^\beta(\mathbf{x}+\frac{1}{2}\boldsymbol{\xi},t+\frac{1}{2}\tau) > \tag{9}$$

(see Marsch and Tu /14/ for more details) and can be expressed by its nondiagonal elements. In the transfer equations these are associated with the turbulence sources related to the large-scale inhomogeneity and to the shear of $\mathbf{B}_o$ and $\mathbf{V}_o$ and with the vorticity $\nabla \times \mathbf{V}_o$ and current $\nabla \times \mathbf{B}_o$. These source terms play, besides the nonlinearities, a prominent role in the evolution of solar wind MHD turbulence and of the turbulent energies

$$e^\pm = \frac{1}{2} < (\delta\mathbf{Z}^\pm)^2 > \tag{10}$$

of the Elsässer fields. Quite comprehensive interplanetary studies (see the above cited reviews) now exist on the phenomenology and spectral characteristics of $e^\pm$, and also numerical simulations (e.g. Grappin et al. /17,18/) have been carried out to understand their nonlinear evolution. Lack of space prohibits us to present a thorough analysis of $\mathcal{E}$ as a function of interplanetary stream structure and magnetic field topology. We shall therefore concentrate on some results obtained in fast solar wind flows.

TURBULENT ELECTRIC FIELD SPECTRA

Tu et al. /19/ and Grappin et al. /20/ have shown that the $e^\pm$ spectra do systematically vary as a function of the solar wind stream structure during periods of solar activity minimum. Similarly, the spectra of electric field fluctuations vary when calculated by means of the Fourier transform of the autocorrelation function of $\mathcal{E}$ according to the definition (3). Generally speaking, the spectra of $\mathcal{E}$ are more fully developed toward a Kolmogorov-type spectrum in the slow wind associated with the heliospheric current sheet. The $\mathcal{E}$ spectra are not positive by definition but may oscillate as a function of frequency or wavevector. However, their positive modular envelopes show evolutionary trends in their shapes and slopes which are similar to the $e^\pm$ spectra.

In Fig. 1 we show representative energy spectra for a fast solar wind (637 km/s) near solar maximum in 1980 from Helios observations at 0.53 AU. Clearly e^+ dominates e^-, indicating Alfvénic fluctuations with a developed turbulent cascade and slope close to the -5/3 value. For pure Alfvén waves the EMF should vanish after the ensemble averages (3) and (8). This is not the case, however, in the in-situ observations, where only a small amount of e^- exists. But this is still sufficient to establish a nonlinear cascade and nonzero turbulent electric field. The spectrum of the z–component of $\mathcal{E}$ in solar-ecliptic coordinates (x-axis radially away from the sun, z-axis to the north) is given in Fig. 2, showing $\mathcal{E}_z$ versus frequency in the range from 10^{-5} to 10^{-2} Hz. Note that crosses indicate positive values and up-side-down triangles negative values. Apparently, there is substantial scatter in the data, but the overall shape of the spectrum is close to a Kolmogorov one. Whereas the $\mathcal{E}_z$-components are predominantly positive this is quite different for the $\mathcal{E}_y$ spectrum (not shown here), which also has a -5/3 spectral envelope but strongly alternates in sign, particularly at frequencies higher than some 10^{-4} Hz. In slow solar wind one usually finds $e^+ \gtrsim e^-$ (see e.g. /14,20/), and the turbulence level and intensity is considerably lower than in fast streams. These features are also reflected in the electric field spectra. The turbulent electric field is smaller in the current-sheet associated low-speed solar wind.

DYNAMO THEORY AND THE ALPHA EFFECT

As emphasized in the introduction the electromotive force $\mathcal{E}$ is of central importance in dynamo theory (Krause and Rädler /6/). One of the key questions is whether $\mathcal{E}$ can be expressed in terms of the mean flow velocity and magnetic field in such a way that closure of the induction equation is obtained in combination with the fluid equation of motion. The linearity in the mean field of the induction equation for the fluctuations then guaranties that $\mathcal{E}$ can only depend linearly on $\mathbf{B}_o$ and its spatial derivatives. A Taylor expansion on the large-scale L may then be written as:

$$\mathcal{E} = \underline{\underline{\alpha}} \cdot \mathbf{B}_o + \underline{\underline{\beta}} : \frac{\partial}{\partial \mathbf{x}} \mathbf{B}_o + \ldots\ldots \tag{11}$$

The expansion coefficients are tensors of different ranks. First the pseudo-tensor $\underline{\underline{\alpha}}$ depends on $\mathbf{V}_o$, $\delta\mathbf{V}$, ℓ and may generally still vary on the large scale L. If the turbulence is statistically isotropic and homogeneous then one has

$$\underline{\underline{\alpha}} = \underline{1}\alpha , \quad \alpha \propto \ <\delta\mathbf{V}\cdot(\boldsymbol{\nabla}\times\delta\mathbf{V})> \tag{12}$$

Obviously, the alpha effect described by the parameter α is built on kinetic helicity and therefore requires that reflexional symmetry is broken, since otherwise (12) would vanish identically. As a result one has $\mathcal{E} = \alpha\mathbf{B}_o$. In this simplest model of the electromotive force also the current is linear in $\mathbf{B}_o$, thus implying $\mathbf{J}_o = \sigma\alpha\mathbf{B}_o$ with the conductivity σ. Consequently, a dynamo cycle is achieved by the alpha effect as this generates toroidal currents, and hence poloidal fields, from toroidal fields. If there really is a turbulent dynamo of this kind acting in the solar wind we should be able to find evidence for the linear relation (11). We therefore analysed the data with respect to correlations between $\mathcal{E}$ and $\mathbf{B}_o$. The results for a typical fast solar wind flow are shown on the opposite page.

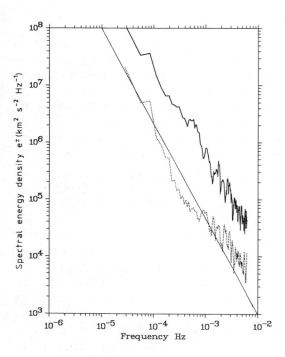
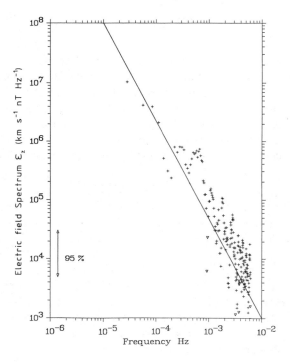

Fig. 1　Spectral energy density of the magneto-hydrodynamic fluctuations in terms of Elsässer variables plotted versus frequency. The dominance of e^+ over e^- indicates that Alfvénic fluctuations prevail for this time period of Helios 1 data from 1980 obtained at 0.53 AU.

Fig. 2　The out-of-ecliptic component $\mathcal{E}_z$ of the ponderomotive electric field versus frequency. The reference line corresponds to a Kolmogorov spectrum with a slope of -5/3.

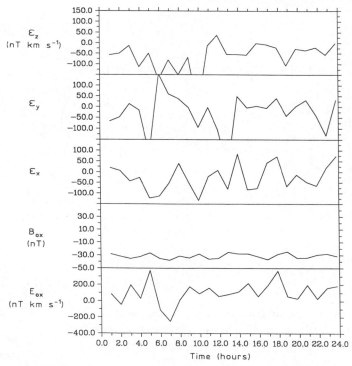

Fig. 3   Helios 2 data obtained near 0.29 AU in a fast stream with a speed of 733 km s^{-1}. From top to bottom the cartesian components of the turbulent electric field $\mathcal{E}$ and the radial components of the mean magnetic field $\mathbf{B}_o$ and the average electric field $\mathbf{E}_o = -\mathbf{V}_o \times \mathbf{B}_o$ are plotted versus time for a 24-hour period. Note the large fluctuations of $\mathcal{E}$ which concur with intense Alfvénic fluctuations. Individual points correspond to 1-h averages.

In Fig. 3 from top to bottom the vector components of $\mathcal{E}$ and then the x–components of the mean magnetic and convection electric field are plotted versus time in hours for a fast wind (733 km/s) as observed by Helios near 0.29 AU. Notice, there are considerable variations in $\mathcal{E}$ at a sizable fraction of $\mathbf{E}_o$, but there is no obvious correlation between $\mathcal{E}$ and $\mathbf{B}_o$. We have investigated quite a few other periods of high as well as low speed wind with the same result: no apparent linear correlation exists between the EMF and B . This negative finding concerning a possible alpha effect is corroborated in the presentation of Fig.4 showing detailed correlation plots between the one-hour averages of the cartesian components of $\mathbf{B}$ and $\mathcal{E}$. Again, there is no indication of a linear relationship. To make shure this is also true for other averaging periods we repeated this study at an 10-hour scale with the same negative conclusion. In this context it is important to note that Marsch and Tu /14/ found in the spectral analysis of the Elsässer fields that MHD turbulence in the solar wind largely seemed to be locally mirror-symmetric, in which case the alpha parameter (12) had to be zero on the average and no linear relation would apply.

CONCLUSIONS

The purpose of this study was to investigate electric field fluctuations in the solar wind and to scrutinize the data with respect to possible turbulent dynamo effects in the interplanetary medium. Electric field fluctuations have been studied in the framework of mean-field electrodynamics. A sizable EMF $\mathcal{E}$ has indeed been found in various types of wind flow, with $\mathcal{E}$ being even particularly strong in Alfvénic fluctuation periods with $e^+ > e^-$, whereas for pure Alfvén waves $\mathcal{E}$ should be identically zero. The $\mathcal{E}$–spectra show signatures of fully developed turbulence, in particular when the fluctuations are intense and in current-sheet associated slow wind. In correlation studies of the α–effect we could not establish the required linear relationship between $\mathcal{E}$ and $\mathbf{B}_o$. No dynamo effects in this sense were found in the data based on 1-hour and 10-hour averages. Although spectra of $\mathcal{E}$ are much smaller than the spectral densities themselves, the EMF is certainly important in the coupling of turbulence to the large-scale gradients of $\mathbf{V}_o$ and $\mathbf{B}_o$, and thus to the inhomogeneity of the interplanetary medium. These couplings are essential in the spatial evolution of MHD turbulence in the solar wind as described by the recent spectral transfer equations and models already mentioned in the introduction.

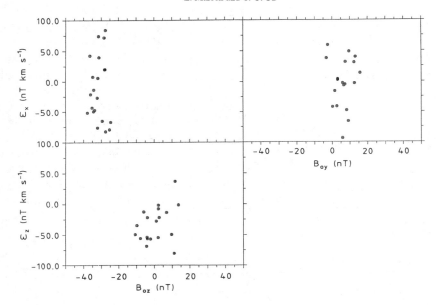

Fig. 4   Correlation plots for the vector components of the turbulent electric field and mean magnetic field in order to indicate a possible linear relationship according to equation (11). Note that no such relation is apparent. The data set is the same as in Fig. 3.

REFERENCES

1.   Marsch, E., in *Physics of the Inner Heliosphere*, vol. 2, ed. by R. Schwenn and E. Marsch, Springer-Verlag, Berlin, Heidelberg, New York, 159 (1990)

2.   Mangeney, A., R. Grappin, and M. Velli, in *Advances in Solar System Magnetohydrodynamics*, ed. by E.R. Priest (1990)

3.   Roberts, D.A., *Reviews of Geophysics*, Supplement, 932 (1991)

4.   Leer, E., T.E. Holzer, and F. Flå, *Space Sci. Rev.*, 33, 161 (1982)

5.   Moffatt, H.K., *Magnetic Field Generation in Electrically Conducting Fluids*, Cambridge University Press, Cambridge (1978)

6.   Krause, F., and K.-H. Rädler, *Mean-field magnetohydrodynamics and dynamo theory*, Akademie-Verlag, Berlin (1980)

7.   Matthaeus, W.H., M.L. Goldstein, and J.H. King, *J. Geophys. Res.*, 91, 59 (1986)

8.   Matthaeus, W.H., and M.L. Goldstein, *J. Geophys. Res.*, 87, 10347 (1982)

9.   Matthaeus, W.H., and M.L. Goldstein, *J. Geophys. Res.*, 87, 6011 (1982)

10. Montgomery, D.C., in *Solar Wind Five*, ed. by M. Neugebauer, NASA CP-2280, 107 (1983)

11. Marsch, E., in *Physics of the Inner Heliosphere*, vol. 2, ed. by R. Schwenn and E. Marsch, Springer-Verlag, Berlin, Heidelberg, New York, 45 (1990)

12. Zhou, Y., and W.H. Matthaeus, *J. Geophys. Res.*, 95, 14881 (1990)

13. Zhou, Y., and W.H. Matthaeus, *J. Geophys. Res.*, 95, 14863 (1990)

14. Marsch, E., and C.-Y. Tu, *J. Plasma Phys.*, 41, 479 (1989)

15. Tu, C.-Y., and E. Marsch, *J. Plasma Phys.*, 44, 103 (1990)

16. Velli, M., R. Grappin, and A. Mangeney, in *Plasma Phenomena in the Solar Atmosphere*, ed. by M. A. Dubois, Editions de Physique, Orsay (1989)

17. Grappin, R., U. Frisch, J. Léorat, and A. Pouquet, Astron. Astrophys., 105, 6 (1982)

18. Grappin, R., A. Pouquet, and J. Léorat, *Astron. Astrophys.*, 126, 51 (1983)

10. Tu, C.-Y., E. Marsch, and K.M. Thieme, *J. Geophys. Res.*, 94, 739 (1989)

20. Grappin, R., A. Mangeney, and E. Marsch, *J. Geophys. Res.*, 95, 8197 (1990)

WEAKLY INHOMOGENEOUS MHD TURBULENCE AND TRANSPORT OF SOLAR WIND FLUCTUATIONS

W. H. Matthaeus,* Y. Zhou,** S. Oughton* and G. P. Zank*

*Bartol Research Institute, The University of Delaware, Newark, DE 19716, U.S.A.
** Center for Turbulence Research, Stanford University, Stanford, CA 94305, U.S.A.

Abstract

Certain observed properties of solar wind fluctuations require descriptions outside the traditional WKB or homogeneous turbulence frameworks. This article presents a brief review of recent theories of transport of small scale MHD turbulence in an inhomogeneous background which show promise for providing more complete explanations of the evolution of solar wind turbulence.

1 Introduction

A useful way of looking at solar wind MHD fluctuations is to first ask how the fluctuations behave locally in space and time, say, on the scale of a few correlation lengths and times, and second, to ask how this characterization changes as the fluctuations evolve under the influence of the slowly varying background. For example, because the mean wind is both supersonic and super-Alfvénic, the radial evolution of the fluctuations corresponds to a temporal evolution of each outward moving blob of magnetofluid. The first of these questions has often been addressed by treating the fluctuations as propagating MHD waves, mostly Alfvén waves, a view supported both by minimum variance arguments and by high correlations of plasma velocity and magnetic field fluctuations in the inner heliosphere /1/. Alternatively, a viewpoint that the fluctuations are a form of actively evolving MHD turbulence affords an appealing mechanism for plasma heating in the solar wind /2/, while also providing a framework for a reasonably consistent interpretation of a variety of detailed statistical properties of the observations /3/. The question of the large scale transport and evolution of the fluctuations has traditionally been addressed in the context of WKB theory /4,5,6,7/. This approach works fairly well to explain the radial evolution of the wave energy density, but fails in a number of important ways. In this paper we review briefly the current status of recently developed theories of the spatial and wavenumber transport of MHD *turbulence* in the weakly inhomogeneous solar wind that may provide explanations of a number of observed phenomena that cannot be treated using wave-based WKB transport.

2 Solar Wind Spectral Evolution in WKB and related theories

The traditional way to write transport equations for the evolution of the fluctuation power spectra in the solar wind begins with the WKB formalism for MHD waves in a weakly inhomogeneous background /4,5,6,7/. In the lowest order WKB approximation one finds that

$$\partial_t P^\pm(k) + L^\pm_{WKB} P^\pm(k) = 0, \tag{1}$$

where $P^\pm$ are the reduced (or one-dimensional) power spectra of inward $(-)$ and outward $(+)$ propagating waves, with argument wavenumber k. We can denote the WKB spatial transport operator as $L^\pm_{WKB} = L^\pm_x + 2L^\pm$, with $L^\pm_x \equiv (\mathbf{U} \mp \mathbf{V}_A) \cdot \nabla$ and $L^\pm \equiv \nabla \cdot (\mathbf{U}/4 \pm \mathbf{V}_A/2)$ The conditions for validity of the WKB expansion (discussed further below) include the enforcement of both the wave dispersion relation, and the scale separation condition $kR \gg 1$.

This WKB formalism has been employed in many solar wind and other space physics applications, and it is widely believed to account reasonably well for certain observed phenomena, such as the radial evolution of the fluctuation level δB^2 in the outer (> 1 AU) heliosphere. However, there are also known shortcomings. For example, it is a consequence of the wave description that there exists no "mixing" of Alfvénic fluctuations at the leading order, i.e., the equations for the development of the $+$ and $-$ fluctuations are not coupled. Although mixing does nevertheless occur at higher orders /8/, this mixing is insufficient /9/ to account for the evolution of cross-helicity with radial distance in the solar wind /10/. A similar statement can be made regarding the behavior of the "Alfvén ratio", i.e. the ratio of kinetic to magnetic energy per unit mass in inertial range fluctuations. Observations /3,10/ indicate that the Alfvén ratio is often somewhat less than unity in the solar wind, but it remains exactly one in leading order WKB theory.

Further shortcomings of WKB theory involve its inability to (a) account for the rapid evolution of the spectral shape of solar wind fluctuations in the inner heliosphere /10,11/ and (b) provide a rapid wave damping mechanism that would address the source of heating of solar wind plasma /12/ between the lower corona and 1 AU. Coleman /2/ had suggested that turbulent heating, at a rate such as that predicted for homogeneous turbulence in Kolmogoroff theory /13/, might provide the required mechanism.

In a pair of important and seminal papers, Tu and coworkers /14,15/, combined Coleman's heating suggestion with WKB transport, opening the way for more complete treatments of the turbulence. The essence of Tu's models is to combine leading order WKB spatial transport with a simple phenomenology for spectral transfer in wavenumber, giving rise to a transport equation

$$\partial_t P^+(k) + L^+_{WKB} P^+(k) = NL^+, \tag{2}$$

for the spectrum of outward propagating fluctuations. The term NL^+ in (2) represents triple correlations arising from nonlinear terms in the MHD equations. The first model /14/ adopted what is essentially the Kraichnan /16/ phenomenology of spectral transfer in the inertial range for modeling NL^+, while the second /15/ utilized the Kolmogoroff /13/ inertial range phenomenology. In each case the nonlinearities are written as $NL = \partial G/\partial k$ where the energy flux in wavenumber space, G, is approximated by dimensional analysis and the usual statistical assumptions applied to the turbulent inertial range. A number of similar approaches can be adopted in modeling nonlinearities responsible for local turbulence /17/. The Tu theories for evolution of the spectrum have enjoyed some success in accounting for modifications to the energy spectrum in the inner heliosphere. However, because the inward and outward waves are still uncoupled, they cannot describe nontrivial dynamics of either the cross helicity or the Alfvén ratio.

An equally important advance in solar wind transport, was made /18,19/ in connection with dynamical transport equations for the total wave energy. Phenomenological theories of *total* turbulent energy decay in homogeneous turbulence /13/ make use of the global eddy turnover time as an estimate of the decay time of the energy-containing eddies. Hollweg /18,19/ adopted this "Kolmogoroff" perspective, along with a WKB approximation for the spatial transport of the total wave energy, in a model for the acceleration and heating of the solar wind in the inner heliosphere. While such a procedure directly addresses the ideas set forth by Coleman /2/, the predictions of this theory /20/ have not yet been able to account simultaneously for both the wave energies and the temperatures at 1 AU, when reasonable parameter values at the coronal critical points are used.

3   Structure of two-scale transport theories

The models of Tu and Hollweg provide ample motivation to develop more comprehensive theories for transport of solar wind turbulence. One would like to describe in such a theory not only the radial evolution of turbulent energy in both the energy-containing and inertial ranges, but also the possibility of variations in the cross helicity content, the interaction of "inward" and "outward" type fluctuations, and unequal values of kinetic and magnetic fluctuation energies. There exists also the possibility of studying transport of other spectral quantities, such as magnetic helicity and induced electric field. Moreover, there is a basic conceptual question that arises in both the Tu and Hollweg models: If the turbulence is "strong" enough to be treated by Kolmogoroff theory, how can the dispersion relation underlying WKB theory also be enforced? In the last several years, these questions have been investigated by appeal to more general formalisms for transport of turbulence /21,22,23,24/ based mainly upon the assumptions of scale separation and local incompressibility. The simplest forms of the transport theory assume the large scale plasma velocity and magnetic fields are specified.

The theory can be cast into two forms, one appropriate to spectral evolution in the inertial range /21,22,23/ and the other to the evolution of the energy-containing eddies near the correlation scale /25/. In each case, dynamical equations based on a two length scale expansion are derived, from which the evolution of various quantities may be computed, including magnetic and kinetic energies, cross helicity, induced electric field, and the corresponding helicities.

By the assumption of scale separation, we mean that the turbulent fluctuations, consisting of fluctuations at scales up to a correlation length λ, admit locally well-defined statistical properties that vary on the scale R that characterizes changes in the inhomogeneous background velocity, magnetic and density fields. Taking R to be the local heliocentric distance, this amounts to the assertion that $\epsilon = \lambda/R$ is a small parameter. Fast and slow-varying space coordinates, are introduced to separate local effects (e.g., turbulent spectral transfer) and effects associated with inhomogeneities. Similarly, fast and slow time scales can also be introduced. Thus, for position r and time t, we let $\mathbf{r} = \mathbf{x}$, $t = \tau$ (slow scale) and $\mathbf{x}' = \mathbf{x}/\epsilon$ and $t' = t/\epsilon$ (fast scale). To facilitate calculations, we introduce an averaging operator $< ... >$ that annihilates fast scale variations. For example, the solar wind (proton) fluid velocity $\mathbf{V}$ may be decomposed into $\mathbf{V} = \mathbf{U} + \mathbf{u}$ where $\mathbf{U} =< \mathbf{V} >$ is the mean large scale flow and $\mathbf{u}$ is the fluctuating velocity. Similarly, the magnetic field can be separated according to $\mathbf{B} = \mathbf{B_0} + \mathbf{b}$, with $\mathbf{B_0} =< \mathbf{B} >$ and turbulent fluctuation $\mathbf{b}$. In the simplest cases, the density ρ is taken to be locally incompressible, $\rho =< \rho >$ and the mean fields $\mathbf{U}$, $\mathbf{B_0}$ and ρ are assume to be specified, time independent functions varying only with heliocentric distance R. It is also convenient to work in Elsässer variables for the fluctuations, $\mathbf{z}^{\pm} = \mathbf{u} \pm \mathbf{b}/\sqrt{4\pi\rho}$.

Transport equations are computed by forming the correlation functions $R_{ij}^{++} =< z_i^+ z_j^{+\prime} >$, $R_{ij}^{--} =< z_i^- z_j^{-\prime} >$ and $R_{ij}^{+-} =< z_i^+ z_j^{-\prime} >$, and making use of the compressible MHD equations and the assumption of statistical homogeneity on the fast scales. In the above the primed ($'$) and unprimed variables are evaluated at distinct spatial positions and the associated separation vector (generally in the radial direction for solar wind spacecraft observations) is the sole dependency of the correlation matrices. The associated spectral tensors, i.e., the Fourier transform of the correlation matrices, are denoted as S_{ij}^{++}, S_{ij}^{--} and S_{ij}^{+-} having wavevector $\mathbf{k}$ as argument. The procedure leads to equations involving the operators introduced earlier, $L_x^{\pm}$ and $L^{\pm}$, which appeared in WKB theory. Also appearing is a new matrix operator $M_{ij}^{\pm}$ that involves only derivatives of the large scale fields, and which gives rise to "mixing" type couplings between inward and outward type fluctuations /21,22,26/. The explicit form of the spectral equations so obtained is quite complex /22,23/ and will not be repeated here. However, the general form of the equations is $(1\partial_t + \mathbf{L}_S) \cdot \mathbf{s} = \mathbf{nl}$ where $\mathbf{L}_S$ is a linear spatial transport operator The elements s_i of the solution vector $\mathbf{s}$ are the wavenumber dependent scalar spectra, and the right hand side is a column vector $\mathbf{nl}$ whose elements are the modeled nonlinear terms associated with the corresponding spectra.

In the most general case of homogeneous turbulence, it can be shown that the maximum number of independent elements s_i is 16. Four of these are identified with the antisymmetric parts of the spectral tensors and can be represented by the magnetic helicity, the helicity of the velocity field, the "helicity of the cross helicity" and the single spectral scalar that generates the induced electric field spectrum /22/. Of the 12 spectra contained in the symmetric parts of the three spectral tensors, six arise from the three independent elements of each of the symmetric energy spectral tensors $S_{ij}^{++} + S_{ji}^{++}$ and $S_{ij}^{--} + S_{ji}^{--}$. The remaining six are accounted for by the three independent elements of the energy difference tensor (kinetic minus magnetic) and the three elements connected with the "helicity of the electric field" /22/.

For a particular symmetry of the turbulence, a natural choice of independent spectral scalars is usually evident. A selection of slab, two-dimensional (2D) or isotropic turbulence forces equality between certain of the scalars, and causes others to vanish. Moreover, it is sometimes convenient to write the spectral equations in terms of scalar spectra that depend upon wave vector **k**, retaining the full three dimensional domain of the vector argument. In other cases it may prove better to write everything in terms of one dimensional, "reduced" spectra that are easily connected with single spacecraft solar wind observations /3/. In the latter case especially, the symmetry of the turbulence enters in a crucial fashion, since various symmetry-dependent relationships between modal, omnidirectional and reduced spectra /13/ need to be used to simplify the final equations. For typical cases, including slab, 2D and isotropic turbulence, it is convenient to write the spectral transport equations in a form including energy spectra of the $\pm$ fields $P^{\pm}(k) = S_{ii}^{\pm\pm}/4$, the difference of kinetic and magnetic energy spectra $F(k) = S_{ii}^{+-}/4$, and so on.

The spatial transport operator $\mathbf{L}_S$ includes familiar effects such as convection and expansion associated with the solar wind velocity, and wave propagation in the direction of (or opposite to) the large scale Alfvén velocity. For the symmetries mentioned above, the transport equation for the $\pm$ energies has the form

$$\partial_t P^{\pm}(k) + L_{WKB}^{\pm} P^{\pm}(k) + M^{\pm} F = NL^{\pm}, \tag{3}$$

where further equations for F, etc. are required to close the model. The new term $M^{\pm}$ represents leading order mixing, and has a form specific to the choice of symmetry /22,26/. Space prohibits an exhaustive treatment of either the structure or solutions of the full transport theory in this paper. The model remains a subject of intense investigation and comparison with solar wind observations. Thus, we restrict our comments here to several pertinent general points.

- The spectral evolution theory consists of up to 16 coupled equations, which can be simplified by approximations in special cases. For isotropic turbulence, the number of coupled equations may be as many as seven, or, with suitable approximations regarding the behavior of F, as few as two. A three equation isotropic model, involving $P^{\pm}$ and F, appears to be physically reasonable /26/.

- One interesting limit /27/ is that of WKB theory, and convergence to that case is demonstrated for strong Alfvén speed and weak nonlinearities, i.e, enforcement of the wave dispersion relation /26/. Strong "mixing" requires nontrivial couplings of $P^{\pm}$ with F. The distinction between strong mixing and recovery of WKB results is explored further below.

- Several simple analytic solutions /21,22/ predict, with increasing heliocentric distance, a decrease of the preponderance of outward-traveling type Alfvénic fluctuations and a lowering of the small-scale kinetic to magnetic energy ratio. Both of these are roughly consistent with Helios and Voyager observations /10/.

- The theory requires, for closure, modeling of the nonlinear terms. For spectral quantities conserved in the inertial range at high Reynolds numbers, such as $P^{\pm}(k)$, correct models will adopt nonlinearities of the form $NL^{\pm} = \partial G^{\pm}(k)/\partial k$, where $G^{\pm}$ is an appropriately defined wavenumber flux of the associated conserved energy. In strong turbulence such terms vanish. For nonconserved spectral quantities such as $F(k)$, this type of model is inappropriate, and in strong turbulence the associated transport effects may not vanish. Turbulence theory has yet to provide us with well-accepted forms for model nonlinearities. Howver, simple approaches, including diffusion in k-space, one point closures and simple relaxation time models may be useful /17,24,28/ for empirical and observational comparisons.

- The spectral transport model can also be developed further into a model for the evolution of the energy containing fluctuations, in analogy with classical quasi-equilibrium range hydrodynamic turbulence /13/ and making use of phenomenologies of homogeneous MHD turbulence. A closed five equation model for evolution of the turbulence has been proposed /25/, including transport and nonlinear evolution of bulk magnetic energy, kinetic energy, cross helicity and two correlation scales. The energy-containing model differs from the spectral transport model above mainly in the way in which the nonlinearities are handled, and is expected to be useful in providing the required low wavenumber boundary data for the spectral transport model that operates in the inertial range. In fact, it is expected /13/ that inertial range energy transfer rates, and thus the overall heating rate, will be regulated mainly by the decay of the energy containing eddies.

4   Relation to WKB theory: Multiple scales analysis

Some of the consequences of the multi-equation, scale separated transport models /21,22,23/ may be disconcerting, especially in that couplings between inward and outward-type fluctuations, and therefore the"mixing" effect, appear in the leading order theory. How is it that such effects are absent in WKB theory but present here? Is an error implied in the classical derivations of WKB transport? Nonlinearities do not easily explain these discrepancies, since the mixing effect occurs in the *linear* transport terms. These concerns /27/ have been discussed in connection with the spectral transport theory, neglecting nonlinearities, the suggestion emerging that within *linear* theory, the mixing effect should be treated in leading order for certain cases. Specifically, when $\mathbf{k} \cdot \mathbf{V}_A \to 0$ or when $\mathbf{V}_A/U \to 0$, the strong mixing effect is present in leading order, owing to the degeneracy of the two solutions to the wave dispersion relation in those limits. It is also possible to reconcile WKB and a non-WKB "mixing theory" entirely in the context of the primitive linear equations for the fluctuating fields. We outline this procedure here.

Let the fluctuating Elsässer fields be expanded as $\mathbf{z}^{\pm}(\mathbf{x}, \mathbf{x}', \tau, \tau') = \mathbf{z}^{\pm 0} + \epsilon \mathbf{z}^{\pm 1} + \epsilon^2 \mathbf{z}^{\pm 2} + \cdots$. The leading order ϵ^{-1} expansion yields

$$\left(\partial_{\tau'} + L_{\mathbf{x}'}^{\pm}\right) z_i^{\pm 0} = 0, \tag{4}$$

where the primed operators involve derivatives with respect to the fast scale. The solution to (4) is written as $\mathbf{z}^{\pm 0} = \tilde{\mathbf{z}}^{\pm 0} e^{iS_{\pm}}$, in which $\tilde{\mathbf{z}}^{\pm 0}$ is slowly varying envelope function and $S_{\pm}$ (the eikonal or phase function) depends on both fast and slow coordinates. By proper choice of the (arbitrary) slowly varying dependence of $S_{\pm}$, one can simplify subsequent manipulations. This is accomplished by demanding that the phase functions themselves obey the ϵ^{-1} equation with the fast derivatives replaced by slow derivatives.

Proceeding to the $O(1)$ expansion and making use of the leading order solution yields

$$\left(\partial_{\tau'} + L_{\mathbf{x}'}^{\pm}\right) z_i^{\pm 1} = -e^{iS_{\pm}} \left[\left(\partial_{\tau} + L_{\mathbf{x}}^{\pm} + L^{\pm}\right) \tilde{z}_i^{\pm 0}\right] - e^{iS_{\mp}} M_{ik}^{\pm} \tilde{z}_k^{\mp 0}. \tag{5}$$

This equation is an inhomogeneous wave equation of the same type as the leading order $O(\epsilon^{-1})$ wave equation, therefore the solution consists of a particular plus homogeneous solution $z^{\pm 1} = z_p^{\pm 1} + z_h^{\pm 1}$. Clearly, $z_h^{\pm 1}$ has the same form (eikonal) as the zeroth order term, hence $z_h^{\pm 1} = \tilde{z}_h^{\pm 1} e^{iS_{\pm}}$. In determining the particular solution, care must be taken to avoid secularities. This is equivalent to avoiding resonances in the inhomogeneous wave equation (5). Such considerations lead to two distinct choices of solvability condition, one of which leads back to well-known classical WKB theory, while the other corresponds to the non-WKB multiple scales approach with strong mixing.

To arrive at the WKB solvability condition, assume that the $\mathbf{z}^+$ inhomogeneity is non-resonant with $\mathbf{z}^-$. At this order, the assumption is equivalent to $S_+ \neq S_-$. Hence the only restriction on the inhomogeneous wave equation (5) is that

$$\left(\partial_{\tau} + L_{\mathbf{x}}^{\pm} + L^{\pm}\right) \tilde{z}_0^{\pm} = 0.$$

This is the leading order WKB equation for the primitive fields $z^{\pm}$, and implies WKB transport for the energy spectra as given in (1).

The non-WKB solvability criterion is obtained through the assumption that z^+ and z^- are nearly resonant in the sense that $S_+ \approx S_-$. Hence, the solvability condition for (5) is that the full RHS vanish. We now have no inhomogeneities and the leading order slowly varying amplitudes obey the equation

$$\left(\partial_\tau + L_x^{\pm} + L^{\pm} \right) \tilde{z}^{\pm 0} + \epsilon_0 M^{\pm} \tilde{z}_0^{\pm} = 0,$$

where $\epsilon_0 = e^{i(S_- - S_+)}$. This represents a non-WKB form of the transport equation with mixing possible at the leading order, and is equivalent to (3) without the nonlinear terms. In general, the size of the mixing term depends on the magnitude of ϵ_0 which in the present case is $O(1)$. It is apparent that it is necessary to choose the non-WKB conditions when $S_+ \approx S_-$, which is equivalent to either of the conditions $\mathbf{k} \cdot \mathbf{V}_A \approx 0$ or $V_A/U \to 0$. These are two of the conditions identified previously /27/ for leading order "mixing" in linear spectral transport equations.

5 Conclusions

We have briefly reviewed the background and basic principles leading to the recent development of transport equations for MHD turbulence in the weakly inhomogeneous background solar wind. Various approximate theories based on this approach show promise in answering important questions in heliospheric physics. Future applications in lower coronal physics, in shock theory and in astrophysics are also feasible. In a new development given in Sec. 4, we have outlined the mathematical connection between WKB and non-WKB theories, further implications of which will be presented at a future opportunity.

Acknowledgements This work has been supported in part by the Bartol NASA Space Physics Theory Program under grant NAGW-2076 and by NSF grant ATM-8913627.

6 References

1. J.W. Belcher and L. Davis, J. Geophys. Res., **76**, 3534 (1971).
2. P.J. Coleman, Ap. J., **153**, 371 (1968).
3. W. Matthaeus and M. Goldstein, J. Geophys. Res., **87** 6011 (1982)
4. S. Weinberg, Phys. Rev. bf 126 1899 (1962)
5. E. Parker, Space Sci. Rev. **4** 666 (1965)
6. A. Barnes, in Solar System Plasma Physics, North-Holland, New York, 249 (1979).
7. J. Hollweg, J. Geophys. Res., **78**, 3643 (1973); **79**, 1539 (1974)
8. M. Heinemann and S. Olbert, J. Geophys. Res. **85** 1311 (1980)
9. J. Hollweg, J. Geophys. Res. **95** 14873 (1990)
10. D.A. Roberts, M.L. Goldstein, L.W. Klein & W.H. Matthaeus, J. Geophys. Res., **92**, 11021 (1987).
11. B. Bavassano, M. Dobrowolny, F. Mariani and N. F. Ness, J. Geophys. Res., **87**, 3617 (1982)
12. A. J. Hundhausen, *Coronal Expansion and Solar Wind*, Springer-Verlag (1972)
13. G. K. Batchelor, *Theory of Homogeneous Turbulence*, Cambridge, 1953.
14. C. Tu, Z. Pu and F. Wei, J. Geophys. Res, **89**, 9695 (1984)
15. C. Tu, J. Geophys. Res. **93**, 7 (1988)
16. R. Kraichnan, Phys. Fluids **8** 1385 (1965)
17. Y. Zhou and W. Matthaeus, J. Geophys. Res. **95** 14881 (1990)
18. J. Hollweg, J. Geophys. Res., **91**, 4111 (1986)
19. J. Hollweg and W. J. Johnson, J. Geophys. Res., **93**, 9547 (1988).
20. P. Isenberg, J. Geophys. Res., **95**, 6437 (1990).
21. Y. Zhou and W. Matthaeus Geophys. Res. Lett. **16** 755 (1989)
22. Y. Zhou and W. Matthaeus, J. Geophys. Res. **95** 10291 (1990)
23. E. Marsch and C. Tu, J. Plasma Phys. **41**, 479 (1989)
24. C. Tu and E. Marsch, J. Plasma Phys. **44** 103 (1990)
25. W. Matthaeus and Y. Zhou, EOS **71** 1509 (1990)
26. S. Oughton and W. Matthaeus, these proceedings.
27. Y. Zhou and W. Matthaeus, J. Geophys. Res. **95** 14863 (1990)
28. A. Mangeney, R. Grappin and M. Velli, in "Advances in Solar System Magnetohydrodynamics", ed. E.R. Priest and A. W. Hood, Cambridge U. Press (1991), p. 344.

ANOMALOUS DIFFUSION IN SOLAR WIND PLASMAS CAUSED BY ELECTROSTATIC TURBULENCE

C. -V. Meister

Central Institute for Astrophysics, Potsdam, Germany

ABSTRACT

In this paper the diffusion of solar wind electron streams by the interaction with electrostatic waves is considered. The solution of the kinetic equation is found for a plasma with weak density gradient, so that the associated diffusion currents disturb the local thermal equilibrium only slightly. In comparison with former works a recalculated anomalous collision term is taken into account obtained within Klimontovich formalism.

1. INTRODUCTION

Ion acoustic waves have been observed in the solar wind for some ten years. They may be driven by drifting electrons or electron heat fluxes and seem to be important for the plasma dynamics at high ratio of electron temperature T_e to ion temperature T_i. From the dispersion relation of the waves, and the values of the frequencies $\omega \approx$ 3-10 Hz observed by *Helios* follows, that wave lengths above the electron Debye-radius exist.

Anomalous heat conduction in solar wind caused by ion acoustic waves is considered in /1/. Here we study the anomalous space diffusion. A first expression is given for the diffusion coefficient perpendicular to the magnetic field, which seems to be very important for magnetic reconnection.

2. KINETIC EQUATION

If in a plasma electrostatic waves are generated, which have wave lengths larger than the Debye-radius and frequencies lower than the plasma frequency, then it is useful to calculate transport coefficients starting with the following form of the kinetic equation /2,3/:

$$\frac{\partial}{\partial t} f_a(\vec{x}, t) + \vec{v} \frac{\partial}{\partial \vec{r}} f_a(\vec{x}, t) + \frac{q_a}{m_a} \left(\vec{E}_o(\vec{r}) + [\vec{v} \times \vec{B}_o(\vec{r})] \right) \frac{\partial}{\partial \vec{v}} f_a(\vec{x}, t) \tag{1}$$

$$= \frac{\partial}{\partial \vec{v}} \left(\gamma_a(\vec{p}) \vec{v} f_a(\vec{v}, t) \right) + \sum_{ij} \frac{\partial}{\partial v_i} D_{ij}^a(\vec{x}) \frac{\partial}{\partial \vec{v}_j} f_a(\vec{x}, t).$$

For plasmas far from the equilibrium state the small spontaneous contribution to the wave-particle collision integral (second term on the right side of eq.(1)) can be neglected, and for electrostatic waves $(\vec{k} \| \widetilde{\delta E})$ the velocity-diffusion tensor D_{ij}^a is given by

$$D_{ij}^a = \frac{q_a^2}{16\pi^3} \int \delta(\omega - \vec{k}\vec{v}) \left(\widetilde{\delta E_i \delta E_j} \right)_{\omega \vec{k}} d\omega d\vec{k}. \tag{2}$$

$\sim$ means the averaging over possible small-scale fluctuations. $\vec{E}_o$ is the mean electric field, $\vec{B}_o$ the outer magnetic field, ω the frequency of the electrostatic waves, and $\vec{k}$ the wave vector.

2. SOLUTION

To find the stationary solution of eq.(1) spherical coordinate systems $\vec{v}(v, \vartheta_{\vec{v}}, \varphi_{\vec{v}})$, $\vec{k}(k, \vartheta_{\vec{k}}, \varphi_{\vec{k}})$ are introduced (the z-axis is parallel to the magnetic field, and the electric field $\vec{E}_o$ is antiparallel to the z-axis):

$$\vec{v}\frac{\partial f_e}{\partial \vec{r}} - \frac{eE}{m_e}\left(\xi\frac{\partial f_e}{\partial v} + \frac{1-\xi^2}{v}\frac{\partial f_e}{\partial \xi}\right) - \omega_{ce}\frac{\partial f_e}{\partial \varphi_{\vec{v}}} + \frac{1}{v}\frac{\partial}{\partial \xi}\left(D_{\xi\xi}\frac{1}{v}(1-\xi^2)\frac{\partial f_e}{\partial \xi}\right); \quad \xi = cos\vartheta_{\vec{v}} \tag{3}$$

$$D_{\xi\xi}(v,\xi) \approx \frac{2\pi m_i}{n^o m_e^2 v}\int_{-\eta}^{\eta}dy\int_0^\infty dk k\omega_{\vec{k}}^2\frac{(\omega\xi/kv - y)^2}{1-\xi^2}\left(\overline{\delta E_\xi \delta E_{\vec{k}}}\right)_{\vec{k}}\left[1-\xi^2-y^2+\frac{2\omega y\xi}{kv}-\frac{\omega^2}{k^2v^2}\right]^{-1/2}, \eta = \sqrt{1-\xi} \tag{4}$$

In (3) was used, that the characteristic relaxation time of the energy $\tau_{eff}(v) = 0.5v_e^4 k^2\omega_{\vec{k}}^{-2}D_{\xi\xi}^{-1}$ is much larger than the characteristic relaxation time of the momentum $\tau_{eff}(\xi) = \nu_{eff}^{-1} = 0.5v_e^2 D_{\xi\xi}^{-1}$, as $\omega_{\vec{k}}/(kv_e) < 1$ for ion-acoustic waves.

In solar wind plasmas at 1 AU the ratio of the electron mean free path $l_{coll} = v_e/\nu_{pp}$ to the density scale length $L = | \partial lnn_e^o/\partial \vec{r} |^{-1}$ is of the order of 1 /1/. But suggesting that the wave-particle collision frequency ν_{wp} /3/ of the microturbulent plasma is much larger than the Coulomb collision frequency ν_{pp}, we find a small parameter l/L $l = v_e/(\nu_{wp}+\nu_{pp})$, and develop the electron distribution function in a series with respect to l/L /4, 5/. The zero-order distribution function is approximated by a Maxwellian one:

$$f_e(\vec{r},\vec{v}) = f_e^M(\vec{r},v^2) + \delta f_e(\vec{r},\vec{v}), \quad f_e^M(\vec{r},v^2) = n_e^o(\vec{r})(\frac{m_e}{2\pi\theta_e^o})^{3/2}exp\{-\frac{m_e v^2}{2\theta_e^o}\}, \quad \delta f_e(\vec{r},\vec{v}) = f_e^M(v^2)\phi^*(\vec{r},\vec{v}), \tag{5}$$

$$\phi^*(\vec{r},\vec{v}) = \vec{v}\vec{\phi}(\vec{r},v^2), \quad \vec{\phi}(\vec{r},v^2) = A(v^2)\frac{\partial ln\, n_e^*}{\partial z}\vec{n}_z + A_1(v^2)\left(\frac{\partial ln\, n_e^*}{\partial x}\vec{n}_x + \frac{\partial ln\, n_e^*}{\partial y}\vec{n}_y\right)A_2(v^2)\left[\vec{\omega}_{ce} \times \frac{\partial lnn_e^*}{\partial \vec{r}}\right] \tag{6}$$

$$\frac{\partial ln\, n_e^*}{\partial \vec{r}} = \frac{\partial ln\, n_e}{\partial \vec{r}} + \frac{e\vec{E}_o\vec{v}}{\theta_e^o}, \quad \vec{\omega}_{ce} = \frac{q_e\vec{B}}{m_e c} \tag{7}$$

Developing further A, A_1, A_2 in series of Laguerre-polynominals

$$L_n^m(z_*) = \frac{z_*^{-m}e^{z*}}{n!}\frac{d^n}{dz_*^n}(z_*^{n+m}e^{-z_*}) \tag{8}$$

$$\int_o^\infty z^m e^{-z_*}L_p^{(m)}(z_*)L_q^{(m)}(z_*)dz_* = \frac{\Gamma(p+m+1)}{\Gamma(p+1)}\delta_{pq}, \quad z_* = \frac{m_e v^2}{2\theta_e^o}, \quad X(z_*) = \sum_{k=0}^\infty x_k L_k^{(3/2)}(z_*) \tag{9}$$

the system of equations

$$-\frac{3\sqrt{\pi}}{4}\delta_{l0} + \frac{\Gamma(l+5/2)}{l!}a_{2l}^* = 6\sqrt{3}G_o^*\sum_{k=0}^\infty\frac{(5/2)_k(2/2)_l}{k!l!}a_{1k}^*\sum_{q=0}^k\frac{(-k)_q(-1/2)_q}{(5/2)_q(-1/2-l)_q}; \quad l = 0,1,2... \tag{10}$$

$$\frac{\Gamma(l+5/2)}{l!}a_{1l}^* = 3\sqrt{3}G_o^*\sum_{k=0}^\infty\frac{(5/2)_k(2/2)_l}{k!}a_{2k}^*\sum_{q=0}^k\frac{(-k)_q(-1/2)_q}{(5/2)_q(-1/2-l)_q}, \tag{11}$$

$$\sum_{k=1}^\infty\frac{(5/2)_k(3/2)_l}{k!l!}a_k\sum_{q=0}^k\frac{(-k)_q(-1/2)_q}{(5/2)_q(-1/2-l)_q} = 0, \tag{12}$$

is obtained, which has to be solved with respect to $a_{1l}^* = \omega_{ce}a_{1l}$, $a_{2l}^* = \omega_{ce}^2 a_{2l}$, a_{1l}. Here $G^* = v\nu_{wp}/(2\omega_{ce}v_e)$, $(x) = x(x+1)...(x+k-1); \quad (x)_o = 1$,
The space-diffusion coefficient of the electrons, e.g. perpendicular to the magnetic field, has to be calculated via the velocity moment,

$$u_{e\perp} = \int\delta f_e v_\perp d\vec{v} \approx \frac{\theta_e^o}{\pi m_e}\frac{a_{10}^*}{\omega_{ce}}\frac{\partial ln\, n_e^*}{\partial \rho} \tag{13}$$

For the space diffusion of the electrons perpendicular to the magnetic field B_o caused by the interaction with ion-acoustic waves excited by the electrons itself it follows

$$D_\perp \approx \frac{\sqrt{6}\pi m_i}{n_e^\circ m_e^{3/2} \sqrt{\theta_e^\circ} \omega_{ce}^2} \int_0^\infty dk \; k\omega_k^2 \left(\widetilde{\delta E_\xi} \widetilde{\delta E_\xi}\right)_{\vec{k}} \tag{14}$$

REFERENCES

1. J. V. Hollweg, Collisionless electron heat conduction in the solar wind, *J. Geophys. Res.* 81, #10, 1649-1658 (1976)

2. Yu. L. Klimontovich, *Statistical Physics* (in russ.), Nauka, Moscow, 1982

3. V. V. Belyi, Yu.L. Klimontovich, and V.P. Nalivaiko, Kinetic theory of the anomalous electrical conductivity for a turbulent plasma (in russ.), *Fizika Plazmy* 8, #5, 1063-1072 (1982)

4. S. I. Braginski, Transport phenomena in the plasma (in russ.), in: *Voprozy Teorii Plazmy* #1, ed. M.A. Leontovich, Atomizdat, Moscow, 1963, pp. 183-272.

5. C.-V. Meister, and E.V. Mishin, Calculation of the diffusion coefficient in weak turbulence theory approximations, in: *Conference Abstracts of the 8. European Sectional Conference on Atomic and Molecular Physics of Ionized Gases*, Greifswald, 26.-29.8.1986, pp. 256-257

STOCHASTIC FORCES ON PARTICLES IN PLASMAS WITH ION-ACOUSTIC INSTABILITY

C. -V. Meister,* B. Nikutowski* and K. Schindler**

* *Central Institute for Astrophysics, Potsdam, Germany*
** *Ruhr-University, Bochum, Germany*

ABSTRACT

A Langevin equation for charged particles in a plasma with electrostatic turbulence is developed from first principles in consistency with kinetic theory. For the case of ion-acoustic turbulence observed in the solar wind detailed results are given for a Gaussian-distributed stochastic force, which has to be expressed by the diagonalized velocity diffusion tensor.

1. INTRODUCTION

Beyond a distance of 10-20 solar radii from the sun the interplanetary medium can be considered as almost collision-free, and microinstabilities seem to play an important role /1,2/. A basic type of electrostatic instabilities, detected by *Helios* and *Voyager* in the solar wind are ion-acoustic and ion-acoustic-like waves /2/. But nevertheless the significance of this waves for macroscopic effects on the solar wind is unclear /2/. Thus here a Langevin equation is developed describing the influence of electrostatic waves on charged particles.

2. LANGEVIN-EQUATION

The Langevin-equation for a particle (of type a) can be found adding in the equation of motion of this particle the time-dependent force $\vec{F}_a(\vec{x}, t) = \langle \vec{F}_a(\vec{x}) \rangle + \delta \vec{F}_a(\vec{x}, t)$ ($\{\vec{x}\} = \{\vec{r}, \vec{v}\}$) of the waves (considered as large-scale fluctuations),

$$\frac{d\vec{r}}{dt} = \vec{v}, \quad \frac{d\vec{v}}{dt} = \frac{q_a}{m_a} \left(\vec{E}_o(\vec{r}) + [\vec{v} \times \vec{B}_o(\vec{r})] \right) + \vec{F}_a(\vec{x}, t), \tag{1}$$

Usually the average of this force over the dynamical distribution function /3-5/ is expressed by a friction term $-\gamma_a(\vec{v})v_i$. Here a second term is added, which has to be found regarding consistency between Langevin-equation and Fokker-Planck-form of the corresponding kinetic equation /3/,

$$\langle F_{ai}(\vec{x}) \rangle = -\gamma_a(\vec{v})v_i + \frac{\partial D_{ii}^a(\vec{v})}{\partial v_i}. \tag{2}$$

For Gaussian stochastic processes, $f(\delta\vec{F}_a) = exp\left\{-(\delta\vec{F}_a)^2/2\sigma^2\right\}/\sqrt{2\pi}\sigma$, all correlation functions of the stochastic force of odd order are zero, correlation functions of even order can be expressed by functions of order $n = 2$, and

$$\langle \delta F_{ai}(\vec{x}, t_1) \delta F_{aj}(\vec{x}, t_2) \rangle = 2\delta_{ij}\delta(t_1 - t_2)D_{ij}^a(\vec{x}). \tag{3}$$

Components of the stochastic force in different directions and at different instants of time are uncorrelated.

Instead of solving the Langevin-equation for given initial conditions $\vec{v}_o = \vec{v}(t_o)$, $\vec{r}_o = \vec{r}(t_o)$ and to determine the probabilities of all its solutions it is also possible to find an equation of motion for the distribution function $f_a(\vec{x})$. If the plasma processes are approximated by Markovian ones, e.g.some $1/\nu_{eff} < \Delta t < t_d$, ($t_d$- dissipation time of the plasma process, ν_{eff}- effective wave-particle collision frequency) one obtains the Fokker-Planck-equation

$$\frac{\partial}{\partial t}f_a(\vec{x},t) + \vec{v}\frac{\partial}{\partial\vec{r}}f_a(\vec{x},t) + \frac{q_a}{m_a}\left(\vec{E}_o(\vec{r}) + [\vec{v}\times\vec{B}_o(\vec{r})]\right)\frac{\partial}{\partial\vec{v}}f_a(\vec{x},t) \tag{4}$$

$$= \frac{\partial}{\partial\vec{v}}\left(\gamma_a(\vec{p})\vec{v}f_a(\vec{v},t)\right) + \sum_i\frac{\partial}{\partial v_i}D_{ii}^a(\vec{x})\frac{\partial}{\partial\vec{v}_i}f_a(\vec{x},t).$$

Transforming the kinetic equation for plasmas with large-scale fluctuations /5/ in an equation of Fokker-Planck-form and comparing it with the obtained Fokker-Planck-equation, we can express the unknown intensity of the stochastic force D^a and the friction coefficient γ_a by the diagonalized velocity-diffusion tensor of particle a, e.g. by the space-time spectral density of the energy of the waves excited in the plasma, and by the dielectric function of the plasma. For electrostatic waves $(\vec{k}\|\delta\vec{E})$ it yields

$$D_{ii}^a = \frac{q_a^2}{16\pi^3}\int\delta(\omega-\vec{k}\vec{v})\left(\widetilde{\delta E_i\delta E_i}\right)_{\omega\vec{k}}d\omega d\vec{k}, \quad \gamma_a\vec{v} = \frac{q_a^2}{2\pi^2}\int\delta(\omega-\vec{k}\vec{v})\frac{\vec{k}}{k^2}\frac{Im\widetilde{\varepsilon}(\omega,\vec{k})}{|\widetilde{\varepsilon}(\omega,\vec{k})|^2}d\omega d\vec{k}. \tag{5}$$

The Gaussian-distributed stochastic force $\delta\vec{F}_a(\vec{x},t)$ may be estimated by its intensity. If the growth rate of the waves is much smaller than the real part of the wave frequency $\omega_o(\vec{k})$, than

$$\sigma = \sqrt{2D_{ii}} = \frac{q_a}{2\sqrt{2}\pi}\left(\sum_{\vec{k}\vec{v}=\pm\omega_o(\vec{k})}\left(\widetilde{\delta E_i\delta E_i}\right)_{\vec{k}}\right)^{1/2} \tag{6}.$$

The tensor of the dielectric function is given by

$$\widetilde{\varepsilon}_{rs}(\omega,\vec{k}) = \delta_{rs} + \sum_a\frac{4\pi q_a^2 n_a}{m_a\omega}\int\frac{v_r\partial f_a/\partial v_s}{\omega-\vec{k}\vec{v}+i\delta I_a(\vec{v})}. \tag{7}$$

$f_a(\vec{x})$ is the solution of the kinetic equation, δI_a the linearized particle-collision operator, $\sim$ means the averaging over possible small-scale fluctuations. A diagonal tensor D_{ii}^a can be found for any electrostatic instability, as D_{ii}^a is symmetric, and any symmetric tensor can be diagonalized. In the case of ion-acoustic instability $(\vec{B}_o$ is directed along the z-axis) one gets

$$\left(\widetilde{\delta E_i\delta E_i}\right)_{\omega\vec{k}} = \begin{pmatrix} 2A+Y & 0 \\ 0 & C-Y \end{pmatrix} \tag{8}$$

$$Y = 2B^2(2d-2A-C)/((d-C)^2+2B^2), \quad d = A + \frac{C}{2} + \sqrt{A^2+C^2/4-AC+2B^2} \tag{9}$$

$$A = (\widetilde{\delta E_x\delta E_x})_{\omega,\vec{k}}, \quad B = (\widetilde{\delta E_x\delta E_z})_{\omega,\vec{k}}, \quad C = (\widetilde{\delta E_z\delta E_z})_{\omega,\vec{k}}. \tag{10}$$

3. CONCLUSIONS

A Langevin-equation (1, 2, 5-7) is found for a charged particle in a plasma with any electrostatic turbulence. The stochastic force has to be expressed by the diagonalized velocity-diffusion tensor. Solving the Langevin-equation one can investigate the influence of electrostatic microturbulence on macroscopic effects, e.g., expressing the space-time spectral density of the wave energy by (8-10) it is possible to estimate the significance of the ion-acoustic instability for solar wind dynamics. If the distribution function $f_a(\vec{x})$ could be measured exactly enough, stochastic forces could be directly obtained from particle observations (taking into account eqs.(5, 7) and the relation between $(\widetilde{\delta E_i\delta E_i})_{\omega\vec{k}}$ and $f_a(\vec{x})$).

REFERENCES

[1] SCHWARZ, S. J., Plasma instabilities in the solar wind: A theoretical review, Rev.Geophys.Space Phys.**18** (1) (1980) 313-336.

[2] GURNETT, D. A., Waves and instabilities, in: Physics and Chemistry in Space — Space and Solar Physics **21**, Physics of the Inner Heliosphere II, eds. R. Schwenn and E. Marsch, Springer-Verlag, Berlin-Heidelberg, 1991, pp. 135-148.

[3] KLIMONTOVICH, Yu. L., Turbulent Motion and Structure of Chaos. A New Approach to the Statistical Theory of Open Systems, Nauka, Moscow, 1990 (in Russ.).

[4] RÖPKE, G., Statistische Mechanik für das Nichtgleichgewicht, Deutscher Verlag der Wissenschaften, Berlin, 1987.

[5] KLIMONTOVICH, Yu. L., Statistical Physics, Nauka, Moscow, 1982 (in Russ.).

EVOLUTION OF SOLAR WIND FLUCTUATIONS AND THE INFLUENCE OF TURBULENT 'MIXING'

S. Oughton and W. H. Matthaeus

Bartol Research Institute, University of Delaware, Newark, DE 19716, U.S.A.

ABSTRACT

We present various numerical and analytical solutions for the transport of solar wind turbulence. The model used takes into account the effects of convection, expansion, and wave propagation, as well as the recently illuminated effects of (non-WKB) 'mixing' terms. The radial evolution of the fluctuating kinetic energy (E_k), magnetic energy (E_b) and normalized cross helicity (σ_c) is computed, and, it is demonstrated that in appropriate limits the solutions converge to the WKB forms. In the general case, solutions which differ substantially from those predicted by WKB theory are obtained. The degree of turbulent 'mixing' shows considerable dependence on the nature of the turbulence, giving rise to varying levels, at 1AU, of the ratio of "inward" and "outward" fluctuation energies and the ratio of kinetic to magnetic energies in the fluctuations. The transport properties described here may provide at least a partial explanation for the observed mixing of cross helicities with increasing heliocentric distance in the solar wind.

INTRODUCTION AND THE MODEL

The problem of how to adequately describe the physics of fluctuations of the interplanetary medium has been present since the earliest spacecraft observations showed that such fluctuations are ubiquitous in the solar wind. Recently developed theories of the transport of MHD scale turbulence in a weakly inhomogeneous background plasma provide a basis for computing both radial and temporal dependence of the spectrum of solar wind fluctuations /1/. Here we report on the results of a numerical and analytic investigation of such a transport model. As a consequence of the length constraints to which this paper is subject we are of necessity concise in our discussion, and we refer the reader to previous and forthcoming publications for further details *e.g.*, /1,2,3,4,5/, and also to the companion articles of Grappin, Mangeney & Velli; Marsch; Tu; Velli, Grappin & Mangeney, and in particular, Matthaeus *et.al.* appearing in this volume.

The interplanetary medium is assumed to be a single component magnetofluid obeying the usual compressible MHD fluid equations. A two length (and time) scale decomposition of these dynamical equations is performed, in which the the fields depend upon both large (**R**) and small (**x**) scale spatial co-ordinates. Thus each field separates into two components: (1) a spatially slowly varying 'mean' part, depending only on the large-scale, and (2) a fluctuating portion which depends on both the large and small spatial scales. On the basis of observational evidence (*e.g.*, /6,7/), and also for simplicity, we assume that the small-scale fluctuations are both incompressible and homogeneous. Hence, the only fluctuating quantities are the velocity (**v**) and magnetic field (**b**). Note that the large scale fields are *not* required to be either homogeneous or incompressible.

Straightforward algebraic manipulations yield a set of coupled transport equations for such physically important correlation functions as $S_{ij}^{vb}(\mathbf{R}, \mathbf{r}) = <v_i(\mathbf{R}, \mathbf{x}) \, b_j(\mathbf{R}, \mathbf{x} + \mathbf{r})>$, where the angle brackets denote averaging over $\mathbf{x}$ /1/. These equations contain terms involving the *large*-scale slowly varying fields, which we take to be specified. As a result of both the symmetries and the simple physical interpretation associated with the Elsässer variables ($\mathbf{z}^\pm = \mathbf{v} \pm \mathbf{b}/\sqrt{4\pi\rho}$), it proves convenient to use this 'inward' and 'outward' propagating modes representation. Finally we Fourier transform the equations with respect to the separation parameter $\mathbf{r}$ (conjugate variable $\mathbf{k}$).

If we denote the energy in the Elsässer fields by $P^\pm(\mathbf{R}, k_r)$, the energy difference (residual energy /8/) by $F \propto E_k - E_b$, and the helicity of the induced electric field by $J \propto Im\{\mathbf{v}^*(\mathbf{k}) \cdot \mathbf{b}(\mathbf{k})\}$, then the final set of equations is:

$$\frac{\partial P^\pm}{\partial t} + (U \mp V_{Ar})\frac{\partial P^\pm}{\partial R} + \left(\frac{U \pm V_{Ar}}{R}\right)P^\pm + M_{ki}^\pm F_{ik} = NL^\pm$$

$$\frac{\partial F}{\partial t} + U\frac{\partial F}{\partial R} + \frac{U}{R}F - (2\mathbf{k}\cdot\mathbf{V}_A)J + 2\left[M_{ki}^+ P_{ik}^- + M_{ki}^- P_{ik}^+\right] = NL^F$$

$$\frac{\partial J}{\partial t} + U\frac{\partial J}{\partial R} + \frac{U}{R}J + (2\mathbf{k}\cdot\mathbf{V}_A)F = NL^J$$

where U is the constant, radially directed mean wind speed, $\mathbf{B_0}$ is the mean magnetic field—taken equal to the standard Parker spiral, $\rho \propto 1/R^2$, is the large-scale fluid density, $\mathbf{V}_A = \mathbf{B_0}/\sqrt{4\pi\rho}$, is the large-scale Alfvén velocity, and we refer to $M_{ki}^\pm$ as the 'mixing' operators. In order to facilitate comparisons with observations we take $\mathbf{r}$ to be in the radial direction, the spectra then being reduced ones (*i.e.*, functions of k_r rather than $\mathbf{k}$). It should be stressed that we have made *no* approximations regarding the relative abundances of the 'inward' and 'outward' modes. In fact the model supports *completely* arbitrary admixtures of these modes. The terms on the left of the equations are all linear and represent the effects of convection, expansion, wave propagation and 'mixing', while those on the right represent non-linear interactions. Note that in contrast to the case of WKB transport *all* of these effects, including 'mixing', are present at *leading* order. In this preliminary study we focus upon the properties of the linear transport operators, dropping hereafter all non-linear terms*.

Before moving on to the results, a few words are in order regarding the nature of the 'mixing' operators. Physically, we can interpret 'mixing' as a scattering of the $\mathbf{z}^\pm$ modes due to large-scale gradients of the mean fields. The operators have the form:

$$M_{ji}^\pm(\mathbf{R}) = \frac{\partial \mathbf{U}_i}{\partial \mathbf{R}_j} \pm \frac{1}{\sqrt{4\pi\rho}}\frac{\partial \mathbf{B}_{0i}}{\partial \mathbf{R}_j} - \frac{1}{2}\delta_{ij}\nabla\cdot\left(\frac{\mathbf{U}}{2} \pm \mathbf{V}_A\right)$$

which is completely determined by the the *large*-scale gradients of the *mean* fields. However, because the operators always appear coupled to *small*-scale spectral tensors (*e.g.*, $Q_{nj}M_{ji}^\pm$), 'mixing' also depends on the nature and rotational symmetry properties of the small-scale turbulence. Assuming that the small-scale turbulence is either isotropic, slab, or two-dimensional (2-D) enables the trace of $Q_{nj}M_{ji}^\pm$ to be evaluated and written as $M^\pm Q$, where $Q = Q_{ii}$, and the $M^\pm$ are effective 'mixing' operators. For values of $R \gtrsim 2AU$ these effective operators are all essentially the same. Inside $1AU$, however, important differences exist between both the plus and minus versions for the same type of turbulence, and also between $M^\pm$ for different types of rotational symmetry.

The impact of the 'mixing' term on the radial evolution of the physical quantities is crucially affected by the size of $\mathbf{k} \cdot \mathbf{V}_A$. This factor is the coupling strength between the F and J fields and may be considered as a 'WKB enforcing' term. If $\mathbf{k} \cdot \mathbf{V}_A \approx 0$, then strong mixing occurs, since the initial dominance of the 'outward' mode causes growth of F, which in turn causes growth of the 'inward' mode. Thus $\sigma_c = (P^- - P^+)/(P^- + P^+)$ decreases significantly with heliocentric distance. However, when F and J are strongly coupled, the energy in the 'inward' fluctuations remains a tiny fraction of that in the 'outward' and WKB-like solutions are obtained.

---

*Note that for fully developed turbulence there is no net spectral transport of $P^\pm$ in the inertial range, *i.e.*, $NL^\pm = 0$.

RESULTS AND DISCUSSION

In order to solve the equations we have, in most cases, resorted to numerical techniques, namely Chebyshev spectral (collocation) methods /9,10/, where the equations are integrated in time to steady-state solutions. We choose to impose the boundary conditions on the fields at R_0 ($\equiv 10 R_{sun}$) so that the fluctuations there are *purely* outwardly propagating. This allows the inner boundary to be interpreted as the Alfvén radius, *i.e.*, the distance at which the (radial) flow velocity becomes equal to the (radial) Alfvén velocity. Chebyshev techniques were chosen to facilitate the subsequent inclusion of non-linear terms, and also because they support arbitrary boundary conditions.

The first case we consider is that of *isotropically* distributed fluctuations. The power-law ($k^{-\alpha}$) inertial range turbulence is characterised by a single parameter, namely its spectral slope α. Typically we choose values corresponding to the Kolmogorov ($\alpha = 5/3$) or Kraichnan ($\alpha = 3/2$) scenarios. It can be shown that for isotropic symmetries J is identically zero. An analytic solution to the $\mathbf{V}_A = 0$ version of these equations was presented by Zhou & Matthaeus /1/, and this provided a useful test of our numerical accuracy. Since J is explicitly zero, the WKB enforcing term cannot come into play, and thus we see a substantial falloff in the normalized cross helicity with increasing heliocentric distance. The solutions show significant dependence on (a) the spectral index α: increasing values causing faster radial decay of σ_c; and (b) the value of $A_0 = V_{Ar}(R_0)/U$, where V_{Ar} is the radial component of the large-scale Alfvén velocity (Figure 1a). This latter dependence decreases the effect of 'mixing' as A_0 is increased from zero to unity. The case of $A_0 = 1$ corresponds to R_0 being the Alfvén (critical) radius, while smaller positive values may be interpreted as R_0 exceeding this radius. In such cases we still enforce purely outward fluctuations at the inner boundary, despite the fact that this is no longer physically necessary. The $A_0 = 0$ case represents the situation in the absence of a large-scale magnetic field, *e.g.*, within the current sheet.

The second case relates to *slab* geometry, where the fluctuations are in the plane perpendicular to the wave-vector $\mathbf{k}$. The direction of $\mathbf{k}$ is taken to be parallel to either $\mathbf{R}$ or $\mathbf{B}_0$, the results being similar for both cases. Strong 'mixing' is seen in the separate cases of $\mathbf{k} = 0$ and $\mathbf{V}_A = 0$. However as the radial component of $\mathbf{k}$ is increased towards a reciprocal correlation scale, the results rapidly return to WKB-like solutions. As mentioned above this is because non-zero $\mathbf{k} \cdot \mathbf{V}_A$ means that F and J are tightly coupled. Such coupling constrains F to oscillate about zero (*i.e.*, remain small), and thus, since P^+ is driven only by F at this order, 'mixing' is strongly inhibited (Figure 1a).

Finally we consider 2-D turbulence, by which we mean (a) $\mathbf{k} \perp \mathbf{B}_0$, (b) fluctuations are perpendicular to both $\mathbf{k}$ and $\mathbf{B}_0$, and (c) fluctuations are distributed isotropically (in the planes normal to $\mathbf{B}_0$) with a power-law inertial range. As a consequence of this geometry $\mathbf{k} \cdot \mathbf{V}_A \equiv 0$, ensuring that

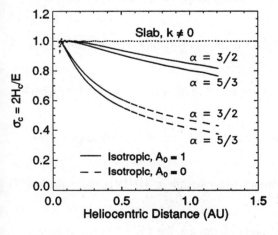
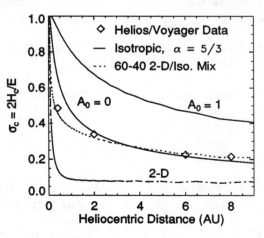

Fig. 1. Plots of the numerical solutions for the normalized cross helicity as a function of distance. (a) isotropic and slab solutions, (b) comparison with data.

F and J are decoupled, and thence that 'mixing' is always strong. In fact, J is identically zero for reasons which are essentially the same as those applying in the fully isotropic case. Significant dependence on the values of the spectral index α, and A_0 is again seen. Simulation results /11/ have shown that in the presence of a mean magnetic field, energy is transferred to the $\perp$ components of the fluctuations much more rapidly than it is to the $\parallel$ ones. In other words the 2-D spectrum 'switches on' first. This suggests that 2-D fluctuations may be particularly relevant to the solar wind system, and indeed there is also some observational evidence that the solar wind can be modeled as an admixture of slab ($\mathbf{k} \parallel \mathbf{B}_0$) and 2-D fluctuations /12/. To this end we show in Figure 1b several numerical solutions for the normalized cross helicity with observational data points from Helios and Voyager data superimposed (3 hour averages). Without attempting to optimise the agreement between theory and data, we note that the 60% 2-D, 40% isotropic mixture provides a remarkably good fit to the data. We are not suggesting that this is in fact the state of affairs in the solar wind, only that the theory clearly shows strong potential to explain the observations.

Acknowledgements. This research has been supported in part by the NASA SPTP at Bartol through grant NAGW-2076. Computations were supported by the NSF San Diego Computing Center.

REFERENCES

1. Y. Zhou and W.H. Matthaeus. Transport and turbulence modeling of solar wind fluctuations. *J. Geophys. Res.*, **95**, 10,291, 1990a.

2. Y. Zhou and W.H. Matthaeus. Remarks on transport theories of interplanetary fluctuations. *J. Geophys. Res.*, **95**, 14,863, 1990b.

3. Y. Zhou and W.H. Matthaeus. Models of inertial range spectra of interplanetary magnetohydrodynamic turbulence. *J. Geophys. Res.*, **95**, 14,881, 1990c.

4. C-Y. Tu, Z-Y. Pu, and F-S. Wei. The power spectrum of interplanetary Alfvénic fluctuations: Derivation of the governing equation and it's solution. *J. Geophys. Res.*, **89**, 9695, 1984.

5. C-Y. Tu. The damping of interplanetary Alfvenic fluctuations and the heating of the solar wind. *J. Geophys. Res.*, **93**, 7, 1988.

6. W. H. Matthaeus and M. L. Goldstein. Measurement of the rugged invariants of magnetohydrodynamic turbulence in the solar wind. *J. Geophys. Res.*, **87**, 6011, 1982a.

7. W. H. Matthaeus and M. L. Goldstein. Stationarity of magnetohydrodynamic fluctuations in the solar wind. *J. Geophys. Res.*, **87**, 10347, 1982b.

8. A. Pouquet, U. Frisch, and J. Léorat. Strong MHD helical turbulence and the nonlinear dynamo effect. *J. Fluid Mech.*, **77**, 321, 1976.

9. D. Gottlieb and S. A. Orszag. *Numerical Analysis of Spectral Methods: Theory and Applications.* SIAM, 1977.

10. C. Canuto, M. Y. Hussaini, A. Quarteroni, and T. A. Zang. *Spectral Methods in Fluid Mechanics.* Springer-Verlag, 1988.

11. J. V. Shebalin, W.H. Matthaeus, and D. Montgomery. Anisotropy in MHD turbulence due to a mean magnetic field. *J. Plasma Phys.*, **29**, 525, 1983.

12. W. H. Matthaeus, M. L. Goldstein, and D. A. Roberts. Evidence for the presence of quasi-two-dimensional nearly incompressible fluctuations in the solar wind. *J. Geophys. Res.*, **95**, 20673, 1990.

SOLAR WIND THERMAL ELECTRON DISTRIBUTIONS

J. L. Phillips and J. T. Gosling

Mail Stop D438, Los Alamos National Laboratory, Los Alamos, NM 87545, U.S.A.

ABSTRACT

Solar wind thermal electron distributions exhibit distinctive trends which suggest that Coulomb collisions and geometric expansion in the interplanetary magnetic field play key roles in electron transport. We introduce a simple numerical model incorporating these mechanisms, discuss the ramifications of model results, and assess the the model in terms of ISEE-3 results and preliminary Ulysses observations. Although the model explains certain observational features, observed gradients in total electron temperature indicate the importance of additional heating mechanisms.

Solar wind electron distributions generally include two populations: a thermal "core" component and a suprathermal "halo" component. While the halo is often strongly anisotropic, the core distributions usually have only modest anisotropies, with the parallel-to-perpendicular temperature ratio most often in the range 1.0 to 1.5. Analysis of a large data base of ISEE-3 electron observations near 1 AU /1/ revealed that the electron core anisotropy, $T_{//}/T_{\perp}$, varies systematically as a function of density, with denser plasmas more nearly isotropic. Further, $T_{\perp}$ is relatively constant, while $T_{//}$ is much more variable, such that the most anisotropic distributions have a high $T_{//}$ and thus a high total temperature. A subsequent study /2/ identified intervals where core $T_{\perp}$ exceeded $T_{//}$ and showed that this condition was associated with slow, dense, plasma and with interplanetary magnetic field (IMF) orientations which were nearly transverse to the flow.

Figure 1 summarizes these observations, for an ISEE-3 data set of 31,000 electron spectra measured near the L1 point. Data are plotted vs. core temperature ratio, $T_{//}/T_{\perp}$, and include (top panel) log number of spectra in each $T_{//}/T_{\perp}$ bin, (second panel) median core density, (third panel) medians of the two temperature components, and (bottom panel) the median departure of the ecliptic IMF direction from the Parker spiral based on the observed proton bulk speed. Note the following: (1) most distributions have a small $T_{//} > T_{\perp}$ anisotropy, with decreasing likelihood for higher anisotropies and for $T_{\perp} > T_{//}$; (2) $T_{//}/T_{\perp}$ varies inversely with density; (3) $T_{//}$ is much more variable than is $T_{\perp}$; (4) radial (transverse) IMF is associated with high (low) temperature ratios. These observations motivated development of a model for core electron transport /1,2,3/. This model is not intended to provide a complete description of electron transport; it explicitly neglects the interplanetary electrostatic potential and the electron heat flux, among other things. Its purpose is to demonstrate the important roles of collisions and adiabatic expansion in the modification of core electron distributions. We use the model to make predictions for the in-ecliptic and high-latitude phases of the Ulysses mission, and show some preliminary Ulysses in-ecliptic results which qualitatively support some of these predictions.

To construct the collision-expansion model for electron transport, we start with the double adiabatic invariants /4/ referred to henceforth as the CGL relations:

$$\frac{d}{dt}\left(\frac{P_\parallel B^2}{\rho^3}\right) = 0 \tag{1}$$

$$\frac{d}{dt}\left(\frac{P_\perp}{\rho B}\right) = 0 \tag{2}$$

where $P_\parallel$ and $P_\perp$ are plasma pressure parallel and perpendicular to the magnetic field B, ρ is mass density, and d/dt is the convective derivative. The expressions are often used to describe proton behaviour in slowly changing magnetic fields in the absence of dissipation and heat flux. By neglecting the ambipolar electric field of the solar wind and by assuming that the electron heat flux is carried by the halo electrons only, one can also apply the CGL relations to thermal electrons. Consider the expansion of a plasma parcel, convecting radially outward as a spherical shell of fixed thickness. For radial IMF, the geometric expansion is everywhere perpendicular to B; such expansion results in cooling only of $T_\perp$. For a spiral IMF, the expansion of the plasma cools both components, with the relative cooling rate of $T_\parallel$ increasing as the field wraps up. Thus one might expect large $T_\parallel/T_\perp$ anisotropies to develop in the nearly radial field of the inner solar system, and then be mitigated by the increasing pitch of the IMF at greater distances from the Sun.

Figure 2 shows the evolution of the electron temperature components and total temperature, for CGL-based transport, starting from 0.1 AU with a scalar temperature of 1.5×10^6 K; density is assumed to vary as r^{-2}. The heavy dot represents a typical electron temperature at 1 AU. For collisionless transport in a radial field (top), the CGL relations reduce to constant $T_\parallel$, $T_\perp \propto r^{-2}$, and an unrealistically high temperature ratio of 100:1 at 1 AU. For a spiral field (middle panel), the wrap-up of the IMF reduces the temperature at 1 AU, but still results in an unrealistic 32:1 temperature ratio. The bottom panel shows the extreme opposite case, one in which we assume that the distributions are collison-dominated such that no anisotropies can develop. In this limit, the CGL relations reduce to the polytropic relation $P \propto n^{5/3}$, which for $n \propto r^{-2}$ yields a temperature gradient of $T \propto r^{-4/3}$. Note that this oft-cited gradient for adiabatic transport applies only in the isotropic case; gradients in total temperature are substantially flatter if anisotropies develop.

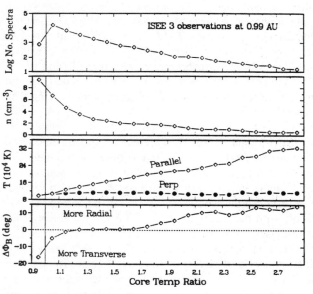

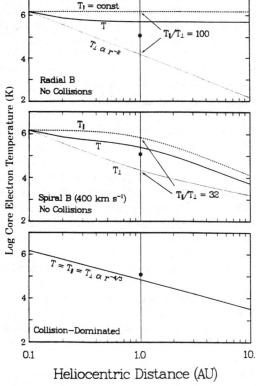

Fig. 1. Overview of ISEE-3 observations. From /3/.

Fig. 2. Evolution of core electron temperature components and total temperature based on the CGL relations for (top) radial IMF and no collisions, (middle) spiral IMF and no collisions, and (bottom) collision-dominated transport. From /3/.

The profiles in Figure 2 show the collisional and collisionless extremes of CGL-based transport. In fact, core electron distributions are marginally collisional, with collisional isotropization serving to mitigate the anisotropies generated by geometric expansion. The collision frequency used in the model, which incorporates electron isotropization by Coulomb collisions with electrons, protons, and alphas, will not be cited explicitly for the sake of brevity; the reader is referred to /3/ for this expression. It resembles more familiar collision frequencies in that it is directly proportional to n and inversely proportional to $T^{3/2}$ (actually to $T_{//}^{3/2}$), but also incorporates terms involving the anisotropy. We implement the collisional relaxation by assuming that total electron temperature is conserved in the relaxation process:

$$\frac{dT_\perp}{dt} = -\frac{1}{2}\frac{dT_{\parallel}}{dt} = 2.55\nu(T_{\parallel} - T_\perp) \tag{3}$$

where v is the electron-electron collision frequency and 2.55 is the scaling factor incorporating electron scattering by protons and a 5% (by number) alpha population /3/. We then prescibe a constant radial convection speed V, require that $n \propto r^{-2}$, and use a standard Parker spiral treatment to calculate B as a function of heliocentric distance r, solar rotation rate Ω, and heliocentric latitude θ. We can then express the variation of the two temperature components with r as follows:

$$\frac{dT_{\parallel}}{dr} = -T_{\parallel}\frac{2r\Omega^2\cos^2\theta}{V^2 + r^2\Omega^2\cos^2\theta} - \frac{2(2.55\nu)(T_{\parallel} - T_\perp)}{V} \tag{4}$$

$$\frac{dT_\perp}{dr} = -T_\perp\frac{2V^2 r^{-1} + r\Omega^2\cos^2\theta}{V^2 + r^2\Omega^2\cos^2\theta} + \frac{2.55\nu(T_{\parallel} - T_\perp)}{V} \tag{5}$$

These equations are integrated numerically from an initial assumed scalar temperature T_o at 0.1 AU.

Figure 3 shows model results, emphasizing the roles of the IMF spiral and convection speed in regulating the electron distributions. T_o is assumed to be 1.5 x 10^6 K and density to be 300 cm^{-3} at 0.1 AU. $T_{//}$ and $T_\perp$ are shown as functions of r for bulk speeds of 250 and 750 km s^{-1}. For radial B (top), bulk speed determines the time available for collisional relaxation. The slower plasma stays nearly isotropic and cools rapidly, while the faster plasma develops a large anisotropy and cools more slowly. The same qualitative effects could be created by changing the plasma density: the denser, more collisional, plasma would be more isotropic and cool more rapidly. For spiral B (bottom), the effects of changing bulk speed are twofold: a higher speed allows less time for collisions, and also creates a less tightly wound IMF spiral. For V = 250 km s^{-1} the spiral and radial IMF results are similar, because the low speed enables collisional isotropization to dominate the distributions such that the IMF orientation matters little. For V = 750 km s^{-1}, the radial and spiral field cases are markedly different; the paucity of collisions allows field orientation to play an important role.

In both the high and low speed, spiral field, cases of Figure 3, $T_\perp$ ultimately exceeds $T_{//}$. When the IMF pitch exceeds 55°, $T_{//}$ cools more rapidly than $T_\perp$ and begins to mitigate the prevailing $T_{//} > T_\perp$ anisotropy of the inner solar system. The point at which the reverse anisotropy ensues depends on a variety of factors, including speed, density, initial temperature, and latitude /2/. Note however, that the $T_\perp > T_{//}$ anisotropy in the outer solar system is never very pronounced, due to the expansion component normal to the plane of the IMF spiral, and that the anisotropy ultimately becomes largest for faster (or less dense) solar wind due to the lesser influence of collisional isotropization.

The collision-expansion model successfully explains some aspects of the ISEE-3 observations at 0.99 AU. For example, the cool and isotropic distributions characteristic of dense, slow interstream flows, despite their origin in the hot coronal streamer belt, can be attributed to the dominance of collisions and the resulting steep temperature gradient. The existence of $T_\perp > T_{//}$ anisotropies during periods of high density, low speed, and transverse IMF are also predicted by the model /2/, though the agreement with observations is far from perfect. Figure 4 shows median observed (by ISEE-3) temperature components (top) and electron density (bottom) at 0.99 AU as functions of temperature ratio. The model incorporated the following initial conditions at 0.1 AU: $V = 400$ km s^{-1}, $T_o = 1.5 \times 10^6$ K. Density was varied to produce the model results shown in Figure 4 as solid traces. This single initial condition reproduces the relative behavior of the two temperature components (top), and is somewhat successful in explaining the dependence of temperature ratio on plasma density (bottom).

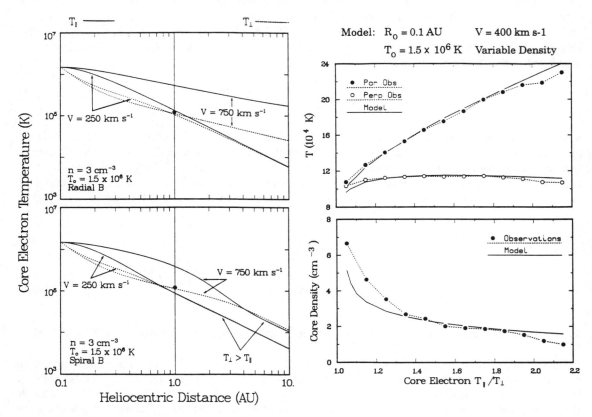

Fig. 3. Radial profiles of core temperature components for marginally collisional transport, for $V = 250$ and 750 km s^{-1}, for radial field (top) and spiral field (bottom). From /3/.

Fig. 4. Temperature components (top) and core density (bottom), as functions of core temperature ratio, showing median ISEE-3 observations and model predictions. From /3/.

The model makes predictions about the variation of electron temperature and anistropy with heliocentric distance and latitude. We chose two sets of initial conditions for purposes of illustration. The first set represents an interstream flow, charcterized by high coronal temperature, high plasma density, and low speed. The initial conditions at 0.1 AU are: $T_o = 2.0 \times 10^6$ K, $V = 350$ km s^{-1}, and $n_o = 1000$ cm^{-3}. The second set of conditions, representing a high-speed stream, is as follows: $T_o = 1.5 \times 10^6$ K, $V = 500$ km s^{-1}, and $n_o = 200$ cm^{-3}. Figure 5 shows model calculations as contours of total temperature (left) and anisotropy (right) for the interstream flow (top) and high-speed stream conditions (bottom). The view is "meridional" in a quarter plane through the solar poles and extending to 5 AU. Note that for the interstream flow, which due to its high density and low speed is highly collision-dominated, there is little variation in temperature with latitude (the contours are nearly circular). The temperature ratio reaches a maximum of only about 1.4 in the inner solar system, then gradually decreases. For the high speed stream, the lower rate of collisional isotropization allows large anisotropies to develop. This is especially pronounced at high latitudes, where the model produces nearly radial magnetic fields and predicts temperature ratios as high as 6:1 at 4-5 AU. The onset of the $T_\perp > T_{//}$ anisotropy occurs relatively close to the Sun (1.2 AU in the equatorial plane) for the interstream flow, but the anisotropy never becomes very pronounced due to the dominance of collisions. For the high-speed stream, the $T_\perp > T_{//}$ anisotropy is delayed to 2.8 AU in the equatorial plane, but ultimately becomes more pronounced due to the lower collision rate.

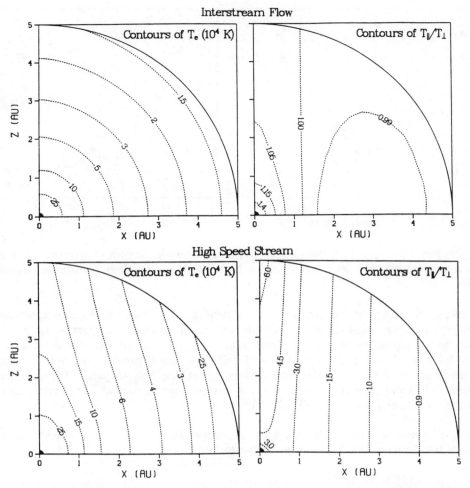

Fig. 5. Meridional view, with Sun at lower left, showing model contours of core electron temperature (left) and temperature ratio (right) for interstream (top) and high-speed (bottom) solar wind conditions. From /3/.

Ulysses in-ecliptic electron temperature results out to 3.76 AU are presented elsewhere in these proceedings /5/. Although the results should be considered preliminary, the Ulysses measurements support an overall core electron temperature gradient which can be expressed as $T \propto r^{-0.70}$. This gradient is considerably flatter than predicted by the collision-expansion model for low-speed, high-density flows, and is slightly flatter than that predicted for high-speed, low-density solar wind. Heating mechanisms not incorporated in the model, such as core-halo interactions and compression due to solar wind stream dynamics, are evidently at work. However, certain aspects of the Ulysses results qualitatively support the importance of collisions and expansion in a spiral IMF. Figure 6 (left) shows the percent occurrence of various core electron maximum-to-minimum temperature ratios (definitive identification of $T_{//}$ and $T_{\perp}$ will require further analysis) for Ulysses 3-d measurements between 1.15 and 3.76 AU. Data are sorted into three subsets, shown in the legend, based on density normalized by r^{-2} to 1 AU. Note that as density increases, the core electrons tend to become more isotropic. This is a distinctive signature of the effect of collisions in isotropizing the distributions. In the right panel, we limit the data to the medium density ($4 < nr^2 < 8$ cm^{-3}) subset, and plot the anisotropy occurrence in three ranges of heliocentric distance. Note that the distributions become more nearly isotropic with increasing distance. Thus the trends in anisotropy qualitatively support the key roles of collisions and geometric expansion in regulating electron distributions, but the observed gradient in total temperature indicates that heating mechanisms, not incorporated in the model, are probably at work.

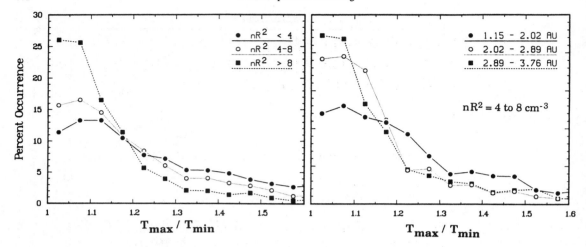

Fig. 6. Ulysses observations from 1.15 to 3.76 AU, showing percent occurrence of maximum-to-minimum temperature ratio for three density subsets (top), and for three heliocentric distance ranges (bottom).

The simple collision-expansion model presented here is not intended to be a comprehensive model for transport of solar wind thermal electrons. Other mechanisms neglected by the model also contribute to the regulation of electron distributions. However, we believe that the model provides a zeroth order understanding of the interplay of collisions and expansion in a spiral magnetic field, which together control the development and mitigation of anisotropies. We anticipate further comparison of model predictions with Ulysses observations at various heliocentric distances and latitudes, as well as exploitation of ISEE 3 - Ulysses lineups.

We thank S.J. Bame for use of Ulysses electron observations. This work was performed under the auspices of the United States Department of Energy.

REFERENCES

1. J.L. Phillips, J.T. Gosling, D.J. McComas, S.J. Bame, S.P. Gary, and E.J. Smith, Anisotropic thermal electron distributions in the solar wind, *J. Geophys. Res.*, *94*, 6453 (1989).

2. J.L. Phillips, J.T. Gosling, D.J. McComas, S.J. Bame, and E.J. Smith, ISEE 3 observations of solar wind thermal electrons with $T_\perp > T_{||}$, *J. Geophys. Res.*, *94*, 13377 (1989).

3. J.L. Phillips and J.T. Gosling, Radial evolution of solar wind thermal electron distributions due to expansion and collisions, *J. Geophys. Res.*, *95*, 4217 (1990).

4. G.F. Chew, M.L. Goldberger, and F.E. Low, The Boltzmann equation and the one-fluid hydrodynamic equations in the absence of particle collisions, *Proc. R. Soc. London, Ser. A.*, *236*, 112, 1956.

5. S.J. Bame, D.J. McComas, J.L. Phillips, J.T. Gosling, and B.E. Goldstein, The Ulysses solar wind plasma experiment: description and in-ecliptic results, these proceedings.

OBSERVATION AND SIMULATION OF THE RADIAL EVOLUTION AND STREAM STRUCTURE OF SOLAR WIND TURBULENCE

D. A. Roberts

NASA, Goddard Space Flight Center, Code 692, Greenbelt, MD 20771, U.S.A.

ABSTRACT

This paper presents a brief overview of the observed evolution in a variety of quantities describing the turbulent evolution of the interplanetary plasma and describes simulation results consistent with many features of the evolution. The turbulence is manifested through a dissipation at small scales in the inner heliosphere with a corresponding evolution in the breakpoint between a relatively flat and a Kolmogoroff spectrum; an evolution from kinetically to (slightly) magnetically dominated energy of the plasma fluctuations; a general decrease in the cross helicity or "Alfvénicity"; changes in the anisotropy of the fluctuations; and the increasing predominance of quasi-pressure-balanced structures in the compressive component of the fluctuations. MHD simulations with shear layers either side of a central current sheet show that even in the absence of compressibility the lack of a mean field along the direction of the main flow in the current sheet leads to rapid nonlinear evolution and the observed characteristics of "Elsässer spectra" of the fields in the inner heliosphere. Adding compressibility to the simulations does not greatly change the "incompressive" quantities but leads in addition to observed correlations between a measure of compression and other quantities.

INTRODUCTION

The interplanetary medium gives us an excellent laboratory for the study of MHD turbulence /1/. While this viewpoint has been bolstered by many recent studies /2,3/, many of the details of this picture have yet to be understood. The answers to such questions as the relative role of compression, shear, and the nearly spherical expansion of the plasma in the evolution of the fluctuations are beginning to become clearer, but are still controversial. This paper summarizes some of the observations that must be explained by theories of the evolution of the solar wind fluctuations, and then describes recent simulations that support the general correctness of the nearly incompressible, shear-driven turbulence viewpoint for giving a unified framework for understanding the observations. The presentation of the simulation results also gives an overview of many recent conclusions based on an "Elsässer variable" analysis of the observations. Due to space limitations, the majority of the references in this paper are to our work; this is not meant to slight the contributions of others.

SUMMARY OF OBSERVATIONS

The tables presented in this section summarize the evolution of the fluctuations in magnetic field, **B**, and velocity, **v**, over much of the range in heliocentric distances (0.3 to ~20 AU) covered by the Helios and Voyager missions. Most of the results shown here are based on the analysis of hour-averaged data, with quantities at the various scales being "ensemble averaged" by time averaging over 30 to 100 days. The term "small" will be used here for scales of 0.1 AU or less implying a period of 10 hours or less in the spacecraft frame and "large" will imply scales of about 0.1 to 1 AU. The basic energy balance is described by the first two tables /4/. The results for dissipationless ("WKB," $\delta\mathbf{B} \propto r^{-3/2}$) and "saturated" ($\delta\mathbf{B} \propto B$) waves are used as benchmarks.

TABLE 1 Magnetic Field Amplitude Evolution

	0.3 AU	1 AU	2 AU	> 8 AU
Small scale	Dissip.	Intermed.	"WKB"	"WKB"
Large scale	"WKB"	Intermed.	Intermed.	"Saturated"

TABLE 2 Alfvén ratio: δKE/δME

	0.3 AU	1 AU	2 AU	> 8 AU
Small scale	≈1	0.5	0.5	0.5
Large scale	5	3	2	0.5
Large scale*	≈1	0.6	0.5	0.5

*Not including the radial stream "fluctuations"

TABLE 3 "Alfvénicity"

	0.3 AU	1 AU	2 AU	> 8 AU
Small scale	0.8	0.4	0.3	≈0
Large scale	0.5	≈0	≈0	≈0

TABLE 4 Anisotropy

	0.3 AU	1 AU	2 AU	> 8 AU
δB	5:1	4:1	3:1	3:1
δV	3:1	3:1	3:1	3:1

TABLE 5 |B|, ρ correlation: % corr < -0.8

	0.3 AU	1 AU	2 AU	> 8 AU
Small scale	15	20	20	50

Tables 1 and 2 show that the magnetic and velocity fields experience strong damping at small scales in the inner heliosphere, but attain a state consistent with equal energy supply from large scales and dissipation from small scales in the outer heliosphere. The large scales are nearly undamped in the inner heliosphere, but the dominant kinetic energy of differential stream flow (δKE) is transferred to the large scale magnetic field (and presumably to an energy cascade as well) as the flow moves outward. The dominance of the relative streaming is shown in Table 2; the third line shows that the transverse components of the flow are nearly equipartitioned. Table 3 shows that the fluctuations become systematically less Alfvénic with increasing heliocentric distance /5/ as measured by the correlation known as the "Alfvénicity" or normalized, reduced cross helicity $\sigma_c = 2<\delta v \cdot \delta b>/<\delta v^2 + \delta b^2>$, where **b** is the magnetic field in Alfvén speed units. The fluctuations are seldom isotropic /6/, as indicated in Table 4 which gives the ratio of the power in fluctuations transverse to the minimum variance direction to that along it for small scales. The velocity is generally more isotropic than the magnetic field, and the minimum variance direction of the velocity becomes more radial in the outer heliosphere rather than following the Parker spiral as does the minimum variance of the magnetic field. Finally, Table 5 provides evidence that nearly pressure balanced structures, as indicated by the correlation between the magnetic field magnitude and the density fluctuations, are a progressively more dominant feature of the compressive component of the fluctuations /7/ although the overall role of density fluctuations remains relatively constant with $\delta\rho/\rho \approx 0.1$ for small scales at any heliocentric distance.

SIMULATION RESULTS AND COMPARISONS TO DATA

Many of the features described above, as well as the properties of the recently measured spectra of the Elsässer variables $z^\pm = v \pm b$ are reproduced in the 2-D MHD spectral method simulations described in this section /8/. The initial conditions consist of a "low-speed" region—a change of frame makes it "fast"—of plasma flowing primarily in the x direction in the center of a square box surrounded by high speed flow on either side. A current sheet, simulating the interplanetary sector bound-

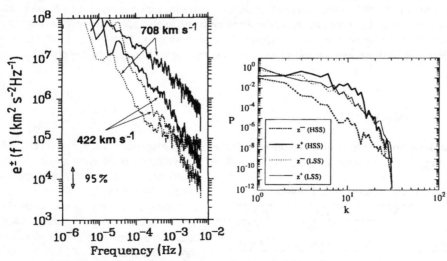

Fig. 1: Elsässer variable spectra for high and low speed streams. Observations from Helios /10/ are on the left.

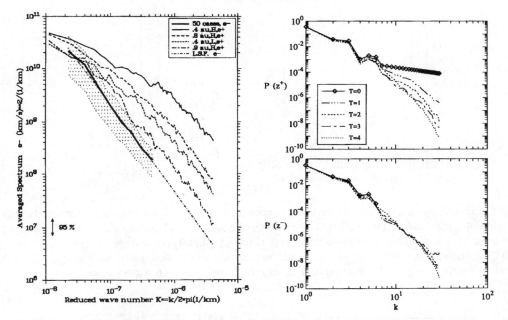

Fig. 2: Observation /11/ and simulation of the "background" z⁻ spectrum and the decay of the
z⁺ spectrum as a function of heliocentric distance (time, in the simulation).

ary, is in the middle of the low speed flow. A population of Alfvénic fluctuations with a relatively flat
spectrum is distributed uniformly throughout the box, but σ_C is small at large scales. Figures 1 through 4 refer
to a purely incompressible MHD run at 64^2 resolution. The left hand sides of all the figures show observed
properties of the solar wind, and the right sides present the simulation results.

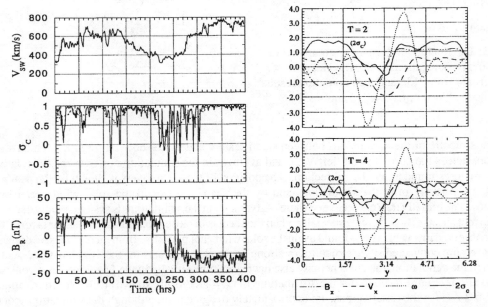

Fig. 3: Cross helicity and other quantities across a current sheet imbedded between shear
layers as seen in Helios data /12/ and in averages over x in the simulations.

Figure 1 shows wave number spectra of z^+ and z^- for plasma in the high and low speed regions, indicating
the "breathing" of the spectra seen in going from one region to the next /9,10/; in the less Alfvénic regions
the spectra collapse toward each other with corresponding changes in spectal slopes. The proximity to the
current sheet is probably the dominant factor in the differences seen here, as spectra like those of the of slow
wind are seen near the borders of the box in high speed flow where a second current sheet occurs, and spectra

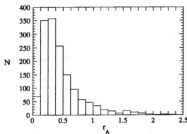

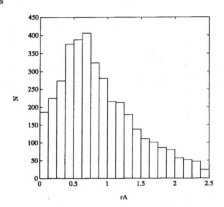

Fig. 4: Distributions of r_A from Voyager observations near 2 AU and the simulation at T = 2.

similar to that for high-speed flow shown here are seen for low-speed wind away from current sheets in the solar wind. Figure 2 shows that the Elsässer spectra in the simulation have the observed /11/ property of a nearly constant z^- spectrum when averaged over the box but a rapidly decreasing z^+ spectrum. This seems to result from equal inputs of energy to the cascade form large scales for both variables, but a greatly enhanced dissipation for the dominant fluctuations due to their high power at high wave number.

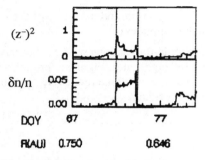

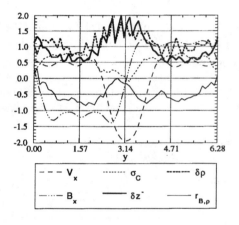

Fig. 5: The fluctuation level of z^- compared to the density fluctuation level. The Helios observations are from /13/, and the simulation results include the "incompressive" quantities.

The spatial structure of the correlations between small scale **v** and **b** are shown in Figure 3 which shows observations across the current sheet (left) /12/, and averages in the simulation box over the the relatively uniform structures found along the x direction. The position of the shear layers is shown by the peaks and valleys of the vorticity ω in the simulations and by the sharp changes in the solar wind speed in the observations. The initially flat ($\sigma_C = 1$ everywhere) Alfvénicity in the simulations is systematically decreased near the current sheets ($B_x = 0$), and especially when the sheet is in close proximity to the shear layers. Thus the similar observed low values of σ_C found using the solar wind data may arise from incompressive dynamics and do not require convected static structures or strong compressive effects for their explanation. As time progresses in the simulation, the cross helicity also decreases substantially in the fast flows, also consistent with observations. Figure 4 shows that the distributions of $r_A = \delta KE/\delta ME$, shown at T = 2 for the simulation and at about 2 AU for the observations, are approximately in agreement, although the simulations contain too many high values of this ratio.

The final set of graphs, in Figure 5, show measures of compressive effects with the same initial conditions but calculated using a polytropic compressive MHD code (see references in /2,8/). The quantities plotted in Figure 3 (right) are nearly the same as before, but we now see that the density fluctuations are correlated with the level of z^- as observed /9,13/ (left panel of the figure), and that the density and magnetic field are anticorrelated, again as observed /7/.

SUMMARY

We conclude that much of the evolution of interplanetary field fluctuations is accounted for by nonlinear incompressive MHD processes. The shear in the velocity field produced near the Sun leads to approximately equal injection of power in z^+ and z^- at large scales. Because z^- is initially small at high wave numbers, this injection leads to a spectrum that grows rapidly until it is balanced by dissipation; both injection and dissipation then remain at a relatively low, nearly constant level. The z^- spectrum thus is well developed, or "old," very near the Sun. The injection of z^+ is inadequate to balance the rapid dissipation of the large high-k fluctuations, and thus the z^+ spectrum initially decreases rapidly. When the z^+ level is nearly equal to the z^- level the two spectra continue to evolve slowly toward each other with their overall amplitudes decreasing at nearly the rate they would if no spectral transfer were occurring. This implies that in the outer heliosphere the fluctuation amplitudes will decay at nearly the "WKB" rate and this is what is observed.

The nonlinear processes are accelerated in the regions where the local mean field has only a small component along the shear layer and especially, as for the heliospheric current sheet at solar minimum, when this occurs near strong shear layers. In the latter regions the rapid dissipation of the flat z^+ spectrum has already occurred by the time it is observed at 0.3 AU; thus the turbulence is "older" and the observed temperature evolution is nearly adiabatic. By contrast, regions experiencing the rapid z^+ decay outside 0.3 AU have a slower than adiabatic temperature decrease because the dissipation heats the plasma. Broad slow speed regions that do not contain a current sheet have properties similar to high-speed streams at solar minimum, and in particular slow speed regions near solar maximum are frequently highly Alfvénic and have relatively flat z^+ spectra at low wave numbers.

Compressive effects are largely a result of the incompressive dynamics, as is to be expected in flows with low turbulent sonic Mach numbers. In particular, the presence of both z^+ and z^- in the "low speed" regions implies variations in the magnitude of the magnetic field, and these variations lead naturally to enhanced density fluctuations. The density and magnetic field magnitude variations will tend to become anticorrelated to establish the quasi-pressure-balanced state associated with "psuedo-sound." This is again consistent with the observations presented in the previous section.

The simulations do not explain how the large scale σ_c becomes small, how the initial population of Alfvénic fluctuations arises, or why the anisotropy of the fluctuations evolves as described above. These questions, along with the further theoretical understanding of the results (see elsewhere in these proceedings), are major topics for further research.

ACKNOWLEDGMENTS

The data for the original observational studies presented here were kindly provided by the principal investigators of the relevant Helios (R. Schwenn, plasma, and F. Neubauer, magnetic field) and Voyager (J. Belcher, plasma, and N. Ness, magnetic field) instruments. I also thank many collegues for helpful discussions on all aspects of this work, and especially M. Goldstein, W. Matthaeus, L. Klein, S. Ghosh, R. Grappin, A. Mangeney, E. Marsch, C.-Y. Tu, R. Bruno, and B. Bavassano. This work was supported in part by the Space Physics Theory Program grant to NASA Goddard Space Flight Center.

REFERENCES

1. P. J. Coleman, Turbulence, viscosity, and dissipation in the solar wind plasma, 153, *Astro. Phys. J.*, 371 (1968).

2. D. A. Roberts and M. L. Goldstein, Turbulence and waves in the solar wind, *Rev. Geophys., Suppl.*, 932 (1991).

3. E. Marsch, MHD turbulence in the solar wind, in *Physics of the Inner Heliosphere II*, edited by R. Schwenn, and E. Marsch, Springer, New York 1991.

4. D. A. Roberts, M. L. Goldstein, and L. W. Klein, The amplitudes of interplanetary fluctuations: Stream structure, heliocentric distance, and frequency dependence, *J. Geophys. Res.*, *95*, 4203, 1990.

5. D. A. Roberts, M. L. Goldstein, L. W. Klein, and W. H. Matthaeus, Origin and evolution of fluctuations in the solar wind: Helios observations and Helios-Voyager comparisons, *J. Geophys. Res.*, *92*, 12023 (1987).

6. L. W. Klein, D. A. Roberts, and M. L. Goldstein, Anisotropy and minimum variance direction of solar wind fluctuations in the outer heliosphere, *J. Geophys. Res.*, *96*, 3779 (1991).

7. D. A. Roberts, Heliocentric distance and temporal dependence of the interplanetary density-magnetic field magnitude correlation, *J. Geophys. Res.*, *95*, 1087, 1990a.

8. D. A. Roberts, S. Ghosh, M. L. Goldstein, and W. H. Matthaeus, MHD simulation of the radial evolution and stream structure of solar wind turbulence, *Phys. Rev. Lett.*, *67*, 3741 (1991).

9. R. Grappin, A. Mangeney, and E. Marsch, On the origin of solar wind MHD turbulence: Helios data revisited, *J. Geophys. Res.*, 95, 8197 (1990).

10. E. Marsch and C.-Y. Tu, On the radial evolution of MHD turbulence in the inner heliosphere, *J. Geophys. Res.*, 95, 8211 (1990).

11. Tu, C.-Y., and E. Marsch, Evidence for a "background" spectrum of solar wind turbulence in the inner heliosphere, *J. Geophys. Res.*, *95*, 433, (1990).

12.. D. A. Roberts, M. L. Goldstein, W. H. Matthaeus, and S. Ghosh, Velocity-shear generation of solar wind turbulence, *J. Geophys. Res.*, submitted (1991).

13. Bruno, R, and B. Bavassano, Origin of low cross-helicity regions in the inner solar wind, *J. Geophys. Res.*, *96*, 7841 (1991).

THE EVOLUTION OF LARGE-AMPLITUDE MHD WAVES NEAR QUASI-PARALLEL SHOCKS IN THE SOLAR WIND

S. R. Spangler

Department of Physics and Astronomy, University of Iowa, Iowa City, IA 52242, U.S.A.

ABSTRACT

We discuss attempts to explain the steepening of magnetohydrodynamic waves upstream of the Earth's bow shock and the formation of short-wavelength "shocklets" in terms of simple analytic models. The effort involves use of the Derivative Nonlinear Schrödinger Equation (DNLS) to describe the wave evolution. We review shortcomings of previous attempts to model foreshock wave phenomena in terms of wave packet evolution according to the DNLS. It is pointed out that: (a) the oblique propagation of the waves and (b) their growth by unstable particle distribution are conducive to the processes of wave steepening and shocklet formation. Numerical solutions of the DNLS with initial conditions corresponding to envelope-modulated, obliquely propagating wave packets yield evolved wave packets with many properties similar to the observed waves.

I. INTRODUCTION AND SUMMARY OF PRIOR INVESTIGATIONS

This paper will deal with the large-amplitude MHD waves observed upstream of the Earth's bow shock and other shock waves in the solar wind. As generated by unstable ion distributions, these waves have wavelengths of 10 to 20 thousand kilometers and they are of right circular polarization. Numerous observational investigations (e.g., Hoppe et al. /1/) have demonstrated well-defined changes in the wave properties as one goes deeper in the ion foreshock and therefore deals with more highly evolved waves.

- The waveforms show steepening.
- There appear high-wavenumber (short-wavelength) wave packets ("shocklets") associated with the waves, obviously generated by them and generally located near the fronts of the wave packets.
- There is a polarization evolution from circular to elliptical.

In the past we (and others) have attempted to understand the evolution of these waves in terms of the Derivative Nonlinear Schrödinger Equation (DNLS), a nonlinear wave equation /2/-/7/. Assume that the waves propagate along a large-scale magnetic field $B_0 \hat{e}_x$. Defining a complex field $\phi = b_y + i b_z$ where b_y and b_z are the components of the wave magnetic field, the evolution of the wave can be described by the DNLS;

$$\frac{\partial \phi}{\partial t} + \frac{1}{2} \left[\frac{1}{8\pi \rho_0 V_A (1 - \beta)} \right] \frac{\partial}{\partial x} \left\{ |\phi|^2 \phi \right\} \pm i \frac{V_A^2}{\Omega_i} \frac{\partial^2 \phi}{\partial x^2} = 0 \qquad (1)$$

In equation (1) V_A, ρ_0, Ω_i, and β represent, respectively, the Alfvén speed, plasma mass density, ion-cyclotron frequency, and plasma beta. The sign on the last (dispersive) term is determined by the wave polarization, either right hand $(-)$ or left hand $(+)$. Equation (1) is appropriate for a frame moving in the x direction with speed V_A. We have previously used the DNLS to model the evolution of foreshock wave packet models, both excluding /2/-/3/ and including /4/ the effects of wave growth and damping. These studies assumed the waves propagate parallel to the large-scale field. A representative result from numerical solutions of the DNLS is shown in Figure 1.

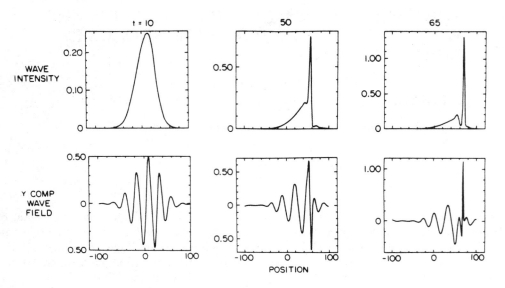

Fig. 1.   Evolution of a circularly polarized, amplitude-modulated model solar wind wave packet, subject to the DNLS (from Spangler /3/, Figure 4). Shown are the wave energy density (top rows) and one component of the transverse wave field (bottom row) at three times in the wave packet evolution. The units of time, spatial coordinate, and magnetic field strength have been de-dimensionalized according to the prescription in /3/.

The evolution in Figure 1 is characteristic of "modulationally unstable" wave evolution. This term owes its origin to the well-known fact that a constant-amplitude, circularly polarized wave packet is a solution (albeit an uninteresting one) to equation (1). One can use the DNLS to study the evolution of phase and amplitude perturbations to this circularly polarized wave. For certain, modulationally stable combinations of wave polarization and plasma β these perturbations oscillate about their initial values, whereas for modulationally unstable conditions they exponentially increase. Modulational instability occurs for RCP waves if $\beta > 1$ and LCP waves if $\beta < 1$. For realistic, heavily modulated wave packets such as shown in Figure 1, modulational instability corresponds to steepening of the wave packet and the formation of a contracted, short-wavelength pulse.

The evolution of modulationally stable wave packets is quite different from that shown in Figure 1. As discussed in Spangler /3/ and illustrated in Figure 6 of that paper, modulated wave packets tend to spread out (demodulate) and become of smaller peak amplitude. These characteristics do not resemble observed upstream wave properties.

A comparison of numerous numerical solutions of the DNLS with observed properties of foreshock waves indicate some points of agreement and others of disagreement. The aspects of foreshock wave evolution that the DNLS can reproduce are (under some modulationally unstable conditions) the steepening of wave packets and the formation of high wavenumber wavelets which can be associated with the shocklets.

However, the simple DNLS theory discussed above encounters three difficulties in being accepted as a theory of upstream waves.

• First, as noted by Spangler /4/, there is a difficulty with the available time for nonlinear development. The lifetime of a wave is the time for convection through the foreshock and into the bow shock. This time is short, or at best only comparable to the time scale for nonlinear development.

• A far more serious difficulty is the seeming unlikelihood that evolution of the form shown in Figure 1 would ever occur for the predominantly RCP waves in the foreshock. For steepening of RCP waves to occur, it is necessary for the coefficient of the nonlinearity in (1) to be negative. This would seem to be the case if $\beta > 1$. A kinetic theory derivation of the DNLS /9/–/10/ shows that this sign reversal does not occur except for extreme electron-to-ion temperature ratios. For $T_e \approx T_i$, the coefficient of nonlinearity remains positive and modulational stability of RCP wave packets would be expected.

• The description above is completely unable to account for polarization evolution. In the case of parallel propagation, the DNLS predicts that a circularly polarized wave packet remains circularly polarized.

II. THE PROSPECT OF AN "AUGMENTED DNLS"

At the end of the last section, we concluded that a "simplified" DNLS theory, employing parallel-propagating waves, with or without wave growth, failed in several important respects in accounting for the properties of upstream waves.

In this paper we explore the possibility that this failure is due to the omission of relevant foreshock processes which can be described within the context of a modified nonlinear wave equation. There are at least three physical processes known to be important in the foreshock which can be easily accommodated within an augmented DNLS theory.

• The real waves are observed to be obliquely propagating. Although the angles of propagation are observed to be small ($\sim 5° - 20°$), a fact which motivated the purely parallel approximation in our earlier work, the small-but-finite obliquity is actually quite important dynamically. The importance of oblique propagation in determining the observed characteristics of foreshock waves has recently been discussed by Malara and Elaofir /6/.

• The waves are subject to growth and damping processes, with growth and damping occurring in different wavenumber regimes. Galinsky et al. /7/ have recently suggested that different types of waves observed near comets can be interpreted in terms of different growth and damping regimes.

• The presence of energetic ions responsible for wave amplification will cause significant modification of the wave dispersion relation.

We now consider the first two of these processes. A discussion of the third process will appear in a future paper.

Oblique Wave Propagation

As has been pointed out by Mjølhus and Wyller /8/–/9/, Kennel et al. /11/, and extensively discussed by Hada et al. /12/, the DNLS also describes the propagation of oblique Alfvén waves. Once again, the direction of wave propagation (and the only direction in which spatial variations occur) is the x direction. In the case of oblique propagation, we take the large-scale field to lie in the $x - y$ plane rather than coinciding with the x axis as was the case for parallel propagation.

If one redefines the wave field by

$$\phi = b_y + ib_z + B_{Y0} \tag{2}$$

where b_y and b_z are the fluctuating wave magnetic fields, the field (2) satisfies equation (1). This is an enormous advantage in that one can use familiar analytic and numerical techniques to study obliquely propagating waves.

It should be emphasized that the unchanged nature of the DNLS for oblique propagation does not at all mean that the characteristics of the solutions are the same. Derivation of the DNLS for obliquely propagating waves illustrates two features which allow a more ready explanation of observed wave properties. First, the equation for b_y contains Korteweg-deVries-like nonlinearities which could explain steepened wave packets. Second, the differential equations for b_y and b_z are different, providing a ready explanation for polarization evolution.

Wave Growth and Damping

The foreshock waves are subject to growth processes responsible for their formation, as well as damping processes. A method for studying wave growth and damping effects was discussed in Spangler /4/. Let $\tilde{\phi}_k$ represent the spatial Fourier transform of the wave field. Equation (1) may then be rewritten

$$\frac{\partial \tilde{\phi}_k}{\partial t} \pm ik^2 \tilde{\phi}_k + \left\{ \text{nonlinearity} \right\}_k = \gamma(k)\tilde{\phi}_k \; , \tag{3}$$

where a subscript k denotes a spatial Fourier transform. The left-hand side of (3) is simply the spatially Fourier transformed DNLS. The right-hand side is a linear growth and damping term.

We make the following suggestion as to why wave growth can enhance steepening and shocklet formation. As discussed in Spangler /4/ and Ghosh and Papadopoulos /5/, the DNLS nonlinearity engenders a transfer of wave energy from the "carrier" wavelength to much shorter scales. Although not discussed in Spangler /4/, this spectral transfer also occurs in the modulationally stable case. This spectral transfer is a nonlinear effect, and therefore proceeds more rapidly for large-amplitude waves. In the case of no wave growth, a modulationally unstable wave packet will show contraction and an increase in the maximum wave amplitude. This will cause an enhancement in the spectral transfer of wave energy from the carrier mode to high wavenumbers. By the same token a modulationally stable wave packet becomes more extended and of smaller wave amplitude. The spectral transfer is slowed as the evolution progresses, leading to an asymptotically linear phase in the wave packet evolution. However, if wave growth is present, a large wave amplitude is maintained through linear processes, so spectral transfer via the DNLS nonlinearity can remain an effective process. This certainly will result in enhanced spectral power at high wavenumbers in the case of modulationally stable evolution with wave growth, and we may speculate that these high wavenumber spectral components correspond to steepened waves and shocklets.

III.   NUMERICAL STUDIES OF WAVE PACKET EVOLUTION WITH OBLIQUITY AND GROWTH

The previous section gave heuristic arguments as to why oblique wave propagation and the presence of wave growth and damping might produce steepening of wave packets and the formation of shocklets, even in the modulationally stable regime. To demonstrate the importance of these processes, we use numerical solutions of the DNLS. The code used is essentially the same as that in Spangler, Sheerin, and Payne /2/ and Spangler /3/, and modified to include the effects of wave growth and damping as described by Spangler /4/. The only difference between the present study and the previous ones is that the wave field (3) is chosen, instead of the parallel-propagating wave packets previously investigated in which $B_{Y0} = 0$.

We have made many computational runs involving a range of initial wave amplitudes, angle of propagation θ with respect to the large-scale magnetic field, wavelength, scale of spatial modulation, and growth rate of the "carrier mode." For reasons of brevity we describe here only one numerical investigation, which illustrates what we feel are the principal results. The initial conditions for this calculation were as follows. The wave amplitude is $0.50 B_0$; we assume here that linear growth has amplified the wave to such a value from an initially much smaller one. The wavelength is 20π in units of V_A/Ω_i, and a Gaussian modulation scale ℓ (defined in /2/) is 70 in these units. The angle of propagation θ was taken to be $11.5°$, characteristic of that observed for ULF waves in the Earth's foreshock. The "carrier mode," i.e., the wave with a wavelength of 20π, was amplified with a growth rate $\gamma_{\max} = 0.005\Omega_i$. Wave damping was also introduced for wavelengths of $2\pi V_A/\Omega_i$ and smaller, but this seems to be unimportant for the evolution presented below.

The results of the calculation are summarized in Figure 2. The top row shows the initial conditions, and the bottom row shows how the DNLS has evolved these initial conditions to a time of $100/\Omega_i$.

The evolved wave packet shows a number of features reminiscent of those observed in upstream wave packets, such as a steepened character (most obvious in the panel showing wave intensity), the generation of a high wavenumber wave packet, the shocklet-like waves being near the front of the wave packet, and the creation of a broadband spatial power spectrum. Examination of plots of b_y and b_z also shows pronounced polarization evolution.

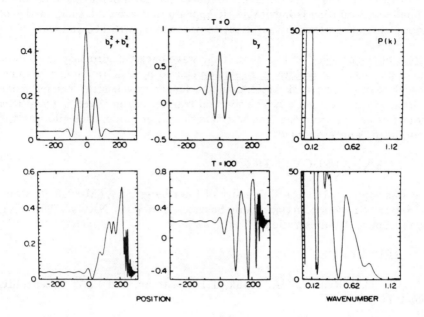

Fig. 2. Evolution of a spatially modulated, obliquely propagating wave packet. Upper row shows initial condition, bottom row shows wave field at $t = 100/\Omega_i$. The columns show, from left to right respectively, wave intensity, one component of the wave field, and the spatial power spectrum. The units of wave intensity and field are B_0^2 and B_0, respectively, while the units of the power spectrum are arbitrary.

IV. CONCLUSIONS

In this paper we have considered the extent to which a relatively simple theory based on the Derivative Nonlinear Schrödinger Equation (DNLS) can account for the nonlinear evolution observed for the MHD waves at the Earth's foreshock. The discussion differs from our earlier papers /3/–/4/ in that we consider the effects of oblique wave propagation and growth via a kinetic plasma instability of modulationally stable waves.

Wave packets evolving subject to a modified DNLS which incorporates these effects show many important features both of the observed waves and of hybrid code simulations which do not have as restricted physics content as the DNLS. The advantage of a simple analytic theory based on a single equation is that it allows identification of the physical process or processes responsible for an observed wave characteristic.

Numerical studies of the DNLS similar to those shown in Figure 2 reveal four important foreshock wave characteristics which are readily explained by the DNLS.

- The most important result is the <u>occurrence of wave steepening</u> for modulationally stable, RCP waves. In agreement with observations, the steepening occurs on the leading part of the wave. We suggest that this facet may be attributed to <u>Korteweg-deVries-like terms</u> which appear in the equation of motion for the wave field when oblique propagation is taken into account.

- The <u>generation of the shocklet wave packets</u> can be associated with <u>spectral transfer</u> of wave energy caused by the <u>DNLS nonlinearity</u>. Linear wave growth is important to this process because it keeps the wave amplitude high, thus counteracting the demodulation natural to modulationally stable wave packets. The large amplitude insures the effective nonlinear spectral transfer necessary for the formation of shocklets.

- Polarization evolution is particularly simple to interpret and is due to the oblique propagation of the wave. In this case there are different nonlinear partial differential equations for b_y and b_z, so different nonlinear evolution is inevitable. It appears that even the small angles of propagation observed for the upstream waves are sufficient to produce important nonlinear effects.

- The location of the shocklet at the front of the wave packet is also easy to explain and involves primarily linear physics. The quadratic dispersion caused by a finite ion gyrofrequency means the short-wavelength waves will travel faster than the longer wavelength "carrier modes," and thus move to the front of the wave packet. The spectral transfer of the DNLS is nonetheless important, because it transfers energy from the long wavelength carrier mode, thus allowing the shocklets to be of large amplitude and prominent.

ACKNOWLEDGMENTS

This research was supported at the University of Iowa by grant ATM-8918190 from the National Science Foundation, Division of Atmospheric Sciences, and grant NAGW-1594 from the National Aeronautics and Space Administration.

REFERENCES

1. M.M. Hoppe, C.T. Russell, L.A. Frank, T.E. Eastman, and E.W. Greenstadt, J. Geophys. Res. 86, 4471 (1981)

2. S.R. Spangler, J.P. Sheerin, and G.L. Payne, Phys. Fluids 28, 104 (1985)

3. S.R. Spangler, Astrophys. J. 299, 122 (1985)

4. S.R. Spangler, Phys. Fluids 29, 2535 (1986)

5. S. Ghosh and K. Papadopoulos, Phys. Fluids 30, 1371 (1987)

6. F. Malara and J. Elaofir, J. Geophys. Res. 96, 7641 (1991)

7. V.I. Galinsky, A.V. Khrabrov, and V.I. Shevchenko, Planet. Space Sci. 38, 1069 (1990)

8. E. Mjølhus and J. Wyller, Phys. Scr. 33, 442 (1986)

9. E. Mjølhus and J. Wyller, J. Plasma Phys. 40, 229 (1988)

10. S.R. Spangler, Phys. Fluids B 2, 407 (1990)

11. C.F. Kennel, B. Buti, T. Hada, and R. Pellat, Phys. Fluids 31, 1949 (1988)

12. T. Hada, C.F. Kennel, and B. Buti, J. Geophys. Res. 94, 65 (1989)

ESTIMATION OF HIGH ENERGY SOLAR PARTICLE TRANSPORT PARAMETERS DURING THE GLE's IN 1989

J. J. Torsti,* T. Eronen,* M. Mähönen,* E. Riihonen,* C. G. Schultz,*
K. Kudela** and H. Kananen***

* *Space Research Laboratory, University of Turku, SF-20520 Turku, Finland*
** *Institute of Experimental Physics, Slovak Akad. Sci., Kosice, Czechoslovakia*
*** *Geophysical Observatory, University of Oulu, Finland*

ABSTRACT

Analysis of five ground level enhancements in 1989 is carried out for the observations of the Lomnicky Stit and Oulu neutron monitors. The objective of the analysis is the estimation of the particle transport parameters in the solar corona and in the interplanetary space at relativistic proton energies. For the September 29 and October 24 flares the observations at both stations reveal fine structures which can be interpreted as a double injection process into IP space.

INTRODUCTION

During 1989 several ground level enhancements (GLE) were observed by neutron monitors. The purpose of this investigation is to analyse the Lomnicky Stit and Oulu observations which represent moderately different energy ranges in the primary radiation, but on the other hand have partly overlapping asymptotic cones. As a consequence some insight on the rigidity behaviour of the transport parameters is obtained. Preliminary results of this investigation were presented in /1,2/ by expanding the analysis to concern all flares in 1989 in which the maximum solar particle intensity exceeds the cosmic ray background by more than 5 %. Special attention is paid to the functional form of the transport model in cases where the observed intensity has two more or less apparent maxima.

In this work five neutron monitor observations of GLE's in 1989 are analysed: August 16, September 29, October 19, October 22, and October 24. Oulu neutron monitor registered maximal count rate increases 12 %, 179 %, 40 %, 16 %, and 97 % respectively. At Lomnicky Stit the observed highest intensities where 149 % (September 29), 8 % (October 19), and 21 % (October 24). During the August 16 and October 22 events the Lomnicky Stit observations did not reveal any significant increase. At least the increase was less than a few percent. At both stations the data collection time is 10 seconds which facilitates accurate estimation of the particle flux profile for arriving particles. The geographical coordinates of Oulu and Lomnicky Stit are (65.06 N, 25.47 E) and (49.20 N, 20.22 E). The cut-off rigidity of the Lomnicky Stit, 4.0 GV, is several times greater than that of Oulu, 0.77 GV.

MODEL FOR PARTICLE PROPAGATION

We separate the model representing particle transport from the source to the Earth into three parts:

1) The particle acceleration and release at the source into the ambient corona (U_S) is usually considered as instantaneous compared to the whole transport time, and is therefore presented as a delta-function. In this work we allow a certain spread-out of this release, represented by an adjustable parameter c_S. We suppose that the release function is gaussian:

$$U_S(t) = \dot{q} \, (2\pi c_S)^{-\frac{1}{2}} \exp(-t^2/2c_S^2)$$

where

$\dot{q}$ = production rate of the released particles.

2) The particle transport in the corona is modelled by a 2-dimensional diffusion. We allow two consecutive particle eruptions. The escape of both particle populations into interplanetary space is exponential:

$$U_c(t) = A_1/t \cdot \exp(-c_c/t - t/\tau_c) + H(t-t_0) \cdot A_2/(t-t_0) \cdot \exp(-m\, c_c/(t-t_0) - (t-t_0)/\tau_c)$$

where

c_c = the coronal diffusion factor depending on the diffusion coefficient and the distance between the source and IP emission point,

τ_c = the loss constant into IP space,

t_0 = time difference between the 1st and 2nd particle injections into the IP space.

m = correction factor of the diffusion factor c_c for the second generation particle diffusion,

$H(t-t_0)$ = function which is unity for $t > t_0$, and vanishes elsewhere.

This model presumes that there is no change in the general diffusion conditions in the solar corona when the two particle populations migrate from the source to the foot of IP field lines guiding particles to the Earth. On the other hand the model allows that the locations of the two sources are not the same. If we denote by r_i the distance from the source to the point where particles escape into IP, m represents the square of these distances, $m = (r_2/r_1)^2$.

3) For the propagation of particles in the interplanetary space we suppose that the relativistic particle transport is free escape in the interplanetary magnetic field:

$$U_i(t) = B \exp(-t/\tau_i)$$

where

τ_i =the interplanetary loss constant of particles.

The diffusion due to magnetic scattering in IP is minute compared to the escape of the relativistic particles during the first hours /1,2/.

The particle flux in interplanetary space near the Earth is then a convolution of U_s, U_c, and U_i:

$$U(t) = \int \int U_s(t_1)\, U_c(t_2 - t_1)\, U_i(t - t_1 - t_2)\, dt_1\, dt_2.$$

The transport parameters were obtained by determing the minimum of the χ^2-function between the observed and estimated fluxes N and U during the observation periods.

TABLE 1. Analysis of the 1989 GLEs

Station (model)	c_c (min)	τ_c (min)	τ_i (min)	c_s (min)	t_c (min)	t_0 (min)	m	χ^2/d.f.	K (10^{16} cm^2/s)
\multicolumn{10}{c}{16 August 1989 (X: 1:18 UT; Loc: S18,W84)}									
Oulu (s)	8	150	190	1.0	8			1.8	69
\multicolumn{10}{c}{29 September 1989 (X: 11:33 UT; Loc: W 105)}									
Oulu (d)	14	42	181	4.5	11	48	9.7	4.5	114
Lomn. (d)	51	15	109	3.5	21	36	4.9	1.8	31
\multicolumn{10}{c}{19 October 1989 (X: 12:58 UT; Loc: S27, E10)}									
Oulu (s)	32	106	1200	9.0	26			1.8	51
Lomn. (s)	150	22	650	12.0	47			3.3	11
\multicolumn{10}{c}{22 October 1989 (X: 18:05; Loc: S27, W31)}									
Oulu (s)	26	13	168	6.0	13			2.0	36
\multicolumn{10}{c}{24 October 1989 (X: 18:31; Loc: S30, W57)}									
Oulu (d)	43	33	310	11.0	25	30	8.7	2.8	13
Lomn. (d)	240	5	170	1.0	32	30	0.9	2.1	2

Model: s=single injection; d=double injection. c_c=coronal diffusion parameter. τ_c=coronal loss parameter. τ_i=interplanetary loss parameter. c_s=half width of the production rate at source. t_c=time of maximum particle intensity at corona. t_0=time difference between the 1st and 2nd particle injections into the IP space. χ^2/d.f.=χ^2-value / degrees of freedom. K=coronal diffusion parameter.

In the Table 1 the $\chi 2$-values/degree of freedom and values of the parameters of the model are given for five and three GLE analysis for the Oulu and Lomnicky Stit stations respectively. In the table s and d represent the transport model where single or double particle eruption processes are present. Coronal diffusion parameter K is calculated from $K = s^2/4c_c$ where s is the distance between the source and IP emission point at the solar surface.

FLUX ESTIMATIONS

The **September 29** flare represents, due to its exceptionally high count rate, a stringent test for transport models. In the case of Oulu, the single eruption model leads to a $\chi 2$-value (23/d.f.) which convincingly shows that the present transport model is ruled out. In all other cases the $\chi 2$-values are between 1.8-4.5/d.f. which can be regarded as a satisfactory agreement when taking into account that the observed intensities will maintain variations caused by changing IP conditions near the Earth, and modifications of the asymptotic cones due to the changing geomagnetic field during the few hours of the analysis.

Therefore we calculated two other model estimations for the Oulu observations allowing two injections. Assuming that both have the same source, m = 1, we obtained 7.3/d.f. for $\chi 2$. In the case of different sources our optimisation gave 40 % smaller $\chi 2$-value with m = 9.7.

Figure 1 shows the comparison between U and N profiles, and decomposition of U into two injection profiles. In Oulu the maximum intensities of the first and second particle pulses are at 1253 UT and 1425 UT. The maximum intensity of the second pulse is about 54 % from the first.

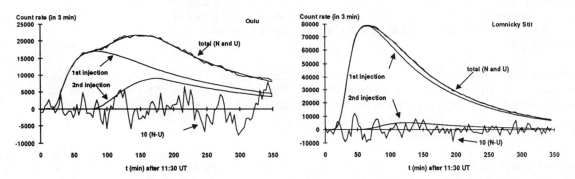

Figure 1. Count rates of the 29 Sep 1989 flare observed at Oulu (left) and Lomnicky Stit (right).

Though the intensity profile observed at Lomnicky Stit does not contain an easily discernible double pulse behaviour the analysis did find the presence of a double peak with a time separation of about 59 min. The amplitude of the second was 7.5 % from the first. The drop in the $\chi 2$-value is about 40 %.

The **October 24** flare indicates the presence of the 2nd maximum with 48 min separation and 20 % amplitude for the Oulu particle profiles, and 30 min separation and 42 % amplitude for the Lomnicky Stit particles. The differences between the injection times into IP space, t_0, are between 30-50 min for the 29 September and 24 October flares.

For **other 1989 GLE's** the transport model with single injection process gives prediction of intensity profiles which are in satisfactory agreements with the observed fluxes. Especially the estimations for the August 16, October 19, and October 22 fluxes are in good agreement with the Oulu observations ($\chi 2$-value less than 2/d.f.).

TRANSPORT PARAMETERS

The values obtained for the half width of the production rate at the source (c_s) are only tentative. They are not very sensitive in the fitting procedure. Therefore the values tell only the general trend of the relativistic particle production and injection: the time span is of the order of a few minutes.

Values obtained for the coronal diffusion parameters K of particles dominating the Oulu fluxes are systematically 3-6 times larger than those for the Lomnicky Stit measurements. As a consequence this leads to a quite stable behaviour between the various coronal flux intensity profiles at the foot of the IP field line incident with the Earth. The maximum of U_C occurs 10-50 minutes after the particle eruption at the source. The domination of low rigidity particles in the Oulu observations will have the maximum intensity 5-20 minutes earlier than the higher rigidity Lomnicky Stit particles. This difference is later compensated by the longer transport times of low rigidity particles in interplanetary space so that there is no large difference in the moments of the start of the GLE observations.

The interplanetary loss parameter is the most sensitive of all the parameters used in the present model in searching for the minimum of $\chi2$. The ratios of these parameters for Oulu and Lomnicky Stit observations are 1.66, 1.85 and 1.82, respectively for the September 29, October 19 and October 24 flares.

DISCUSSION

Smart et. al. /3/ showed in their analysis using the world-wide network of cosmic ray stations that at low energies (1 to 3 GV) the "sunward" viewing neutron monitors recorded two distinct injections. The first maximum occurred between 1200 and 1230 UT, depending on the station cutoff rigidity and the orientation of the asymptotic cone. The second maximum occurred between 1315 UT and 1410UT. The "backward viewing" monitors (in the 1 to 3 GV cutoff range) observed a single maximum at about 1345±25 UT.

The asymptotic cones of the Oulu and Lomnicky Stit stations are directed "vertically" to the IP field lines during the first hours of the September 29 GLE. They however are able to register the two injection processes in the same time intervals as the forward viewing stations.

CONCLUSIONS

The analysis of relativistic solar particle observations of both the Oulu and Lomnicky Stit neutron monitors reveal the presence of a second particle injection process in the September 29 and October 24 flares. The time separation in injection times into IP space is of the order of 30-50 min. The amplitude of the second injection is several tens of percent from the first.

ACKNOWLEDGEMENTS

The Academy of Finland and J. and A. Wihuri foundation are thanked for financial support.

REFERENCES

1. J.J. Torsti, T. Eronen, M. Mähönen, E. Riihonen, C.G. Schultz, K. Kudela and H. Kananen, Estimation of Transport Parameters for the Solar Cosmic Ray Events in 1989, 22nd International Cosmic Ray Conference, SH 3.1.15, Dublin, 1991.

2. J.J. Torsti, T. Eronen, M. Mähönen, E. Riihonen, C.G. Schultz, K. Kudela and H. Kananen, Search of Peculiarities in the Flux Profiles of GLE's in 1989, 22nd ICRC, SH 3.1.16, Dublin, 1991.

3. D.F. Smart, M.A. Shea, M.D. Wilson and L.C. Gentile, Solar Cosmic Rays on 29 September 1989; an Analysis Using the World-Wide Network of Cosmic ray Stations, 22nd ICRC, SH 3.1.2, Dublin 1991.

THE EVOLUTION OF MHD TURBULENCE IN THE SOLAR WIND

C. Y. Tu*,** and E. Marsch**

* Department of Geophysics, Peking University, Beijing 100871, China
** Max-Planck-Institut für Aeronomie, W-3411 Katlenburg-Lindau, Germany

ABSTRACT

Based on the previous work by Marsch and Tu /1/ and Tu and Marsch /2,3/, an extended theoretical model is presented in this paper in order to explain the radial evolution of the solar wind fluctuations. The idea is to combine systematically in a single model the three basic observed features: Alfvén waves, turbulence and convective structures. Part of the small-scale variations are Alfvén waves that are believed to be created near the coronal base and to propagate outward along the magnetic field lines. The large-scale variations along the magnetic field lines may be considered as convective structures that are static during the solar wind expansion time. The variations perpendicular to the convected magnetic field lines can also have small scales and contribute considerably to the solar wind fluctuations when the data sampling conditions are favourable. Nonlinear interactions will take place between inward and outward propagating Alfvén waves, between the cross-field variations themselves (2-D-turbulence) and between the cross-field fluctuations and the Alfvén waves. These nonlinear interactions then make the fluctuations to be turbulent in nature. With this idea some empirical trends of the radial evolution of solar wind turbulence can easily be explained. An increase of the relative amount of convective magnetic fluctuations against propagating Alfvén waves in the sampled data will result in a decrease of the cross-helicity, σ_c, and Alfvén ratio r_A. Under some reasonable assumptions, the spectral transfer equations for the fluctuations composed of propagating Alfvén waves and convective structures can be derived. For a very simple model an analytical solution of these transfer equations is given and found to be in qualitative agreement with the observations.

INTRODUCTION

Since the beginning of the space epoch, solar wind fluctuations have been studied for more than 30 years. However the nature of the fluctuations and their evolution have not yet been fully understood. In a pioneering paper, Coleman /4/ suggested that the solar wind fluctuations are akin to the MHD turbulence described by Kraichnan /5/. This turbulence was assumed to be driven by the velocity shear in the solar wind. In another seminal paper, Belcher and Davis /6/ suggested that the fluctuations in high-speed streams, especially in their trailing edges, are just outward propagating Alfvén waves, while the fluctuations in low-speed wind may be strongly intermixed with structures which are non-Alfvénic and possibly static in nature. The turbulence picture and wave picture apparently are not consistent with each other. Both of them are either not obviously connected with convective structures. To make a unique picture explaining these three features (or two of them) systematically has been a major theoretical goal in the passed years (see reviews /7–9/). Considerable progress has been made in this direction, however the key observational results still remain to be understood theoretically.

To facilitate a comparison with theoretical predictions, we present some statistical observational results in Fig. 1. The original data were obtained by Helios 2 from days 19 to 109 in 1976. The spectra $e^+(k')$, $e^-(k')$, $\sigma_c(k')$, and $r_A(k')$ were calculated with data sets corresponding to one-day periods, which were taken from the original data by shifting a one-day-interval successively in one-hour steps. For the detailed definitions see Tu et al. /10/. The dotted lines are the model results; see the subsequent sections for more details. We can see that the general trends of these parameters are consistent with previous studies. However, the format of our presentation is more convenient for a comparison with theoretical results.

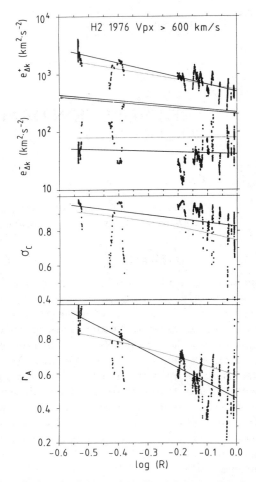

Fig. 1. The radial evolution of the energies of the Elsässer variables $e^+_{\Delta k}$ and $e^-_{\Delta k}$, the normalized cross-helicity σ_c and the Alfvén ratio r_A as observed by Helios in 1976 for solar wind velocities greater that 600 km/s. The solid lines are the least-squares fits to these data points. The dotted lines are the theoretical curves (see equations (12) and (13)).

AN EXTENDED MODEL FOR THE RADIAL EVOLUTION OF SOLAR WIND FLUCTUATIONS

We have suggested a new model incorporating the properties of Alfvén waves, convective structures and turbulence /3/. The basic idea is shown in Fig. 2. The solar wind fluctuations are conceived to be composed of outward and inward propagating Alfvén waves and convective structures. The inhomogeneity of the background fluid and the nonlinear interactions are not so strong as to destroy the dispersion relation of Alfvén waves, but they modify and couple the Alfvén waves and convective structures. Thus both Alfvén waves and convective structures together determine the radial evolution of the observed solar wind fluctuations. This idea is consistent with the earlier suggestion by Matthaeus et al. /11/ that the solar wind at 1 AU contains a population of Alfvénic fluctuations, mostly originating in the solar corona, along with a quasi-two-dimensional component that could be evidence of turbulent evolution between the sun and 1 AU. This quasi-two-dimensional component is just identical with the convective structures mentioned above.

Fig. 3 is a cartoon which shows how the Alfvén waves and the convective structures could be combined together to contribute to the observed solar wind fluctuations. In Figure 3a, each sheet represents a family of correlated field lines. There are large scale variations along the magnetic field lines. For these large-scale variations the Alfvénic transit time $T_A = \ell_\ell/V_A$ is much larger than the solar wind expansion time $T_S = r/V_S$, where ℓ_ℓ is the large scale of the variations along the magnetic field lines, V_A is the Alfvén speed, r the heliocentric distance and V_S the solar wind speed. Since the magnetic stress tensor is too small to drive the fluid to move with a considerable speed during the expansion time (Jokippii and Kota /12/; Hollweg and Lee /13/), the large-scale variations are just convected by the solar wind. However, these convective structures could result in small-scale variations in the perpendicular direction.

The small-scale variations along the magnetic field lines are Alfvén waves propagating both outwardly and inwardly. For them we have $\ell_m/V_A << L/V_S$, where ℓ_m is the small-scale length and L is the large scale. WKB theory can be used to describe the evolution of the amplitude of these waves. If we make small-scale measurements along the magnetic field line, we sample preferentially Alfvén waves. If we take measurements perpendicular to the sheets, we sample the variations caused by the convective structures. These perpendicular variations can have the same scale as the Alfvén waves. For compressible fluctuations the sheets may not be parallel to each other. However, the fluctuations of the magnetic field components may not differ a lot from the ones in the parallel case.

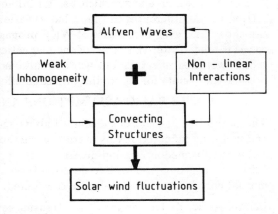

Fig. 2. The composition of the solar wind fluctuations.

With this cartoon in mind, we can easily explain qualitatively the radial evolution of the solar wind fluctuations. We simply assume that the observed solar wind fluctuations are composed of outward-propagating Alfvén waves, for which $\sigma_c = 1$ and $r_A = 1$, and of convective magnetic structures, for which $\sigma_c = 0$ and $r_A = 0$. The relative contributions of Alfvén waves and convective structures to the composed fluctuations will determine the value of σ_c and r_A. If the Alfvén waves trains are limited to narrow areas between the sheets, then the angles between radial and magnetic field direction decide on the relative importance of the Alfvén waves in the sampling of data, since due to the spacecraft orbits the sampling is always approximately along the radial direction. According to the Parker spiral, the angle between the magnetic field and the radial direction increases with increasing heliocentric distance; see Fig. 3b. As a consequence the spacecraft samples more Alfvén waves near 0.3 AU, while it samples more convective magnetic structures at large heliocentric distances. For the same reason the observed fluctuations are composed of more Alfvén waves in high-speed wind, while they are composed of more convective structures in low-speed wind at the same heliocentric distance. This angular effect is one of the possible causes for the decrease of σ_c and r_A with increasing heliocentric distance or with decreasing velocity of the streams.

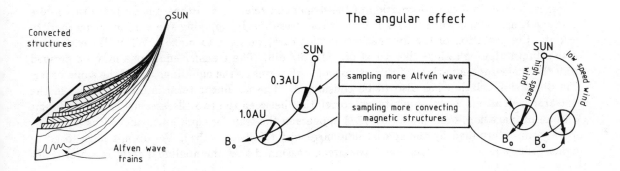

Fig. 3a. (left) Cartoon to show the idea about the occurence of Alfvén waves and convective structures in the solar wind. 3b. (right) Cartoon to show our ideas about the angular effect in the sampling of solar wind fluctuations.

The other possible effect, which has an influence on the radial evolution of the fluctuations, is related to the differences in the radial variations of Alfvén waves and convective structures. One can show that, even in the case of WKB propagation, the decrease of the energy density of Alfvén waves is much faster than that of the magnetic structures. The angular effect and the differential radial evolution effect should work simultaneously. However, in this short paper we shall only present the results for the radial evolution effect.

THE BASIC ASSUMPTIONS AND APPROXIMATIONS OF THE MODEL

The convective fluctuations discussed above resemble two-dimensional turbulence in their nature and can thus be described by statistical methods without any problem. However, for the Alfvén waves the dispersion relation dominates the properties of the fluctuations. Since we want to use statistical concepts, we have to describe these two kinds of fluctuations separately, although their mutual correlations should also be considered.

Before we can derive the combined transfer equations for the two kinds fluctuations, we have to separate them in the MHD equations. The basic assumption for this separation is that the terms related to inhomogeneity and nonlinear interactions in the fluctuation equations are of first order ($O(\ell/L)$ and small terms. This is different from the assumptions made by Zhou and Matthaeus /14,16/. They assumed that the nonlinear interaction is strong enough to destroy the dispersion relation. Since there is clear evidence of outward propagating Alfvén waves in the solar wind, we think that the nonlinear interactions are in fact weak, at least in the inner solar system. Therefore, the dispersion relation of Alfvén waves is assumed to be valid.

The wave length of the magnetic field variations along the field lines ($\ell_{\parallel}$) will determine the nature of the fluctuations, being Alfvén waves or convective structures. If $\ell_{\parallel}/V_A >> r/V_S (T_A >> T_S)$, the fluctuations can be considered as convective structures during the expansion time (Jokipii and Kota /12/; Hollweg and Lee /13/). Yet, the perpendicular scale $\ell_{\perp}$ of these convective variations may be small. These perpendicular small-scale fluctuations are considered as convected inhomogeneous two-dimensional turbulence. The transfer equations can be derived from the general transfer equations (presented by Marsch and Tu /1/, Tu and Marsch /2/ and Zhou and Matthaeus /14/) by neglecting the term $\mathbf{V}_A \cdot \mathbf{k}$. Since in this case there are no propagating Alfvén waves in the fluctuations, the use of Elsässer variables is not advantageous any more. It may be more convenient to use the correlations of $\delta\mathbf{B}$ and $\delta\mathbf{V}$ to describe the fluctuations, because evidently $\delta\mathbf{B}$ and $\delta\mathbf{V}$ have different properties (Tu and Marsch /3/; Klein et al. /17/). The large-scale variation along the magnetic field lines will be described by the equations governing the background flow and will not be discussed any further in this paper.

For fluctuations with $\ell_{\parallel}/V_A \ll L/V_S (T_A \ll T_S)$ the wave features are most prominent. If one derives the correlation equations directly from the original fluctuation equation by use of ensemble averages, he will lose the information of the dispersion relation and thus mis-evaluate the orders of the various terms, which could lead to wrong results. In applying the statistical method to describe the evolution of the fluctuations in this case, we have to make a WKB-like expansion of the fluctuation equations like Tu et al. /18,19/ did. The transfer equations may be derived from the first-order equations of this expansion. All terms in the equations are of the same order. The dispersion relation has also to be considered. The nonlinear term is included also in the first-order equation. Then no e^R–term (correlation between the two Elsässer fields with different sense of propagation, Marsch and Tu /1/) appears in the transfer equations, since e^R is a second-order term determined by the spatial inhomogeneity (Hollweg, /15/). We do not have to go to higher-order equations, because we have already assumed that the nonlinear term is a first-order term.

From the above discussion we see that for modelling solar wind fluctuations we only have to consider two kinds of contributions, the perpendicular small-scale variations produced by large-scale convective structures and the small-scale Alfvénic fluctuations propagating along the magnetic field lines. The scale-separation method prevents us from considering also intermediate-scale fluctuations either parallel or perpendicular. Such fluctuations could result in oscillations of the amplitude of Alfvén waves (Heinemann and Olbert, /20/). So we can not describe the transition from small-scale Alfvénic fluctuations to large-scale convective structures.

Under the assumption that the solar wind fluctuations only consist of small-scale Alfvén waves and large-scale convective structures, the separation of these two modes can be achieved by averaging the fluctuations over a length scale $\ell_c(\ell_\ell > \ell_c > \ell_m)$ along the magnetic field lines. The perpendicular variations of these averages are expected to represent 2-D-turbulence, while the fluctuations along the magnetic field lines remain small-scale Alfvén waves. Even if with single spacecraft observations one can generally not seperate the two modes, for comparison with the observations the models should include contributions from both Alfvén waves and convective structures.

THE TRANSFER EQUATIONS

Based on the assumptions of the previous section, two sets of transfer equations have been obtained separately for both Alfvén waves (denoted by $\delta\mathbf{V}^A$ and $\delta\mathbf{B}^A$) and convective structures ($\delta\mathbf{V}^C$ and $\delta\mathbf{B}^C$). However, the equations are interrelated by nonlinear terms. For Alfvén waves the equations are the same as the equations presented by Marsch and Tu /1/ and Tu and Marsch /2/, but with $e^R = 0$ and an additional term describing nonlinear interactions between Alfvén waves and structures. For structures we assume $\nabla \cdot \delta\mathbf{V}^C = 0$, and that there are no large-scale Alfvén waves. We define the following correlation tensors:

$$\mathsf{P}_{ij} = < \delta V_{A,i} \delta V_{B,j} >, \quad \mathsf{Q}_{ij} = < \delta B_{A,i} \delta B_{B,j} >, \quad \mathsf{C}_{ij} = \tfrac{1}{2} < \delta B_{B,i} \delta V_{A,j} + \delta B_{A,i} \delta V_{B,j} >$$

The symbols p, q, c are the traces of the three tensors, respectively. The corresponding transfer equations read as follows:

$$\frac{\partial}{\partial t}p + \mathbf{U} \cdot \nabla p + (\mathsf{P} + \mathsf{P}^\mathsf{T}) : \nabla\mathbf{U} - \frac{2}{4\pi\rho}\mathsf{C}^\mathsf{T} : \nabla\mathbf{B}_o = W_C^p + W_{C,A}^p \tag{1}$$

$$\frac{\partial}{\partial t}q + \mathbf{U} \cdot \nabla q + 2(\nabla \cdot \mathbf{U})q - 2(\mathsf{Q} + \mathsf{Q}^\mathsf{T}) : \nabla\mathbf{U} + 2\mathsf{C} : \nabla\mathbf{B}_o = W_C^q + W_{C,A}^q \tag{2}$$

$$\frac{\partial}{\partial t}c + \mathbf{U} \cdot \nabla c + (\nabla \cdot \mathbf{U})c + (\mathsf{C} - \mathsf{C}^\mathsf{T}) : \nabla\mathbf{U} + \frac{1}{2}(\mathsf{P} + \mathsf{P}^\mathsf{T} - \frac{(\mathsf{Q} + \mathsf{Q}^\mathsf{T})}{4\pi\rho}) : \nabla\mathbf{B}_o = W_C^c + W_{C,A}^c \tag{3}$$

The superscript T indicates the transposed matrix, W_C describes the nonlinear interactions between convective structures, and $W_{C,A}$, the nonlinear interactions between the convective structures and Alfvén waves.

COMPARISON OF THE MODEL RESULTS WITH OBSERVATIONS

As a first step to solve the transfer equations we consider a simple case under the following assumptions: 1. Spherical symmetry; 2. $\mathbf{U} = V_S\mathbf{e}_r$, $\mathbf{B}_o = B_o\mathbf{e}_r$ and $V_S = const.$; 3. $V_A \ll V_S$; 4. A -5/3 power law, for which all nonlinear terms disappear; 5. $\delta\mathbf{B}^c = (0, \delta B_\theta^c, \delta B_\phi^c)$; 6. $\delta\mathbf{V}^c = \mathbf{0}$. Under these conditions we found the following approximate solutions of the transfer equations:

$$e^\pm = e_o^\pm(\frac{r}{r_o})^{-1} , \ e^m \equiv \frac{q}{4\pi\rho} = \frac{< \delta\mathbf{B} \cdot \delta\mathbf{B} >}{4\pi\rho} = const. \tag{4}$$

With the assumption of isotropy of these spectra in k–space, we can evaluate the observed spectra. Considering that $e^- \ll 0.5e_o^m$, we have

$$e_{ob}^+ = e_o^m(a(r/r_0)^{-1} + 0.5) , \ e_{ob}^- = 0.5e_o^m \tag{5}$$

$$\sigma_c = a(r/r_o)^{-1}/(a(r/r_o)^{-1} + 1) , \ r_A = 0.5a(r/r_o)^{-1}/(0.5a(r/r_o)^{-1} + 1) \tag{6}$$

For high-speed wind near 0.3 AU, $a = e_o^+/e_o^m = 10$ and $e_o^m = 10^{2.2}(km/sec)^2$. The results are shown in Fig. 1 by dotted lines. We see that the model results can describe some major evolutionary trends of $e^\pm$, σ_c, and r_A. It should be pointed out that, as we will see in our subsequent paper, considering the anisotropic spectra, the Parker spiral geometry, and the sampling angular effect does not change the basic evolution trend presented here. Although the assumption about the isotropy of the spectra is not consistent with the observations /11/, the comparison presented here is meaningful. The basic evolution trend is determined by the differential radial evolution effect of these two components, while e^+ decreases like WKB theory describes it, e^m remains a constant. It should also be pointed out that equation (6) predicts that $r_A \to 0$ with $r \to \infty$. This is not consistent with the observational "saturated" value of about 0.5 in the outer heliosphere.

CONCLUSIONS AND DISCUSSIONS

A theoretical model has been suggested to explain the observed radial evolution of the solar wind fluctuations. It is assumed that the fluctuations are composed of two components: Alfvén waves and convective structures. The nonlinear interactions and spatial inhomogeneity are assumed to be weak. Yet, they can modify and cause couplings of the Alfvén waves and the structures, but can not destroy the dispersion relation. This assumption allows outward propagating Alfvén waves to exist, which is consistent with the observations. A way to separate these two components has been suggested and the transfer equations for the structures have been given. The model results of a simple example have been presented and found to be qualitatively consistent with the in situ measurements. These results suggest that the basic evolutionary trends of the solar wind fluctuations may be explained by a radial decrease in the ratio of the fluctuation energy of the Alfvén waves and convective structures.

ACKNOWLEDGEMENTS

The authors thank H. Rosenbauer and F. M. Neubauer for use of the Helios plasma and magnetic field data. Discussions with B. Inhester and J. F. McKenzie are gratefully acknowledged. Part of Tu's work is supported by the National Natural Science Foundation of China.

REFERENCES

1. Marsch, E., and C.-Y. Tu, *J. Plasma Phys.*, 41, 479 (1989)
2. Tu, C.-Y., and E. Marsch, *J. Plasma Phys.*, 44, 103 (1990)
3. Tu, C.-Y., and E. Marsch, *Ann. Geophysicae*, 9, 319 (1991)
4. Coleman, P.J., *Astrophys. J.*, 153, 371 (1968)
5. Kraichnan, R.H., *Phys. Fluids*, 8, 1385 (1965)
6. Belcher, J.W., and L. Davis, *J. Geophys. Res.*, 76, 3534 (1971)
7. Marsch, E., in *Physics of the Inner Heliosphere*, vol. 2, ed. by R. Schwenn and E. Marsch, Springer-Verlag, Berlin, Heidelberg, New York, 159 (1990)
8. Roberts, D.A., and M. Goldstein, *Reviews of Geophysics, Supplement*, 932 (1991)
9. Mangeney, A., R. Grappin, and M. Velli, in *Advances in Solar System Magnetohydrodynamics*, ed. by E.R. Priest (1990)
10. Tu, C.-Y., E. Marsch, and K.M. Thieme, *J. Geophys. Res.*, 94, 739 (1989)
11. Matthaeus, W.H., M.L. Goldstein, and D.A. Roberts, *J. Geophys. Res.*, 95, 20673 (1990)
12. Jokipii, J.R., and J. Kóta, *Geophys. Res. Lett.*, 16, 1 (1989)
13. Hollweg, J.V., and M.A. Lee, *Geophys. Res. Lett.*, 16, 919 (1989)
14. Zhou, Y., and W.H. Matthaeus, *J. Geophys. Res.*, 95, 14881 (1990)
15. Hollweg, J.V., *J. Geophys. Res.*, 95, 14873 (1990)
16. Zhou, Y., and W.H. Matthaeus, *J. Geophys. Res.*, 95, 14863 (1990)
17. Klein, L., D.A. Roberts, and M. Goldstein, *J. Geophys. Res.*, 96, 3779 (1991)
18. Tu, C.-Y., Z.-Y. Pu, and F.-S. Wei, *J. Geophys. Res.*, 89, 9695 (1984)
19. Tu, C.-Y., *J. Geophys. Res.*, 93, 7 (1988)
20. Heinemann, M., and S. Olbert, *J. Geophys. Res.*, 85, 1311 (1980)

CORRELATIONS BETWEEN THE LEVEL OF MHD FLUCTUATIONS AND THE BULK SPEED AND MASS FLUX IN THE SOLAR WIND

C. Y. Tu,*,** E. Marsch** and H. Rosenbauer**

* Department of Geophysics, Peking University, Beijing 100871, China
** Max-Planck-Institut für Aeronomie, W-3411 Katlenburg-Lindau, Germany

ABSTRACT

A new type of solar wind speed and mass-flux diagram has been used to study the correlations of the flow speed and mass flux with the energy of outward propagating fluctuations (e^+), their normalized cross-helicity (σ_c) and the Alfvén ratio (r_A). The data were obtained by Helios 1 and 2 near 0.3 AU in solar activity maximum (1979–1980). The results show that the variations of these correlations, as obtained by changing the amplitude and Alfvénicity of the fluctuations, can not be described by using the single parameter V_p. But they can be described satisfactorily by means of the velocity and mass flux diagram. The statistical results also show that the well known correlations between e^+ and proton velocity V_p and temperature T_p do not exist in the solar wind streams with a low energy flux (0.65–0.87 erg cm^{-2} s^{-1} at 1 AU) for the periods we studied. This low-energy-flux wind is a new kind of solar wind that appears only near activity maximum. The correlations of the fluctuations with the speed of this type of solar wind are more complicated than the correlations found near the minimum (75–76).

INTRODUCTION

Marsch et al. /1/ found a data period in 1978 showing that relatively pure Alfvén waves can also appear in low-speed wind. Roberts et al. /2/ further pointed out that for some periods in 1980 the purest Alfvénic fluctuations are not found in the trailing edges of high-speed streams but are associated with low-speed flows. These results stimulated the following questions for the (79–80) solar maximum: Is there a correlation between the properties of the fluctuations and the background flow, and if it exists, how do we describe the correlation?

We present here a new way to describe the correlations between the fluctuations and the streaming velocity by means of ordering the data according to both the solar wind speed V_p and the mass flux $\dot{M} = n_p V_p r^2$. With these two parameters, we can calculate the total energy flux $F = (\mathcal{E} + 0.5 V_\infty^2)\dot{M}$, where the escape speed $V_\infty = 618$ km/s, and the solar wind energy per amu $\mathcal{E} \cong 0.5 V_p^2$. For the purposes of this paper the enthalpy and other terms in the energy equation can be neglected in the Helios orbital range. F and $\dot{M}$ are the two solar wind constants of motion indicative of coronal boundary conditions. In the solar minimum (1975–1976), F was found to be nearly a constant for all wind speeds (Schwenn /3,4/, Marsch /5/). Given the value of V_p, both F and $\dot{M}$ are fixed. However, in solar maximum F (one-day-averages) varies by more than a factor of 3. The simple concept of two states of the solar wind, high-speed and low-speed wind, can not sufficiently describe the observed properties of the fluctuations in these cases. It is necessary to use both V_p and $\dot{M}$ (or F and $\dot{M}$).

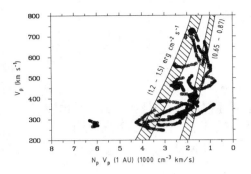

Fig. 1   The distribution of the data points of the investigated periods in the $V_p - \dot{M}$ diagram. Each point represents a set consisting of proton velocity (V_p) and number density flux ($N_p V_p$, normalized to 1 AU), which both have been averaged over a one-day period.

DATA SOURCE AND ANALYSIS

We analysed 39 one-day to two-days periods of Helios 1 and 2 data near solar maximum (113,1979 – 180,1980) for the heliocentric distance range 0.29–0.52 AU. In all these periods, the solar wind velocities did not change dramatically and did not reveal strong trends. No shock events and current-sheet crossings were contained in these periods. The detailed description of the data source and the method of the spectrum analysis were given in Tu et al. /6/. Original data gaps were filled in by averaging the adjacent data. Then the data were further averaged over 32-min periods in order to create new time series with less data gaps. Spectral analyses were made of the resulting time series of $\delta\mathbf{V}, \delta\mathbf{B}, \delta\mathbf{Z}^{\pm} = \delta\mathbf{V} \pm \delta\mathbf{V}_A$ for one-day-periods, which were created by shifting a one-day window in one-hour steps successively over the periods we had chosen. We kept all the resulting spectra with data gaps less than 12 %. Finally, 689 spectra were thus obtained. The spectra of $e^{\pm} = \frac{1}{2}(\delta Z^{\pm})^2$, $r_A = \delta V^2 / \delta V_A^2$, $\sigma_c = (e^+ - e^-)/(e^+ + e^-)$ were further averaged over a frequency band between $f = 1 \times 10^{-4}$ Hz and $f = 2 \times 10^{-4}$ Hz. For each one-day-period, we also calculated the averages of proton velocity, V_p, density, n_p, and temperature, T_p. The mass flux $\dot{M} = n_p V_p r^2$ was then calculated, where r is in AU. The results are shown in Figs. 1–3.

THE LOW–ENERGY–FLUX WIND

Fig. 1 shows the distribution of all the 689 one-day-periods in the $V_p - \dot{M}$ diagram. A prominent feature is that about half of the points are distributed in the area with $n_p V_p$ (1AU) $< 2 \times 10^8$ cm^{-2}/sec, which is the lower limit set by Leer and Holzer /7/ for modeling the solar wind. Most of these low-$\dot{M}$ cases correspond to low-energy flux wind. The two shaded belts represent high-energy flux wind (1.2 – 1.5 erg cm^{-2} s^{-1}) and low-energy flux wind (0.65 – 0.87 erg cm^{-2} s^{-1}), respectively. Schwenn /3/ showed that one-solar-rotation averages of F are around 1 erg cm^{-2} s^{-1} in 1979–1980. These averages are consistent with our results, however, we found that our one-day averages of F have wider distributions. F changes by at least a factor of 3 in Fig. 1. It is clear that the state of the solar wind can not be fully described by using only the wind velocity as an order parameter. We also made a similar calculation for the solar minimum (1975–1976) and found that the data points of V_p and $\dot{M}$ were then merely distributed around the high-energy-flux belt. The low-energy-flux wind presented in Figure 1 therefore is a new kind of the solar wind which has not yet been studied systematically.

Fig. 2 shows the distributions of data points for high, medium and low values of e^+, σ_c and r_A, respectively. We see that while these parameters decrease the patterns of the distributions evolve systematically. The relative numbers of points distributed in the range $V_p < 350$ km/sec and $n_p V_p < 2 \times 10^8$ cm^{-2} s^{-1} increase from left to right in each sub-figure. At last, for low e^+ and r_A almost all the points correspond to $V_p < 350$ km/s, while for low σ_c, all the data points are limited to the range determined by $V_p < 350$ km/s and $n_p V_p < 2 \times 10^8$ cm^{-2} s^{-1}. The low-energy-flux wind contains fluctuations with both high and low Alfvénicity, while the high-energy-flux wind seems to contain fluctuations with relatively higher Alfvénicity. An individual case study shows that the standard high-speed-flow fluctuations and standard low-speed-flow fluctuations, which typically appear in solar minimum (Tu et al. /6/), do also appear in the solar maximum (1979–1980). However, the characteristic properties of the fast-stream fluctuations,

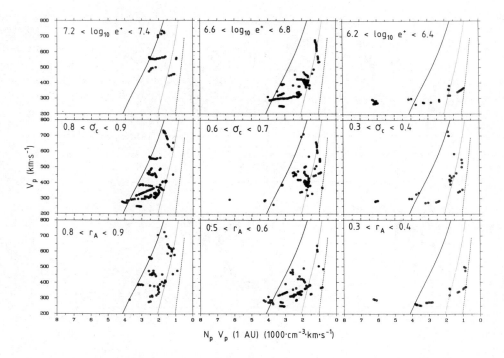

Fig. 2 The frequency distributions of data points in the $V_p - \dot{M}$ diagram for high, medium and low values of e^+, σ_c and r_A, respectively, as indicated in each box.

such as high Alfvénicity, high Alfvén ratio, a flatter spectrum, and high proton temperature, also occur in some low-speed flows, as pointed by Marsch et al. /1/ and Roberts /2/. Here we found that most of the points indicating such fluctuations are limited to the low−$\dot{M}$ range (in the low-energy-flux belt), while the points for the standard low-speed-wind fluctuations are distributed mostly in the high−$\dot{M}$ region.

Fig. 3 shows the variations of e^+ as a function of V_p and T_p for the high- and low-energy flux belts respectively. We see that in the high-energy-flux case there are clear correlations between e^+ and V_p and T_p, which are consistent with the earlier results of Feldman et al. /8/, Tu et al. /9/, and Grappin et al. /10/. The new finding here is that there are no clear correlations between e^+ and V_p and T_p in low-energy-flux wind. Therefore the Alfvénic fluctuations seem to be not related to the acceleration and heating of this low-energy flux wind.

CONCLUSIONS

We have shown that the solar wind as observed during solar maximum (1979–1980) had in many cases a comparatively low energy flux. The relative number of such cases, in which the wind has a low energy flux or has a low speed, increases with decreasing e^+, σ_c, and r_A. In the high-energy flux wind, the observed fluctuations have properties which are similar to the ones found in solar minimum. However, the fluctuations in low-energy-flux wind do have new features. The turbulent energy of these fluctuations does not at all correlate with the flow speed and temperature. This low-energy-flux wind appears to be a different kind of solar wind. The explanation of its dynamical behaviour and acceleration mechanism represents a new theoretical problem for solar wind theorists. This analysis is based on one-day periods. We have also made similar analyses based on two-day periods. The conclusions are similar to the ones reported here.

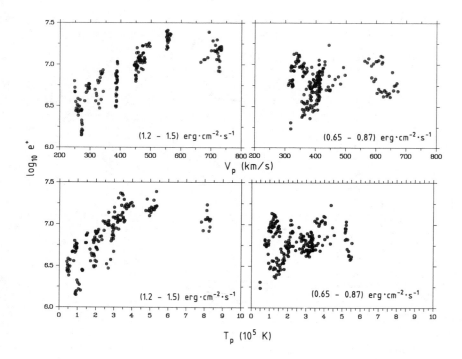

Fig. 3   The correlations of e^+ with V_p and T_p for the high- and low-energy flux solar wind.

ACKNOWLEDGEMENTS

The authors thank F. M. Neubauer for use of the Helios magnetic field data. The authors also like to thank R. Schwenn for discussion and for providing the list of possible shock-wave crossings in the Helios data. Discussions with A. Roberts are gratefully acknowledged. Part of Tu's work is a project supported by the National Natural Science Foundation of China.

REFERENCES

1.   Marsch, E., K.-H. Mühlhäuser, H. Rosenbauer, R. Schwenn, K.U. Denskat, *J. Geophys. Res.*, 86, 9199 (1981)
2.   Roberts D.A., M.L. Goldstein, L.W. Klein, W.H. Matthaeus, *J. Geophys. Res.*, 92, 12023 (1987)
3.   Schwenn, R., *Solar Wind Five*, ed. by M. Neugebauer, NASA Conf. Publ., CP-2280, 489 (1983)
4.   Schwenn, R., in *Physics of the Inner Heliosphere*, Vol. I, ed. by R. Schwenn and E. Marsch, Springer Verlag, Berlin, Heidelberg, New York, 99 (1990)
5.   Marsch, E., in *Physics of the Inner Heliosphere*, Vol. II, ed. by R. Schwenn and E. Marsch, Springer Verlag, Berlin, Heidelberg, New York, 45 (1991)
6.   Tu, C.-Y., E. Marsch, K.M. Thieme, *J. Geophys. Res.*, 95, 11739 (1989)
7.   Leer, E. and T.E. Holzer, *Solar Phys.*, 63, 143 (1979)
8.   Feldman, W.C., B. Abraham-Shrauner, J.R. Asbridge, S.J. Bame, in *Physics of Solar Planetary Environments*, Vol. I, ed. by D.A. Williams, AGU, 413 (1976)
9.   Tu, C.-Y., J.W. Freeman, R.E. Lopez, *Solar Phys.*, 119, 197 (1989)
10. Grappin, R., A. Mangeney, E. Marsch, *J. Geophys. Res.*, 95, 8197 (1990)

CORRELATION, ANISOTROPY AND COMPRESSIBILITY OF LOW FREQUENCY FLUCTUATIONS IN SOLAR WIND

P. Veltri, F. Malara and L. Primavera

Dipartimento di Fisica, Università della Calabria, 87030 - Roges di Rende (CS), Italy

ABSTRACT

The properties of low frequency fluctuations in Solar Wind, have often been studied in the framework of incompressible, statistically homogeneous MHD turbulence. In this framework, the problem of the low level of compressible fluctuations is decoupled from that of the high degree of velocity and magnetic field correlation. The former is usually ignored while relaxation processes of MHD turbulence account for the latter. Using the results of numerical simulations we analyze the effect of both Solar Wind compressibility and spatial inhomogeneity on the propagation of an initially pure Alfvenic fluctuation and we show that during the propagation the normalized cross-helicity is reduced while the level of compressible fluctuations is increased. The existence of Alfvenic fluctuations up to 1 AU, in the high speed streams, requires to take into account a physical mechanism, which is not described by the MHD equations.

INTRODUCTION

Low frequency fluctuations in Solar Wind extend over a wide (about 5 decades) frequency spectrum according to a power law: this fact seems to be the signature of a turbulent nonlinear cascade. In spite of that, these fluctuations display, mainly in the trailing edges of high speed streams up to 1 AU and over a more limited range of frequencies (about 2 decades), a number of striking properties, which have motivated their name (Alfvenic fluctuations) and have been the origin of a lot of theoretical work:
(i) A high degree of correlation between velocity and magnetic field, which corresponds to an excess of outward propagating Alfven waves [1]

$$\delta v \approx \frac{\delta B}{\sqrt{4 \pi \rho}} \qquad \text{or} \qquad \delta z^- = \delta v - \frac{\delta B}{\sqrt{4 \pi \rho}} \approx 0 \qquad (1)$$

(ii) A very low level of fluctuations in mass density and magnetic field intensity [1]

$$\delta \rho \approx 0 \qquad \text{and} \qquad \delta |B| \approx 0 \qquad (2)$$

(iii) A considerable anisotropy revealed by minimum variance analysis of magnetic fluctuations [2]

$$\lambda_1 : \lambda_2 : \lambda_3 \approx 10 : 3 : 1 \qquad (3)$$

(λ_i are the eigenvalues of the magnetic field fluctuations autocorrelation matrix).
These characteristics seem to be less common in slow streams, where there is almost no excess of outward propagating waves, a considerable level of fluctuations in mass density an magnetic field intensity and a less pronounced anisotropy: without any restriction to Alfvenic fluctuations Belcher and Davis found [1]

$$\lambda_1 : \lambda_2 : \lambda_3 \approx 5 : 4 : 1 \qquad (4)$$

In the outer heliosphere, the correlation is progressively destroyed and the anisotropy reduces to /3/

$$\lambda_1 : \lambda_3 \approx 3 : 1 \tag{5}$$

Since in fast streams $\delta\rho \approx 0$ the behavior of low frequency "Alfvenic" fluctuations, has often been studied in terms of incompressible and statistically homogeneous MHD turbulence. The main result of such studies has been to show that this turbulence displays a strong tendency to develop self-organized states where the value of the normalized cross-helicity is maximum and the energy is distributed on the different wavevectors according a power law spectrum /4-10/.

ANISOTROPY AND INCOMPRESSIBILITY

The Solar Wind is by no means an incompressible fluid, since this would require $\beta = (8\pi n\, k_B T)/B^2 >> 1$, it is however worth discussing if incompressible MHD turbulence can be assumed at least as a zero order approximation to the low frequency fluctuations behavior in the "Alfvenic periods". In this case the evolution of the MHD incompressible turbulence should account for the observed anisotropy.

In a statistically homogeneous medium there are two possible polarizations of the magnetic field fluctuations for each wave vector $\mathbf{k}$ with respect to the mean magnetic field $\mathbf{B}_0$

$$\mathbf{B}(\mathbf{k}) = b_1(\mathbf{k})\, e^{(1)}(\mathbf{k}) + b_2(\mathbf{k})\, e^{(2)}(\mathbf{k})$$

with $\qquad\qquad\qquad\qquad\qquad\qquad\qquad\qquad\qquad\qquad\qquad\qquad\qquad$ (6)

$$e^{(1)}(\mathbf{k}) = \frac{i\,\mathbf{k} \times \mathbf{B}_0}{|\,\mathbf{k} \times \mathbf{B}_0\,|} \qquad \text{and} \qquad e^{(2)}(\mathbf{k}) = \frac{i\,\mathbf{k}}{|\mathbf{k}|} \times e^{(1)}(\mathbf{k})$$

(the first polarization is that of the Alfven waves, while the second is that of both fast and slow magnetosonic waves).

Carbone and Veltri /11/ have numerically integrated a set of statistical equations /12/, which describe the behavior of incompressible MHD turbulence in the presence of a background magnetic field and have shown that in the relaxation of an initially isotropic state nonlinear interactions produce a considerable anisotropy in the wave vector distribution simultaneously with a growth of the correlation between velocity and magnetic field fluctuations, but almost no anisotropy in the polarization: in the final state, the two polarizations have the same wavevectors distributions and their energy contents differ by less than 10%. This result is clearly due to the fact that anisotropy is produced by the different rates of transfer towards small scales of parallel and perpendicular wavevectors (Alfven effect) /11-13/. In incompressible turbulence "Alfven effect" is the same for both polarizations since both have the same dispersion relation $\omega = \mathbf{k} \cdot \mathbf{v}_A$

By analyzing the Solar Wind data Carbone et al. have shown /14/ that the properties of the two polarizations are quite different, in particular: (i) the energy in the "magnetosonic" polarization is between 1/2 and 1/3 of the energy in the "Alfvenic" polarization; (ii) the spectral index of the "magnetosonic" polarization is always higher than that of the "Alfvenic" polarization.

This analysis then shows that also during the Alfvenic periods the characteristics of low frequency fluctuations cannot be directly explained in terms of incompressible statistically homogeneous MHD turbulence theory.

CORRELATION, COMPRESSIBILITY AND INHOMOGENEITY

The most recent analysis of Solar Wind fluctuations /15-17/ suggest that low frequency fluctuations consist of outward propagating Alfven waves near the Sun. At larger radial distances a "mixed" state is progressively formed, in which the normalized cross-helicity is lowered, the level of mass density and magnetic field intensity fluctuations is increased and the anisotropy is reduced. It has been suggested /18/ that this state is possibly due to local nonlinear shear instabilities.

We want to explore the possibility that mixing is only due to the propagation of initially pure Alfven waves in a "compressible" and statistically "inhomogeneous" medium like the Solar Wind and does not requires the free energy contained in the shears. We have then set up a solution of the MHD equations, representing pure Alfven waves

$$\mathbf{B} = B_0\, \mathbf{e}_y + \delta\mathbf{B} \qquad\qquad \delta\rho \approx 0 \qquad\qquad \delta\mathbf{v} \approx \frac{\delta\mathbf{B}}{\sqrt{4\,\pi\,\rho}}$$

and

$$\delta\mathbf{B} = B_1 \{ \cos[\ \Phi(y)\ ]\, \mathbf{e}_x + \sin[\ \Phi(y)\ ]\, \mathbf{e}_z \}$$

(7)

and we have separately studied the effects of "compressibility" and "inhomogeneity" on its propagation by numerical integration of the following set of equations

$$\frac{\partial\rho}{\partial t} + \nabla\cdot(\rho\,\mathbf{v}) = 0$$

$$\frac{\partial\mathbf{v}}{\partial t} + (\mathbf{v}\cdot\nabla)\,\mathbf{v} = -\frac{1}{\rho}\nabla p + \frac{1}{4\pi\rho}(\nabla\times\mathbf{B})\times\mathbf{B} + \frac{1}{\rho\,S_v}\nabla^2\mathbf{v}$$

(8)

$$\frac{\partial\mathbf{B}}{\partial t} = \nabla\times(\mathbf{v}\times\mathbf{B}) + \frac{1}{S_R}\nabla^2\mathbf{B}$$

$$\frac{\partial T}{\partial t} + (\mathbf{v}\cdot\nabla)\,T + (\gamma-1)\,T\,(\nabla\cdot\mathbf{v}) = \frac{1}{\rho S_k}\nabla^2 T + (\gamma-1)\frac{1}{\rho S_v}\left(\frac{\partial v_i}{\partial x_j}\frac{\partial v_j}{\partial x_i}\right) + (\gamma-1)\frac{1}{4\pi\rho S_R}(\nabla\times\mathbf{B})^2$$

Compressibility Effects

It is well known that a circularly polarized Alfven wave can decay in a sound wave and a backward propagating Alfven wave (parametric instability). Using a 1D code we have let propagate our simple wave, with $B_1 = 0.5\, B_0$, in a homogeneous magnetic field B_0 with a value of $\beta = 0.02$. By an appropriate choice of the function $\Phi(y)$, we have studied two different initial spectra for the magnetic

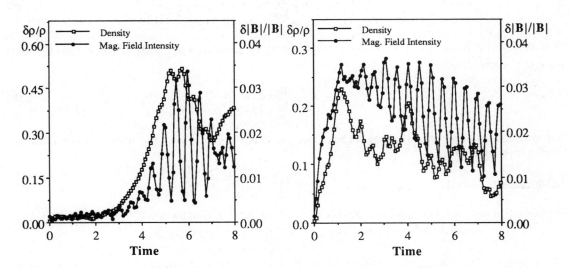

Fig.1. Homogeneous simulation: time evolution of the density and magnetic field intensity fluctuations for an initial narrow band spectrum. The time unit is the fundamental wave period.

Fig.2.Homogeneous simulation: time evolution of the density and magnetic field intensity fluctuations for an initial broad band spectrum. The time unit is the fundamental wave period.

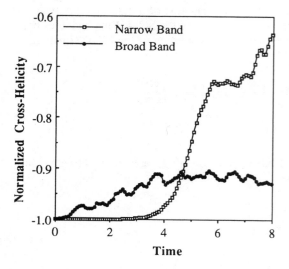

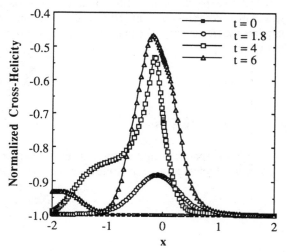

Fig.3. Homogeneous simulation: time evolution of the normalized cross-helicity. The time unit is the fundamental wave period.

Fig.4. Inhomogeneous simulation: profiles of the the normalized cross-helicity at different times. Times are normalized to a / V_A .

energy density: (i) a narrow band spectrum, i.e. a spectrum where only few wave numbers were initially excited (ii) a broad band spectrum, i.e. a power law spectrum extending over all the allowed wave numbers. In both cases, in the wave propagation, a parametric instability develops and density and magnetic intensity fluctuations grow up (fig. 1 and fig.2), while the normalized cross-helicity is progressively reduced (fig. 3). The density and magnetic intensity fluctuations, produced by the instability, are always anticorrelated as often observed in Solar Wind data. In the case of the broad band spectrum the overall process is accelerated but it saturates at a lower level: saturation requires 1 or 5 wave periods respectively in the broad band and in the narrow band spectrum case. To decide if this instability has time sufficient to develop in the Solar Wind we have compared the saturation times with the transit time, i.e. the time the Solar Wind needs to arrive at a distance R from the Sun:

$$\tau_{tr} \approx \frac{R}{V_{SW}} \tag{9}$$

(V_{SW} is the Solar Wind velocity). It can be easily shown the instability develops for those waves whose periods T satisfy

$$T < \frac{2\pi}{5} \frac{R}{V_{SW}} \frac{V_A}{V_{SW}} \tag{10}$$

Introducing typical numerical values in (10), we find that for R = 1 AU, the wave periods must be smaller than 10 hours .

<u>Inhomogeneity Effects</u>

We have then assumed an inhomogeneous background magnetic field of the form

$$B_0(x) = B_0^0 [1 + \Delta \, tgh \, (x/L)] \tag{11}$$

To avoid resistive instabilities we have chosen $\Delta = 0.157$, which ensures the fact that the magnetic field is nowhere zero. The typical shear length associated to (11) is given by $a = L/\Delta$. The equilibrium magnetic pressure related to (11) has been balanced by an appropriate choice of the temperature profile. In correspondence to the variations of B_0 and T, β varies from 1 to 0.1 across the shear. We have studied

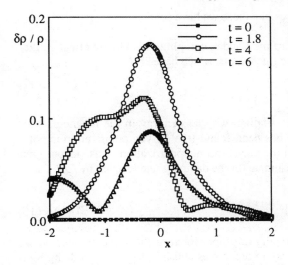

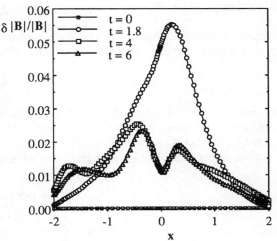

Fig.5. Inhomogeneous simulation: profiles of the normalized density fluctuations at different times. Times are normalized to a / V_A.

Fig.6. Inhomogeneous simulation: profiles of the magnetic field intensity fluctuations at different times. Times are normalized to a / V_A.

the propagation in this magnetic field of the simple wave described by equations (7), with $\phi(y) = ky$ and $ka = 21$. We have used a spectral compressible code, where a Fourier expansion in the y-direction and a double Chebyshev expansion in the x-direction are performed /19/. Looking at figs 4, 5 and 6 it can be seen that in the shear region: (i) the normalized cross-helicity is progressively reduced (fig. 4), thus indicating that backward propagating Alven waves are being produced, (ii) the energy of the Alfven wave is progressively converted into density (fig. 5) and magnetic field intensity (fig. 6) fluctuations. It is also worth noting that in the high magnetic field region $\delta\rho$ and $\delta|B|$ are correlated while they are anticorrelated in the low magnetic field region. In order to decide whether the modification in the Alfven wave due to the inhomogeneity have time sufficient to develop, we have compared the characteristic evolution time

$$\tau_{inh} \approx \frac{a}{V_A} \tag{12}$$

(V_A being the Alfven velocity) with the transit time (9) and we have found that the following condition must be satisfied

$$\frac{a}{R} < \frac{V_A}{V_{SW}} \tag{13}$$

which means that transverse inhomogeneities with length scales smaller than about 1/6 of the radial distance from the Sun are effective in modifying the initial Alfven waves.

CONCLUSIONS

At variance with the behavior of an incompressible statistically homogeneous medium, where correlation between velocity and magnetic field fluctuations are naturally produced, in a compressible, statistically inhomogeneous medium there are many processes which tend to destroy such correlation and the associated properties. These processes then furnish a natural explanation for the characteristics of low frequency fluctuations in slow streams and more generally for the observed evolution of these characteristics in the outer heliosphere.

The main physical problem which remain to be solved is then to explain how it is possible to conserve the characteristics of Alfvenic fluctuations in fast streams up to 1 AU /20/, i.e. to determine the nature of the feedback which inhibits the efficiency of processes like those we have studied.

A tentative explanation can perhaps be given in terms of the Landau damping: this kinetic effect could represent a good candidate, in that it tends to damp those fluctuations which have /21/

$$\delta |\mathbf{B}| \neq 0 \tag{14}$$

With respect to the anisotropy analysis /13/ this could explain why less energy in the magnetosonic polarization with respect to the Alfvenic polarization has been found (the Alfven polarization is not affected by Landau damping) and why the spectrum of the magnetosonic polarization was steeper than that of the Alfvenic one. In fact an order of magnitude estimate for the damping rate is given by /20/

$$\gamma_L \approx \exp\{-\frac{1}{\beta}\} \omega \tag{15}$$

(ω being the wave frequency). Since the Landau damping is proportional to the wave frequency, we expect that it is more effective at higher wave frequencies). In order to establish if the Landau damping is sufficient to stop the growth of the magnetic intensity fluctuations produced, for example, by the background magnetic field shear, we must compare t_{inh} with γ_L^{-1}. The comparison shows that magnetic intensity fluctuations with periods smaller than T_{cut}

$$T_{cut} \approx \frac{2\pi a}{V_{SW}} \exp\{-\frac{1}{\beta}\} \tag{16}$$

are efficiently damped. Assuming $a/V_{SW} \approx 1$ day we find for $\beta = 0.5$ $T_{cut} \approx 0.8$ days and for $\beta = 0.2$ $T_{cut} \approx 4.2 \cdot 10^{-2}$ days: very small variations of β give rise to considerable differences in the Landau damping effectiveness, a fact which could be related with the completely different properties of low frequency fluctuations in fast streams and slow streams. However only an extended search for a correlation between local β values and the presence of Alfvenic fluctuations could furnish a definitive answer to the problem.

REFERENCES

1. J.W. Belcher and L. Davis, *J. Geophys. Res.*, 76, 3534 (1971).
2. B. Bavassano, M. Dobrowolny, G. Fanfoni, F. Mariani, and N.F. Ness, *Sol. Phys.*, 78, 373 (1982).
3. L.W. Klein, D.A. Roberts, and M. Goldstein, *J. Geophys. Res.*, 96, 3779 (1991).
4. M. Dobrowolny, A. Mangeney, and P. Veltri, *Phys. Rev. Lett.*, 45, 144 (1980).
5. R. Grappin, U. Frisch, J. Leorat, and A. Pouquet, *Astron. Astrophys.*, 105, 6 (1982).
6. R. Grappin, A. Pouquet, and J. Leorat, *Astron. Astrophys.*, 126, 51 (1983).
7. W.H. Matthaeus, M.L. Goldstein, and D. Montgomery, *Phys. Rev. Lett.*, 51, 1484 (1983).
8. A. Ting, W.H. Matthaeus, and D. Montgomery, *Phys. Fluids*, 29, 3261 (1986).
9. A. Pouquet, M. Meneguzzi. and U. Frisch, *Phys. Rev.*, A33, 4266 (1986).
10. R. Grappin, *Phys. Fluids*, 29, 2433 (1986).
11. V. Carbone, and P. Veltri, *Geophys. Astrophys. Fluid Dynamics*, 52, 153 (1990).
12. P. Veltri, A. Mangeney, and M. Dobrowolny, *Nuovo Cimento*, 68B, 235 (1982).
13. J.D. Shebalin, W.H. Matthaeus, and D. Montgomery, *J.Plasma Phys.*, 29, 525 (1983).
14. V. Carbone, F. Malara, and P.Veltri, this issue (1991).
15. B. Bavassano, and R. Bruno, *J. Geophys. Res.*, 94, 11977 (1989).
16. Roberts et al., J. Geophys. Res., 95, 4203 (1990).
17. R. Grappin, A. Mangeney, and E. Marsch, *J. Geophys. Res.*, 95, 8197 (1990).
18. D.A. Roberts, and M.L. Goldstein, in: *Proceedings of the Third International Conference on Supercomputing*, Vol. I, ed. L.P. Kartashev and S.I. Kartashev, International Supercomputing Institute, p.310 (1988).
19. F. Malara, P. Veltri, and V. Carbone, *Phys. Fluids*, submitted (1992).
20. R. Grappin, M. Velli, and A. Mangeney, Ann. Geophys., 9, 416 (1991).
21. A. Barnes, *Phys. Fluids*, 9, 1483 (1966).

WHISTLER INSTABILITY AND MAGNETIC MOMENT DIFFUSION OF ELECTRON DISTRIBUTION FUNCTIONS AT THE EARTH'S BOW SHOCK

P. Veltri and G. Zimbardo

Università della Calabria, Dipartimento di Fisica, I-87030 Arcavacata di Rende, Italy

ABSTRACT

The wave-particle interaction at the Earth's bow shock is considered. The electron distribution function obtained from a previous Monte Carlo simulation is found to be strongly unstable with respect to whistler modes. In turn, the whistlers interact with the electrons by changing their pitch angle and magnetic moment. Hence, new diffusion terms are introduced in the Monte Carlo simulation, and a better fit of the perpendicular temperature is obtained.

INTRODUCTION

The interaction of the solar wind with a planetary magnetosphere leads to the formation of a collisionless bow shock. In particular, the November 7, 1977, bow shock crossing of ISEE 1 at $\sim$ 2251 UT is a quasi-perpendicular shock, in which the ions are heated by the reflection process /1/, and the electrons by the electrostatic potential in the deHoffman-Teller frame /2–4/. Recent numerical work /5/ has shown that the electron distribution function f_e can be obtained by means of a Monte Carlo simulation where the electrons are evolving, in the drift approximation, into the shock electric and magnetic fields. Although many features of the shock are well reproduced, the perpendicular temperature $T_\perp$ is systematically lower than observed, and the resulting distribution functions are strongly anisotropic, with a clear loss-cone structure, Figure 1. This suggests that a wave particle interaction capable of changing the electron magnetic moment, such as that due to whistler waves, is needed in order to smooth out the loss cone and to allow the transfer from parallel to perpendicular energy /4–6/.

In this paper, first we study the stability of the simulated electron distribution function f_e with respect to whistlers /7–9/, and compare the obtained growth rates with the data. The diffusion coefficient due to the whistler waves is evaluated, and a model diffusion profile for the electron magnetic moment and parallel velocity is inferred: This is used to introduce a diffusion operator in the simulation equations. Then, we carry out a second set of Monte Carlo simulations: The preliminary results show that an improved fit of the perpendicular and parallel temperatures is obtained.

WHISTLER INSTABILITY ANALYSIS

In the Monte Carlo simulations performed in /5/, a random force term, which models the electrostatic wave-particle interaction in the drift approximation, is added to the electron equations of motion. In a numerical run, 50,000 particles are injected in the shock layer, using the measured electric and magnetic field profiles of the November 7, 1977, bow shock crossing /2/, to which we refer in what follows. Finally, $f_e(x, v_\perp, v_\parallel)$ is obtained from the occupation time of each phase space cell (a sample is shown in Figure 1a for 2251:12 UT), and the results for $T_\perp$ are shown in Figure 1b.

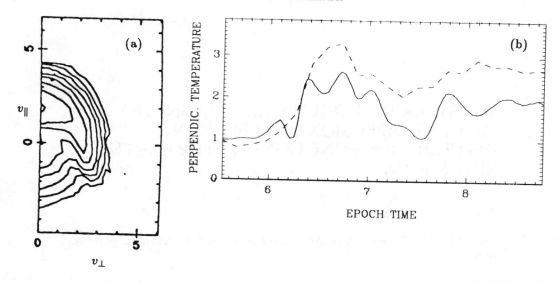

Fig. 1. (a) Contour plots, logarithmically spaced, of the electron distribution functions obtained from the Monte Carlo simulation, corresponding to 2251:12 UT, i.e., to the shock foot. (b) Perpendicular temperature: computed $T_\perp$ (solid lines) compared with the measured one (dashed lines) versus time. The Epoch Time is computed in minutes after 2245:00 UT.

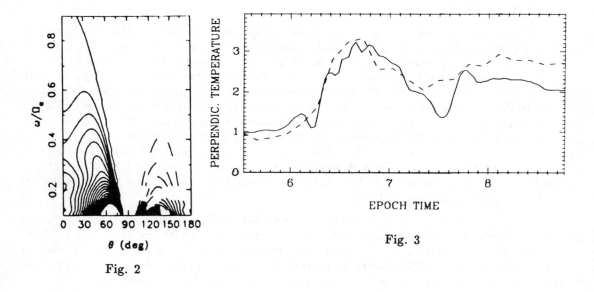

Fig. 2

Fig. 3

Fig. 2. Whistler growth rate contours, obtained with the distribution function of Figure 1a. The growth rate is normalized to Ω_e, and the contours are spaced by 0.075 Ω_e. Solid lines: positive values, implying instability. Dashed lines: negative values, implying damping of the waves. The "outermost" solid line corresponds to $\gamma = 10^{-3}\Omega_e$.

Fig. 3. Perpendicular temperature: Same format as Figure 1b, but for the Monte Carlo simulation including the random terms due to the whistler waves.

The growth rate of an electromagnetic wave in a plasma can be obtained by the semi-classical treatment of Melrose /10–12/, and for a non relativistic plasma can be written as

$$\gamma_\sigma(\mathbf{k}) = \sum_{\nu=-\infty}^{\infty} \frac{2\pi^2 e^2}{m\Omega_e^2} \int d^3v\ v_\perp^2 \frac{[C_\sigma J_\nu + J_\nu']^2}{\frac{1}{2}[\frac{\partial}{\partial\xi}(\mu_\sigma^2 \xi^2)](1+T_\sigma^2)}$$

$$\times \delta\left[\xi\left(1 - \frac{\mu_\sigma v_\parallel \cos\theta}{c}\right) - \nu\right]\left\{\frac{\nu\Omega_e}{v_\perp}\frac{\partial}{\partial v_\perp} + k_\parallel \frac{\partial}{\partial v_\parallel}\right\}f_e \tag{1}$$

where $\mathbf{k}$ is the wavevector, Ω_e is the electron gyrofrequency, $v_\perp$ ($v_\parallel$) is the velocity perpendicular (parallel) to the magnetic field direction, C_σ and T_σ are coefficients related to the wave mode polarization σ /11/, J_ν are the Bessel function of order ν, whose argument is $k_\perp v_\perp/\Omega_e$, ξ is the wave frequency normalized to Ω_e, μ_σ is the refraction index, θ the angle between $\mathbf{k}$ and the magnetic field, and f_e is normalized to number density. In the following we adopt the whistler mode refraction index

$$\mu_W = \left[1 + \frac{\omega_{pe}^2/\Omega_e^2}{\xi(|\cos\theta| - \xi)}\right]^{1/2} \tag{2}$$

which implies that $\sqrt{m_e/M_i} < \xi < 1$, since the ion contribution is neglected. The growth rate is computed by making use of the electron distribution functions obtained by the Monte Carlo simulations, and the results for the same f_e as in Figure 1a are shown in Figure 2. It can be seen that γ/Ω_e is fairly large in some regions of the parameter space (ω, θ), in particular for low frequencies. In general, γ is large and positive for $\theta < 90°$ between 2250:36 UT and 2251:36 UT, corresponding to the shock foot and ramp, and where f_e has a marked loss cone. Substantial emission in the direction of $\mathbf{B}$, $\theta \sim 0°$, at an increased frequency $\omega \sim 0.5\ \Omega_e$, is observed at some locations, as at 2253:00 UT (not shown). These results are in good agreement with the observation of magnetic noise at the bow shock, in the frequency channels from 5.6 Hz to 100 Hz /2/; note that the upstream gyrofrequency is $\Omega_e/2\pi = 150$ Hz, and that the observed Doppler shifted frequency is within a factor two of the emission frequency /8/. In particular, the magnetic noise is most intense in the foot and the ramp, corresponding to the regions where the whistler growth rate is the largest, whereas the attenuated magnetic noise downstream of the ramp can be understood as damping in the regions where $\gamma < 0$.

MAGNETIC MOMENT DIFFUSION AND MONTE CARLO SIMULATION

The excited whistler modes interact with the electrons by changing their magnetic moment and parallel velocity: The diffusion coefficient in magnetic moment μ (and similarly in $v_\parallel$) is obtained as /10/

$$D_{\mu\mu} = \sum_{\nu=-\infty}^{\infty} \int \frac{d^3k}{(2\pi)^3}\ \nu^2 \left(\frac{\hbar e}{mc}\right)^2 \frac{w^W(\nu, \mathbf{p}, \mathbf{k})E^W(\mathbf{k})}{\hbar\omega} \tag{3}$$

where $w^W(\nu, \mathbf{p}, \mathbf{k})$ is the transition probability per unit time, and $E^W(\mathbf{k})$ the energy density of whistler waves with wavevector $\mathbf{k}$. Note that the magnetic moment elementary change $\hbar e/mc$ is twice Bohr's magneton. An estimate of the diffusion time, made with the measured electromagnetic energy density, shows that this is short enough to influence the electron motion when traversing the shock layer, but longer than the electrostatic wave diffusion time considered in the previous simulation /5/.

The electron interaction with the whistlers can be introduced in the Monte Carlo simulation by adding proper random terms in the evolution equation of the magnetic moment and the parallel velocity used earlier:

$$\frac{dx}{dt} = v_\parallel b_x \tag{4}$$

$$\frac{dv_\parallel}{dt} = b_x\left[-\frac{e}{m}E - \frac{\mu}{m}\frac{dB}{dx}\right] + F_r^{e.s.}(v_\parallel, x, t) + F_{r\parallel}^W(v_\parallel, x, t) \tag{5}$$

$$\frac{d\mu}{dt} = F_{r,\mu}^W(v_\parallel, x, t) \tag{6}$$

where x is along the shock normal, $b_x = \mathbf{B} \cdot \mathbf{e}_x / B$, B is the magnetic field intensity, E the cross shock electric field in the deHoffman-Teller frame, and $F_r^{e.s.}$, $F_{r\parallel}^W$, and $F_{r,\mu}^W$ are random force terms. These are taken to be of the form

$$F_r(v_\parallel, x, t) = A(v_\parallel, x)\eta_r(t), \tag{7}$$

where $\eta_r(t)$ is a random function with zero correlation time (white noise), while the amplitude A is related to the diffusion coefficients given above, with proper spatial and velocity profiles. The terms $F_{r\parallel}^W$ and $F_{r,\mu}^W$, are taken to be statistically independent of $F_r^{e.s.}$, which is the only random term in the first Monte Carlo simulation, and is related to the electrostatic noise in the lower hybrid regime. The distribution function and its moments are obtained as before, and $T_\perp$ is shown in Figure 3. When comparing with the perpendicular temperature obtained without magnetic moment diffusion (Figure 1b), one can notice that the fit of $T_\perp$ is much improved. Also improved is the fit of $T_\parallel$. A better fit of all of the moments should be obtained in the forthcoming runs by optimizing the parameters of the simulation.

CONCLUSIONS

The present study further clarifies the role of steady electric and magnetic fields, and of electrostatic (adiabatic) and whistler (non-adiabatic) wave-particle interactions in collisionless shock waves:

1) The steady electric and magnetic fields energize the electrons by means of a reversible heating in the deHoffman-Teller frame /2–4/.

2) The electrostatic (lower hybrid) wave-particle interactions in the drift approximation distribute the energy over the whole electron population in an irreversible process, but this type of interaction is not effective in increasing $T_\perp$ /5/. The resulting distribution functions are strongly unstable with respect to whistler modes.

3) Wave-particle interactions which change the adiabatic invariant μ, such as that due to whistler waves, are effective in increasing the perpendicular temperature $T_\perp$. Also, they further smooth out the electron distribution function and allow a better fit of its moments.

REFERENCES

1. Kennel, C.F., J.P. Edmiston, T. Hada, in *Collisionless Shocks in the Heliosphere: Reviews of Current Research*, ed. B.T. Tsurutani and R.G. Stone, pp. 1–36, AGU, Washington, D.C., 1985.

2. Scudder, J.D., A. Mangeney, C. Lacombe, C.C. Harvey, T.L. Aggson, R.R. Anderson, J.T. Gosling, G. Paschmann, and C.T. Russell, *J. Geophys. Res. 91*, 11019 (1986)

3. Scudder, J.D., A. Mangeney, C. Lacombe, C.C. Harvey, and T.L. Aggson, *J. Geophys. Res. 91*, 11053 (1986)

4. Scudder, J.D., A. Mangeney, C. Lacombe, C.C. Harvey, C.S. Wu, and R.R. Anderson, *J. Geophys. Res. 91*, 11075 (1986)

5. Veltri, P., A. Mangeney, J.D. Scudder, *J. Geophys. Res. 95*, 14939 (1990)

6. Kennel, C.F. and H.E. Petschek, *J. Geophys. Res. 71*, 1 (1966)

7. Kennel, C.F. and H.V. Wong, *J. Plasma Phys. 1*, 75 (1967)

8. Tokar, R.L., D.A. Gurnett, W.C. Feldman, *J. Geophys. Res. 89*, 105 (1984)

9. Tokar, R.L., D.A. Gurnett, *J. Geophys. Res. 90*, 105 (1985)

10. Melrose, D.B., *Astrophys. Space Sci. 2*, 171 (1968); Melrose, D.B., in *Plasma Astrophysics*, vol. 1, The emission, absorption and transfer of waves in plasmas, pp. 144–178, Gordon and Breach, New York, 1980.

11. Mangeney, A. and P. Veltri, *Astron. Astrophys. 47*, 165 (1976)

12. Goldstein, M.L. and C.K. Goertz, in *Physics of the Jovian Magnetosphere*, ed. A.J. Dessler, pp. 317–352, Cambridge University Press, New York, 1983.

ALFVÉN WAVE PROPAGATION IN THE SOLAR ATMOSPHERE AND MODELS OF MHD TURBULENCE IN THE SOLAR WIND

M. Velli,* R. Grappin and A. Mangeney

Observatoire de Paris-Meudon, 92195 Meudon Cedex, France

ABSTRACT

The propagation of Alfvén waves along a purely radial magnetic field in the solar atmosphere is discussed, with particular emphasis on the role of the Alfvénic critical point in determining the transmission of the waves into the wind. Models for the evolution of Alfvénic turbulence are compared to the low-frequency limit of the linear equations.

INTRODUCTION

The basic equations for magnetic field ($\vec{b}$) and incompressible velocity ($\vec{v}$) fluctuations, in a plasma of density ρ, may be conveniently expressed in terms of Elsässer variables $\vec{z}^{\pm} = \vec{v} \mp \text{sign}(\vec{B}_0)\vec{b}/\sqrt{(4\pi\rho)}$, describing Alfvén waves propagating in opposite directions along the average magnetic field $\vec{B}_0$:

$$\frac{\partial \vec{z}^{\pm}}{\partial t} + (\vec{U} \pm \vec{V}_a) \cdot \vec{\nabla}\vec{z}^{\pm} + \vec{z}^{\mp} \cdot \vec{\nabla}(\vec{U} \mp \vec{V}_a) + \frac{1}{2}(\vec{z}^- - \vec{z}^+)\vec{\nabla} \cdot (\vec{V}_a \mp \frac{1}{2}\vec{U}) = 0, \qquad (1)$$

where U is the average wind velocity, and V_a the average Alfvén velocity. The first two terms in (1) describe wave propagation; the third term describes the reflection of waves by the gradients of the equilibrium fields along the fluctuations (and depend explicitly on the relative directions of the backround field and fluctuation polarisation); the fourth term describes the WKB losses and the isotropic part of the reflection. Nonlinear terms, of the form $(\nabla z^{\mp} z^{\pm})$, have been neglected. An easily derived consequence of (1) is the conservation of net upward wave-action /1/ which may be written as

$$S^+ - S^- = S_{\infty}, \qquad S^{\pm} = F\frac{(U \pm V_a)^2}{U\,V_a}|\vec{z}^{\pm}|^2, \qquad (2)$$

where S_{∞} is constant and F ($= \rho\,U R^2$ for a spherical expansion) is the solar wind mass flux. In the following we will consider only radial Alfvén speed and wind profiles. First we discuss the transmission problem in the static case: $S^{\pm}$ then reduce to the upward and downward propagating energy flux integrated across a flux tube cross-section, and (2) is a statement of the conservation of the net upward wave power.

PROPAGATION IN THE STATIC CORONA

Considering waves of frequency ω and wavevector $k = \omega/V_a$, (1) becomes, after elimination of the systematic amplitude variation of $z^{\pm}$ through the normalization $z^{\pm} = \rho^{1/4}z^{\pm}$,

$$z^{\pm\prime} \mp ikz^{\pm} - \frac{1}{2}\frac{k'}{k}z^{\mp} = 0. \qquad (3)$$

---

* NATO-CNR Advanced research fellow

The solution may be written formally /2/ as (heliocentric distance is normalized to the solar radius $R_\odot$ i.e., $r = R/R_\odot$ throughout)

$$z^\pm = \mathrm{Texp}\left(\int_0^r dr\ \left[ik\hat\sigma_3 + \frac{1}{2}\frac{k'}{k}\hat\sigma_1\right]\right)z^\pm{}_0, \tag{4}$$

where $\hat\sigma_{1,3}$ are the Pauli matrices, and the time-ordered propagator is quasi-unitary. The first order WKB expansion of the solution is given by oppositely propagating waves, and the region where this solution breaks down may be found by applying the Liouville-Green transformation/3,4/ to the equations for the velocity and magnetic field fluctuations (also in velocity units):

$$v'' + k^2\ v = 0$$
$$b'' + (k^2 + k''/k - 2k'^2/k^2)\ b = 0. \tag{5}$$

This procedure allows one to define a local critical frequency

$$\omega_c^{2\pm} = V_a'^2/4 \pm V_a V_a''/2. \tag{6}$$

At frequencies above the greater of $\omega_c^\pm$, we have propagating WKB solutions, while at frequencies below the smallest value the anticlassical expansion (decaying WKB) is valid. In the solar context, only two situations have been considered: a) a static multilayer planar isothermal atmosphere /5,6,7,8/, in which case calculations show that the transmission is very low, except for resonant peaks at the eigenfrequencies of the cavity considered /7/; b) a spherically or supraspherically diverging flux-tube/4/, where calculations have been limited to the transition frequencies (6). The conclusion of these studies must be taken *cum grano salis*: planar models generally underestimate transmission through the lower atmospheric layers, in particular because flux tube expansion at the chromospheric level is neglected /9/. In the case of a spherically symmetric isothermal atmosphere, the Alfvén speed, after an initial exponential increase, decreases as r^{-2} and a significant amount of tunneling should occur.

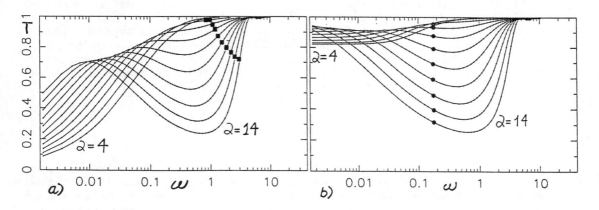

Fig. 1.   Transmission coefficient as a function of frequency for an isothermal atmosphere a), and b) for an isothermal wind. Frequency is normalized so that $\omega = 1$ corresponds to a period of 1 hour.

To calculate the transmission coefficient, we integrate (3) backwards starting at a heliocentric distance large enough such that an outwardly propagating WKB solution is adequate ($\omega \gg \omega_c^\pm$); the Alfvén speed profile is given by

$$V_a = \frac{V_{a0}}{r^2} * \exp\left(\left(\frac{\alpha}{2}\left(1 - \frac{1}{r}\right)\right)\right) \qquad \alpha = \frac{GM_\odot}{R_\odot C_s^2},$$

where C_s is the thermal speed and the parameter α typically lies in the range $4 \leq \alpha \leq 15$ for coronal temperatures between $8.0\ 10^5 - 3.0\ 10^6$ °K . In fig. 1a) we show the transmission coefficient $T = S_\infty/S_\odot^+$ (S_∞ is the energy flux carried by the asymptotic WKB wave, while $S_\odot^+$ is the outward propagating energy flux at the coronal base) as α is varied from 4.0 to 14.0 but the ratio β of kinetic to magnetic pressure at the coronal base is held constant at a value of 4% (the curves shown are for integer values of α between 4 and 14). A strong dependence of transmission on temperature for waves of periods greater than one hour is apparent, while waves with periods less than about 15 mins. are completely transmitted by the solar atmosphere. Notice that the greatest critical frequency (6), denoted by the dark squares on the transmission curves, yields only a very crude estimation of the actual transmission, because of the significant tunneling effect.

PROPAGATION IN THE SOLAR WIND AND ALFVENIC TURBULENCE

The problem of the transmission of Alfvén waves cannot be solved correctly unless we take into account the solar wind which after some 10 to 20 solar radii replaces the static gravitational stratification. When the wind is included, the asymptotic boundary condition is replaced by the condition of regularity /1,10/ at the Alfvénic critical point where $U = V_a$. The transmission coefficient is again defined as $T = S_\infty/S_\odot^+$, where now however the wave action flux is determined by the amplitude of the outwardly propagating wave at the critical point as $S_\infty = S_c^+ = 4F|\vec{z}^+|_c^2$. The transmission coefficient as a function of frequency is shown in fig. 1b) for a family of isothermal wind models of varying temperature. Again, the coronal base β is fixed at a value of 4%: the models are then defined completely by the single parameter α. The shape of the transmission curves is similar to that of the static case, except for a general increase in transmission at low frequencies, due to the presence of the wind. An interesting feature to remark again is the effect of the background temperature on the transmission, which may be related to the strong observed correlation,(at least at solar minimum), between the level of the turbulence and the stream temperature /11/ at periods around one hour; of course, the isothermal wind models described above are rather crude, not taking into account important effects such as flux tube expansion. It would be interesting to develop more realistic models to test whether the observed correlation may be attributed to the transmission properties of the ambient medium.

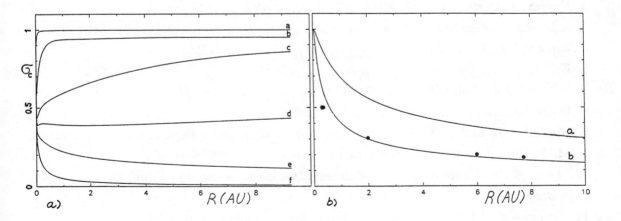

Fig. 2. Variation of the normalized cross helicity with distance a) for different frequencies (decreasing from curves a to f) in the linear case, and b) as observed (dark circles) and for the "turbulent" models /14/ defined in (8).

There is a major difference between the propagation properties in the static and expanding case, and in the significance of figs. 1a) and 1b): in the static case, waves of all frequencies are asymptotically outwardly propagating, a fact which may conveniently be rephrased in terms of the normalized cross-helicity σ_c as

$$\sigma_c = \frac{z^{+2} - z^{-2}}{z^{+2} + z^{-2}} \to 1, \ r \to \infty. \tag{7}$$

In the spherically expanding case, fig. 1b) shows the proportion of outward waves that reach the superalfvénic wind, but the asymptotic behaviour of σ_c depends on the frequency/10,12/. For all frequencies greater than a critical one, given by $\omega_* = U^2/2V_{a\infty}R_\odot$ and denoted by the dark circles in fig. 1b), σ_c increases with distance from the critical point to a frequency-dependent limiting value which tends to one at high frequencies as $1 - O(1/\omega^2)$. At lower frequencies however σ_c decreases with distance, as is shown in fig. 2a) (this calculation was done using an empirical wind model /10/), and tends asymptotically to 0, i.e., we have total reflection at infinity. The decrease in σ_c with distance from the sun is a feature which is observed in Alfvénic solar-wind turbulence [/13/ and fig. 2b)] and it has been modeled recently /14/ in terms of the second order moments of (1). These models are based on an isotropized version of the system (1), written in terms of the energies $E^\pm$ in the $\pm$ modes and the correlation $E^R = <z^+ \cdot z^->$:

$$(U \pm V_a)E^{\pm'} \mp E^\pm \vec{\nabla} \cdot (\vec{V}_a \mp \vec{U}/2) \pm E^R \xi \vec{\nabla} \cdot (\vec{V}_a \pm \vec{U}/2) = 0$$

$$UE^{R'} + \vec{\nabla} \cdot \vec{U}/2E^R + \xi(E^+ + E^-)\vec{\nabla} \cdot \vec{U}/4 - \xi(E^+ - E^-)\vec{\nabla} \cdot \vec{V}_a/2 = 0, \tag{8}$$

where ξ is a parameter which depends on the technique used to isotropize the moments of (1). Integrating these equations starting from the Alfvénic critical point yields a decrease in the σ_c, essentially because the system (8) is structurally similar to the low frequency limit of (1). Integration neglecting the Alfvén speed yields an agreement with observations [Fig. 2b), curve b and dark circles]. This result however is coincidental, as inclusion of the Alfvén speed leads to a higher asymptotic value of σ_c (curve a).

REFERENCES

1. M. Heinemann, and S. Olbert, J. Geophys. Res., 85, 1311 (1980).

2. Kholkunov, V. A., Theor. Math. Phys., 2, 169 (1970).

3. Nayfeh, A.,Perturbation Methods, (J. Wiley, 1973), p. 314.

4. R.L. Moore, Z.E. Musielak, S.T Suess and An C.-H., Astrophys. J., 378, 347 (1991).

5. Ferraro,V.C.A. and Plumpton, Astrophys. J., 127, 459 (1958).

6. Hollweg, J.V., Cosmic Electrodynamics, 2, 423 (1972).

7. Hollweg, J.V., Solar Phys., 56, 305 (1978).

8. Leroy B. , Astron. Astrophys., 97, 245 (1980).

9. Similon P.L. and S. Zargham in, Mechanisms of chromospheric and coronal heating, ed. P. Ulmschneider E.R. Priest R. Rosner (Springer, 1991), p. 438.

10. Velli, M., R. Grappin, and A. Mangeney, Geophys. Astrophys. Fluid Dyn., , (1991, in press).

11. Grappin R., M. Velli and A. Mangeney, Ann. Geophys., 9, 416 (1991).

12. Barkhudarov,M. N., Solar Phys., 135, 131 (1991).

13. Roberts, D. A., M. L. Goldstein, L. W. Klein, and W. H. Matthaeus, J. Geophys. Res., 92, 12023 (1987).

14. Zhou Y. and W. H. Matthaeus, J. Geophys. Res., 95A, 10291 (1990).

MAGNETOSPHERIC LOW-FREQUENCY NONRESONANT ION-BEAM TURBULENCE

F. Verheest and G. S. Lakhina*

*Instituut voor Theoretische Mechanika, Universiteit Gent, Krijgslaan 281,
B-9000 Gent, Belgium*

ABSTRACT

A theoretical framework is established for estimating the upper bound on magnetic field fluctuations produced in multi–ion beam systems, due to the excitation of short–wavelength nonresonant modes. In the region upstream of Earth's bow shock such modes can be excited due to reflected (or diffuse) ion beams. The estimates of the present theory show that $\delta B/B_0$ can be of order unity, in agreement with numerical simulations and observational evidence.

INTRODUCTION

Under solar wind conditions where the relative drift between two or more ion components is large in comparison with the Alfvén speed, unstable modes belong to the magnetosonic branch and have their maximum growth rates when k parallels $\mathbf{B}_0$. In certain regions of parameter space, the most unstable right–hand/left–hand polarized mode is not resonant with either ion component and propagates antiparallel/parallel to the magnetic field. Simulations show that for denser beams nonresonant modes grow and reach higher amplitudes than resonant modes which saturate at smaller amplitudes /1,2/.

In a previous paper we used a version of Fowler's theorem for unstable plasmas in order to estimate the magnetic field turbulence levels for nonresonant, linearly polarized firehose–like unstable modes in a multi-beam plasma, for any number of beams /3/. Our predictions, however, were for longer wavelengths than covered by existing simulations in the literature, so we now address the shorter–wavelength modes, to find that Fowler's theorem predicts higher levels than for the firehose–like modes.

Our formalism could be of importance for the diffuse ion beams and shocklets upstream of Earth's bow shock and analogous applications /4/, modelled by a two–beam proton system or by a system consisting of a background plasma and a beam, both consisting of protons. The theoretical levels of low–frequency turbulence agree reasonably well with observations and/or numerical simulations.

THEORETICAL FRAMEWORK

To be fully general, we start with a homogeneous, charge neutral multispecies beam–plasma immersed in a uniform magnetic field $\mathbf{B}_0$. For every ion species (with label s) we define the cyclotron frequencies $\Omega_s = e_s B_0/m_s$, the other symbols having their usual meaning. All constituents can in principle have drift velocities $\mathbf{U}_s$ parallel to $\mathbf{B}_0$. Perpendicular equilibrium drifts are eliminated by going to a deHoffmann–Teller frame. We assume charge and current neutrality for every beam, consisting of ions with a given set of characteristics (N_s, e_s, m_s, U_s) plus the electrons necessary to provide the return current. This differs slightly from the usual treatment, where all the electrons are taken together as one single fluid, but the electrons do not really contribute anyway to all important quantities, which are mass density averages.

We restrict ourselves to parallel wave propagation and look at low–frequency modes, such that for all beams *except one* the Doppler-shifted frequencies are small compared to their respective gyrofrequencies:

$$|\omega - kU_s| \ll \Omega_s \qquad (s \neq b). \tag{1}$$

The label b will be used to denote *the* energetic beam for which this condition is not satisfied. In this way we will generalize the nonresonant ion beam instability of *Winske and Leroy* /1/, characterized by a wavelength

---

*Permanent address: Indian Institute of Geomagnetism, Colaba, Bombay 400005, India

much larger than that corresponding to resonance with the main constituents $(s \neq b)$. This allows one to approximate the dispersion law (see *e.g.* /5/) as

$$(\omega - k\overline{U})^2 = k^2(V_A^2 - W) + \frac{\rho(\omega - kU_b)^3}{\omega - kU_b \pm \Omega_b}, \tag{2}$$

up to second order in the small quantities following from (1). Various global quantities have been introduced, the mean bulk velocity $\overline{U}$ of the plasma as a whole, the global Alfvén velocity V_A and the normalized kinetic energy in the relative drift motions /3/

$$W = \left(\sum_{s=1}^{n} \rho_s (U_s - \overline{U})^2\right) \left(\sum_{s=1}^{n} \rho_s\right)^{-1} = \frac{1}{2} \left(\sum_{s=1}^{n} \sum_{s'=1}^{n} \rho_s \rho_{s'} (U_s - U_{s'})^2\right) \left(\sum_{s=1}^{n} \rho_s\right)^{-2}. \tag{3}$$

Finally ρ is the mass density of the energetic beam which violates condition (1), normalized with respect to the total mass density.

In order to compute an upper estimate for the ultimate levels of magnetic turbulence in the case of unstable modes, we will make use of Fowler's equipartition ideas /6/. Neglecting the polarization drift because of the low–frequency modes, we compute from Maxwell's equations and the linearized equations of motion the electric field and fluid velocity fluctuations in terms of the magnetic field fluctuations

$$\delta \mathbf{E} = \frac{\omega}{k} \delta \mathbf{B} \times \mathbf{e}_B,$$

$$\delta \mathbf{v}_s = \frac{(\omega - kU_s)q_s}{m_s k} \frac{\Omega_s \delta \mathbf{B} + i(\omega - kU_s)\delta \mathbf{B} \times \mathbf{e}_B}{(\omega - kU_s)^2 - \Omega_s^2} \qquad (s = 1, \cdots, n). \tag{4}$$

Denoting $\delta B^2 = \delta \mathbf{B} \cdot \delta \mathbf{B}^\star$, we use Fowler's thermodynamics theorem /6/ in the form

$$\frac{\delta B^2}{2\mu_0} + \frac{\varepsilon_0 \delta E^2}{2} + \frac{1}{2}\sum_{s=1}^{n} \rho_s \delta v_s^2 \leq \frac{1}{2}\left(\sum_{s=1}^{n} \rho_s\right) F. \tag{5}$$

The terms on the left–hand side respectively describe the fluctuation energies in the magnetic and electric fields and in the plasma particles, whereas the right–hand side expresses the normalized free energy F accessible to the instabilities to be considered here. Inserting (4) into (5) and performing the same low–frequency approach as in deriving the dispersion law (2), one gets

$$\left\{ V_A^2 + W + \left(\frac{\text{Re}\,\omega}{k} - \overline{U}\right)^2 - \rho\left(\frac{\text{Re}\,\omega}{k} - U_b\right)^2 + (1-\rho)\left(\frac{\text{Im}\,\omega}{k}\right)^2 \right.$$

$$\left. + \rho \frac{|\omega - kU_b|^2}{k^2} \frac{\Omega_b^4 + \Omega_b^2|\omega - kU_b|^2}{[\Omega_b^2 - (\omega - kU_b)^2][\Omega_b^2 - (\omega^\star - kU_b)^2]} \right\} \frac{\delta B^2}{B_0^2} \leq F. \tag{6}$$

SPECIAL CASES

If we were to suppose that (1) also holds for the energetic beam, then we are in the long–wavelength limit of the nonresonant firehose–like mode, discussed in our earlier paper /3/. For wavelengths which are not as long, we will have a closer look at the regime defined by *e.g.* *Winske and Leroy* /1/, where ω is small in the sense that

$$\frac{\omega}{|k|U_b + \Omega_b} \ll 1. \tag{7}$$

Here we have taken $k = \mp|k|$ for the RH, respectively LH mode, so as to avoid the wavenumber regime where $-kU_b \pm \Omega_b$ could become very small. Defining now $u = |k|U_b/\Omega_b$, we can approximate (2) as

$$\left[1 - \rho + \frac{\rho}{(1+u)^3}\right]\left(\frac{\omega}{\Omega_b}\right)^2 \pm \left[2\frac{|k|\overline{U}}{\Omega_b} - \rho\frac{u^2(3+2u)}{(1+u)^2}\right]\left(\frac{\omega}{\Omega_b}\right) + \frac{k^2(\overline{U}^2 + W - V_A^2)}{\Omega_b^2} - \rho\frac{u^3}{1+u} = 0. \tag{8}$$

This quadratic equation can have unstable roots provided its discriminant Δ is negative, in which case one finds

$$\frac{\text{Re}\,\omega}{\Omega_b} = \mp \frac{\dfrac{k\overline{U}}{\Omega_b} - \dfrac{\rho u^2(3+2u)}{2(1+u)^2}}{1 - \rho + \dfrac{\rho}{(1+u)^3}}, \qquad \frac{\gamma}{\Omega_b} = \frac{\text{Im}\,\omega}{\Omega_b} = \frac{\sqrt{|\Delta|}}{2 - 2\rho + \dfrac{2\rho}{(1+u)^3}}. \tag{9}$$

One can substitute the above expressions in (6) in order to get an estimate for the saturated magnetic field fluctuations for any number of beams, provided a suitable expression for the free energy F is given.

We will specify the above results to *a two-beam model*, however, in order to show the relevance to the region upstream of Earth's bow shock. Then $\overline{U} = \rho U_b$ and $W = \rho(1 - \rho)U_b^2$, evaluated in a frame where the main constituents are at rest, and $\rho = \rho_b/(\rho_b + \rho_m)$. The indices b and m refer to the energetic beam and the main plasma, respectively. Hence the discriminant becomes $\Delta = 4\rho u^2 H(u, \rho)$, with

$$H(u, \rho) = \rho \frac{\left(1 + \frac{u}{2}\right)^2 - 1}{(1 + u)^4} - (1 - \rho)\left[\frac{1}{1 + u} - \frac{1}{M^2(1 + u)^3}\right] + \frac{(1 - \rho)^2}{\rho M^2}. \tag{10}$$

The Alfvénic Mach number $M = U_b/V_{AR}$ has been expressed with regard to the *main* plasma. One can show that

$$H(0, \rho) = (1 - \rho)\left(\frac{1}{\rho M^2} - 1\right), \tag{11}$$

and this is negative, provided initially $W/V_A^2 = \rho M^2 > 1$. On the other hand $H(u, \rho)$ is positive for $u \to \infty$, and has at least one positive root. There is thus at given M and ρ a range of unstable wavenumbers k.

We will suppose that $(1 - \rho)^2/\rho M^2$ is small, and look for values of u (or k), namely u_m (or k_m), which maximise the growth rate $\gamma = \mathrm{Im}\,\omega$ of the unstable modes, with $\gamma_m = \gamma(u_m)$. This results in

$$u_m = \frac{|k_m|U_b}{\Omega_b} \approx \frac{\rho M^2}{2(1 - \rho)}, \tag{12}$$

with the corresponding values

$$\frac{\mathrm{Re}\,\omega_m}{\Omega_b} = \mp \frac{\rho}{2(1 - \rho)}, \qquad \frac{\gamma_m}{\Omega_b} = \frac{\rho M}{2(1 - \rho)} = M|\mathrm{Re}\,\omega_m|. \tag{13}$$

We are thus dealing with an almost purely growing mode. For the application of *Fowler's* theorem (6) we also need for this fastest growing mode

$$\frac{\mathrm{Re}\,\omega_m}{k} = \frac{U_b}{M^2} = \frac{V_{AR}}{M}, \qquad \frac{\gamma_m}{|k|} = \frac{U_b}{M} = V_{AR}. \tag{14}$$

Hence, always up to small quantities in $(1 - \rho)^2/\rho M^2$, (6) is rewritten as

$$\frac{\delta B^2}{B_0^2} \leq \frac{F}{2(1 - \rho)V_{AR}^2}. \tag{15}$$

The level of turbulence crucially depends on the estimate of the available free energy F. First of all, since the instability demands that $\rho M^2 > 1$, an absolute upper bound would be $F = W - V_A^2 = (1 - \rho)V_{AR}^2(\rho M^2 - 1)$, because all nonresonant unstable modes are quenched when $W \leq V_A^2$. This yields

$$\frac{\delta B^2}{B_0^2} \leq \frac{1}{2}(\rho M^2 - 1). \tag{16}$$

Hence the saturation level can be quite large, $|\delta B/B_0| \geq 1$, as found in simulations for the shorter wavelengths nonresonant modes. However, the present estimate supposes that *all* the available free energy is pumped in the most unstable mode. This is maybe not quite reasonable. We want to emphasize here that, although the most unstable mode would grow the fastest, yet the other modes also take away a portion of the available free energy. Thus, the upper bound on the saturation levels given in (16) is certainly on the high side.

So a more conservative approach would be to assume that the most unstable mode takes up free energy in proportion to its growth rate squared, and then *Fowler's* theorem (15) yields

$$\frac{\delta B^2}{B_0^2} \leq \frac{1}{2(1 - \rho)}. \tag{17}$$

This is independent of M since the maximum growth rate is kV_{AR}, always provided $(1 - \rho)^2/\rho M^2$ is small compared to 1.

DISCUSSION

Satellite observations show that upstream of Earth's bow shock one encounters low–frequency electromagnetic turbulence associated with reflected and diffuse ion beams, streaming away from the bow shock /4,7/. One could model this by a two–beam system, one for the solar wind protons and one for the reflected (or diffuse) ion beams. In this case $\rho = N_b/(N_m + N_b)$, where N_m and N_b represent the number densities of the main solar wind and of the reflected beam protons. Assuming a Mach number of $M = 8$ and taking $\rho \simeq 0.1$, the higher estimate for the fluctuations follows from (16) as $|\delta B/B_0| \leq 1.65$, whereas the lower estimate in (17) would give $|\delta B/B_0| \leq 0.75$. This shows that the shorter wavelength nonresonant modes, defined by the inequality (7), can lead to higher levels of turbulence than the longer wavelength firehose–like modes discussed in our earlier paper /3/. This is in agreement with numerical simulations for a similar situation /8,9/.

Since the formulation given in the preceding sections is very general, one can also apply it to similar situations occurring elsewhere in space plasmas. In the case e.g. of the outer coma of comets, ρ could be as high as 0.5 closer to the nucleus, since the pick–up ions are mostly of the water group with an effective mass of 16.8 m_p, with mass densities which could thus become comparable to that of the undisturbed solar wind, even at relatively small number densities for the cometary ions. In that case $|\delta B/B_0| \leq 1$, even from (17).

It has to be remarked, however, that our estimates would hold true provided all the free energy goes into nonresonant modes. As a nonresonant mode grows, it could become modulationally unstable and pump energy into new modes (which need not be nonresonant!). Indeed it is rather hard to conceive of a true final stationary state with fluctuations as large as the background.

As a final remark, we might mention that we have also looked at resonant modes, which need $W < V_A^2$ to be excited, but much smaller levels of turbulence occur.

ACKOWLEDGMENTS

It is a pleasure to thank the (Belgian) National Fund for Scientific Research for a special grant which made the stay of GSL at Gent possible. GSL also wants to thank the International Centre for Theoretical Physics (Trieste, Italy) for a travel grant.

REFERENCES

1. D. Winske and M. M. Leroy, Diffuse ions produced by electromagnetic ion beam instabilities, *J. Geophys. Res.* **89**, 2673–2688 (1984)

2. S. P. Gary and D. Winske, Computer simulations of electromagnetic instabilities in the plasma sheet boundary layer, *J. Geophys. Res.* **95**, 8085–8094 (1990)

3. F. Verheest and G. S. Lakhina, Nonresonant low–frequency instabilities in multibeam plasmas: Applications to cometary environments and plasma sheet boundary layers, *J. Geophys. Res.* **96**, 7905–7910 (1991)

4. M. M. Hoppe, C. T. Russell, L. A. Frank, T. E. Eastman and E. W. Greenstadt, Upstream hydromagnetic waves and their association with backstreaming ion populations: ISEE 1 and 2 observations, *J. Geophys. Res.* **86**, 4471–4492 (1981)

5. G. S. Lakhina and F. Verheest, Alfvén wave instabilities and ring current during solar wind–comet interaction, *Astrophys. Space Sci.* **143**, 329–338 (1988)

6. T. K. Fowler, Thermodynamics of unstable plasmas, in : *Advances in Plasma Physics* (Ed. A. Simon and W. B. Thompson, Wiley Interscience) **1**, 201–225 (1968)

7. D. D. Sentman, C. F. Kennel and L. A. Frank, Plasma rest frame distributions of suprathermal ions in the Earth's foreshock region, *J. Geophys. Res.* **86**, 4365–4373 (1981)

8. S. P. Gary, C. D. Madland, D. Schriver and D. Winske, Computer simulation of electromagnetic cool ion beam instabilities, *J. Geophys. Res.* **91**, 4188–4200 (1986)

9. D. Winske and S. P. Gary, Electromagnetic instabilities driven by cool heavy ion beams, *J. Geophys. Res.* **91**, 6825–6832 (1986)

PARAMETRIC INSTABILITIES OF LARGE AMPLITUDE ALFVÉN WAVES WITH OBLIQUELY PROPAGATING SIDEBANDS

A. F. Viñas and M. L. Goldstein

NASA/Goddard Space Flight Center, Code 692, Greenbelt, MD 20771, U.S.A.

ABSTRACT

This paper presents a brief report on properties of the parametric decay, modulational, filamentation and magnetoacoustic instabilities of a large amplitude, circularly polarized Alfvén wave. We allow the daughter and sideband waves to propagate at an arbitrary angle to the background magnetic field so that the electrostatic and electromagnetic characteristics of these waves are coupled. We investigate the dependance of these instabilities on dispersion, plasma β, pump wave amplitude and propagation angle. Analytical and numerical results are compared with numerical simulations to investigate the full nonlinear evolution of these instabilities.

INTRODUCTION

Large amplitude Alfvén waves have been observed directly in the solar wind /1–3/, in the vicinity of planetary bow shocks /4/ and of interplanetary shocks /5/. These waves are of great theoretical interest because of their ubiquity in astrophysical magnetized systems and because they are easily excited via different plasma instabilities. Although Alfvén waves are relatively stable, especially against kinetic (i.e. wave-particle) interactions, they are, parametrically unstable /6–7/ to the generation of plasma density fluctuations and other Alfvénic fluctuations through nonlinear wave-wave couplings. Linearized theory of the parametric instabilities of Alfvén waves has been studied by various authors /6–13/ under a variety of limits and approximations. All these studies made use of either single or two fluid MHD to describe the process and considered either linear or circularly polarized waves. Others assumed small or finite amplitude pump waves. Other work included dispersive effects but still assumed that the Alfvén pump wave, the daughter and the sideband waves propagated parallel to the background magnetic field in which case the electrostatic and electromagnetic nature of these instabilities decouples.

Alfvén waves undergo two basic processes for parallel propagation, namely parametric decay ($k/k_o \geq 1$) and modulational instabilities ($k/k_o < 1$). Here k and k_o represent the wavenumbers of the daughter and pump waves respectively. Another independent decay process, namely the 'beat wave' decay instability ($k/k_o \approx 1$), distinct from the usual decay process, has a smaller growth rate and a narrower wavenumber bandwidth was also found /10/. Another parametric process, the parametric filamentation instability, was also investigated at perpendicular propagations /14–15/. The study of this instability was restricted to non-oscillatory ($Re(\omega) = 0$), purely growing (spatially and temporally) modes strictly propagating orthogonal to the incoming pump wave and background magnetic field.

Recently, Viñas and Goldstein /16,17/ generalized the theory of parametric instabilities of large amplitude, circularly polarized, dispersive Alfvén waves by allowing the daughter (compressive) and Alfvénic sideband waves to propagate at arbitrary angle to the background DC magnetic field thus coupling to both electrostatic and electromagnetic waves. In the case of parallel propagation ($\theta=0°$) a new parametric modulational channel in which a right or left circularly polarized Alfvén pump wave couples directly to electromagnetic daughter waves was found. Their analysis showed that the electromagnetic coupling is strictly a three-wave process which can only excite one electrostatic (ion acoustic) sideband while the other sideband wave vanishes exactly. Viñas and Goldstein also found besides the filamentation instability another parametric process at oblique propagation angles which has not been described before, namely a parametric magnetoacoustic instability. This instability has a distinct signature compared to the filamentation instability because it is characterized by propagating density fluctuations with large real frequencies that satisfy the condition $Re(\omega) \gg \gamma$ and it extends over a broad angular region.

MATHEMATICAL DESCRIPTION: LINEAR THEORY

The mathematical derivation of the general dispersion equation for obliquely propagating daughter and sideband waves is described by Viñas and Goldstein /16,17/. Here we only mention the basic assumptions that went into the derivation. The analysis is based on the two-fluid plasma model which describes the conservation of mass and momentum, together with Maxwell's equations. We neglect electron inertia effects since our objective is to study waves with frequencies below the ion gyrofrequency. Several effects such as Landau and cyclotron damping of the daughter waves, particularly in high β plasmas, are overlooked because they are strictly kinetic in nature. However, recent results /18–19/ suggest the possible importance of kinetic effects which cannot be addressed in a fluid treatment.

The model assumes a large amplitude, parallel propagating (i.e. in the z-direction), circularly polarized dispersive Alfvén pump wave which can couple to obliquely propagating (i.e. in the x-z plane) daughter and sideband waves. The coupling conditions $\mathbf{k}^{\pm} = \mathbf{k} \pm \mathbf{k_o}$, $\omega_{\pm} = \omega \pm \omega_o$ give the resonance wave-wave interactions representing conservation of momentum and energy. The general dispersion equation that describes the parametric instability of large amplitude Alfvén wave for upper and lower obliquely propagating sideband waves is the determinant of a six-by-six matrix which depends on

$$D = D\left(\widehat{\omega}, \widehat{k}, \theta, \beta, \eta, \kappa\right) = 0 \tag{1}$$

(see /16/) which depends therefore on six independent parameters written in normalized variables $\widehat{\omega} = \omega/k_o C_a$, $\widehat{k} = k/k_o$, $\beta = C_s^2/C_a^2$, $\kappa = k_o C_a/\Omega_i$ where C_a and C_s are the Alfvén and sound speeds respectively, k_o is the wavenumber of the Alfvén pump wave which propagates strictly along the mean field and Ω_i is the proton gyrofrequency. The parameters represent the frequency and growth rate of the mode, the magnitude of the wavevector $\widehat{k}$, the angle of propagation (θ) with respect to the background magnetic field, the thermal characteristic (β) of the plasma, the amplitude of the Alfvén pump wave ($\eta = \widetilde{B}/B_o$) and the amount of dispersion (κ).

The linear dispersion analysis in equation (1) is restricted only to the excitation of the fundamental sideband waves. The excitation of harmonics while observed in the simulations, is beyond the linear analysis.

NUMERICAL RESULTS

The numerical solutions of equation (1) are illustrated in Figures (1)-(2) in terms of the normalized variables. Recall from /16/ that the parameter κ plays a dual role in that its magnitude represents the amount of dispersion and its sign indicates the left ($\kappa > 0$) and right ($\kappa < 0$) hand sense of polarization of the pump wave. For a given value of κ, the corresponding value of $\widehat{\omega}_o$ is computed using equation (20) of /16/ in normalized form. We use high precision numerical techniques to find the unstable roots of (1) without writing the determinant as a polynomial.

Figure (1a) show the dispersion characteristics of both Alfvén decay ($k/k_o \geq 1$ and $\kappa = -0.3$) and modulational ($k/k_o < 1$ and $\kappa = 0.3$) instabilities respectively, as a function of $\widehat{k}$ for various propagation angles and for $\eta = 0.2$ and $\beta = 0.5$. The top curves on this figure represent the real frequency and the bottom curve the growth rate. The real frequencies and growth rate are shown for propagation angles $\theta = 0°$, $5°$, $10°$, $15°$ and $20°$ for the Alfvén decay instability and $\theta = 0°$, $10°$ and $20°$ for modulation. The principal characteristic is that the growth rate of these instabilities decreases and narrows their bandwidth as the propagation angle increases. Also note that as the propagation angle increases the maximum growth rate of the decay instability shifts toward $\hat{k} \approx 1$. The results at $\theta = 0°$ are similar to those found by Wong and Goldstein /10/. A different independent instability namely, the 'beat' wave decay instability that grows as the propagation angle increases was also found and the reader is referred to /16/ for a discussion of its properties.

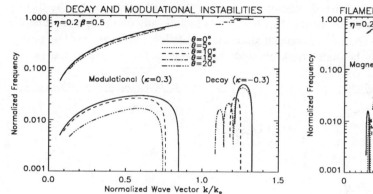
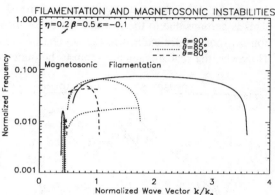

Fig. 1 Dispersion characteristics of a) decay and modulational instabilities and b) filamentation and magnetoacoustic instabilities, as a function of wavevector at different propagation angles.

Figure (1b) presents the dispersion curves of the filamentation and magnetoacoustic instabilities at various angles of propagation ($\theta = 90°$, $85°$ and $80°$) for parameters $\eta = 0.2$, $\kappa = -0.1$ and $\beta = 0.5$. It is found that as the propagation angle decreases from $90°$, the real part of the frequency of the filamentation instability becomes finite and by $80°$ actually compares to the imaginary part. This indicates that near $\theta = 90°$ the instability is essentially non-propagating but as the propagation angle decreases the real part of the frequency of the filamentation instability increases rapidly, turning the mode into a propagating one with a narrow bandwidth in $\hat{k}$ and a smaller growth rate. Therefore the non-propagating feature of the filamentation instability suggested by Kuo et al. /14–15/ is confined to a very narrow quasi-perpendicular region near $\theta = 90°$. In contrast to the filamentation instability, the magnetoacoustic instability shown in Figure (1b) has a sizeble real frequency ($Re(\hat{\omega}) \gg \hat{\gamma}$). Also note that this instability has a narrow bandwidth compared to the filamentation instability. Its maximum growth rate occurs at $\theta = 90°$ and decreases as the propagation angle decreases shifting its maximum peak to larger $\hat{k}$ values. A comparison of the filamentation and magnetoacoustic instabilities for the same set of parameters, as shown in Figure (1b) indicates that the filamentation instability is a more dominant process at least for perpendicular propagation. But as the propagation angle decreases from $90°$, the magnetoacoustic instability dominates since the non-propagating filamentation instability is highly localized.

We have also investigated the nonlinear evolution of these parametric instabilities using a two-dimensional hybrid simulation to compare the linear stages of their evolution with the analytical and numerical results presented here and in /16,17/. The simulation was carried out in a rectangular grid $(\hat{x}, \hat{y})$ of 128×64 mesh with grid spacing $\triangle\hat{x} = \triangle\hat{y} = 1.96$, and for $\beta = 0.2$. The spatial $(\hat{x}, \hat{y})$ coordinates of the simulation are normalized to the ion inertial length (C_a/Ω_i). The simulation was initiated by setting a large amplitude, right-handed, dispersive ($\kappa = -0.1$) Alfvén wave at T=0 propagating along a constant background magnetic field along the x-axis with an amplitude of $\eta = 0.5$. The pump wave was set at modenumber $m_o = 4$ which corresponds to the wavevector $|\kappa| = 0.1$ defined in normalized units by $2\pi m_o/(128 \times \triangle\hat{x})$. Figure (2a) shows the evolution of the density power spectrum (in units of Hz^{-1}) for parallel propagation. Note that the evolution of m_x-modes 6, 13 and 15 where m_x is the wave modenumber of the daugther wavevector $\hat{k}_x = 2\pi m_x/(128 \times \Delta\hat{x})$ (corresponding in the normalization used above for the linear analysis to wavenumbers $\hat{k}$= 1.5, 3.3 and 3.8 respectively) undergo an exponential growth in the time interval T=120–320 (time is normalized to the proton gyrofrequency) for the three modes before they reach saturation. From the slopes of these curves we obtained their linear growth rates and compare it with the analytical/numerical results. A similar analysis is performed for the real frequencies of the modes. The comparison of these results are presented in Figure (2b). Because of the coarse resolution of the spatial grid along the background magnetic field we can only resolve two modes (6 and 7 corresponding to wavenumbers $\hat{k}$= 1.5 and 1.8 respectively) that can be compared with the linear theory of the decay instability. Note that the growth rates of the simulation are smaller than that predicted by the theory. This is perhaps due to the fact that the linear theory /16,17/ does not include any dissipation and that the ion-acosutic modes excited in the simulation are damped since the simulation treats the ions kinetically. However, the real frequencies seem to be quite consistent with those predicted by the linear theory.

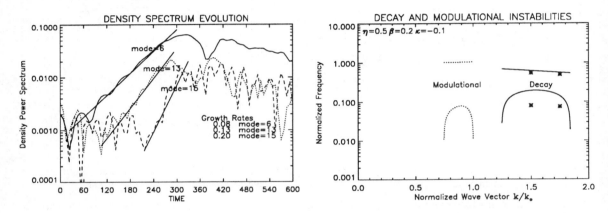

Fig. 2 a) Density spectrum evolution for the decay instability, b) Growth rates and real frequencies of linear theory (lines) and simulation (asterisks).

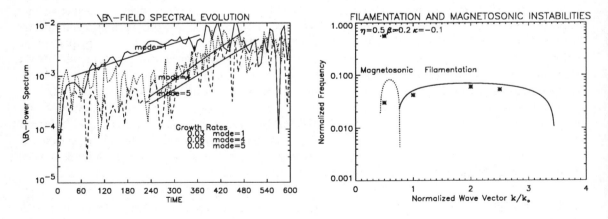

Fig. 3 a) Magnetic field spectrum evolution for filamentation and magnetoacoustic instabilities, b) Growth rates and real frequencies of linear theory (lines) and simulation (asterisks).

Another feature of the simulation results (Figure 2a) is the excitation of higher order modes whose growth rates increses with harmonic number n $(\hat{\gamma}(n\hat{k})/\hat{\gamma}(\hat{k})) \approx n$, the growth rate of modes 13 and 15 are approximately 2 and 3 times the growth rate of mode 6). This is consistent with the results obtained in /20/ using a one dimensional MHD simulation. A similar study is shown in Figure (3a) for the evolution of the magnetic field magnitude power spectrum (in units of Hz^{-1}) at perpendicular propagation and for n_y-modes 1, 2, 4 and 5 where n_y is the wave modenumber of the daugther wavevector $\widehat{k}_y = 2\pi n_y/(64 \times \triangle \widehat{y})$ (corresponding in the linear analysis to wavenumbers $\widehat{k}$= 0.5, 1.0, 2.0 and 2.5 respectively). Here we show the excitation of the parametric magnetoacoustic instability (mode 1) and the filamentation instability (modes 2, 4 and 5) respectively. Because of the coarse resolution of the spatial grid perpendicular to the background magnetic field we can only resolve one mode for the magnetoacoustic instability and four modes for the filamentation process. A comparison of these results with linear theory are presented in Figure (3b). As in the decay process, the simulation shows that the growth rates of the magnetoacoustic process are smaller than those predicted by linear theory but the real frequencies show very good agreement with the predicted values. In general the growth rates of the filamentation instability seems in good agreement as well as the real frequencies with the linear results.

SUMMARY

We have presented a general treatment of the parametric instabilities of a large amplitude, circularly polarized, dispersive Alfvén wave by allowing the unstable daughter and sideband waves to propagate obliquely to the background magnetic field. Four parametric processes can be excited by an Alfvén wave: parametric decay, modulational, filamentation and magnetoacosutic instabilities. A comparison of the analytical/numerical results with a numerical hybrid simulations indicates good agreement between the linear stages of the modes evolution and the predicted linear instabilities.

REFERENCES

1. J. Belcher and L. Davies, J. Geophys. Res., 76, 3534 (1971).

2. T. W. J. Unti and M. Neugebauer, Phys. Fluids, 11, 563 (1968).

3. B. Abraham-Shrauner and W. C. Feldman, J. Geophys. Res., 82, 618 (1977).

4. M. L. Goldstein, H. K. Wong, A. F. Viñas and C. W. Smith, J. Geophys. Res., 90, 302 (1985).

5. A. F. Viñas, M. L. Goldstein and M. H. Acuña, J. Geophys. Res., 89, 6813 (1984).

6. A. A. Galeev and V. N. Oraevskii, Sov. Phys. Dokl., Engl. Transl., 7, 988 (1963).

7. R. Z. Sagdeev and A. A. Galeev, Nonlinear Plasma Theory, edited by T. O'Neil and D. Book, W. A. Benjamin, New York, p 7, 1969.

8. C. N. Lashmore-Davies, Phys. Fluids, 19, 587 (1976).

9. M. L. Goldstein, Astrophys. J., 219, 700 (1978).

10. H. K. Wong and M. L. Goldstein, J. Geophys. Res., 91, 5617 (1986).

11. J.-I. Sakai and B. U. Ö. Sonnerup, J. Geophys. Res., 88, 9069 (1983).

12. C. R. Ovenden, H. A. Shah and S. J. Schwartz, J. Geophys. Res., 88, 6095 (1983).

13. M. Longtin and B. U. Ö. Sonnerup, J. Geophys. Res., 91, 6816 (1986).

14. S. P. Kuo, M. H. Whang and M. C. Lee, J. Geophys. Res., 93, 9621 (1988).

15. S. P. Kuo, M. H. Whang and G. Schmidt, Phys. Fluids B, 1, 734 (1989).

16. A. F. Viñas and M. L. Goldstein, J. Plasma Physics, 46, 107 (1991a).

17. A. F. Viñas and M. L. Goldstein, J. Plasma Physics, 46, 129 (1991b).

18. B. Inhester, J. Geophys. Res., 95, 10525 (1990).

19. E. Mjølhus and J. Wyller, Physica Scripta, 33, 442 (1986).

20. M. Hoshino and M. L. Goldstein, Phys. Fluids B, 1, 1405 (1989).

OBSERVATIONS OF THE ELECTRON DENSITY SPECTRUM NEAR THE SUN DURING SPECTRAL BROADENING TRANSIENTS

R. Woo and J. W. Armstrong

Jet Propulsion Laboratory, California Institute of Technology, Pasadena, CA 91109, U.S.A.

ABSTRACT

We present high-time resolution spectral broadening data for several transients caused by propagating interplanetary disturbances observed inside 20 R_o. The results show that the shape and level of the ambient near-Sun electron density spectrum undergoes abrupt and substantial change during passage of the disturbance. The steepening of the spectrum to near the Kolmogorov value suggests fully-developed inertial-range turbulence in the wake of the interplanetary disturbance.

INTRODUCTION

The spatial wavenumber spectrum of solar wind density fluctuations is of interest because it provides a statistical description of the fluctuations as well as clues to their nature and evolution. Since in situ measurements have so far only been available beyond 0.3 AU /1–2/, we must rely on remote sensing radio scintillation and scattering measurements using spacecraft and natural radio sources to study this region. Scintillation measurements have yielded information on the spectrum of the background electron density fluctuations covering a very wide range of scale sizes and radial distances /3–4/. The purpose of this paper is to investigate the detailed evolution of the electron density spectrum near the Sun during rare occurrences of transients caused by propagating interplanetary disturbances.

SPECTRAL BROADENING OBSERVATIONS

When a spacecraft flies behind the Sun, its initially monochromatic radio signal is broadened by scattering from moving electron density irregularities in the solar wind. The observed spectral broadening depends on the velocity of the scatterers and the shape and level of their turbulence spectrum. Spectral broadening measurements are line-integrated measurements of electron density fluctuations, but they essentially probe the region of closest approach since the density fluctuations fall off with heliocentric distance R as $\sim 1/R^2$.

Transients caused by propagating interplanetary disturbances occasionally appear in intensity scintillations /5–6/, Doppler scintillations /7/, and spectral broadening measurements /6,8/. These transients are observed simultaneously in all data types that are available /6/. The frequency of occurrence of transients near solar maximum has been investigated based on Doppler scintillation measurements in 1979–1983 /9/. More recently, Doppler scintillation measurements observed in 1981–1982 have been compared with in situ plasma measurements. Most of the transients observed in that study were produced by interplanetary shocks /10/.

In this paper, we examine the electron density spectrum of several transients observed within 20 R_o. We do this with spectral broadening measurements (which are sensitive to spatial scales smaller than ≈ 100 km) because there is a simple and robust connection between the observable and the electron density spectrum. Interpretation of the observations depends only on the "second moment" wave propagation theory, and is independent of strength of scintillations /3/.

TRANSIENT RESULTS

Spectral broadening bandwidth B is defined as the frequency range containing half the RF power. The radial variation of B at 2.3 GHz (13 cm wavelength) for the background solar wind near the Sun has been investigated earlier (see Figure 10 in /3/). Shown in Figure 1 is the time history of B (for 2-min. integrations) for the 23 April 1981 transient as it passed across the radio path of Pioneer Venus at 15.8 R_o. The pre-transient value of B is within the range of values of B observed in the background solar wind /3/. This transient has recently been identified as an interplanetary shock /10/ based on comparisons with radially-aligned in situ plasma measurements made by Helios 1. The rapid increase in bandwidth following the onset of the transient corresponds to passage of the shock front. The shock speed estimated from the Helios measurements at 174 R_o is 610 km/s while the shock speed based on the transit time between 15.8 and 174 R_o is 615 km/s /10/.

Shown in Figure 2 in log-linear plots are the RF spectra for 2-min. integrations corresponding to times before (top panel), near the peak (middle panel), and following the peak of the transient (lower panel). Following /3/, we have constructed model spectral broadening curves assuming a power law (with one-dimensional index α) for the spectrum of the electron density fluctuations. The noise level of the observed spectra is estimated from the power near the edges of the band and removed. The observed spectra are then compared with model spectra constrained to have the same bandwidth, center frequency, and total power, with α as a varied parameter. Examples of fits, represented by the dashed curves, are included in Figure 2. As can be seen, the model spectra represent good fits over a wide ($\approx 1000{:}1$) dynamic range. In this study, we fit the observations in the frequency domain. Alternatively, they could have been fit in the spatial domain as was done in /3-4/, and as will be done in future work when we will additionally analyze the simultaneous phase scintillation measurements.

The time history of the inferred spectral index α is shown in the upper panel of Figure 1. Ahead of the shock, $\alpha \approx 1.0$, a value consistent with other measurements of the background solar wind at this distance from the Sun /3-4/. The striking result of Figure 2 is that the abrupt and substantial increase in bandwidth during passage of the interplanetary shock front is accompanied by a correspondingly rapid rise in spectral index α. Although the spectrum of the electron density fluctuations remains approximated by a power law throughout the event, it steepens quickly to near the Kolmogorov value ($\alpha = 5/3$). It remains steeper than the pre-shock values for considerable distance in the wake of the shock, where the spectral broadening level is still enhanced because of increased turbulence and solar wind speed.

Spectral broadening measurements near 7 R_o on 15–16 November 1979 using Helios 1 are shown in Figure 3. The pre-event values of B and α are consistent with those of the background solar wind /3/. Although the results are dominated by a large transient starting around 2340 UT, there is also a smaller one starting around 2240 UT. Both transients (especially the large one) are probably shocks because of their rapid rise times /10/, and are associated with two CMEs observed in Solwind white-light coronagraph pictures /11/. The important result in Figure 3 is that the events display behavior similar to that of the

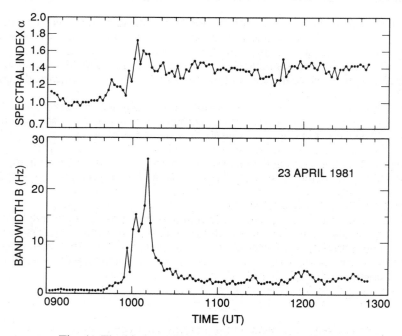

Fig. 1. The history of B and α for 23 April 1981 event.

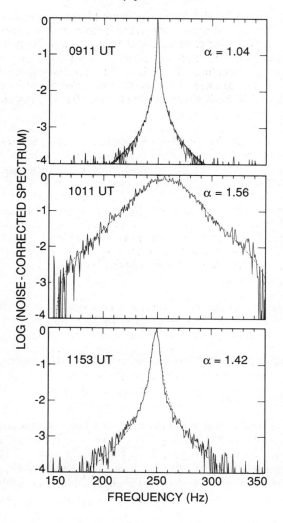

Fig. 2. RF spectra for 23 April 1981 event.

23 April 1981 shock, i.e. rapid rise-time and consistently steeper spectra after shock passage. This result is consistent with the steepening of the electron density spectrum observed in high-time resolution Voyager spectral broadening measurements of the 18 August 1979 shock at 13.1 R_o /6/, and evidence for occasional steeper spectra seen in the Voyager and Arecibo spectral broadening measurements analyzed in /4/.

CONCLUSIONS

Measurements of spectral broadening represent a robust means for studying the spatial wavenumber spectrum of small-scale (smaller than $\approx$100 km) electron density fluctuations in the inner heliosphere. Unfortunately, observations that capture complete transients (start and recovery) are rare. This stems not only from the infrequent occurrence of the transients /9/, but also the paucity of tracking time near the Sun.

We have presented detailed, time-resolved, S-band data for three transients observed inside 20 R_o, one of which is an interplanetary shock and the other two of which are probable shocks. A fourth transient, also confirmed as a shock, has been analyzed in /6/. In all four cases, the general morphology of the event is the same. The spectral broadening bandwidth increases rapidly from its pre-event value, then decays to a post-event level that is substantially above the "background" for that solar distance. Throughout the event the functional form of the small-scale density spectrum is well-approximated by a power-law. However, the power-law index increases (spectrum steepens) at the onset of the event, and persists even after the disturbance passes the radio path. The steepening of the spectrum to near the Kolmogorov value suggests fully-developed inertial-range turbulence /12/ in the wake of the interplanetary disturbance.

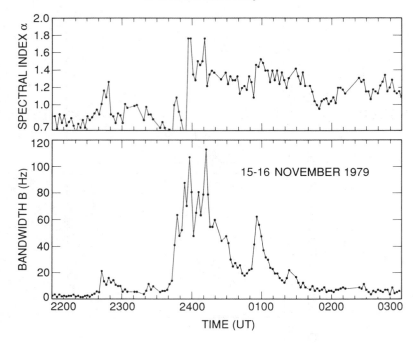

Fig. 3.   Time history of B and α on 15–16 November 1979.

ACKNOWLEDGMENTS

It is a pleasure to thank C. Chang for assistance in computing. We wish to acknowledge the support received from the NASA Deep Space Network and the Pioneer Venus and Helios Projects, especially A. Beers, A. Bouck, S. Borutzki, L. Colin, R. Fimmel, K. Heftman, R. Jackson, E.R. Kursinski, D. Lozier, and R. Ryan. This paper describes research carried out at the Jet Propulsion Laboratory, California Institute of Technology, under contract with the National Aeronautics and Space Administration.

REFERENCES

1.   B. Goldstein and G.L. Siscoe, in: Solar Wind Three, ed. C.P. Sonett, P.J. Coleman, and J.M. Wilcox, NASA Special Publ. SP-308, 506 (1972).

2.   E. Marsch and C.Y. Tu, J. Geophys. Res., 95, 11945 (1990).

3.   R. Woo and J.W. Armstrong, J. Geophys. Res., 84, 7288 (1979).

4.   W.A. Coles, W. Liu, J.K. Harmon, and C.L. Martin, J. Geophys. Res., 96, 1745 (1991).

5.   B.J. Rickett, Solar Phys., 43, 237 (1975).

6.   R. Woo and J.W. Armstrong, Nature, 292, 608 (1981).

7.   R. Woo, J.W. Armstrong, N.R. Sheeley, Jr., R.A. Howard, M.J. Koomen, and D.J. Michels, J. Geophys. Res., 90, 154 (1985).

8.   R.M. Goldstein, Science, 166, 598 (1969).

9.   R. Woo, J. Geophys. Res., 93, 3919 (1988).

10. R. Woo and R. Schwenn, J. Geophys. Res., 96, 21227 (1991).

11. M.K. Bird, H. Volland, R.A. Howard, M.J. Koomen, D.J. Michels, N.R. Sheeley, Jr., J.W. Armstrong, B.L. Seidel, C.T. Stelzried, and R. Woo, Solar Phys., 98, 341 (1985).

12. D. Montgomery, M. Brown, and W. Matthaeus, J. Geophys. Res., 92, 282 (1987).

NEARLY INCOMPRESSIBLE FLUID DYNAMICS

G. P. Zank and W. H. Matthaeus

Bartol Research Institute, The University of Delaware, Newark, DE 19716, U.S.A.

Abstract

The theory of nearly incompressible magnetohydrodynamics is reviewed in the context of solar wind observations.

1 Introduction

The solar wind is an interesting medium, characterized on the one hand by properties typical of a fully compressible magnetofluid (shocks, magnetosonic waves, etc.) yet, on the other hand, many solar wind phenomena are best understood in terms of an incompressible MHD description. This apparent dichotomy is further complicated by the presence of an admixture of small scale density fluctuations in the solar wind whose characteristic spectra are of the "Kolmogorov" type. Similar density fluctuation spectra have been inferred for the interstellar medium /1/. Although it has long been known that (proton) density fluctuations in the solar wind possess a $k^{-5/3}$ omni-directional wavenumber spectrum /2/, only recently has this topic begun to be addressed in detail. While of obvious interest to both the solar wind and the ISM, the properties and nature of density fluctuations in a fluid which is in a nearly incompressible state are of fundamental importance to the general theory of (subsonic) fluid dynamics, representing as it does the interface between the compressible and incompressible descriptions. Given the nature of this paper, we do not attempt to develop in any detail the full theory of nearly incompressible fluid dynamics, but present instead a synopsis of the observational results most pertinent to models of nearly incompressible MHD together with an overview of the general theoretical developments /3,4,5,6,7,8,9,10,11/.

2 Clues and constraints from the solar wind

Turbulence in the solar wind represents one of the few examples of a high Reynolds number turbulent flow which can be measured *in situ*. Although the collisional mean free path λ is about 1AU, virtually the size of the system, wave-particle interactions reduce the effective λ to the order of the Larmor radius. Thus, one may reasonably use a fluid description of the solar wind on these scales /12,13/.

An important parameter characterizing a magnetofluid is the plasma beta β_p. Classical 3D incompressible MHD is valid only in the $\beta_p \gg 1$ regime /14/, a condition that is rarely met in the inner Heliosphere although perhaps in the polar regions /15/. Instead, we might expect that $\beta_p \sim O(1)$ throughout the supersonic, equatorial region of the solar wind /16,17/. Within about 10–$20\,R_\odot$, we can expect $\beta_p \ll 1$. Clearly, caution must be exercised in applying the incompressible 3D MHD description to the solar wind.

Using high resolution Voyager magnetometer and plasma analyser data, Matthaeus *et al.* /18/ have presented the most detailed set of observations appropriate to a nearly incompressible solar wind state. In Fig. 1a, a histogram of rms density fluctuations $\delta\rho/\rho_0$ for two hour intervals is presented for the region 1–3AU. Since the distribution is strongly peaked at $\delta\rho/\rho_0 \sim 10^{-1}$, a large fraction of the sampled plasma admits only weakly compressive fluctuations.

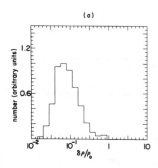

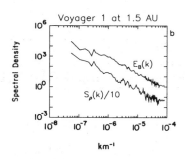

Figure 1: (a) Distribution of $\delta\rho/\rho_0$. (b) Density and magnetic energy power spectra.

Montgomery et al. /3/ suggested that since the incompressible pressure fluctuations δp^∞ were, in a sense, enslaved to the magnetic field fluctuations δB, then, provided the pressure and density fluctuations were related via a "pseudosound" relation $\delta p^\infty = C_s^2 \delta\rho$ (C_s the sound speed), the density fluctuation spectrum should follow that of δB. Thus, if the magnetic energy spectrum has a $k^{-5/3}$ Kolmogorov dependence, then so too does the density fluctuation spectrum (see also /19,20/). Examples of density and magnetic fluctuation spectra /18,21/ using Helios and Voyager data reveal the close correspondence between the two, with both spectra often following a $k^{-5/3}$ power law closely (Fig. 1b).

Micro-scale pressure-balanced-structures (PBS's) /22/ reveal various correlations that exist between different fluctuating fluid quantities. Recently, Zank et al. /7/ identified a second form of PBS in which density temperature fluctuations were anti-correlated. The various correlations imply that a theory of weakly compressible fluctuations must explain the possibility of at least two distinct fluid descriptions /6,7/.

Investigation of temperature spectra in the inner Heliosphere /23/ indicate that the spectral slope in the low frequency regime follows closely the Kolmogorov index, and in particular, the density and temperature spectra also follow each other closely.

Finally, mention should be made of the scintillation measurements of density and velocity fluctuations in the solar wind. These measurements reveal that within $20R_\odot$, density fluctuation spectra are Kolmogorov-like in the scale range of 10^3–10^6km followed by a local flattening on smaller scales and then a cut-off at an inner scale of ~ 10km /24,25/. A surprising result, however, was the discovery of a highly anisotropic density microstructure inside of 10–$15R_\odot$, a structure which Coles (these proceedings) argues is consistent with a 2D microstructure. Beyond $15R_\odot$, the density microstructure becomes quite isotropic. As noted by Harmon & Coles (these proceedings), the "recent measurements of high spatial anisotropy at small scales provide strong observational constraints for any theory which purports to explain the nature of plasma turbulence ... near the sun".

Two perspectives have come to dominate our view of MHD fluctuations in the solar wind—the first is Coleman's /26/ picture of a highly turbulent, non-linear medium, and the second is the Belcher & Davis /27/ perspective of a superposition of non-interacting Alfvénic modes. As noted by Roberts et al. /28/, however, it seems that both the wave and the turbulence viewpoints have some validity. Owing to the difficulty in obtaining fully 3D information about fluctuations in the solar wind, simplified/idealized theoretical models have been postulated. The wave perspective leads to a model of MHD fluctuations in which non-zero spectral amplitudes are associated with wave vectors parallel to the ambient dc magnetic field (the "slab" model). Alternatively, turbulence descriptions often invoke an "isotropic" model in which all wave numbers with the same magnitude have equal spectral amplitudes. Thus, for the slab model, $\mathbf{k} \parallel \mathbf{B_0}$ ($\mathbf{B_0}$ the average magnetic field) whereas for the isotropic model, $\mathbf{k}$ is arbitrary.

Unfortunately, real spectra are more complex than suggested by the idealized models and ques-

$$
\begin{array}{rcl}
\partial_t \rho + \nabla \cdot (\rho \mathbf{u}) &=& 0 \\
\rho \left(\partial_t \mathbf{u} + \mathbf{u} \cdot \nabla \mathbf{u} \right) &=& -\frac{1}{\varepsilon^2} \nabla p \\
\partial_t p + \mathbf{u} \cdot \nabla p + \gamma p \nabla \cdot \mathbf{u} &=& 0
\end{array}
\rightsquigarrow
\begin{array}{rcl}
\partial_t \mathbf{u}^\infty + \mathbf{u}^\infty \cdot \nabla \mathbf{u}^\infty &=& -\nabla p^\infty \\
\nabla \cdot \mathbf{u}^\infty &=& 0
\end{array}
$$

Figure 2:

tions that need to be addressed include (1) is the isotropic turbulence model appropriate to a magnetofluid suffused by a strong dc magnetic field? (2) can a "turbulence" and a "wave" viewpoint be reconciled? and (3) is the observed power anisotropy of the magnetic field fluctuations /27/ consistent with a turbulence description?

Some observational results concerning the symmetry properties of the magnetic and velocity fluctuation fields have been presented by Matthaeus *et al.* /29/. A contour plot of the 2D correlation function was presented—the "Maltese Cross" of their Fig. 3, plotted as a function of distance parallel $(r_\parallel)$ and perpendicular $(r_\perp)$ to $\mathbf{B}_0$. To interpret their 2D correlation, we note that (i) for Alfvén waves with $\mathbf{k} \parallel \mathbf{B}_0$, the contours are elongated parallel to $r_\perp$; (ii) for isotropic turbulence, the contour plot should consist of concentric circles, and (iii) for 2D turbulence orthogonal to $\mathbf{B}_0$, the correlation function contours are elongated parallel to $r_\parallel$. The "Maltese Cross" correlation reveals MHD fluctuations to be neither slab nor isotropic in the solar wind. Instead, one interpretation is that the correlation function identifies two populations—one with large correlation lengths transverse to $\mathbf{B}_0$ (slab turbulence) and the other with large correlation lengths parallel to $\mathbf{B}_0$ (2D turbulence).

3 Theoretical description

The impetus for the development of many of the ideas below came from two important papers by Montgomery and co-workers /3,30/. In developing a theory to describe density fluctuations in the solar wind successfully, we need to understand how the compressible and incompressible fluid descriptions are related. We follow here the developments of Zank & Matthaeus /6,7,8,9,10,11/. They have constructed *nearly incompressible* (NI) fluid models which converge, as an appropriately defined Mach number tends to zero, to a suitable incompressible description. To remind the reader, it should be noted that we are working in the zero momentum frame and that the fluctuations of interest are on scales much smaller than the scale of the system.

3.1 Derivation of the incompressible hydrodynamic equations (Example)

To illustrate the "bounded derivative" technique of deriving an "incompressible" set of equations from a compressible hyperbolic system, consider the highly subsonic hydrodynamic equations. The problem is illustrated most simply in Fig. 2—in the left box are the normalized fluid equations, $\varepsilon^2 \equiv \gamma u_0^2 / C_s^2 = \gamma M_s^2 \ll 1$ (u_0 the "convective" fluid velocity), and in the right box are the quite different incompressible fluid equations (non-singular, no acoustic effects, etc.). How then are the two systems related?

The answer is to be found in a Kreiss's bounded derivative principle /31,8/ which asserts that for solutions of a singular hyperbolic system of PDE's to vary on slow time scales only, it is necessary that several time derivatives of the solution be $O(1)$. Then, if $\partial_t \mathbf{u}$ is to be bounded independently of ε, the functional form of the normalized pressure must be $p = 1 + \varepsilon^2 p_1$. Similarly, since $\partial_t^2 \mathbf{u} =$ bounded terms $+\varepsilon^{-2} \nabla (\gamma p \nabla \cdot \mathbf{u})$, we can ensure the boundedness of the second order time derivative only if $\nabla \cdot \mathbf{u} = 0$. Thus, application of Kreiss's theorem yields the incompressible equations as constraints on the subsonic compressible fluid equations which eliminate all solutions that vary on fast time scales. Our derivation of the incompressible equations is exactly analogous to the standard physical arguments used typically to justify their use /8/. This analysis is, however, of considerable value because it provides a powerful and flexible approach to the more complex problem of MHD.

3.2 NI MHD for $\beta_p \sim 1$

Consider a magnetofluid suffused by a large applied magnetic field $\mathbf{B}_0 = B_0 \mathbf{z}$ such that the plasma beta $\beta_p \sim 1$. This is a reasonable approximation for the supersonic solar wind in the equatorial regions. The normalized continuity, induction and gas pressure equations are standard. For $\beta_p \sim 1$, it is necessary to use the following normalization for the total momentum equation

$$\rho \left(\partial_t + \mathbf{u} \cdot \nabla \right) \mathbf{u} = -\varepsilon^{-2} \nabla p + \varepsilon^{-2} \left(\nabla \times \mathbf{B} \right) \times \mathbf{B}. \tag{1}$$

Here, we have identified ε with the Alfvénic Mach number M_A and $|C_s| < |V_A|$ and $M_A^2 \ll 1$. On this basis, we can derive the following system of core equations

$$\nabla_\perp \cdot \mathbf{u}_\perp^\infty = 0; \qquad \nabla_\perp \cdot \mathbf{B}_\perp^\infty = 0;$$
$$(\partial_t + \mathbf{u}_\perp^\infty \cdot \nabla_\perp) \mathbf{u}_\perp^\infty = -\nabla_\perp p^\infty - \frac{1}{2} \nabla_\perp \left(B_\perp^{\infty 2} + B_z^{\infty 2} \right) + (\mathbf{B}_\perp^\infty \cdot \nabla_\perp) \mathbf{B}_\perp^\infty; \tag{2}$$
$$(\partial_t + \mathbf{u}_\perp^\infty \cdot \nabla_\perp) \mathbf{B}_\perp^\infty = (\mathbf{B}_\perp^\infty \cdot \nabla_\perp) \mathbf{u}_\perp^\infty,$$

together with

$$(\partial_t + \mathbf{u}_\perp^\infty \cdot \nabla_\perp) u_z^\infty = (\mathbf{B}_\perp^\infty \cdot \nabla_\perp) B_z^\infty.$$
$$(\partial_t + \mathbf{u}_\perp^\infty \cdot \nabla_\perp) B_z^\infty = (\mathbf{B}_\perp^\infty \cdot \nabla_\perp) u_z^\infty, \tag{3}$$

where $\mathbf{B}_0 = B_0 \mathbf{z}$, $\nabla_\perp \equiv (\partial_x, \partial_y)$, $\mathbf{u}_\perp^\infty \equiv (u_x^\infty, u_y^\infty)$, etc. The spatially 3D compressible system is reduced to two spatial dimensions when one eliminates high frequency solutions from the $\beta_p \sim 1$ MHD equations subject to an anisotropic applied magnetic field. Equations (2)–(3) are often described as $2\frac{1}{2}$D MHD.

Since various time scales are widely separated in a highly subsonic flow, a multiple scales analysis of the compressible $\beta_p \sim 1$ MHD equations is possible. Since solutions to the incompressible MHD equations vary on slow convective time scales only, convective modes can interact with acoustic and Alfvénic modes on long wavelength scales /8/. We therefore introduce the fast and slow time scales $\tau' = t/\varepsilon$ and $\tau = t$ together with the short and long wavelength scales $\eta = \mathbf{x}$ and $\xi = \varepsilon \mathbf{x}$. Kreiss's theorem can then be utilized to develop an *Ansatz* for the functional form of the NI solutions

$$p = p_0 + \varepsilon p_1 + \varepsilon^2 (p^\infty + p^*); \qquad \mathbf{u} = \mathbf{u}^\infty + \varepsilon \mathbf{u}_1;$$
$$\rho = 1 + \varepsilon \rho_1 + \varepsilon^2 \rho^*; \qquad \mathbf{B} = \mathbf{B}_0 + \varepsilon (\mathbf{B}^\infty + \mathbf{B}_1) + \varepsilon^2 \mathbf{B}^*, \tag{4}$$

where the $\cdot^\infty$ variables are solutions to the incompressible model. One can show that we must have $p_1 = 0 = \rho_1$ at the NI level, from which it follows that $B_z^\infty = 0$. Hence the incompressible model must instead be properly 2D (*i.e.*, equations (2)) when one considers the NI corrections. This rather remarkable result shows that a $\beta_p \sim 1$ magnetofluid with an applied magnetic field $\mathbf{B}_0$ admits a 2D incompressible MHD description in planes orthogonal to $\mathbf{B}_0$.

On considering further variables in the *Ansatz* (4), we find, for example, that high frequency magnetosonic waves are associated with p^*, ρ^* and $\mathbf{B}^*$, *i.e*,

$$\left[\frac{\partial^4}{\partial \tau'^4} - \left(\gamma p_0 + B_0^2 \right) \nabla_\eta^2 \frac{\partial^2}{\partial \tau'^2} + \gamma p_0 B_0^2 \frac{\partial^2}{\partial \eta_z^2} \nabla_\eta^2 \right] \begin{pmatrix} p^* \\ \rho^* \\ \mathbf{B}^* \end{pmatrix} = 0, \tag{5}$$

$$
\begin{aligned}
\nabla_\perp \cdot \mathbf{u}_\perp^\infty &= 0 \qquad \nabla_\perp \cdot \mathbf{B}_\perp^\infty = 0 \\
(\partial_t + \mathbf{u}_\perp^\infty \cdot \nabla_\perp)\,\mathbf{u}_\perp^\infty &= -\nabla_\perp p^\infty - \tfrac{1}{2}\nabla_\perp \left(B_\perp^{\infty\,2} + B_z^{\infty\,2} \right) \\
&\quad + (\mathbf{B}_\perp^\infty \cdot \nabla_\perp)\,\mathbf{B}_\perp^\infty \\
(\partial_t + \mathbf{u}_\perp^\infty \cdot \nabla_\perp)\,\mathbf{B}_\perp^\infty &= (\mathbf{B}_\perp^\infty \cdot \nabla_\perp)\,\mathbf{u}_\perp^\infty
\end{aligned}
$$

$$
\begin{aligned}
D_t \rho^* + \varepsilon^{-1}\nabla \cdot \mathbf{u}_1 &= 0 \\
D_t \mathbf{u}_1 + \mathbf{u}_1 \cdot \nabla \mathbf{u}^\infty &= -\varepsilon^{-1}\nabla p^* - \varepsilon^{-1}\left[\nabla(\mathbf{B}_0 \cdot \mathbf{B}^*) - (\mathbf{B}_0 \cdot \nabla)\mathbf{B}^* \right] \\
- \nabla(\mathbf{B}_\perp^\infty \cdot \mathbf{B}^*) &+ (\mathbf{B}_\perp^\infty \cdot \nabla)\mathbf{B}^* + (\mathbf{B}^* \cdot \nabla)\mathbf{B}_\perp^\infty \\
D_t p^* + \varepsilon^{-1}\gamma p_0 \nabla \cdot \mathbf{u}_1 &= -D_t p^\infty \\
D_t \mathbf{B}^* + \mathbf{B}^* \cdot \nabla \mathbf{u}_\perp^\infty &- \nabla \times (\mathbf{u}_1 \times \mathbf{B}_\perp^\infty) = \varepsilon^{-1}\left[(\mathbf{B}_0 \cdot \nabla)\mathbf{u}_1 - \mathbf{B}_0(\nabla \cdot \mathbf{u}_1) \right]
\end{aligned}
$$

Figure 3: Schematic of $\beta_p \sim 1$ NI MHD

which is recognized to be the normalized magnetosonic wave equation. The NI variables p^*, etc. are functions of (x, y, z, t) and $B_z^* \neq 0$, so one has isotropic magnetosonic wave propagation in this plasma beta limit.

On following /8/, one can derive the full NI model for $\beta_p \sim 1$ MHD, illustrated in Fig. 3. Note that $D/Dt \equiv \partial_t + \mathbf{u}_\perp^\infty \cdot \nabla_\perp$ is the 2D incompressible convective derivative. Fig. 3 illustrates a number of important features: (i) compressible effects ride parasitically on the back of the 2D incompressible flow field; (ii) compressible fluctuations are properly 3D; (iii) high frequency effects enter through the ε^{-1} terms; (iv) besides acting as non-constant coefficients, the incompressible fluid variables also act as density fluctuation source terms through Dp^∞/Dt, this representing a generalization of the Lighthill /32/ mechanism for the generation of sound by turbulence; (v) the compressible equations form a linear system. In non-normalized form, one derives self-consistently the pseudosound relation

$$
p^* + p^\infty = C_s^2 \rho^*, \tag{6}
$$

suggesting that the result of Montgomery et al. /3/ may require some modification since there is no guarantee that the magnetosonic pressure contribution is significantly smaller than that of the incompressible pressure. Another important solution is the long wavelength 1D Alfvén wave equation for B_x^* or u_{x1},

$$
\frac{\partial^2 u_{x1}}{\partial t^2} = V_A^2 \frac{\partial^2 u_{x1}}{\partial \xi_z^2}, \tag{7}
$$

V_A the Alfvén speed. Thus, long wavelength Alfvén modes propagate along the applied magnetic field $\mathbf{B}_0$. Therefore, the $\beta_p \sim 1$ NI MHD picture is one of fluctuations convected in 2D planes orthogonal to the directed mean magnetic field interacting non-linearly with higher-order Alfvén waves which propagate along the mean field, as well as an isotropically propagating magnetosonic wave component.

3.3 NI MHD for $\beta_p \ll 1$

In view of the perceived importance of the $\beta_p \ll 1$ case /24,25/, this case is discussed briefly. As one might expect, the incompressible core description is 2D incompressible MHD. However, the large applied field $\mathbf{B}_0 = B_0 \mathbf{z}$ also imposes quasi-2D characteristics on the compressible fluctuations /9/. Although a simple pseudosound relation again relates pressure and density fluctuations, the density fluctuations satisfy in addition a quasi-2D acoustic wave equation with source term

$$\frac{D^2\rho^*}{Dt^2} - \frac{C_s^2}{\delta^2}\frac{\partial^2\rho^*}{\partial z^2} = S(\mathbf{x}, t), \tag{8}$$

where D/Dt is the 2D incompressible convective derivative. From (8), short wavelength density fluctuations are generated and convected by the incompressible 2D fluid, thereby leading to long correlation lengths along the magnetic field. Long wavelength modes tend to propagate along the $\mathbf{B}_0$, so giving the magnetofluid a compressible "reduced MHD" character.

One might therefore expect the nature of density fluctuations in the solar wind to undergo a marked change in going from the subsonic $\beta_p \ll 1$ region of the solar wind (highly anisotropic fluctuations) to the $\beta_p \sim 1$ supersonic region (isotropic fluctuations).

3.4 Thermal effects

The above considerations have been based on a polytropic gas description. However, use of the ideal gas law leads to interesting modifications /6,7,8,10/ to NI MHD. The medium is assumed to be ideal so $p = RT\rho$. The principle of bounded derivatives and the principle of least degeneracy yield two possibilities for the NI *Ansatz*: (i) the heat-fluctuation-dominated (HFD) expansion with $\rho = 1 + \varepsilon\rho_1$ and $T = T_0 + \varepsilon T_1$; and (ii) the heat-fluctuation-modified (HFM) expansion $\rho = 1 + \varepsilon^2\rho_1$ and $T = T_0 + \varepsilon^2 T_1$. The HFD and HFM expansions reveal that two distinct approaches to incompressibility are possible, although the solar wind appears to favour the latter.

For HFD fluids, density fluctuations are transported as a passive scalar on the incompressible flow field

$$\partial_t\rho_1 + \mathbf{u}^\infty \cdot \nabla\rho_1 = Pr^{-1}\nabla^2\rho_1, \tag{9}$$

with Pr the Prandtl number. Also, density and temperature fluctuations are anti-correlated *i.e.*, $\rho_1 \propto -T_1$, a relationship that has been verified observationally in the solar wind recently /7/. The power spectrum of passive scalar fluctuations can be proportional to $k^{-5/3}$ in the inertial range /33/ and a high k flattening may be expected. It is noteworthy that (9) represents a possible origin of the observed density fluctuation spectra different from that of the pseudosound theory /6,7,8,34/.

Alternatively, for the HFM *Ansatz*, the thermal transport equation now admits acoustic effects, and a generalized form of the pseudosound relation (6) now holds, with

$$p^* + p^\infty - [(\gamma - 1)/\gamma]\,T_1 = \rho_1, \tag{10}$$

in normalized form. If the total pressure fluctuation $\delta P \equiv p^* + p^\infty$ is correlated with the temperature fluctuation, it then follows that $p^* + p^\infty = C_s^2\rho_1$ in non-normalized form. This, together with the correlation $\rho_1 \propto T_1$, forms the basis for the pseudosound theory.

Since $\rho_1 \propto -T_1$ (HFD) or $\rho_1 \propto T_1$ (HFM) can both occur in the solar wind /7/, we can expect temperature fluctuation spectra in the solar wind to follow the density fluctuation spectra. Tu *et al.* /23/ have used Helios data to investigate temperature fluctuation spectra in the solar wind finding (i) that the spectra are Kolmogorov-like; and (ii) that density and temperature spectra track each other closely, to the extent that each have the same stream-structure dependence.

3.5 Geometric effects of flux tubes

If a magnetofluid with an applied magnetic field is characterized by a large aspect ratio (*i.e.*, with a major z axis and minor x and y axes), then the system simplifies to that of reduced MHD

(RMHD). Such models have been utilized widely in the fusion community and only recently has it been suggested that RMHD may be a useful paradigm for the solar wind /18,10/. Zank & Matthaeus /10/ show that RMHD represents a purely incompressible description which describes a class of exact slowly varying solutions to the compressible MHD equations, a result which clarifies a number of issues in earlier treatments. Thus, the RMHD description admits no compressible magnetosonic modes, yet a number of interesting consequences for the solar wind emerge. The incompressible fluctuations consist of a 2D family which is convected in planes orthogonal to the applied magnetic field, together with an Alfvén wave population propagating parallel to $\mathbf{B}_0$. One therefore has a picture not unlike that of the quasi-2D $\beta_p \sim 1$ NI MHD description except that compressible effects are now completely absent. However, the RMHD description emerges from the imposition of a strong geometric constraint on the fluid, and it is by no means clear that the large-aspect-ratio constraint is adequately fulfilled in the solar wind except perhaps very close to the sun. It is conceivable that a combination of geometric and NI effects are observed in the solar wind.

4 Conclusions

A brief overview has been presented of some general theoretical models which may be useful in investigating both incompressible and compressible MHD turbulence in the solar wind. In writing this paper, we have tried to present solar wind observations which are difficult to understand in terms of previous theoretical descriptions but which seem to admit a natural explanation in terms of our theory of nearly incompressible fluid dynamics. No attempt has been made (and indeed no attempt can be made within MHD) to explain why the solar wind should, on suitably small scales, behave in an incompressible or nearly incompressible fashion and for this, a kinetic treatment may be necessary. A natural candidate to suppress fast acoustic motions in the solar wind is e.g., Landau damping /36/.

To summarize, we can list our major conclusions for the solar wind as follows.

(i) The underlying incompressible description for investigating turbulence in the solar wind should be 2D incompressible MHD rather than the more usual 3D description (since $\beta_p \sim 1$ and a mean magnetic field exists).

(ii) The NI $\beta_p \sim 1$ MHD model suggests that fluctuations in the solar wind are either convected in 2D planes orthogonal to the mean magnetic field or propagate as Alfvén waves along $\mathbf{B}_0$. Both populations, of course, interact non-linearly with each other.

(iii) NI MHD can account for the observed density, temperature and magnetic fluctuation spectra. The density fluctuations are found to be partially slaved to the magnetic field fluctuations and also generated by the incompressible flow field and so NI fluid dynamics represents a generalization of the Lighthill mechanism for the generation of aerodynamic sound.

(iv) NI fluid dynamics predicts the existence of two possible classes—the HFD and HFM descriptions. Both models show that the density and temperature fluctuation spectra should track one another.

(v) Finally, NI MHD predicts that density fluctuations in a $\beta_p \ll 1$ plasma should be quasi-2D convecting with the flow yet properly isotropic in a $\beta_p \sim 1$ magnetofluid.

We believe that significant observational evidence exists to support the general framework of nearly incompressible fluid dynamics and that this represents a new and developing field, rich in possibilities, both theoretically and observationally.

Acknowledgements This work has been supported in part by the Bartol NASA Space Physics Theory Program under grant NAGW-2076 at the BRI and by NSF grant ATM-8913627. We thank D.C. Montgomery for his initial suggestions and formative contributions to this field of research.

1. J.W. Armstrong, J.M. Cordes & B.J. Rickett, Nature **291**, 561 (1981).

2. B. Goldstein & G.L. Siscoe, NASA Spec. Publ. SP-308, 506 (1972).

3. D.C. Montgomery, M.R. Brown & W.H. Matthaeus, J. Geophys. Res., **92**, 282 (1987).

4. W.H. Matthaeus & M.R. Brown, Phys. Fluids, **31**, 3634 (1988).

5.J. Shebalin & D.C. Montgomery, J. Plasma Phys., **39**, 339 (1988).

6. G.P. Zank & W.H. Matthaeus, Phys. Rev. Lett., **64**, 1243 (1990).

7.G.P. Zank, W.H. Matthaeus & L.W. Klein, Geophys. Res. Lett., **17**, 1239 (1990).

8.G.P. Zank & W.H. Matthaeus, Phys. Fluids A, **3**, 69 (1991).

9. G.P. Zank & W.H. Matthaeus, submitted Phys. Fluids A (1991).

10. G.P. Zank & W.H. Matthaeus, submitted J. Plasma Phys. (1991).

11. G.P. Zank & W.H. Matthaeus, submitted J. Geophys. Res. (1991).

12. D.C. Montgomery, NASA Conf. Pub. **2260**. 107 (1983).

13. A. Mangeney, R. Grappin & M. Velli, in "Advances in Solar System Magnetohydrodynamics", ed.'s E.R. Priest, A.W. Hood, CUP, 327 (1991).

14. M.J. Lighthill, Waves in Fluids, CUP (1979).

15. D.A. Roberts, Geophys. Res. Lett., **17**, 567 (1990).

16. L.F. Burlaga & K.W. Ogilvie, Sol. Phys., **15**, 61 (1970).

17. A. Richter & E. Marsch, Ann. Geophys., **6**, 319 (1988).

18. W.H. Matthaeus, L.W. Klein, S. Ghosh & M.R. Brown, J. Geophys. Res., **96**, 5421 (1991).

19. R. Grappin, M. Velli & A. Mangeney, Ann. Geophys., **9**, 416 (1991).

20. V.I. Klaitskin, Izv. Atoms. Oceanic Phys., **2**, 474 (1966).

21. E. Marsch & C.-Y. Tu, J. Geophys. Res., **95**, 8211 (1990).

22. M. Vellante & A.J. Lazarus, J. Geophys. Res., **92**, 9893 (1987).

23. C.-Y. Tu, E. Marsch & H. Rosenbauer, Ann. Geophys., in press (1991).

24. W.A. Coles & J.K. Harmon, Ap. J., **337**, 1023 (1989).

25. W.A. Coles, W. Liu, J.K. Harmon & C.L. Martin, J. Geophys. Res., **96**, 1745 (1991).

26. P.J. Coleman, Ap. J., **153**, 371 (1968).

27. J.W. Belcher & L. Davis, J. Geophys. Res., **76**, 3534 (1971).

28. D.A. Roberts, M.L. Goldstein, L.W. Klein & W.H. Matthaeus, J. Geophys. Res., **92**, 11021 (1987).

29. W.H. Matthaeus, M.L. Goldstein & D.A. Roberts, J. Geophys. Res., **95**, 20673 (1990).

30. D.C. Montgomery, Phys. Scri. T, T2/1, 83 (1982).

31. H.-O. Kreiss, Commun. Pure Appl. Math., **33**, 399 (1980).

32. M.J. Lighthill, Proc. R. Soc. London Ser. A, **211**, 564 (1952).

33. G.K. Batchelor, J. Fluid Mech., **5**, 113 (1959).

34. B.J. Bayly, C.D. Levermore & T. Passot, Phys. Fluids A, in press (1991).

35. H.R. Strauss, Phys. Fluids, **19**, 134 (1976).

36. A. Barnes, in Solar System Plasma Physics, North-Holland, New York, 249 (1979).

Heliospheric Dynamic Phenomena

Conveners: Len Burlaga
Ernie Hildner
Richard Steinolfson

EVOLUTION PROPERTIES OF THE MAGNETIC FIELD STRUCTURE OF THE FLARING ACTIVE REGION HR 16 631 AND ASSOCIATED SOLAR WIND PHENOMENA ON FEBRUARY 2-6, 1980

V. N. Borovik,* N. A. Drake** and N. G. Ptitsyna***

Pilkovo Observatory, St. Petersburg, 196140, Russia
*** University Observatory, St Petersburg, 198904, Russia*
*** LOIZMIRAN, 2 Liniya 23, St Petersburg, 199053, Russia*

ABSTRACT

Data obtained at Helios 1,2, ISEE-3 and interplanetary scintillations are examined in order to reveal the physical characteristics and 3D-structure of the extreme disturbance of the solar wind on 5-6 Feb 1980. The disturbance and associated comet and Earth's magnetosphere phenomena was shown to be due to the solar flare on 3 Feb at 13^h38^m UT in active region HR 16 631. The magnetographic and radioastronomical investigations have revealed unusual evolution properties of this region before the flare: rapid evolution and complex configuration of the magnetic field, an existence of a radio source with peculiar spectral-polarization characteristics testifying to the abnormal local release of energy in the corona.

INTRODUCTION

The unique series of events in the interplanetary space and on the Sun was observed in the beginning of February 1980. The extreme solar wind disturbance was registered with Helios 2. It was responsible for the very rapid tail-turning of the comet Bradfield X 1979 on 6 Feb and for sharp Earth's magnetosphere compression accompanied by ssc events and geomagnetic pulsations. The results of the interplanetary events timing information and solar data analysis indicate that the most plausible candidate for the solar source is the flare in the active region HR 16 631 (S15° E15°) on 3 Feb at 13^h38^m UT, which is associated with a type IV radio emission, CME and outstanding radio bursts in the centimeter, decimeter and meter wavelength ranges.

The magnetographic and radioastronomical data obtained by Panoramic magnetograph and radio telescope RATAN-600 with high angular resolution was investigated in order to reveal properties that could be associated with the extreme phenomena in the interplanetary space.

DATA AND RESULTS

Helios 2 (Fig. 1a) data show solar wind disturbance on February 5 at about 16^h UT: the solar wind velocity increases rapidly from 350 km/s to 880 km/s. This disturbance proceeded at very low solar wind density $n<1$ cm^{-3}. The density increases to $n \approx 8$ cm^{-3} only after 6-7 hours, remains almost constant during 20 hours and then rapidly falls. There is a drastic cooling from $8 \cdot 10^5$K to $5 \cdot 10^4$K in the region with dense material. The He^{++}/H^+ ratio observed with ISEE-3 shows an increase after the velocity enhancement at about 7^h UT, February 6. This V, n, T profiles could be identified as the signature of flare ejecta driving a shock.

Radial size of the region $L=0.28$ AU, latitudinal extent $l=0.3$ AU (Fig.2). Approximating the transient as a tore segment, M could be estimated as $M \approx 4 \cdot 10^{15}g$.

This event is unusual since the disturbance propagates through the corotating low density ($n<1$ cm^{-3}) stream which can be traced with Helios 1-2 data. Appearance of a sudden commencement of geomagnetic storm ssc on 6 Feb at 3^h20^m UT testifies to a sharp compression of magnetosphere at the moment of the interplanetary shock arrival. ssc was accompanied by geomagnetic pulsations Pc5.

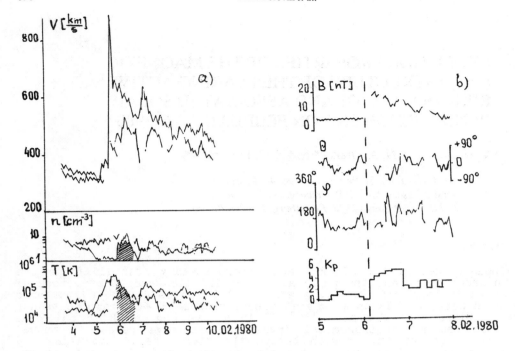

Fig.1. a) Helios 2 (solid line) and ISEE-3 solar wind data between 4 Feb and 10 Feb 1980 . The data presented are: the solar wind velocity V, the solar wind density n and the temperature T, the ISEE-3 data being shifted by corotating and radial transit time.

b). Magnetic field intensity B, the direction angles θ and φ near the Earth (ISEE-3) and Kp-index of geomagnetic activity.

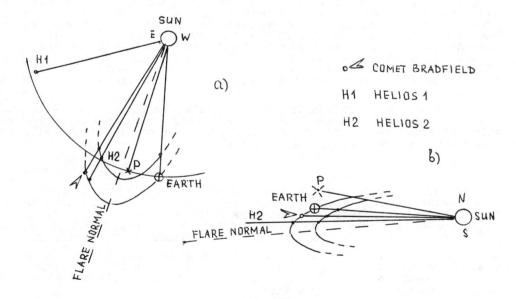

Fig.2. The positions of the Earth, the Sun, comet Bradfield and spacecrafts. P-the effective position of IPS 3C 48 (the position of the peak of the scattering weighing function) in the ecliptic plane (a) and in the plane normal to the ecliptic and containing the flare normal (b). The normal of the 3 Feb flare is indicated by an arrow. Solid lines illustrate the interplanetary disturbance 3D-speed model. The location of the transient leading edges are obtained for the moments T_1 on 5 Feb at 15^h UT (encounter with Helios 2) and T_2 on 6 Feb at 3^h UT (encounter with the Earth which is indicated by *ssc*).

Fig.1b shows the enhancement of the magnetic field ($B \approx 20$ nT) after ssc during about a day. After 12^h UT on February 6 one can see rapid alteration of the magnetic field direction ($\approx 180°$) in the ecliptic plane (φ). During this period the magnetic field latitude direction (θ) changes from large southern direction to large northern direction. Being comparatively cold (Fig.1a) the crosshatched region seems to be a "magnetic cloud".

A sequence of photographs show the unusually rapid turning of the plasma tail of comet Bradfield on 6 Feb 1980 at $2^h 20^m$ UT /1/. Exceptionally apt mutual position of the Earth, comet and spacecrafts (Fig.2) allow to associate these events and to determine the solar source of the disturbance as 1B flare on 3 Feb 1980 at $13^h 38^m$ UT in the active region HR 16 631.

The propagation speed of the disturbance is highly anisotropic: the highest speed is along the flare normal, the speed rapidly decreases as the angular distance increases. The propagation speed could be approximated by the formula: $V = V_0 (cos 2\Delta l)^{3/2}$ in the ecliptic plane and $V = V_0 (cos 2\Delta l)^{5/2}$ in the plane perpendicular to the ecliptic and containing flare normal.

Such angular distribution of the velocities explains the lack of disturbance in the solar wind data measured with Helios 1.

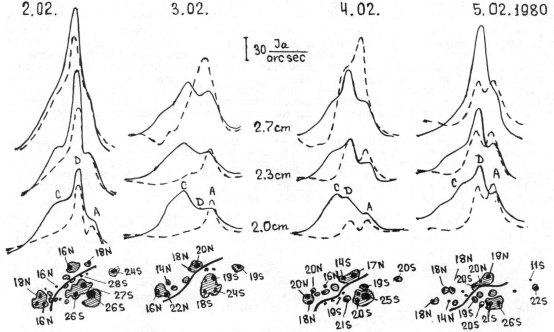

Fig.3. One-dimensional fan beam scans of HR 16 631 at wavelengths of 2.0, 2.3 and 2.7 cm with the RATAN-600 radio telescope on 2-5 Feb 1980 (I-total intensity, V-circular polarization). They are compared with a sunspot picture (umbra black and penumbra crosshatched). The neutral line is designated by the solid line in the sunspot picture.

One can see in Fig.3 a complex multipolar configuration of sunspots. On 2 Feb many small sunspots appear in the central part of the group (shown in a squares) which testify to the emergence of a new magnetic flux. On 3 Feb the majority of these newly emerged sunspots disappear.   At the shortest wavelengths 2.0 and 2.3 cm one can see a number of components in the structure of local source: "A"-core associated source connected with the largest leading umbra in the group. This source has a spectrum of fluxes sharply growing with wavelength and a high degree of circular polarization (60-80%). It has a flat spectrum of fluxes in wavelength 2-4 cm and low degree of circular polarization. "A" and "C" sources are conventional sunspot-associated sources and "D" is a new type source with peculiar characteristics. This source does not coincide in position with any big sunspot in the group, but is associated with the region of the new magnetic flux. Its position is related to the neutral line of photolspheric magnetic polarity in the region of the steepest magnetic field gradient as revealed from magnetograms obtained with Panoramic magnetograph of SibIZMIR. This peculiar source has a spectrum of fluxes sharply growing with wavelength while the degree of circular polarization is moderate (20-40%)/2/. An important characteristics of the peculiar source is its high brightness temperature $\approx 2.5 \cdot 10^6$ K ($\lambda=3.2$ *cm*). This value has been obtained under the assumption of circular symmetry of the radio source, so the real brightness temperature may be even higher.

This peculiar component of radio emission is a new type of source, which generation mechanism is not quite understood. The possible mechanisms are /3/: 1)Thermal cyclotron radiation; 2)Thermal bremsstrahlung of a hot ($T\approx10^7\div10^8$K) and very dense ($Ne\approx10^{12}\div10^{13}cm^{-3}$) condensation; 3)Nonthermal (but incoherent) generation mechanism. However all these mechanisms met difficulties. In any case abnormal local long-lasting release of energy is involved. A strong current (probably current sheets) is, almost certainly, present.

SUMMARY

1. The extreme disturbance of the solar wind and the Earth's magnetosphere on 5-6 Feb, 1980 was shown to be due to the solar flare on 3 Feb at 13^h38^m UT in active region HR 16 631. This region during 2-5 Feb was characterized by unusual properties of radio emission and magnetic field evolution:a) The emergence of the new magnetic flux with rapid evolution ($5 \cdot 10^{19}$ *Mx/hour*) and the high gradient of the photospheric magnetic field ($dH_\parallel$ / dx> 0.1 *G/km*) is revealed. b) In the local radio source structure the peculiar detail, the position of which is related to the neutral line of the photospheric magnetic field in the region of the steepest magnetic field gradient, appears. Its location coincides with the region of the new magnetic flux with rapid evolution. This peculiar source has a spectrum of fluxes sharply growing with wavelength in the range of 2 - 4 *cm*, the moderate degree of circular polarization (20-40%) and high brightness temperature about 10^7K. This spectral-polarization characteristics of the peculiar radio source provide evidence for the long-lasting abnormal local release of energy in the corona.

2. We have determined *3D*-structure and mass of the disturbance in the interplanetary space and on the Earth in the beginning of Feb 1980 at *r*=1 *AU*: the radial size *L*=0.28 *AU*, latitudinal extent *l*=0.3 *AU*, half width in the ecliptic plane <56°. The mass of the disturbance has been estimated to be $\approx 4 \cdot 10^{15}$ *g*. We associate the comparatively dense and cold material after a shock front with *CME* following the 3 Feb flare.

We assume that the interplanetary disturbance under discussion could be associated with the revealed peculiarities of the active region HR 16 631 before the flare (rapid evolution and complex configuration of the magnetic field, long-lasting local release of energy).

REFERENCES

1. J.C.Brandt, J.D.Hawley and M.B.Niedner, A very rapid turning of the plasma-tail axis of comet Bradfield 1979 1 on 1980 Feb.6, Astrophys.J.241, L51-L54 (1980)

2. V.N.Borovik, N.A.Drake and A.A.Golovko, Evolution of the magnetic flux in the flaring active region based on optical and radio observations, in: Solar Magnetic Fields and Corona, Novosibirsk, "Nauka", Vol.2, 162-166 (1989)

3. A.N.Korzhavin, G.B.Gelfreikh and S.M.Vatrushin, Peculiar sources of solar radio emission and their possible interpretation, in: Solar Magnetic Fields and Corona, Novosibirsk, "Nauka", Vol.2, 119-124 (1989)

MAGNETIC CLOUD OBSERVATIONS BY THE HELIOS SPACECRAFT

V. Bothmer and R. Schwenn

Max-Planck-Institut für Aeronomie D-3411 Katlenburg-Lindau, Germany

ABSTRACT

A possible interpretation for the observed characteristics of an interplanetary magnetic cloud is the passage of a magnetic flux rope. For simplification the flux rope might be considered as a cylindrically symmetric structure with the magnetic field lines being directed parallel to the axis at its center and circular at its outer edges. Near the center of this flux rope the magnetic field strength would be strongest. The minimum variance technique was applied to several magnetic clouds observed by the Helios spacecraft between 0.3 and 1 AU in order to determine the orientations of the magnetic flux rope axis. The calculated orientations are examined with respect to the global solar wind stream structure, the surrounding solar wind flow, the radial distance to the sun and their solar origin.

INTRODUCTION

Magnetic clouds are transient phenomena in the solar wind with peculiar plasma and magnetic field properties compared to the ambient solar wind (Klein and Burlaga [1981]). A good association was found between coronal mass ejections (CMEs) and magnetic clouds (Wilson and Hildner [1984]), and there is strong evidence that all magnetic clouds are related to CMEs at the sun. One of their striking features discovered by spacecraft measurements (Burlaga et al. [1981]) is the rotation of the magnetic field direction parallel to a plane suggesting a looplike magnetic field topology. An idealized sketch that could explain the changes of the magnetic field direction is presented in figure 1, in reference to Goldstein [1983]. The Helios spacecraft is passed by an idealized cylindrically symmetric magnetic cloud that moves radially away from the sun. First the cloud's outer circular oriented magnetic field lines are encountered and a negative Bz-component is observed for the cloud sketched in figure 1. At the center of the cloud the magnetic field is directed only in the By-direction, representing the cloud's axis. Finally the cloud's outer circular field lines are encountered again yielding a positive Bz-component in our case. The time profiles for the change of the individual magnetic field components during the clouds passage are included in figure 1. If By is plotted versus Bz the result should be an arc in the Bx-By plane. Goldstein [1983] has solved the MHD-equations for a cylindrically symmetric magnetic cloud with a force free magnetic field. Under these assumptions the magnetic field strength should be strongest at the cloud's axis. If a minimum variance analysis (Sonnerup and Cahill [1969]) is applied to the magnetic field data the cloud's axis should lie along the direction of intermediate variance. The reliability of the calculated orientation clearly depends on the space-craft's trajectory through the magnetic cloud, i.e. the distance from the spacecraft to the cloud's axis. More sophisticated model simulations for magnetic clouds accounting for this fact have

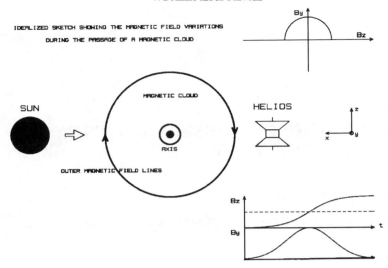

Figure 1. Idealized sketch showing a cross-section through a cylindrically symmetric magnetic cloud that moves towards the Helios spacecraft. The time profiles for the changes of the individual magnetic field components during the cloud's passage are given in the right part of the picture. It cannot be uniquely decided whether the field lines of the magnetic cloud remain connected to the sun or if they are entirely disconnected from the sun.

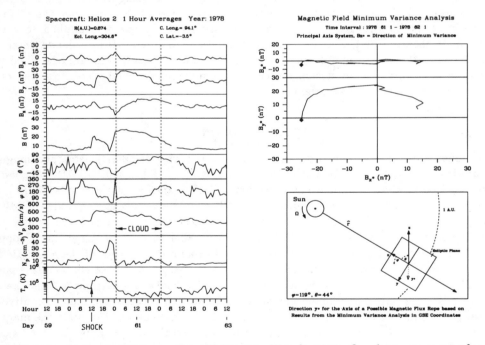

Figure 2. A magnetic cloud observed by Helios 1 in March, 1978. One hour averages of magnetic field and plasma parameters : Bx, By, Bz in GSE-coordinates, magnetic field magnitude B, magnetic field latitude angle θ and azimuthal angle ϕ, proton bulk speed V_p, proton number density N_p and proton temperature T_p. The magnetic cloud is marked by dashed lines, and the interplanetary shock preceeding the cloud is depicted by an arrow. The right upper part of the figure shows results from the minimum variance analysis of the magnetic field. The magnetic field component Bz* along the direction of minimum variance is very small and the smooth rotation of the magnetic field in the plane of maximum variance resembles figure 1. The calculated orientation for the cloud's axis is given in the lower right portion. For this particular case the axis is inclined by 44° to the ecliptic plane and its azimuthal angle ϕ is 119°. The GSE-coordinate system is centered at the spacecraft position.

been established and applied to spacecraft data by Burlaga [1988], Lepping et al. [1990], Burlaga and Lepping [1990].

MAGNETIC CLOUDS OBSERVED BY THE HELIOS SPACECRAFT

A magnetic cloud observed by Helios 1 on day 61 in 1978 at 0.87 AU is shown in figure 2. The magnetic cloud indicated by dashed vertical lines is following the interplanetary shock at 12:15 UT on day 60. In the right part of figure 2 the results of the minimum variance analysis are displayed. The component of the magnetic field along the direction of minimum variance, $Bz*$ is small ($\langle B \rangle / \langle |Bz*| \rangle = 0.09$), the ratio of the intermediate to minimum eigenvalue $\lambda 2 / \lambda 3$ is 16.8, and the angle between the first and the last hourly averaged magnetic field values is 150°. The direction of minimum variance is therefore well determined (Lepping and Behannon [1980]). The calculated direction for the cloud's axis in GSE longitude and latitude angles ϕ and θ is $\phi = 119°$ and $\theta = 44°$. The minimum variance normal, $\phi = 189°$, $\theta = -20°$ is pointing to the south of the ecliptic plane in the anti-sunward direction. Unique criteria to determine the exact start and end times for a magnetic cloud have not been established. Uncertainties resulting from the minimum variance method itself also have to be taken into account, so that the calculated axis orientation should be considered an approximation.

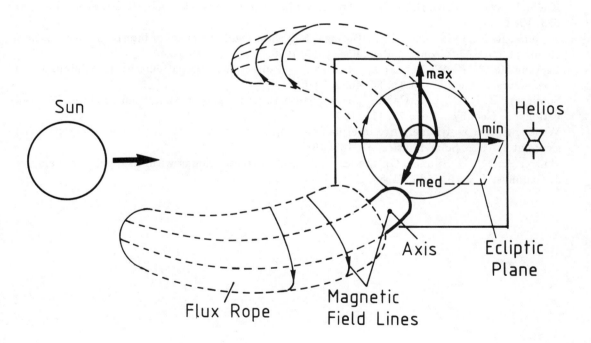

Figure 3. Sketch of a magnetic cloud as a large magnetic flux rope. The orientation of the flux rope shown in the figure was the most frequent one according to our analysis of several magnetic clouds observed by the Helios spacecraft between 1974 and 1981. The directions of minimum, intermediate and maximum variance are indicated.

RESULTS AND CONCLUSIONS

For the years 1974 to 1981, 32 magnetic clouds have been selected from the plasma and magnetic field data of the Helios spacecraft between 0.3 and 1 AU. 26 of these clouds were associated with interplanetary shocks. The minimum variance method was used to calculate the axis orientations for the magnetic clouds, yielding $\langle \phi \rangle = 100°$ with $rms = 38°$, and $\langle \theta \rangle = +1°$ with $rms = 30°$. The axes orientations scattered about the east-west direction , with mostly small inclinations to the ecliptic plane. A sample cloud is schematically illustrated in figure 3. The results are similar to those of Lepping et al. [1990] which were derived from model simulations for a set of magnetic clouds observed at 1 AU. The lack of clouds having axes nearly normal to the ecliptic plane may be an artifact of the selection procedure, certainly the orientation of filaments on the sun's disk plays an important role. This will be the subject of further investigations.

REFERENCES

Burlaga, L. F., R. P. Lepping, J. A. Jones, Global configuration of a magnetic cloud, in Physics of Magnetic Flux Ropes, ed. by E. R. Priest, L. C. Lee and C. T. Russell, AGU Geophysical Monograph 58, 1990.

Burlaga, L. F., Magnetic clouds: constant alpha force-free configurations, J. Geophys. Res., 93, 7217, 1988.

Burlaga, L. F., E. Sittler, F. Mariani, and R. Schwenn, Magnetic loop behind an interplanetary shock: Voyager, Helios, and Imp 8 observations, J. Geophys. Res., 86, 6673, 1981.

Goldstein, H., On The Field Configuration in Magnetic clouds, Solar Wind Five, NASA Conference Publ. 2280, 731, 1983.

Klein, L. W., L. F. Burlaga, Interplanetary Magnetic Clouds At 1 AU, J. Geophys. Res., 87, 613, 1982.

Lepping, R. P., J. A. Jones, L. F. Burlaga, Magnetic Field structure of interplanetary magnetic clouds at 1 AU, J. Geophys. Res., 95, 11957, 1990.

Lepping, R. P., K. W. Behannon, Magnetic field directional discontinuities: 1. Minimum variance errors, J. Geophys. Res., 85, 4695, 1980.

Sonnerup, B. U. O., L. J. Cahill, Magnetopause structure and attitude from Explorer 12 observations, J. Geophys. Res., 72, 171, 1967.

Wilson, R. M., E. Hildner, Are interplanetary magnetic clouds manifestations of coronal transients at 1 AU ?, Solar Physics, 91, 168, 1984.

Wilson, R. M., E. Hildner, On the association of magnetic clouds with disappearing filaments, J. Geophys. Res., 91, 5867, 1986.

THE MODELING OF DYNAMICAL PROCESSES IN THE SOLAR WIND BY THE METHOD OF MACROPARTICLES

A. D. Chertkov* and V. A. Anisimov**

*Institute of Physics, St Petersburg State University, Petrodvorets,
St. Petersburg, 198904, Russia
** Institute for Informatics and Automation of Russian Academy of Sciences,
14th line 39, St Petersburg, 199178, Russia

ABSTRACT

A computer code "SOLAR WIND DYNAMICS" for MHD simulation of non-stationary developing 3D processes in the solar wind is suggested. The solution is constructed in Lagrange's approach. The code calculates and shows in a clear form (as a movie) the well-known dynamical processes in the solar wind: high-speed streams of different origin, quasi-shock fronts, rarefaction regions, magnetic field spirals. Some empirical regressions are introduced. This code could be used for the real time forecast of the solar wind disturbances and the level of the geomagnetic activity on the base of the solar magnetic field measurements. The code could be a useful tool for persons, who are interested in the solar-terrestrial predictions and the interplanetary dynamics modeling of the inner and middle heliosphere. THE CODE IS AVAILABLE UPON REQUEST.

INTRODUCTION

The solar wind models operating in terms of MHD approximation /1,2/ satisfactorily describe the observed basic properties of solar coronal expansion /3/. But they meet difficulties with explanation of internal correspondence of important parameters of this description /4/. For example, semi-empirical Parker's model of the interplanetary dynamical processes permitted to forecast the existence of the solar wind /1/, despite the internal contradictions of the model: the hydrodynamical description requires the small value of the mean free path l while the extraordinarily high magnitude of the thermal conductivity requires big l /4/. In the model of the work /1/ the important processes of thermal energy transfer were described by simple polytrope law, and it occurred very fruitful /2-4/. Sometimes only empirical and simulating models are suitable in case when the fully adequate theory is still absent /4/. The constructing of an effective algorithm for solution of 3D space MHD problems is possible in framework of the macroparticle method /3,5,6/. The simulation methods of different kinds are intensively developed nowadays /7,8/. The proposed code was chosen for its flexibility and economy of computer resources.

CALCULATION TECHNIQUE

The description of method. The suggested code constructs the solution of MHD problem in Lagrange's approach (see /5/). The final results can be given in Lagrange's and Euler's representation. The ideal non-dissipative MHD approximation is supposed. The macroparticle method provides the fulfillment of the mass conservation law. The macroparticles collide moving out of the Sun.

The result of this interaction (in accordance with the impulse conservation law) is the formation of high-speed stream structures, shocks, compression and rarefaction regions of plasma and magnetic field. The streams are adiabatic: there exists no energy influx into the stream during its motion out of the Sun. The kinetic energy of bulk movement of the streams is partially converted into the thermal form in the stream interaction regions (according to energy conservation law). The magnetic field plays only passive role and it follows the macroparticle motions. The values of solar wind parameters (velocity v, density n, temperature T, GSM Cartesian components of the interplanetary magnetic field B_x, B_y, B_z), averaged over Euler's volumes, can be interpreted as solutions of MHD set /5/. The model contains a certain number of free parameters which have clear sense. These parameters are optimized in the process of the inverse problem solution (best fit procedure). The main model parameters are: the masses of macroparticles, their sizes, initial velocities, the collision parameters (non-elasticity ratios). The best fit procedure lets us satisfactorily describe the empirical and theoretical dependences between solar wind parameters reported in /2,4/. Hence, our model can reproduce (whithin the limits of its accuracy) the interaction of corotating streams, including the steepening of moving fronts of evolving disturbances, quasi-shock fronts, rarefaction regions, the moving spirals of magnetic field. The number of macroparticles, the temporal discretization step and the size of the modeled region should be chosen from the computer memory and productiveness and from time required for simulations. The code creates the computer movie, calculates and displays on the color screen the parameters of the solar wind. They are: v, n, T, B_x, B_y, B_z and vector standard deviation of magnetic field σ_B. Then $\sum K_p$ index of the geomagnetic activity is evaluated according to formula from /9/: $\sum K_p = 5 + (\sigma_B - 0.5 B_z) v$. It is suggested that the variations in σ_B are proportional to n /4/. The geomagnetic activity forecast based on these ideas was relatively successful /4,6/.

The initial and boundary conditions. At first the model space should be filled up with macroparticles moving uniformly with velocity $v_o = 400$km/s along the radius from the rotating Sun outside the initial surface. The "source surface" of the model by K.H.Schatten et al. /10/ was chosen as an initial surface for our calculation. The magnetic field in space is simulated by Parker's spirals: each separate magnetic field line represents the line connecting the particles emitted from the same place of rotating Sun. The corresponding value of measured (or artificially set up) radial component of the solar magnetic field is ascribed to this field line. This magnetic field value must be constant in frame of suggested concept during all the process of modeled movement. Then the high-speed (or low-speed) macroparticles start from the "source surface". Their initial velocities should be specified from experimental data or artificially generated. Unlike the magnetic field boundary conditions, the hydrodynamical boundary parameters (masses and initial velocities) can vary from one temporal step to another.

The menu system. The menu system allows to load and save the parameter and data block from and to the disk in interactive regime; to choose the simulation modes; to set up (from real data) or artificially create the initial condition vector; to set up the form of representation. The model has some changeable parameters, which can be established from special sub-menu. They are: the size of modeled region, the number of macroparticles along the longitude and latitude, the temporal discretization step, the value of "primordial" wind velocity and density, the radius of macroparticle, modification of coordinate system and others.

The modes. "Forecast in pictures" mode can be used for obtaining the dynami cal picture of the solar wind and for graphic representation of the velocity, density and temperature near the Earth over the time. This variant lets show the evolving process in the visual form. It is the easiest way for user to get the whole notion about all the situation in space, leading to the consequence of events observed near the Earth (or near any other place specified in good time). The "forecast in numbers" mode can be used for obtaining some fundamental solar wind parameters (v, n, T, B_x, B_y, B_z, σ_B) near the Earth and $\sum K_p$ index of the geomagnetic activityin digital form . This way is convenient for further digital analysis and report. The "demonstration in pictures" mode can be used for clear representation of some interesting specific effects: recurrent flows, solar flare flows and high speed streams of different origin. The initial and boundary conditions should be generated artificially. The developing disturbances can be seen from different positions of an observer in space. An additional interest represents the case when the modeling is continued up to distances $2 \div 5 A.U.$ The co-rotating spirals, stemming from different solar turns, interact at these distances. This mode can be used for educational purposes. The "demonstration in numbers" mode is displayed as "forecast in numbers" mode but its other features are the same as ones of the "demonstration in pictures" mode. An additional possibility of this mode is obtaining of Euler's averaged solar wind parameters near artificial spacecraft, the trajectory of which can be set up.

The required experimental data. The "forecast" modes require the magnetogram data files, which may be prepared from the real magnetogram data received at Stanford (USA) and Sayan (Russia) observatories. The special computer service program is designed for this preparation. The profile of the solar wind velocity at the source surface v_s is calculated on the base of these data.

According to the work /4/ the regression was introduced:
$v_s = v_o (1 + \frac{a}{\pi} \arctan((|B_r| - cB_\perp)/B))$, where v_o is the given mean velocity (e.g. 400km/s), "a" and "c" are the regression coefficients obtained from the best fit, B_r is the radial component, $B_\perp$ is the transversal component, and B is the magnitude of the magnetic field at the source surface. The profile of the velocity may be changed by the user (when the user has the additional information about these initial velocities).

THE CAPABILITIES AND DRAWBACKS OF THE CODE

This code could be a good scientific tool: a user could perform rather complicated scenarios of the interplanetary disturbance development, specifying the initial conditions on the initial surface. He can change the highlight for different populations of macroparticles (united by common condition), marking the most interesting structures. The marked surfaces could serve as tracers for spatial positions of peculiarities in the solar wind. This can help to find the right reconstruction of the spatial distribution and temporal evolution of magnetic field and plasma structures when at least some of the initial conditions are known. Drawbacks of the code are determined by ones of the "ideal" MHD approximation /2,4/ and macroparticle method /5/. The solar wind is non-adiabatic in reality /2,3/. The varying in time solar magnetic fields are not allowed in frame of ideal MHD concept. But according to work /11/ this variability plays the principal role in the creation of the solar wind. Hence, these non-adiabatic and non-stationary processes work mostly inside the Alfven surface. Therefore the variant of the code suggested here should give reliable results outside this surface. The macroparticle method has a restricted spatial resolution which cannot be changed in the process of calculation. That is why we can see only quasi-shock fronts and not the shocks. The dynamical range of calculation is restricted also from the same reason. Therefore we cannot simulate extremely strong shocks. The increase in spatial resolution is connected with huge increase in required

computer time and memory. Hence, this code is suitable for preliminary quantitative estimation of space dynamical processes. The incorporation of non-
adiabatic processes and non-stationarity of magnetic field requires more
complicated programming. It is the subject of the forthcoming papers.

DISCUSSION AND CONCLUSION

The code is designed for the solution of the "direct" problem: the initial
conditions on the source surface are known, the situation in space should be
calculated. But to solve this problem one should have good (best fitted)
free parameters of the model and good initial data. The model parameters
used in this variant of the code are fitted, of course. But this fitting
probably is not the best. This originates from lack of good initial data: as
a rule we know only the situation close to the Earth, and some of necessary
data on the source surface (usually the magnetic field). It should be noted
that the direct information about the hydrodynamical parameters (v, n) on
the initial surface is absent. Observing, e.g., a coronal mass ejection
event, we cannot derive the quantitative values of the velocity and the mass
of this disturbance. Just the same is for the solar flares in optical and
all the other frequency bands: all these events are only the qualitative indicators of energy influx into the solar wind. The possibility to use the
quantitative information about the energy influx contains the model of induction coronal heating /11/. Nevertheless, the progress in this field requires further attempts to solve the inverse problem. Using this code one might
perform a great deal of different scenarios of interplanetary disturbance
behavior. THE CODE IS AVAILABLE UPON REQUEST. The E-mail address is:
adchert@space.phys.lgu.spb.su (for A.D.Chertkov).

Acknowledgements. This work was partially sponsored by the private Scientific Innovation Firm "Nienschanz".

REFERENCES

1. E.N.Parker, Interplanetary Dynamical Processes Interscience (1963).

2. A.J.Hundhausen, Coronal Expansion and Solar Wind. Springer Verl. (1972).

3. S.Cuperman, Space Sci.Rev. 26, 277 (1980).

4. A.D.Chertkov, Solnechnyi Veter i Vnutrennee Stroenie Solnca (The Solar
Wind and the Internal Structure of the Sun) Nauka Publ., Moscow (1985).

5. R.W.Hockney and J.W.Eastwood, Computer Simulation Using Particles. McGraw
Hill Int. Book Comp. (1981).

6. A.D.Chertkov, in: Magnitosfernye Issledovanija (Magnetospheric Research),
#13, 97, VINITI Publ., Moscow (1989).

7. C.D.Olmsted and S.-I.Akasofu, Planet.Space Sci. 33, 831 (1983).

8. S.T.Wu, M.Dryer and S.M.Hun, Solar Physics 84, 395 (1983).

9. M.I.Pudovkin, D.I.Poniavin, M.A.Shukhtina, and S.A.Zaitseva, in: Geomag-
nitnye Issledovanija (Geomagnetic Research), Sov.Radio Publ.,Moscow, #27, 69
(1980).

10. K.H.Schatten, J.M.Wilcox and N.F.Ness, Solar Physics 6, 442 (1969).

11. A.D.Chertkov and Yu.V.Arkhipov, The induction coronal heating and solar
wind acceleration (this issue) (1992).

RELATIONS BETWEEN PARAMETERS OF CORONAL MASS EJECTIONS AND SOLAR FLARE MICROWAVE AND SOFT X-RAY BURSTS

I. M. Chertok, A. A. Gnezdilov and E. P. Zaborova

Institute of Terrestrial Magnetism, Ionosphere, and Radio Wave Propagation (IZMIRAN), Troitsk, Moscow Region, 142092, Russia

ABSTRACT

We have analyzed the distribution of flare events without CMEs and with different CMEs on plots of the peak intensity and effective duration of microwave and soft X-ray bursts. Several zones are distinguished where CMEs with definite angular span, speed, mass, shape, as well as events without CMEs are concentrated. The results are discussed in terms of the flare particle acceleration and the long-duration, high-coronal energy release during a restoration of the coronal magnetic field, disturbed by CMEs, to its initial state.

INTRODUCTION

The Coronal Mass Ejections (CMEs) CMEs and their relationships with the flare activity are of great interest. Up to now associations of CMEs with soft X-ray bursts have been mainly analyzed (see /1/ and references therein). The principal result of these investigations is a conclusion that the duration of the burst is the most important characteristic for the appearance of CMEs.

In the present report, the relationships of CMEs and microwave bursts are primarily discussed. The soft X-ray emission is also considered. The important distinctive point of our analysis is that the combination of the intensity and duration is taken as a main characteristic of the burst. Besides, in our consideration, the relations of microwave and soft X-ray bursts to all the main parameters of CMEs (angular size, speed, mass, and shape) are analyzed. We used the data on about 60 P78-1 SOLWIND CMEs that according to /2-5/ and others, are identified with flares on the near-limb heliolongitudes $|\ell| \geq 45°$, as well as data on about 20 flares which are not accompanied by CMEs.

DISTRIBUTION OF THE CME EVENTS ON THE INTENSITY-DURATION PLOTS

For microwave radio bursts, the combination of the absolute peak flux density (S) at frequencies above 3 GHz and the effective duration (d) at the half peak flux level is considered. The analysis reveals that the location of the events on the S-d-plot allows us not only to separate the flares with CMEs and ones without CMEs but also to determine the relations between the characteristics of microwave bursts and such parameters of CMEs as angular sizes, speed, mass, shape. As an illustration the S-d-plot is shown in Figure 1 where the events with large, medium, and small angular sizes, as well as the events without CMEs are marked by different signs. One can see that the microwave events of two qualitatively different types are associated with CMEs: (1) large non-impulsive bursts with $S \sim 10^2 - 4 \times 10^4$ s.f.u. and $d \sim 2$-20 min; (2) small-intensive prolonged "gradual rise and fall" (GRF) bursts with $S \sim$ 6-100 s.f.u. and $d \geq 20$ min. Moreover, the definite zones may be distinguished on the

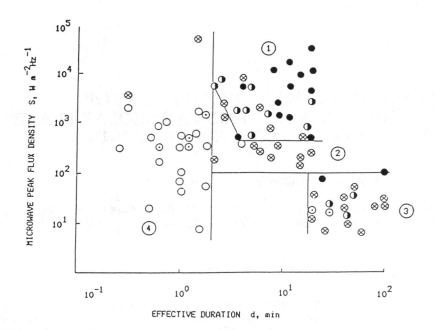

Fig 1. Scattered S-d-plot showing the relation between the microwave burst
properties and angular width of corresponding CMEs: ● – very large (≥ 90°),
◑ – large (60-90°), ⊗ – middle (30-60°), ⊙ – small (≤ 30°) CMEs;
○ – events without CMEs. Lines mark the conditional zones of intensive
(1), moderate (2), GRF (3), and impulsive (4) radio bursts.

S-d-plot in which the CMEs with different angular sizes ($\Delta\Theta$) are concentrated. The
majority (80 %) of large CMEs with $\Delta\Theta \geq 60°$ are observed in a combination with
the most intensive ($S \geq 4 \times 10^2 - 5 \times 10^3$ s.f.u.) and long-duration ($d \geq 2-4$ min) radio
bursts. In the intermediate burst zone ($S \sim 10^2 - 5 \times 10^3$ s.f.u. for $d \sim 2-4$ min and
$S \sim 10^2 - 4 \times 10^2$ s.f.u. for $d > 4$ min) and GRF-zone, the CMEs of medium sizes
($\Delta\Theta \sim 30-60°$) predominate. The impulsive burst zone ($d \leq 2$ min) may be characterized
as one of events either without CMEs or with CMEs of small sizes ($\Delta\Theta \leq 30-40°$).

Besides angular sizes, the analogous separation takes place also for other parameters
of CMEs on the S-d-plot (Figure 2). In the zone of the intensive, non-impulsive
bursts, the CMEs of complex shapes (filled loops, curved front, remnants), high speed
($V \geq 1000$ km/s), and big mass ($m \geq (8-12) \times 10^{15}$ g) are mainly observed. For the
intermediate burst and GRF zones, the CMEs of simpler shapes (spikes, fan), medium
speed ($V \sim 400-1000$ km/s), and moderate mass ($m \sim (1-10) \times 10^{15}$ g) are typical. As for
as several CMEs associated with sufficiently intensive impulsive bursts are
concerned, they have as a rule the simple shape, moderate speed ($V \sim 400-900$ km/s)
and mass ($m < 8 \times 10^{15}$ g).

The similar (but slightly different) regularities are obtained when the combination
of the intensity (F) and whole duration (d) of soft X-ray bursts is used instead of
the radio parameters S and d (Figure 3).

CONCLUSION AND DISCUSSION

Thus, the results described above mean that there is a wide spectrum of events with
the different correlation between the eruption of CMEs and the flare or flare-like
energy release, but this correlation discovers the definite reflection in
characteristics of the microwave and soft X-ray emission. Such reflection is possible
both when the ascending CME initiates a flare and when on the contrary the eruption
of the CME is stimulated by the flare energy release. In the majority of confined
(impulsive) flares, the energy release happens without visible CMEs in general. The
magnetic field reconnection may be initiated by other phenomena such as the emergence

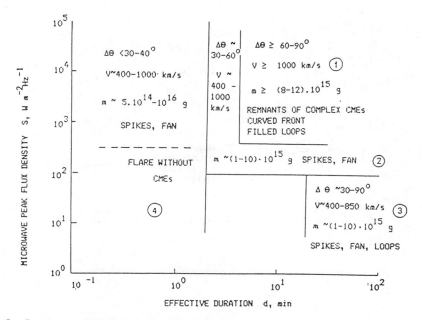

Fig. 2. Summary of the correlation between the microwave bursts and the characteristics of CMEs. The zones of intensive (1), intermediate (2), GRF (3), and impulsive (4) radio bursts are indicated.

of a new magnetic flux, the interaction of magnetic loops and so on.

On the other hand, the CME eruption may be a consequence of a disappearance of the equilibrium of existing coronal structures /6,7/. According to /8/, the eruption of the CME starts often before the impulsive phase of the flare which takes place then near one of legs of the transient loop. However, such a situation seems to be typical for relatively small flares. In large flares with the developed spatial-time structure, the more complicated relationship between the CME and flare takes place accompanied by the multiple energy release in different parts of the magnetosphere above the active region. In such cases, the observed dependence between parameters of microwave bursts and CMEs reflects the fact that the eruption of large, massive and fast CME combines with the powerful, long energy release including the numerous particle acceleration.

However, the eruption of CMEs including average or even large ones not always results in the explosive energy release. When there are not suitable conditions in the active region, the eruption of a prominence and ascent of a CME lead rather to the prolonged heating of a large plasma volume in the low corona and to the generation of GRF radio emission than to the particle acceleration.

In the different eruptive events, the prolonged energy release during the late phase of the flare is very important and can give a remarkable additional contribution to the relationships between the CMEs and electromagnetic emission. On this stage, which must take place in any eruptive events, including GRF-bursts, the magnetic field disturbed by the CME recovers to its initial state. The restoration happens via the magnetic reconnection in the vertical current sheet located in the corona. In large flares, it is accompanied by a formation of the post-flare loops, prolonged particle acceleration, and generation of the long-duration soft X-ray emission, as well as the so-called delayed or extended microwave bursts with soft radio spectra of (see /9-12/). It should be noted that just such soft radio spectrum is observed in the majority of flares associated with large CMEs.

The data indicated above reveal also that the various shapes of CMEs observed with the white-light coronagraphs, reflect the real physical differences of events but are

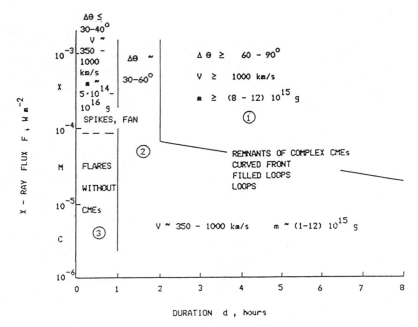

Fig. 3. Summary of the correlation between the soft X-ray burst parameters and characteristics of CMEs. The zones of intensive and long-duration (1), moderate (2) and impulsive (3) soft X-ray bursts are indicated.

not only consequence of geometrical or other analogous factors. The close dependence which according to the present analysis exists between parameters of microwave (soft X-ray) bursts and CMEs can be used for the electromagnetic diagnostics of flares causing interplanetary disturbances and geomagnetic storms.

REFERENCES

1. S.W.Kahler, N.R.Sheeley Jr., and M.Liggett, Astrophys. J. 344, 1026 (1989).

2. N.R.Sheeley Jr., R.T.Stewart, R.D.Robinson, R.A.Howard, M.J.Koomen, and D.J.Michels, Astrophys. J. 279, 839 (1984).

3. S.W.Kahler, N.R.Sheeley Jr., R.A.Howard, M.J.Koomen, D.J.Michels, R.E.McGuire, T.T.von Rosenvinge, and D.V.Reames, J. Geophys. Res. 89, 9683 (1984).

4. N.R.Sheeley Jr., R.A.Howard, M.J.Koomen, D.J.Michels, R.Schwenn, K.H.Muhlhauser, and H.Rosenbauer, J. Jeophys. Res. 90, 163 (1985).

5. H.V.Cane, N.R.Sheeley Jr., and R.A.Howard, J.Geophys. Res. 92, 9869 (1987).

6. R.Wolfson and S.A.Gould, Astrophys. J. 296, 287 (1985).

7. B.C.Low, Highlights of Astron. 7, 743 (1986).

8. R.A.Harrison, E.Hildner, A.J.Hundhausen, D.G.Sime, and G.M.Simnett, J.Geophys. Res. 95, 917 (1990).

9. U.Anzer, and G.W.Pneuman, Solar Phys. 79, 129 (1982).

10. S.W.Kahler, Solar Phys. 90, 133 (1984).

11. E.W.Cliver, B.R.Dennis, A.L.Kiplinger, S.R.Kane, D.F. Neidig, N.R.Sheeley Jr, and M.J.Koomen, Astrophys. J. 305, 920 (1986).

12. K.Kai, H.Nakajima, T.Kosugi, R.T.Stewart, G.J.Nelson, and S.R.Kane, Solar Phys. 105, 383 (1986).

A COMPARATIVE STUDY OF DYNAMICALLY EXPANDING FORCE-FREE, CONSTANT-ALPHA MAGNETIC CONFIGURATIONS WITH APPLICATIONS TO MAGNETIC CLOUDS

C. J. Farrugia,* L. F. Burlaga,* V. A. Osherovich** and R. P. Lepping*

Goddard Space Flight Center, Greenbelt, MD 20771, U.S.A.
**NASA/GSFC, Hughes STX, 4400 Forles Blvd., Lanham, MD 20706, U.S.A.*

ABSTRACT

We contrast two different solutions of the constant alpha, force-free MHD equation, both of which have been suggested as models for magnetic clouds: a solution in cylindrical coordinates and one in spherical polar coordinates. In line with the observation that magnetic clouds expand, we generalize these static models and construct their expanding counterparts. We find that expansion introduces in both cases a large asymmetry in the field strength signature which is in the same sense as that seen in the data, i.e. towards the leading edge of the cloud. We then do a least squares fit of the respective models to one-spacecraft data on a magnetic cloud. We find that the fitting routine converges in both cases. However, while purely formally we cannot distinguish between the two models using data from one spacecraft, the field components in the 'spherical' model have features not compatible with data on magnetic clouds.

INTRODUCTION

Magnetic clouds /1/ have been suggested to be force-free configurations /2/. Burlaga /3/ showed that the solution in cylindrical coordinates of the force-free MHD equation, curl $\mathbf{B} = \alpha \mathbf{B}$, with α a real number, first obtained by Lundquist /4/ , describes the signatures of magnetic clouds well. (See also /5/.). The field line geometry thus envisaged has helical field lines on the surfaces of coaxial cylinders with pitch decreasing with distance from the axis. By contrast, Vandas et al. /6/ suggested that the solutions of the force-free equation with constant α in spherical polar coordinates, first derived by Chandrasekhar and Kendall /7/, or variations thereof (e.g. spheroid, /8/) are sometimes more appropriate to describe magnetic clouds. This claim was based mainly, but not exclusively, on the grounds that observed asymmetries in the field strength profile are explicable on a spherical model, which has inherent 'geometrical' asymmetries, but not on the straight 'flux rope' model, which has none. The field line geometry of this second model (in its main variant) is similar to the streamlines in Hill's spherical vortex. The field lines wind on tight tori in the azimuthal direction. Roughly, the issue is therefore: Is the magnetic cloud more like a flux rope or more like a spherical plasmoid ?

We propose to contrast these two models by carrying out a least-squares fit of both to a specific data example. Departing from the practice used hitherto of contrasting static models, we shall incorporate expansion in both models, since magnetic clouds expand /9/. It turns out that expansion introduces large asymmetries in the field strength profile of both models. We shall conclude that, purely on the basis of a least-squares minimization algorithm both models describe the data well. However, the field components of the spherical solution exhibit polarity changes at the rear edge of the cloud which are at variance with the observations, and which are not seen in the data of other examples either. For reasons of space we can but outline the main points here. For a complete description the reader is referred to /10/

DATA EXAMPLE

Figure 1 (after /11/) shows interplanetary magnetic field and plasma data at 1 min resolution. From top to bottom: flow speed (V; km s^{-1}), density (N; cm^{-3}), temperature (T; $^{\circ}$K), B$_x$, B$_y$, B$_z$ (GSE; nT), and total field (B; nT). As described in /11/, the magnetic cloud is identified as the data shown between vertical guidelines. An interplanetary shock precedes the cloud by ~ 5 h. Of relevance here are the last four

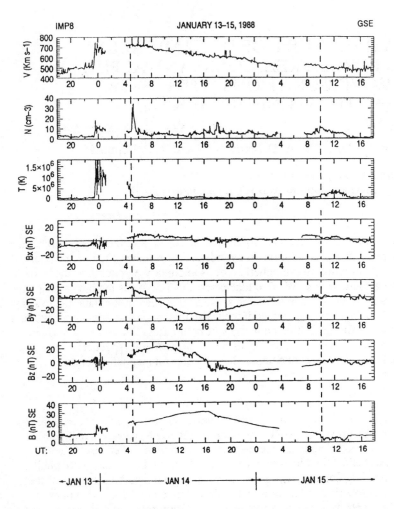

panels, showing the large rotation of the field components (mainly in y and z) and the enhanced field strength, well above pre-shock values, peaking where B_z changes polarity and B_y maximizes negative. Note that there is an asymmetry in the field strength, with the field maximum well displaced towards the leading edge of the cloud. Another important feature of the data is the falling V-profile, (panel 1) symptomatic of expansion. In our fitting procedure we use information gained from the flow in the first panel of Figure 1 to fit the magnetic field in the last four panels

Fig. 1 IMP-8 plasma and magnetic field data relating to an interplanetary magnetic cloud observation. The latter is shown within vertical guidelines. For details, see text.

THE EXPANDING MAGNETIC FIELD CONFIGURATIONS

Farrugia et al. /11/ investigated expanding magnetic configurations with the aim of deriving a velocity law consistent with the equations of ideal MHD. They treated the case of a radially expanding, self-similar magnetic configuration. The last assumption means that the magnetic flux function, ψ, rather than being a function of the two variables r and t separately, is a function of a combination of them given by the similarity parameter η: $\psi = \psi(\eta, z)$, where $\eta = r\,\xi(t)$. One seeks to determine $\xi(t)$. For free expansion $\xi = -1/(t + t_0)$, t_0 constant, with a consequent (unique) velocity law of expansion $v = r/(t + t_0)$. In brief, this says that the speed at a fixed distance from the axis (or origin) drops inversely with time. From Figure 1, top panel, one can see qualitatively that the flow speed is behaving like this.

With the help of ξ, we are now able to construct the two expanding configurations which initially coincide with the static configurations of constant α. (We note that during the expansion the Hill vortex-type configuration remains force-free, whereas the flux rope solution deviates from strict force-freeness. A detailed asssessment of this is given in /12/.) For instance, accepting for simplicity of analysis the above value of ξ as approximate solution, the expanding Lundquist solution, which does not depend on the z coordinate, is

$$B_r = 0$$
$$B_\phi = (B_0 / \tau) J_1 (\alpha r / \tau)$$
$$B_z = (B_0 / \tau^2) J_0 (\alpha r / \tau),$$

where $\tau = (t + t_0)/t_0$. A self-similar solution can be obtained for the expanding spherical magnetic cloud /10/. Both these expanding configurations have a field strength profile which is strongly asymmetric by virtue of the expansion, and the maximum field strength comes always toward the leading edge of the cloud /10/, as are the field strength maxima seen in the data.

LEAST SQUARES FITTING: RESULTS

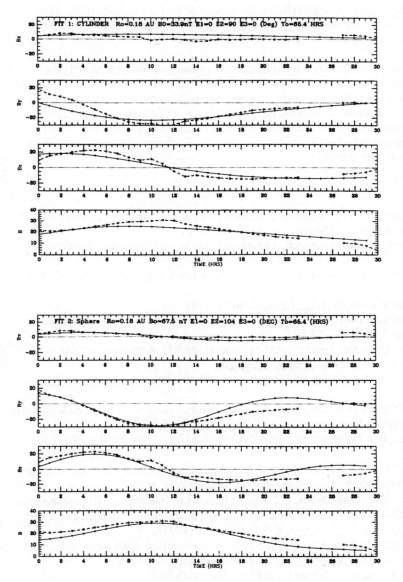

The least-squares fitting proceeds in two steps. In the first step we fit the theoretical V-profile to the data, taking care to include the bulk motion of the cloud. This yields two quantities, the first (t_0) is the time the cloud had been expanding when first observed (its 'age'), and the second (r_0) gives the radius it attained at that time. In our case, $t_0 = 65.40$ h and $r_0 = 0.18$ AU /11/.

The second step is a least-squares fit to the GSE components, using the above derived values of t_0 and r_0. The free parameters are B_0, the orientation of the magnetic cloud (specified by Euler angles), and, in the case of the spherical model, also the impact parameter. We refer the reader to /10/ for details.

As already anticipated, the algorithm converged in both cases and the variances were comparable. Figures 2a (cylinder) and 2b (sphere) show the quality of the fits and the results obtained. The dashed lines are the data points (hour averages) and the continuous curves are the fits. It is clear that, while from the point of view of the fitting routine both models do a good job by

Fig. 2a, b. Results of the fits to (a) the model in cylindrical coordinates and (b) the model in spherical polar coordinates. The data are hourly averages and are joined by dashed lines.

minimizing the variance and obtaining comparable values for its minimum, an inspection of the fit to the spherical model (Figure 2b) shows why this model does an inferior job and thus why the expanding flux rope solution should be preferred. Towards the rear edge of the cloud the B_y and B_z components both change sign. No observational magnetic cloud signature has more than one 'cycle' in its components. The reason why this model has these 'wings' is, heuristically speaking, because the geometry is similar to that of a very tight torus, the tightest you can get, i.e. the radius of the small arm is equal to the radius of the big arm. Thus traversals near the centre will yield more than one cycle of the components.

CONCLUSIONS

We have constructed the two freely-expanding counterparts of two force-free, constant-alpha configurations, one in cylindrical and one in spherical coordinates. In contradistinction to the respective static counterparts, where only the spherical solution has field strength asymmetries, the expanding models both exhibit a large asymmetry stemming from the effect of the expansion on the magnetic field. This expansion shifts the maximum value of the field strength towards the leading edge of the cloud in both cases. The direction of the asymmetry is, then, in agreement with observations.

A least-squares fit to the data from one spacecraft was insufficient to uniquely identify the underlying geometry. We noted, however, essential qualitative differences in the fitted field direction. In particular, in the spherical solution the modelled components develop a second 'wing'. This must be considered a serious shortcoming in as much as magnetic cloud data, such as the event studied here, do not share this feature. Furthermore, for this cloud in particular, we have elsewhere /11, 13/ shown evidence which argues for connection of the magnetic field lines of the cloud to the sun. Such connection is hard to reconcile with a spherical plasmoid-type topology.

REFERENCES

1. Burlaga, L.F., E. Sittler, F. Mariani, and R. Schwenn, Magnetic loop behind an interplanetary shock: Voyager, Helios and IMP-8 observations, J. Geophys. Res., 86, 6673 (1981)

2. Goldstein, H, On the field configuration in magnetic clouds, in Solar Wind Five, *NASA Conference Publ. 2280*, p.731 (1983)

3. Burlaga, L.F., Magnetic clouds: constant alpha force-free configurations, J. Geophys. Res., 93, 7217 (1988)

4. Lundquist, S., Magnetohydrostatic fields, Arkiv Fysik, 2, 361 (1950).

5. Lepping, R.P., J.A. Jones, and L.F. Burlaga, Magnetic field structure of interplanetary magnetic clouds at 1 AU, J. Geophys. Res., 95, 11957, 1990

6. Vandas, M., S. Fischer, and A. Geranios, Spherical and cylindrical models of magnetized plasma clouds and their comparison with spacecraft data, Planet. Space Sci., 39, 1147 (1991)

7. Chandrasekhar, S. and K.H. Prendergast, The equilibrium of magnetic stars, Proc. Nat. Acad. Sci., Washington, 42, 5, (1956)

8. Vandas, M., S. Fischer, P. Pelant, and A. Geranios, Magnetic clouds: comparison between spacecraft measurements and theoretical magnetic force-free solutions, These Proceedings, 1992

9. Klein, L.W., and L.F. Burlaga, Interplanetary magnetic fields at 1 AU, J. Geophys. Res., 87, 613 (1982)

10. Farrugia, C.J., L.F. Burlaga, V.A. Osherovich, and L.P.Lepping, A study of the topology and time evolution of interplanetary magnetic clouds, J Geophys. Res., to be submitted, 1992.

11. Farrugia, C.J., L.F. Burlaga, V.A. Osherovich, I.G. Richardson, M.P. Freeman, R.P. Lepping, and A. Lazarus, A study of an expanding interplanetary magnetic cloud and its interaction with the Earth's magnetosphere: The interplanetary aspect, J. Geophys. Res., submitted (1991)

12 Osherovich, V.A., C.J. Farrugia, and L. F. Burlaga, The non-linear evolution of a cylindrical magnetic flux rope. I. The low beta case, J. Geophys. Res., to be submitted, 1992

13. Richardson, I.G., C.J. Farrugia, and L.F. Burlaga, Energetic observations in the magnetic cloud of 14-15 January 1988 and their implications for the magnetic field topology, Proc. 22nd International Cosmic Ray Conf. (Dublin), Paper SH 7.8, 1991

RADIAL EXPANSION OF AN IDEAL MHD CONFIGURATION AND THE TEMPORAL DEVELOPMENT OF THE MAGNETIC FIELD

C. J. Farrugia,* V. A. Osherovich,** L. F. Burlaga,* R. P. Lepping* and
M. P. Freeman***

Goddard Space Flight Center, Greenbelt, MD 20771, U.S.A.
**NASA/GSFC, Hughes STX, 4400 Forles Blvd., Lanham, MD 20706, U.S.A.*
***The Blackett Laboratory, Imperial College, London, SW7 2BZ, U.K.*

ABSTRACT

We study the free radial expansion of a 3-component magnetic configuration. The emphasis of this paper is on the behaviour of a field undergoing non-self-similar expansion. Comparing our results with the evolution of a magnetic configuration expanding self-similarly, we find that self-similar expansion appears as the asymptotic limit (with time) of the general case. Using a model field we show that a non-self-similar velocity profile need not have a strict monotonic decrease with time.

INTRODUCTION

Magnetic clouds /1/ are large-scale MHD configurations which expand as they propagate antisunwards /2/. In an earlier work /3/ we modelled this expansion using ideal MHD. In that study we assumed self-similarity. The resulting velocity law for free, radial expansion of an axisymmetric field structure was

$$v = r/t$$

where r is the radial distance from the cloud's axis. In this paper we relax self-similarity and derive new expressions, in two coordinate systems, for the temporal evolution of an axisymmetric magnetic field. We find that, as time tends to infinity, the field behaviour approaches that of self-similar expansion. We show that for a non-self-similar expansion, the speed departs from the $1/t$ variation. In so far as magnetic clouds expand freely, our theoretical results might help understand the non-monotonic behaviour of some observed magnetic cloud speed profiles.

BASIC EQUATIONS AND ASSUMPTIONS

For the description of the magnetic field we employ the equations of ideal MHD, i.e.(Gaussian units) momentum equation:

$$\rho\, \partial v/\partial t + \rho\, (v.\ \mathrm{grad})v = 1/4\pi\ (\mathrm{curl}\ \mathbf{B}) \times \mathbf{B} - \mathrm{grad}\ P$$

divergence-free condition for **B**:

$$\mathrm{div}\ \mathbf{B} = 0$$

induction equation:

$$\partial \mathbf{B}/\partial t = \mathrm{curl}\ (v \times \mathbf{B})$$

continuity equation:

$$\partial \rho/\partial t + \mathrm{div}\ (\rho\ v) = 0$$

where p is the gas pressure, ρ is the mass density, v is the velocity of the flow, and B is the magnetic field intensity. A possible equation for energy transport is a polytropic law. We do not, however, discuss the energetics of magnetic clouds here.

We define free radial expansion as the MHD configuration obeying the momentum equation

$$\partial v/\partial t + v\, \partial v/\partial r = 0$$

Free expansion is a justifiable assumption when the inertial force dominates the total body force. The combined body force may be neglected if either (i) the body forces are in balance, or (ii) the body forces are separately small. An example of the latter category is where the magnetic $j \times B$ force is zero and the pressure gradient is small. In magnetic clouds there is evidence to suggest that the magnetic field is approximately "force-free", i.e. the magnetic pressure and tension forces are in balance /4/. As a particular example, magnetic field data from the magnetic cloud observed at 1 AU on January 14-15, 1988 have been successfully fitted to force-free models /5/. Figure 1 shows the thermal pressure measured in this magnetic cloud (see /3/). It is evident that the thermal pressure inside the cloud rarely exceeds 0.005 nPa, except at its boundaries. From simple arguments relating to the magnetic field and velocity asymmetries in this cloud we can estimate the expansion velocity to be of order 100 km/s. Thus the kinetic energy density of expansion is found to be of order 0.05 nPa, greatly in excess of the thermal energy density. This suggests that free expansion can be an appropriate approximation for magnetic clouds.

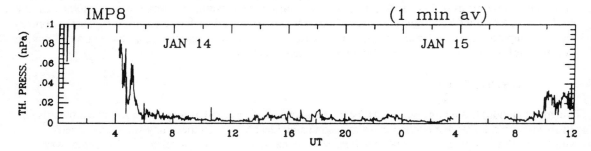

Fig. 1 The thermal pressure measured in the magnetic cloud on January 14 - 15, 1988.

TIME EVOLUTION OF THE MAGNETIC FIELD

(A) *Cylindrical coordinates* (r, ϕ, z) The magnetic field is derived from the magnetic flux function, $\psi(r, z, t)$ and an arbitrary function G according to

$$B_z = (1/r)\partial\psi/\partial r\,;\ B_r = -(1/r)\,\partial\psi/\partial z;\ B_\phi = -\partial G/\partial r$$

where ψ = constant are the magnetic surfaces /6/. The integral of motion which satisfies both momentum equation and induction equation is

$$r = v(\psi, z)\, t + F(\psi, z)$$

where $F(\psi, z)$ is an arbitrary function. Thus the magnetic field affects the velocity profile through the function F, which depends on the magnetic flux function, ψ. Our goal is to express the field components as functions of time on the magnetic surfaces at fixed z. A straightforward analysis shows that the evolution of the magnetic field on a magnetic surface (for fixed z) is given by :

$$B_r = [(\partial v/\partial z)t + \partial F/\partial z]\,[(\partial v/\partial\psi)t + \partial F/\partial\psi)\,(vt + F)]^{-1}$$
$$B_\phi = -(\partial G/\partial\psi).\,[(\partial v/\partial\psi)t + \partial F/\partial\psi]^{-1}$$
$$B_z = [vt + F]^{-1}[(\partial v/\partial\psi)t + \partial F/\partial\psi]^{-1}$$

(B) *Spherical polar coordinates* (r, θ, ϕ). The field can in this case be expressed through the Chandrasekhar potential $A(r, \theta, t)$ and another function $S(r, \theta, t)$ as

$$B_r = (1/r^2\sin\theta)\,(\partial A/\partial\theta);\ B_\theta = -(1/r\sin\theta)\,\partial A/\partial r;\ B_\phi = -(1/r\sin\theta)\,\partial S/\partial r$$

This time, the integral of motion and the components of the magnetic field are as follows

$$r = v(A, \theta)\, t + F(A, \theta)$$

$$B_r = -[\sin\theta\,(vt+F)^2]^{-1}[t\,\partial v/\partial\theta + \partial F/\partial\theta][t\,\partial v/\partial A + \partial F/\partial A]^{-1}$$
$$B_\theta = -[(\sin\theta\,(vt+F)]^{-1}[t\,\partial v/\partial A + \partial F/\partial A]^{-1}$$
$$B_\phi = -\partial S/\partial A\,[t\,\partial v/\partial A + \partial F/\partial A]^{-1}$$

ASYMPTOTIC BEHAVIOUR

As time goes to infinity, the field on a magnetic surface approaches *(cylindrical coordinates)*

$$B_r \longrightarrow (1/t)\,\partial v/\partial z\,[\partial/\partial\psi\,(v^2/2)]^{-1}$$
$$B_\phi \longrightarrow (1/t)\,\partial G/\partial\psi\,(\partial v/\partial\psi)^{-1}$$
$$B_z \longrightarrow 1/t^2[\partial/\partial\psi\,(v^2/2)]^{-1}$$

Thus, for instance, for fixed ψ and z, $B_z(t_1)/B_z(t_2) = (t_2/t_1)^2$. The corresponding expressions in spherical polar coordinates are

$$B_r \longrightarrow -(1/t^2)(1/v^2)\,[\partial v/\partial\theta][\partial v/\partial A]^{-1}$$
$$B_\theta \longrightarrow -(1/t^2)\,[\sin\theta\,\partial/\partial A\,(v^2/2)]^{-1}$$
$$B_\phi \longrightarrow -(1/t)\,\partial S/\partial A\,[\partial v/\partial A]^{-1}$$

For example, if A and θ are fixed then $B_\phi(t_1)/B_\phi(t_2) = t_2/t_1$

Note that the expressions for the field components given in this section are the same as those formally obtained from the general formulae by setting $F = 0$.

SPECIAL CASE: SELF-SIMILAR EXPANSION

From above, the integral of the motion (cylindrical coordinates) is

$$r = v(\psi, z)\, t + F(\psi, z)$$

For $F = 0$ we have

$$v(\psi, z) = r/t$$

This means that, after inverting,

$$\psi = \psi(r/t, z)$$

In general, self-similar, radial expansion is given by $\psi = \psi(r/f(t), z)$, where $f(t)$ is an arbitrary function. Thus the $F = 0$ case corresponds to self-similar expansion, the particularly simple $f(t) = t$ here resulting from the free expansion approximation. We therefore see that the time evolution of the magnetic field for self-similar expansion coincides with the asymptotic limit of a non-self-similar expansion.

WEAKLY NON-SELF-SIMILAR VELOCITY PROFILE

Let us consider the velocity profile of a weakly non-self-similarly expanding magnetic cloud. Suppose in the integral of motion the function F is given by

$$F(\psi, z) = K\,\psi, \quad K \text{ constant}$$

where $F(\psi, z)$ is only a small perturbation, i.e. $|K\psi| \ll r$. In this case we can assume self-similarity in the small correction, i.e.

$$v(\psi, z) = r/t - [K\psi(r/t, z)] / t$$

Following /5/, let the field of the magnetic cloud be given by

$$B_r = 0; \quad B_\phi = B_O J_1(\alpha r); \quad B_z = B_O J_O(\alpha r)$$

where α and B_O are constants and J_O and J_1 are Bessel functions. Then

$$\psi = - (1/\alpha) B_O (r/t) J_1 (\alpha r/t)$$

and

$$v(\psi, z) = r/t + (1/t) (1/\alpha) [KB_O r/t J_1(\alpha r/t)].$$

We illustrate this profile in Figure 2. Figure 2a, top curve, shows the temporal evolution of the velocity appropriate to self-similar expansion. The perturbation is shown by the bottom curve. The resultant non-self-similar profile is shown in Figure 2b, where it is noted that the velocity is no longer strictly monotonically decreasing throughout.

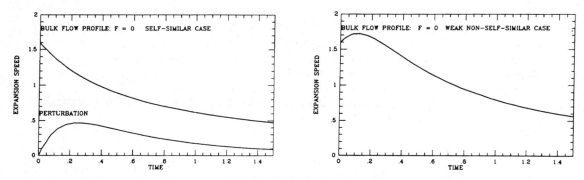

Fig. 2a, b. A weakly non-self-similar profile. For details, see text

CONCLUDING REMARKS

The relevance of the above to studies of magnetic clouds lies in the following. The radial, self-similar expansion of an axially symmetric magnetic cloud ($v = r/t$) suggests a universal velocity profile, one which is valid independent of the details of the magnetic structure of the cloud. Comparison of the observed velocity profiles with the 1/t law confirms the expectation that clouds expand self-similarly in a first approximation. Within this approximation we successfully modelled the time evolution of a particular magnetic cloud, and its size and age were inferred from this expansion law / 3 /. Nonetheless, in general, one should expect the velocity profile to be affected by the magnetic field. Such departures from self-similarity might be understood in a non-self-similar treatment along the lines given here.

REFERENCES

1. Burlaga, L.F., E. Sittler, F. Mariani, and R. Schwenn, J. Geophys. Res., 86, 6673 (1981)
2. Klein, L.W., and L.F. Burlaga, J. Geophys. Res., 87, 613 (1982)
3. Farrugia, C.J., L.F. Burlaga, V.A. Osherovich, I.G. Richardson, M.P. Freeman, R.P. Lepping, and A. Lazarus, J. Geophys. Res., submitted (1991)
4. Burlaga, L.F., J. Geophys. Res., 93, 7217 (1988)
5. Farrugia, C.J., L.F. Burlaga, V.A. Osherovich, R.P. Lepping, This Issue (1992)
6. Chandrasekhar, S., and K.H. Prendergast, Proc. Nat. Acad. Sci., Washington, 42, 5 (1956)

COUNTERSTREAMING SOLAR WIND HALO ELECTRON EVENTS ON OPEN FIELD LINES?

J. T. Gosling, D. J. McComas and J. L. Phillips

MS D438, Los Alamos National Laboratory, Los Alamos, NM 87545, U.S.A.

ABSTRACT

Counterstreaming solar wind halo electron events have been identified as a common 1 AU signature of coronal mass ejection events, and have generally been interpreted as indicative of closed magnetic field topologies, i.e., magnetic loops or flux ropes rooted at both ends in the Sun, or detached plasmoids. In this paper we examine the possibility that these events may instead occur preferentially on open field lines, and that counterstreaming results from reflection or injection behind interplanetary shocks or from mirroring from regions of compressed magnetic field farther out in the heliosphere. We conclude that neither of these suggested sources of counterstreaming electron beams is viable and that the best interpretation of observed counterstreaming electron events in the solar wind remains that of passage of closed field structures.

INTRODUCTION

Solar wind halo electrons (above about 80 eV at 1 AU) are nearly collisionless and are generally beamed outward from the Sun along the interplanetary magnetic field, IMF. This unidirectional flux of halo electrons arises because field lines in the solar wind usually are "open" and are effectively connected to the hot solar corona at only one end. However, discrete interplanetary events in which halo electrons are observed streaming in both directions along the IMF are common at times of high solar activity /1/. These events typically have durations of the order of 12 - 18 hours (although considerably shorter and longer events are also observed), and appear to be reliable signatures of coronal mass ejections, CMEs, in the solar wind at 1 AU. Counterstreaming has generally been interpreted as a signature of passage of closed field structures, i.e., magnetic loops or magnetic flux ropes rooted at both ends in the Sun, or detached plasmoids. In magnetic loops or flux ropes the counterstreaming fluxes of hot halo electrons arise because both ends of field lines threading these structures are rooted in the hot solar corona. Presumably the counterstreaming fluxes are trapped on the field lines and continue to circulate if the field lines disconnect to form plasmoids. All three of these field topologies are consistent with the observation that CMEs generally originate in closed field regions in the solar corona.

Recently Kahler and Reames have noted that at least some counterstreaming electron events are nearly transparent to energetic solar electrons (0.2 - 2.0 MeV) and protons (22 - 27 MeV) and to cosmic rays /2/. On the basis of this observation they have concluded that counterstreaming electron events occur on open, rather than closed, field lines, and that counterstreaming results from reflection or injection behind a shock or from mirroring from a region of compressed field beyond 1 AU. Our purpose here is to demonstrate that counterstreaming electron events in the solar wind at 1 AU do not originate in the manner they suggest, and that it is thus unlikely that the counterstreaming events which we identify as CMEs occur on open field lines.

OBSERVATIONS

Let us first consider the possibility that counterstreaming results from reflection of the solar wind

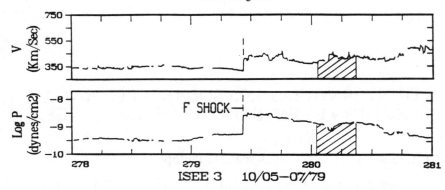

Fig. 1. Solar wind speed and pressure (total kinetic gas plus field) measured by ISEE 3 for a typical transient interplanetary shock (dashed line) and associated counterstreaming solar wind electron event (cross-hatched). The latter is identified as the CME driving the shock.

electron heat flux at an interplanetary shock beyond 1AU or from the production of hot electrons at the shock that then travel back to the spacecraft. In either case, if the shock were the source of an additional beam of electrons streaming back toward the Sun along the IMF then one would expect to observe counterstreaming on all field lines connected to the shock. In particular, one would expect to observe counterstreaming immediately following shock passage, for that is the one time when it is certain that a spacecraft is magnetically connected to the shock. However, while counterstreaming electron events often are observed behind interplanetary shocks, the counterstreaming never begins immediately following shock passage. Rather, as illustrated by the event shown in Figure 1, counterstreaming always lags the shock by a number of hours /1/. This aspect of the observations argues strongly against the possibility that beams of halo electrons streaming back toward the Sun along the IMF are commonly produced at interplanetary shocks.

There are several reasons why mirroring from regions of compressed magnetic fields beyond 1 AU is an unlikely source of counterstreaming electron beams. The prime reason is that increases in field magnitude beyond 1 AU are inadequate to mirror large fractions of the highly field-aligned,

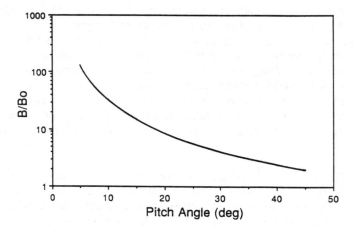

Fig. 2. Field increase required to mirror a given pitch angle particle. Bo is field strength at the measurement point (1 AU) and B is the field strength required at the mirror point on the same field line.

antisunward-directed, solar wind electron heat flux. Figure 2 shows the field increases required beyond 1 AU to mirror electrons of various pitch angles. As an example, in order to mirror a typical heat flux electron with a pitch angle of 20 degrees at 1 AU, the field strength must increase by a factor of ~8.5 over its 1 AU value. Considerably larger increases are required to mirror particles with smaller pitch angles. Typical field strengths within counterstreaming events at 1 AU are ~10 nT, and considerably stronger fields are not uncommon /1/. Thus field strengths of the order of 85 nT and greater are required beyond 1 AU to reflect electrons with pitch angles less than or equal to 20 degrees at 1 AU.

The left panel of Figure 3 compares 25-day averages of the IMF magnitude observed by Voyager 1 with the variation with distance predicted by Parker's spiral model (the solid curve), while the right panel shows variations in the measured field strength relative to the Parker value at various heliocentric distances. Although the contrast in field magnitude between compression regions and rarefactions is greater at larger distances, the absolute value of the field magnitude within compression regions is generally less than at 1 AU. For example, at 5 AU, where Bp is about 1 nT, the maximum (10-hr average) field strength is about 4.5 nT. Clearly, the field increases required beyond 1 AU to mirror substantial portions of the highly field-aligned electron heat flux are not observed. The reason, of course, is that the overall field magnitude decreases with increasing heliocentric distance, even within compression regions, owing to the 3-dimensional divergence of the solar wind flow.

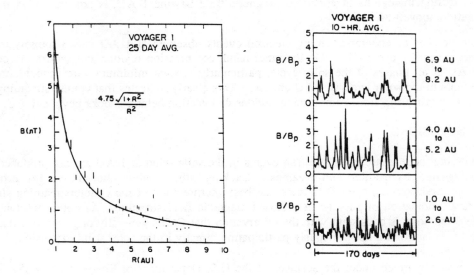

Fig. 3. Heliocentric variation of 25-day averages of the strength of the IMF measured by Voyager 1 (left panel) and 10-hr averages of the field for selected 170-day intervals at different heliocentric distances (right panel) (from /4/). Bp is the average field value predicted by Parker's spiral field model.

In addition, if mirroring were the source of counterstreaming electron beams at 1 AU, then field-aligned loss cone "holes" should be apparent in the mirrored beams since virtually an infinite field increase is required to mirror particles with very small pitch angles. However, our examination of measured halo electron angular distributions within counterstreaming events has not yet revealed any evidence for such "holes". The distributions shown in Figure 4 are representative of those observed within these events. Note that the halo beams peak parallel and antiparallel to the field direction, in contrast with what would be observed if one of the beams were primarily a result of imperfect mirroring of the other. Although we have examined in detail only one of the events specifically discussed by Kahler and Reames, that of April 3,4, 1979, distributions similar to that shown in Figure 4 were observed at that time.

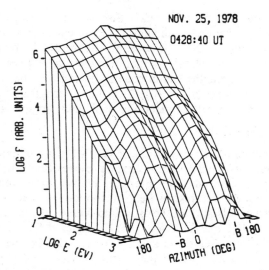

Fig. 4. Three-dimensional view of a representative two-dimensional solar wind electron distribution obtained during a counterstreaming solar wind halo electron event (from /1/). Note that the field-aligned loss cone "hole", expected if counterstreaming were associated with mirroring in regions of enhanced magnetic field beyond 1 AU, is not present in the distribution above ~80 eV.

Finally, the number of counterstreaming electron events observed at 1 AU varies roughly in phase with the solar activity cycle /3/.   On the other hand, compression regions and shocks are common beyond 1 AU at all phases of the solar cycle, particularly at solar minimum when counterstreaming electron events in the solar wind at 1 AU are rare. This clearly indicates that counterstreaming is not associated with mirroring from compression regions or injection behind shocks beyond 1 AU.

CONCLUSIONS

We conclude that counterstreaming electron events in the solar wind at 1 AU are not produced in the manner suggested by Kahler and Reames.   Lacking other viable alternatives for producing counterstreaming beams on open field lines, the best interpretation of the counterstreaming signature is that it is an indication of passage of a closed magnetic field structure. As noted previously, this type of field topology is consistent with the observation that CMEs generally originate in closed field regions in the solar corona not previously participating directly in the solar wind expansion.

This work was performed under the auspices of the U.S. Department of Energy with NASA support under S-04039-D.

REFERENCES

1. J. T. Gosling, D. N. Baker, S. J. Bame, W. C. Feldman, R. D. Zwickl, and E. J. Smith, Bidirectional solar wind electron heat flux events, *J. Geophys. Res.*, *92*, 8519, (1987)

2. S. W. Kahler and D. V. Reames, Probing the magnetic topologies of magnetic clouds by means of solar energetic particles, *J. Geophys. Res.*, *96*, 9419, (1991)

3. J. T. Gosling, D. J. McComas, J. L. Phillips, and S. J. Bame, Counterstreaming solar wind halo electron events: Solar cycle variations, *J. Geophys. Res.*, in press, (1992)

4. L. F. Burlaga, L. W. Klein, R. P. Lepping, and K. W. Behannon, Large-scale interplanetary magnetic fields: Voyager 1 and 2 observations between 1 AU and 9.5 AU, *J. Geophys. Res.*, *89*, 10659, (1984)

SOLAR-GENERATED DISTURBANCES IN THE HELIOSPHERE

B. V. Jackson

Center for Astrophysics and Space Sciences, University of California at San Diego, La Jolla, CA 92093, U.S.A.

ABSTRACT

It has long been known that disturbances can propagate from Sun to Earth with periods of a few days following large solar flares. Other disturbances re-occur with the solar rotation rate implying that they are more or less stably generated by a specific region on the solar surface. At the Sun some of these disturbances are readily observed in coronagraphs and against the solar disk. Several techniques have been used to remotely detect and follow different disturbances in the interplanetary medium as they propagate outward from the Sun. These techniques include interplanetary scintillation, kilometric radio and Helios photometer observations. *In situ*, spacecraft can mark the passage of disturbances by direct measurement along the column convected past the observation point. As probes of the heliospheric magnetic field and disturbances in themselves, particles above the energy of the thermal plasma traverse the heliosphere and indicate the extent of its structures. Both the basic physics, as well as the spatial and temporal evolution of disturbances, can be confused as they propagate through the interplanetary medium largely because of the data coverage limitations. However, as their basic physics becomes better known through more complete observations and theory, the extent and accuracy of these disturbances can be better described.

INTRODUCTION

The outermost parts of the solar atmosphere - the corona and solar wind - experience dramatic perturbations in the form of disturbances associated with discrete ejections of mass and energy from the Sun. These disturbances extend to the magnetosphere of Earth and to Earth itself. In the past, remote-sensing observations of the origins of these disturbances on the Sun have been restricted to coronal emission-line observations /1, 2/ and the meter-wave and shorter radio wavelengths /3/, but since the late 1960's we have seen the addition of powerful new tools for observation. Our observation tools now include a wide variety of techniques from white-light observations of coronal and heliospheric structures /4, 5, 6, 7/ and kilometric radio bursts /8, 9/ associated with these, to interplanetary scintillation (IPS) observations /10, 11/. Unfortunately, the observational techniques most commonly used are each sensitive to a specific type of disturbance and more sensitive or more limited at one portion of the interplanetary medium than another. In addition, none of the remote sensing techniques has complete spatial or temporal coverage /12/. *In situ* observations /13, 14, 15/ have the disadvantage that they do not observe the whole. Thus the basic physics, as well as the spatial and temporal evolution of heliospheric structures, can be misinterpreted as they evolve away from the Sun.

This review is limited to those solar-generated disturbances directly associated with mass ejections and shock waves observed in the interplanetary medium. I give a description of the techniques used and show a limited number of examples of the data which trace each heliospheric structure. In the next two sections examples of the techniques used are given as much as possible in the order of the distance from the Sun at which they are sensitive. In the first of these, I describe the observations used to follow mass ejections as they propagate outward from the Sun. In the second, measurements of shocks in the interplanetary medium are described. Concluding remarks are found in the last section.

MASS EJECTIONS

Coronagraph Observations

Near the Sun, coronal mass ejections (CMEs) observed by coronagraphs /4, 5, 6/ (Figure 1) mark the location and the approximate coronal involvement of major disturbances. The speed and masses of these ejecta can be shown to involve a significant and, in some instances, a major portion of the mass and energy present in the solar wind /16, 17/. To date Hα instrumentation and coronagraph techniques can image ejections and directly view the outward motion of these disturbances in the lower corona from 1 - 10R$_S$.

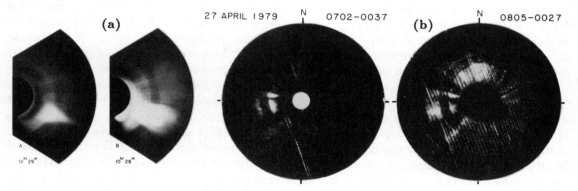

Fig. 1. Examples of coronal mass ejections. **(a)** the 21 January 1974 CME as observed by the Skylab coronagraph (from /18/). The outer edge of the field-of-view extends to 6R$_S$. **(b)** the 27 April 1979 CME as observed by the Solwind coronagraph (from /19/). The outer edge of the field-of-view extends to 8R$_S$ in these images.

Type IV and Other Coronal Radio Signatures

Dynamic spectra were first used with metric-wave observations to distinguish various disturbances in the corona within a few solar radii of the Sun /20/. Figure 2 gives a stylized version of these three-dimensional plots which in essence depict the intensity of radiation simultaneously with time and height above the solar surface (/21/). Four distinct types of metric disturbances are commonly found (Figure 2) which often have a high temporal correlation with solar flares observed in Hα light on the solar surface /22/. The only disturbances (marked in Figure 2) which have the proper characteristics and average speeds to be directly associated with CMEs in the low corona are indicated by type IV radiation.

Although moving type IV metric radiation is considered indicative of an outward-moving plasmoid of material somehow associated with mass ejection events /23/, the exact details of the association are unknown. It has been suggested that these events are evidence for a closed or confining magnetic field topology /24/ associated with an expulsion of mass. However, these events are even less

good correspondences with coronagraph CMEs have been shown /25/, it is not at all certain what radiation process or CME topology is necessary for the radiation to occur.

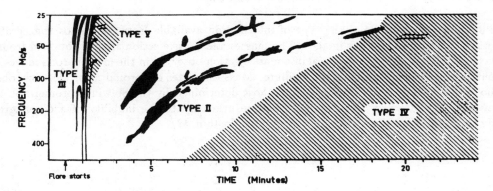

Fig. 2. A stylized dynamic high-frequency radio spectra showing various metric radiation components. Each component is indicative of a specific type of disturbance or high-energy particle distribution in the lower corona, and each can be generally temporally associated with solar flares (adapted from /21/).

Faraday Rotation and Doppler Scintillation Observations

Beyond the region normally viewed by a coronagraph and metric radiation sensors there is a region of the heliosphere from approximately 10 to 20 R_S that has been little-observed by any technique. Two experiments which have probed this region deserve mention, however. These are Faraday rotation measurements /26/ and doppler scintillation measurements /27, 28/. Both techniques usually require the fortuitous placement of a spacecraft behind the Sun. Several mass ejections have been observed by the Faraday rotation technique /26/. Significant in the observations is the fact that while a simple loop of current-carrying plasma should show first a one-way directed and then an oppositely-directed excursion of the response from the ambient, perhaps only one of the CMEs does. Thus, an explanation of all CMEs as current driven loops is not valid. Scintillation observations in the same height range can provide extremely high signal to noise detection of the onset of solar disturbances. These data have been mapped to both CMEs observed by the Solwind coronagraph and solar surface flares /29/. Significant in these measurements are the short times often observed for the propagation of a CME to heights many solar radii above the Sun following its surface or near-surface manifestation. Two scenarios present themselves from this aspect of the data – either some events have been accelerated to extremely high speed very close to the Sun or else the event actually extends outward to a greater height at its onset than is evident from usual observations near the solar surface. In addition, these doppler scintillation observations are most clearly associated with shocks observed *in situ* at the Helios spacecraft /30/ and not with the density enhancement often associated with a CME. This has implications for the intensity IPS observations discussed in the following subsections.

Helios Photometer Observations

The Helios spacecraft, the first of which was launched into heliocentric orbit in 1974, had on board three sensitive zodiacal-light photometers for the study of the zodiacal-light distribution /31/. These photometers swept the celestial sphere at 16°, 31° and 90° ecliptic latitude to obtain data fixed with respect to the solar direction, with a sample interval of about five hours. The two spacecraft were placed in heliocentric orbits with perihelia of about 0.3 AU. The Helios photometer data were first identified as a valid source of information for mapping mass ejections by /32/ in 1982. Since then, the Helios photometers have been used to image the interplanetary medium from

20 R_S out to 1 AU /7, 33/ and, in particular, disturbances in it /34/. Before this, the only other ways to remotely sense disturbances in the inner heliosphere were by kilometric radio radiation from space and by interplanetary scintillation techniques.

Figure 3, shows an example of the type of information available from the Helios spacecraft photometers in the form of contour images. The masses and mass ejection shapes obtained from these observations indicate that the material of a mass ejection observed in the lower corona moves coherently outward into the interplanetary medium. Mass estimates of coronal mass ejections observed by Helios are generally approximately twice those determined by the Solwind coronagraph for the same events. We interpret this difference as due primarily to the inability of a coronagraph to measure the total mass of an ejection at any given instant /7/.

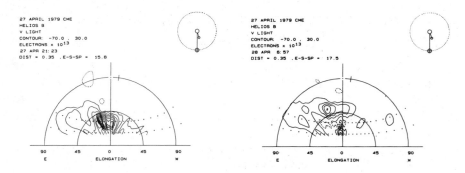

Fig. 3. 27 April 1979 CME from Helios 2 data contour-imaged in levels of electron columnar density. Image parameters are given in the figure captions to the upper left. The photometer data in these images are interpolated at the time intervals (indicated) of the data obtained from the 31° photometer. The diagram in the upper right of each figure depicts the relative locations of the Sun, Earth and Helios 2 (from /35/).

The orbits of the Helios spacecraft can give them a perspective view of mass ejections that are 90° orthogonal to that from Earth. The shapes of three of these CMEs, which appeared as loop-like mass ejections as observed by Earth-based coronagraphs, were measured in order to determine their edge-on thicknesses as they moved past the Helios photometers. The extents of these CMEs as obtained from the Helios data was nearly the same as in the coronagraph view /36/ for each event studied. From this, the implication is that mass ejections (even those of loop-like appearance) have considerable thicknesses as they move outward into the interplanetary medium.

For several CMEs, the motion of the ejection can be traced far out into the heliosphere. One CME measured by the Solwind coronagraph as a "halo" event on 27 November 1979 and observed in Helios data to move outward, can be observed several days later *in situ* at Earth /37/. Another CME is traced outward using these two techniques until an enhancement of speed and the scintillation level during passage of the excess mass is observed perpendicular to the line-of-sight to 3C48 /38/. The small-scale inhomogeneties (~200km) within the ejection to which the IPS is sensitive show a speed enhancement to about 500 km s^{-1} for two days following the event and compare favorably with the speed of bulk motion obtained from Helios data. Still another event observed both by Solwind and the Helios photometers can be observed finally in the low-frequency IPS observations obtained near Cambridge, England /19/.

Low–Frequency Intensity IPS observations

Intensity IPS has been used since 1962 to observe the interplanetary medium /39/. The technique relies on measuring the rapidly fluctuating intensity level from point-like radio sources. Whether a single-site array is used or a multiple one where velocities can be measured /40/, the technique relies on the presence of density inhomogeneties in the interplanetary medium to move across the line-of-sight to the source. The presence of the disturbance is indicated by either an enhancement of the scintillation level /41/, or an increase of the IPS-determined speed /42/, or both /43/.

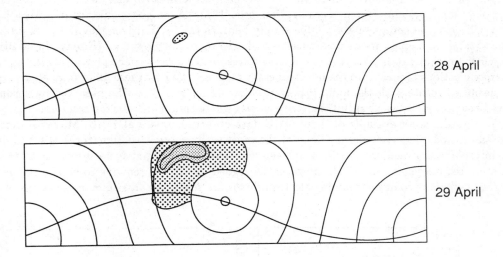

Fig. 4. Schematic of 81.5 MHz IPS data of the mass ejection imaged in Figures 1b and 3 as observed by the Cambridge, England IPS array in April 1979. Sky maps in right ascension and declination in this image show the CME as a region of enhanced IPS level (stippled) relative to average values. In the figure the Sun is centered with smooth lines marking each 30° elongation increment. Not shown in this schematic are other disturbances present in the same image (from /19/).

An example of the analysis from the Cambridge group using 1979 data is shown in Figure 4. Many different IPS disturbances are observed in the data each day. Over a 24-hour time interval it is possible to determine an approximate size and shape of the structures which move past Earth and to classify them as in /11/. These features, presumably modified significantly by their passage through the interplanetary medium on the way to 1 AU, are generally classified into structures that are either co-rotating or cut-off from the Sun. Some of these structures certainly are remnants of mass ejections as discussed in /10/ or even /41/. However, most structures observed by /41/ appear to be disturbances which are best associated with solar co-rotating features. Often, *in situ* measurements (*e.g.*, /44/) can help to elucidate the differences between these two forms of disturbances.

Most of the experience with mass ejections comes from coronagraph observations in white light from disturbances close to the solar limb. Unfortunately, although intensity IPS observations measure disturbances in the solar wind, they do not have a straightforward interpretation as mass unless the scintillation level enhancement proportionality to a line-of-sight density enhancement is known. While this proportionality has been determined in several ways, albeit with different results /45, 46/, the technique maps a disturbance well at 81.5 MHz only beyond approximately 50° (>0.75 AU) from the Sun, and thus close to Earth. Not only is the extent of an IPS disturbance difficult to discern given the signal to noise present in the observations, but at this solar distance most mass ejections have evolved significantly /47/. IPS simulation studies by /48/ and others have shown that it is often possible to confuse a co-rotating structure from one that moves radially outward from the Sun. Add this to the observations discussed earlier by /30/ which throw into question the validity of the determined proportionality, and the relation of intensity IPS disturbances with differently shaped mass structures (*e.g.*, mass ejections) becomes very difficult.

In Situ Observations

As mass ejections traverse the heliosphere, they evolve with time so that *in situ* even at 1 AU they often become difficult to distinguish from other structures (or enhanced mass) present in the interplanetary medium (for a recent review see /15/). A unique signature of CMEs in *in situ* data will never be completely available simply because the CME driver may not fully intersect the spacecraft even though its effects (such as a shock) surrounding the CME may. In addition, it is not clear to what extent slowly-evolving structures on the solar surface play a role in the release of discrete ejections of mass into the interplanetary medium and how these differ from structures which primarily co-rotate with the Sun. However, as mentioned, some CMEs observed by coronagraphs have been traced outward until they arrive *in situ* where they show varying responses (*e.g.*, /29, 47, 49/). Figure 5 is an example of these CME data observed *in situ* at 1 AU. Magnetic clouds /14/ and associated effects are perhaps one of the most easily recognized signatures of CME drivers in the interplanetary medium. Perhaps an even more easily-recognized signature of the CME driver is counterstreaming electrons /50/. However, as argued in /51/, it remains to be shown (perhaps as by /52/) whether counterstreaming electron fluxes are solely the *in situ* manifestations of CMEs.

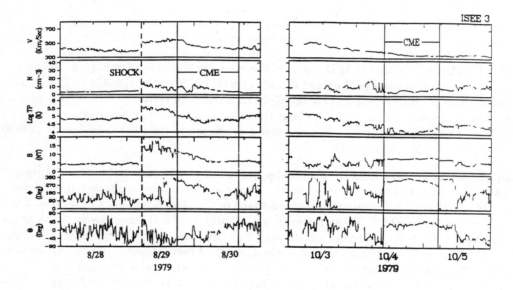

Fig. 5. (Left) Selected solar wind plasma and field measurements encompassing a shock wave disturbance driven by a fast CME. (Right) Selected solar wind plasma and field measurements encompassing a slow CME (adapted from /50/).

SHOCKS

In situ, fast shocks are noted by abrupt increases in solar wind density, speed, temperature, and magnetic field strength. The magnetic field changes can be interpreted to give directions of the shock normal /53 , 54/, and thus the directional properties of the shock at the position of the spacecraft. Examples of a shock in the interplanetary medium at 1 AU have been given previously in Figure 5 in conjunction with a CME. While it is clear that not all interplanetary shock waves are caused by mass ejections even in the inner heliosphere (see /55/ for a review of shocks from co-rotating structures in this region of space), most co-rotating shocks are formed far beyond 1 AU. Because inner heliospheric shocks can arise from very near the Sun, they are usually first observed by remote-sensing techniques.

Metric Remote Sensing Observations

Shock waves, thought to be highlighted by the type II metric radiation associated with them, can be observed remotely to move outward in the lower corona associated with some flares and CMEs. An example of a metric type II burst dynamic spectrum is shown in Figure 6.

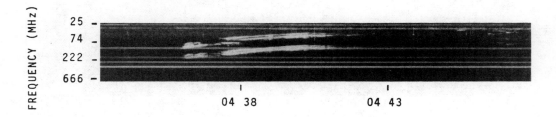

Fig. 6. Dynamic spectrum of a metric type II radio burst as observed by the Culgoora radio spectrograph on 9 October 1969.

The mass ejections associated with type II radiation are commonly those with the highest speeds /56/. While large, bright Hα flares, type II radio bursts implying fast coronal shocks, and mass ejections are all indicative of a common phenomena, the presence of such features is not at all in one-to-one correspondence. Furthermore, the location of type II radiation relative to CMEs in the lower corona often does not coincide with the top or fastest portion of the CME (reviewed in /57/). Figure 7 gives an example from SMM of the location of a CME and the corresponding type II metric radiation. Clearly the type II radiation is not present in conjunction with the outer portion of the CME and, in this case, the type II radiation appears to arise from two distinct locations to either side of the CME /58/. Modeling of the shock response to CMEs /59/ has shown that the formation of a fast shock can be highly dependent on the magnetic field direction and plasma density relative to the propagation direction of the shock.

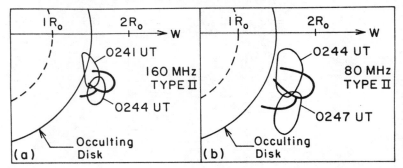

Fig. 7. Schematic diagrams comparing 160 and 80 MHz type II burst positions with a CME at two times (a) 024410 UT and (b) 024730 UT, (from /58/).

Kilometric Remote Sensing Observations

Kilometric type II radio bursts often appear as the interplanetary counterpart of their lower metric cousins. Figure 8 is the presentation of the dynamic spectrum of one such burst as observed from the Ulysses spacecraft in 1991 /60/. These unprecedented observations have such good frequency resolution that they have the potential of giving the three-dimensional shape of shock fronts in the interplanetary medium. In general, these remotely-sensed features are associated with the most massive and highest speed CMEs observed earlier in the corona by coronagraphs /61/. Unfortunately, very few simultaneous observations of mass ejections and kilometric type II radiation of the same event exist, and none will probably be available during the Ulysses fly-by of the Sun from

observed in the interplanetary medium is thus only a matter of speculation. I suspect that most of the positional qualifiers for lower coronal shocks, and their placement relative to CMEs, behave in a similar manner when dealing with the formation of kilometric type II radiation.

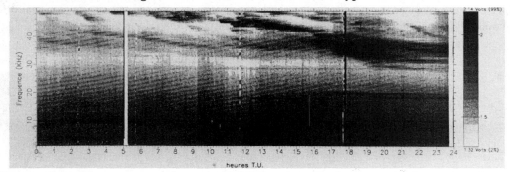

Fig. 8. Dynamic spectrum of a type II kilometric radio burst on 21 January 1991 as observed from the Ulysses spacecraft.

In Situ Shock Measurements

To date, five density enhancements behind shocks observed *in situ* have also been remotely measured by the Helios spacecraft photometers /62/. As expected from kilometric radio observations and *in situ* density measurements, brightness increases observed by Helios indicate that interplanetary density enhancements associated with shocks are extensive heliospheric features. Figure 9 gives one such comparison. These observations show that the shock-associated density enhancement can be remotely sensed at less than 90° elongation a few hours prior to its arrival at the spacecraft. The brightness registered in the photometers indicates a density enhancement, perpendicular to the shock front, which is significantly greater (~0.5 AU) than the thickness of the density enhancement measured *in situ* (~0.1 AU).

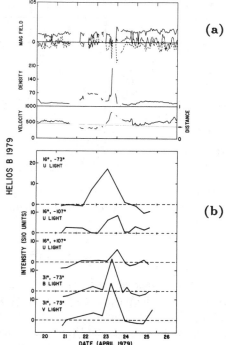

Fig. 9. Helios 2 observations of the shock of 23 April 1979. (a) *In situ* magnetic field, proton density and velocity plasma data. (b) Photometer time series from photometer ecliptic latitudes and longitudes of the sector centers relative to the Sun (–to the west)

Accelerated Particles

High energy particles (to ~1 GeV), accelerated at approximately the times of solar flares, propagate through the heliosphere as disturbances and can be used to trace the extents of heliospheric structures. Recent work by /63/ argues against a mechanism in which cross field particle transport of the particles takes place to any significant extent. It is thus generally concluded that the shock process associated with the flare, or its associated mass ejection, accelerates particles in the interplanetary medium along nearby magnetic field lines connected with the shock /64/. These important discoveries provide even more emphasis for study and observation of the propagation of heliospheric disturbances, since some of the particles accelerated by these shocks can pose significant danger to future space travellers.

CONCLUSIONS

Near the solar surface, observations of the corona at the approximate time of a flare or a mass ejection show that a significant fraction of the energy involved is manifest in the form of coronal motions. As this disturbance moves outward into the interplanetary medium it can be observed remotely by a variety of techniques. The shapes and speeds of different structures evolve somewhat as they move outward through the heliosphere, and since none of the techniques that trace the events overlap well temporally or spatially, there are usually ambiguities in the observations. Thus, none of the techniques is complete in itself, and all must operate in conjunction with additional information to describe these disturbances in the most complete fashion.

As more events are observed, general rules to describe their behavior will become well-accepted and their physics understood. In addition, new and better observations such as the better data from the Ulysses spacecraft /60/ will certainly help in the description, and thus our understanding of the physics involved. At some future date an instrument (or set of instruments) such as a combination of the SOHO coronagraphs (e.g., /65/) and the SMEI /66, 67/ that can simultaneously view the whole heliosphere in white light from extremely near the Sun to 1 AU, will be in operation. It can be hoped that these instruments will open a new era in our understanding of solar-generated disturbances in the heliosphere.

Acknowledgements: I am grateful for the many helpful comments and encouragement from my colleagues during the course of this review and to Paul Hick for critically reading this manuscript. My work has been supported by grants NASA NAGW-2002 and AFOSR-91-0091 to the University of California at San Diego.

REFERENCES

1. Menzel, D.H., *Our Sun*, Harvard University Press, Cambridge, Massachusetts, 1959.

2. Abetti, G., *The Sun*, Faber and Faber, London, 1963.

3. Kundu, M.R., *Solar Radio Astronomy*, Interscience, New York, 1965.

4. MacQueen, R.M., J.A. Eddy, J.T. Gosling, E. Hildner, R.H. Munro, G.A. Newkirk, Jr., A.I. Poland and C.L. Ross, "The Outer Solar Corona as Observed from Skylab: Preliminary Results", *Astrophys. J. Lett.*, **187**, 85, 1974.

5. Koomen, M.J., C.R. Detwiler, G.E. Brueckner, H.W. Cooper and R. Tousey, "White Light Coronagraph in OSO-7", *Appl. Opt.*, **14**, 743, 1975.

6. MacQueen, R.M., A. Csoeke-Poeckh, E. Hildner, L. House, R. Reynolds, A. Stanger, H.

Tepoel and W. Wagner, "The High Altitude Observatory Coronagraph/Polarimeter on the Solar Maximum Mission", *Sol. Phys.*, **65**, 91, 1980.

7. Jackson, B.V., "Imaging of Coronal Mass Ejections by the Helios Spacecraft", *Sol. Phys.*, **100**, 563, 1985.

8. Steinberg, J.-L., "Satellite Observations of Solar Radio Bursts", in *Radio Physics of the Sun*, M.R. Kundu and T.E. Gergeley eds., 387, 1980.

9. Bougeret, J.-L., "Physics of Low Frequency Radio Emissions", in *AIP Conference Proceedings*, **207**, 139, 1990.

10. B.J. Rickett, "Disturbances in the Solar Wind from IPS Measurements", *Solar Phys.*, **43**, 237, 1975.

11. Gapper, G.R., A. Hewish, A. Purvis, P.J. Duffett-Smith, "Observing Interplanetary Disturbances from the Ground", *Nature* **296**, 633, 1982.

12. Bird, M.K. and P. Edenhofer, "Remote-Sensing Observations of the Solar Corona", in *Physics of the Inner Heliosphere*, R. Schwenn and E. Marsch, eds., 13, 1990.

13. Hundhausen, A.J., S.J. Bame and M.D. Montgomery, "The Large-Scale Characteristics of Flare-Associated Solar Wind Disturbances", *J. Geophys. Res.*, **75**, 4631, 1970.

14. Klein, L.W. and L.F. Burlaga, "Interplanetary Magnetic Clouds at 1 AU", *J. Geophys. Res.*, **87**, 613, 1982.

15. Gosling, J.T., "Coronal Mass Ejections and Magnetic Flux Ropes in Interplanetary Space", in *Physics of Magnetic Flux Ropes*, C.T. Russell, L.C. Lee and E. Priest, eds., 343, 1990.

16. Howard, R.A., N.R. Sheeley, Jr., M.J. Koomen, D.J. Michels, "Coronal Mass Ejections: 1979-1981", *J. Geophys. Res.*, **90**, 8173, 1985.

17. Jackson, B.V., "Remote Sensing Observations of Mass Ejections and Shocks in Interplanetary Space", in *Eruptive Solar Flares*, Z. Svestka, B.V. Jackson and M.E. Machado, eds., Springer-Verlag, Heidelberg, in press, 1991.

18. Sheridan, K.V., B.V. Jackson, D.J. McLean and G.A. Dulk, "Radio Observations of a Massive Slow–Moving Ejection of Coronal Material", *Proc. Astron. Soc. Australia*, **3**, 249, 1978.

19. Jackson, B.V., "Comparison of Helios Photometer and Interplanetary Scintillation Observations", in the proceedings of the First Soltip Symposium held in Liblice, Czechoslovakia, 1991.

20. Wild, J.P. and L.L. McCready, "Observations of the Spectrum of High-Intensity Solar Radiation at Metre Wavelengths – I. The Apparatus and Spectral Types of Solar Burst Observed", *Australian J. Sci. Res.*, **3**, 387, 1950.

21. Aller, L.A., *Astrophysics the Atmospheres of the Sun and Stars*, Ronald, New York, 1963.

22. Dulk, G.A., D.J. McLean and G.J. Nelson, "Solar Flares", in *Solar Radiophysics*, D.J. McLean and N.R. Labrum eds., Cambridge University Press, Sydney, 53, 1985.

23. Wild, J.P., "Some Investigations of the Solar Corona: The First Two Years of Observation with the Culgoora Radioheliograph", *Proc. Astron. Soc. Australia*, **1**, 365, 1970.

24. Dulk, G.A., "The Gyro-Synchrotron Radiation from Moving Type IV Sources in the Solar Corona", *Solar Phys.*, **32**, 491, 1973.

25. Stewart, R.T., G.A. Dulk, K.V. Sheridan, L.L. House, W.J. Wagner, C. Sawyer and R. Illing, "Visible Light Observations of a Dense Plasmoid Associated with a Moving Type IV Solar Radio Burst", *Astron. Astrophys.*, **116**, 217, 1982.

26. Bird, M.K., H. Volland, R.A. Howard, M.J. Koomen, D.J. Michels, N.R. Sheeley, Jr., J.W. Armstrong, B.L. Seidel, C.T. Stelzried and R. Woo, "White-Light and Radio Sounding Ob-

27. Woo, R., "Radial Dependence of Solar Wind Properties Deduced from Helios 1/2 and Pioneer 10/11 Radio Scattering Observations", *Astrophys. J.*, **219**, 727, 1978.

28. Woo, R., "A Synoptic Study of Doppler Scintillation Transients in the Solar Wind", *J. Geophys. Res.*, **93**, 3919, 1988.

29. Woo, R., J.W. Armstrong, N.R. Sheeley, Jr., R.A. Howard, D.J. Michels and M.J. Koomen, "Doppler Scintillation Observations of Interplanetary Shocks within 0.3 AU", *J. Geophys. Res.*, **90**, 154, 1985.

30. Woo, R. and R. Schwenn, "Comparison of Doppler Scintillation and In Situ Spacecraft Plasma Measurement of Interplanetary Disturbances", *J. Geophys. Res.*, in press, 1991.

31. Leinert, C., H. Link, E. Pitz, N. Salm, and D. Kluppelberg, "Helios Zodiacal Light Experiment", *Raumfahrtforschung*, **19**, 264, 1975.

32. Richter, I., C. Leinert, and B. Planck, "Search for Short Term Variations of Zodiacal Light and Optical Detection of Interplanetary Plasma Clouds", *Astron. Astrophys.*, **110**, 115, 1982.

33. Jackson, B.V. and C. Leinert, "Helios Images of Solar Mass Ejections", *J. Geophys. Res.*, **90**, 10759, 1985.

34. Webb, D.F. and B.V. Jackson, "The Identification and Characteristics of Solar Mass Ejections Observed by the Helios 2 Photometers", *J. Geophys. Res.*, **95**, 20641, 1990.

35. Jackson, B.V., "The Dynamics of Mass Ejections in the Heliosphere Observed Using Helios Photometer Data", in the Proceedings of Solar Wind 7 (this Issue), 1992.

36. Jackson, B.V., R.A. Howard, N.R. Sheeley, Jr., D.J. Michels, M.J. Koomen and R.M.E. Illing, "Helios Spacecraft and Earth Perspective Observations of Three Looplike Solar Mass Ejections", *J. Geophys. Res.*, **90**, 5075, 1985.

37. Jackson, B.V., "Helios Observations of the Earthward-Directed Mass Ejection of 27 November, 1979", *Sol. Phys.*, **95**, 363, 1985.

38. Jackson, B.V., B. Rompolt and Z. Svestka, "Solar and Interplanetary Observations of the Mass Ejection on 7 May 1979", *Solar Phys.*, **115**, 327, 1988.

39. Hewish, A., P.F. Scott and D. Wills, "Interplanetary Scintillation of Small Diameter Radio Sources", *Nature*, **203**, 1214, 1964.

40. Coles, W.A. and J.J. Kaufman, "Solar Wind Velocity Estimation From Multi-Station IPS", *Radio Science*, **13**, 591, 1978.

41. Hewish, A. and S. Bravo, "The Sources of Large-Scale Heliospheric Disturbances", *Solar Phys.*, **106**, 185, 1986.

42. Watanabe, T. and T. Kakinuma, "Radio-Scintillation Observations of Interplanetary Disturbances", *Adv. Space Res.*, **4**, 331, 1984.

43. Jackson, B.V., "IPS Observations of the 14 August 1979 Mass Ejection Transient", in *Proceedings of the Maynooth, Ireland Symposium on Solar/Interplanetary Intervals*, M.A. Shea, D.F. Smart and S.M.P. McKenna-Lawlor, eds., 169, 1984.

44. Behannon, K.W., L.F. Burlaga and A. Hewish, "Structure and Evolution of Compound Streams at ≤ 1 AU", *J. Geophys. Res.*, in press, 1991.

45. Tappin, S.J., "Interplanetary Scintillation and Plasma Density", *Planet. Space Sci.*, **35**, 271, 1987.

46. Zwickl, R.D., "The Study of Fluxuations in the Solar Wind Density and Their Impact on IPS Measurements", private communication (Presentation at the SEIIM conference held in Colorado Springs, Colorado, 1988), 1988.

47. Sheeley, N.R., Jr., R.A. Howard, M.J. Koomen, D.J. Michels, R. Schwenn, K.H. Muhlhauser and H. Rosenbauer, "Coronal Mass Ejections and Interplanetary Shocks", *J. Geophys. Res.*, **90**, 163, 1985.

48. Pizzo, V.J. and D.G. Sime, "Interpretation of Interplanetary Scintillation Observations of Large Scale Solar Wind Structures", in the Proceedings of Solar Wind 7 (this Issue), 1992.

49. Burlaga, L.G., L. Klein, N.R. Sheeley, Jr., D.J. Michels, R.A. Howard, M.J. Koomen, R. Schwenn and H. Rosenbauer, "A Magnetic Cloud and a Coronal Mass Ejection", *Geophys. Res. Lett.*, **9**, 1317, 1982.

50. Gosling, J.T., D.N. Baker, S.J. Bame, W.C. Feldman, R.D. Zwickl and E.J. Smith, "Bidirectional Solar Wind Electron Heat Flux Events", *J. Geophys. Res.*, **92**, 8519, 1987.

51. Kahler, S.W and D.V. Reames, "Probing the Magnetic Topologies of Magnetic Clouds by Means of Solar Energetic Particles", *J. Geophys. Res.*, **96**, 9419, 1991.

52. Gosling, J.T., D.J. McComas and J.L. Phillips, "Counterstreaming Electron Events on Open Field Lines?", in the Proceedings of Solar Wind 7 (this Issue), 1992.

53. Hundhausen, A.J., *Coronal Expansion and the Solar Wind*, Springer-Verlag, New York, 1972.

54. Colburn, D.S. and C.P. Sonnett, "Discontinuities in the Solar Wind", *Space Sci. Revs.* **5**, 439, 1966.

55. Schwenn, R., "Large-Scale Structure of the Interplanetary Medium", in *Physics of the Inner Heliosphere*, R. Schwenn and E. Marsch, eds., 99, 1990.

56. Gosling, J.T., E. Hildner, R.M. MacQueen, R.H. Munro, A.I. Poland and C.L. Ross, "The Speeds of Coronal Mass Ejection Events", *Solar Phys.*, **48**, 389, 1976.

57. Wagner, W.J., "SERF Studies of Mass Motions Arising in Flares", *Adv. Space Res.*, **2**, 203, 1983.

58. Gary, D.E., G.A. Dulk, L. House, R. Illing, C. Sawyer, W.J. Wagner, D.J. McLean and E. Hildner, "Type II Bursts, Shock Waves and Coronal Transients: The Event of 1980 June 29, 0233 UT", *Astron. Astrophys.*, **134**, 222, 1984.

59. Steinolfson, R.S., "Type II Radio Emission in Coronal Transients", *Solar Phys.*, **94**, 193, 1984.

60. Stone, R.G. and 26 co-authors, "The ISPM Unified Radio and Plasma Wave Experiment", *ESA Special Publication*, **SP-1050**, 1983.

61. Cane, H.V., N.R. Sheeley, Jr. and R.A. Howard, "Energetic Interplanetary Shocks, Radio Emission, and Coronal Mass Ejections", *J. Geophys. Res.*, **92**, 9869, 1987.

62. Jackson, B.V., "Helios Photometer Measurement of In-Situ Density Enhancements", *Adv. Space Res.*, **6**, 307, 1986.

63. Reames, D.V., "Acceleration of Energetic Particles by Shock Waves from Large Solar Flares", *Astrophys. J.*, **358**, L63, 1990.

64. Cane, H.V. D.V. Reames, and T.T. von Rosenvinge, "The Role of Interplanetary Shocks in the Longitude Distribution of Solar Energetic Particles", *J. Geophys. Res.*, **93**, 9555, 1988.

65. Schwenn, R., P. Hemmerich, R. Kramm, B.Podlipnik, K. Pahlke and R. Roll, "The LASCO-C1 Coronagraph: First Coronal Images Taken with the Bergmodell at Sacramento Peak and Mauna Loa", in the Proceedings of Solar Wind 7 (this Issue), 1992.

66. Jackson, B., R. Gold and R. Altrock, "The Solar Mass Ejection Imager", *Adv. Space Res.*, **11**, 377, 1991.

67. Jackson, B.V., D.F. Webb, R.C. Altrock and R. Gold, "Considerations of a Solar Mass Ejection Imager in a Low-Earth Orbit", in *Eruptive Solar Flares*, Z. Svestka, B.V. Jackson and M.E. Machado, eds., Springer-Verlag, Heidelberg, in press, 1991.

RADIO OBSERVATIONS OF ENERGETIC ELECTRONS ASSOCIATED WITH CORONAL SHOCK WAVES: EVIDENCE FOR SHOCK ACCELERATION?

K. -L. Klein

Observatoire de Paris, DASOP and URA 324, F-92195 Meudon, France

ABSTRACT

Intense hectometric radio emission has been considered as evidence for the acceleration of electrons upstream of flare-associated coronal shock waves. The analysis presented here of the temporal evolution of the hard X-ray and radio emission between the low corona and $\simeq 10 R_\odot$ above the photosphere does not confirm the hypothesis of shock acceleration, but suggests that the hectometric emission traces electrons escaping from a height-extended region of acceleration/injection in the low and middle corona.

INTRODUCTION

Evidence for different mechanisms of particle acceleration during the impulsive and gradual phase of solar flares has been inferred from the analysis of energetic particles in space. The particle populations differ from each other in terms of electron-to-proton ratio, abundances and charge states of heavy ions [1,2,3] and electron spectra [4]. A large-scale coronal shock wave, revealed by a characteristic slowly drifting, narrow-band metric to kilometric radio emission ("type II burst" [5]), is considered as evidence for shock acceleration in the gradual phase. It has been shown earlier that this perturbation does not significantly contribute to the production of mildly relativistic electrons downstream, in the low solar atmosphere: Metric continuum emission ("type IV" [6]) and gradual hard X-ray bursts cannot be explained by electrons accelerated in the shock wave, but are likely energized by a continually (10 min-1 hr) acting process in the low and middle corona, downstream of the coronal shock [7,8,9,10].

In the upstream region the existence of a specific class of hectometric to kilometric radio emission (observed at $\nu < 2$ MHz) has been suggested, consisting of intense fast-drift bursts which cannot be tracked to coronal type III bursts in the impulsive phase, but are found to accompany type II bursts [11,12,13]. Because of this association, the hectometric bursts were termed "shock associated events" and tentatively attributed to electrons accelerated in the coronal shock. However, as type II bursts are often accompanied by metric continua and gradual hard X-ray emission in the low and middle corona, the association of another phenomenon with type II emission will not favour a causal link with the coronal shock wave, irrespective of the statistical significance of association, unless a common origin with the electrons producing continuum and gradual hard X-ray emission can be excluded. The present work reexamines the origin of "shock associated events" by comparing the hectometric emission with hard X-ray and metric emission in the low and middle corona.

"SHOCK ASSOCIATED EVENTS" AND METRIC CONTINUA

Data Analysis

Criteria to identify "shock associated events" in the hectometric and kilometric data (high flux density, complex time structure, prolonged duration with respect to metric type III bursts) have

TABLE 1 End Times of the Radio and X-ray Emissions

Date	1.98 MHz	Type II (WEIS)	Hard X	408 MHz	169 MHz
1 oct 1978	8:10	? (160-46)	8:06	<8:19	8:10
13 oct 1978	13:12	none	13:00	no data	12:47/13:20/> 16
1 mar 1979	<10:50	10:28 (130-50)	10:38	10:24/10:51	10:25/10:45
27 apr 1979	7:06/7:17	6:54 (218-86)	7:00/7:30	7:03/7:29	7:05/7:29
2 may 1979	17:28	17:23 (60-64)	17:19	17:35	17:36
14 aug 1979	13:23	none	13:18	no data	13:20
21 aug 1979	6:47	6:39 (123-40)	6:16	6:34/6:57	6:20/6:54
26 aug 1979	18:40	none	18:34	18:38	18:35
15 apr 1980	11:24	10:52 (66-60)	none	11:44	11:45
3 may 1980	13:40	13:13 (70-32)	13:22	14:18	14:12
28 may 1980	17:33	17:35 (140-30)	17:31	17:20/17:32	17:30
1 jul 1980	16:48	16:48 (240-55)	16:34	16:38	16:48

been established in [12], where a list of 28 candidate events is compiled. They are accompanied by metric type II bursts *or continuum emission*. In an attempt to look for distinguishing criteria between the association with either type II bursts or continua, the 12 events of this list also observed with the Nançay Observatory's single dish and imaging instruments (169, 408 MHz) were studied. The time histories of the flux densities at 1.98 MHz were compared with those of the metre and decimetre emission, as well as with the count rate time histories in hard X-rays (ISEE 3 [14]).

Results

In a first step the end times in the different spectral ranges were compared (Table 1). The end time of the hectometric, metric and hard X-ray (26-43 keV) events was read from the plotted time histories by estimating the instant when the emission has fallen back to the pre-event level. This gives a different end time of the hectometric emission with respect to [12, Table II] who considered the time of the last significant peak at 1 MHz (Flux density > 500 sfu) as the end time of the "shock associated" event. When the metric or hard X-ray emission had a pronounced minimum in the course of an event, this instant is given by the first number in columns 4 to 6. The third column refers to the type II burst, with the highest and lowest frequency of the drifting bands in parentheses [15]. The type II burst occurs at frequencies lower than 169 MHz in most and lower than 408 MHz in all cases, so that no confusion with the metric continuum can occur. In three cases no type II burst was reported (Weissenau, Solar Geophysical Data), and in four events the type II ends long before the continuum.

Figs. 1 show comparative time histories at radio and

X-ray wavelengths for two events. On 3 may 1980 (Fig. 1.a) a 1N/M2 flare occurred at S26 E 43. At cm-waves a moderately intense "great burst" and long-lasting GRF emission are reported in Solar Geophysical Data. The "shock-associated" event at 1.98 MHz is accompanied by hard and soft X-rays and a metric storm continuum ("stationary type IV burst" [6]). This implies that electrons are accelerated in the low and middle corona from $\simeq$ 12:50 UT to at least 14 UT, i.e. 20 minutes after the end of the event at 1.98 MHz. The earlier end of the hard X-ray burst with respect to the metric continuum is probably due to the acceleration of less and less energetic electrons in the later phase of the flare (e.g. [7]). It is not clear whether this also explains the longer duration of the metric continuum with respect to the emission at 1.98 MHz, or whether the energetic electrons lose access to the higher altitudes during the late phase of the event (e.g. as a consequence of the changing magnetic topology during a coronal mass ejection).

Fig. 1.b shows a flare continuum from centimetric to metric waves, which is accompanied by a gradual hard X-ray burst (from a 1B flare at S22 E73 [16]). The gross features of the temporal

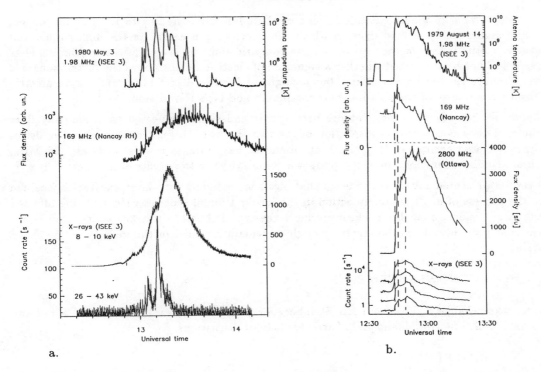

Figure 1: Evolution of radio emission ("shock associated" event at 1.98 MHz, metric continuum at 169 MHz) in the high and middle corona and in X rays: **a.** metric storm continuum ("stationary type IV"); **b.** flare continuum (X-ray channels, from top to bottom: 26-43, 43-78, 78-154, 154-398 keV).

evolution are similar in all spectral ranges, indicating that during the whole duration of particle injection in the low and middle corona electrons escape into the high corona where they generate the hectometric emission.

In summary, the 12 events with common observations at hectometric and metric wavelengths show the presence of energetic electrons in the low and middle corona during the whole duration of the hectometric emission in the high corona. While in general no detailed correspondence of fine structure can be found between the metric and hectometric data, gross features of the temporal evolution are similar in some, end times in nearly all cases. No event is found where the hectometric emission ends more than five minutes after the metric continuum. In the case of storm continua, the emission may last longer at metric than at hectometric wavelengths.

DISCUSSION

The association between hectometric fast-drift bursts and metric continua suggests that if the hectometric emission were due to electrons accelerated in the type II-associated coronal shock wave, the same mechanism should account for the continuum and X-ray emission in the downstream region. At least two reasons argue against this:

- The type II-associated shock wave is neither a sufficient [10] nor a necessary (e.g. [7,9]) condition for metric continua or gradual hard X-ray emission in the post-impulsive phase of a flare.
- The gradual hard X-ray emission comes from structures in the low corona ($\simeq 1 - 5 \times 10^4$ km [17,8]), which is hard to reconcile with particles being accelerated in the shock at several $R_\odot$ above the photosphere.

The observed correspondence of temporal evolution rather points to the acceleration/injection of electrons over a probably extended range of altitudes, from the low to the middle corona, including regions of open field lines on which the electrons can escape to the high corona and eventually to the interplanetary space. The hectometric emission then traces the escaping electrons. It has not been checked whether a signature of electron beams above the metric source is indeed found in the events considered, but it might be hard to detect in metric and decametric radio spectra because of the simultaneous continuum and type II emission.

If the above scenario is correct, intense hectometric emission is expected to accompany flare continua. This was checked by inspection of the two events of [16] not contained in the list of "shock associated events" of [12]. Both are found to have intense counterparts at 1.98 MHz, which are similar (duration, intensity, temporal complexity) to the events of Table 1.

The results presented here give evidence that electrons escaping into interplanetary space can trace the process of acceleration/injection in the low and middle corona without being affected by the large-scale coronal shock which produces the type II burst. Electrons accelerated in the shock front probably are not energetic enough or remain confined in the vicinity of the shock (cf. [18]).

ACKNOWLEDGEMENTS

K.-L.K. acknowledges S. Kane and J.L. Steinberg for giving him access to the ISEE 3 X-ray and radio data, R.J. MacDowall and E.T. Sarris for helpful discussions.

REFERENCES

[1] R. Ramaty, R.J. Murphy, and J.A. Miller, in: *Particle Astrophysics*, ed. W.V. Jones, F.J. Kerr, and J.F. Ormes, AIP Conf. Proc. 203, 143 (1990)

[2] H.V. Cane, R.E. McGuire, and T.T. von Rosenvinge, Ap.J. 301, 448 (1986)

[3] H.V. Cane, D.V. Reames, and T.T. von Rosenvinge, Ap.J. 373, 675 (1991)

[4] D. Moses, W. Dröge, P. Meyer, and P. Evenson, Ap. J. 346, 523 (1989)

[5] G.J. Nelson and D.B. Melrose, in: *Radiophysics of the Sun*, ed. D.J. McLean and N.R. Labrum, The University Press, Cambridge (1985)

[6] M. Pick, Solar Phys. 104, 19 (1986)

[7] G. Trottet, Solar Phys. 104, 145 (1986)

[8] S. Kahler, Solar Phys. 90, 133 (1984)

[9] E.W. Cliver, B.R. Dennis, A.L. Kiplinger, S.R. Kane, D.F. Neidig, N.R. Sheeley, and M.J. Koomen, Ap.J. 305, 920 (1986)

[10] K.-L. Klein, G. Trottet, A.O. Benz, and S.R. Kane, ESA SP-285 Vol. 1, 157 (1988)

[11] H.V. Cane, R.G. Stone, J. Fainberg, J.L. Steinberg, and S. Hoang, Geophys. Res. Letters 8, 1285 (1981)

[12] R.J. MacDowall, R.G. Stone, and M.R. Kundu, Solar Phys. 111, 397 (1987)

[13] S.W. Kahler, E.W. Cliver, and H.V. Cane, Solar Phys. 120, 393 (1989)

[14] K.A. Anderson, S.R. Kane, J.H. Primbsch, R.H. Weitzmann, W.D. Evans, R.W. Klebesadel, and W.P. Aiello, IEEE Trans. GE-16, 157 (1978)

[15] H.W. Urbarz, Report UAG-98, NOAA Boulder 1990.

[16] K.-L. Klein, K. Anderson, M. Pick, G. Trottet, N. Vilmer, and S. Kane, Solar Phys. 84, 295 (1983)

[17] S.R. Kane, Solar Phys. 86, 355 (1983)

[18] E.T. Sarris and S.M. Krimigis, Ap.J. 298, 676 (1985)

RELATION BETWEEN CORONAL mm-WAVE SOURCES, NOISE STORMS, AND CMEs

A. Krüger,* A. Böhme,* J. Hildebrandt* and S. Urpo**

*Central Institute for Astrophysics, Potsdam, Germany
** University of Technology, Helsinki, Finland

ABSTRACT

Coronal millimeter-wave sources (CMMSs) centered at heights in the solar atmosphere of the order of 100000 km above the photospheric level have been observed for more than ten years at the Metsähovi Radio Research Station of the Helsinki University of Technology. The CMMSs appear to form a special class of long-duration post-flare emission although their relationship to single flares is not always clearly established. There is an interesting relation to gradual microwave bursts and noise storms, and also to soft X-ray long-duration events (LDEs) and coronal mass ejections (CMEs). Results of these findings are discussed and consequences for the physical interpretation outlined.

1. INTRODUCTION

Active phenomena in the solar corona indicate the presence of quite different characteristic time scales. In particular one can distinguish
- long lasting (hours, days), slowly developing features in "closed" magnetic loop or arch-like structures, and
- dynamical processes of shorter time scales (seconds, minutes) occurring in "closed" or "open" field structures.

Both groups appear to be closely related to each other, but the detailed physical processes responsible are not yet well clarified.

As a typical example of the first type of phenomena we quote long-duration soft X-ray events (LDEs), hard X-ray (HXR) post-flare arches, coronal mm-wave sources (CMMSs), and radio noise storms (NSs). Dynamical processes are in particular flare-related, associated with, e.g., different types of radio bursts, shock waves, electron and proton beams, and further filament eruptions and coronal mass ejections (CMEs). In the following we study the question of connection between the above two large groups of processes mainly by consideration of CMMSs, NSs, LDEs, and CMEs.

2. GENERAL CHARACTERISTICS OF CMMSs

Millimeter waves are rarely detected in the solar corona. Usually the solar mm-wave emission is confined to levels close to the chromosphere. It was shown in /1/ however, that in some cases the sources of *coronal mm-wave radiation* can be observed at the solar limb at altitudes of about 100000 km (and more) above the photosphere. These sources are related to solar flares and microwave bursts and last much longer than the impulsive phase of the related flare and burst events. A detailed inspection shows that the coronal mm-wave sources (CMMSs) signify the presence of infrequent, extended, complex, and flare-rich solar active regions or complexes of activity. CMMSs form a special class of long living post- and inter-flare events which can be considered as a "super-gradual" background underlying a series of individual flare events. A review on CMMSs can be found in /2/.

The main information hitherto available on CMMSs is derived from solar maps of the Metsähovi Radio Research Station of the Helsinki University of Technology. The observations have been made at 37 and 22 GHz (8 and 13.5 mm wavelength, respectively) with an angular resolution of 2.5 and 4 arc minutes, respectively (cf. Figure 1). Other (but less numerous) observations are available also from other stations.

The observations of mm-wave off-limb sources comprise
- quasi-stationary post- (and intermediate-) flare sources,
- huge burst sources during the main phase of flares,
- sources associated with erupting filaments.

Comparisons with flare reports show that most of the CMMSs of the first group are related to flares starting a few hours earlier in the same active region.

Some cases of missing flare association are to be attributed to behind-limb events. The post-burst CMMSs differ from the main-phase burst sources by the decreasing brightness. Post-flare CMMSs can be related to weak flare events occurring with shorter time delay between the observation and flare start or to larger flare events at larger delay times. Repeated flare activity in the same active region can lead to revivals of CMMSs. The information about the temporal development of the source height in the early flare phase is still rather poor, it appears that the source height grows rapidly during the initial phase of a flare and then remains approximately constant.

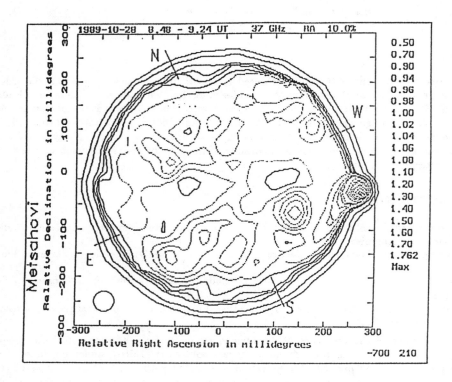

Fig. 1 Example of a CMMS on 28 October 1989, a LDE with a lifetime of about 20 hours occurred 14 hours before the time of the mm-wave observation.

CMMSs are related to microwave bursts in the cm-wave range, but the lifetime of the CMMSs is much greater than that of these bursts measured by small radiotelescopes. The spectrum of a CMMS on 22 September 1980 was measured by the large radio telescope RATAN 600 at cm-waves. This spectrum is in good agreement with optically thin bremsstrahlung ($T \sim 10^7 K, N \sim 8 \times 10^9 cm^{-3}, \Delta s \sim 5 \times 10^9 cm, B \sim 300G$) /3/. Outside the frequency range measured in the above case, model calculations predict different contributions of high-temperature gyromagnetic emission and low- or medium-temperature bremsstrahlung. In essence, the analysis of the whole spectrum leaves open an interpretation by an inhomogeneous source composition: Optically thick, cold prominence-like plasma may be responsible for the radiation at higher frequencies f > 37 GHz while hotter components are capable of generating gyro-synchrotron radiation at lower frequencies f < 8 GHz.

3. RELATION BETWEEN CMMSs, X-RAYS, NOISE STORMS, AND CMEs

It was shown in /4/ that CMMSs are closely related to flares, LDEs in soft X-rays, and NSs. Recent observations of CMMSs qualitatively confirm the earlier results. The study of /4/ was based on 59 observations of CMMSs related to a total of 39 active regions during the period 1977 - 1983. The new material consists

of CMMSs observed on 26 days during the period 1987 - 1991 related to 19 active regions. Nevertheless, there is still a need for more observations. NSs are of special interest because they signify the existence of a persistent release of nonthermal energy in extended coronal regions. NSs are known to be asociated with activity in underlying loop-like magnetic configurations, e.g.
- the S-component of microwave emission /5/
- CMMSs /6/
- SXR-LDEs /7/
- gradual HXR-arches /8/
- disappearing prominences /9/.

Because there is a good correlation between LDEs and CMEs /10/, the CMMSs and NSs also coexist together with CMEs. Further associations with CMEs are known, e.g., with
- giant HXR arches /11/
- hydrodynamic shocks /12/
- flares and erupting filaments, but there is a disparity in sizes between these phenomena /13/.
Simplifying we put the two large classes of physical processes, viz. the quasi-stationary and dynamical processes as indicated by the various phenomena denoted above into the two upper boxes of Figure 2. Then the main questions concern the coupling of the processes and phenomena within each box and the interaction between the two boxes. All these processes must be guided by the evolution of magnetic stuctures during the lifetime of an active region.

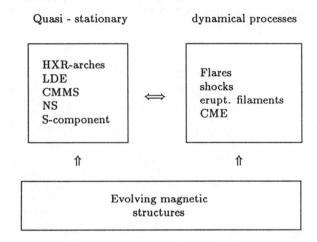

Fig. 2 Outline of two classes of coronal processes

4. CONCLUSIONS

- CMMSs are an useful complementary diagnostic tool of quasi-stationary post- (and inter-)flare coronal loop or arch structures.
- CMMSs are related to HXR-arches and LDEs and form the base of noise storm emitting regions.
- CMMSs are indicative for large, flare-rich active regions.
- the physical link between microwave and X-ray features and NSs is still poorly understood.
- There are characteristic statistical connections of CMMSs with LDEs, and of LDEs with CMEs.

REFERENCES

1. S. Urpo, A. Krüger, and Hildebrandt, Millimetre wave sources in the solar corona, *Astron. Astrophys.*, 163, 340-342 (1986)

2. A. Krüger and S. Urpo, Coronal millimeter sources associated with eruptive flares, *Proceedings of the IAU-Colloquium 133 on Eruptive Flares*, in press (1991)

3. V.N. Borovik, G.B. Gelfreikh, V.M. Bogod, A.N. Korzhavin, A. Krüger, J. Hildebrandt, and S. Urpo, The spectrum of coronal sources at millimetre and centimetre waves, *Solar Phys.* 124, 157-166 (1989)

4. A. Krüger, J. Hildebrandt, S. Urpo, and S. Pohjolainen, On the Association between coronal mm-wave sources, LDE X-ray flares, and coronal mass ejections *Adv. Space Res.* 9, (4), 47-51 (1989)

5. T.A. Bratova, G.B. Gelfreikh, and A.A. Gnezdilov, On the relation of local sources of centimeter solar radioemission with noise storms based on the RATAN-600 observations, (in Russian), *Pisma v Astron. Zhurn.* 9, 495-499 (1983)

6. A. Böhme and A. Krüger, Characteristics of coronal loop structures indicated by Proton Flares, LDE X-ray bursts, and radio noise storms, in *Solar Magnetic Fields and Corona*, ed. R.B. Teplitskaya, Novosibirsk, Nauka, Part II, 313-315 (1989)

7. P. Lantos, A. Kerdraon, G.G. Rapley, and R.D. Bently, Relationship between a soft X-ray long duration event and an intense metric noise storm, *Astron. Astrophys.* 101, 33-38 (1981)

8. L. Klein, K. Anderson, M. Pick, G. Trottet, and V. Vilmer, Association between gradual hard X-ray emission and metric continua during large flares, *Solar Phys.* 84, 295-310 (1983)

9. P. Lantos, Relationship between noise storms and X-ray emissions (a review), in *Solar Radio Storms*, ed. A.O. Benz and P. Zlobec, Trieste, 247-254 (1982)

10. D.F. Webb and A.J. Hundhausen, Activity associated with the solar origin of coronal mass ejections, *Solar Phys.* 108, 383-401 (1987)

11. Z.F. Švestka, B.V. Jackson, R.A. Howard, N.R. Sheeley, Jr., Giant solar arches and coronal mass ejections in November 1980, *Solar Phys.* 122, 131-143 (1989)

12. S. Kahler, N.R. Sheeley, Jr., R.A. Howard, M.J. Koomen, and D.J. Michels, Characteristics of flares producing metric type II bursts and coronal mass ejections, *Solar Phys.* 93, 133-141 (1984)

13. S.W. Kahler, N.R. Sheeley, Jr., and M. Liggett, Coronal mass ejections and associated X-ray flare durations, *Astrophys. J.* 344, 1026-1033 (1989)

REGULATION OF THE INTERPLANETARY MAGNETIC FIELD

D. J. McComas, J. T. Gosling and J. L. Phillips

Space Plasma Physics Group, MS-D438, Los Alamos National Laboratory, Los Alamos, NM 87545, U.S.A.

ABSTRACT

In this study we use a recently developed technique for measuring the combined magnitudes of inward and outward (sunward and antisunward) pointing 2-D magnetic flux in the ecliptic plane to examine 1) the long term variation of the amount of magnetic field open to interplanetary space and 2) the apparent rate at which coronal mass ejections (CMEs) may be opening new magnetic field from the Sun. Since there is a substantial variation (~50%) of these combined fluxes in the ecliptic plane over solar cycle 21, we conclude that there must be some means whereby new field can be opened from the Sun and previously open magnetic field can be closed off. We briefly describe recently discovered coronal disconnection events which could serve to close off previously open magnetic field. CMEs appear to retain at least partial magnetic connection to the Sun and hence open up new field, while disconnections appear to be likely signatures of the process that returns closed field to the Sun. The combination of these processes could regulate the amount of inward and outward magnetic flux open to interplanetary space.

INTRODUCTION

One of the most stunning comparisons to be made with eclipse and coronagraph images of the solar corona is that between solar minimum and maximum conditions. At solar minimum the corona generally displays a relatively simple structure with a coronal streamer belt at low heliolatitudes and large coronal holes at the solar poles. At solar maximum the corona has a much more complicated configuration with streamers and coronal holes found at all heliolatitudes. On shorter time scales of several days to several solar rotations, coronal holes and streamers typically appear and disappear in the solar corona. Since the "open" magnetic field observed in the solar corona connects out into interplanetary space to form the interplanetary magnetic field (IMF), both long and short term modifications of the coronal magnetic field map into the IMF configuration.

Figure 1 schematically portrays various magnetic topologies possible in interplanetary space (above the dashed line). Typical interplanetary field lines (A) are referred to as "open" although they clearly must close somewhere in the outer heliosphere in order for the field to be divergence free. Coronal mass ejections (CMEs) are observed in sequences of coronagraph pictures as bright (high density) structures rising through the solar corona and expanding into interplanetary space. Classic CMEs appear as closed loops of newly expanding magnetic field. The outward progression of these loops ultimately produces new open field lines. The magnetic bottle geometry (B) is the natural extension of CMEs which maintain simple magnetic connection to the Sun.

One long-standing question of heliospheric physics is why doesn't the IMF magnitude grow without bound if CMEs open new magnetic flux into interplanetary space? Possible solutions to this problem include 1) CMEs falling back into the Sun and 2) CMEs being released as detached "plasmoid" structures (C in Figure 1). One possible intermediate configuration for CMEs is that of magnetic flux ropes (D in Figure 1) which would form when reconnection occurs across some fraction of a CME's sheared magnetic loops /1,2/. With one unpublished exception that we are aware of, CME material never appears to fall back sunward. Similarly, few coronagraph observations support the hypothesis of magnetic detachment of CMEs as plasmoids. Instead, most CMEs appear to open previously magnetically closed regions of the Sun.

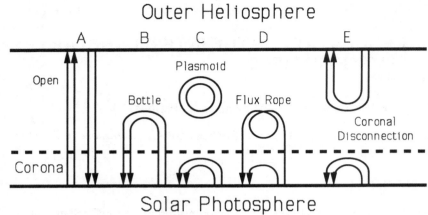

Fig. 1. Schematic diagram of possible magnetic topologies in interplanetary space: (A) "open" magnetic field; (B-D) possible CME geometries; and (E) a coronal disconnection event topology resulting from reconnection above a helmet streamer.

An alternate solution to the build-up of the interplanetary magnetic field has been suggested recently by McComas et al. /3,4/. These authors observed coronal disconnection events, which appear as the separation and release of U-shaped magnetic structures above helmet streamers in the corona. Since helmet streamers are the coronal footprints of the heliospheric current sheet, we believe that coronal disconnections arise from reconnection between oppositely directed fields across the current sheet. This reconnection process converts open magnetic field (A in Figure 1) to closed field arches (E in figure 1), thus reducing the amount of sunward and antisunward magnetic flux open to interplanetary space.

OBSERVATIONS

The magnetic flux, Φ, integrated over a surface completely bounding the Sun must be identically equal to zero since $\nabla \cdot B = 0$. However, the amounts of inward and outward pointing magnetic flux open to interplanetary space could vary greatly. These inward and outward fluxes would be found, for example, by integrating the radially inward and outward magnetic field components over a spherical surface centered on the Sun. Unfortunately, such complete measurements are not available. By way of contrast, only single point spacecraft measurements have generally been available in interplanetary space and the standard proxy for the amount of open field has simply been long-term averages of the IMF magnitude. Using this proxy, solar cycle 20 displayed little long term variation while solar cycle 21 displayed a substantial and systematic variation /5/. There are a number of problems with inferring the global variation in the amount of open magnetic field from simple averages of the IMF magnitude measured at a single point in space. Obviously, single point measurements do not provide information about the field elsewhere. In addition, averages of the IMF magnitude do not correct for the rate at which magnetic flux is convected past the measurement point.

Recently a somewhat more sophisticated method for measuring the amount of magnetic flux passing a single observing spacecraft was developed /2/. While this technique still suffers from the unavoidable undersampling of single point spacecraft measurements, at least the rate at which flux is convected past the spacecraft is taken into account. These authors defined a pseudo-flux which combines the magnitudes of the inward and outward pointing fluxes as a measure of the the total amount of magnetic field open to interplanetary space; the 2-D pseudo-flux, ϕ^*, is then given by

$$\phi^* = \int |B_y| \, v_x \, dt \tag{1}$$

in solar ecliptic (SE) coordinates under the simplifying assumption that the magnetic flux in the ecliptic plane is confined to that X-Y plane /2/. This 2-D assumption is not precisely correct, however, McComas et al. /2/ demonstrated that it is a reasonable approximation for these purposes.

Figure 2 displays annual averages of ϕ^* calculated using the NSSDC's OMNI data file of one hour averaged solar wind and IMF parameters from 1973 - 1988. This interval spans solar cycle 21 which extends from 1976 through 1987. Clearly there is a significant (~50%) variation in the amount of pseudo-flux encountered in the ecliptic plane at 1 AU over solar cycle 21, as discussed previously by Slavin et al. for field magnitude /5/. We have not tried to separate normal open magnetic field from those which are newly opening (bottle geometry) or simply convecting past (plasmoid geometry) over the entire interval shown in Figure 2. However, such a significant variation in ϕ^* over the solar cycle suggests there must be some means by which 1) new field can be opened (such as by CMEs which remain at least partially magnetically attached to the Sun) and 2) open magnetic field can be closed off (such as by reconnection above helmet streamers).

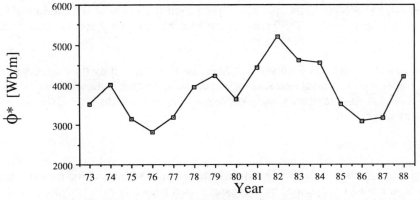

Fig. 2. Annual averages of the pseudo-flux in the ecliptic plane from 1973 - 1988.

While the method for calculating ϕ^* discussed above is an improvement for examining the long term variations in the magnetic field open to interplanetary space, its real advantage is that it can be applied to short duration, transient structures in the solar wind. This is because the combination of magnetic field and flow speed in equation (1) provides a measurement independent of the rate of the structure's motion past the spacecraft. Coronal mass ejections can be most readily identified in the solar wind by the streaming of suprathermal halo electrons in both directions along the local magnetic field /1/. We have used the counterstreaming signature to identify 58 CMEs over the 18 month period from mid-August 1987 through mid-February 1980 and have calculated ϕ^* for each event /2/.

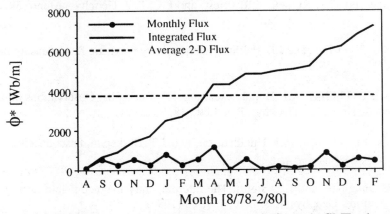

Fig. 3. Monthly and integrated ecliptic plane pseudo-fluxes in CMEs observed by ISEE-3 at 1 AU. The dashed line indicates the average ϕ^* from Figure 2.

Figure 3 displays the monthly observed 2-D pseudo-fluxes (solid curve with dots) and the cumulative pseudo-flux (solid trace). The latter gives the rate at which 2-D ϕ^* would build up in interplanetary space if all CMEs remained simply connected to the Sun in bottle-type geometries. The dotted horizontal line indicates the average ϕ^* in the ecliptic plane from Figure 2. At the rate of build-up shown in Figure 3, the amount of open magnetic field in the ecliptic plane would be expected to double over only ~9 months. This rate is slower than the doubling time of ~100 days estimated by MacQueen /6/ from coronagraph observations of CMEs prior to 1980. Of course if CMEs remain only partially magnetically attached to the Sun, as in flux ropes, the rate at which ϕ^* would build up in interplanetary space would be slower.

SUMMARY

1. Coronal mass ejections appear to remain at least partially magnetically attached to the Sun and, hence, probably open new magnetic field into interplanetary space.

2. A recently developed technique for measuring the combination of inward and outward 2-D magnetic flux passing an observing spacecraft in interplanetary space confirms that there is a substantial variation in the ecliptic plane over solar cycle 21. This variation suggests that there must also be some means for closing off field previously open to interplanetary space.

3. Integration of the pseudo-flux involved in CMEs, as distinguished by the counterstreaming electron signature, over an 18 month interval near solar maximum indicates that in the absence of other processes the 2-D pseudo-flux in the ecliptic plane would double over only ~9 months if all CMEs represented simple "bottle" geometry magnetic fields.

CONCLUSION

We believe that the evidence strongly suggests that the amounts of inward and outward magnetic flux in interplanetary space must be regulated by processes which open and close magnetic field from the Sun. CMEs appear to be the obvious choice for opening new magnetic field while reconnection across the heliospheric current sheet above helmet streamers is the most likely method for closing it off.

We thank E.J. Smith and the JPL magnetometer group for ISEE-3 data and the NSSDC for OMNI data. This work was conducted under the auspices of the United States Department of Energy with support from NASA under S-04039-D.

REFERENCES

1. J.T. Gosling, Coronal mass ejections and magnetic flux ropes in interplanetary space, Physics of Magnetic Flux Ropes, ed. C.T. Russell, E.R. Priest, and L.C. Lee, Geophys. Mono. 58, AGU, 1990, p. 343.

2. D.J. McComas, J.T. Gosling, and J.L. Phillips, Interplanetary magnetic flux: measurement and balance, J. Geophys. Res., 97, 171 (1992)

3. D.J. McComas, J.L. Phillips, A.J. Hundhausen, and J.T. Burkepile, Observations of disconnection of open coronal magnetic structures, Geophys. Res. Lett., 18, 73 (1991)

4. D.J. McComas, J.L. Phillips, A.J. Hundhausen, and J.T. Burkepile, Disconnection of open coronal magnetic structures, this issue.

5. J.A. Slavin, G. Jungman, and E.J. Smith, The interplanetary magnetic field during solar cycle 21: ISEE-3/ICE observations, Geophys. Res. Lett., 13, 513 (1986)

6. R.M. MacQueen, Coronal transients: a summary, Phil. Trans. R. Soc. Lond., 297, 605 (1980)

REMOTE RADIO OBSERVATIONS OF SOLAR WIND PARAMETERS UPSTREAM OF PLANETARY BOW SHOCKS

R. J. MacDowall, R. G. Stone and J. D. Gaffey Jr

Laboratory for Extraterrestrial Physics, NASA/Goddard Space Flight Center, Greenbelt, MD 20771, U.S.A.

Abstract: Radio emission is frequently produced at twice the electron plasma frequency $2f_p$ in the foreshock region upstream of the terrestrial bow shock. Observations of this emission provide a remote diagnostic of solar wind parameters in the foreshock. Using ISEE-3 radio data, we present the first evidence that the radio intensity is proportional to the kinetic energy flux and to other parameters correlated with solar wind density. We provide a qualitative explanation of this intensity behavior and predict the detection of similar emission at Jupiter by the Ulysses spacecraft.

INTRODUCTION

Observations of radio emissions are frequently used to provide remote analyses of source regions in the interplanetary medium; examples include type II and type III solar radio bursts. Another source, which is the subject of this paper, is radio emission produced at the leading edge of the terrestrial foreshock. A schematic of this region is shown in Figure 1. This region is particularly interesting for the study of plasma radio emission because it is more readily accessible to spacecraft than the larger interplanetary medium, because the bow shock is quasistationary, and because the magnetic field geometry is readily modeled.

The $2f_p$ emission from the terrestrial foreshock was first reported by Dunckel /1/ and subsequently studied by many other authors. Lacombe *et al.* /2/ analyzed the dimensions of the radio source using data from ISEE-1 and ISEE-3 and obtained source dimensions of 120–150 R_E in diameter and only 1 R_E thick. Their paper also gives a concise summary of previous work on this emission.

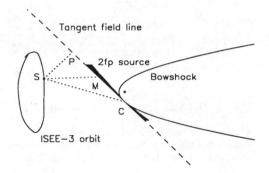

Fig. 1. Schematic in the ecliptic plane showing terrestrial bow shock, orbit of ISEE-3, tangent field line, $2f_p$ source region, and distances, as discussed in this paper.

Filbert and Kellogg /3/ associated this emission with solar wind electrons reflected and accelerated at the bow shock, i.e., the electrons that form the electron foreshock. These electrons are accelerated by a fast Fermi process described in detail by Wu /4/ and Leroy and Mangeney /5/. Adiabatic arguments /5/ alone yield the average energy parallel to the magnetic field of a reflected electron, where v is solar wind velocity.

$$\langle E_{r\parallel} \rangle \sim 2m_e v^2 / \cos^2 \theta_{Bn} \tag{1}$$

Adiabatic considerations also permit determining the density of electrons reflected, which (depending on the formulation of the problem) is proportional to the solar wind density or the density of suprathermal electrons /4,5/. Numerical simulations /6/ confirm the validity of these adiabatic approximations. The total energy of reflected electrons (parallel to the magnetic field) is a maximum for θ_{Bn} near 88°. This effect, in conjunction with time of flight, results in the most energetic electrons forming a beam at the leading edge of the foreshock.

The maximum magnetic field at the bow shock plays an important role in determining the fraction of electrons reflected. The terrestrial bow shock is usually supercritical and, therefore, B usually undergoes a considerable overshoot before settling to a downstream value. Observations by Mellott and Livesey /7/ show that the magnitude of the overshoot in B is correlated with the shock Mach number and electron beta. Consequently, solar wind magnetic field, density, and temperature enter into the determination of the density and energy of the reflected electron beam.

It is generally believed that the electrons produce $2f_p$ radio emission by a two-stage process. As the electrons move away from the point of reflection, time of flight effects produce an unstable distribution function. Langmuir waves are excited by the electron beam. Subsequently, they coalesce with oppositely-directed waves, produced by a back-scattering process involving ion-acoustic waves, to produce $2f_p$ electromagnetic waves (see, for example, Cairns and Melrose, /8/).

To greatly simplify consideration of the emission mechanism, one might assume that the radio power output is proportional to the power of the reflected electrons. This is approximately the product of the flux of electrons entering the acceleration region times the fraction of electrons reflected times the average energy of a reflected electron (given by equation (1)), yielding $P(2f_p) \propto nm_ev^3$. We will show that this is indeed the case.

This introduction has concentrated on terrestrial foreshock $2f_p$ emission, which is the only planetary bow shock at which such radio emission has been detected. At other planets such as Jupiter /9/, Langmuir waves strongly suggest the presence of a similar electron foreshock. The Voyager radio instrument did not provide observations at frequencies low enough to detect $2f_p$ emission at the outer planets. Type II solar bursts are also believed to be produced by similar processes.

OBSERVATIONS AND RESULTS

In this paper, we present the first analysis of statistical relations between the intensity of terrestrial $2f_p$ emission and observed solar wind parameters. The radio data used for this analysis come from the ISEE-3 Radio Mapping Experiment /10/. The receiver covers the frequency range from 30 kHz–2 MHz with 23 discrete frequencies. Figure 2 is a dynamic spectrum representation of 6 hours of ISEE-3 data showing thermal noise immediately above the plasma frequency and terrestrial radio emission at $2f_p$. Note that the density enhancement observed in the *in situ* thermal noise at 12:25 UT appears about 20 minutes later in the $2f_p$ emission. This is the time required for the density enhancement to propagate from the spacecraft to the terrestrial foreshock. (At this time, ISEE-3 had not yet arrived at its halo orbit position.) See /11/ for additional events from ISEE-3 data.

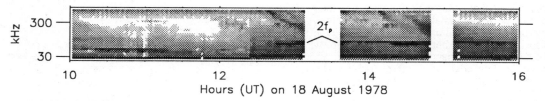

Fig. 2. Dynamic spectrum for 18 Aug 1978, showing remote $2f_p$ emission.

Thirty-three one-hour intervals when $2f_p$ emission was observed were selected and averaged to obtain the emission frequency f_{em} and the peak flux density S of the $2f_p$ emission. An example of the data prior to averaging is shown in Figure 3 (where $f_p \approx 33$ kHz). Note the narrow bandwidth of the $2f_p$ emission.

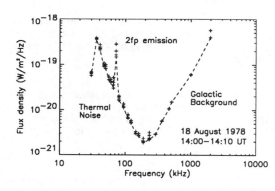

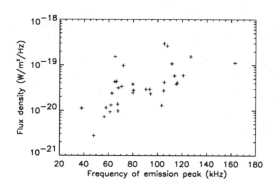

Fig. 3. Spectrum showing $2f_p$ emission line and thermal noise continuum.

Fig. 4. Plot of radio flux density S versus f_{em}.

As shown in Figure 1, the ISEE-3 spacecraft position relative to the $2f_p$ source region varies considerably throughout its halo orbit. The position of the nearest field line tangent to the bow shock also undergoes large changes in position as the field direction (θ_B, ϕ_B) varies. To study the intrinsic variations in $2f_p$ intensity, it is necessary to determine the distance and direction to the source region. We calculate the point of tangency by assuming that the bow shock is a paraboloid, described by $X - A + (Y^2 + Z^2)/B = 0$, where A = 13.8 R_E and B = 45 R_E.

We solve for the coordinates (X_c, Y_c, Z_c) of the contact point in the same manner as Lacombe *et al.* /2/, after correcting the errors in their cubic equation for Z_c. Because of the large dimensions of the source region, we choose the midpoint (labelled M) between the tangent (contact) point C of the closest field line and the perpendicular P from the spacecraft to this line as the centroid of the radio source. (See Figure 1.) Using the distance SM, we normalize the flux density to a standard distance of 1.5×10^6 km.

The solar wind density, temperature, velocity, and vector magnetic field used for this study were obtained from the National Space Science Data Center /12/. These hour–averaged data have been extrapolated from the position of the spacecraft that acquired them to that of the Earth. We use these data values as if they correspond to data acquired at the radio source position (marked M).

The radio data alone show that the flux density varies directly with the emission frequency f_{em} and, therefore, with the density of the source region. This correlation is shown in Figure 4; the linear correlation coefficient is 0.43 (linear flux density versus frequency) for 33 data values.

These observations also demonstrate the accuracy with which the radio emission provides the solar wind density upstream of the bow shock. Figure 5 shows the excellent agreement; the dashed line is $f_{em} = 2f_p = 18\sqrt{n}$ kHz, with a linear correlation coefficient of 0.95.

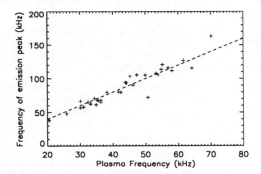

Fig. 5. Plot of f_{em} versus solar wind plasma frequency f_p.

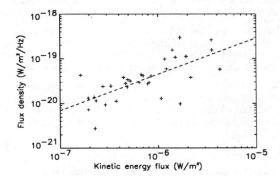

Fig. 6. Plot of radio flux density S versus kinetic energy flux, $1/2\ nm_e v^3$.

Table 1 lists the parameters that have been correlated with flux density. The largest correlation coefficient, 0.67, occurs for the product of the magnetic field magnitude and velocity. It should be noted that the magnetic field is correlated with density; the correlation coefficient is 0.78. On the other hand, velocity and density are uncorrelated, not anti-correlated as is generally the case for the solar wind. These relationships arise, in part, because the dense regions with which the $2f_p$ emission is associated are frequently compressed material located downstream of interplanetary shocks; they are not unperturbed solar wind.

As discussed in the Introduction, simple arguments suggest that the $2f_p$ intensity will be correlated with the kinetic energy flux ($1/2\ m_e nv^3$). This is indeed the case, with a correlation coefficient of 0.61. Noting that the relationship is approximately linear, we fit a trend to the data, as shown in Figure 6, and remove the trend. Recalculating the correlation coefficients, we obtain the final column of results shown in Table 1. The recalculated correlations are small, with the exception of T_i and related parameters. The large correlation for T_i results from a single outlying point; removing this point reduces the correlation to 0.29. The small residual correlations indicate no additional trends in this dataset for the parameters shown.

TABLE 1 Linear Correlation Coefficients for Flux Density of $2f_p$ Emission vs. Solar Wind Parameters

Solar Wind Parameter	Linear Correlation coefficient	Corr. coeff after nv^3 trend removed
Bv	0.67	0.11
B	0.64	0.14
T_i	0.64	0.49
nv^2	0.62	0.06
nv^3	0.61	0.04
nv	0.59	0.08
n	0.48	0.10
v	0.47	-.0.05
f_{em}	0.43	0.06
β_i	0.38	0.46
θ_B	-0.01	-0.14
M_A	-0.11	-0.22
ϕ_B	-0.13	-0.04
M_s	-0.43	-0.40
f_p/f_c	-0.43	-0.08

DISCUSSION

We have demonstrated that the power of $2f_p$ radiation varies directly with parameters that depend on solar wind density. Because density varies over a much broader range than solar wind velocity, the relevant parameters include particle flux, ram pressure and kinetic energy flux. We phrase the discussion in terms of the kinetic energy flux because of its simple relationship to the energy of the reflected electron beam. The trend indicates that the emitted $2f_p$ power is proportional to the energy flux of the electrons entering the acceleration region.

From the trend shown in Figure 6, we derive the following relation. The numerical coefficient is derived by assuming that the $2f_p$ emission is emitted isotropically and has a bandwidth of 3 kHz /11/.

$$P(terrestrial\ 2f_p) \approx 10^3 \left(\frac{n}{10\ cm^{-3}}\right) \left(\frac{v}{400\ km/sec}\right)^3 watts \tag{2}$$

It is interesting to apply the same concept to the Jovian foreshock. If the area of the acceleration region is set by a fixed range of θ_{Bn}, then the area will be determined by the radius of curvature r_{BS} of the bow shock. Then, the following scaling law should apply to any bow shock.

$$P(\ 2f_p) \approx 10^3 \left(\frac{n}{10\ cm^{-3}}\right) \left(\frac{v}{400\ km/sec}\right)^3 \left(\frac{r_{BS}}{r_{BS,Earth}}\right)^2 watts \tag{3}$$

This scaling law suggests that the power of $2f_p$ emission from the Jovian foreshock should be about 200 times more intense than terrestrial $2f_p$ emission. Such emission will be readily detected by the Ulysses Radio Astronomy experiment /13/ prior to the encounter with Jupiter on 8 Feb 1992.

Many details of the reflection, beam propagation, and emission processes have been neglected in this discussion. A complete model of the emission process would be complicated and contain too many parameters to be constrained by this data. It might be possible, however, to use a model like that of Fitzenreiter *et al.* /14/ to compute the extent of unstable electron distribution functions throughout the foreshock.

The next analysis involving these observations should include more of the available data. There are at least a thousand hours of ISEE-3 data where $2f_p$ emission is observed. A larger dataset would serve to constrain the correlations better and permit testing for additional correlations. In addition, it should include analysis of the nature of the overdense regions of solar wind, such as coronal mass ejections, in which the $2f_p$ emission occurs.

SUMMARY

Analysis of the emission frequency and flux density of $2f_p$ radio emission from the terrestrial foreshock indicates that the radio flux density varies directly with the emission frequency. We ascribe this behavior to a dependence on the solar wind kinetic energy flux entering the acceleration region. This assumption provides a scaling law, which predicts that the Ulysses spacecraft will observe $2f_p$ emission from the Jovian foreshock with an intensity 2 orders of magnitude greater than terrestrial $2f_p$ emission.

Acknowledgments: The ISEE-3 Radio Mapping Experiment is a joint project of the Department de Recherche Spatiale of the Observatoire de Paris-Meudon and the Laboratory for Extraterrestrial Physics, NASA/Goddard Space Flight Center. R.J.M. acknowledges helpful discussions with Iver Cairns, Gordon Holman, Denise Lengyel-Frey, Dennis Papadopoulos, and Ron Zwickl.

References

1. N. Dunckel, *Ph.D. Thesis*, Stanford University (1974)

2. C. Lacombe, C.C. Harvey, S. Hoang, A. Mangeney, J.-L. Steinberg, and D. Burgess, *Ann. Geophys.* 6, 113 (1988)

3. P.C. Filbert and P.J. Kellogg, *J. Geophys. Res.* 84, 1369 (1979)

4. C.S. Wu, *J. Geophys. Res.* 89, 8857 (1984)

5. M.M. Leroy and A. Mangeney, *Ann. Geophys.* 2, 449 (1984)

6. D. Krauss-Varban, D. Burgess, and C.S. Wu, *J. Geophys. Res.* 94, 15,089 (1989)

7. M.M. Mellott and W.A. Livesey, *J. Geophys. Res.* 92, 13,661 (1987)

8. I.H. Cairns and D.B. Melrose, *J. Geophys. Res.* 90, 6637 (1985)

9. D.A. Gurnett, J.E. Maggs, D.L. Gallagher, W.S. Kurth, and F.L. Scarf, *J. Geophys. Res.* 86, 8833 (1981)

10. R. Knoll, G. Epstein, S. Hoang, G. Huntzinger, J.-L. Steinberg, F. Grena, S.R. Mosier, and R.G. Stone, *IEEE Trans. Geosci. Electr.* GE–16, 231 (1979)

11. S. Hoang, J. Fainberg, J.-L. Steinberg, R.G. Stone, and R.H. Zwickl, *J. Geophys. Res.* 86, 4531 (1981)

12. D.A. Couzens and J.H. King, Interplanetary Medium Data Book, 1977–1985, *Rep. NSSDC 86–04*, National Space Science Data Center, NASA/GSFC, Greenbelt, Maryland, U.S.A. (1986)

13. R.G. Stone and 31 coauthors, *Astr. and Astrophys. Sup.* 92, in press (1992)

14. R.J. Fitzenreiter, J.D. Scudder, and A.J. Klimas, *J. Geophys. Res.* 95, 4155 (1990)

QUANTITATIVE ANALYSIS OF BIDIRECTIONAL ELECTRON FLUXES WITHIN CORONAL MASS EJECTIONS AT 1 AU

J. L. Phillips, J. T. Gosling, D. J. McComas, S. J. Bame and W. C. Feldman

Mail Stop D438, Los Alamos National Laboratory, Los Alamos, NM 87545, U.S.A.

ABSTRACT

The solar wind electron heat flux is carried primarily by suprathermal electrons beamed antisunward along the interplanetary magnetic field. However, analysis of electron observations at 1 AU has shown that counterstreaming electron beams, suggesting closed magnetic structures, prevail within coronal mass ejections (CMEs). These structures might be magnetic "tongues", magnetically detached plasmoids, or complex flux ropes. Here we show results of analysis of ISEE-3 observations within 39 CMEs, including the asymmetry between the two beams, its control by magnetic field orientation, and the variation of the electron distributions as CMEs convect past the spacecraft. We find that some CMEs are strongly asymmetric, with the antisunward beam generally dominant, while others contain nearly symmetric beams. The beam asymmetries, and the magnetic field orientations, exhibit characteristic trends as CMEs pass over the spacecraft. We present an example of a distinctive "strahl-on-strahl" distribution, suggesting continued magnetic connection to the corona, in which a narrow antisunward beam is superimposed on a broader beam. Our results favor continuing magnetic connection to the Sun in a tongue or flux rope geometry rather than a fully detached plasmoid.

INTRODUCTION

Coronal mass ejections (CMEs) are phenomena in which large volumes of material from the solar atmosphere are released. We believe that solar wind "halo" electrons, that is electrons with energies of ~50 to 80 eV and greater (at 1 AU) provide the most reliable signatures of CMEs in interplanetary space. In the typical solar wind, halo electrons are beamed along the interplanetary magnetic field (IMF) and carry an antisunward heat flux, indicating effective magnetic connection to the hot corona in only one direction. Spacecraft observations have shown that the electron distributions within CMEs are characterized by halo fluxes streaming in both directions along the IMF /1/. This suggests a closed field topology, with the spacecraft connected to a heat source in both directions. Various closed field topologies have been proposed: (1) a magnetic "tongue" or "bottle" rooted in the Sun at both ends; (2) complete magnetic disconnection as a closed "plasmoid"; (3) a helical "flux rope" structure, which suggests a plasmoid in its magnetic signature but which remains connected to the Sun /2/. These topologies, and the prevailing "open" IMF, are shown schematically in Figure 1. The purpose of this study is to assess these models using evidence contained in the halo beams.

OBSERVATIONS

ISEE-3 was launched in 1978 and was initially stationed in orbit about the L1 libration point, 0.01 AU sunward from Earth. The primary measurements used here are from the electron analyzer /3/, which in its normal mode accepts electrons in an energy range of 8.5 to 1140 eV and a polar angle range of ±67.5° centered on the spacecraft spin plane (roughly the ecliptic plane). A two-dimensional electron spectrum representing one 3-s spacecraft spin is made every 84 s, and includes

651

16 spectral scans separated in azimuth by 22.5°. Calculations were performed in seven energy bands covering the halo distribution, from 116-161 eV through 822-1140 eV; results are qualitatively similar for each. Additionally, measurements from a separate ion analyzer were used for solar wind bulk speed and density. Our study also uses one-minute averaged magnetic field measurements from the Jet Propulsion Laboratory magnetometer. The data base for this work includes 39 CMEs observed from August 1978 through January 1980. The events are the same as those listed in /1/, with some of the smaller CMEs combined into single larger events. The CMEs used ranged from 1.5 hours to 36 hours in transit time past the spacecraft, and in all contain nearly 18,000 electron spectra.

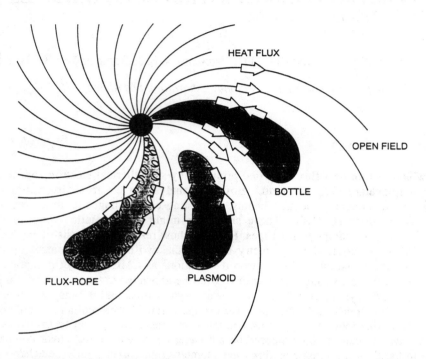

Fig. 1. Schematic diagram of normal open magnetic flux and possible CME topologies. The large arrows show the direction of the electron heat flux.

The shapes and intensities of the halo electron beams are highly variable. Figure 2 contains a sampling of various beam configurations. Each panel includes four traces, representing ISEE-3 measurements in passbands centered at the following energies: 37.3, 71.4, 136.7, and 261.6 eV, ordered from top to bottom. Each trace shows log counts on an arbitrary scale, with horizontal grid lines separated by one decade, vs. ecliptic look angle with the sunward look direction shown as 0°. In all panels, the 37.3 eV (top) trace shows core electrons, with the single broad hump centered at 0° representing antisunward convection (note, however, that for two of the panels there is some evidence of counterstreaming halo electrons even at this low energy). The second (71.4 eV) trace shows various degrees of core-halo superposition, while the two highest energy traces show purely halo distributions. The top left panel represents a typical "open" field halo, with a single beam centered on the IMF. The top right panel shows a CME spectrum for which the two beams are quite symmetric, while the bottom left panel represents a CME measurement for which the beams are strongly asymmetric in intensity. Finally, the bottom right panel illustrates the variability in beam angular width; this CME spectrum contains one broad beam and one narrow beam.

To assess the symmetry of the counterstreaming electron fluxes, the measurement fan (centered roughly on the ecliptic plane) was divided into "outward" and "inward" hemispheres based on the IMF orientation for each spectrum. Each hemisphere is centered on the ecliptic projection of B, with the outward hemisphere being the one directed more nearly antisunward, and contains spectral scans at 8 different azimuths. Interhemisphere comparisons for each energy band were performed for the peak electron count rates observed in a hemisphere and for the counts integrated over the hemispheres. Only hemisphere-integrated results will be shown in this study; however, peak and integrated results were qualitatively similar in all cases. Figure 3 displays results of the interhemisphere comparison for the energy range 223-309 eV; results are similar for all halo energies. The plots show the number of spectra for which the outward hemisphere contained more counts than the inward hemisphere (solid shading), and for which the inward hemisphere dominated (hatching). The data are binned by the ratio of dominant to lesser hemisphere counts. Notable trends are: (1) the outward electron fluxes are dominant roughly 75% of the time; (2) strong outward > inward asymmetries are common, while strong inward > outward asymmetries are rare.

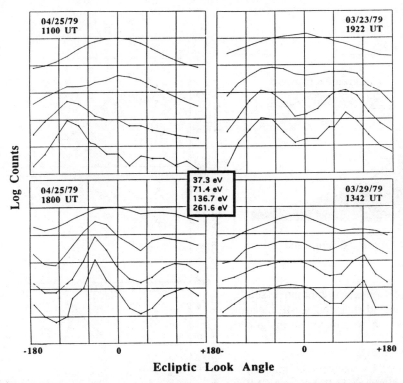

Fig. 2. Four ISEE-3 log count vs. angle electron spectra, each including measurements centered at the four energies indicated in the legend, illustrating the variability in the halo electron beams.

In Figure 4 the data are binned by IMF ecliptic orientation, with inward and outward field sectors folded into a 90-270 degree range; a nominal spiral field of either 135° or 315° would map to 135° in this system. The top panel shows the frequency distribution for the CME spectra and for all solar wind spectra observed by ISEE-3 from August 1978 through January 1980. Note that the overall solar wind data are dominated by the peak near the Parker spiral at 120-140 degrees, while the CME measurements are more uniformly distributed (see also /1/). The CME spectra often have IMF nearly transverse to the solar wind flow (90° or 270°), but also exhibit a skew toward the nominal spiral orientation. The middle panel shows the percentage of CME measurements for which the outward electron fluxes were dominant in the energy range 223-309 eV. Note that spectra with nearly radial IMF (180°) are nearly always dominated by the outward beam, while for nearly transverse IMF (90° or 270°) this trend is much less pronounced. The bottom panel shows the median ratio of dominant-to-lesser beam counts, again binned by IMF azimuth. While the distributions of count ratios for inward-dominated spectra (open dots) show little modulation, for outward-dominated spectra (solid dots) the asymmetries are much larger for nearly radial fields than for transverse fields.

A superposed epoch analysis was performed to identify prevailing trends in electron counts rates as CMEs pass over the spacecraft, under the assumption that the bidirectional electron intervals exactly match the CME intervals. CMEs of less than 8 hours of transit time were disregarded, which reduced the data set to 29 events. This was a somewhat arbitrary step designed to minimize the contribution of glancing encounters with the edges of CMEs and to maximize the likelihood that the analyzed CME observations represented radial transits from leading edge to trailing edge. Measurements were binned by fractional transit time for a given CME, and were weighted so each event contributed equally to the aggregate data set. Medians of various parameters were then calculated, binning by fractional time. Certain trends, not shown here, were noted which are probably attributable to compressive effects at the leading edges of fast CMEs. For example, count rates in both inward and outward beams tend to increase quickly after CME arrival, reach maxima roughly 20% through the CME transit, then decline slowly. A similar trend is seen in ion bulk speed, while ion density tends to peak near the leading edge.

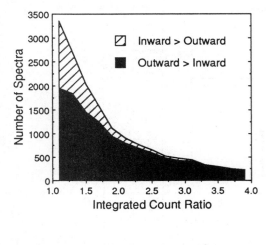

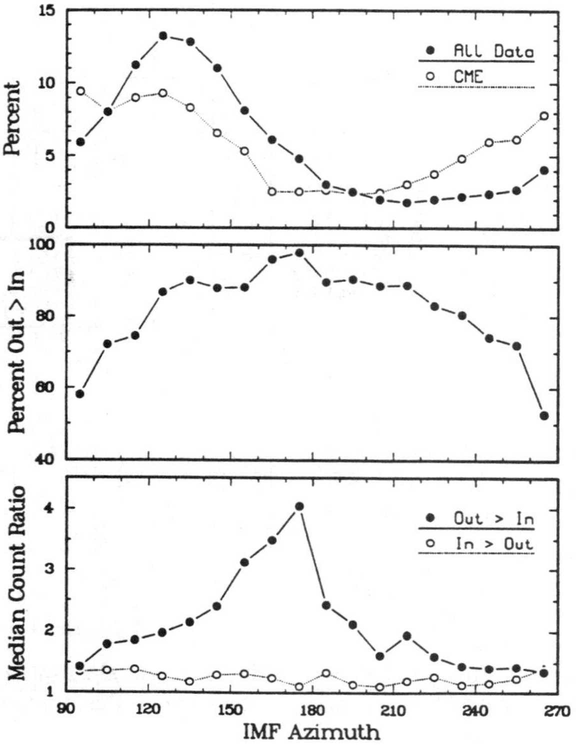

Fig. 3. Number of outward-dominated (solid) and inward dominated (hatched) CME electron spectra, as a function of hemisphere-integrated count ratio.

Fig. 4. Data binned by magnetic ecliptic azimuth: (top) distribution of all data and CME spectra; (middle) percent of spectra dominated by outward beam; (bottom) median ratio of dominant-to-lesser beam integrated counts.

We noted two transit-time effects which may provide insight into likely topologies. The left panel of Figure 5 shows median 161-223 eV integrated count ratios for spectra in which the outward beam was dominant. Note the pronounced trend for high outward/inward count ratios in the last 20% of the aggregate event. This effect typically results from the gradual decay of the inward beam, while the outward beam continues roughly unchanged. The right panel shows the median angle from radial of the ecliptic projection of the IMF for the aggregate CME data set. Note that, on average, this angle declines as CMEs pass over the spacecraft, and that the median angle near the trailing edge roughly matches the nominal spiral angle of 45 degrees.

Halo electrons, although less affected by scattering than are the slower core electrons, are still subject to a variety of scattering processes. Additionally, the various magnetic topologies suggested for CMEs may result in mirroring and multiple transits of the CME structure. The importance of these factors in modifying the highly variable halo distributions is not completely understood. In Figure 6 we present an example of previously unreported compound electron beams, which we will refer to as "strahl-on-strahl" (SOS) spectra. To improve the azimuthal resolution of the measurements, we combined 9 to 12 adjacent spectra during periods in which the IMF was nearly steady in azimuth and had a very small polar angle. Because the measurement timing is asynchronous with spacecraft rotation, this enables aasembly of spectra with fine angular spacing. Figure 6 shows electron counts at 161-223 eV for a 20-min interval within a CME; the vertical lines indicate the range of observed

IMF azimuths for this interval. Note the presence of a narrow electron beam superimposed on the broader outward (azimuth -90° to +90°) halo beam.

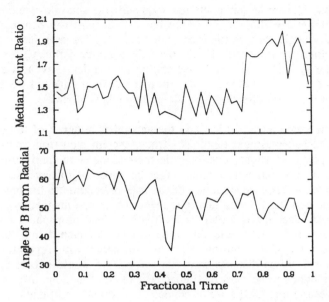

Fig. 5. Medians binned by fractional time of CME transit: (top) integrated count ratios for spectra with dominant outward beams; (bottom) ecliptic angle of **B** from radial for all CME data.

We emphasize that such distributions are a recent, preliminary finding. However, a cursory survey indicates that SOS distributions are present in roughly half of the CMEs surveyed, and that they are not present on open field lines. We suggest the following hypothesis. In a magnetic tongue or flux rope the counterstreaming halo electrons may mirror repeatedly, enabling broadening of the beams through various scattering processes. However, halo electrons making their first outward transit are subject to scattering for a much shorter time and thus may be observable as a narrower beam superimposed on the broader, scattered beam.

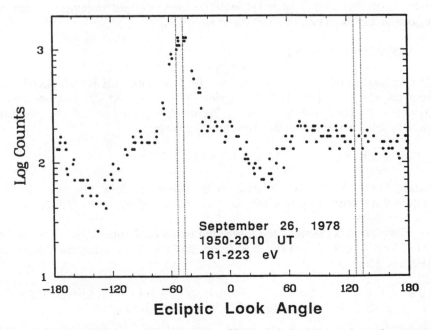

Fig. 6. Strahl-on-Strahl count vs. angle spectrum. Vertical dashed lines show the range of ecliptic **B** azimuth during the spectrum interval.

SUMMARY AND CONCLUSIONS

The observations shown here are an initial attempt at quantitative analysis of the electron observations within CMEs. We summarize the statistical results: (1) the electron beam moving more nearly antisunward is generally larger than its sunward counterpart; (2) The asymmetry tends to be maximal for radial B, and minimal for transverse B; (3) the IMF azimuth within CMEs is distributed more uniformly than for the typical solar wind, but exhibits a similar skew toward the Parker spiral direction (4) beam anisotropies tend to maximize near the trailing edges of CMEs; (5) the magnetic field tends to approach the spiral direction near the trailing edges of CMEs.

The existence of strong beam asymmetries, and the general dominance of the outward beam, seem inconsistent with a detached plasmoid topology. Once disconnected from the coronal heat source, halo beams within a plasmoid should exhibit little modulation with field direction. Furthermore, while either a plasmoid or a magnetic tongue might be expected to have the observed transverse fields near the leading edge, the ultimate return of the IMF orientation to the prevailing spiral implies the continued influence of solar rotation and thus favors continued magnetic connection to the Sun. However, a CME which disconnected after its leading edge had propagated far into interplanetary space might have a magnetic field profile similar to that of an attached tongue. The dominance of the outward beam near the CME trailing edges is consistent with regulation of the beams by scattering proportionate to travel distance from the corona; we would expect the field-aligned path lengths in opposite directions to be most dissimilar near the trailing edge of a magnetic tongue. We find no evidence suggesting that CMEs are disconnected plasmoids, and tentatively conclude that the statistical results best support a tongue or connected flux rope model.

The strahl-on-strahl distributions discovered in this study offer a new opportunity for use of electrons as tracers of magnetic structure. Our tentative hypothesis is that the SOS spectra represent a superposition of a narrow, pristine halo beam upon a broader, scattered halo beam, requiring continued magnetic connection to the Sun. More extensive analysis of beam widths in CMEs, the "open" solar wind, and SOS distributions, will be necessary for assessment of this hypothesis.

We note in closing that if CMEs remain fully magnetically attached to the corona, as in a simple tongue, or partially attached in a flux rope, they open new magnetic flux in interplanetary space. To avoid a monotonic increase in the magnitude of the IMF, some mechanism must return flux to the Sun. One candidate for this mechanism is reconnection above coronal streamers, which may lead to coronal disconnection and the return of closed field arches /4/.

ACKNOWLEDGMENTS

Magnetic field data used in this study were provided by E.J. Smith, via NASA/NSSDC. Work was performed under the auspices of the U.S. Department of Energy, and was supported by the NASA Guest Investigator and Supporting Research & Technology programs and by the Los Alamos National Laboratory Directed Research and Development program.

REFERENCES

1.   J.T. Gosling, D.N. Baker, S.J. Bame, W.C. Feldman, R.D. Zwickl, and E.J. Smith Bidirectional solar wind electron heat flux events, *J. Geophys. Res.* 92, 8519 (1987).

2. J.T. Gosling, Coronal mass ejections and magnetic flux ropes in interplanetary space, in: *Physics of Magnetic Flux Ropes*, ed. C.T. Russell, E.R. Priest, and L.C. Lee, American Geophysical Union, Washington, 1990, p. 343.

3. S.J. Bame, J.R. Asbridge, H.E. Felthauser, J.P. Glore, H.L. Hawk, and J. Chavez, ISEE-C solar wind plasma experiment, *IEEE Trans. Geosci. Electron.* GE-16, 160 (1978).

4. D.J. McComas, J.L. Phillips, A.J. Hundhausen, and J.T. Burkepile, Disconnection of open coronal magnetic structures, these proceedings.

OBSERVATION OF LOCAL RADIO EMISSION ASSOCIATED WITH TYPE III RADIO BURSTS AND LANGMUIR WAVES

M. J. Reiner, R. G. Stone and J. Fainberg

NASA/GSFC, Greenbelt, MD 20771, U.S.A.

ABSTRACT

The first clear detection of fundamental and harmonic radiation from the type III radio source region is presented. This radiation is characterized by its lack of frequency drift, its short rise and decay times, its relative weakness compared to the remotely observed radiation and its temporal coincidence with observed Langmuir waves. The observations were made with the radio and plasma frequency (URAP) receivers on the Ulysses spacecraft between about 1 and 2 AU from the Sun.

INTRODUCTION

Type III solar radio bursts are remotely observed radio emissions generated by suprathermal electrons which are injected into the interplanetary medium above active regions on the Sun, often in association with solar flares or subflares, and which propagate through the interplanetary medium along interplanetary magnetic field lines. We analyze here type III radio emissions observed by the Ulysses radio receiver /1/ in the frequency range between 1.25 kHz and 940 kHz, which corresponds to heliocentric distances from about 20 $R_\odot$ to about 2.5 AU.

Type III radio emission is characterized by a frequency drift, i.e., the onset of the radio emission occurs later for lower frequencies. The frequency drift rate varies from about 10 kHz/s near 1 MHz to about 0.01 kHz/s at 50 kHz /2/. The radio emission for most type III bursts does not extend to frequencies below about 50 kHz. However, the Ulysses observations have shown that the radio emission for many intense type III bursts extends to frequencies as low as 20 or 30 kHz, and in rare cases even lower.

It is now well established that type III radiation is generated from a plasma emission process. This is already indirectly indicated by the observed frequency drift. As the beam electrons move out from the Sun, they encounter decreasing plasma densities and therefore emit at lower frequencies.

The plasma emission mechanism is a two step process /3/. First, as the faster electrons outpace the slower ones, a bump-on-tail is formed in the electron parallel velocity distribution function. This bump-on-tail instability leads to the growth of Langmuir waves. A second process then converts the Langmuir wave energy into the observed transverse radio waves. Fundamental emission may be produced by the scattering of the Langmuir waves off thermal ions or low frequency waves. Second harmonic radiation may be produced by the coalescence of two Langmuir waves; third harmonic emission, if it exists, by the coalescence of a transverse second harmonic wave with a Langmuir wave /4/.

Although this theory is not currently in a completely satisfactory state, a number of investigators have attempted to check the basic validity of the mechanism. The idea is to simultaneously observe the electrons (including the formation of the bump-on-tail in the electron velocity distribution), the Langmuir waves, and the radio waves in the type III source region. Spacecraft observations have established the causal relationship between observed electron bursts and Langmuir waves, as well as the time coincidence between the formation of the bump-on-tail in the electron parallel velocity distribution function and the onset of the Langmuir waves /5,6/. However, there has always been a discrepancy between the Langmuir waves and radio emission observed at the frequency of the second harmonic. Second harmonic radio emission has been observed to precede the onset of the Langmuir waves by some 20 to 30 minutes, suggesting that this radio emission is not produced in the vicinity of the spacecraft. However, because of the relatively high ambient densities during these observations, weak local radio emission generated at the second harmonic may have been overwhelmed by the radiation generated remotely.

ULYSSES OBSERVATIONS OF LOCAL TYPE III EMISSION

Observations of radio emission made when the Ulysses spacecraft was in the radio source region are better able to distinguish local radiation for two reasons.

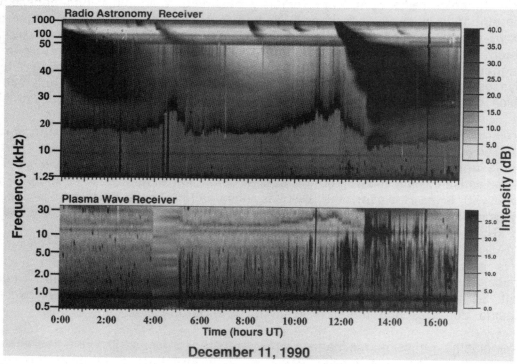

Fig. 1.   Dynamic spectra of the Ulysses URAP data for 11 December 1990. The upper dynamic spectrum is the radio receiver data covering the frequency range from 1.25 kHz to 940 kHz. The lower dynamic spectrum is the plasma frequency receiver data covering the frequency range from 0.57 kHz to 35 kHz.

First, the low frequency receiver on Ulysses provides continuous frequency coverage below 50 kHz, with a bandwidth of only 750 Hz. Secondly, the Ulysses spacecraft, as it is moves farther from the Sun, encounters regions of lower plasma density (frequency) where the intensity of the remote type III radio emission is greatly reduced, thereby exposing potentially weaker locally generated radio emission.

This is illustrated by the observed type III radio burst that commenced at 11:35 UT on 11 December 1990, shown by the dark drifting feature in the dynamic spectrum in Fig. 1. This intense type III was associated with a 1N flare which was located about 53º to the west of the Ulysses-Sun line, so that the interplanetary magnetic field lines from above this flare site should have passed over the Ulysses spacecraft. Thus the burst electrons should have eventually arrived at Ulysses. Indeed, an intense burst of Langmuir waves was observed in the plasma frequency receiver data from about 13:00 UT to 14:10 UT, signaling the arrival of the burst electrons. These Langmuir waves were centered near 13.6 kHz, the ambient plasma frequency at the spacecraft at the time of the observation.

This type III radio burst exhibited a normal frequency drift down to about 30 kHz. This drift is more visible in the intensity-time plots shown in Fig. 2 for selected observing frequencies.

An abrupt change is evident in the character of the observed radiation for frequencies below 30 kHz. There is no evident frequency drift in the observed radiation between 30 kHz and the plasma frequency near 14 kHz. Furthermore, the rise and decay times are much shorter than for the radiation at the higher frequencies, contrary to what is expected /2/. The rise time of only about 20 minutes is independent of frequency. Finally, this non-drifting radiation is in precise time coincidence with the intense Langmuir wave bursts observed in the plasma frequency receiver at 13.8 kHz (shown in the bottom line plot).

We interpret this non-drifting radiation as locally generated radio emission associated with the type III burst observed at higher frequencies and the Langmuir waves observed *in situ*. The time profiles of this local emission presumably represent the basic type III emission profiles from an elemental spatial volume /7/.

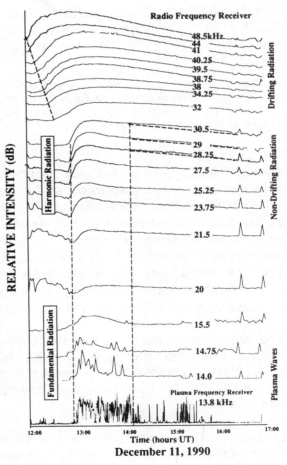

Fig. 2. Intensity versus time plots for selected frequencies from the radio data shown in Fig. 1. The values of the frequency are given on the right. The lowest line plot shows the peak intensities in the plasma frequency receiver at 13.8 kHz.

The fundamental/harmonic structure of this radio emission is clearly revealed in the frequency spectrum at 13:20 UT shown in Fig. 3. The error bars shown on the data points represent an estimated 10% uncertainty in the measurement of the radio flux.

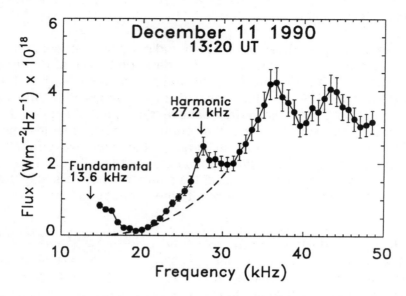

Fig. 3. The frequency spectrum at 13:20 UT for the radio burst on 11 December 1990. The fall off of the remote emission is indicated by the dashed line. Enhancements in the radio emission are evident near the fundamental (13.6 kHz) and the harmonic (27.2 kHz) of the local plasma frequency.

As the intense remote emission rapidly falls off below 30 kHz, clear peaks are observed in the radio emission at the plasma frequency of 13.6 kHz and at 27.2 kHz, corresponding to the second harmonic of the plasma frequency. The intensity of the harmonic emission in this example is about 1.2×10^{-18} $Wm^{-2}Hz^{-1}$ above the remote emission component; the intensity of the fundamental is estimated to be 0.8×10^{-18} $Wm^{-2}Hz^{-1}$. The ratio of harmonic to fundamental intensity is about 1.5. The intensity of the remote radio emission observed at higher frequencies is 4×10^{-18} $Wm^{-2}Hz^{-1}$, so the intensity of the local fundamental emission represents about 1/5 of the remote emission.

The rather large bandwidth of the observed fundamental and harmonic emission results from the bandwidth of the receiver, which samples radiation from plasma densities over an extended spatial region, but more significantly from the small scale fluctuations of the plasma density in the vicinity of the spacecraft.

Clearly, the lower the ambient plasma density, the more potential for observing the harmonic structure of locally generated radio emission. For a type III burst that was observed on 22-23 December 1990, the *in-situ* plasma frequency decreased from about 10 kHz at 00:50 UT to about 8.7 kHz at 01:30 UT (as determined from the frequency spectrum of the observed Langmuir waves). The frequency spectra of the local radio emission observed at these times showed a corresponding shift in the harmonic structure to lower frequencies. The frequency spectrum of the radio emission observed at 01:30 UT, shown in Fig. 4, shows two peaks, centered at about 17.5 kHz and at 26 kHz, below the cut off of the intense remote radio emission.

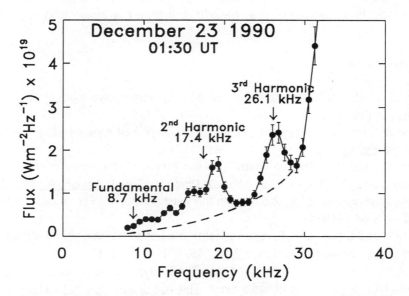

Fig. 4. The frequency spectrum at 01:30 UT for the radio burst on 23 December 1990. The fall off of the remote emission is indicated by the dashed line. Enhancements in the radio emission are evident near the second harmonic (17.4 kHz) and the third harmonic (26.1 kHz) of the local plasma frequency (8.7 kHz).

Since the frequencies of these two peaks are in the ratio 2:3, it is natural to interpret them as corresponding to local emission at the second and third harmonic of the *in situ* plasma frequency. This is the first evidence for third harmonic emission associated with kilometric type III radio bursts. Alternatively, the peak at 26 kHz could represent second harmonic emission produced in a density enhanced region near the spacecraft. Since the *in-situ* plasma frequency is not observed to go significantly above 10 kHz during this entire period, we believe that this interpretation is less likely.

Discussion

After examining hundreds of type III radio bursts observed by the Ulysses URAP experiment from November 1990 through April 1991, we found that even though many can be established to lie far to the west of the Ulysses-Sun line, clear evidence of locally produced radio emission and associated Langmuir waves are rare. During this time, we found only three good examples. This suggests that type III radio emission is generated in localized regions in the interplanetary medium, rather than uniformly along the entire extent of the electron exciter beam. It is only in rare cases, like those illustrated here, when the spacecraft happens to lie in or very near such a region that local Langmuir and radio waves are simultaneously observed.

Conclusion

We have presented evidence for the observation of local radio emission at the fundamental and harmonic of the *in situ* plasma frequency in direct time association with observed Langmuir waves. These observations provide the missing causal link between the *in-situ* Langmuir waves and the type III radio emission.

References

1. R.G. Stone, *et al.*, The Ulysses unified radio and plasma wave investigation, Astron. Astrophys., In Press.
2. J. Fainberg and R.G. Stone, Satellite observations of type III solar radio bursts at low frequencies, Space Sci. Rev. **16**, 145-188 (1974).
3. V.L. Ginzburg and V.V. Zheleznyakov , On the Possible Mechanisms of Sporadic Solar Radio Emissions (Radiation in an Isotropic Plasma), Soviet Astron.-AJ **35**, 653 (1958).
4. V.V. Zheleznyakov and E.Ya. Zlotnik, On the third harmonic in solar radio bursts, Solar Phys. **36**, 443-449 (1974).
5. D.A. Gurnett and L.A. Frank, Electron plasma oscillations associated with type III radio emission and solar electrons, Solar Phys. **45**, 477-493 (1975).
6. R.P.Lin, D.W. Potter, D.A. Gurnett, and F.L. Scarf, Energetic electrons and plasma waves associated with a solar type III radio burst, The Astrophys. Journal **251**, 364-373 (1981).
7. M.J. Reiner and R.G. Stone, Model interpretation of type III radio burst characteristics II. Temporal aspects, Astron. Astrophys. **217**, 251-269 (1989).

A QUANTITATIVE ASSESSMENT OF THE ROLE OF THE POST-SHOCK TURBULENT REGION IN THE FORMATION OF FORBUSH DECREASES

T. R. Sanderson,* A. M. Heras,* R. G. Marsden* and D. Winterhalter**

* Space Science Dept of ESA, ESTEC, Noordwijk, The Netherlands
** Jet Propulsion Laboratory, California Institute of Technology, Pasadena, CA, U.S.A.

ABSTRACT

We present results of a survey of the relation between Forbush decreases, magnetic clouds, and interplanetary shocks during the period August 1978 to November 1982. We have used data from the ISEE-3 study of bi-directional ions associated with magnetic structures or clouds of Marsden et al. /1/, and ground based observations of Forbush decreases from several neutron monitors. We use the two step model of a Forbush decrease. We assume that the first step is due to the passage of the post-shock turbulent region, and that the second is due to the passage of the magnetic cloud or structure which usually follows the post-shock turbulent region. To determine the effectiveness of the post-shock turbulent region in causing a Forbush decrease, we have evaluated the radial diffusion coefficient of the post-shock turbulent region for the eight largest events during the above period using observations of the magnetic field. We have made a quantitative assessment of the relative importance of the post-shock turbulent region in the formation of the Forbush decrease, concluding that the post-shock turbulent region alone is not sufficient to cause a Forbush decrease.

INTRODUCTION

Although there has been much discussion over the years as to the cause of Forbush decreases, it is only recently that satellite data have been used to study in-situ the interplanetary medium at the time of occurrence of the Forbush decreases observed in earth-based neutron monitors. Two candidates have been studied extensively, namely (1) the turbulent post-shock region which often follows the passage of an interplanetary shock, and (2) the magnetic cloud or coronal mass ejection which often follows the passage of the post-shock region. In a previous analysis we selected the eight most intense Forbush decreases from a data set of 31 events /2/. All eight of these Forbush decreases could be considered as having two steps, the first step commencing at the time of passage of the start of the post-shock region, and the second step commencing at the time of passage of the start of the magnetic cloud or structure. In some cases the first step was essentially absent since the post-shock region was not sufficiently turbulent to affect the propagation of the cosmic rays, whilst in some other cases the second step was essentially absent, since the magnetic cloud was not effective enough to cause a decrease.

In this paper we examine in detail the scattering characteristics of the post-shock region by evaluating the diffusion coefficient of the interplanetary medium during the time of passage of the post-shock region for the eight largest Forbush decreases associated with interplanetary shocks observed during the period August 1978 to November 1982.

MAGNETIC FIELD AND SOLAR WIND CHANGES

We first examined the relevant upstream and downstream solar wind velocities and magnetic field magnitudes, which according to Nishida /3/ and Kadokura and Nishida /4/ are important parameters in the determination of the size of a Forbush decrease. Table 1 shows the relevant upstream (suffix 1) and downstream (suffix 2) solar wind velocities (in km/s) and magnetic field magnitudes (in nT) in descending order of the size of the Forbush decrease, d, for the eight events.

Table 1. *Upstream and downstream solar wind and magnetic field parameters for the shocks and the size of the Forbush decrease, d, for the eight events.*

Event	Day	Year	V_1	V_2	V_2/V_1	δV	B_1	B_2	B_2/B_1	δB	d (%)
1	031	1982	350	550	1.7	200	8.4	15.8	1.9	7.4	8.0
2	206	1981	375	500	1.33	125	6.5	19.5	3.0	13	7.5
3	042	1982	350	500	1.4	150	8	15	1.9	7	6.1
4	114	1982	440	650	2.2	110	4	9.5	2.4	5.5	6.0
5	272	1978	640	690	1.1	50	8	10	1.2	4	5.7.
6	095	1979	410	410	1.46	180	13	28	2.2	15	5.0
7	316	1978	400	570	1.43	170	8	18	2.3	10	4.5
8	079	1980	gap	-	-	-	-	-	-	-	4.5

We found no significant correlation between the size of the Forbush decrease and either (1) the ratio of the downstream to upstream solar wind velocity, (2) the jump in solar wind velocity at the shock, (3) the ratio of downstream to upstream magnetic field, or (4) the magnetic field jump at the shock. This suggests that none of the above are important parameters which are related to the magnitude of the decrease.

RADIAL DIFFUSION COEFFICIENT

Using 3-second resolution magnetic field data from the magnetometer on ISEE-3 we have evaluated the radial diffusion coefficient of the interplanetary medium as a function of time for each of the eight events and compared it with the intensity profile of the corresponding Forbush decrease. We have derived 1-hour values of the vector sum, P_0 of the noise power at 1 Hz of the three components of the magnetic field, and the slope q of the noise power spectra. We have evaluated the diffusion coefficient parallel to the magnetic field $K_\parallel$ for a 1 GeV proton from quasi-linear theory using the slab model for magnetic field fluctuations /5/

$$K_\parallel = v^{3-q} V^{q-1} B^2 \omega^{q-2} / (2\pi)^q P_0 (4-q)(2-q)$$

where v is the particle velocity, V is the solar wind velocity, B is the magnitude of the magnetic field, and ω is the gyro frequency of the particle. We evaluate the perpendicular diffusion coefficient, $K_\perp$ from

$$K_\perp = v^4 / 9\omega^2 K_\parallel$$

For the purpose of determining the effectiveness of different regions in producing a Forbush decrease, we assume that the Forbush decrease is produced by a radially outward moving slab, and therefore we have evaluated the radial diffusion coefficient

$$K_r = K_\parallel cos^2 \psi + K_\perp sin^2 \psi$$

where ψ is the angle between the radial direction and the direction of the magnetic field. Although these formulae do not give an accurate absolute measure of the diffusion coefficient when the level of turbulence is high, they are still able to be used to give a reasonable estimate of the changes in the diffusion coefficient.

Figure 1 shows one-hour averages of K_r as a function of time for the eight events, for a 48-hour sample which includes both the passage of interplanetary shock and the start of the magnetic cloud. In each panel vertical lines are drawn to show the time of arrival of the shock and the time of arrival of the magnetic cloud. The gaps are periods when either no data were available from ISEE-3, or the calculation of the diffusion coefficient gave a negative value (i.e. when the slope of the magnetic field power spectrum $q \geq 2$).

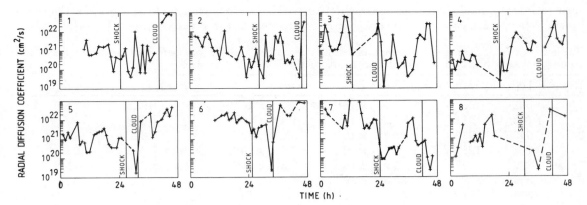

Figure 1. *One hour averages of the radial diffusion coefficient plotted as a function of time for the eight events in Table 1. Also shown, with vertical lines, are the times of arrival of the interplanetary shock, and the magnetic cloud.*

We find a reasonable agreement between the behaviour of the radial diffusion coefficient as a function of time, and the Forbush decrease profiles. Events 1, 2, 3, 5, and 6 show little or no change in the neutron monitor counting rate when the shock or the post-shock region passes the earth. There is little or no change in the radial diffusion coefficient at the time of arrival or passage of the post-shock turbulent region. Event 3, Figure 2a, is a typical example.

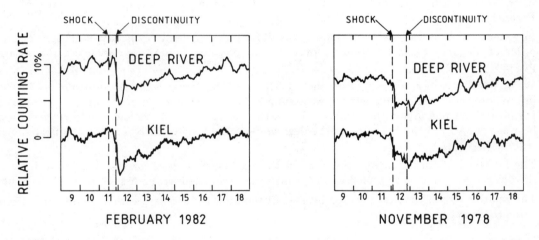

Figure 2. *(a) Neutron monitor counting rates for event 3, February 11 1982. The Forbush decrease is observed only when the magnetic cloud passes the earth. (b) Neutron monitor counting rates for event 7, November 12, 1978. The Forbush decrease is observed when the post-shock region passes the earth, and continues whilst the magnetic cloud passes the earth.*

In Event 7 a Forbush decrease commences around the time of passage of the shock, as shown in Figure 2b, and at the same time a large change is observed in the radial diffusion coefficient. In this event, (November 12, 1978) a large and sudden change in the diffusion coefficient at the time of arrival of the shock is observed. A Forbush decrease commences rapidly almost at the time of arrival of the interplanetary shock. The increase in turbulence, which we measure by a large change in the radial diffusion coefficient, is most likely responsible for the first step of this Forbush decrease. All other events have only small changes in the diffusion coefficient, and small onsets at the time of shock passage. In event 7, the ratio of the upstream to downstream diffusion coefficient changes by a factor of 77, and corresponds to a initial drop in the Deep River counting rate of 3.2 %.

Table 2 shows the values of the upstream and downstream diffusion coefficients, K_{r1} and K_{r2}, and their ratio, calculated for the six-hour period upstream and downstream of the shock, for the eight events. Where data gaps exist the six-hour period nearest to the shock is taken.

Table 2. *Six-hour averaged upstream and downstream radial diffusion coefficients, ratio of upstream to downstream diffusion coefficients, and size of the Forbush decrease.*

Event	Day	Year	K_{r1} (cm^2 s^{-1})		K_{r2} (cm^2 s^{-1})		K_{r1}/K_{r2}	d (%)
1	031	1982	1.9	x 10^{20}	3.6	x 10^{20}	0.5	8.0
2	206	1981	5.3	x 10^{20}	1.4	x 10^{20}	3.8	7.5
3	042	1982	36.8	x 10^{20}	2.6	x 10^{20}	14.1	6.1
4	114	1982	3.0	x 10^{20}	0.9	x 10^{20}	3.5	6.0
5	272	1978	6.5	x 10^{20}	3.3	x 10^{20}	2.0	5.7
6	095	1979	50.3	x 10^{20}	24.4	x 10^{20}	2.1	5.0
7	316	1978	38.9	x 10^{20}	.5	x 10^{20}	77.0	4.5
8	079	1980	23.4	x 10^{20}	1.3	x 10^{20}	17.8	4.5

The events in Table 2 are ordered in descending order of the size of the Forbush decrease, *d*. We do not observe a similar ordering of the radial diffusion coefficient in the post-shock turbulent region for these events, which leads us to conclude that the turbulence in the post-shock region alone is not sufficient to produce a Forbush decrease.

CONCLUSIONS

1. We find no obvious correlation between the change in the solar wind velocity or the change in the magnetic field magnitude at the time of the passage of the shock and the size of the Forbush decrease.

2. In general we observe large variations in the diffusion coefficient. Typically, variations of one order of magnitude can be observed over periods of a few hours, even when no shocks are present.

3. We do not find a correlation between the overall size of the decrease and the diffusion coefficient in the post-shock region. However, if we assume a 2-step decrease, where the first step is due to the post-shock region and the second is due to the magnetic cloud, we find a better correlation between the size of the first step of the decrease and the level of turbulence in the post-shock region.

4. The highest level of turbulence observed in the post-shock region corresponded to a drop in the diffusion coefficient by a factor of 77, which was associated with a first step of 3.2 per cent.

5. The relative size of the first and second step of the Forbush decrease varies considerably from event to event.

REFERENCES

1. Marsden, R. G., T. R. Sanderson, C. Tranquille, K.-P. Wenzel, and E. J. Smith, ISEE-3 observations of low energy proton bidirectional events and their relation to isolated interplanetary magnetic structures, *J. Geophys. Res.*, 92, 11009 (1987)

2. Sanderson, T. R., J. Beeck, R. G. Marsden, C. Tranquille, K.-P. Wenzel, R. B. McKibben, and E. J. Smith, A study of the relation between magnetic clouds and Forbush decreases, *Proc 21st ICRC*, Adelaide, 6, 251 (1990)

3. Nishida, A, Numerical evaluation of the precursory increase to the Forbush decrease expected from the diffusion convection model, *J. Geophys. Res., 87*, 6003, (1982)

4. Kadokura, A., and A. Nishida, Two-dimensional numerical modelling of the cosmic ray storm, *J. Geophys. Res.* 91, 13-29 (1986)

5. Hasselmann, K., and G. Wibberenz, Scattering of charged particles by random magnetic fields, *Z. Geophys. 34*, 353 (1968)

COMPARISON OF 2 1/2 D AND 3D SIMULATIONS OF PROPAGATING INTERPLANETARY SHOCKS

Z. K. Smith, T. R. Detman and M. Dryer

NOAA, Space Environment Laboratory, Boulder, CO 80302, U.S.A.

ABSTRACT

We present results from a parametric study of interplanetary shocks made using a 3D, time-dependent MHD numerical simulation code. The results are compared to those obtained from a 2½D code and also to spacecraft data. The simulations model the propagation within 1 AU of interplanetary shocks that are produced by relatively short-duration and modest-width inputs at 18 $R_\odot$. The 3D results are qualitatively the same as those of the 2½D study but quantitatively different. When simulations are compared to Prognoz spacecraft data it is seen that both cover the same range of parameters.

INTRODUCTION

As computer capabilities have expanded, numerical models have increased in complexity. In this study, models were used to gain a better understanding of the evolution of solar–generated interplanetary (IP) shocks as they propagate out to 1 AU. The usefulness of the results is verified by comparing them to both *in situ* and ground based observational data. We also investigate the effect of increasing the model dimensions from 2½D to 3D, and compare simulation results to data.

SIMULATIONS

The 3D code described by Han *et al.* /1/ is used to perform the parametric study. The evolution of IP shocks as they propagate out to 1 AU is modeled for a variety of inputs. These inputs are initiated at the inner boundary of the numerical grid at 18 $R_\odot$ in a manner described below.

This study is analogous to that described in Smith and Dryer /2/, who used a 2½D code. The initial steady state required by the code is similar to that used in /1/. It has an axially symmetric magnetic field in the form of a rotating unipole and, thus, has the form of an archimedian spiral in the equatorial plane. The values of the parameters used in this study are given in Table 1.

The computational domain used is a sector of 90° in latitude and longitude; the grid size is 2 $R_\odot$ in radius, 3° in meridional and azimuthal angles. The properties of the input pulse and the range of input values used are as described in /2/ for *modest* shocks, namely shocks with relatively short durations (< 2.5 hr) and narrow widths, (< 60°). In the 3D case, the latitudinal shape must also

TABLE 1. Initial Steady-state Solar Wind Parameters

Parameter		In the equatorial plane 18 $R_\odot$	1 AU	45 deg above the equatorial plane 18 $R_\odot$	1 AU
V_r	(km/s)	250	362	250	362
V_θ	(km/s)	0	0	−.8	−.1
V_ϕ	(km/s)	13	1	9.4	0.7
B_r	(nT)	150	1	150	1.0
B_θ	(nT)	0	0	−.5	−.0
B_ϕ	(nT)	−8	−.9	-5.6	−.65
n	(cm^{-3})	1400	6.6	1400	6.4
T	(deg kK)	1100	30.3	1100	30.4

be specified. It is defined analogously to the longitudinal shape as a sine curve of full width ω (= ω_θ = ω_ϕ). The radius vector of the pulse center, marked as the Pulse Central Meridian (PCM), is chosen to lie in the ecliptic plane in the center of the grid. The total net energy added by each input pulse to the solar wind is again computed as the sum of the kinetic, thermal, and magnetic energies. We show that the exact nature of the input pulse is of minor importance, by plotting the 1AU properties of the shocks as a function of the added energy. The properties of the shocks at 1 AU are examined analogously to /2/ in the meridional and ecliptic planes.

RESULTS

Figure 1 shows the dependence of the 1 AU shock properties as a function of energy for five angular distances of the observer from PCM in the ecliptic plane. These properties are transit time to 1 AU and jump in dynamic pressure (which is used to characterize shock strength) at 1 AU. These plots show that the net input energy is the organizing factor for this set of shocks, both for transit time and strength. For transit time, except for the weakest of these shocks, all the points fall close to a straight line; that is, the transit time has a power-law dependence with input energy. The exponent of –¼ is close to the theoretical value of –1/3 for a spherical, constant velocity, piston driven shock. The bend towards the horizontal for low energies is due to the fact that no outward-propagating disturbance will travel more slowly than the sum of the wave and convective speeds of the solar wind. This energy relationship also holds at angles other than PCM, with decreasing scatter for transit time and increasing scatter for the dynamic pressure jump. These shocks are nearly symmetric about PCM. The slight asymmetry (shocks being faster and stronger when observed to the west of PCM) is due to the archimedian spiral of the background magnetic field.

The 3D shock simulations in the meridional plane were analyzed analogously to those for the ecliptic plane, with almost identical results. The difference between the results in the two planes is that in the meridional case there is symmetry about the equator. This is to be expected because of the choice of initial conditions, which are symmetric about the equatorial plane and in which the main departure from symmetry is due to solar rotation. This situation may apply to conditions near solar maximum as suggested by IPS velocity, K-coronameter, and Stanford source surface maps /3/.

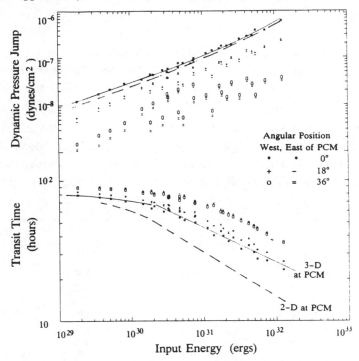

FIG. 1. 3D Forward Shock Properties in the Equatorial Plane at 1 AU.

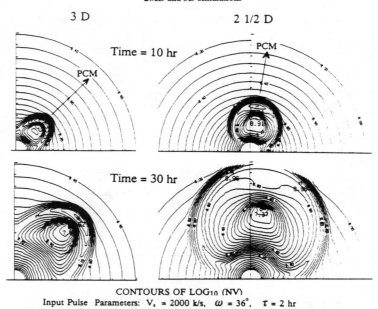

CONTOURS OF LOG₁₀ (NV)
Input Pulse Parameters: V, = 2000 k/s, ω = 36°, τ = 2 hr

FIG. 2. Comparison of 2½D and 3D Shock Evolution in the Equatorial Plane.

The 3D results were then compared to those of the 2½D study. Figure 2 shows the evolution of a typical shock in the ecliptic plane as modeled by the two codes. The IP shocks are qualitatively the same in their spatial development as they propagate out to 1 AU. Their 1 AU properties are also qualitatively similar in that they both depend primarily upon input energy and angle of the observer from PCM.

Quantitatively, the results of the two studies differ in two ways. Figure 2 shows that the 3D IP shocks are narrower in angular extent. This is primarily due to the geometric effect due to the fact that, in the 2½D code, quantities (including the input pulse) have no latitudinal dependence (the latitudinal derivatives are all essentially zero). The more-collimated shock shapes of the 3D code are also in agreement with the IPS results of Wei and Dryer /5/ and Pinter /6/. Note also the difference in the background steady state used in the 2½D and 3D studies. The most influential difference is that the density of the 3D steady state is about double that of the 2½D. However, the density ratio is identical across a given shock. We do not expect the effect of the background density difference to be large because the retardation due to the increased background density is compensated for by the increased shock strength, and because the energy density in the shock front far exceeds that of the background for all but the weakest shocks modeled here.

The second quantitative difference is illustrated in Figure 1 for the PCM direction. Here the dashed lines show PCM values from the 2½D results; the 3D results are represented by the solid curves. This emphasizes that the dependence of transit time and strength on energy obtained from the two models differs somewhat in slope and intercept. These differences can be partially explained by the methods used to calculate input–pulse area in the codes. In the 3D code the area is computed directly, using the longitudinal and latitudinal widths, ω_θ, ω_ϕ. In the 2½D code the area is directly proportional to longitudinal width ω_ϕ, as ω_θ is taken as constant with a value of 0.5 radian (~30°). However, the fundamental difference is the additional dimension into which energy flux diverges in the 3D code.

The 3D results were also compared to 10 data points published by Zastenker et al. /4/, who obtained values for dynamic pressure jumps (DPJ) for interplanetary shocks from the Prognoz 7, 8, and 10 spacecraft. They identified the solar sources for these shocks and thus obtained estimates

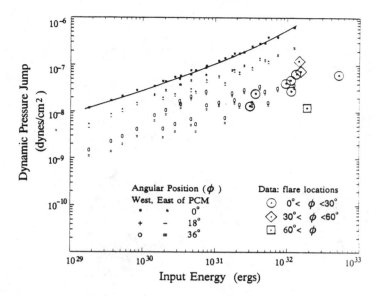

FIG. 3. Comparison of 3D Code Results to Prognoz Data.

of the source energy and angular distance ϕ between the observer and the source. Their energies, however, included the potential energy contribution.

The points are superimposed on the 3D simulation results of DPJ vs. energy in Figure 3. All but 2 of the 10 data cases are well ordered in ϕ. The two points in the range $30° < \phi < 60°$ lie above the points for the smaller ϕ values. The solid curve drawn through the simulated points for PCM direction represents the theoretical upper limit for DPJ (see Smith and Dryer [2]). Thus we expect all data to fall below this curve, and indeed they all do. That they fall well to the lower right of the simulations can be explained in part by the difference in energy calculation in simulation and data.

CONCLUSIONS

The results of the 3D MHD code indicate that input pulses with the range of input parameters used here produce fast forward shocks with a large angular extent and nearly axial symmetry about the PCM direction. The departure from symmetry in longitude is a result of the background magnetic field's archimedian spiral.For our range of input ω and τ, the net input energy, and not the exact nature of the input pulse, determines forward shock properties such as transit time, strength, and angular extent at 1 AU.

Comparison of 2½D and 3D results show that they are qualitatively the same. Quantitatively they differ in that the 3D simulated interplanetary shocks are more collimated than are the 2½D shocks. They also differ in the power-law dependence of the 1 AU shock properties on energy. These differences may be explained by the divergent geometric considerations inherent in the 2½D approximation and by the assumptions made in computing the area of pulse input for the 2½D code.Comparison of the 3D simulation to Prognoz spacecraft data shows that the parameter ranges obtained from the numerical simulations cover the same range as those observed.

REFERENCES

1. Han, S. M., S. T. Wu and M. Dryer, *Computers and Fluids*, **16**, 81, 1987
2. Smith, Z. and M. Dryer, *Solar Phys.* **129**, 387, 1990
3. Rickett, B. J. and W. A. Coles, *JGR* **96**, 1715, 1991.
4. Zastenker, G., Yu. I. Yermolaev, V. I. Zhuravlev, N. L. Borodkova, Z. Nemecek and Ya. Safrankova, *Adv. Space Res.*, **9** (4), 117, 1989
5. Wei, F. S. and M. Dryer, *Solar Phys.* **132**, 373, 1991
6. Pinter, S., *Space Sci. Rev.* **32**, 145, 1982

MAGNETIC CLOUDS: COMPARISON BETWEEN SPACECRAFT MEASUREMENTS AND THEORETICAL MAGNETIC FORCE-FREE SOLUTIONS

M. Vandas,* S. Fischer,* P. Pelant* and A. Geranios **

* Astronomical Institute, CSAS, 120 23 Prague, Czechoslovakia
** University of Athens, Physics Department, 157 71 Athens, Greece

ABSTRACT

Several theoretical magnetic force-free solutions of magnetic clouds (cylindrical, spheroidal) and their differences are discussed. The cloud diameters following from the fit are generally larger for spherical model than for cylindrical one. From the plasma data we try to estimate cloud boundaries as an independent check and compare them with model cloud boundaries in the magnetic field data.

INTRODUCTION

Magnetic clouds are specific regions in the solar wind which exhibit the following properties: (1) the magnetic field direction rotates smoothly through a large angle during an interval of the order of one day; (2) the magnetic field strength is higher than in surroundings; and (3) the temperature is lower than average /1, 2/. The rotation of magnetic field and the temperature decrease inside clouds indicate that magnetic clouds represent a closed magnetic field configuration which isolates them from ambient plasma to some extent and thus suppresses the heat transport. The topology of the magnetic field inside clouds is under intensive examinations now. There are two basic concepts of a magnetic cloud shape, a loop maybe rooted at the Sun, and closed plasmoid (bubble).

Burlaga et al. /2, 3/ have suggested a cylindrical model as a basis for the comparison with measured data. The physical reason for this was that a magnetic cloud as a loop can locally be described with the cylindrical solution. They have brought several indices in favour of a loop configuration for a few cases /3, 4/ and have reached quite good results in matching the angles of the magnetic field vector (ϑ to the ecliptic plane, φ in the ecliptic plane) but fits of magnetic field magnitude (B) are not very good. Magnetic field profiles are often non-symmetric and the ratio of maximum to boundary magnitude sometimes exceeds 2 contrary to the cylindrical solution. These facts have led us to examine other solutions, namely spheroidal ones (representing plasmoid). Spheroidal models are based on a solution in /5/ and are described in a detail in /6/ or in an extended version of this paper submitted to *Journal of Geophysical Research*. Spheroidal clouds have an z-axial symmetry with the corresponding semi-axis c and the other semi-axes a, the oblateness is $\epsilon = (a - c)/a$; we speak about oblate ($\epsilon > 0$), spherical ($\epsilon = 0$), or prolate ($\epsilon < 0$) clouds.

COMPARISON BETWEEN MEASURED MAGNETIC FIELD PROFILES OF CLOUDS AND MODELS

We have shown /7/ that the spherical solution gives the larger ratio of maximum to boundary magnitude and slightly non-symmetric profile. The course of the angles ϑ, φ is similar to the cylindrical case but the fit of B is sometimes better. The cloud diameters are generally larger than for the cylindrical case (Fig. 1).

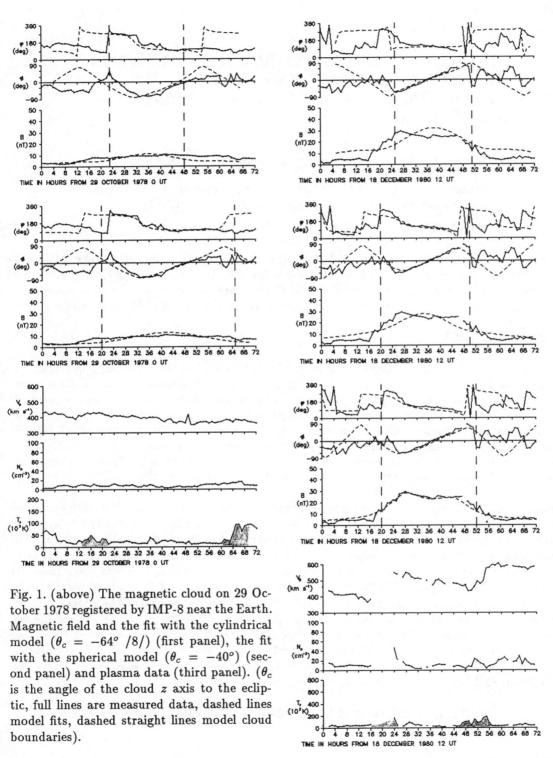

Fig. 1. (above) The magnetic cloud on 29 October 1978 registered by IMP-8 near the Earth. Magnetic field and the fit with the cylindrical model ($\theta_c = -64^\circ$ /8/) (first panel), the fit with the spherical model ($\theta_c = -40^\circ$) (second panel) and plasma data (third panel). (θ_c is the angle of the cloud z axis to the ecliptic, full lines are measured data, dashed lines model fits, dashed straight lines model cloud boundaries).

Fig. 2. (right) The magnetic cloud on 18 December 1980 registered by IMP-8 near the Earth. Magnetic field and the fit with the cylindrical model ($\theta_c = -10^\circ$ /8/) (first panel), the fit with the spherical model ($\theta_c = -20^\circ$) (second panel), the fit with the oblate spheroidal model ($\theta_c = -20^\circ$, $\epsilon = 0.1$, i.e. $a : c = 10 : 9$) (third panel) and plasma data (fourth panel).

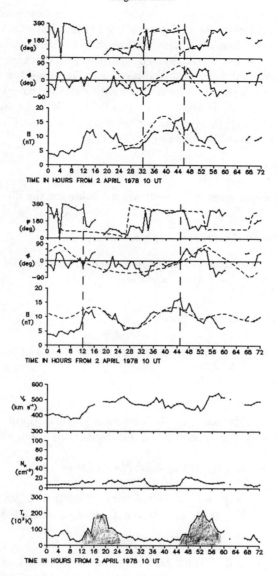

Fig. 3. The magnetic cloud on 2 April 1978 registered by IMP-8 near the Earth. Magnetic field and the fit with the cylindrical model ($\theta_c = 33°$ /8/) (first panel), the fit with the prolate spheroidal model ($\theta_c = 85°$, $\epsilon = -0.43$, i.e. $a : c \doteq 2 : 3$) (second panel) and plasma data (third panel). — A candidate for a prolate spheroidal cloud?

Introducing an oblateness we get a more pronounced non-symmetric profile. Fig. 2 shows the cylindrical and spherical fits. An oblate spheroid gives the best fit of B, even the non-symmetric profile of B with two maxima is reproduced.

For a prolate case the magnetic field can give a very distinct double-peak profile. Fig. 3 shows the first attempt to fit the data with a prolate spheroid. The fit is not ideal but double-peak structure is plausibly matched.

Lepping et al. /8/ have published a list of magnetic clouds and fits with the cylindrical model. We have tried to fit the listed clouds with the spherical model. This larger set confirm our conclusions given above /7/. Because the cloud diameters are quite different in many cases it would be good to have some independent determination of cloud boundaries, e.g. from plasma data, to distinguish between models. In some cases the cylidrical model only chooses a part of the magnetic field increase, apparently shorter then one would put boundaries by "eye and hand", i.e. to cover the full increase of magnetic field over the background level (Fig. 1). Defining the boundaries in such a way we noticed that these often coincide with irregular temperature increases. In the leading edge of clouds the increase is usually connected with the postshock region but an analogical or smaller increase can also be found near the trailing edge. We suggested to use these temperature increases for cloud physical boundaries because they indicate that "something is happening there". The spherical model includes mostly whole magnetic field peak and temperature increases are in a number of cases near or coincide with the model boundaries (see Figs. 1–3 where the related temperature increases are stressed in shadow).

CONCLUSIONS

- The spheroidal models of magnetic clouds are available for the direct comparison with measured data.

- Probably we have found signatures of spheroidal clouds.

- We suggest to take irregular temperature increases near the ends of the magnetic field increase as an indicator of the cloud boundaries.

ACKNOWLEDGEMENT

This work was supported by the grant no. 30309 of the Czechoslovak Academy of Sciences.

REFERENCES

1. L. Burlaga, E. Sittler, F. Mariani and R. Schwenn, Magnetic loop behind an interplanetary shock: Voyager, Helios and IMP-8 observations, *J. Geophys. Res.* **86**, 6673 (1981)

2. L. Burlaga, Magnetic clouds, *Preprint of LEP NASA* (June 1989)

3. L. Burlaga, R. Lepping and J. Jones, Global configuration of a magnetic cloud, in: *Physics of Magnetic Flux Ropes,* ed. E. Priest, L. Lee and C. Russel, AGU Geophysical Monograph 58, 1990, p. 373

4. I. Richardson, C. Farrugia and L. Burlaga, Energetic ion observations in the magnetic cloud of 14-15 January 1988 and their implications for the magnetic field topology, in: *Proc. 22nd ICRC,* Dublin 1991, in press

5. K. Ivanov and A. Harshiladze, Interplanetary spheromaks, *Geomag. Aeronomy* **25**, 377 (in Russian) (1985)

6. M. Vandas, S. Fischer and A. Geranios, Spheroidal models of magnetic clouds, in: *Proc. 1st SOLTIP Symposium,* Liblice 1991, in press

7. M. Vandas, S. Fischer and A. Geranios, Spherical and cylindrical models of magnetic clouds and their comparison with spacecraft data, *Planet. Space Sci.* **39**, 1147 (1991)

8. R. Lepping, J. Jones and L. Burlaga, Magnetic field structure of interplanetary magnetic clouds at 1 AU, *J. Geophys. Res.* **95**, 11957 (1990)

PARAMETRIC MODEL OF INTERPLANETARY SHOCK WAVE PROPAGATION

Z. Vörös* and M. Karlicky**

* Geophysical Institute, 947 01 Hurbanovo, Czechoslovakia
** Astronomical Institute, 251 65 Ondrejov, Czechoslovakia

ABSTRACT

In the paper a simple model relation is proposed for propagation of driven interplanetary shock wave. The free parameters of the model are discussed in a retrospective manner. On the basis of 2D MHD numerical model the "physically meaningful range" of the input parameters is also discussed.

INTRODUCTION

The most fascinating task of the science is to move ahead the front of our knowledge, enlarge the horizon where scientific predictions are possible to make. Quantum physics, statistical physics, nonlinear science are the special fields where our prediction capabilities have its natural limitations.

The need to forecast geomagnetic storms and substorms, ionospheric effects, plasma structure formation within the external geospace, etc. has led to the appearance of cosmical geophysics, solar-terrestrial physics, and has motivated numerous experimental investigations in outer space. The obtained experience shows that,the transition from micro -scale to meso-scale or from meso-scale to macro-scale physics and prediction is far from straightforward. Nevertheless, a wide range of solar, interplanetary and magnetospheric phenomena can be understood using a macroscopic MHD approach. At the same time MHD equilibrium solutions relevant for physical systems often have the feature of non-uniqueness, when sudden transitions from one equilibrium state to another equilibrium state are possible. At the transition or bifurcation points the physically possible solutions are selected at random which limit the reliability of forecasting a later state of the system considered /1/.Such bifurcations can lead to sudden changes of the velocities (e.g. decelerations or accelerations) of coronal shock waves /2/.

A special task of solar-terrestrial physics is to determine the transit time required by shock waves or coronal mass ejections to cover a distance of 1 AU. The modern approach to this task is represented by various MHD numerical simulation models /3,4,5,6,7/. Simple parametric models of shock wave propagation were also constructed /8,9,10/ which allow to make transit time predictions with smaller or greater errors. The errors appear as a consequence of model simplicity and pure measurement precision of input parameters introduced to model. The errors are caused to some extent also by nonlinearities of the physical problem or by non-uniqueness of the possible solutions of MHD equations. It is expected that, after changing the values of input parameters within a "physically meaningful range" (it will be discussed later) , better agreement should be achieved between experimentally obtained transit times and the predicted values.

DRIVEN PHASE OF THE SHOCK WAVE PROPAGATION

On the basis of the analysis of the 2 1/2 dimensional non-planar numerical MHD simulation Smart and Shea /8/ proposed their simple model by which the flare generated shock wave in the initial phase of its development is driven and in the following phase turns into a blast wave. The velocity of the driven shock wave can be estimated from frequency drift of the type II radio bursts, we denote it as v_{II}.The velocity of the blast wave in the system of the solar wind varies according to the power law.Experiments were made to determine the duration of the driven phase (T_D) on the basis of the soft X-radiation and/or long-term stationary type IVm meter bursts /8, 10/.

Driven phase durations as obtained from softX-ray and type-IVm data are compared in Fig.1. We studied the time interval from September 1, 1978 to July 30, 1982 with the total number of 70 cases.The data were obtained from Solar Geophysical Data, Solnechnije Dannije, from measurements carried out on SMS-GOES and PROGNOZ-8 satellites /11/. The time interval, when the soft X-ray power has decreased to 0.5 of the log of the maximum flux above the preevent background was considered as the indicator of driven phase.Fig.1 shows that the relationship between the time intervals of long duration type IV bursts and long-decay soft X-ray events is not simple.That indicates the piston driven shock arises not only as a consequence of thermal

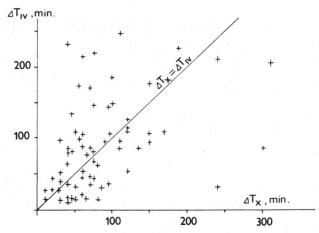

Fig. 1. Comparison of time-duration of soft X-ray and type IVm radiation with total number of 70 cases

explosion. It is a well known fact, however, simple prediction schemes which use a time interval of the soft X-ray events or the type IVm bursts as indicators of driven phase, give rather near to observed transit time estimates. By way of illustration we refer to results obtained by Pinter and Dryer /10/. They suggested to use type IVm radio bursts duration as indicator of the driven phase. Power-law dependence of the blast wave radial velocity was supposed after the driven phase. Fig.2 shows a histogram of shock time-of-arrival predictions. The total number of cases is 39.Their simple prediction scheme yields results within a few hours of the actual shock arrival time. The results of Smart and Shea /8/ obtained from soft X-ray data are similar.

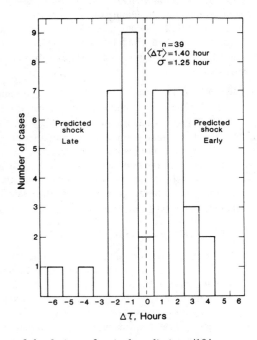

Fig. 2. Histogram of shock time-of-arrival predictions /10/

In the following sections we outline our simple model of propagation of piston driven shock waves and analyse the influence of changes of the input parameters to the transit time values. More detailed discussion is to be found in /9/.

SIMPLE MODEL OF PROPAGATION OF THE PISTON DRIVEN SHOCK WAVES

In our model the propagation velocity of the disturbance $v_d(R)$ relative to an observer is equal to the sum of the velocity of the solar wind and of the shock front. We have proposed the following formula /9/

$$v_d(R, \mathcal{V}) = (v_{II} - v_{sw}(D))(D/R)^N g(\mathcal{V}) + v_{sw}(R) \quad \text{for } R \geq D \tag{1}$$

$$v_d(R) = v_{II} = \text{const.} \quad \text{for } R < D \tag{2}$$

$$\partial v_{sw}/\partial \mathcal{V} \sim 0 \tag{3}$$

where D is the driving distance for the flare generated shock wave, the power N gives the pure blast wave in terms of similarity theory for N=0.5, $g(\mathcal{V})$ is the geometrical factor corresponding to the shape of the shock front, given by

$$g(\mathcal{V}) = (0.5 \times (R/D)(\cos\mathcal{V} + 1))/(0.5 \times (D/R)^k(\cos\mathcal{V} - 1) + R/D) \tag{4}$$

For small values of k (k<1) and large values of D (R>D>0.1AU) function $g(\mathcal{V})$ is asymmetrical. For k>10 $g(\mathcal{V})$ is a quasi-spherical curve. The velocities $v_{sw}(R)$ and $v_{sw}(D)$ can be calculated on the basis of Parker's isothermal model of the solar wind given by

$$v_{sw}(R) = f_p(T_o, R) \tag{5}$$

where T_o is the coronal temperature. Then the model transit time is given by

$$T_{mod} = D/v_{II} + \int_D^{R_E} dR / v_d(R, \mathcal{V}) \tag{6}$$

where R_E is the distance of the Earth from the Sun. In the actual solving of Eq.(6) we first determined the values D and $v_{sw}(D)$. The system of equations (1,2,5) for $v_{sw}(D)$ and D was solved by means of the iteration method with $v_{sw}(D) = 0$ in the first step. The result of the iteration satisfies Lipschitz's convergence condition. The integral in Eq.6 was calculated numerically. At different points of the numerical integration, the velocity of solar wind $v_{sw}(R)$ was varied according to Parker's isothermal model. In a previous paper /9/ , we have analysed four well-documented cases of shock wave propagation in a retrospective manner being used the input parameters of $v_{II}, T_o, N, v_d(R_E, \mathcal{V})$. The following physical conditions were chosen

$$T_{mod} = T_{Tr} = |T_{SSC} - T_{H\alpha}| \tag{7}$$

$$[v_d(R_E, \mathcal{V})]_{mod} = [v_d(R_E, \mathcal{V})]_{exp} = [v_B(R_E, \mathcal{V}) + v_{sw}(R_E)]_{exp} \tag{8}$$

where T_{Tr} is the total transit time, T_{SSC} is the time of the sudden commencement of geomagnmetic storm, $T_{H\alpha}$ is the time of the beginning of the flare in line H_α, $[v_d(R_E, \mathcal{V})]_{exp}$ is the measured local velocity of the disturbance.

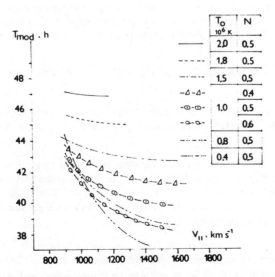

Fig. 3. Dependence of the total transit time on velocity v_{II}, coronal temperature T_0, and power N

For the illustration we show how the total transit time calculated from Eq.(6) depends on coronal temperature To, power N and on coronal shock wave velocity vII (Fig.3).Different values of these parameters, which seem to be within a physically meaningful region (see later), lead to different transit times. We suggest the difference of several hours could explain the forecasting error of Pinter and Dryer /10/ that is an equal order (see Fig.2).

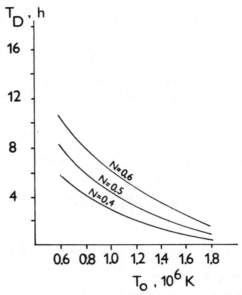

Fig. 4. Duration of the driven phase influenced by coronal temperature T_0 and power N

In Fig. 4 we show another example; the influence of coronal temperature T_o and power N on the duration of the driven phase T_D. A few hour differences in T_D are allowed to achieve changing T_o and N within relatively narrow interval. The discrepancy between the measured and compared values of type IVm and soft X-ray radiation durations is approximately the same order (see Fig.1).On this basis the important conclusion should be drawn: within the frame of the prediction schemes given by simple propagation models containing input parameters as $T_D=T_{IVm}$ or $T_D=T_{softX}$, v_{II},T_o,N ,no additional improvement is possible to achieve, however, it is possible to make rough estimates and forecast sudden commencements of geomagnetic storms by error of a few hours. The retrospective analysis proposed by Vörös and Karlický /9/ give another interesting results. Namely, different combinations of parameters v_{II}, T_o, N, $v_d(R_E,\vartheta)$ and shock front geometry given by $g(\vartheta)$ satisfy the conditions of Eqs. (7) and (8), it means that, in principle, different realisations of shock waves are allowed. Unfortunately, due to our incapability to establish the accurate duration of the driven phase, it is essentially not possible to decide whether the distinct physical realisations indicated by simple model correspond to physical non-uniqueness of solutions of MHD equations.

COMPARISON OF SIMPLE MODEL AND 2D MHD MODEL OF SHOCK WAVE PROPAGATION

We have examined in some detail the two-dimensional MHD calculations of Karlický et al. /12/. Their numerical program solves the interplanetary shock wave propagation as an initial-boundary value problem in one-half of the solar equatorial plane.

In the comparison the MHD model was considered as the standard, and the parameters of the simple model were varied in order to find the best fit between both models. Three runs of the MHD numerical program were made with three different shock wave velocities; v_{II}=1000; 1500; and 2000 kms[-1]. The duration of piston driven phase of these shocks was chosen to be T_D=1h. The range of other input parameters correspond to typical observational values. We came to the following conclusions:

The values of exponent N are from the interval (0.4,0.7) and, the best fit between the simple model and MHD model is achieved when N≈0.5. Experimental results obtained by Volkmer and Neubauer /13/ also indicate that N=0.54±0.07. So,the "physically meaningful range" for the exponent N is well-defined.

At the distance of 1 AU, as it is indicated by MHD model, the shock-front shape for directions of about 45°E and 45°W from pulse central meridian deviates from quasi-spherical behaviour.This deviation leads to the uncertainity of transit time prediction by 3-5 h.The factor g($\smile$) defined by Eq.4 allows such geometrical asymmetry, however, experimental data are necessary to model the actual anisotropy of shock wave propagation.

A small E-W asymmetry of shock wave propagation was found by the MHD numerical simulation. The velocities in the eastern directions are smaller than those in the western directions. The transit time uncertainity is 1h in this case.

The assumed constant velocity of shock wave during the piston-driven phase of the simple model was not confirmed by the MHD numerical simulations.Really, it has also been experimentally discovered and theoretically studied that, coronal shock waves may be both accelerated and decelerated /2,14,15/. We suppose that an average velocity of piston driven shock wave should has some physical sense. The variance of this average due to the use of different models of coronal density and due to the noise of recordings with the radio emission of type II as well as due to the change of shock wave velocity is several hundred kms^{-1}.

CONCLUSIONS

In this paper we compared the experimentally obtained time intervals of long duration soft X-ray radiation and type IVm radiation, as well as the histogram of shock time-of-arrival predictions by the characteristic time intervals of driven phase and total Sun-Earth transit time.The simple model equations (Eqs.1-8) used in the analysis allowed to evaluate the influence of changes of input parameters in retrospective manner. Examination of the results of 2D MHD model calculations of Karlický et al./12/ made possible to span the "physically meaningful range" of considered input parameters. It was also possible to evaluate transit time uncertainities caused by geometrical asymmetry of shock front shape and by E-W asymmetry of shock wave propagation. In summary: a.) simple forecasting models of shock wave propagation permit to predict SSC by a few hour errors. It is possible to remove these errors changing the input parameters within the "physically meaningful range"; b.) it seems to be possible to improve the forecasting efficiency of blast wave propagation by consideration the actual anisotropy features of shock wave propagation; c.) due to the complex and nonlinear dynamics of solar explosions, within the frame of the simple models cited, it seems to be hard to improve the forecasting efficiency of driven wave propagation. For these reasons we suggest to consider more parameters, especially those, which characterize solar explosions as complex dynamical events.

REFERENCES

1. Z.Vörös, Synergetic approach to substorm phenomenon, in:AGU monograph on magnetospheric substorms, in press(1991)

2. Z.Vörös, Bifurcation of the velocities of the propagation of coronal shock waves, Bull. Astron. Inst. Czech., 41,82,1990

3. A.J.Hundhausen, and R.A.Gentry, Numerical simulation of flare generated disturbances in the solar wind, J.Geophys.Res., 74, 2908, 1969

4. M.Dryer, Interplanetary shock waves generated by solar flares, Space Sci. Rev., 15, 403, 1974

5. M.Dryer, Interplanetary shock waves: Recent developments, Space Sci. Rev., 17, 403, 1974

6. S.T.Wu, M.Dryer, and S.M.Han, Non-planar MHD model for solar-flare generated disturbances in the heliospheric equatorial plane, Sol.Phys., 84, 395, 1983

7. D.Odstrčil, Numerical simulation of the April 24, 1981 interplanetary shock wave, Bull. Astr. Inst. Czech., 42, 181, 1991

8. M.A.Smart, and D.F.Shea, A simplified model for timing the arrival of solar flare-initiated shocks, J.Geophys.Res., 90, 183, 1985

9. Z.Vörös, and M.Karlický, A simple model of propagation of driven interplanetary shock waves and prediction of SSC, Bull. Astron. Inst. Czech., 39, 250, 1988

10. S.Pinter, and M.Dryer, Conversion of piston driven shocks from powerful solar flares to blast wave shocks in the solar wind, Bull.Astron.Inst.Czech. 41, 137, 1990

11. B.Valníček, F.Fárnik, J.Sudova, B.Komárek, O.Likin, and N.Pisarenko, Solar X-ray flux in the 6.5-10 and 2.2-7 keV band, Publ.Astron.Inst.Ondrejov, Prognoz data, 1983

12. M.Karlický, Z.Smith, and M.Dryer, Comparison of 2-dimensional MHD and semiempirical models of interplanetary shock wave propagation Bull. Astron. Inst. Czech., in press (1991)

13. P.M.Volkmer, and F.M.Neubauer, Statistical properties of fast magnetoacoustic shock waves in the solar wind between 0.3 AU and 1 AU:Helios-1, 2 observations, Ann. Geophys., 3, 1, 1985

14. R.D.Robinson, Velocities of type II solar radio events, Sol.Phys., 95, 343, 1985

15. M.Karlický,K.Jiřička, O.Kepka, L.Křivský, and A.Tlamicha, Bull. Astr. Inst. Czech., 33, 72, 1982

CHARACTERISTICS OF CMEs OBSERVED IN THE HELIOSPHERE USING HELIOS PHOTOMETER DATA

D. F. Webb* and B. V. Jackson**

* Institute for Space Research, Boston College, Newton Center, MA 02159, U.S.A.
** CASS, University of California at San Diego, La Jolla, CA 92093, U.S.A.

ABSTRACT

The zodiacal light photometers on the two Helios spacecraft have been used to detect and study mass ejections and other phenomena emanating from the Sun and traversing the heliosphere within 1 AU. We have recently compiled a complete list of all of the significant white light transient events detected from the 90° photometers on both the Helios spacecraft. This is a preliminary report on the long-term frequency of occurrence of these events; it emphasizes newly processed data from Helios-1 from 1975 through 1982 and viewed south of the ecliptic. With the large Helios photometer data base, we will be able to identify the fraction of the 90° events which are heliospheric CMEs and determine their characteristics.

We are using the Helios white light photometer data to identify, classify and determine the characteristics of transient disturbances and corotating structures observed in the inner heliosphere. Most of the brightest transient events have been identified as coronal mass ejections (CMEs) and their characteristics in the inner heliosphere studied (/1, 2/). Until now most of the Helios observations of CMEs have involved data from only the Helios-2 zodiacal light photometers, which operated for four years from 1976 through 1979 and viewed north of the ecliptic plane (e.g., /2/). The 90° photometers on Helios-2 and Helios-1 pointed at the north and south ecliptic poles, respectively. All of the photometer data has now been collected on optical disk, greatly facilitating manipulation of the data. We used this data base to identify all significant transient electron plasma events passing to the north or south of or enveloping the spacecraft.

Webb and Jackson /2/ used orbital time-series plots of the Helios-2 photometer data to identify transient white light events. They describe in detail the method of selecting and analyzing these

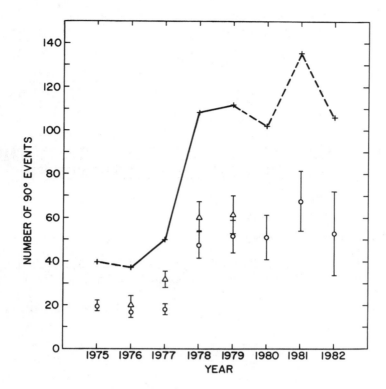

Figure 1: Plot of the annual occurrence rates of transient white light events detected in the Helios 90° photometers. Circled points and triangles denote the number of events at Helios-1 and Helios-2, respectively. Error bars are proportional to statistical counting rates and the amount of data coverage, which decreased markedly after 1977. Crosses are an estimate of the total number of events per year as follows: for years of overlapping coverage (1976–1979) the total is the sum of the duty-cycle corrected counts from both spacecraft; for 1975 and 1980–1982 the total is twice the corrected Helios-1 count.

events. Following the same procedure, we selected only events which met the following criteria: 1) They had to have distinct rise and fall flux profiles consisting of two or more data points (no single-data point enhancements were included). The typical time resolution of these data was 5.2 hrs. 2) The events had to have peak brightnesses above background (normalized to 1 AU) of ≥ 0.5 S10 units. 3) The profiles had to be detectable with similar shapes in separate time-series plots of data obtained in both unpolarized and polarized visible light.

The long term variation of the 90° events is presented in Figure 1. These are shown in the form of their annual rates of occurrence corrected for the Helios spacecraft duty-cycles. During the period of overlapping coverage, the Helios-1 photometers detected an average of 18% fewer events per year than Helios-2. Therefore, this suggests that the Helios-1 90° photometer was less sensitive and detected fewer events than Helios-2.

By classifying the 90° events through examination of the lower-latitude (16 and 31°) photometer data to determine their temporal evolution and spatial extent, Webb and Jackson /2/ found that most (80%) of the initially defined Helios-2 events were CMEs moving outward from the Sun. During the period of overlapping coverage in 1979, the Helios CMEs were found to be well associated with major Solwind CMEs observed near the Sun. In a similar manner we are presently examining the lower-latitude photometer data to classify the final list of 90° events. However, by analogy, we believe it likely that an approximately similar fraction of the final list of events selected by the above criteria will be identified as the heliospheric manifestation of CMEs. The photometers also detected elongated corotating structures, probably streamers and compression regions (e.g., /3/). However, such structures constitute only a small fraction of the total number of identified white light events /2/.

Webb /4/ has determined the annual occurrence rates of CMEs observed near the Sun by Earth-orbiting coronagraphs over the last solar cycle. By comparing the data in Figure 1 with Webb's figure, it is evident that the total annual number of Helios 90° events tracks well with the CME rate curve, including a peak rate in 1981. Counterstreaming particle events are generally considered to be one of the better signatures of mass ejections in *in-situ* solar wind data (e.g., /5/). Elsewhere in these proceedings, Gosling presents a comparison of the annual occurrence rates of counterstreaming electron events from the ISEE/ICE *in-situ* data with Webb's /4/ figure. The reader will note that the annual rates of the 1 AU electron events track reasonably well with those of the Helios 90° events.

The complete Helios photometer data set will permit us to study important characteristics of the heliospheric CMEs, such as their durations, brightnesses, speeds, scale sizes and masses, and to better understand their signatures in the solar wind from *in-situ* data. These results will be published in due course.

We wish to thank S. Kahler and E. Cliver for helpful comments. The work of D. Webb was supported at Boston College by the AF Phillips Laboratory/GPS under contract AF19628-90-K-0006. B. Jackson was supported at UCSD by NASA grant NAGW-2002 and AF contract AFOSR 91-0091.

REFERENCES

1. B.V. Jackson, Solar Phys. 100, 563 (1985)
2. D.F. Webb and B.V. Jackson, J. Geophys. Res. 95, 20641 (1990)
3. B.V. Jackson, J. Geophys. Res. 96, 11307 (1991)
4. D.F. Webb, Adv. Space Res. 11(1), 37 (1991)
5. J.T. Gosling, D.N. Baker, S.J. Bame, W.C. Feldman, R.D. Zwickl, and E.J. Smith, J. Geophys. Res. 92, 8519 (1987)

COMPARISON OF DOPPLER SCINTILLATION AND *IN SITU* SPACECRAFT PLASMA MEASUREMENTS OF INTERPLANETARY DISTURBANCES

R. Woo* and R. Schwenn**

* *Jet Propulsion Laboratory, California Institute of Technology, Pasadena, CA 91109, U.S.A.*
** *Max-Planck-Institut für Aeronomie, W-3411 Katlenburg-Lindau, Germany*

ABSTRACT

Comparisons of extended Doppler scintillation and in situ plasma measurements of the same solar wind are possible for the first time, and have been made in order to improve our understanding of the transients observed in Doppler scintillation measurements. Although the results show that most of the transients are interplanetary shocks, particular attention is given to those that are not.

INTRODUCTION

Ever since Doppler scintillations of the solar wind were first observed using coherent radio signals received from planetary space probes, it was recognized that transients appearing in the observations showed promise for detecting and studying disturbances in the solar wind. While it was soon evident that major Doppler scintillation transients were unmistakably interplanetary shocks /1/, the identities of less pronounced features remained uncertain. Since 1979, the radio path of the Pioneer Venus Orbiter (PVO) spacecraft has spent considerable time probing the near-Sun solar wind off the limbs of the Sun. On occasion, Helios 1 has also been in position above the Sun's limb at the same solar longitude to simultaneously observe in situ the solar wind plasma. Shown in Figure 1 are the trajectories of the PVO and Helios spacecraft in the ecliptic plane and referenced to a fixed Sun-Earth system during the two periods of this study. The PVO Doppler scintillations essentially probe the solar wind near P, the closest approach point of PVO's radio path. The fortuitous circumstances depicted in Figure 1, along with the availability of near-continuous measurements, have made it possible for the first time to carry out detailed comparisons between Doppler scintillation and in situ plasma measurements, and to improve our understanding of Doppler scintillation transients. Since the details of these comparisons will be published in /2/, only a few examples will be shown and a summary of the results presented here.

EXAMPLES OF COMPARISONS

Shown in Figure 2 are the time histories of Helios 1 measurements of proton T_p, bulk solar wind speed v_p, proton density n_p, proton mass flux density $n_p v_p$, and PVO rms Doppler scintillations σ_D for 23–28 April 1981. The three Doppler scintillation transients T2, T3, T4 detected by PVO at 15-18 R_o were observed later at 170-174 R_o by Helios 1 as shocks S2, S3, S4, respectively. Note that while the profiles of $n_p v_p$ and σ_D are similar, the relative magnitudes of the scintillation transients are different from the relative magnitudes $n_p v_p$.

Data for 17–26 May 1981 are shown in Figure 3. A mass flux density event, corresponding to an non-compressive density enhancement NCDE and/or magnetic cloud, is observed by Helios 1 near 127 R_o on May 18. The solar wind probed by PVO Doppler scintillations near 39 R_o is void of any corresponding transient signature.

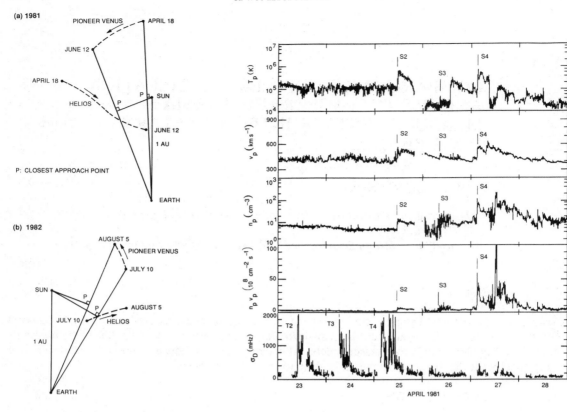

Fig. 1. PVO and Helios trajectories
in the ecliptic plane.

Fig. 2. Time histories of Helios in situ plasma and PVO
Doppler scintillation measurements: 23–28 April 1981.

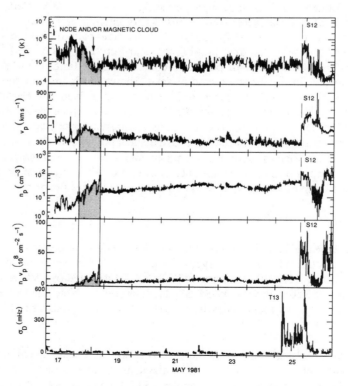

Fig. 3. Similar to Figure 2: 17–26 May 1981.

Data for 5–9 May 1981 when PVO remotely probed at 27-32 R_o and Helios 1 observed directly at 144-153 R_o are shown in Figure 4. There are no fast-mode shocks associated with transients T6 and T7. The two disturbances marked SS and D1 observed by Helios 1 probably correspond to T6 and T7, respectively. SS is a slow shock (suggested by Helios 1 magnetic field measurements) and D1 a minor mass flux density enhancement.

Displayed in Figure 5 are the shock speeds of the 18 transient-shock pairs identified in this study for which $|\Delta\phi| < 16°$, where $|\Delta\phi|$ is the difference in solar longitude between the closest approach point of the PVO radio path and the Helios 1 spacecraft location. The transit speed v_T between the closest approach point of the PVO radio scintillation path and Helios 1 is represented by a horizontal line indicating distance over which it is observed. In situ speeds observed by Helios 1 V_{sh} are displayed as circled points and connected to the transit speeds by a vertical line to reveal deceleration or acceleration of the shock. Two of the three unattached in situ points correspond to shocks for which $|\Delta\phi| < 29°$, and the third is associated with a shock for which only a range of transit speeds was available due to a data gap during the start of the scintillation transient.

SUMMARY OF RESULTS

- During a combined observing period of nearly three months in 1981–1982 near solar maximum (see Figure 1), 22 transients were observed by PVO and 23 shocks by Helios 1.

- Based on a comparison of mass flux density and rms Doppler scintillations, at least 84% of the transients were shocks, while at least 90% of the shocks were transients.

- Although the temporal profiles of Doppler scintillation and mass flux density are similar, the magnitudes of the the Doppler scintillation transients may not simply reflect those of mass flux density (see e.g., Figure 2).

- Only one pronounced solar wind event observed in the mass flux density measurements showed no signature in the scintillation data (see Figure 3). Fields and particles measurements by Helios 1 suggest an NCDE and/or magnetic cloud.

- One of the scintillation transients that is not a fast-mode shock appears to correspond to a slow shock (see Figure 4); however, when scintillations alone are available, slow shocks may be difficult to identify.

- Shock speeds based on transit times between the PVO radio scintillation path and the Helios 1 spacecraft are consistent with those from the in situ plasma measurements, and indicate shock deceleration in essentially all cases (see Figure 5).

- A significant consequence of the improved understanding of Doppler scintillation transients resulting from this investigation is that they can now be used alone to detect and locate interplanetary shocks near the Sun with a relatively high degree of confidence, and hence conduct useful correlative studies with other solar and interplanetary observations in the future.

ACKNOWLEDGMENTS

This paper describes research carried out at the Jet Propulsion Laboratory, California Institute of Technology, under contract with the National Aeronautics and Space Administration.

REFERENCES

1. R. Woo, J.W. Armstrong, N.R. Sheeley, Jr., R.A. Howard, M.J. Koomen, and D.J. Michels, Doppler scintillation observations of interplanetary shocks within 0.3 AU, J. Geophys. Res., 90, 154 (1985).

2. R. Woo and R. Schwenn, Comparison of Doppler scintillation and in situ spacecraft plasma measurements of interplanetary disturbances, J. Geophys. Res., in press (1991).

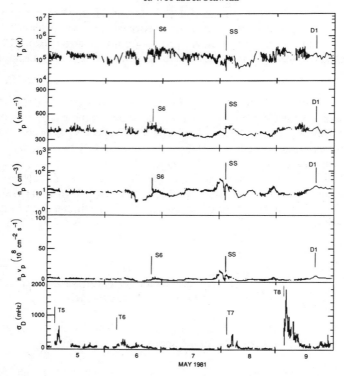

Fig. 4. Similar to Figure 2: 5–9 May 1981.

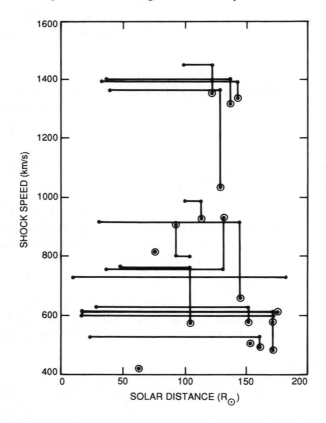

Fig. 5. Shock speeds of the 18 transient-shock
pairs of this study.

INTENSE MAGNETIC CLOUDS AND THEIR INTERACTIONS WITH AMBIENT SOLAR WIND STREAMS*

G. -L. Zhang

Center for Space Science and Applied Research, Academia Sinica, PO Box 8701, Beijing 100080, China

ABSTRACT

Structure characteristics of typical intense magnetic clouds are discussed for different conditions of ambient solar wind streams, with emphasis on the dynamic processes that govern the interaction between cloud and ambient stream. It is suggested that large magnetic pressure gradient forces at the expanding cloud boundaries are able to accelerate the stream ahead of cloud and decelerate that behind, building up double saw-tooth speed profiles and driving shocks.

INTRODUCTION

Magnetic cloud is an important interplanetary transient structure /1,2/. On the solar side, it is the plasma output of coronal mass ejection /3,4/; and on the interplanetary and geospace sides, it can generate drastic cosmic ray decreases and geomagnetic storms /5-7/. Most of the cloud studies so far published emphasized on its magnetic topology, and force free magnetic field model with constant α was proposed to fit the internal magnetic field configuration of cloud /8/. However, little was known about the boundary that plays an important role in the interaction of cloud with ambient stream. Also, less attention was paid to the dynamic effects of cloud, although Burlaga had pointed out that a cloud can expand /9/. It is not clear that whether the expansion at 1 au is dynamical or geometrical.

Stream interaction and shock formation have been extensively studied since the beginning of space era /10-12/. Most of the studies of interplanetary shock so far published restricted themselves on the discussion of shocks driven by the kinetic momentum or thermal impulse. Shocks were observed in association with magnetic cloud /6/, and a possible mechanism for shock driven by expanding magnetic cloud was suggested by Zhang /13/.

This paper intends to present observational evidences showing the dynamic effects of an expanding intense magnetic cloud including driving shock. The discussions involve solar wind speed (V), number density (N), thermal (p) and magnetic (p_m) pressures, rms fluctuation of IMF (σ_c), inclination of IMF vector with respect to ecliptic plane (θ), thermal sonic (C_s) and Alfvenic (C_a) speeds, and ratio of magnetic to kinetic energy density (μ_k), and that of magnetic to thermal pressure (μ_p, i.e. the inverse β parameter). The latter two dimensionless ratios are important quantities for characterizing the strong transient interplanetary disturbances /14-16/, which were used for identifying an intense magnetic cloud that was defined as a cloud having high values of these energy ratios /13/.

---

* The Project supported by the National Natural Science Foundation of China

MAGNETIC CLOUD IN VERY-LOW-SPEED SOLAR WIND

On March 19-22, 1980, a magnetic cloud was moving in an ambient solar wind with very low speed about 300 km/s (Fig.1). Also very low was the rms fluctuation of magnetic field components σ_c, indicative of a nearly structureless quiet solar wind condition. The cloud can be regarded as a simple one free from interaction with large solar wind streams and preserving its own basic structure. The interplanetary magnetic field strength increased from a background value of about 5 nT to 16 nT, then recovered symmetrically, corresponding to one order of magnitude increment of magnetic pressure. The main structure of cloud was characterized by low temperature and density, i.e. large enhancements of dimensionless energy ratios μ_p and μ_k, (30 and 0.37 respectively), and Alfven speed (210 km/s). Magnetic pressure and Alfven speed dominated respectively the dynamic processes and wave propagations inside the cloud.

A high density structure was observed ahead of the cloud with a maximum number density about 45 particles/cm^3, inside which the magnetic fluctuations, temperature as well as thermal pressure were highly enhanced. Such turbulent high density structure can be called a cloud sheath. A forward shock was observed at 6 UT, March 19. Both the magnetic field strength and temperature jumped suddenly at the shock front, and a geomagnetic sudden commencement was recorded /7/. A peak speed of 390 km/s appeared quickly after the shock, nearly in coincidence with the thermal pressure peak but led that of magnetic pressure. Most likely, the shock was driven by the speed hump. The next question was what drove the speed hump? Comparison of the profiles described above ruled out the possibility for speed hump either driven by interplanetary stream interaction or resulted from solar thermal pressure impulse. It is worth noting that the speed peak was located in the rising part of magnetic field profile, a remarkable feature also noted even in the average curve for clouds with a shock ahead /6/. The speed peak located in front of cloud would imply that it resulted from acceleration due to expanding cloud. The unique observable dynamic force was the negative gradient force of total pressure near the cloud boundary. At present, we do not know whether the cloud boundary can be described in terms of a simplified force free model. However, the gradient force of total pressure near the boundary of cloud should be strong enough at least in balance to the dynamic pressure of ambient stream, otherwise, cloud would be crashed by interaction. Because of the very-low thermal pressure inside the cloud, the main driving force should be magnetic origin.

Another important dynamic factor is the distribution of Alfven speed. Higher Alfven, so magnetosonic speed would help quick propagation of magnetic fluctuations across the cloud, and negative gradient of the characteristic speed will result in intersection of characteristics, speeding up the formation of a forward shock driven by velocity gradient only /11/. The fluctuations will then concentrate into the cloud sheath, resulting in dissipation, heating, piling up plasma, increasing thermal pressure in cloud sheath and finally accelerating the ambient solar wind. Therefore, the forward shock was magnetically driven in nature.

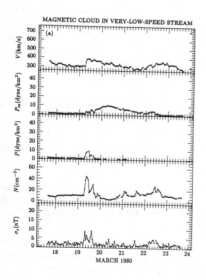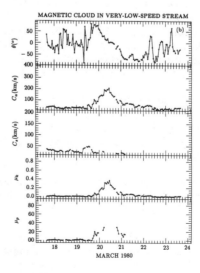

Fig.1 The 19 March 1980 event, a magnetic cloud moving in very-low-speed structureless solar wind.

MAGNETIC CLOUD IN A SMALL STREAM WITH MEDIUM SPEED

In the December 19, 1980 event (Fig.2), a very strong magnetic cloud was moving in the declining part of a small stream with medium speed between 360-520 km/s. Inside the cloud, the peak value of magnetic field strength reached a very high value of 34 nT. The maximum magnetic pressure was about 46 dyne/km^2, nearly 90 times the thermal pressure, and magnetic energy density was as high as 60% of kinetic energy density. Also very high was Alfven speed with a maximum value of 390 km/s, i.e. about 80% of wind speed. Thus the main structure of cloud was characterized by large enhancements of dimensionless energy ratios μ_p and μ_k, and Alfven speed. Magnetic force dominated the dynamic processes inside the main structure of cloud, which should have much more magnetic energy in driving the ambient solar wind without changing its basic characteristics.

A high density sheath was observed ahead of cloud with a maximum number density about 53 particles/cm^3. Magnetic fluctuations, temperature, and thermal pressure were greatly enhanced respectively to peak values of 18 nT, $\geq 10^5$ K, and ≥ 21 dyne/km^2. A strong forward shock was observed in front of cloud sheath at 6 UT, December 19. The bulk speed increased to about 550 km/s quickly after the shock, nearly in phase with the density peak (also the temperature peak as estimated), but led that of magnetic pressure several hours. Like the above example, the shock should be driven by strong magnetic pressure gradient force in the front part of cloud. In the present case, as the shock was moving up to the top speed of ahead small stream, the speed profile after shock followed a same trend as that before, displaying a saw-tooth like profile similar to that observed in stream evolution /11/. A second saw-tooth speed profile appeared by 20 UT, close to the rear boundary of cloud, and the trough was located in the central part of positive magnetic pressure gradient. Such double saw-tooth speed profile was very similar to those observed in high stream interaction /10/, and shock pair moving in the rarefaction part of solar wind stream /11/. The example given here proved such profile driven by magnetic gradient force. However, for the low speed cloud with enlarged space extent shown in Fig.1, the second saw-tooth speed profile was not so obvious as compared to that in Fig.2.

A strong magnetic field structure should expand, besides the forward acceleration offered by large negative gradient force of magnetic pressure, the large positive gradient force of magnetic pressure in the rear part of cloud can decelerate the solar wind behind the cloud, resulting in the first and second saw-tooth speed profiles near the front and rear boundaries of cloud respectively. Also, the negative gradient of Alfven speed in the first part of cloud contributed the formation of front turbulent sheath as discussed above. On the other hand, the positive gradient of Alfven speed can help the forward fluctuations to propagate quickly across the cloud, and make the reverse fluctuations stagnating onto the rear boundary, building up a back sheath of cloud with enhanced turbulences and thermal pressure, which can also contribute to decelerate the solar wind behind the cloud. A reverse shock might be expected, however, based on the hourly value data only, it is difficult to identify such a reverse shock that may be weaker than the forward one.

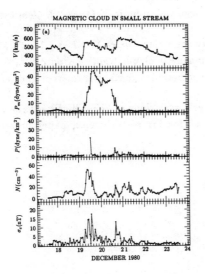
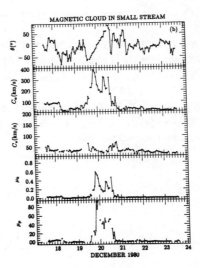

Fig.2 The 19 December 1980 event, a magnetic cloud overtaking a small medium-speed solar wind stream.

CONCLUSIONS

1. An intense magnetic cloud is characterized by large enhancement of dimensionless energy ratios of magnetic to kinetic energy density and that of magnetic to thermal. Magnetic field dominates the dynamic processes of cloud.

2. Large magnetic pressure gradient forces at an intense magnetic cloud boundaries are able to accelerate the stream ahead of cloud and decelerate that behind the cloud, building up the observed double saw-tooth speed profiles.

3. Inside the cloud, greatly enhanced Alfven speed dominates the characteristic speeds, which would help MHD fluctuations to propagate quickly across the cloud. Stagnation of the fluctuations raises temperature in the high density sheaths bounding the cloud.

4. The cloud-stream interaction eventually results in shock formation driven by expanding structure with strong magnetic field, which is different from those generated from thermal impulse or streams interaction.

REFERENCES

1. L. F. Burlaga, E. C. Sittler, Jr., F. Mariani, and R. Schwenn, Magnetic loop behind an interplanetary shock: Voyager, Helios, and IMP 8 observations, *J. Geophys. Res.* *86*, 6673, (1981).

2. L. W. Klein, and L. F. Burlaga, Magnetic clouds at 1 AU, *J. Geophys. Res.* *87*, 613, (1982).

3. L. F. Burlaga, L. W. Klein, N. R. Sheeley, Jr., D. J. Michels, R. A. Howard, M. J. Koomen, R. Schwenn, and H. Rosenbauer, A magnetic cloud and a coronal mass ejection, *Geophys. Res. Lett.* *9*, 1317, (1982).

4. J. T. Gosling, CME and magnetic flux rope in interplanetary space, in *Phys. of Magnetic Flux Rope*, *Geophys. Monog. Ser. 58*, ed. C. T. Russel, E. R. Priest, and L. C. Lee, AGU, Washington D.C., 1988, p.343.

5. R. M. Wilson, Geomagnetic response to magnetic cloud, *Planet. Space Sci.*, *35*, 329, 1987.

6. G.-L. Zhang, and L. F. Burlaga, magnetic clouds, cosmic ray decreases, and geomagnetic disturbances, *J. Geophys. Res.* *93*, 2511, (1988).

7. G.-L. Zhang, Morphology category of geomagnetic storms and characteristics of interplanetary magnetic clouds, *Sci. in China, Ser. A*, *34*, 717, 1991.

8. L. F. Burlaga, Magnetic clouds: Constant alpha force-free configurations, *J. Geophys. Res.* *93*, 7217, (1988).

9. L. F. Burlaga, and K. W. Behannon, Magnetic clouds: Voyager observations between 2 and 4 AU, *Sol. Phys.*, *81*, 181, 1982.

10. A. J. Hundhausen, and J. T. Gosling, Solar wind structure at large heliospheric distances: An interpretation of Pioneer 10 observations, *J. Geophys. Res.* *81*, 1436, (1976).

11. Y. C. Whang, The forward-reverse shock pair at large heliospheric distances, *J. Geophys. Res.* *89*, 7367, (1984).

12. M. Dryer, Interplanetary shock waves: Recent developments, *Space Sci. Rev.*, *17*, 277, (1975).

13. G.-L. Zhang, Structure characteristics of simple magnetic cloud, *Chin. J. Space Sci.*, *8*, 261, 1988.

14. G.-L. Zhang, Y.-F. Xu, Y.-F. Gao, C. Lu, Heliospheric magnetic parameters and solar activity, *Sci. Sinica (Series A)*, *28*,, 1081, (1985).

15. G.-L. Zhang, and Y.-F. Xu, Two kinds of solar cycle variations of heliospheric parameters, this issue.

16. G.-L. Zhang, Distinguishing the transient from corotating disturbances, paper presented at SOLTIP Symposium, Czechoslovakia, 1991.

A SHOCKED –Bz EVENT CAUSED BY FAST STEADY FLOW–SLOW TRANSIENT FLOW INTERACTION

X. Zhao

Center for Space Science and Astrophysics, Stanford University, Stanford, CA 94305, U.S.A.

Abstract

We show that the 25 November 1978 shock pair was caused by the interaction of a fast steady flow with a slow coronal mass ejection in interplanetary space (ICME). It is suggested that the slow ICME may be disconnected from the Sun. In addition, a new method to infer the shock angle and Mach number from the observed upstream plasma β and the jump ratios of proton density and magnetic flux density across a shock is described.

1. INTRODUCTION

By using the ISEE 3 data a shock pair was identified with a forward shock at 11:40 UT day 329 1978, a reverse shock at 0140 UT day 330 and a discontinuity at 1714 UT day 329 which separates the shocked fast plasma from the shocked slow plasma (see Figure 6 of Tsurutani et al., 1988). A $-B_z$ event occurred in the shocked slow plasma. This shock pair has been suggested to be caused by fast steady flow-slow steady flow (SS) interaction or by fast transient flow-slow steady flow (TS) interaction [Tsurutani et al., 1988]. For the case of either SS or TS shock pairs the ambient plasma is the slow steady flow and the magnetic field should be along the Parker spiral. However the existence of a large southward IMF component ahead of the forward shock is hard to explain with the spiral field. The purpose of the present paper is to search for the cause of the observed shock pair and the $-Bz$ event.

2. MAGNETIC CONFIGURATION IN THE FAST AND SLOW FLOW

The interplanetary magnetic field configuration on long time scales can be divided into spiral and nonspiral fields. The expected angular distribution of the spiral field should cluster tightly around $(\phi, \theta) = (-45°, 0°)$ or $(135°, 0°)$ if the MHD fluctuations are weak; the distributions will be broader if the fluctuations are strong as in fast steady flows. A nonspiral field configuration indicates the field in a transient flow, i.e., a interplanetary counterpart of the coronal mass ejection (ICME). Both rotational and nonrotational fields in ICMEs [Gosling, 1990] should show significant departures from the Parker spiral.

We use hourly-average values of the azimuthal and latitudinal angle of the IMF in the solar ecliptic coordinate system from November 24, 00 UT to November 26, 23 UT, 1978 to determine the angular distributions shown in Figure 1. The angular distribution of the field right behind the reverse shock (see panel 1 of Figure 1) is around $(-45°, 0°)$ with rather wide diffusion, as expected for fast steady flows with strong fluctuations. The angular distribution for the field ahead of the forward shock shows multiple component (see panels 1 and 2 of Figure 1). The panel 1 shows spiral field with polarity away from the Sun, the same as that behind the reverse shock but having weak fluctuations. The structure between the spiral field and the forward shock (panel 2) is certainly not spiral field shown by significant departures from the directions $(-45°, 0°)$ or $(135°, 0°)$, implying that it is field in a transient flow. panels 3 and 4 display the angular distributions for fields within the interaction region separated by the discontinuity. The differing distributions between the two panels indicate the discontinuity being a flow interface.

Therefore, the magnetic characteristics show that the field in the slow flow around the forward shock is nonspiral, the field behind the reverse shock is spiral, and the discontinuity within the shock pair is a flow interface.

3. PLASMA PROPERTIES OF THE SLOW FLOW

694   X. Zhao

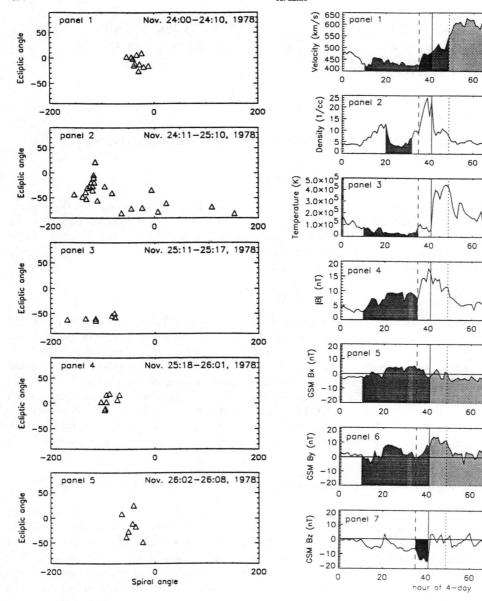

Figure 1. Angular distributions of various plasma flows in the solar wind. Panels from 1 to 5 correspond to (see the five intervals in Panel 1 of Figure 2) the field direction for ambient (0-10), preshock (11-34), post shock before (35-41) and after (42-49) the discontinuity, and after the reverse shock (50-56).

Figure 2. Profiles of magnetic field and plasma properties between November 24 and November 27, 1978. Panels from top to bottom correspond to flow velocity, proton density, proton temperature, magnetic flux density, and the X, Y and Z components of the magnetic field in the geocentric-solar-magnetospheric coordinate system. The dashed, solid and dotted lines correspond to the forward shock, discontinuity and reverse shock, respectively.

    The panel 1 of Figure 2, the proton velocity profile, displays the five intervals of time from left
to right, corresponding to the five panels in Figure 1. The −Bz event studied here is shown in
the shading part of panel 7. Panels 1–3 of Figure 2 show that in addition to a velocity increase, a
substantial proton temperature increase and a abrupt proton density decrease occurred at the flow
interface, which is similar to the stream interface.

    The shading regions in Panels 3 and 4 indicate that the slow flow with nonspiral field men-
tioned above is just a "Cold Magnetic Enhancement", a characteristic of ICMEs. The temperature
remaining low even in the shocked plasma between the forward shock and the interface suggests
that it probably extends through the forward shock to the discontinuity. The high helium density
in the shocked plasma (see the top panel of Figure 6 of Tsurutani et al., 1988) also suggests that
the slow flow is a ICME.

    Gosling found that bidirectional electron flux events, or BDEs, may be one of the most char-
acteristic signatures of CMEs in interplanetary space. A BDE is observed between November 24
20:00 UT and November 25 09:10 UT, 1978 [Gosling et al., 1987] though the time interval is a bit
narrower than the "Cold Magnetic Enhancement" (see the shading parts of panel 2 and panel 3
of Figure 2). Crooker et al. [1990] recently analyzed ICME geometry and found that the radial
boundaries of a ICME were determined based on low temperature, with BDE and counterstreaming proton events occurring within those boundaries. Thus all plasma properties of this slow flow
indicate that it is a slow ICME.

4. CONTRIBUTION OF FAST SHOCK TO −Bz EVENT

    By averaging on the time scale of the shocked ICME, it can be estimated that the southward IMF
component, total magnetic flux and proton density jumped to 13 nT, 17 nT and 14 protons/cm^3,
from 7 nT, 10 nT and 8 protons/cm^3, respectively, and the jump ratios across the shock are about
1.85, 1.70 and 1.75.

    For cases when the plasma β ahead of the fast forward shock is 0.5, 1.0 and 5.0 [Zhao, et al.,
1991], we calculate the jump ratios of proton density and total magnetic flux across shocks when
the shock angle increases from 0° to 90° and the Mach number increases from 1.1 to 6.0. Figure 3

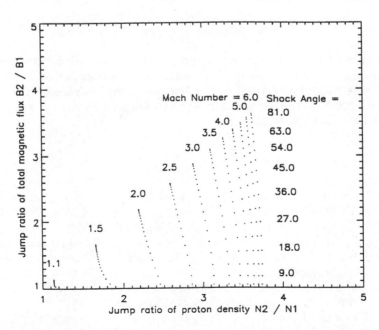

Figure 3. Dependence of the jump ratio of magnetic flux density on the jump ratio of proton density. The
shock angle ranges from 4.5° to 85.5°, and the Mach number ranges from 1.1 to 6.0.

shows the dependence of the total magnetic flux ratio on the proton density ratio for $\beta = 1.0$. It
is found that for shock angles greater than 80°, the total magnetic flux ratio approximately equals
the proton density ratio no matter what the values of the Mach number and plasma β are. If

the proton density ratio is less than or equal to 1.8, the magnetic flux ratio is nearly proportional to the proton density ratio when the shock angle is greater than 50°. This inference holds true approximately for β values between 0.5 and 5.0. The shock normal seems nearly parallel to the ecliptic because the gas driving the shock is a high speed stream, and the shock angle is about 50°. Based on Figure 3 the total magnetic flux ratio can be inferred to be close to or slightly less than proton density ratio. The consistency between the prediction and observations shows that shock compression alone appears to be adequate to explain the large value of southward field component observed within the shocked ICME plasma.

5. Conclusion and Discussion

In analyzing the event of 25 Nov. 1978, we come to the following conclusions:

1. The discontinuity between the shock pair is a flow interface. Its existence suggests that the shock pair is not a TS shock pair associated with a ICME centered far away from Sun-Earth line.

2. The slow flow ahead of the interface is a ICME.

3. The shock pair associated with the 25 November 1981 $-Bz$ event is caused by fast steady flow-slow ICME (ST) interaction (see the shading parts of panels 5 and 6 in Figure 2).

4. The compression alone appears to be adequate to explain the large value of $-Bz$ observed within the shocked ICME because of the strong upstream field which is the internal field of the ICME.

5. The fact that the leading slow ICME has a planar magnetic structure because it is not a magnetic cloud and the fact that a large southward field component exists in the slow ICME suggest that the planar structure is not parallel the solar equator. Thus existence of the ST interaction suggests that the slow ICME may be disconnected from the Sun.

It may be interesting to note that Figure 3 provides a new way to estimate Mach number and shock angle if the upstream plasma β and the proton density and total magnetic flux ratios across a shock can be specified by observations. For instance, by using the ratio values specified above we can grossly infer from Figure 3 that for the forward shock studied here the Mach number is between 1.5 and 1.6, and the shock angle is about 70°. More accurate values may be estimated if the observational values can be specified more accurately.

Acknowledgement.

The author thanks Phil H. Scherrer and J. Todd Hoeksema for helpful discussions and careful reading and commenting on the manuscript, This work was supported in part by the Office of Naval Research under Grant N00014-89-J-1024, by the National Aeronautics and Space Administration under Grant NGR5-020-559, and by the Atmospheric Sciences Section of the National Science Foundation under Grant ATM90-22249.

REFERENCES

Crooker, N. U., J. T. Gosling, E. J. Smith and C. T. Russell, A bubblelike coronal mass ejection flux rope in the solar wind, *Physics of Magnetic Flux Ropes* ed. by C. T. Russell, E. R. Priest, L. C. Lee, pp. 365–372, American Geophysical Union, Washington DC, 1990.

Gosling, J. T., Coronal mass ejections and magnetic flux ropes in interplanetary space, in *Physics of Magnetic Flux Ropes* ed. by C. T. Russell, E. R. Priest, L. C. Lee, pp. 343–364, American Geophysical Union, Washington DC, 1990.

Gosling, J. T., D. N. Baker, S. J. Bame, W. C. Feldman, and R. D. Zwickl, Bidirectional solar wind electron heat flux events, *J. Geophys. Res., 92*, 8519–8535, 1987.

Gosling, J. T., S. J. Bame, E. J. Smith and M. E. Burton, Forward-reverse shock pairs associated with transient disturbances in the solar wind at 1 AU, *J. Geophys. Res., 93*, 8741–8748, 1988.

Tsurutani, B. T., W. D. Gonzalez, F. Tang, S.-I. Akasofu, and E. J. smith, Origin of interplanetary southward magnetic field responsible for major magnetic storms near solar maximum (1978–1979), *J. Geophys. Res., 93*, 8519–8531, 1988.

Zhao, X., K. W. Ogilvie and Y. C. Whang, Modeling the effects of fast shocks on solar wind minor ions, *J. Geophys. Res., 96*, 5437–5445, 1991.

PREDICTION OF LARGE NORTH–SOUTH IMF COMPONENT EVENTS OCCURRING IN DRIVER GAS

X. Zhao and J. T. Hoeksema

Center for Space Science and Astrophysics, Stanford University, Stanford, CA 94305, U.S.A.

Abstract

An approach for identifying a driver gas-associated Bz event and its solar source is introduced. Seven newly identified events are used to further test the model developed by Hoeksema and Zhao [1991] for prediction of the magnetic orientation of some driver gas-associated Bz events from photospheric field observations. Comparison of the model predictions with observations confirms that the model may be appropriate only for driver gas-associated Bz events which have magnetic structures lacking large internal field rotation and are associated with active region-CMEs.

1. Introduction

We have developed a model [Hoeksema and Zhao, 1991, hereafter referred to as paper 1] to predict the direction of driver gas-associated Bz events from the computed coronal magnetic-field orientation above the site of solar parent events which can be classified as active region(AR)-CMEs or quiescent prominence(QP)-CMEs. The comparison of predictions of the model with *in situ* observations of four −Bz events showed that the magnetic field orientations only in driver gas associated with the AR-CMEs can be predicted by the model. The purpose of the present work is to identify more events and to further test the model.

2. Identifying A Driver Gas-Associated Bz Event and Its Solar Source

We define a Bz event as an interval when the magnitude of the north-south component of the interplanetary magnetic field observed at 1 AU is greater than 10 nT for more than 3 hours. 54 such events have been identified using the Omnitape data set between 1978 and 1982. Of these, 28 events were preceded by shocks and 26 events lacked a shock association.

We explicitly assume that a Bz event preceded by a shock is driver gas-associated if it occurred in an interval when a bidirectional electron heat flux event (BDE) or a bidirectional low-energy proton event (BDP) was observed. The tabulations of BDEs [Gosling et al., 1987, 1990] and BDPs [Marsden et al., 1987] show that sixteen of the events were associated with driver gas. The four −Bz events observed between 1978 and 1979 are the same as those detected by Tsurutani et al. [1988] on the basis of a full complement of solar wind and magnetic field data from ISEE-3, and have been analyzed in paper 1.

To identify the solar source of a driver gas-associated Bz event, we compared the Sun-Earth travel time computed from the observed velocity of the driver gas with the elapsed time between the associated solar parent event and the onset of the driver gas for each of the five events analyzed in paper 1. The variations in these values suggest uncertainties of ±12 hours in the computed Sun-Earth travel time for AR-CME associated events and ±24 hours for QP-CME associated events. By examing the SMS-GOES X-ray data (Solar–Geophysical Data comprehensive reports) and the catalog of solar (quiescent) filament disappearances [Wright, 1991] we identified, of the twelve driver gas-associated Bz events not previously analyzed, seven having one or more candidate solar parent events within the time window.

3. Model and Method of Calculation

As pointed out in paper 1, if a magnetic structure lacking large internal field rotation is convected by the solar wind into interplanetary space with its configuration maintained, the internal field of the associated CME is probably potential during the late phase. This may validate the use of the potential field model for predicting the field configuration in AR-CMEs.

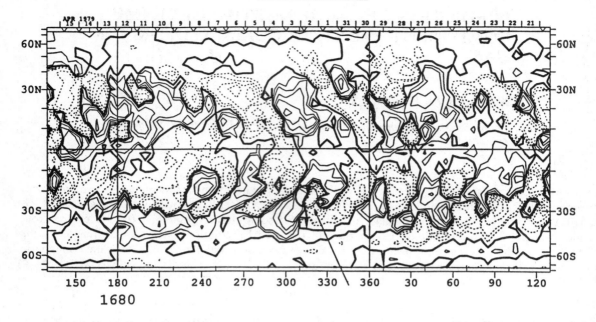

Figure 1. The NSO synoptic chart of the photospheric magnetic field with the 1979/04/03 01:09 (S26 W14) flare-associated CME centered. The field has been averaged into 30 bins of sine latitude from north to south each 5 Carrington degrees. The photosphere is shown in an equal area projection (equal steps in sine latitude). The bottom axis shows the Carrington longitude, while the top axis is labeled by central meridian passage date. The small circle denotes the location of the flare associated with a CME and Bz event. The thick arrow denotes the orientation of Earth's rotation axis when the flare occurred.

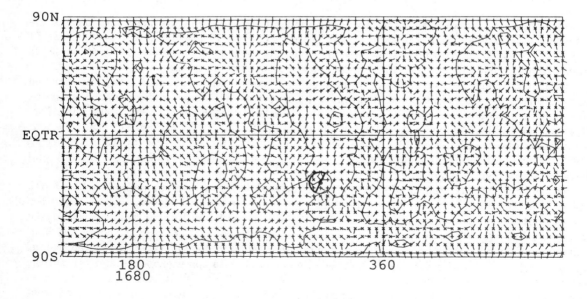

Figure 2. Predicted magnetic orientation at 1.03 $R_\odot$. The small circle and the thick arrow have the same meaning as in Figure 1. The '+' symbols mark grid points having a radial field component directed away from the Sun, 'x' symbols indicate field directed towards the Sun the Sun. The contour of zero radial field is drawn on the chart. The arrow drawn from each grid point shows the direction of the transverse field computed from the θ and ϕ components. This plot gives no indication of the magnitude of the field.

The field at the "release height" (the height where the front of a CME ceases to accelerate) is calculated using a potential field-source surface model [Hoeksema, 1984]. The contour map in Figure 1 shows the photospheric line-of-sight magnetic field over the entire Sun used as the inner boundary condition. The center of the small circle shows the location of the flare that occurred at S26° W14° on April 4, 1979, 01:09 UT. This flare, accompanied by a M4 soft X-ray event, is associated with the CME that produced the driver gas which caused the northward Bz event observed near Earth by ISEE-3 on April 5,1979.

We selected 1.03 $R_\odot$ and 1.20 $R_\odot$ as the candidate release heights (see paper 1). Figure 2 shows the field direction at 1.03 $R_\odot$ computed from the photospheric field displayed in Figure 1. The field above the parent event site has a northward component at 1.03 $R_\odot$.

Table 1: Observation and Prediction

ID Number	Bz Event UT	Driver Gas UT	NERA	Solar Source Onset Location	Field Orientation 1.03	Field Orientation 1.20
#1	2908-2920 March 1979 (southward)	2908-2915 March 1979 (BDE)	-26°	1979/03/25 21:19 S29 E48	SW	W
				1979/03/26 11:50 N05 W78	N	NW
#2	0510-0514 April 1979 (northward)	0510-0605 April 1979 (BDE) 0518-0613 April 1979 (BDP)	-26°	1979/04/03 01:09 S26 W14	N	NE
#3	1916-2005 March 1980 (northward)	1921-2002 March 1980 (BDP) 1916-2119 March 1980 (Cloud)	-25°	1980/03/16 06:30 N36 E04	NW	N
#4	2523-2604 July 1981 (southward) 2607-2610 July 1981 (northward)	2519-2618 July 1981 (BDE) 2522-2605 July 1981 (BDP)	8°	1981/07/23 23:06 RS10 E55	SW	W
#5	2210-2221 Oct. 1981 (southward)	2207-2214 Oct. 1981 (BDE)	26°	1981/10/19 13:18 N20 E20	SW	W
#6	1200-1210 Feb. 1982 (northward)	1200-1206 Feb. 1982 (BDP)	-16°	1982/02/09 04:10 S12 W90	SE	S
	1222-1317 Feb. 1982 (southward)	1203-1317 Feb. 1982 (Cloud)		1982/02/09 13:15 S10 E06	N	NW
#7	2508-2511 April 1982 (southward)	2510-2709 April 1982 (BDE) 2600- > 2618 April 1982 (BDP)	-25°	1982/04/22 14:54 N14 E48	SE	E

4. PREDICTION AND CONCLUSION

The computed magnetic-field orientation at the location of the candidate parent events for each of the seven Bz events at the altitudes 1.03 $R_\odot$, and 1.20 $R_\odot$ are shown in Table 1.

The first and second columns of Table 1 list the time intervals during which the Bz events and the driver gas were observed. The magnetic orientation of the observed Bz events in GSM coordinate system and the characteristic signature of the driver gas are shown in parentheses in the first and second columns, respectively. The next columns under the title of Solar Source show the candidate solar sources of the Bz events and the predicted magnetic orientations at two possible release heights. We assigned each event to one of eight 45 degree sectors, N, NW, etc., relative to the projection of Earth's rotation axis onto the synoptic chart.

Five Bz events in Table 1 are associated with single candidate AR-CMEs. The remaining two events (#1 and #6) have two candidate AR-CMEs. It is seen from Table 1 that the prediction of magnetic orientation at 1.03 $R_\odot$ for events #2, #3, #5 and #7 is consistent with that of the observed Bz events. As shown in paper 1, because the release height for AR-CMEs is usually less than 1.20 $R_\odot$, not all predictions at 1.20 $R_\odot$ are consistent with interplanetary observations. Event

#4 consists of a southward and a northward Bz event separated by an interval of 3 hours. The associated BDE covers both southward and northward subevents. Thus it probably is a structure with large internal field rotation and cannot be described by the potential model, though the southward solar orientation does agree with the direction of the leading part of the event. For the remaining two events we can infer nothing because both northward and southward field components are predicted and we cannot determine which is the real solar source of the Bz event. However, if our hypothesis is true, we can predict which event is the solar source.

In summary, we have now identified 7 driver gas-associated Bz events for which the solar source can relatively confidently be identified as an AR-CME. In each case the north-south component of the coronal field direction computed at 1.03 $R_\odot$ using the potential field model agrees with the B_z component of the IMF observed by spacecraft at 1 AU. These results show that the assumption made in paper 1, that the excess magnetic energy associated with coronal currents is transferred totally to other forms of energy during the late phase of magnetic energy release, may be proper for the AR-CMEs. The model developed in paper 1 may be used to predict the magnetic-field orientation of interplanetary north-south IMF events associated with with AR-CMEs.

Acknowledgement.

We thank P. Scherrer for helpful discussions, T. Bai for providing his catalog of major flares and J. Harvey for providing data from the National Solar Observatory. The Omnitape for interplanetary medium data is provided by the National Space Science Data Center. This work was supported in part by the Office of Naval Research under Grant N00014-89-J-1024, by the National Aeronautics and Space Administration under Grant NGR5-020-559, and by the Atmospheric Sciences Section of the National Science Foundation under Grant ATM90-22249.

REFERENCES

Gosling, J. T., D. N. Baker, S. J. Bame, W. C. Feldman, R. D. Zwickl and E. J. Smith, Bidirectional solar wind electron heat flux events, *J. Geophys. Res.*, *92*, 8519–8535, 1987.

Gosling, J. T., S. J. Bame, D. J. McComas, and J. L. Phillips, Coronal mass ejections and large geomagnetic storms, *Geophys. Res. Lett.*, *17*, 901–904, 1990.

Hoeksema, J. T., Structure and Evolution of the Large Scale Solar and Heliospheric Magnetic Fields, Thesis, CSSA-ASTRO-84-07, Stanford University, 1984.

Hoeksema, J. T. and Xuepu Zhao, Prediction of magnetic orientation in driver gas-associated −Bz events, *J. Geophys. Res.*, in press, 1991.

Marsden, R. G., T. R. Sanderson, C. Tranquille, K.-P. Wenzel, and E. J. Smith, ISEE 3 observations of low-energy proton bidirectional events and their relation to isolated interplanetary magnetic structures, *J. Geophys. Res.*, *92*, 11,009–11,019, 1987.

Tsurutani, B. T., W. D. Gonzalez, F. Tang, S.-I. Akasofu, and E. J. Smith, Origin of interplanetary southward magnetic fields responsible for major magnetic storms near solar maximum (1978–1979), *J. Geophys. Res.*, *93*, 8519–8531, 1988.

Wright, C. S., *Catalogue of Solar Filament Disappearances 1964–1980*, REPORT UAG–100, National Geophysical Data Center, 1991.

AUTHOR INDEX

3rd COSPAR Colloquium
SOLAR WIND SEVEN
List of Participants

Astudillo Hernan, Dr., Chile
Aurass Henry, Dr., Germany
Axford W. Ian, Prof. Dr., Germany
Bagenal Fran, Dr., U.S.A.
Balogh André, Dr., United Kingdom
Bame Samuel J., Dr., U.S.A.
Banaszkiewicz Marek, Dr., Poland
Barnes Aaron, Dr., U.S.A.
Baumgärtel Klaus, Dr., Germany
Belcher John W., Dr., U.S.A.
Bertaux Jean Loup, Dr., France
Bird Michael, Dr., Germany
Bochsler Peter, Dr., Switzerland
Bodmer Roland, Switzerland
Bohlin J. David, Dr., U.S.A.
Bothmer Volker, Germany
Brueckner Guenter E., Dr., U.S.A.
Bruno Roberto, Dr., Italy
Bürgi Alfred, Dr., Switzerland
Burton Marcia, Dr., U.S.A.
Campos L.M.B.C., Dr., Portugal
Carbone Vincenzo, Dr., Italy
Chashei Igor V., Dr., U.S.S.R.
Cheng Chung-Chieh, Dr., U.S.A.
Chertkov Alexandr D., Dr., U.S.S.R.
Chertok Ilya, Dr., U.S.S.R.
Coles W.A., Dr., U.S.A.
Coplan Michael, Dr., U.S.A.
Cuperman Sami, Prof. Dr., Israel
Curdt Werner, Dr., Germany
Czechowski Andrzej, Dr., Poland
Dere Kenneth P., Dr., U.S.A.
Detman Thomas R., Dr., U.S.A.
Domingo Vincente, Dr., Netherlands
Dougherty Michele .K., Dr., United Kingdom
Drillia Georgia Athanasia, Dr., Greece
Dryer Murray, Dr., U.S.A.
Eselevich Viktor G., Dr., U.S.S.R.
Esser Ruth, Dr., Norway
Fahr H.J., Prof. Dr., Germany
Fainshtein Victor G., Dr., U.S.S.R.
Farrugia Charles J., Dr., U.S.A.
Feldman U., Dr., U.S.A.
Fichtner Horst, Dr., Germany
Filippov Michael Adolph, Dr., U.S.S.R.
Fillius W., Dr., U.S.A.

Gaffey John D., Dr., U.S.A.
Galvin Antoinette B., Dr., U.S.A.
Geiss Johannes, Prof. Dr., Switzerland
Gloeckler George, Dr., U.S.A.
Gold Robert E., Dr., U.S.A.
Goldstein Bruce E., Dr., U.S.A.
Gosling John T., Dr., U.S.A.
Grall R., Dr., Norway
Grappin R., Dr., France
Grib Sergei A., Dr., U.S.S.R.
Grünwaldt Heiner, Germany
Grzedzielski Stanislaw, Prof. Dr.; Poland
Gubchenko Vladimir M., Dr., U.S.S.R.
Gubler Lukas, Dr., Switzerland
Habbal Shadia, Dr., U.S.A.
Hansteen Viggo, Dr., Norway
Harmon John K., Dr., U.S.A.
Heras Ana M., Dr., Netherlands
Hewish Anthony, Prof. Dr., United Kingdom
Hick Paul, U.S.A.
Hilchenbach Martin, Dr., Germany
Hildner Ernest, Dr., U.S.A.
Hoang Sang, Dr., France
Hoeksema J. Todd, Dr., U.S.A.
Hollweg Joseph V., Prof. Dr., U.S.A.
Holzer Thomas E., Dr., U.S.A.
Hovestadt Dieter, Dr., Germany
Hsieh K.C., Prof. Dr., U.S.A.
Hubert D., Dr., France
Ipavich Fred M., Dr., U.S.A.
Isenberg Philip, Dr., U.S.A.
Jackson Bernard V., Dr., U.S.A.
Jokipii J.R., Dr, U.S.A.
Kayser Susan E., Dr., U.S.A.
Kecskemety Karoly, Dr., Hungary
Keppler Erhard, Dr., Germany
Kiraly Peter, Dr., Hungary
Klein Karl-Ludwig, Dr., France
Klein Larry, Dr., U.S.A.
Kliem Bernhard, Dr., Germany
Kohl John L., Dr., U.S.A.
Kojima Masayoshi, Dr., Japan
Kota Jozsef, Dr., U.S.A.
Krüger Albrecht, Dr., Germany
Kryshtal Alexander N., Dr., U.S.S.R.
Kulcár Ladislav, Dr., Czecho-Slowakia
Kyrölä Erkki, Dr., Finnland

Lallement Rosine, Dr., France
Lantos Pierre, Dr., France
Larson Davin, Dr., U.S.A.
Lazarus Alan J., Prof. Dr., U.S.A.
Lee Martin A., Dr., U.S.A.
Leer Egil, Dr., Norway
Lengyel-Frey Denise, Dr., U.S.A.
Leubner Manfred, Dr., Österrreich
Liewer Paulett C., Dr., U.S.A.
Lin Robert P., Prof. Dr., U.S.A.
Ling James C., Dr., U.S.A.
Liu Shuhui, China
Livi Stefano, Dr., Germany
Lotova Natalya Andreevna, Dr., U.S.S.R.
MacDowall Robert J., Dr., U.S.A.
Malara Francesco, Dr., Italy
Mangeney Andre, Dr., France
Mann Gottfried, Dr., Germany
Markkanen Jussi, Dr., Finland
Marsch Eckart, Dr., Germany
Marsden Richard G., Dr., Netherlands
Marubashi Katsuhide, Dr., Japan
Matthaeus William H., Dr., U.S.A.
McComas David J., Dr., U.S.A.
McDonald Frank B., Dr., U.S.A.
McKenzie Jim F., Dr., Germany
McNutt, Jr. Ralph L., Dr., U.S.A.
Meister Claudia-Veronika, Dr., Germany
Michels Donald J., Dr., U.S.A.
Mihalov John D., Dr., U.S.A.
Möbius Eberhard, Dr., U.S.A.
Montgomery David, Prof.Dr., U.S.A.
Moussas Xenophon, Dr., Greece
Nash Ana G., Dr., U.S.A.
Neugebauer Marcia, Dr., U.S.A.
Nozawa Satoshi, Dr., Japan
Oetliker Michael, Dr., Switzerland
Ogilvie K.W., Dr., U.S.A.
Oughton Sean, Dr., U.S.A.
Pantellini Filippo, Dr., France
Parker Eugene N., Prof. Dr., U.S.A.
Parker Gary, Dr., U.S.A.
Pätzold Martin, Dr., Germany
Peres Giovanni, Dr., Italia
Pizzo Victor, Dr., U.S.A.
Porsche Herbert, Dr., Germany
Prigancová Alla, Dr., Czecho-Slovakia
Ptitsyna Natalya G., Dr., U.S.S.R.
Quemerais Eric, Dr., France
Ratkiewicz Romana, Dr., Poland
Reiner Michael J., Dr., U.S.A.
Rickett Barney, Dr., U.S.A.
Roatsch Thomas, Dr., Germany
Roberts D. Aaron, Dr., U.S.A.
Roelof Edmond C., Dr., U.S.A.
Rosenbauer Helmut, Dr., Germany
Rucinski Daniel, Dr., Poland
Rusin Vojtech, Dr., Czecho-Slovakia
Sanderson T.R., Dr., Netherlands
Sauer Konrad, Dr., Germany
Schmidt Joachim, Germany

Schmidt Wolfgang K.H., Dr., Germany
Scholer Manfred, Dr., Germany
Schultz C. Göran, Dr., Finland
Schulz Michael, Dr., U.S.A.
Schwenn Rainer, Dr., Germany
Scudder Jack D., Dr., U.S.A.
Shagiev Khalil Rimovich, Dr., UdSSR
Sheeley, Jr. Neil R., Dr., U.S.A.
Shishov V.I., Dr., U.S.S.R.
Shoub Edward C., Dr., U.S.A.
Sime David G., Dr., U.S.A.
Simnett George, England
Smith Charles W., Dr., U.S.A.
Smith Edward J., Dr., U.S.A.
Smith Zdenka, Dr., U.S.A.
Spadaro Daniele, Dr., Italia
Spangler Steven R., Dr., U.S.A.
Steinberg John T., Dr., U.S.A.
Steinitz Raphael, Prof. Dr., Israel
Steinolfson Richard S., Dr., U.S.A.
Stepanova Tatyana Vladimirovna, Dr., UdSSR
Stiller Wolfgang, Dr., Germany
Suess Steve T., Dr., U.S.A.
Summanen Tuula, Dr., Finnland
Sykora Julius, Dr., Czecho-Slovakia
Tatrallyay Mariella, Dr., Hungary
Thiessen Jerold P., Dr., U.S.A.
Tokumaru Munetoshi, Dr., Japan
Torsti Jarmo J., Dr., Finland
Triskova Ludmila, Dr., Czecho-Slowakia
Tu Chuan-yi, Prof. Dr., China
Urbarz H.W., Dr., Germany
Vahia M.N., Dr., India
Valenzuela Arnoldo, Dr., Germany
Vandas Marek, Dr., Czecho-Slovakia
Velli Marco, Dr., France
Veltri Pierluigi, Dr., Italy
Ventura Rita, Dr., Italia
Verheest Frank, Prof. Dr., Belgium
Veselovsky Igor Stanislavovich, Dr., U.S.S.R.
Villanueva Louis A., Dr., U.S.A.
Vinas Adolfo F., Dr., U.S.A.
Vlasov Vladimir I., Dr., U.S.S.R.
von Steiger Rudolf, Dr., Switzerland
Vörös Zoltan, Dr., Czecho-Slovakia
Wang Yi-Ming, Dr., U.S.A.
Washimi Haruichi, Dr., Japan
Webb David F., Dr., U.S.A.
Weigel Andreas, Switzerland
Wenzel K.-P., Dr., Netherlands
Widing Kenneth G., Dr., U.S.A.
Wimmer Robert, Switzerland
Winterhalter Daniel, Dr., U.S.A.
Witte Manfred, Dr., Germany
Woo Richard, Dr., U.S.A.
Woodward Tim, Germany
Wüest Martin, Dr., Switzerland
Yermolaev Yuri I., Dr., U.S.S.R.
Zank Gary P., Dr., U.S.A.
Zhao Xuepu, Dr., U.S.A.
Zimbardo Gaetano, Dr., Italy
Zurbuchen Thomas, Switzerland

3rd COSPAR Colloquium
SOLAR WIND SEVEN
List of Unpublished Papers

S. Arya and J. Freeman
Solar wind velocity gradients inside 1 AU

H. Aurass
The February 1986 active region complex – radio burst signatures of flare related coronal disturbances

Badruddin
High speed solar wind streams and their effects on modulation of cosmic rays

A. Balogh, D. J. Southwood, E. J. Smith and B. T. Tsurutani
Ulysses observation of the heliospheric magnetic field

M. K. Bird, H. Volland, M. Pätzold, P. Edenhofer, S. W. Asmar and J. P. Brenkle
Coronal radio sounding with Ulysses

V. N. Borovik and M. S. Kurbanov
Coronal holes as based on radio observations used RATAN–600 radio telescope

R. Bruno and B. Bavassano
Solar wind turbulence at large scale

M. E. Burton, E. J. Smith and G. L. Siscoe
ICE observations of heliospheric current sheet crossings: Magnetic field and plasma observations

L.M.B.C. Campos
Exact and approximate theories of dissipation of Alfvén waves in atmospheres

I. V. Chashei
The model of solar wind turbulence

C.-C. Cheng, D. J. Michels, A.-H. Wang and S. T. Wu
MHD simulation of coronal mass ejection: Effect of the driving force

A. D. Chertkov
The microscopic structure of the solar wind plasma

A. D. Chertkov
A critical analysis of reconnection theory (Petschek's model)

A. D. Chertkov
The induction coronal heating and the solar wind acceleration

A. D. Chertkov
The boundary conditions for space magnetohydrodynamic problems

A. D. Chertkov and A. E. Shkrebets
The coronal holes and the solar wind acceleration by coronal active regions

A. I. Efimov and O. I. Yakovlev
Acceleration features of the solar wind deduced from radio occultation data

V. G. Eselevich
Relationships of quasi-stationary solar wind flows with their sources on the sun

V. G. Fainshtein
The interaction effect of fast and slow solar wind flows in interplanetary space on solar wind characteristics at the earth's orbit

V. G. Fainshtein
Magnetic field structure and dynamics in the solar corona and the formation of high speed solar wind streams

M. A. Filippov and V. G. Eselevich
Concerning to operational forecast of cosmic disturbances at the earth orbit

J. D. Gaffey, Jr., R. G. Stone and R. J. Mac Dowall
Induced emission of 2 fpe radiation in the terrestrial foreshock by the electron cyclotron maser instability

J. Geiss
Ulysses: new results related to the FIP effect

G. Gloeckler
New results from the solar wind ion composition spectrometer (SWICS) on Ulysses: coronal and interplanetary processes

G. Gloeckler, A.B. Galvin, J. Geiss, K.W. Ogilvie and B. Wilken
Solar wind Ne/O abundance ratios measured with SWICS on Ulysses

B. E. Goldstein
Waves involving minor ion species in the solar wind

B. E. Goldstein, B. T. Tsurutani and W. C. Feldman
The solar probe mission

J. T. Gosling, D. J. McComas, J. L. Phillips and S. J. Bame
Coronal mass ejections in the solar wind at 1 AU: Solar cycle variations

H. Grünwaldt
Slow wind streams merged from multiple coronal sources

S. Grzedzielski
The large-scale structure in the outer solar system

V. M. Gubchenko
Generation of inductive fields by a magnetic dipole submerged into a flowing hot collisionless plasma and formation of a 3-D coronal streamer

T. E. Holzer
> Short summary of the session and what kind of observations do we need in the future

B. V. Jackson
> The dynamics of mass ejections in the heliosphere observed using Helios photometer data

A. Kakouris and X. Moussas
> 3–D wind models: solar wind and jets

L.J. Lanzerotti, E.C. Roelof, R.E. Gold, K.A. Anderson, T.P. Armstrong, S.M. Krimigis, R.P. Lin, M. Pick, E.T. Sarris, G.M. Simnett and W.E. Frain
> Low energy ion and electron measurements beyond 2 AU on the Ulysses spacecraft

L.J. Lanzerotti, T.P. Armstrong, R.E. Gold, C.G. Maclennan, P. White, E. Hawkins, K.A. Anderson, S.M. Krimigis, R.P. Lin, M. Pick, E. C. Roelof, E.T. Sarris and G.M. Simnett
> Low energy ion composition during solar events

D. E. Larson and R. P. Lin
> Evidence for heating of solar wind electrons upstream of the comet Halley bowshock

A. J. Lazarus, J. W. Belcher, G. S. Gordon, Jr., R. L. McNutt, Jr., A. Barnes, J. D. Mihalov and P. R. Gazis
> Global properties of the plasma in the outer heliosphere: II. Detailed structure and transient phenomena

R. P. Lin, D. E. Larson and S. W. Kahler
> Evidence for solar connection during heat flux dropout events

S. Livi, B. Wilken, T. Roatsch, J. Geiss and G. Gloeckler
> Proton and Helium parameter near interplanetary shocks observed by the SWICS experiment

Y. I. Logachev, V. G. Stolpovski, M. A. Zeldovich, K. Kecskeméty, P. Király, A. J. Somogyi, M. Tátrallyay and V. Varga
> Background fluxes of low-energy particles in interplanetary space

U. Mall, J. Geiss, H. Balsiger, R. von Steiger, F. Ipavich and B. Wilken
> Determination of He/O abundances with SWICS – Ulysses

F. B. McDonald and J. A. le Roux
> Studies of cosmic ray modulation in the outer heliosphere over successive solar cycles (21,22)

F.B. McDonald, M.A. Lee, L.F. Burlaga, A.J. Lazarus, J.D. Mihalov and A. Barnes
> Studies of solar/interplanetary energetic particle increases in the distant heliosphere

R. L. McNutt, Jr., A. J. Lazarus, L. Villanueva and S. A. Rappaport
> Periodicities in observable solar wind parameters

A. G. Nash, Y.-M. Wang and N. R. Sheeley, Jr.
> Estimating solar wind speed at latitudes beyond the ecliptic

V. J. Pizzo and D. G. Sime
Interpretation of interplanetary scintillation observations of large scale solar wind structures

V. J. Pizzo
Three-dimensional dynamics of tilted-dipole corotating interaction regions

T. R. Sanderson, R. G. Marsden, A. M. Heras and K.-P. Wenzel
Observations of 1-75 MeV/nucleon particles on the Ulysses spacecraft: First results

M. Schulz and J. T. Hoeksema
Harmonic spectra of solar magnetic fields and their interpretation

R. Schwenn, P. Hemmerich, R. Kramm, B. Podlipnik, K. Pahlke and R. Roll
The LASCO–C1 coronograph: first coronal images taken with the Bergmodell at Sacramento Peak and Mauna Loa

E. C. Shoub
Role of close encounters in Coulomb scattering

D. G. Sime
Large scale structure of the solar corona over the last solar cycle

E. J. Smith
The heliospheric magnetic field

R. S. Steinolfson
Modeling the quiescent and dynamic corona

R. S. Steinolfson
A numerical study of the interaction between the solar wind and the interstellar medium

R. Steinitz
Diamagnetic diffusion and FIP abundance differentiation

M. Tátrallyay, I. Apáthy, I. Szemerey, A. P. Remizov, M. I. Verigin and M. Delva
Large scale structure of the interplanetary medium at the time of spacecraft encounters with comet Halley

J. P. Thiessen, P. J. Kellogg, S. J. Monson and K. Goetz
Observations of low frequency plasma waves by Ulysses

V. I. Vlasov
Heliospheric dynamic phenomena

Y.M. Voytenko, A.N. Kryshtal, S.V. Kootz, P.P. Malovichko and A.K. Yukhimuck
Kinetic Alfvén waves in transition region of solar wind

D. F. Webb, S. W. Kahler, K. L. Harvey and E. W. Cliver
The scale sizes of CMEs and associated surface activity

D. F. Webb and N. U. Crooker
Heliospheric evolution of the coronal streamer belt

Y. C. Whang and L. F. Burlaga
Radial evolution of a recurrent solar wind structure between 1 and 80 AU

D. Winterhalter and E. J. Smith
The identification of solar wind/coronal mass ejection boundaries

M. Witte and H. Rosenbauer
First results of the Ulysses neutral gas experiment